# Applied Mathematical Sciences
Volume 112





Springer Science+Business Media, LLC

# Applied Mathematical Sciences



*(continued following index)*

Yuri A. Kuznetsov

# Elements of Applied Bifurcation Theory

## Third Edition

With 251 Illustrations

Springer

Yuri A. Kuznetsov
Department of Mathematics
Utrecht University
Budapestlaan 6
3584 CD Utrecht
The Netherlands
and
Institute of Mathematical Problems of Biology
Russian Academy of Sciences
142290 Pushchino, Moscow Region
Russia



Mathematics Subject Classification (2000): 34C23, 37Gxx, 37M20, 3704

Library of Congress Cataloging-in-Publication Data
Kuznet͡sov, I͡U. A. (I͡Uriĭ Aleksandrovich)
Elements of applied bifurcation theory/Yuri A. Kuznetsov.—3rd ed.

[on file]

DOI 10.1007/978-1-4757-3978-7

Printed on acid-free paper.



9 8 7 6 5 4 3 2 1 SPIN 10952337

*springeronline.com*
www.springer.com/mycopy

*To my family*

# Preface to the Third Edition

The years that have passed since the publication of the first edition of this book proved that the basic principles used to select and present the material made sense. The idea was to write a simple text that could serve as a serious introduction to the subject. Of course, the meaning of "simplicity" varies from person to person and from country to country. The word "introduction" contains even more ambiguity. To start reading this book, only a moderate knowledge of linear algebra and calculus is required. Other preliminaries, qualified as "elementary" in modern mathematics, are explicitly formulated in the book. These include the Fredholm Alternative for linear systems and the multidimensional Implicit Function Theorem. Using these very limited tools, a framework of notions, results, and methods is gradually built that allows one to read (and possibly write) scientific papers on bifurcations of nonlinear dynamical systems. Among other things, progress in the sciences means that mathematical results and methods that once were new become standard and routinely used by the research and development community. Hopefully, this edition of the book will contribute to this process.

The book's structure has been kept intact. Most of the changes introduced reflect recent theoretical and software developments in which the author was involved. Important changes in the third edition can be summarized as follows. A new section devoted to the fold-flip bifurcation for maps has appeared in Chapter 9. We derive there a parameter-dependent normal form for the map having at critical parameter values a fixed point with eigenvalues $\pm 1$, approximate this normal form by a flow, present generic bifurcation diagrams, and discuss their relationships with the original map. The treatment of the strong resonance 1:1 is extended considerably along similar lines.

A modern technique, due to Coullet and Spiegel, of the simultaneous center manifold reduction and normalization on the center manifold is introduced in Chapters 8 and 9. Using this technique, explicit formulas for the normal form coefficients for codim 2 local bifurcations of $n$-dimensional ODEs and maps are derived, including the fold-flip bifurcation. The resulting formulas are independent of $n$ and equally suitable for both symbolic and numerical evaluation in the original basis. The center manifold computations for codim 1 local bifurcations in Chapter 5 were also rectified. The reader should be warned that the scaling of some coefficients differs slightly from that in the previous editions.

The Hénon map now plays a prominent role in the book. It is used to illustrate the Smale horseshoe in Chapter 1 and the codim 1 bifurcation curves in Chapter 4. In Chapter 7, it is shown that this map approximates rescaled return maps defined near a nontransversal homoclinic orbit to a saddle limit cycle. Using these facts, we sketch a proof of the classical results by Gavrilov and Shil'nikov concerning the accumulation of folds and flips on this bifurcation.

Following numerous suggestions, two new appendices are included. Appendix B to Chapter 3 gives a rather standard introduction to the theory of the Poincaré normal forms of ODEs near an equilibrium. This provides the reader with a uniform point of view on various normal form computations in the book. Appendix B to Chapter 10 is devoted to an elementary treatment of the bialternate matrix product. This product has proved to be an important tool in the numerical analysis of Hopf and Neimark-Sacker bifurcations but it is not considered in the standard courses on Linear Algebra.

While preparing this edition, I have corrected misprints and errors found in the second edition. The bibliography was extended and checked against the AMS *MathSciNet* reference database; if known, the references to English translations are added. All figures were revised, and in some cases relevant data were recomputed.

I am thankful to all colleagues for comments, suggestions, and discussions that led to text improvements. Finally, may the constant support by my wife, Liodmila, and my daughters, Elena and Ouliana, be acknowledged.

Yuri A. Kuznetsov
Utrecht
April 2004

# Preface to the Second Edition

The favorable reaction to the first edition of this book confirmed that the publication of such an application-oriented text on bifurcation theory of dynamical systems was well timed. The selected topics indeed cover major practical issues of applying the bifurcation theory to finite-dimensional problems. This new edition preserves the structure of the first edition while updating the context to incorporate recent theoretical developments, in particular, new and improved numerical methods for bifurcation analysis. The treatment of some topics has been clarified.

Major additions can be summarized as follows: In Chapter 3, an elementary proof of the topological equivalence of the original and truncated normal forms for the fold bifurcation is given. This makes the analysis of codimension-one equilibrium bifurcations of ODEs in the book complete. This chapter also includes an example of the Hopf bifurcation analysis in a planar system using MAPLE, a symbolic manipulation software. Chapter 4 includes a detailed normal form analysis of the Neimark-Sacker bifurcation in the delayed logistic map. In Chapter 5, we derive explicit formulas for the critical normal form coefficients of all codim 1 bifurcations of $n$-dimensional iterated maps (i.e., fold, flip, and Neimark-Sacker bifurcations). The section on homoclinic bifurcations in $n$-dimensional ODEs in Chapter 6 is completely rewritten and introduces the Melnikov integral that allows us to verify the regularity of the manifold splitting under parameter variations. Recently proved results on the existence of center manifolds near homoclinic bifurcations are also included. By their means the study of generic codim 1 homoclinic bifurcations in $n$-dimensional systems is reduced to that in some two-, three-, or four-dimensional systems. Two- and three-dimensional cases are treated in the main text, while the analysis of bifurcations in four-dimensional systems with a homoclinic orbit to a focus-focus is outlined in the new appendix. In Chapter 7, an explicit example of the "blue sky" bifurcation is discussed. Chapter 10, devoted to the numerical analysis of bifurcations, has been changed most substantially. We have introduced bordering methods to continue fold and Hopf bifurcations in two parameters. In this approach, the defining function for the bifurcation used in the minimal augmented system is computed by solving a bordered linear system. It allows for explicit computation of the gradient of this function, contrary to the approach when determinants are used as the defining functions. The main text now includes BVP methods to continue codim 1 homoclinic bifurcations in two parameters, as well as all codim 1 limit cycle bifurcations. A new appendix to this chapter provides test functions to detect all codim 2 homoclinic bifurcations involving a single homoclinic orbit to an equilibrium. The software review in the last appendix to this chapter is updated to present recently developed programs, including AUTO97 with HomCont, DsTool, and CONTENT providing the information on their availability by ftp.

A number of misprints and minor errors have been corrected while preparing this edition. I would like to thank many colleagues who have sent comments and suggestions, including E. Doedel (Concordia University, Montreal), B. Krauskopf (VU, Amsterdam), S. van Gils (TU Twente, Enschede), B. Sandstede (WIAS, Berlin), W.-J. Beyn (Bielefeld University), F.S. Berezovskaya (Center for Ecological Problems and Forest Productivity, Moscow), E. Nikolaev and E.E. Shnoll (IMPB, Pushchino, Moscow Region), W. Langford (University of Guelph), O. Diekmann (Utrecht University), and A. Champneys (University of Bristol).

I am thankful to my wife, Lioudmila, and to my daughters, Elena and Ouliana, for their understanding, support, and patience, while I was working on this book and developing the bifurcation software CONTENT.

Finally, I would like to acknowledge the Research Institute for Applications of Computer Algebra (RIACA, Eindhoven) for the financial support of my work at CWI (Amsterdam) in 1995–1997.

Yuri A. Kuznetsov
Amsterdam
September 1997

# Preface to the First Edition

During the last few years, several good textbooks on nonlinear dynamics have appeared for graduate students in applied mathematics. It seems, however, that the majority of such books are still too theoretically oriented and leave many practical issues unclear for people intending to apply the theory to particular research problems. This book is designed for advanced undergraduate or graduate students in mathematics who will participate in applied research. It is also addressed to professional researchers in physics, biology, engineering, and economics who use dynamical systems as modeling tools in their studies. Therefore, only a moderate mathematical background in geometry, linear algebra, analysis, and differential equations is required. A brief summary of general mathematical terms and results, which are assumed to be known in the main text, appears at the end of the book. Whenever possible, only elementary mathematical tools are used. For example, we do not try to present normal form theory in full generality, instead developing only the portion of the technique sufficient for our purposes.

The book aims to provide the student (or researcher) with both a solid basis in dynamical systems theory and the necessary understanding of the approaches, methods, results, and terminology used in the modern applied mathematics literature. A key theme is that of *topological equivalence* and *codimension*, or "what one may expect to occur in the dynamics with a given number of parameters allowed to vary." Actually, the material covered is sufficient to perform quite complex bifurcation analysis of dynamical systems arising in applications. The book examines the basic topics of bifurcation theory and could be used to compose a course on nonlinear dynamical systems or systems theory. Certain classical results, such as Andronov-Hopf and homoclinic bifurcation in two-dimensional systems, are presented in great detail, including self-contained proofs. For more complex topics of the theory, such as homoclinic bifurcations in more than two dimensions and two-parameter local bifurcations, we try to make clear the relevant geometrical ideas behind the proofs but only sketch them or, sometimes, discuss and illustrate the results but give only references of where to find the proofs. This approach, we hope, makes the book readable for a wide audience and keeps it relatively short and able to be browsed. We also present several recent theoretical results concerning, in particular, bifurcations of homoclinic orbits to nonhyperbolic equilibria and one-parameter bifurcations of limit cycles in systems with reflectional symmetry. These results are hardly covered in standard graduate-level textbooks but seem to be important in applications.

In this book we try to provide the reader with explicit procedures for application of general mathematical theorems to particular research problems. Special attention is given to numerical implementation of the developed techniques. Several examples, mainly from mathematical biology, are used as illustrations.

The present text originated in a graduate course on nonlinear systems taught by the author at the Politecnico di Milano in the Spring of 1991. A similar postgraduate course was given at the Centrum voor Wiskunde en Informatica (CWI, Amsterdam) in February, 1993. Many of the examples and approaches used in the book were first presented at the seminars held at the Research Computing Centre[1] of the Russian Academy of Sciences (Pushchino, Moscow Region).

Let us briefly characterize the content of each chapter.

**Chapter 1. Introduction to dynamical systems.** In this chapter we introduce basic terminology. A *dynamical system* is defined geometrically as a family of evolution operators $\varphi^t$ acting in some state space $X$ and parametrized by continuous or discrete time $t$. Some examples, including symbolic dynamics, are presented. Orbits, phase portraits, and invariant sets appear before any differential equations, which are treated as one of the ways to define a dynamical system. The Smale horseshoe is used to illustrate the existence of very complex invariant sets having fractal structure. Stability criteria for the simplest invariant sets (equilibria and periodic orbits) are formulated. An example of infinite-dimensional continuous-time dynamical systems is discussed, namely, reaction-diffusion systems.

**Chapter 2. Topological equivalence, bifurcations, and structural stability of dynamical systems.** Two dynamical systems are called *topologically equivalent* if their phase portraits are homeomorphic. This notion is then used to define structurally stable systems and bifurcations. The topological classification of generic (hyperbolic) equilibria and fixed points of dynamical systems defined by autonomous ordinary differential equations (ODEs) and iterated maps is given, and the geometry of the phase portrait near such points is studied. A *bifurcation diagram* of a parameter-dependent system is introduced as a partitioning of its parameter space induced by the topological equivalence of corresponding phase portraits. We introduce the notion of *codimension* (codim for short) in a rather naive way as the number of conditions defining the bifurcation. *Topological normal forms* (universal unfoldings of nondegenerate parameter-dependent systems) for bifurcations are defined, and an example of such a normal form is demonstrated for the Hopf bifurcation.

[1] Renamed in 1992 as the Institute of Mathematical Problems of Biology (IMPB).

**Chapter 3. One-parameter bifurcations of equilibria in continuous-time dynamical systems.** Two generic codim 1 bifurcations – *tangent (fold)* and *Andronov-Hopf* – are studied in detail following the same general approach: (1) formulation of the corresponding topological normal form and analysis of its bifurcations; (2) reduction of a generic parameter-dependent system to the normal form up to terms of a certain order; and (3) demonstration that higher-order terms do not affect the local bifurcation diagram. Step 2 (finite normalization) is performed by means of polynomial changes of variables with unknown coefficients that are then fixed at particular values to simplify the equations. Relevant normal form and nondegeneracy (genericity) conditions for a bifurcation appear naturally at this step. An example of the Hopf bifurcation in a predator-prey system is analyzed.

**Chapter 4. One-parameter bifurcations of fixed points in discrete-time dynamical systems.** The approach formulated in Chapter 3 is applied to study *tangent* (*fold*), *flip* (*period-doubling*), and *Hopf* (*Neimark-Sacker*) bifurcations of discrete-time dynamical systems. For the Neimark-Sacker bifurcation, as is known, a normal form so obtained captures only the appearance of a closed invariant curve but does not describe the orbit structure on this curve. Feigenbaum's universality in the cascade of period doublings is explained geometrically using saddle properties of the period-doubling map in an appropriate function space.

**Chapter 5. Bifurcations of equilibria and periodic orbits in $n$-dimensional dynamical systems.** This chapter explains how the results on codim 1 bifurcations from the two previous chapters can be applied to multidimensional systems. A geometrical construction is presented upon which a proof of the Center Manifold Theorem is based. Explicit formulas are derived for the quadratic coefficients of the Taylor approximations to the center manifold for all codim 1 bifurcations in both continuous and discrete time. An example is discussed where the linear approximation of the center manifold leads to the wrong stability analysis of an equilibrium. We present in detail a projection method for center manifold computation that avoids the transformation of the system into its eigenbasis. Using this method, we derive a compact formula to determine the direction of a Hopf bifurcation in multidimensional systems. Finally, we consider a reaction-diffusion system on an interval to illustrate the necessary modifications of the technique to handle the Hopf bifurcation in some infinite-dimensional systems.

**Chapter 6. Bifurcations of orbits homoclinic and heteroclinic to hyperbolic equilibria.** This chapter is devoted to the generation of periodic orbits via homoclinic bifurcations. A theorem due to Andronov and Leontovich describing homoclinic bifurcation in planar continuous-time systems is formulated. A simple proof is given which uses a constructive $C^1$-linearization of a system near its saddle point. All codim 1 bifurcations of homoclinic orbits to saddle and saddle-focus equilibrium points in three-dimensional ODEs are then studied. The relevant theorems by Shil'nikov are formulated together with the main geometrical constructions involved in their proofs. The role of the orientability of invariant manifolds is emphasized. Generalizations to more dimensions are also discussed. An application of Shil'nikov's results to nerve impulse modeling is given.

**Chapter 7. Other one-parameter bifurcations in continuous-time dynamical systems.** This chapter treats some bifurcations of homoclinic orbits to nonhyperbolic equilibrium points, including the case of several homoclinic orbits to a saddle-saddle point, which provides one of the simplest mechanisms for the generation of an infinite number of periodic orbits. Bifurcations leading to a change in the rotation number on an invariant torus and some other global bifurcations are also reviewed. All codim 1 bifurcations of equilibria and limit cycles in $\mathbb{Z}_2$-symmetric systems are described together with their normal forms.

**Chapter 8. Two-parameter bifurcations of equilibria in continuous-time dynamical systems.** One-dimensional manifolds in the direct product of phase and parameter spaces corresponding to the tangent and Hopf bifurcations are defined and used to specify all possible codim 2 bifurcations of equilibria in generic continuous-time systems. Topological normal forms are presented and discussed in detail for the cusp, Bogdanov-Takens, and generalized Andronov-Hopf (Bautin) bifurcations. An example of a two-parameter analysis of Bazykin's predator-prey model is considered in detail. Approximating symmetric normal forms for zero-Hopf and Hopf-Hopf bifurcations are derived and studied, and their relationship with the original problems is discussed. Explicit formulas for the critical normal form coefficients are given for the majority of the codim 2 cases.

**Chapter 9. Two-parameter bifurcations of fixed points in discrete-time dynamical systems.** A list of all possible codim 2 bifurcations of fixed points in generic discrete-time systems is presented. Topological normal forms are obtained for the cusp and degenerate flip bifurcations with explicit formulas for their coefficients. An approximate normal form is presented for the Neimark-Sacker bifurcation with cubic degeneracy (Chenciner bifurcation). Approximating normal forms are expressed in terms of continuous-time planar dynamical systems for all strong resonances (1:1, 1:2, 1:3, and 1:4). The Taylor coefficients of these continuous-time systems are explicitly given in terms of those of the maps in question. A periodically forced predator-prey model is used to illustrate resonant phenomena.

**Chapter 10. Numerical analysis of bifurcations.** This final chapter deals with numerical analysis of bifurcations, which in most cases is the only tool to attack real problems. Numerical procedures are presented for the location and stability analysis of equilibria and the local approximation of their invariant manifolds as well as methods for the location of limit cycles (including orthogonal collocation). Several methods are discussed for equilibrium continuation and detection of codim 1 bifurcations based on predictor-corrector schemes. Numerical methods for continuation and analysis of homoclinic bifurcations are also formulated.

Each chapter contains exercises, and we have provided hints for the most difficult of them. The references and comments to the literature are summarized at the end of each chapter as separate bibliographical notes. The aim of these notes is mainly to provide a reader with information on further reading. The end of a theorem's proof (or its absence) is marked by the symbol $\Box$, while that of a remark (example) is denoted by $\Diamond$ ($\Diamond$), respectively.

As is clear from this Preface, there are many important issues this book does not touch. In fact, we study only the first bifurcations on a route to chaos and try to avoid the detailed treatment of chaotic dynamics, which requires more sophisticated mathematical tools. We do not consider important classes of dynamical systems such as Hamiltonian systems (e.g., KAM-theory and Melnikov methods are left outside the scope of this book). Only introductory information is provided on bifurcations in systems with symmetries. The list of omissions can easily be extended. Nevertheless, we hope the reader will find the book useful, especially as an interface between undergraduate and postgraduate studies.

This book would have never appeared without the encouragement and help from many friends and colleagues to whom I am very much indebted. The idea of such an application-oriented book on bifurcations emerged in discussions and joint work with A.M. Molchanov, A.D. Bazykin, E.E. Shnol, and A.I. Khibnik at the former Research Computing Centre of the USSR Academy of Sciences (Pushchino). S. Rinaldi asked me to prepare and give a course on nonlinear systems at the Politecnico di Milano that would be useful for applied scientists and engineers. O. Diekmann (CWI, Amsterdam) was the first to propose the conversion of these brief lecture notes into a book. He also commented on some of the chapters and gave friendly support during the whole project. S. van Gils (TU Twente, Enschede) read the manuscript and gave some very useful suggestions that allowed me to improve the content and style. I am particularly thankful to A.R. Champneys of the University of Bristol, who reviewed the whole text and not only corrected the language but also proposed many improvements in the selection and presentation of the material. Certain topics have been discussed with J. Sanders (VU/RIACA/CWI, Amsterdam), B. Werner (University of Hamburg), E. Nikolaev (IMPB, Pushchino), E. Doedel (Concordia University, Montreal), B. Sandstede (IAAS, Berlin), M. Kirkilonis (CWI, Amsterdam), J. de Vries (CWI, Amsterdam), and others, whom I would like to thank. Of course, the responsibility for all remaining mistakes is mine. I would also like to thank A. Heck (CAN, Amsterdam) and V.V. Levitin (IMPB, Pushchino/CWI, Amsterdam) for computer assistance. Finally, I thank the Nederlandse Organisatie voor Wetenschappelijk Onderzoek (NWO) for providing financial support during my stay at CWI, Amsterdam.

Yuri A. Kuznetsov
Amsterdam
December 1994

# Contents

# 1

# Introduction to Dynamical Systems

This chapter introduces some basic terminology. First, we define a *dynamical system* and give several examples, including symbolic dynamics. Then we introduce the notions of *orbits, invariant sets*, and their *stability*. As we shall see while analyzing the *Smale horseshoe*, invariant sets can have very complex structures. This is closely related to the fact discovered in the 1960s that rather simple dynamical systems may behave "randomly," or "chaotically." Finally, we discuss how differential equations can define dynamical systems in both finite- and infinite-dimensional spaces.

## 1.1 Definition of a dynamical system

The notion of a dynamical system is the mathematical formalization of the general scientific concept of a *deterministic process*. The future and past states of many physical, chemical, biological, ecological, economical, and even social systems can be predicted to a certain extent by knowing their present state and the laws governing their evolution. Provided these laws do not change in time, the behavior of such a system could be considered as completely defined by its initial state. Thus, the notion of a dynamical system includes a set of its possible states (*state space*) and a law of the *evolution* of the state in *time*. Let us discuss these ingredients separately and then give a formal definition of a dynamical system.

### 1.1.1 State space

All possible states of a system are characterized by the points of some set $X$. This set is called the *state space* of the system. Actually, the specification of a point $x \in X$ must be sufficient not only to describe the current "position" of the system but also to determine its evolution. Different branches of science provide us with appropriate state spaces. Often, the state space is called a *phase space*, following a tradition from classical mechanics.

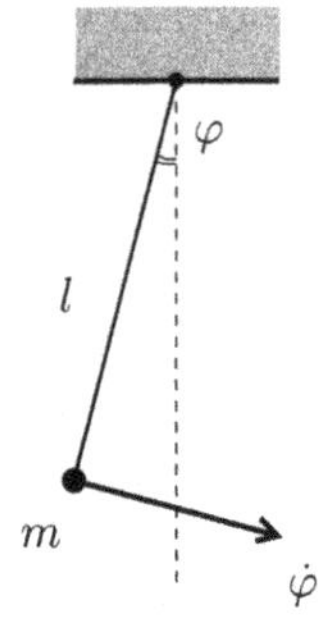


**Fig. 1.1.** Classical pendulum.

**Example 1.1 (Pendulum)** The state of an ideal pendulum is completely characterized by defining its angular *displacement* $\varphi \pmod{2\pi}$ from the vertical position and the corresponding angular *velocity* $\dot{\varphi}$ (see **Fig. 1.1**). Notice that the angle $\varphi$ alone is insufficient to determine the future state of the pendulum. Therefore, for this simple mechanical system, the state space is $X = \mathbb{S}^1 \times \mathbb{R}^1$, where $\mathbb{S}^1$ is the unit circle parametrized by the angle, and $\mathbb{R}^1$ is the real axis corresponding to the set of all possible velocities. The set $X$ can be considered as a smooth two-dimensional *manifold* (cylinder) in $\mathbb{R}^3$. $\Diamond$

**Example 1.2 (General mechanical system)** In classical mechanics, the state of an isolated system with $s$ degrees of freedom is characterized by a $2s$-dimensional real vector:

$$(q_1, q_2, \ldots, q_s, p_1, p_2, \ldots, p_s)^T,$$

where $q_i$ are the *generalized coordinates*, while $p_i$ are the corresponding *generalized momenta*. Therefore, in this case, $X = \mathbb{R}^{2s}$. If $k$ coordinates are *cyclic*, $X = \mathbb{S}^k \times \mathbb{R}^{2s-k}$. In the case of the pendulum, $s = k = 1$, $q_1 = \varphi$, and we can take $p_1 = \dot{\varphi}$. $\Diamond$

**Example 1.3 (Quantum system)** In quantum mechanics, the state of a system with *two observable states* is characterized by a vector

$$\psi = \begin{pmatrix} a_1 \\ a_2 \end{pmatrix} \in \mathbb{C}^2,$$

where $a_i, i = 1, 2$, are complex numbers called *amplitudes*, satisfying the condition

$$|a_1|^2 + |a_2|^2 = 1.$$

The probability of finding the system in the $i$th state is equal to $p_i = |a_i|^2, i = 1, 2$. $\Diamond$

**Example 1.4 (Chemical reactor)** The state of a well-mixed isothermic chemical reactor is defined by specifying the volume *concentrations* of the $n$ reacting chemical substances

$$c = (c_1, c_2, \ldots, c_n)^T.$$

Clearly, the concentrations $c_i$ must be nonnegative. Thus,

$$X = \{c : c = (c_1, c_2, \ldots, c_n)^T \in \mathbb{R}^n, c_i \geq 0\}.$$

If the concentrations change from point to point, the state of the reactor is defined by the reagent *distributions* $c_i(x), i = 1, 2, \ldots, n$. These functions are defined in a bounded spatial domain $\Omega$, the reactor interior, and characterize the local concentrations of the substances near a point $x$. Therefore, the state space $X$ in this case is a *function space* composed of vector-valued functions $c(x)$, satisfying certain smoothness and boundary conditions. $\diamondsuit$

**Example 1.5 (Ecological system)** Similar to the previous example, the state of an ecological community within a certain domain $\Omega$ can be described by a vector with nonnegative components

$$N = (N_1, N_2, \ldots, N_n)^T \in \mathbb{R}^n,$$

or by a vector function

$$N(x) = (N_1(x), N_2(x), \ldots, N_n(x))^T, \quad x \in \Omega,$$

depending on whether the spatial distribution is essential for an adequate description of the dynamics. Here $N_i$ is the number (or density) of the $i$th species or other group (e.g., predators or prey). $\diamondsuit$

**Example 1.6 (Symbolic dynamics)** To complete our list of state spaces, consider a set $\Omega_2$ of all possible bi-infinite *sequences* of two symbols, say $\{1, 2\}$. A point $\omega \in X$ is the sequence

$$\omega = \{\ldots, \omega_{-2}, \omega_{-1}, \omega_0, \omega_1, \omega_2, \ldots\},$$

where $\omega_i \in \{1, 2\}$. Note that the zero position in a sequence must be pointed out; for example, there are *two* distinct periodic sequences that can be written as

$$\omega = \{\ldots, 1, 2, 1, 2, 1, 2, \ldots\},$$

one with $\omega_0 = 1$, and the other with $\omega_0 = 2$. The space $\Omega_2$ will play an important role in the following.

Sometimes, it is useful to identify two sequences that differ only by a shift of the origin. Such sequences are called *equivalent*. The classes of equivalent sequences form a set denoted by $\tilde{\Omega}_2$. The two periodic sequences mentioned above represent the same point in $\tilde{\Omega}_2$. $\diamondsuit$

In all the above examples, the state space has a certain natural structure, allowing for comparison between different states. More specifically, a *distance* $\rho$ between two states is defined, making these sets *metric spaces*.

In the examples from mechanics and in the simplest examples from chemistry and ecology, the state space was a real vector space $\mathbb{R}^n$ of some finite dimension $n$, or a (sub-)manifold (hypersurface) in this space. The *Euclidean norm* can be used to measure the distance between two states parametrized by the points $x, y \in \mathbb{R}^n$, namely

$$\rho(x,y) = \|x-y\| = \sqrt{\langle x-y, x-y\rangle} = \sqrt{\sum_{i=1}^{n}(x_i-y_i)^2}, \tag{1.1}$$

where $\langle \cdot, \cdot \rangle$ is the standard scalar product in $\mathbb{R}^n$,

$$\langle x, y\rangle = x^T y = \sum_{i=1}^{n} x_i y_i.$$

If necessary, the distance between two (close) points on a manifold can be measured as the minimal length of a curve connecting these points within the manifold. Similarly, the distance between two states $\psi, \varphi$ of the quantum system from Example 1.3 can be defined using the standard scalar product in $\mathbb{C}^n$,

$$\langle \psi, \varphi\rangle = \bar{\psi}^T \varphi = \sum_{i=1}^{n} \bar{\psi}_i \varphi_i,$$

with $n = 2$. Meanwhile, $\langle \psi, \psi\rangle = \langle \varphi, \varphi\rangle = 1$.

When the state space is a function space, there is a variety of possible distances, depending on the *smoothness* (differentiability) of the functions allowed. For example, we can introduce a distance between two continuous vector-valued real functions $u(x)$ and $v(x)$ defined in a bounded closed domain $\Omega \in \mathbb{R}^m$ by

$$\rho(u,v) = \|u-v\| = \max_{i=1,\ldots,n} \sup_{x\in\Omega} |u_i(x) - v_i(x)|.$$

Finally, in Example 1.6 the distance between two sequences $\omega, \theta \in \Omega_2$ can be measured by

$$\rho(\omega, \theta) = \sum_{k=-\infty}^{+\infty} \delta_{\omega_k \theta_k} 2^{-|k|}, \tag{1.2}$$

where

$$\delta_{\omega_k \theta_k} = \begin{cases} 0 & \text{if} \quad \omega_k = \theta_k, \\ 1 & \text{if} \quad \omega_k \neq \theta_k. \end{cases}$$

According to this formula, two sequences are considered to be close if they have a long block of coinciding elements centered at position zero (check!).

Using the previously defined distances, the introduced state spaces $X$ are *complete metric spaces*. Loosely speaking, this means that any sequence of states, all of whose sufficiently future elements are separated by an arbitrarily small distance, is convergent (the space has no "holes").

According to the dimension of the underlying state space $X$, the dynamical system is called either *finite-* or *infinite-dimensional.* Usually, one distinguishes finite-dimensional systems defined in $X = \mathbb{R}^n$ from those defined on manifolds.

### 1.1.2 Time

The evolution of a dynamical system means a change in the state of the system with time $t \in T$, where $T$ is a number set. We will consider two types of dynamical systems: those with continuous (real) time $T = \mathbb{R}^1$, and those with discrete (integer) time $T = \mathbb{Z}$. Systems of the first type are called *continuous-time* dynamical systems, while those of the second are termed *discrete-time* dynamical systems. Discrete-time systems appear naturally in ecology and economics when the state of a system at a certain moment of time $t$ completely determines its state after a year, say at $t + 1$.

### 1.1.3 Evolution operator

The main component of a dynamical system is an evolution law that determines the state $x_t$ of the system at time $t$, provided the *initial state* $x_0$ is known. The most general way to specify the evolution is to assume that for given $t \in T$ a map $\varphi^t$ is defined in the state space $X$,

$$\varphi^t : X \to X,$$

which transforms an initial state $x_0 \in X$ into some state $x_t \in X$ at time $t$:

$$x_t = \varphi^t x_0.$$

The map $\varphi^t$ is often called the *evolution operator* of the dynamical system. It might be known explicitly; however, in most cases, it is defined *indirectly* and can be computed only approximately. In the continuous-time case, the family $\{\varphi^t\}_{t \in T}$ of evolution operators is called a *flow.*

Note that $\varphi^t x$ may not be defined for all pairs $(x, t) \in X \times T$. Dynamical systems with evolution operator $\varphi^t$ defined for both $t \geq 0$ and $t < 0$ are called *invertible.* In such systems the initial state $x_0$ completely defines not only the *future* states of the system, but its *past* behavior as well. However, it is useful to consider also dynamical systems whose future behavior for $t > 0$ is completely determined by their initial state $x_0$ at $t = 0$, but the history for $t < 0$ can not be unambigously reconstructed. Such (*noninvertible*) dynamical systems are described by evolution operators defined only for $t \geq 0$ (i.e., for $t \in \mathbb{R}^1_+$ or $\mathbb{Z}_+$). In the continuous-time case, they are called *semiflows.*

It is also possible that $\varphi^t x_0$ is defined only *locally* in time, for example, for $0 \leq t < t_0$, where $t_0$ depends on $x_0 \in X$. An important example of such a behavior is a "blow-up," when a continuous-time system in $X = \mathbb{R}^n$ approaches infinity within a finite time, i.e.,

$$\|\varphi^t x_0\| \to +\infty,$$

for $t \to t_0$.

The evolution operators have two natural properties that reflect the deterministic character of the behavior of dynamical systems. First of all,

$$\text{(DS.0)} \qquad \varphi^0 = \text{id},$$

where id is the identity map on $X$, $\text{id}\,(x) = x$ for all $x \in X$. The property (DS.0) implies that the system does not change its state "spontaneously." The second property of the evolution operators reads

$$\text{(DS.1)} \qquad \varphi^{t+s} = \varphi^t \circ \varphi^s.$$

It means that

$$\varphi^{t+s}x = \varphi^t(\varphi^s x)$$

for all $x \in X$ and $t, s \in T$, *such that both sides of the last equation are defined.*[1] Essentially, the property (DS.1) states that the result of the evolution of the system in the course of $t + s$ units of time, starting at a point $x \in X$, is the same as if the system were first allowed to change from the state $x$ over only $s$ units of time and then evolved over the next $t$ units of time from the resulting state $\varphi^s x$ (see **Fig. 1.2**). This property means that the law governing the

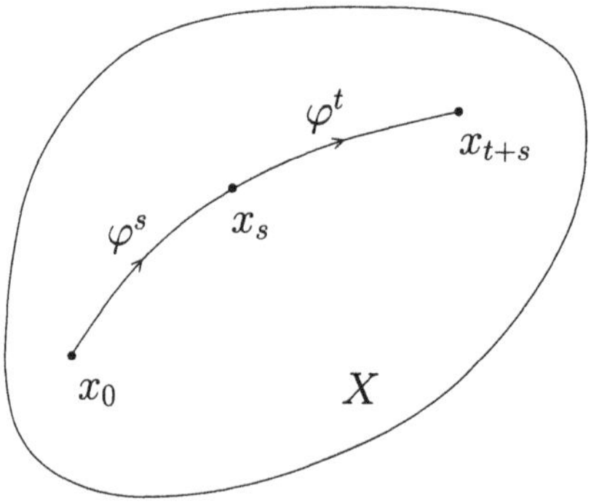


**Fig. 1.2.** Evolution operator.

behavior of the system does not change in time: The system is "autonomous."

For invertible systems, the evolution operator $\varphi^t$ satisfies the property (DS.1) for $t$ and $s$ both negative and nonnegative. In such systems, the operator $\varphi^{-t}$ is the inverse to $\varphi^t$, $(\varphi^t)^{-1} = \varphi^{-t}$ since

[1] Whenever possible, we will avoid explicit statements on the domain of definition of $\varphi^t x$.

$$\varphi^{-t} \circ \varphi^t = \text{id}.$$

A discrete-time dynamical system with integer $t$ is fully specified by defining only one map $f = \varphi^1$, called *"time-one map."* Indeed, using (DS.1), we obtain

$$\varphi^2 = \varphi^1 \circ \varphi^1 = f \circ f = f^2,$$

where $f^2$ is the *second iterate* of the map $f$. Similarly,

$$\varphi^k = f^k$$

for all $k > 0$. If the discrete-time system is invertible, the above equation holds for $k \leq 0$, where $f^0 = \text{id}$.

Finally, let us point out that, for many systems, $\varphi^t x$ is a continuous function of $x \in X$, and if $t \in \mathbb{R}^1$, it is also continuous in time. Here, the continuity is supposed to be defined with respect to the corresponding metric or norm in $X$. Furthermore, many systems defined on $\mathbb{R}^n$, or on smooth manifolds in $\mathbb{R}^n$, are such that $\varphi^t x$ is smooth as a function of $(x, t)$. Such systems are called *smooth dynamical systems.*

### 1.1.4 Definition of a dynamical system

Now we are able to give a formal definition of a dynamical system.

**Definition 1.1** *A* dynamical system *is a triple* $\{T, X, \varphi^t\}$, *where* $T$ *is a time set,* $X$ *is a state space, and* $\varphi^t : X \to X$ *is a family of evolution operators parametrized by* $t \in T$ *and satisfying the properties* (DS.0) *and* (DS.1).

Let us illustrate the definition by two explicit examples.

**Example 1.7 (A linear planar system)** Consider the plane $X = \mathbb{R}^2$ and a family of linear nonsingular transformations on $X$ given by the matrix depending on $t \in \mathbb{R}^1$:

$$\varphi^t = \begin{pmatrix} e^{\lambda t} & 0 \\ 0 & e^{\mu t} \end{pmatrix},$$

where $\lambda, \mu \neq 0$ are real numbers. Obviously, it specifies a continuous-time dynamical system on $X$. The system is invertible and is defined for all $(x, t)$. The map $\varphi^t$ is continuous (and smooth) in $x$, as well as in $t$. $\Diamond$

**Example 1.8 (Symbolic dynamics revisited)** Take the space $X = \Omega_2$ of all bi-infinite sequences of two symbols $\{1, 2\}$ introduced in Example 1.6. Consider a map $\sigma : X \to X$, which transforms the sequence

$$\omega = \{\ldots, \omega_{-2}, \omega_{-1}, \omega_0, \omega_1, \omega_2, \ldots\} \in X$$

into the sequence $\theta = \sigma(\omega)$,

$$\theta = \{\ldots, \theta_{-2}, \theta_{-1}, \theta_0, \theta_1, \theta_2, \ldots\} \in X,$$

where

$$\theta_k = \omega_{k+1}, \quad k \in \mathbb{Z}.$$

The map $\sigma$ merely shifts the sequence by one position to the left. It is called a *shift map*. The shift map defines a discrete-time dynamical system $\{\mathbb{Z}, \Omega_2, \sigma^k\}$ called the *symbolic dynamics*, that is invertible (find $\sigma^{-1}$). Notice that two sequences, $\theta$ and $\omega$, are equivalent if and only if $\theta = \sigma^{k_0}(\omega)$ for some $k_0 \in \mathbb{Z}$. ◊

Later on in the book, we will encounter many different examples of dynamical systems and will study them in detail.

## 1.2 Orbits and phase portraits

Throughout the book we use a geometrical point of view on dynamical systems. We shall always try to present their properties in geometrical images since this facilitates their understanding. The basic geometrical objects associated with a dynamical system $\{T, X, \varphi^t\}$ are its *orbits* in the state space and the *phase portrait* composed of these orbits.

**Definition 1.2** *An* orbit starting at $x_0$ *is an ordered subset of the state space* $X$*,*

$$Or(x_0) = \{x \in X : x = \varphi^t x_0, \text{ for all } t \in T \text{ such that } \varphi^t x_0 \text{ is defined}\}.$$

Orbits of a continuous-time system with a continuous evolution operator are *curves* in the state space $X$ parametrized by the time $t$ and oriented by its direction of increase (see **Fig. 1.3**). Orbits of a discrete-time system are

**Fig. 1.3.** Orbits of a continuous-time system.

*sequences of points* in the state space $X$ enumerated by increasing integers. Orbits are often also called *trajectories*. If $y_0 = \varphi^{t_0} x_0$ for some $t_0$, the sets $Or(x_0)$ and $Or(y_0)$ coincide. For example, two equivalent sequences $\theta, \omega \in \Omega_2$ generate the same orbit of the symbolic dynamics $\{\mathbb{Z}, \Omega_2, \sigma^k\}$. Thus, all different orbits of the symbolic dynamics are represented by points in the set $\widetilde{\Omega}_2$ introduced in Example 1.6.

The simplest orbits are *equilibria.*

**Definition 1.3** *A point $x^0 \in X$ is called an* equilibrium (fixed point) *if $\varphi^t x^0 = x^0$ for all $t \in T$.*

The evolution operator maps an equilibrium onto itself. Equivalently, a system placed at an equilibrium remains there forever. Thus, equilibria represent the simplest mode of behavior of the system. We will reserve the name "equilibrium" for continuous-time dynamical systems, while using the term "fixed point" for corresponding objects of discrete-time systems. The system from Example 1.7 obviously has a single equilibrium at the origin, $x^0 = (0,0)^T$. If $\lambda, \mu < 0$, all orbits converge to $x^0$ as $t \to +\infty$ (this is the simplest mode of *asymptotic* behavior for large time). The symbolic dynamics from Example 1.7 have only two fixed points, represented by the sequences

$$\omega^1 = \{\ldots, 1, 1, 1, \ldots\}$$

and

$$\omega^2 = \{\ldots, 2, 2, 2, \ldots\}.$$

Clearly, the shift $\sigma$ does not change these sequences: $\sigma(\omega^{1,2}) = \omega^{1,2}$.

Another relatively simple type of orbit is a *cycle.*

**Definition 1.4** *A* cycle *is a periodic orbit, namely a nonequilibrium orbit $L_0$, such that each point $x_0 \in L_0$ satisfies $\varphi^{t+T_0} x_0 = \varphi^t x_0$ with some $T_0 > 0$, for all $t \in T$.*

The minimal $T_0$ with this property is called the *period* of the cycle $L_0$. If a system starts its evolution at a point $x_0$ on the cycle, it will return exactly to this point after every $T_0$ units of time. The system exhibits *periodic oscillations.* In the continuous-time case a cycle $L_0$ is a closed curve (see **Fig. 1.4**(a)).

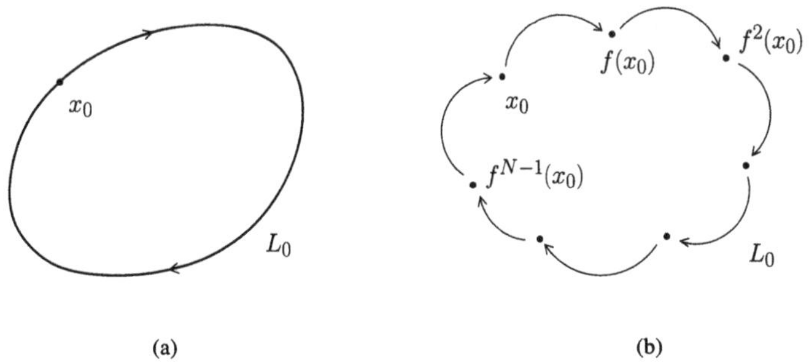


**Fig. 1.4.** Periodic orbits in (a) a continuous-time and (b) a discrete-time system.

**Definition 1.5** *A cycle of a continuous-time dynamical system, in a neighborhood of which there are no other cycles, is called a* limit cycle.

In the discrete-time case a cycle is a (finite) set of points

$$x_0, f(x_0), f^2(x_0), \ldots, f^{N_0}(x_0) = x_0,$$

where $f = \varphi^1$ and the period $T_0 = N_0$ is obviously an integer (**Fig. 1.4**(b)). Notice that each point of this set is a *fixed point* of the $N_0$th iterate $f^{N_0}$ of the map $f$. The system from Example 1.7 has no cycles. In contrast, the symbolic dynamics (Example 1.8) have an *infinite* number of cycles. Indeed, any *periodic sequence* composed of repeating blocks of length $N_0 > 1$ represents a cycle of period $N_0$ since we need to apply the shift $\sigma$ exactly $N_0$ times to transform such a sequence into itself. Clearly, there is an infinite (though countable) number of such periodic sequences. Equivalent periodic sequences define the same periodic orbit.

We can roughly classify all possible orbits in dynamical systems into fixed points, cycles, and "all others."

**Definition 1.6** *The* phase portrait *of a dynamical system is a partitioning of the state space into orbits.*

The phase portrait contains a lot of information on the behavior of a dynamical system. By looking at the phase portrait, we can determine the number and types of *asymptotic states* to which the system tends as $t \to +\infty$ (and as $t \to -\infty$ if the system is invertible). Of course, it is impossible to draw all orbits in a figure. In practice, only several key orbits are depicted in the diagrams to present phase portraits schematically (as we did in **Fig. 1.3**). A phase portrait of a continuous-time dynamical system could be interpreted as an image of the flow of some fluid, where the orbits show the paths of "liquid particles" as they follow the current. This analogy explains the use of the term "flow" for the evolution operator in the continuous-time case.

## 1.3 Invariant sets

### 1.3.1 Definition and types

To further classify elements of a phase portrait – in particular, possible asymptotic states of the system – the following definition is useful.

**Definition 1.7** *An* invariant set *of a dynamical system* $\{T, X, \varphi^t\}$ *is a subset* $S \subset X$ *such that* $x_0 \in S$ *implies* $\varphi^t x_0 \in S$ *for all* $t \in T$.

The definition means that $\varphi^t S \subseteq S$ for all $t \in T$. Clearly, an invariant set $S$ consists of orbits of the dynamical system. Any individual orbit $Or(x_0)$ is obviously an invariant set. We always can *restrict* the evolution operator $\varphi^t$ of the system to its invariant set $S$ and consider a dynamical system $\{T, S, \psi^t\}$, where $\psi^t : S \to S$ is the map induced by $\varphi^t$ in $S$. We will use the symbol $\varphi^t$ for the restriction, instead of $\psi^t$.

If the state space $X$ is endowed with a metric $\rho$, we could consider *closed invariant sets* in $X$. Equilibria (fixed points) and cycles are clearly the simplest examples of closed invariant sets. There are other types of closed invariant sets. The next more complex are *invariant manifolds*, that is, finite-dimensional hypersurfaces in some space $\mathbb{R}^K$. **Fig. 1.5** sketches an invariant two-dimensional *torus* $\mathbb{T}^2$ of a continuous-time dynamical system in $\mathbb{R}^3$ and a typical orbit on that manifold. One of the major discoveries in dynamical systems theory was the recognition that very simple, invertible, differentiable dynamical systems can have extremely complex closed invariant sets containing an infinite number of periodic and nonperiodic orbits. Smale constructed the most famous example of such a system. It provides an invertible discrete-time dynamical system on the plane possessing an invariant set $\Lambda$, whose points are in one-to-one correspondence with all the bi-infinite sequences of two symbols. The invariant set $\Lambda$ is not a manifold. Moreover, the restriction of the system to this invariant set behaves, in a certain sense, as the symbolic dynamics specified in Example 1.8. That is, how we can verify that it has an infinite number of cycles. Let us explore Smale's example in some detail.

### 1.3.2 Smale horseshoe

Consider the geometrical construction in **Fig. 1.6**. Take a square $S$ on the plane (**Fig. 1.6**(a)). Contract it in the horizontal direction and expand it in the vertical direction (**Fig. 1.6**(b)). Fold it in the middle (**Fig. 1.6**(c)) and place it so that it intersects the original square $S$ along two vertical strips (**Fig. 1.6**(d)). This procedure defines a map $f : \mathbb{R}^2 \to \mathbb{R}^2$. The image $f(S)$ of the square $S$ under this transformation resembles a horseshoe. That is why it is called a *horseshoe map*. The exact shape of the image $f(S)$ is irrelevant; however, let us assume for simplicity that both the contraction and expansion are linear and that the vertical strips in the intersection are rectangles. The map $f$ can be made invertible and smooth together with its inverse. The inverse map $f^{-1}$ transforms the horseshoe $f(S)$ back into the square $S$ through stages (d)–(a). This inverse transformation maps the dotted square $S$ shown

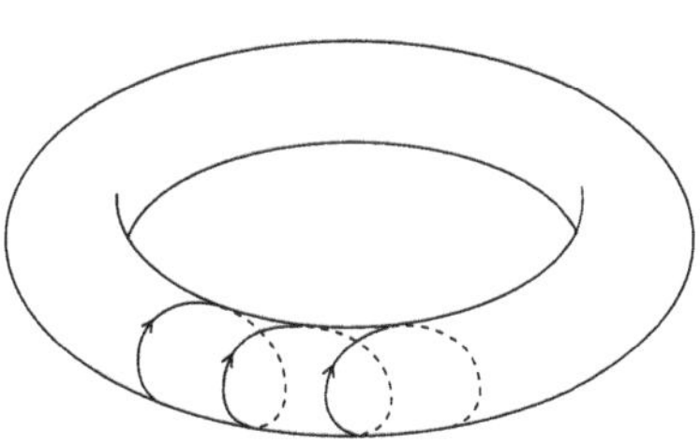

**Fig. 1.5.** Invariant torus.

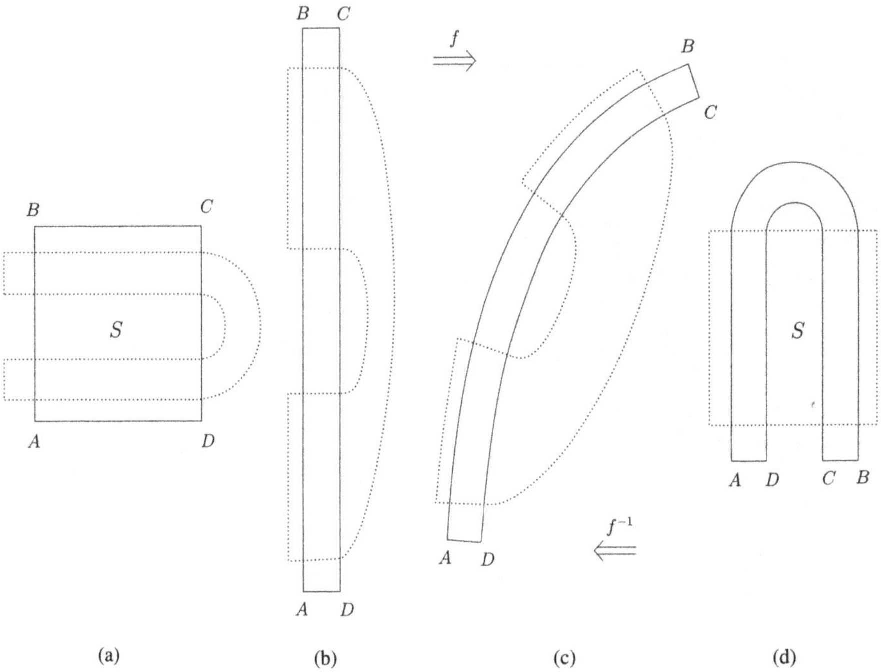


**Fig. 1.6.** Construction of the horseshoe map.

in **Fig. 1.6**(d) into the dotted horizontal horseshoe in **Fig. 1.6**(a), which we assume intersects the original square $S$ along two horizontal rectangles.

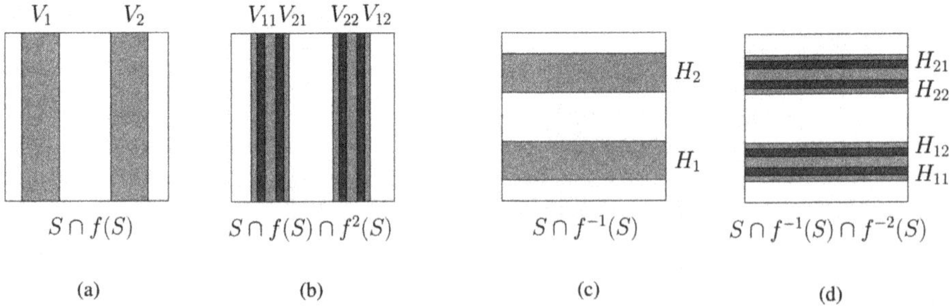


**Fig. 1.7.** Vertical and horizontal strips.

Denote the vertical strips in the intersection $S \cap f(S)$ by $V_1$ and $V_2$,

$$S \cap f(S) = V_1 \cup V_1$$

(see **Fig. 1.7**(a)). Now make the most important step: Perform the *second iteration* of the map $f$. Under this iteration, the vertical strips $V_{1,2}$ will be transformed into two "thin horseshoes" that intersect the square $S$ along four narrow vertical strips: $V_{11}, V_{21}, V_{22}$, and $V_{12}$ (see **Fig. 1.7**(b)). We write this as

$$S \cap f(S) \cap f^2(S) = V_{11} \cup V_{21} \cup V_{22} \cup V_{12}.$$

Similarly,

$$S \cap f^{-1}(S) = H_1 \cup H_2,$$

where $H_{1,2}$ are the horizontal strips shown in **Fig. 1.7**(c), and

$$S \cap f^{-1}(S) \cap f^{-2}(S) = H_{11} \cup H_{12} \cup H_{22} \cup H_{21},$$

with four narrow horizontal strips $H_{ij}$ (**Fig. 1.7**(d)). Notice that $f(H_i) = V_i,\ i = 1,2$, as well as $f^2(H_{ij}) = V_{ij},\ i,j = 1,2$ (**Fig. 1.8**).

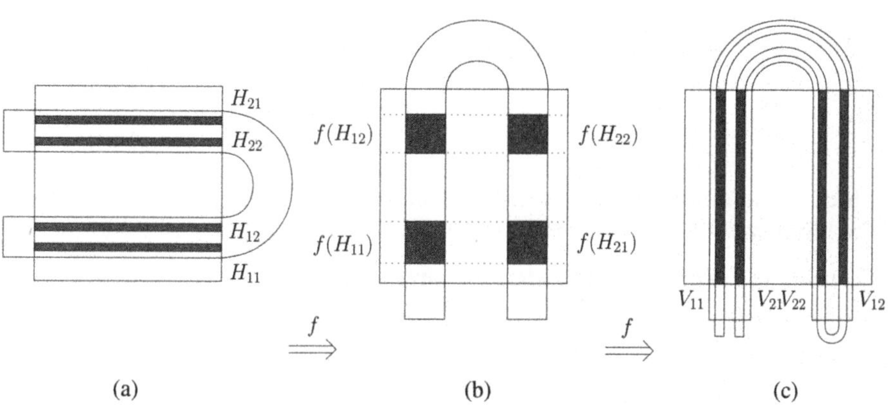


**Fig. 1.8.** Transformation $f^2(H_{ij}) = V_{ij},\ i,j = 1,2.$

Iterating the map $f$ further, we obtain $2^k$ vertical strips in the intersection $S \cap f^k(S),\ k = 1,2,\ldots$. Similarly, iteration of $f^{-1}$ gives $2^k$ horizontal strips in the intersection $S \cap f^{-k}(S),\ k = 1,2,\ldots$.

Most points leave the square $S$ under iteration of $f$ or $f^{-1}$. Forget about such points, and instead consider a set composed of all points in the plane that remain in the square $S$ under all iterations of $f$ and $f^{-1}$:

$$\Lambda = \{x \in S : f^k(x) \in S, \text{ for all } k \in \mathbb{Z}\}.$$

Clearly, if the set $\Lambda$ is nonempty, it is an *invariant set* of the discrete-time dynamical system defined by $f$. This set can be alternatively presented as an infinite intersection,

$$\Lambda = \cdots \cap f^{-k}(S) \cap \cdots \cap f^{-2}(S) \cap f^{-1}(S) \cap S \cap f(S) \cap f^2(S) \cap \cdots f^k(S) \cap \cdots$$

(any point $x \in \Lambda$ must belong to each of the involved sets). It is clear from this representation that the set $\Lambda$ has a peculiar shape. Indeed, it should be located within

$f^{-1}(S) \cap S \cap f(S)$ $f^{-2}(S) \cap f^{-1}(S) \cap S \cap f(S) \cap f^2(S)$

(a) (b)

**Fig. 1.9.** Location of the invariant set.

$$f^{-1}(S) \cap S \cap f(S),$$

which is formed by *four* small squares (see **Fig. 1.9**(a)). Furthermore, it should be located inside

$$f^{-2}(S) \cap f^{-1}(S) \cap S \cap f(S) \cap f^2(S),$$

which is the union of *sixteen* smaller squares (**Fig. 1.9**(b)), and so forth. In the limit, we obtain a *Cantor (fractal) set.*

**Lemma 1.1** *There is a one-to-one correspondence $h : \Lambda \to \Omega_2$, between points of $\Lambda$ and all bi-infinite sequences of two symbols.*

**Proof:**

For any point $x \in \Lambda$, define a sequence of two symbols $\{1, 2\}$

$$\omega = \{\dots, \omega_{-2}, \omega_{-1}, \omega_0, \omega_1, \omega_2, \dots\}$$

by the formula

$$\omega_k = \begin{cases} 1 & \text{if} \quad f^k(x) \in H_1, \\ 2 & \text{if} \quad f^k(x) \in H_2, \end{cases} \tag{1.3}$$

for $k = 0, \pm 1, \pm 2, \dots$. Here, $f^0 = \text{id}$, the identity map. Clearly, this formula defines a map $h : \Lambda \to \Omega_2$, which assigns a sequence to each point of the invariant set.

To verify that this map is invertible, take a sequence $\omega \in \Omega_2$, fix $m > 0$, and consider a set $R_m(\omega)$ of all points $x \in S$, not necessarily belonging to $\Lambda$, such that

$$f^k(x) \in H_{\omega_k},$$

for $-m \le k \le m - 1$. For example, if $m = 1$, the set $R_1$ is one of the four intersections $V_j \cap H_k$. In general, $R_m$ belongs to the intersection of a vertical and a horizontal strip. These strips are getting thinner and thinner as $m \to +\infty$, approaching in the limit a vertical and a horizontal segment, respectively. Such segments obviously intersect at a single point $x$ with $h(x) = \omega$. Thus, $h : \Lambda \to \Omega_2$ is a one-to-one map. It implies that $\Lambda$ is nonempty. $\square$

**Remark:**

The map $h : \Lambda \to \Omega_2$ is continuous together with its inverse (a *homeomorphism*) if we use the standard metric (1.1) in $S \subset \mathbb{R}^2$ and the metric given by (1.2) in $\Omega_2$. $\diamondsuit$

Consider now a point $x \in \Lambda$ and its corresponding sequence $\omega = h(x)$, where $h$ is the map previously constructed. Next, consider a point $y = f(x)$, that is, the image of $x$ under the horseshoe map $f$. Since $y \in \Lambda$ by definition, there is a sequence that corresponds to $y : \theta = h(y)$. Is there a relation between these sequences $\omega$ and $\theta$? As one can easily see from (1.3), such a relation exists and is very simple. Namely,

$$\theta_k = \omega_{k+1}, \quad k \in \mathbb{Z},$$

since $f^k(f(x)) = f^{k+1}(x)$. In other words, the sequence $\theta$ can be obtained from the sequence $\omega$ by the *shift map* $\sigma$, defined in Example 1.8:

$$\theta = \sigma(\omega).$$

Therefore, the restriction of $f$ to its invariant set $\Lambda \subset \mathbb{R}^2$ is *equivalent* to the shift map $\sigma$ on the set of sequences $\Omega_2$. Let us formulate this result as the following short lemma.

**Lemma 1.2** *$h(f(x)) = \sigma(h(x))$, for all $x \in \Lambda$.*

We can write an even shorter one:

$$f|_\Lambda = h^{-1} \circ \sigma \circ h.$$

Combining Lemmas 1.1 and 1.2 with obvious properties of the shift dynamics on $\Omega_2$, we get a theorem giving a rather complete description of the behavior of the horseshoe map.

**Theorem 1.1 (Smale [1963])** *The horseshoe map $f$ has a closed invariant set $\Lambda$ that contains a countable set of periodic orbits of arbitrarily long period, and an uncountable set of nonperiodic orbits, among which there are orbits passing arbitrarily close to any point of $\Lambda$.* $\square$

The dynamics on $\Lambda$ have certain features of "random motion." Indeed, for any sequence of two symbols we generate "randomly," thus prescribing the phase point to visit the horizontal strips $H_1$ and $H_2$ in a certain order, there is an orbit showing this feature among those composing $\Lambda$.

The next important feature of the horseshoe example is that we can slightly perturb the constructed map $f$ without qualitative changes to its dynamics. Clearly, Smale's construction is based on a sufficiently strong contraction/expansion, combined with a folding. Thus, a (smooth) perturbation $\tilde{f}$ will have similar vertical and horizontal strips, which are no longer rectangles but curvilinear regions. However, provided the perturbation is sufficiently small (see the next chapter for precise definitions), these strips will shrink to

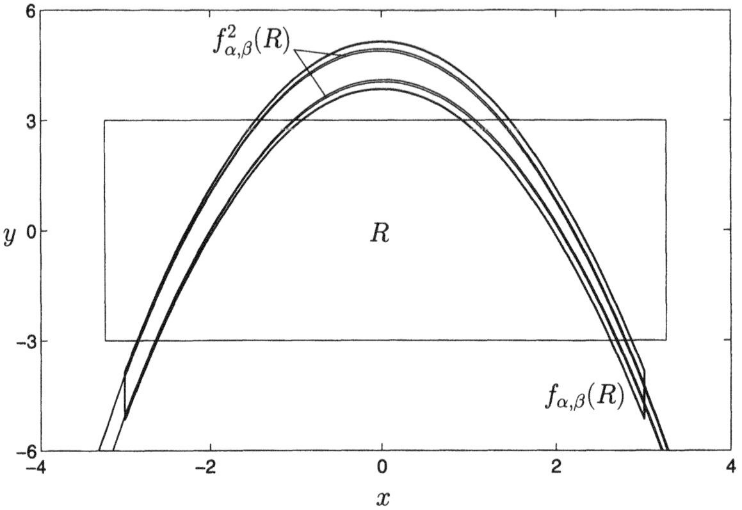


**Fig. 1.10.** Smale horseshoe in the Hénon map.

*curves* that deviate only slightly from vertical and horizontal lines. Thus, the construction can be carried through verbatim, and the perturbed map $\tilde{f}$ will have an invariant set $\tilde{\Lambda}$ on which the dynamics are completely described by the shift map $\sigma$ on the sequence space $\Omega_2$. As we will discuss in Chapter 2, this is an example of *structurally stable* behavior.

**Remark:**
One can precisely specify the contraction/expansion properties required by the horseshoe map in terms of *expanding* and *contracting cones* of the Jacobian matrix $f_x$ (see the literature cited in the bibliographical notes in Appendix B to this chapter). $\diamondsuit$

**Example 1.9 (Hénon map)** Consider the following planar quadratic map depending on two parameters:

$$f_{\alpha,\beta} : \begin{pmatrix} x \\ y \end{pmatrix} \mapsto \begin{pmatrix} y \\ \alpha - \beta x - y^2 \end{pmatrix}. \tag{1.4}$$

An equivalent map was introduced by Hénon [1976] as the simplest map with "random dynamics." The map $f_{\alpha,\beta}$ is invertible if $\beta \neq 0$ and has the essential stretching and folding properties of the horseshoe map in a certain parameter range.

For example, fix $\alpha = 4.5$, $\beta = 0.2$, and consider the first two images $f_{\alpha,\beta}(R)$ and $f^2_{\alpha,\beta}(R)$ of a rectangle $R$, shown in **Fig. 1.10**. The similarity to **Fig. 1.8**(c) is obvious. $\diamondsuit$

### 1.3.3 Stability of invariant sets

To represent an observable asymptotic state of a dynamical system, an invariant set $S_0$ must be stable; in other words, it should "attract" nearby orbits. Suppose we have a dynamical system $\{T, X, \varphi^t\}$ with a complete metric state space $X$. Let $S_0$ be a closed invariant set.

**Definition 1.8** *An invariant set $S_0$ is called* stable *if*

(i) *for any sufficiently small neighborhood $U \supset S_0$ there exists a neighborhood $V \supset S_0$ such that $\varphi^t x \in U$ for all $x \in V$ and all $t > 0$;*
(ii) *there exists a neighborhood $U_0 \supset S_0$ such that $\varphi^t x \to S_0$ for all $x \in U_0$, as $t \to +\infty$.*

If $S_0$ is an equilibrium or a cycle, this definition turns into the standard definition of stable equilibria or cycles. Property (i) of the definition is called

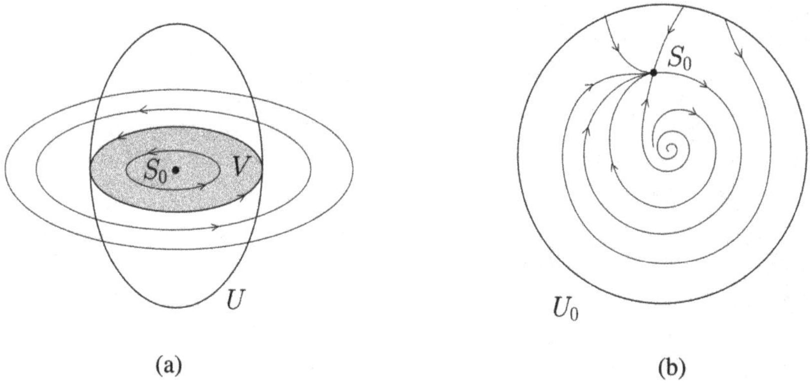


**Fig. 1.11.** (a) Lyapunov versus (b) asymptotic stability.

*Lyapunov stability.* If a set $S_0$ is Lyapunov stable, nearby orbits do not leave its neighborhood. Property (ii) is sometimes called *asymptotic stability.* There are invariant sets that are Lyapunov stable but not asymptotically stable (see **Fig. 1.11**(a)). In contrast, there are invariant sets that are attracting but not Lyapunov stable since some orbits starting near $S_0$ eventually approach $S_0$, but only after an excursion outside a small but fixed neighborhood of this set (see **Fig. 1.11**(b)).

If $x^0$ is a fixed point of a finite-dimensional, smooth, discrete-time dynamical system, then sufficient conditions for its stability can be formulated in terms of the Jacobian matrix evaluated at $x^0$.

**Theorem 1.2** *Consider a discrete-time dynamical system*

$$x \mapsto f(x), \quad x \in \mathbb{R}^n,$$

*where $f$ is a smooth map. Suppose it has a fixed point $x^0$, namely $f(x^0) = x^0$, and denote by $A$ the Jacobian matrix of $f(x)$ evaluated at $x^0, A = f_x(x^0)$. Then the fixed point is stable if all eigenvalues $\mu_1, \mu_2, \dots, \mu_n$ of $A$ satisfy $|\mu| < 1$.* $\square$

Recall that the eigenvalues are roots of the *characteristic equation*

$$\det(A - \mu I_n) = 0,$$

where $I_n$ is the $n \times n$ unit matrix.

The eigenvalues of a fixed point are usually called *multipliers*. In the linear case the theorem is obvious from the Jordan normal form. Theorem 1.2, being applied to the $N_0$th iterate $f^{N_0}$ of the map $f$ at any point of the periodic orbit, also gives a sufficient condition for the stability of an $N_0$-cycle.

Another important case where we can establish the stability of a fixed point of a discrete-time dynamical system is provided by the following theorem.

**Theorem 1.3 (Contraction Mapping Principle)** *Let $X$ be a complete metric space with distance defined by $\rho$. Assume that there is a map $f : X \to X$ that is continuous and that satisfies, for all $x, y \in X$,*

$$\rho(f(x), f(y)) \leq \lambda \rho(x, y),$$

*with some $0 < \lambda < 1$. Then the discrete-time dynamical system $\{\mathbb{Z}_+, X, f^k\}$ has a stable fixed point $x^0 \in X$. Moreover, $f^k(x) \to x^0$ as $k \to +\infty$, starting from any point $x \in X$.* $\square$

The proof of this fundamental theorem can be found in any text on mathematical analysis or differential equations. Notice that there is no restriction on the dimension of the space $X$: It can be, for example, an infinite-dimensional function space. Another important difference from Theorem 1.2 is that Theorem 1.3 *guarantees* the existence and uniqueness of the fixed point $x^0$, whereas this has to be *assumed* in Theorem 1.2. Actually, the map $f$ from Theorem 1.2 is a contraction near $x^0$, provided an appropriate metric (norm) in $\mathbb{R}^n$ is introduced. The Contraction Mapping Principle is a powerful tool: Using this principle, we can prove the Implicit Function Theorem, the Inverse Function Theorem, as well as Theorem 1.4 ahead. We will apply the Contraction Mapping Principle in Chapter 4 to prove the existence, uniqueness, and stability of a closed invariant curve that appears under parameter variation from a fixed point of a generic planar map. Notice also that Theorem 1.3 gives *global* asymptotic stability: Any orbit of $\{\mathbb{Z}_+, X, f^k\}$ converges to $x^0$.

Finally, let us point out that the invariant set $\Lambda$ of the horseshoe map is *not* stable. However, there are invariant fractal sets that are stable. Such objects are called *strange attractors*.

## 1.4 Differential equations and dynamical systems

The most common way to define a continuous-time dynamical system is by *differential equations*. Suppose that the state space of a system is $X = \mathbb{R}^n$

with coordinates $(x_1, x_2, \ldots, x_n)$. If the system is defined on a manifold, these can be considered as local coordinates on it. Very often the law of evolution of the system is given implicitly, in terms of the velocities $\dot{x}_i$ as functions of the coordinates $(x_1, x_2, \ldots, x_n)$:

$$\dot{x}_i = f_i(x_1, x_2, \ldots, x_n), \quad i = 1, 2, \ldots, n,$$

or in the vector form

$$\dot{x} = f(x), \tag{1.5}$$

where the vector-valued function $f : \mathbb{R}^n \to \mathbb{R}^n$ is supposed to be sufficiently differentiable (smooth). The function in the right-hand side of (1.5) is referred to as a *vector field* since it assigns a vector $f(x)$ to each point $x$. Equation (1.5) represents a system of $n$ *autonomous ordinary differential equations*, ODEs for short. Let us revisit some of the examples introduced earlier by presenting differential equations governing the evolution of the corresponding systems.

**Example 1.1 (revisited)** The dynamics of an ideal pendulum are described by Newton's second law,

$$\ddot{\varphi} = -k^2 \sin \varphi,$$

with

$$k^2 = \frac{g}{l},$$

where $l$ is the pendulum length, and $g$ is the gravity acceleration constant. If we introduce $\psi = \dot{\varphi}$, so that $(\varphi, \psi)$ represents a point in the state space $X = \mathbb{S}^1 \times \mathbb{R}^1$, the above differential equation can be rewritten in the form of equation (1.5):

$$\begin{cases} \dot{\varphi} = \psi, \\ \dot{\psi} = -k^2 \sin \varphi. \end{cases} \tag{1.6}$$

Here

$$x = \begin{pmatrix} \varphi \\ \psi \end{pmatrix},$$

while

$$f\begin{pmatrix} \varphi \\ \psi \end{pmatrix} = \begin{pmatrix} \psi \\ -k^2 \sin \varphi \end{pmatrix}. \ \diamond$$

**Example 1.2 (revisited)** The behavior of an isolated energy-conserving mechanical system with $s$ degrees of freedom is determined by $2s$ *Hamiltonian equations*:

$$\dot{q}_i = \frac{\partial H}{\partial p_i}, \quad \dot{p}_i = -\frac{\partial H}{\partial q_i}, \tag{1.7}$$

for $i = 1, 2, \ldots, s$. Here the scalar function $H = H(q, p)$ is the *Hamilton function*. The equations of motion of the pendulum (1.6) are Hamiltonian equations with $(q, p) = (\varphi, \psi)$ and

$$H(\varphi, \psi) = \frac{\psi^2}{2} - k^2 \cos \varphi. \ \Diamond$$

**Example 1.3 (revisited)** The behavior of a quantum system with two states having different energies can be described between "observations" by the *Heisenberg equation,*

$$i\hbar \frac{d\psi}{dt} = H\psi,$$

where $i^2 = -1$,

$$\psi = \begin{pmatrix} a_1 \\ a_2 \end{pmatrix}, \quad a_i \in \mathbb{C}^1.$$

The symmetric real matrix

$$H = \begin{pmatrix} E_0 & -A \\ -A & E_0 \end{pmatrix}, \quad E_0, A > 0,$$

is the Hamiltonian matrix of the system, and $\hbar$ is Plank's constant divided by $2\pi$. The Heisenberg equation can be written as the following system of two linear *complex* equations for the amplitudes

$$\begin{cases} \dot{a}_1 = \dfrac{1}{i\hbar}(E_0 a_1 - A a_2), \\ \dot{a}_2 = \dfrac{1}{i\hbar}(-A a_1 + E_0 a_2). \ \Diamond \end{cases} \tag{1.8}$$

**Example 1.4 (revisited)** As an example of a chemical system, let us consider the *Brusselator* [Lefever & Prigogine 1968]. This hypothetical system is composed of substances that react through the following irreversible stages:

$$\begin{aligned} A &\xrightarrow{k_1} X \\ B + X &\xrightarrow{k_2} Y + D \\ 2X + Y &\xrightarrow{k_3} 3X \\ X &\xrightarrow{k_4} E. \end{aligned}$$

Here capital letters denote reagents, while the constants $k_i$ over the arrows indicate the corresponding reaction rates. The substances $D$ and $E$ do not re-enter the reaction, while $A$ and $B$ are assumed to remain constant. Thus, the *law of mass action* gives the following system of two nonlinear equations for the concentrations $[X]$ and $[Y]$:

$$\begin{cases} \dfrac{d[X]}{dt} = k_1[A] - k_2[B][X] - k_4[X] + k_3[X]^2[Y], \\ \dfrac{d[Y]}{dt} = k_2[B][X] - k_3[X]^2[Y]. \end{cases}$$

Linear scaling of the variables and time yields the *Brusselator equations*,

$$\begin{cases} \dot{x} = a - (b+1)x + x^2 y, \\ \dot{y} = bx - x^2 y. \end{cases} \quad \Diamond \tag{1.9}$$

**Example 1.5 (revisited)** One of the earliest models of ecosystems was the system of two nonlinear differential equations proposed by Volterra [1931]:

$$\begin{cases} \dot{N}_1 = \alpha N_1 - \beta N_1 N_2, \\ \dot{N}_2 = -\gamma N_2 + \delta N_1 N_2. \end{cases} \tag{1.10}$$

Here $N_1$ and $N_2$ are the numbers of prey and predators, respectively, in an ecological community, $\alpha$ is the prey growth rate, $\gamma$ is the predator mortality, while $\beta$ and $\delta$ describe the predators' efficiency of consumption of the prey. $\Diamond$

Under very general conditions, solutions of ODEs define smooth continuous-time dynamical systems. Few types of differential equations can be solved analytically (in terms of elementary functions). However, for smooth right-hand sides, the solutions are guaranteed to exist according to the following theorem. This theorem can be found in any textbook on ordinary differential equations. We formulate it without proof.

**Theorem 1.4 (Existence, uniqueness, and smooth dependence)** *Consider a system of ordinary differential equations*

$$\dot{x} = f(x), \quad x \in \mathbb{R}^n,$$

*where $f : \mathbb{R}^n \to \mathbb{R}^n$ is smooth in an open region $U \subset \mathbb{R}^n$. Then there is a unique function $x = x(t, x_0)$, $x : \mathbb{R}^1 \times \mathbb{R}^n \to \mathbb{R}^n$, that is smooth in $(t, x_0)$, and satisfies, for each $x_0 \in U$, the following conditions:*

(i) $x(0, x_0) = x_0$;

(ii) *there is an interval $\mathcal{J} = (-\delta_1, \delta_2)$, where $\delta_{1,2} = \delta_{1,2}(x_0) > 0$, such that, for all $t \in \mathcal{J}$,*

$$y(t) = x(t, x_0) \in U,$$

*and*

$$\dot{y}(t) = f(y(t)). \ \square$$

The degree of smoothness of $x(t, x_0)$ with respect to $x_0$ in Theorem 1.4 is the same as that of $f$ as a function of $x$. The function $x = x(t, x_0)$, considered as a function of time $t$, is called a *solution starting at* $x_0$. It defines, for each $x_0 \in U$, two objects: a *solution curve*

$$Cr(x_0) = \{(t, x) : x = x(t, x_0), t \in \mathcal{J}\} \subset \mathbb{R}^1 \times \mathbb{R}^n$$

and an *orbit*, which is the projection of $Cr(x_0)$ onto the state space,

$$Or(x_0) = \{x : x = x(t, x_0), t \in \mathcal{J}\} \subset \mathbb{R}^n$$

(see **Fig. 1.12**). Both curves are parametrized by time $t$ and oriented by the direction of time advance. A nonzero vector $f(x_0)$ is tangent to the orbit $Or(x_0)$ at $x_0$. There is a *unique* orbit passing through a point $x_0 \in U$.

Under the conditions of the theorem, the orbit either leaves $U$ at $t = -\delta_1$ (and/or $t = \delta_2$), or stays in $U$ forever; in the latter case, we can take $\mathcal{J} = (-\infty, +\infty)$.

Now we can define the evolution operator $\varphi^t : \mathbb{R}^n \to \mathbb{R}^n$ by the formula

$$\varphi^t x_0 = x(t, x_0),$$

which assigns to $x_0$ a point on the orbit through $x_0$ that is passed $t$ time units later. Obviously, $\{\mathbb{R}^1, \mathbb{R}^n, \varphi^t\}$ is a continuous-time dynamical system (check!). This system is *invertible*. Each evolution operator $\varphi^t$ is defined for $x \in U$ and $t \in \mathcal{J}$, where $\mathcal{J}$ depends on $x_0$ and is smooth in $x$. In practice, the evolution operator $\varphi^t$ corresponding to a smooth system of ODEs can be found numerically on fixed time intervals to within desired accuracy. One of the standard ODE solvers can be used to accomplish this.

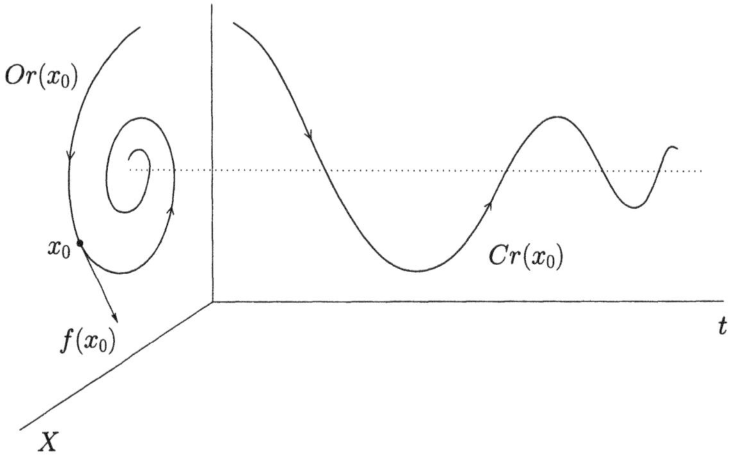


**Fig. 1.12.** Solution curve and orbit.

One of the major tasks of dynamical systems theory is to analyze the behavior of a dynamical system defined by ODEs. Of course, one might try to solve this problem by "brute force," merely computing many orbits numerically (by "simulations"). However, the most useful aspect of the theory is that we can predict some features of the phase portrait of a system defined by ODEs *without* actually solving the system. The simplest example of such information is the number and positions of equilibria. Indeed, the equilibria of a system defined by (1.5) are zeros of the vector field given by its right-hand side:

$$f(x) = 0. \tag{1.11}$$

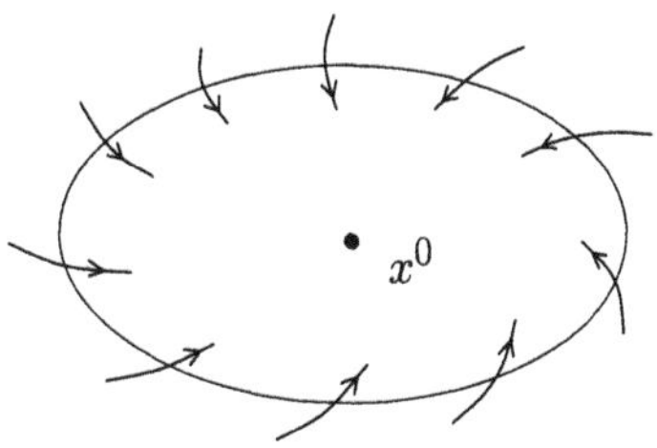

**Fig. 1.13.** Lyapunov function.

Clearly, if $f(x^0) = 0$, then $\varphi^t x_0 = x_0$ for all $t \in \mathbb{R}^1$. The stability of an equilibrium can also be detected without solving the system. For example, sufficient conditions for an equilibrium $x^0$ to be stable are provided by the following classical theorem.

**Theorem 1.5 (Lyapunov [1892])** *Consider a dynamical system defined by*

$$\dot{x} = f(x), \quad x \in \mathbb{R}^n,$$

*where $f$ is smooth. Suppose that it has an equilibrium $x^0$ (i.e., $f(x^0) = 0$), and denote by $A$ the Jacobian matrix of $f(x)$ evaluated at the equilibrium, $A = f_x(x^0)$. Then $x^0$ is stable if all eigenvalues $\lambda_1, \lambda_2, \dots, \lambda_n$ of $A$ satisfy* Re $\lambda < 0$. □

The theorem can easily be proved for a linear system

$$\dot{x} = Ax, \quad x \in \mathbb{R}^n,$$

by its explicit solution in a basis where $A$ has Jordan normal form, as well as for a general nonlinear system by constructing a *Lyapunov function* $L(x)$ near the equilibrium. More precisely, by a shift of coordinates, one can place the equilibrium at the origin, $x^0 = 0$, and find a certain quadratic form $L(x)$ whose level surfaces $L(x) = L_0$ surround the origin and are such that the vector field points strictly inside each level surface, sufficiently close to the equilibrium $x^0$ (see **Fig. 1.13**). Actually, the Lyapunov function $L(x)$ is the same for both linear and nonlinear systems and is fully determined by the Jacobian matrix $A$. The details can be found in any standard text on differential equations (see the bibliographical notes in Appendix B). Note also that the theorem can also be derived from Theorem 1.2 (see Exercise 7).

Unfortunately, in general it is impossible to tell by looking at the right-hand side of (1.5), whether this system has cycles (periodic solutions). Later on in the book we will formulate some efficient methods to prove the appearance of cycles under small perturbation of the system (e.g., by variation of parameters on which the system depends).

If the system has a smooth invariant manifold $M$, then its defining vector field $f(x)$ is *tangent* to $M$ at any point $x \in M$, where $f(x) \neq 0$. For an $(n-1)$-dimensional smooth manifold $M \subset \mathbb{R}^n$, which is locally defined by $g(x) = 0$ for some scalar function $g : \mathbb{R}^n \to \mathbb{R}^1$, the invariance means

$$\langle \nabla g(x), f(x) \rangle = 0.$$

Here $\nabla g(x)$ denotes the *gradient*

$$\nabla g(x) = \left( \frac{\partial g(x)}{\partial x_1}, \frac{\partial g(x)}{\partial x_2}, \ldots, \frac{\partial g(x)}{\partial x_n} \right)^T,$$

which is orthogonal to $M$ at $x$.

## 1.5 Poincaré maps

There are many cases where discrete-time dynamical systems (maps) naturally appear in the study of continuous-time dynamical systems defined by differential equations. The introduction of such maps allows us to apply the results concerning maps to differential equations. This is particularly efficient if the resulting map is defined in a lower-dimensional space than the original system. We will call maps arising from ODEs *Poincaré maps.*

### 1.5.1 Time-shift maps

The simplest way to extract a discrete-time dynamical system from a continuous-time system $\{\mathbb{R}^1, X, \varphi^t\}$ is to fix some $T_0 > 0$ and consider a system on $X$ that is generated by iteration of the map $f = \varphi^{T_0}$. This map is called a $T_0$-*shift map* along orbits of $\{\mathbb{R}^1, X, \varphi^t\}$. Any invariant set of $\{\mathbb{R}^1, X, \varphi^t\}$ is an invariant set of the map $f$. For example, isolated fixed points of $f$ are located at those positions where $\{\mathbb{R}^1, X, \varphi^t\}$ has isolated equilibria.

In this context, the *inverse* problem is more interesting: Is it possible to construct a system of ODEs whose $T_0$-shift map $\varphi^{T_0}$ reproduces a given smooth and invertible map $f$? If we require the discrete-time system to have the same dimension as the continuous-time one, the answer is negative. The simplest counterexample is provided by the linear scalar map

$$x \mapsto f(x) = -\frac{1}{2}x, \quad x \in \mathbb{R}^1. \tag{1.12}$$

The map in (1.12) has a single fixed point $x^0 = 0$ that is stable. Clearly, there is no scalar ODE

$$\dot{x} = F(x), \quad x \in \mathbb{R}^1, \tag{1.13}$$

such that its evolution operator $\varphi^{T_0} = f$. Indeed, $x^0 = 0$ must be an equilibrium of (1.13), thus none of its orbits can "jump" over the origin like those of (1.12). We will return to this inverse problem in Chapter 9, where we explicitly construct ODE systems approximating certain maps.

**Remark:**

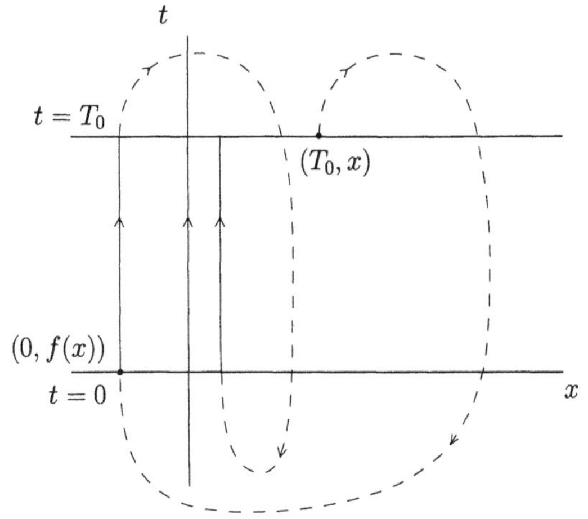


**Fig. 1.14.** Suspension flow.

If we allow for ODEs on *manifolds*, the inverse problem can always be solved. Specifically, consider a map $f : \mathbb{R}^n \to \mathbb{R}^n$ that is assumed to be smooth, together with its inverse. Take a layer

$$\{(t, x) \in \mathbb{R}^1 \times \mathbb{R}^n : t \in [0, T_0]\}$$

(see **Fig. 1.14**) and identify ("glue") a point $(T_0, x)$ on the "top" face of $X$ with the point $(0, f(x))$ on the "bottom" face. Thus, the constructed space $X$ is an $(n+1)$-dimensional manifold with coordinates $(t \bmod T_0, x)$. Specify now an autonomous system of ODEs on this manifold, called the *suspension*, by the equations

$$\begin{cases} \dot{t} = 1, \\ \dot{x} = 0. \end{cases} \tag{1.14}$$

The orbits of (1.14) (viewed as subsets of $\mathbb{R}^1 \times \mathbb{R}^n$) are straight lines inside the layer interrupted by "jumps" from its "top" face to the "bottom" face. Obviously, the $T_0$-shift along orbits of (1.14) $\varphi^{T_0}$ coincides on its invariant hyperplane $\{t = 0\}$ with the map $f$.

Let $k > 0$ satisfy the equation $e^{kT_0} = 2$. The suspension system corresponding to the map (1.12) has the same orbit structure as the system

$$\begin{cases} \dot{t} = 1, \\ \dot{x} = -kx, \end{cases}$$

defined on an (infinitely wide) *Möbius strip* obtained by identifying the points $(T_0, x)$ and $(0, -x)$ (see **Fig. 1.15**). In both systems, $x = 0$ corresponds to a stable limit cycle of period $T_0$ with the multiplier $\mu = -\frac{1}{2}$. $\diamondsuit$

### 1.5.2 Poincaré map and stability of cycles

Consider a continuous-time dynamical system defined by

$$\dot{x} = f(x), \quad x \in \mathbb{R}^n, \tag{1.15}$$

with smooth $f$. Assume, that (1.15) has a periodic orbit $L_0$. Take a point $x_0 \in L_0$ and introduce a *cross-section* $\Sigma$ to the cycle at this point (see **Fig. 1.16**). The cross-section $\Sigma$ is a smooth hypersurface of dimension $n-1$, intersecting $L_0$ at a nonzero angle. Since the dimension of $\Sigma$ is one less than the dimension of the state space, we say that the hypersurface $\Sigma$ is of "codimension" one, codim $\Sigma = 1$. Suppose that $\Sigma$ is defined near the point $x_0$ by the zero-level set of a smooth scalar function $g : \mathbb{R}^n \to \mathbb{R}^1$, $g(x_0) = 0$,

$$\Sigma = \{x \in \mathbb{R}^n : g(x) = 0\}.$$

A nonzero intersection angle ("transversality") means that the gradient

$$\nabla g(x) = \left(\frac{\partial g(x)}{\partial x_1}, \frac{\partial g(x)}{\partial x_2}, \ldots, \frac{\partial g(x)}{\partial x_n}\right)^T$$

is not orthogonal to $L_0$ at $x_0$, that is,

$$\langle \nabla g(x_0), f(x_0) \rangle \neq 0,$$

where $\langle \cdot, \cdot \rangle$ is the standard scalar product in $\mathbb{R}^n$. The simplest choice of $\Sigma$ is a hyperplane orthogonal to the cycle $L_0$ at $x_0$. Such a cross-section is obviously given by the zero-level set of the linear function

$$g(x) = \langle f(x_0), x - x_0 \rangle.$$

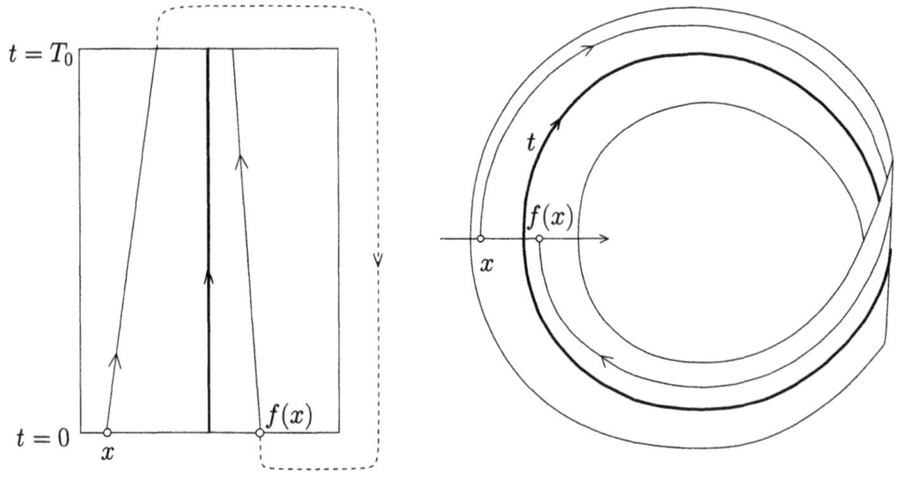


**Fig. 1.15.** Stable limit cycle on the Möbius strip.

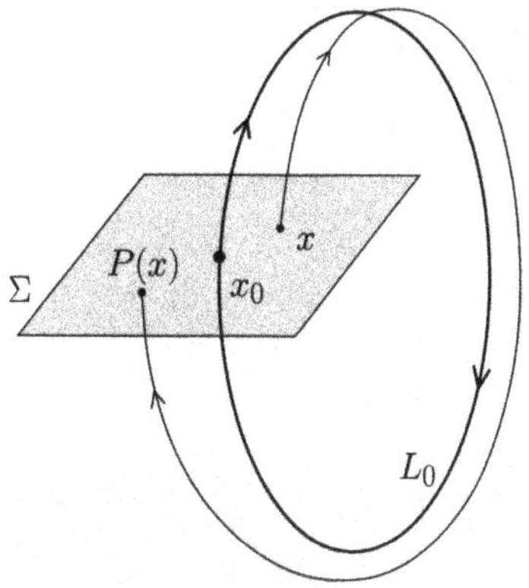


**Fig. 1.16.** The Poincaré map associated with a cycle.

Consider now orbits of (1.15) near the cycle $L_0$. The cycle itself is an orbit that starts at a point on $\Sigma$ and returns to $\Sigma$ at the same point ($x_0 \in \Sigma$). Since the solutions of (1.15) depend smoothly on their initial points (Theorem 1.4), an orbit starting at a point $x \in \Sigma$ sufficiently close to $x_0$ also returns to $\Sigma$ at some point $\tilde{x} \in \Sigma$ near $x_0$. Moreover, nearby orbits will also intersect $\Sigma$ transversally. Thus, a map $P : \Sigma \to \Sigma$,

$$x \mapsto \tilde{x} = P(x),$$

is constructed.

**Definition 1.9** *The map $P$ is called a* Poincaré map *associated with the cycle $L_0$.*

The Poincaré map $P$ is locally defined, is as smooth as the right-hand side of (1.15), and is invertible near $x_0$. The invertibility follows from the invertibility of the dynamical system defined by (1.15). The inverse map $P^{-1} : \Sigma \to \Sigma$ can be constructed by extending the orbits crossing $\Sigma$ backward in time until reaching their previous intersection with the cross-section. The intersection point $x_0$ is a *fixed point* of the Poincaré map: $P(x_0) = x_0$.

Let us introduce local coordinates $\xi = (\xi_1, \xi_2, \ldots, \xi_{n-1})$ on $\Sigma$ such that $\xi = 0$ corresponds to $x_0$. Then the Poincaré map will be characterized by a locally defined map $P : \mathbb{R}^{n-1} \to \mathbb{R}^{n-1}$, which transforms $\xi$ corresponding to $x$ into $\tilde{\xi}$ corresponding to $\tilde{x}$,

$$P(\xi) = \tilde{\xi}.$$

The origin $\xi = 0$ of $\mathbb{R}^{n-1}$ is a *fixed point* of the map $P : P(0) = 0$. The stability of the cycle $L_0$ is equivalent to the stability of the fixed point $\xi_0 = 0$ of the Poincaré map. Thus, the cycle is stable if all eigenvalues (multipliers) $\mu_1, \mu_2, \ldots, \mu_{n-1}$ of the $(n-1) \times (n-1)$ Jacobian matrix of $P$,

$$A = \left. \frac{dP}{d\xi} \right|_{\xi=0},$$

are located inside the unit circle $|\mu| = 1$ (see Theorem 1.2).

One may ask whether the multipliers depend on the choice of the point $x_0$ on $L_0$, the cross-section $\Sigma$, or the coordinates $\xi$ on it. If this were the case, determining stability using multipliers would be confusing or even impossible.

**Lemma 1.3** *The multipliers* $\mu_1, \mu_2, \dots, \mu_{n-1}$ *of the Jacobian matrix* $A$ *of the Poincaré map* $P$ *associated with a cycle* $L_0$ *are independent of the point* $x_0$ *on* $L_0$*, the cross-section* $\Sigma$*, and local coordinates on it.*

**Proof:**

Let $\Sigma_1$ and $\Sigma_2$ be two cross-sections to the same cycle $L_0$ at points $x^1$ and $x^2$, respectively (see **Fig. 1.17**, where the planar case is presented for simplicity). We allow the points $x^{1,2}$ to coincide, and we let the cross-sections $\Sigma_{1,2}$ represent identical surfaces in $\mathbb{R}^n$ that differ only in parametrization. Denote by $P_1 : \Sigma_1 \to \Sigma_1$ and $P_2 : \Sigma_2 \to \Sigma_2$ corresponding Poincaré maps. Let $\xi = (\xi_1, \xi_2, \dots, \xi_{n-1})$ be coordinates on $\Sigma_1$, and let $\eta = (\eta_1, \eta_2, \dots, \eta_{n-1})$ be coordinates on $\Sigma_2$, such that $\xi = 0$ corresponds to $x^1$ while $\eta = 0$ gives $x^2$. Finally, denote by $A_1$ and $A_2$ the associated Jacobian matrices of $P_1$ and $P_2$, respectively.

Due to the same arguments as those we used to construct the Poincaré map, there exists a locally defined, smooth, and invertible *correspondence map* $Q : \Sigma_1 \to \Sigma_2$ along orbits of (1.15):

$$\eta = Q(\xi).$$

Obviously, we have

$$P_2 \circ Q = Q \circ P_1,$$

or, in coordinates,

$$P_2(Q(\xi)) = Q(P_1(\xi)),$$

for all sufficiently small $\|\xi\|$ (see **Fig. 1.17**). Since $Q$ is invertible, we obtain the following relation between $P_1$ and $P_2$:

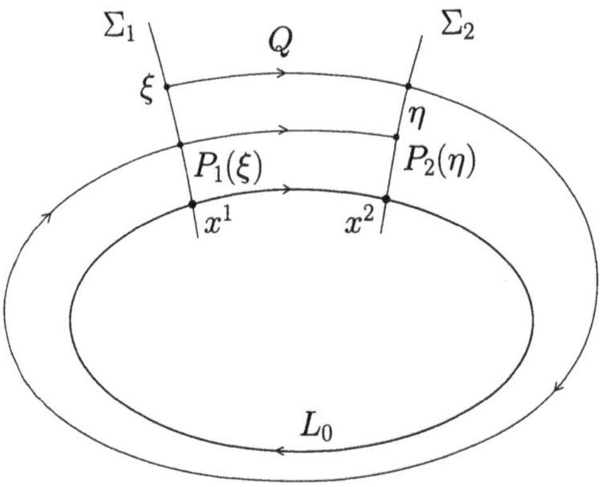


**Fig. 1.17.** Two cross-sections to the cycle $L_0$.

$$P_1 = Q^{-1} \circ P_2 \circ Q.$$

Differentiating this equation with respect to $\xi$, and using the chain rule, we find

$$\frac{dP_1}{d\xi} = \frac{dQ^{-1}}{d\eta}\frac{dP_2}{d\eta}\frac{dQ}{d\xi}.$$

Evaluating the result at $\xi = 0$ gives the matrix equation

$$A_1 = B^{-1}A_2B,$$

where

$$B = \left.\frac{dQ}{d\xi}\right|_{\xi=0}$$

is nonsingular (i.e., $\det B \neq 0$). Thus, the characteristic equations for $A_1$ and $A_2$ coincide, as do the multipliers. Indeed,

$$\det(A_1 - \mu I_n) = \det(B^{-1})\det(A_2 - \mu I_n)\det(B) = \det(A_2 - \mu I_n)$$

since the determinant of the matrix product is equal to the product of the the determinants of the matrices involved, and $\det(B^{-1})\det(B) = 1$. □

According to Lemma 1.3, we can use *any* cross-section $\Sigma$ to compute the multipliers of the cycle: The result will be the same.

The next problem to be addressed is the relationship between the multipliers of a cycle and the differential equations (1.15) defining the dynamical system that has this cycle. Let $x^0(t)$ denote a periodic solution of (1.15), $x^0(t+T_0) = x^0(t)$, corresponding to a cycle $L_0$. Represent a solution of (1.15) in the form

$$x(t) = x^0(t) + u(t),$$

where $u(t)$ is a deviation from the periodic solution. Then,

$$\dot{u}(t) = \dot{x}(t) - \dot{x}^0(t) = f(x^0(t) + u(t)) - f(x^0(t)) = A(t)u(t) + O(\|u(t)\|^2).$$

Truncating $O(\|u\|^2)$ terms results in the linear $T_0$-periodic system

$$\dot{u} = A(t)u, \quad u \in \mathbb{R}^n, \tag{1.16}$$

where $A(t) = f_x(x^0(t)),\ A(t+T_0) = A(t)$.

**Definition 1.10** *System* (1.16) *is called the* variational equation *about the cycle* $L_0$.

The variational equation is the main (linear) part of the system governing the evolution of *perturbations* near the cycle. Naturally, the stability of the cycle depends on the properties of the variational equation.

**Definition 1.11** *The time-dependent matrix* $M(t)$ *is called the* fundamental matrix solution *of* (1.15) *if it satisfies*

$$\dot{M} = A(t)M,$$

*with the initial condition* $M(0) = I_n$*, the unit* $n \times n$ *matrix.*

Any solution $u(t)$ to (1.16) satisfies

$$u(T_0) = M(T_0)u(0)$$

(prove!). The matrix $M(T_0)$ is called a *monodromy matrix* of the cycle $L_0$. The following *Liouville formula* expresses the determinant of the monodromy matrix in terms of the matrix $A(t)$:

$$\det M(T_0) = \exp\left\{\int_0^{T_0} \operatorname{tr} A(t)\, dt\right\}. \tag{1.17}$$

**Theorem 1.6** *The monodromy matrix* $M(T_0)$ *has eigenvalues*

$$1, \mu_1, \mu_2, \ldots, \mu_{n-1},$$

*where* $\mu_i$ *are the multipliers of the Poincaré map associated with the cycle* $L_0$.

**Sketch of the proof:**

Let $\varphi^t$ be the evolution operator (flow) defined by system (1.15) near the cycle $L_0$. Consider the map

$$\varphi^{T_0} : \mathbb{R}^n \to \mathbb{R}^n.$$

Clearly, $\varphi^{T_0}x_0 = x_0$, where $x_0$ is an initial point on the cycle, which we assume to be located at the origin, $x_0 = 0$. The map is smooth, and its Jacobian matrix at $x_0$ coincides with the monodromy matrix:

$$\left.\frac{\partial \varphi^{T_0} x}{\partial x}\right|_{x=x_0} = M(T_0).$$

We claim that the matrix $M(T_0)$ has an eigenvalue $\mu_0 = 1$. Indeed, $v(t) = \dot{x}^0(t)$ is a solution to (1.16). Therefore, $q = v(0) = f(x_0)$ is transformed by $M(T_0)$ into itself:

$$M(T_0)q = q.$$

If no generalized eigenvectors are associated to $q$, then the monodromy matrix $M(T_0)$ has a one-dimensional invariant subspace spanned by $q$ and a complementary $(n-1)$-dimensional invariant subspace $\Sigma : M(T_0)\Sigma = \Sigma$. Take the subspace $\Sigma$ as a cross-section to the cycle at $x_0 = 0$. One can see that the restriction of the linear transformation defined by $M(T_0)$ to $\Sigma$ is the linearization of the Poincaré map $P$ defined by system (1.15) on $\Sigma$. Therefore, their eigenvalues $\mu_1, \mu_2, \ldots, \mu_{n-1}$ coincide.

If generalized eigenvectors are associated to $q$, the theorem remains valid, however, the proof becomes more involved and is omitted here. □

According to (1.17), the product of all eigenvalues of $M(T_0)$ can be expressed as

$$\mu_1\mu_2\cdots\mu_{n-1} = \exp\left\{\int_0^{T_0} (\text{div } f)(x^0(t))\, dt\right\}, \tag{1.18}$$

where, by definition, the *divergence* of a vector field $f(x)$ is given by

$$(\text{div } f)(x) = \sum_{i=1}^{n} \frac{\partial f_i(x)}{\partial x_i}.$$

Thus, the product of all multipliers of any cycle is *positive*. Notice that in the planar case ($n = 2$) formula (1.18) allows us to compute the only multiplier $\mu_1$, provided the periodic solution corresponding to the cycle is known explicitly. However, this is mainly a theoretical tool since periodic solutions of nonlinear systems are rarely known analytically.

### 1.5.3 Poincaré map for periodically forced systems

In several applications the behavior of a system subjected to an external periodic forcing is described by *time-periodic* differential equations

$$\dot{x} = f(t, x), \quad (t, x) \in \mathbb{R}^1 \times \mathbb{R}^n, \tag{1.19}$$

where $f(t + T_0, x) = f(t, x)$. System (1.19) defines an autonomous system on the cylindrical manifold $X = \mathbb{S}^1 \times \mathbb{R}^n$, with coordinates $(t(\text{mod } T_0), x)$, namely

$$\begin{cases} \dot{t} = 1, \\ \dot{x} = f(t, x). \end{cases} \tag{1.20}$$

In this space $X$, take the $n$-dimensional cross-section $\Sigma = \{(x, t) \in X : t = 0\}$. We can use $x^T = (x_1, x_2, \dots, x_n)$ as coordinates on $\Sigma$. Clearly, all orbits of (1.20) intersect $\Sigma$ transversally. Assuming that the solution $x(t, x_0)$ of (1.20) exists on the interval $t \in [0, T_0]$, we can introduce the *Poincaré map*

$$x_0 \mapsto P(x_0) = x(T_0, x_0).$$

In other words, we have to take an initial point $x_0$ and integrate system (1.19) over its period $T_0$ to obtain $P(x_0)$. By this construction, the discrete-time dynamical system $\{\mathbb{Z}, \mathbb{R}^n, P^k\}$ is defined. Fixed points of $P$ obviously correspond to $T_0$-periodic solutions of (1.19). An $N_0$-cycle of $P$ represents an $N_0T_0$-periodic solution (*subharmonic*) of (1.19). The stability of these periodic solutions is clearly determined by that of the corresponding fixed points and cycles. More complicated solutions of (1.19) can also be studied via the Poincaré map. In Chapter 9 we will analyze in detail a model of a periodically (seasonally) forced predator-prey system exhibiting various subharmonic and chaotic solutions.

## 1.6 Exercises

**(1) (Symbolic dynamics and the Smale horseshoe revisited)**

(a) Compute the number $N(k)$ of period-$k$ cycles in the symbolic dynamics $\{\mathbb{Z}, \Omega_2, \sigma^k\}$.

(b) Explain how to find the coordinates of the two fixed points of the horseshoe map $f$ in $S$. Prove that each point has one multiplier inside and one multiplier outside the unit circle $|\mu| = 1$.

**(2) (Hamiltonian systems)**

(a) Prove that the Hamilton function is constant along orbits of a Hamiltonian system: $\dot{H} = 0$.

(b) Prove that the equilibrium $(\varphi, \psi) = (0, 0)$ of a pendulum described by (1.6) is Lyapunov stable. (*Hint*: System (1.6) is Hamiltonian with closed level curves $H(\varphi, \psi) = \text{const}$ near $(0, 0)$.) Is this equilibrium asymptotically stable?

**(3) (Quantum oscillations)**

(a) Integrate the linear system (1.8), describing the simplest quantum system with two states, and show that the probability $p_i = |a_i|^2$ of finding the system in a given state oscillates periodically in time.

(b) How does $p_1 + p_2$ behave?

**(4) (Brusselator revisited)**

(a) Derive the Brusselator system (1.9) from the system written in terms of the concentrations $[X], [Y]$.

(b) Compute an equilibrium position $(x_0, y_0)$ and find a sufficient condition on the parameters $(a, b)$ for it to be stable.

**(5) (Volterra system revisited)**

(a) Show that (1.10) can be reduced by a linear scaling of variables and time to the following system with only one parameter $\gamma$:

$$\begin{cases} \dot{x} = x - xy, \\ \dot{y} = -\gamma y + xy. \end{cases}$$

(b) Find all equilibria of the scaled system.

(c) Verify that the orbits of the scaled system in the *positive quadrant* $\{(x, y) : x, y > 0\}$ coincide with those of the Hamiltonian system

$$\begin{cases} \dot{x} = \dfrac{1}{y} - 1, \\ \\ \dot{y} = -\dfrac{\gamma}{x} + 1. \end{cases}$$

(*Hint*: Vector fields defining these two systems differ by the factor $\mu(x, y) = xy$, which is positive in the first quadrant.) Find the Hamilton function.

(d) Taking into account steps (a) to (c), prove that all nonequilibrium orbits of the Volterra system in the positive quadrant are closed, thus describing periodic oscillations of the numbers of prey and predators.

**(6) (Explicit Poincaré map)**

(a) Show that for $\alpha > 0$ the planar system in polar coordinates

$$\begin{cases} \dot{\rho} = \rho(\alpha - \rho^2), \\ \dot{\varphi} = 1, \end{cases}$$

has the explicit solution

$$\rho(t) = \left(\frac{1}{\alpha} + \left(\frac{1}{\rho_0^2} - \frac{1}{\alpha}\right) e^{-2\alpha t}\right)^{-1/2}, \ \varphi(t) = \varphi_0 + t.$$

(b) Draw the phase portrait of the system and prove that it has a unique limit cycle for each $\alpha > 0$.

(c) Compute the multiplier $\mu_1$ of the limit cycle:

(i) by explicit construction of the Poincaré map $\rho \mapsto P(\rho)$ using the solution above and evaluating its derivative with respect to $\rho$ at the fixed point $\rho_0 = \sqrt{\alpha}$ (*Hint:* See Wiggins [1990, pp. 66-67].);

(ii) using formula (1.18), expressing $\mu_1$ in terms of the integral of the divergence over the cycle. (*Hint:* Use polar coordinates; the divergence is invariant.)

**(7) (Lyapunov's theorem)** Prove Theorem 1.5 using Theorem 1.2.

(a) Write the system near the equilibrium as

$$\dot{x} = Ax + F(x),$$

where $F(x) = O(\|x\|^2)$ is a smooth nonlinear function.

(b) Using the variation-of-constants formula for the evolution operator $\varphi^t$,

$$\varphi^t x = e^{At}x + \int_0^t e^{A(t-\tau)} F(\varphi^\tau x) \, d\tau,$$

show that the unit-time shift along the orbits has the expansion

$$\varphi^1 x = Bx + O(\|x\|^2),$$

where $B = e^A$.

(c) Conclude the proof, taking into account that $\mu_k = e^{\lambda_k}$, where $\mu_k$ and $\lambda_k$ are the eigenvalues of the matrices $B$ and $A$, respectively.

## 1.7 Appendix A: Infinite-dimensional dynamical systems defined by reaction-diffusion equations

As we have seen in Examples 1.4 and 1.5, the state of a spatially distributed system is characterized by a function from a *function space* $X$. The dimension of such spaces is *infinite*. A function $u \in X$ satisfies certain boundary and smoothness conditions, while its evolution is usually determined by a system of equations with *partial derivatives* (PDEs). In this appendix we briefly discuss how a particular type of such equations, namely *reaction-diffusion systems*, defines infinite-dimensional dynamical systems.

The state of a chemical reactor at time $t$ can be specified by defining a vector function $c(x,t) = (c_1(x,t), c_2(x,t), \dots, c_n(x,t))^T$, where the $c_i$ are concentrations of reacting substances near the point $x$ in the reactor domain $\Omega \subset \mathbb{R}^m$. Here $m = 1,2,3$, depending on the geometry of the reactor, and $\Omega$ is assumed to be closed and bounded by a smooth boundary $\partial\Omega$. The concentrations $c_i(x,t)$ satisfy certain problem-dependent *boundary conditions.* For example, if the concentrations of all the reagents are kept constant at the boundary, we have

$$c(x,t) = c_0, \quad x \in \partial\Omega.$$

Defining a deviation from the boundary value, $s(x,t) = c(x,t) - c_0$, we can reduce to the case of *zero Dirichlet boundary conditions*:

$$s(x,t) = 0, \quad x \in \partial\Omega.$$

If the reagents cannot penetrate the reactor boundary, *zero Neumann* (*zero flux*) *conditions* are applicable:

$$\frac{\partial c(x,t)}{\partial n} = 0, \quad x \in \partial\Omega,$$

where the left-hand side is the inward-pointing normal derivative at the boundary.

The evolution of a chemical system can be modeled by a system of *reaction-diffusion equations* written in the vector form for $u(x,t)$ ($u = s$ or $c$):

$$\frac{\partial u(x,t)}{\partial t} = D(\Delta u)(x,t) + f(u(x,t)), \tag{A.1}$$

where $f : \mathbb{R}^n \to \mathbb{R}^n$ is smooth and $D$ is a diagonal *diffusion matrix* with positive coefficients, and $\Delta$ is known as the *Laplacian,*

$$\Delta u = \sum_{i=1}^{m} \frac{\partial^2 u}{\partial x_i^2}.$$

The first term of the right-hand side of (A.1) describes diffusion of the reagents, while the second term specifies their local interaction. The function $u(x,t)$ satisfies one of the boundary conditions listed above, for example, the Dirichlet conditions:

$$u(x,t) = 0, \quad x \in \partial\Omega. \tag{A.2}$$

**Definition 1.12** *A function $u = u(x,t)$, $u : \Omega \times \mathbb{R}^1 \to \mathbb{R}^n$, is called a* classical solution *to the problem* (A.1),(A.2) *if it is continuously differentiable, at least once with respect to $t$ and twice with respect to $x$, and satisfies* (A.1),(A.2) *in the domain of its definition.*

For any twice continuously differentiable *initial function* $u_0(x)$,

$$u_0(x) = 0, \quad x \in \partial\Omega, \tag{A.3}$$

the problem (A.1),(A.2) has a unique classical solution $u(x,t)$, defined for $x \in \Omega$ and $t \in [0, \delta_0)$, where $\delta_0$ depends on $u_0$, and such that $u(x,0) = u_0(x)$. Moreover, this classical solution is actually infinitely many times differentiable in $(x,t)$ for $0 < t < \delta_0$. The same properties are valid if one replaces (A.2) by Neumann boundary conditions.

Introduce the space $X = C_0^2(\Omega, \mathbb{R}^n)$ of all twice continuously differentiable vector functions in $\Omega$ satisfying the Dirichlet condition (A.3) at the boundary $\partial\Omega$. The preceeding results mean that the reaction-diffusion system (A.1),(A.2) defines a continuous-time dynamical system $\{\mathbb{R}^1_+, X, \varphi^t\}$, with the evolution operator

$$(\varphi^t u_0)(x) = u(x,t), \tag{A.4}$$

where $u(x,t)$ is the classical solution to (A.1),(A.2) satisfying $u(x,0) = u_0(x)$. It also defines a dynamical system on $X_1 = C_0^\infty(\Omega, \mathbb{R}^n)$ composed of all infinitely continuously differentiable vector functions in $\Omega$ satisfying the Dirichlet condition (A.3) at the boundary $\partial\Omega$.

The notions of equilibria and cycles are, therefore, applicable to the reaction-diffusion system (A.1). Clearly, equilibria of the system are described by time-independent vector functions satisfying

$$D(\Delta u)(x) + f(u(x)) = 0 \tag{A.5}$$

and the corresponding boundary conditions. A trivial, spatially homogeneous solutions to (A.5) satisfying (A.2), for example, is an equilibrium of the *local system*

$$\dot{u} = f(u), \quad u \in \mathbb{R}^n. \tag{A.6}$$

Nontrivial, spatially nonhomogeneous solutions to (A.5) are often called *dissipative structures*. Spatially homogeneous and nonhomogeneous equilibria can be stable or unstable. In the stable case, all (smooth) small perturbations $v(x)$ of an equilibrium solution decay in time. Cycles (i.e., time-periodic solutions of (A.1) satisfying the appropriate boundary conditions) are also possible; they can be stable or unstable. Standing and rotating *waves* in reaction-diffusion systems in planar circular domains $\Omega$ are examples of such periodic solutions.

Up to now, the situation seems to be rather simple and is parallel to the finite-dimensional case. However, one runs into certain difficulties when trying to introduce a distance in $X = C_0^2(\Omega, \mathbb{R}^n)$. For example, this space is *incomplete* in the "integral norm"

$$\|u\|^2 = \int_\Omega \sum_{\substack{j=1,2,\ldots,n \\ |i| \le 2}} \left| \frac{\partial^{|i|} u_j(x)}{\partial x_1^{i_1} \partial x_2^{i_2} \cdots \partial x_m^{i_m}} \right|^2 d\Omega, \tag{A.7}$$

where $|i| = i_1 + i_2 + \ldots + i_m$. In other words, a Cauchy sequence in this norm can approach a function that is not twice continuously differentiable (it may have no derivatives at all) and thus does not belong to $X$. Since this property is important in many respects, a method called *completion* has been developed that allows us to construct a complete space, given any normed one. Loosely speaking, we add the limits of all Cauchy sequences to $X$. More precisely, we call two Cauchy sequences *equivalent* if the distance between their corresponding elements tends to zero. Classes of equivalent Cauchy sequences are considered as points of a new space $H$. The original norm can be extended to $H$, thus making it a complete normed space. Such spaces are called *Banach spaces.* The space $X$ can then be interpreted as a subset of $H$. It is also useful if the obtained space is a *Hilbert space*, meaning that the norm in it is generated by a certain *scalar product.*

Therefore, we can try to use one of the completed spaces $H$ as a new state space for our reaction-diffusion system. However, since $H$ includes functions on which the diffusion part of (A.1) is undefined, extra work is required. One should also take care that the reaction part $f(u)$ of the system defines a smooth map on $H$. Without going into details, we merely state that it is possible to prove the existence of a dynamical system $\{\mathbb{R}^1_+, H, \psi^t\}$ such that $\psi^t u$ is defined and continuous in $u$ for all $u \in H$ and $t \in [0, \delta(u))$, and, if $u_0 \in X \subset H$, then $\psi^t u_0 = \varphi^t u_0$, where $\varphi^t u_0$ is a classical solution to (A.1),(A.2).

The stability of equilibria and other solutions can be studied in the space $H$. If an equilibrium is stable in $H$, it will also be stable with respect to smooth perturbations. One can derive sufficient conditions for an equilibrium to be stable in $H$ (or $X$) in terms of the linear part of the reaction-diffusion system (A.1). For example, let us formulate sufficient stability conditions (an analogue of Theorem 1.5) for a trivial (homogeneous) equilibrium of a reaction-diffusion system on the interval $\Omega = [0, \pi]$ with Dirichlet boundary conditions.

**Theorem 1.7** *Consider a reaction-diffusion system*

$$\frac{\partial u}{\partial t} = D\frac{\partial^2 u}{\partial x^2} + f(u), \tag{A.8}$$

*where $f$ is smooth, $x \in [0, \pi]$, with the boundary conditions*

$$u(0) = u(\pi) = 0. \tag{A.9}$$

*Assume that $u^0 = 0$ is a homogeneous equilibrium, $f(0) = 0$, and $A$ is the Jacobian matrix of the corresponding equilibrium of the local system, $A = f_u(0)$. Suppose that the eigenvalues of the $n \times n$ matrix*

$$M_k = A - k^2 D$$

*have negative real parts for all $k = 0, 1, 2, \ldots$.*

*Then $u^0 = 0$ is a stable equilibrium of the dynamical system $\{\mathbb{R}^1_+, H, \psi^t\}$ generated by the system* (A.8), (A.9) *in the completion $H$ of the space $C_0^2([0, \pi], \mathbb{R}^n)$ in the norm* (A.7). □

A similar theorem can be proved for the system in $\Omega \subset \mathbb{R}^m, m = 2, 3$, with Dirichlet boundary conditions. The only modification is that $k^2$ should be replaced by $\kappa_k$, where $\{\kappa_k\}$ are all positive numbers for which

$$(\Delta v_k)(x) = -\kappa_k v_k(x),$$

with $v_k = v_k(x)$ satisfying Dirichlet boundary conditions. The modification to the Neumann boundary condition case is rather straightforward.

## 1.8 Appendix B: Bibliographical notes

Originally, the term "dynamical system" meant only mechanical systems whose motion is described by differential equations derived in classical mechanics. Basic results on such dynamical systems were obtained by Lyapunov and Poincaré at the end of the nineteenth century. Their studies have been continued by Dulac [1923] and Birkhoff [1966], among others. The books by Nemytskii & Stepanov [1949] and

Coddington & Levinson [1955] contain detailed treatments of the then-known properties of dynamical systems defined by differential equations. Later on, it became clear that this notion is useful for the analysis of various evolutionary processes studied in different branches of science and described by ODEs, PDEs, or explicitly defined iterated maps. The modern period in dynamical system theory started from the work of Kolmogorov [1957], Smale [1963, 1966, 1967], and Anosov [1967]. A general introduction to the modern theory of dynamical systems can be found in [Katok & Hasselblatt 1995]. However, it is advisable to read [Hasselblatt & Katok 2003] first. Today, the literature on dynamical systems is huge. We do not attempt to survey it here, giving only a few remarks in the bibliographical notes to each chapter.

The horseshoe diffeomorphism proposed by Smale [1963, 1967] is treated in many books; for example, in Nitecki [1971], Guckenheimer & Holmes [1983], Wiggins [1990], and Arrowsmith & Place [1990]. However, the best presentation of this and related topics is still due to Moser [1973]. A map equivalent to (1.4) was introduced by Hénon [1976] as the simplest map with a strange attractor. The Hénon map is probably the best-studied planar map with complicated dynamics.

General properties of ordinary differential equations and their relation to dynamical systems are presented in the cited book by Nemytskii and Stepanov and notably in the texts by Pontryagin [1962], Arnol'd [1973], and Hirsch & Smale [1974]. The latter three books contain a comprehensive analysis of linear differential equations with constant and time-dependent coefficients. The book by Hartman [1964] treats the relation between Poincaré maps, multipliers, and stability of limit cycles.

The study of infinite-dimensional dynamical systems has been stimulated by hydro- and aerodynamics and by chemical and nuclear engineering. Linear infinite-dimensional dynamical systems, known as "continuous (analytical) semigroups," are studied in functional analysis (see, e.g., Hille & Phillips [1957], Balakrishnan [1976], or the more physically oriented texts by Richtmyer [1978, 1981]). Infinite-dimensional dynamical systems also arise naturally in studying differential equations with delays (see Hale [1971], Hale & Verduyn Lunel [1993], and Diekmann, van Gils, Verduyn Lunel & Walther [1995]). The theory of nonlinear infinite-dimensional systems is a rapidly developing field. Early results are presented in the relevant chapters of the books by Marsden & McCracken [1976], Carr [1981], and Henry [1981]. Modern treatments and further references can be found in the books by Temam [1997] and Robinson [2001].

# 2

# Topological Equivalence, Bifurcations, and Structural Stability of Dynamical Systems

In this chapter we introduce and discuss the following fundamental notions that will be used throughout the book: topological equivalence of dynamical systems and their classification, bifurcations and bifurcation diagrams, and topological normal forms for bifurcations. The last section is devoted to the more abstract notion of structural stability. In this chapter we will be dealing only with dynamical systems in the state space $X = \mathbb{R}^n$.

## 2.1 Equivalence of dynamical systems

We would like to study general (qualitative) features of the behavior of dynamical systems, in particular, to classify possible types of their behavior and compare the behavior of different dynamical systems. The comparison of any objects is based on an *equivalence relation*,[1] allowing us to define classes of equivalent objects and to study transitions between these classes. Thus, we have to specify when we define two dynamical systems as being "qualitatively similar" or equivalent. Such a definition must meet some general intuitive criteria. For instance, it is natural to expect that two equivalent systems have the same number of equilibria and cycles of the same stability types. The "relative position" of these invariant sets and the shape of their regions of attraction should also be similar for equivalent systems. In other words, we consider two dynamical systems as equivalent if their phase portraits are "qualitatively similar," namely, if one portrait can be obtained from another by a continuous transformation (see **Fig. 2.1**).

**Definition 2.1** *A dynamical system* $\{T, \mathbb{R}^n, \varphi^t\}$ *is called* topologically equivalent *to a dynamical system* $\{T, \mathbb{R}^n, \psi^t\}$ *if there is a homeomorphism* $h$ :

[1] Recall that a relation between two objects ($a \sim b$) is called *equivalence* if it is reflexive ($a \sim a$), symmetric ($a \sim b$ implies $b \sim a$), and transitive ($a \sim b$ and $b \sim c$ imply $a \sim c$).

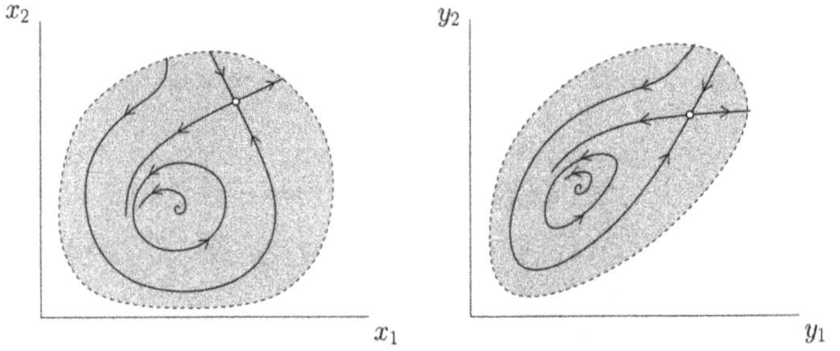


**Fig. 2.1.** Topological equivalence.

*$\mathbb{R}^n \to \mathbb{R}^n$ mapping orbits of the first system onto orbits of the second system, preserving the direction of time.*

A *homeomorphism* is an invertible map such that both the map and its inverse are continuous. The definition of the topological equivalence can be generalized to cover more general cases when the state space is a *complete metric* or, in particular, is a *Banach* space. The definition also remains meaningful when the state space is a smooth finite-dimensional *manifold* in $\mathbb{R}^n$, for example, a two-dimensional torus $\mathbb{T}^2$ or sphere $\mathbb{S}^2$. The phase portraits of topologically equivalent systems are often also called topologically equivalent.

The above definition applies to both continuous- and discrete-time systems. However, in the discrete-time case we can obtain an *explicit* relation between the corresponding maps of the equivalent systems. Indeed, let

$$x \mapsto f(x), \quad x \in \mathbb{R}^n, \tag{2.1}$$

and

$$y \mapsto g(y), \quad y \in \mathbb{R}^n, \tag{2.2}$$

be two topologically equivalent, discrete-time invertible dynamical systems ($f = \varphi^1, g = \psi^1$ are smooth invertible maps). Consider an orbit of system (2.1) starting at some point $x$:

$$\ldots, \; f^{-1}(x), \; x, \; f(x), \; f^2(x), \; \ldots$$

and an orbit of system (2.2) starting at a point $y$:

$$\ldots, \; g^{-1}(y), \; y, \; g(y), \; g^2(y), \; \ldots.$$

Topological equivalence implies that if $x$ and $y$ are related by the homeomorphism $h$, $y = h(x)$, then the first orbit is mapped onto the second one by this map $h$. Symbolically,

$$\begin{array}{ccc} x & \xrightarrow{f} & f(x) \\ {\scriptstyle h}\downarrow & & {\scriptstyle h}\downarrow \\ y & \xrightarrow{g} & g(y). \end{array}$$

Therefore, $g(y) = h(f(x))$ or $g(h(x)) = h(f(x))$ for all $x \in \mathbb{R}^n$, which can be written as

$$f(x) = h^{-1}(g(h(x)))$$

since $h$ is invertible. We can write the last equation in a more compact form using the symbol of map composition:

$$f = h^{-1} \circ g \circ h. \tag{2.3}$$

**Definition 2.2** *Two maps $f$ and $g$ satisfying* (2.3) *for some homeomorphism $h$ are called* conjugate.

Consequently, topologically equivalent, discrete-time systems are often called conjugate systems. If both $h$ and $h^{-1}$ are $C^k$ maps, the maps $f$ and $g$ are called *$C^k$-conjugate.* For $k \geq 1$, *$C^k$-conjugate* maps (and the corresponding systems) are called *smoothly conjugate* or *diffeomorphic.* Two diffeomorphic maps (2.1) and (2.2) can be considered as the same map written in two different coordinate systems with coordinates $x$ and $y$, while $y = h(x)$ can be treated as a smooth *change of coordinates.* Consequently, diffeomorphic discrete-time dynamical systems are practically indistinguishable.

Now consider two continuous-time topologically equivalent systems:

$$\dot{x} = f(x), \quad x \in \mathbb{R}^n, \tag{2.4}$$

and

$$\dot{y} = g(y), \quad y \in \mathbb{R}^n, \tag{2.5}$$

with smooth right-hand sides. Let $\varphi^t$ and $\psi^t$ denote the corresponding flows. In this case, there is no simple relation between $f$ and $g$ analogous to formula (2.3). Nevertheless, there are two particular cases of topological equivalence between (2.4) and (2.5) that can be expressed analytically, as we now explain.

Suppose that $y = h(x)$ is an invertible map $h : \mathbb{R}^n \to \mathbb{R}^n$, which is *smooth* together with its inverse ($h$ is a *diffeomorphism*) and such that, for all $x \in \mathbb{R}^n$,

$$f(x) = M^{-1}(x)g(h(x)), \tag{2.6}$$

where

$$M(x) = \frac{dh(x)}{dx}$$

is the Jacobian matrix of $h(x)$ evaluated at the point $x$. Then, system (2.4) is topologically equivalent to system (2.5). Indeed, system (2.5) is obtained from system (2.4) by the smooth change of coordinates $y = h(x)$. Thus, $h$ maps solutions of (2.4) into solutions of (2.5),

$$h(\varphi^t x) = \psi^t h(x),$$

and can play the role of the homeomorphism in Definition 2.1.

**Definition 2.3** *Two systems* (2.4) *and* (2.5) *satisfying* (2.6) *for some diffeomorphism $h$ are called* smoothly equivalent (*or* diffeomorphic).

**Remark:**
If the degree of smoothness of $h$ is of interest, one writes: $C^k$*-equivalent* or $C^k$*-diffeomorphic* in Definition 2.3. ◇

Two diffeomorphic systems are practically identical and can be viewed as the same system written using different coordinates. For example, the eigenvalues of corresponding equilibria are the same. Let $x_0$ and $y_0 = h(x_0)$ be such equilibria and let $A(x_0)$ and $B(y_0)$ denote corresponding Jacobian matrices. Then, differentiation of (2.6) yields

$$A(x_0) = M^{-1}(x_0)B(y_0)M(x_0).$$

Therefore, the characteristic polynomials for the matrices $A(x_0)$ and $B(y_0)$ coincide. In addition, diffeomorphic limit cycles have the same multipliers and period (see Exercise 4). This last property calls for more careful analysis of different *time parametrizations.*

Suppose that $\mu = \mu(x) > 0$ is a smooth *scalar positive* function and that the right-hand sides of (2.4) and (2.5) are related by

$$f(x) = \mu(x)g(x) \tag{2.7}$$

for all $x \in \mathbb{R}^n$. Then, obviously, systems (2.4) and (2.5) are topologically equivalent since their orbits are identical and it is the velocity of the motion that makes them different. (The ratio of the velocities at a point $x$ is exactly $\mu(x)$.) Thus, the homeomorphism $h$ in Definition 2.1 is the *identity* map $h(x) = x$. In other words, the systems are distinguished only by the time parametrization along the orbits.

**Definition 2.4** *Two systems* (2.4) *and* (2.5) *satisfying* (2.7) *for a smooth positive function $\mu$ are called* orbitally equivalent.

Clearly, two orbitally equivalent systems can be nondiffeomorphic, having cycles that look like the same closed curve in the phase space but have different periods.

Very often we study system dynamics *locally,* e.g., not in the whole state space $\mathbb{R}^n$ but in some region $U \subset \mathbb{R}^n$. Such a region may be, for example, a neighborhood of an equilibrium (fixed point) or a cycle. The above definitions of topological, smooth, and orbital equivalences can be easily "localized" by introducing appropriate regions. For example, in the topological classification of the phase portraits near equilibrium points, the following modification of Definition 2.1 is useful.

**Definition 2.5** *A dynamical system* $\{T, \mathbb{R}^n, \varphi^t\}$ *is called* locally topologically equivalent *near an equilibrium $x_0$ to a dynamical system* $\{T, \mathbb{R}^n, \psi^t\}$ *near an equilibrium $y_0$ if there exists a homeomorphism $h : \mathbb{R}^n \to \mathbb{R}^n$ that is*

(i) *defined in a small neighborhood* $U \subset \mathbb{R}^n$ *of* $x_0$;
(ii) *satisfies* $y_0 = h(x_0)$;
(iii) *maps orbits of the first system in* $U$ *onto orbits of the second system in* $V = h(U) \subset \mathbb{R}^n$, *preserving the direction of time.*

If $U$ is an open neighborhood of $x_0$, then $V$ is an open neighborhood of $y_0$. Let us also remark that equilibrium positions $x_0$ and $y_0$, as well as regions $U$ and $V$, might coincide.

Let us compare the above introduced equivalences in the following example.

**Example 2.1 (Node-focus equivalence)** Consider two linear planar dynamical systems:

$$\begin{cases} \dot{x}_1 = -x_1, \\ \dot{x}_2 = -x_2, \end{cases} \tag{2.8}$$

and

$$\begin{cases} \dot{x}_1 = -x_1 - x_2, \\ \dot{x}_2 = x_1 - x_2. \end{cases} \tag{2.9}$$

In the polar coordinates $(\rho, \theta)$ these systems can be written as

$$\begin{cases} \dot{\rho} = -\rho, \\ \dot{\theta} = 0, \end{cases}$$

and

$$\begin{cases} \dot{\rho} = -\rho, \\ \dot{\theta} = 1, \end{cases}$$

respectively. Thus,

$$\begin{aligned} \rho(t) &= \rho_0 e^{-t}, \\ \theta(t) &= \theta_0, \end{aligned}$$

for the first system, while

$$\begin{aligned} \rho(t) &= \rho_0 e^{-t}, \\ \theta(t) &= \theta_0 + t, \end{aligned}$$

for the second. Clearly, the origin is a stable equilibrium in both systems since $\rho(t) \to 0$ as $t \to \infty$. All other orbits of (2.8) are straight lines, while those of (2.9) are spirals. The phase portraits of the systems are presented in **Fig. 2.2**. The equilibrium of the first system is a *node* (**Fig. 2.2**(a)), while in the second system it is a *focus* (**Fig. 2.2**(b)). The difference in behavior of the systems can also be perceived by saying that perturbations decay near the origin monotonously in the first case and oscillatorily in the second case.

The systems are neither orbitally nor smoothly equivalent. The first fact is obvious, while the second follows from the observation that the eigenvalues of the equilibrium in the first system ($\lambda_1 = \lambda_2 = -1$) differ from those of the second ($\lambda_{1,2} = -1 \pm i$). Nevertheless, systems (2.8) and (2.9) are *topologically equivalent*, for example, in a closed unit disc

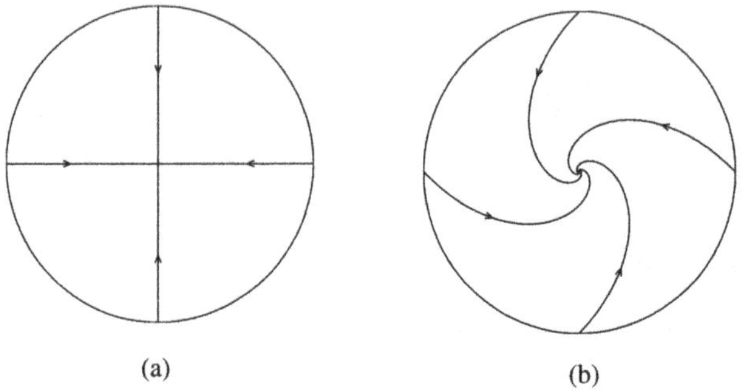


**Fig. 2.2.** Node-focus equivalence.

$$U = \{(x_1, x_2) : x_1^2 + x_2^2 \leq 1\} = \{(\rho, \theta) : \rho \leq 1\},$$

centered at the origin. Let us prove this explicitly by constructing a homeomorphism $h : U \to U$ as follows (see **Fig. 2.3**). Take a point $x \neq 0$ in $U$ with polar coordinates $(\rho_0, \theta_0)$ and consider the time $\tau$ required to move, along an orbit of system (2.8), from the point $(1, \theta_0)$ on the boundary to the point $x$. This time depends only on $\rho_0$ and can easily be computed:

$$\tau(\rho_0) = -\ln \rho_0.$$

Now consider an orbit of system (2.9) starting at the boundary point $(1, \theta_0)$, and let $y = (\rho_1, \theta_1)$ be the point at which this orbit arrives after $\tau(\rho_0)$ units of time. Thus, a map $y = h(x)$ that transforms $x = (\rho_0, \theta_0) \neq 0$ into $y = (\rho_1, \theta_1)$ is obtained and is explicitly given by

$$h : \begin{cases} \rho_1 = \rho_0, \\ \theta_1 = \theta_0 - \ln \rho_0. \end{cases}$$

For $x = 0$, set $y = 0$, that is, $h(0) = 0$. Thus the constructed map transforms $U$ into itself by rotating each circle $\rho_0 = \text{const}$ by a $\rho_0$-dependent angle. This

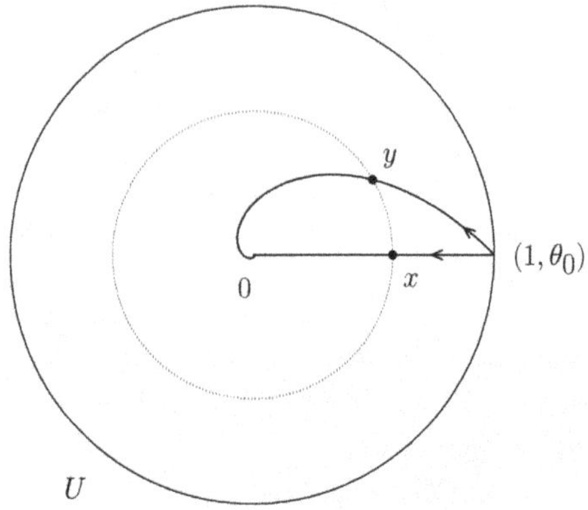


**Fig. 2.3.** The construction of the homeomorphism.

angle equals zero at $\rho_0 = 1$ and increases as $\rho_0 \to 0$. The map is obviously continuous and invertible and maps orbits of (2.8) onto orbits of (2.9), preserving time direction. Thus, the two systems are topologically equivalent within $U$.

However, the homeomorphism $h$ is not differentiable in $U$. More precisely, it is smooth away from the origin but not differentiable at $x = 0$. To see this, one should evaluate the Jacobian matrix $\frac{dy}{dx}$ in $(x_1, x_2)$-coordinates. For example, the difference quotient corresponding to the derivative

$$\left.\frac{\partial y_1}{\partial x_1}\right|_{x_1=x_2=0}$$

is given for $x_1 > 0$ by

$$\frac{x_1 \cos(\ln x_1) - 0}{x_1 - 0} = \cos(\ln x_1),$$

which has no limit as $x_1 \to 0$. $\diamondsuit$

Therefore, considering continuous-time systems modulo topological equivalence, we preserve information on the number, stability, and topology of invariant sets, while losing information relating transient and time-dependent behavior. Such information may be important in some applications. In these cases, stronger equivalences (such as orbital or smooth) have to be applied.

A combination of smooth and orbital equivalences gives a useful equivalence relation, which will be used frequently in this book.

**Definition 2.6** *Two systems* (2.4) *and* (2.5) *are called* smoothly orbitally equivalent *if* (2.5) *is smoothly equivalent to a system that is orbitally equivalent to* (2.4).

According to this definition, two systems are equivalent (in $\mathbb{R}^n$ or in some region $U \subset \mathbb{R}^n$) if we can transform one of them into the other by a smooth invertible change of coordinates and multiplication by a positive smooth function of the coordinates. Clearly, two smoothly orbitally equivalent systems are topologically equivalent, while the inverse is not true.

## 2.2 Topological classification of generic equilibria and fixed points

In this section we study the geometry of the phase portrait near *generic*, namely *hyperbolic*, equilibrium points in continuous- and discrete-time dynamical systems and present their topological classification.

### 2.2.1 Hyperbolic equilibria in continuous-time systems

Consider a continuous-time dynamical system defined by

$$\dot{x} = f(x), \quad x \in \mathbb{R}^n, \tag{2.10}$$

where $f$ is smooth. Let $x_0 = 0$ be an equilibrium of the system (i.e., $f(x_0) = 0$) and let $A$ denote the Jacobian matrix $\frac{df}{dx}$ evaluated at $x_0$. Let $n_-, n_0$, and $n_+$ be the numbers of eigenvalues of $A$ (counting multiplicities) with negative, zero, and positive real part, respectively.

**Definition 2.7** *An equilibrium is called* hyperbolic *if $n_0 = 0$, that is, if there are no eigenvalues on the imaginary axis. A hyperbolic equilibrium is called a* hyperbolic saddle *if $n_- n_+ \neq 0$.*

Since a generic matrix has no eigenvalues on the imaginary axis ($n_0 = 0$), hyperbolicity is a typical property and an equilibrium in a generic system (i.e., one not satisfying certain special conditions) is hyperbolic. We will not try to formalize these intuitively obvious properties, though it is possible using measure theory and transversality arguments (see the bibliographical notes). Instead, let us study the geometry of the phase portrait near a hyperbolic equilibrium in detail. For an equilibrium (not necessarily a hyperbolic one), we introduce two invariant sets:

$$W^s(x_0) = \{x : \varphi^t x \to x_0, t \to +\infty\}, W^u(x_0) = \{x : \varphi^t x \to x_0, t \to -\infty\},$$

where $\varphi^t$ is the flow associated with (2.10).

**Definition 2.8** *$W^s(x_0)$ is called the* stable set *of $x_0$, while $W^u(x_0)$ is called the* unstable set *of $x_0$.*

**Theorem 2.1 (Local Stable Manifold)** *Let $x_0$ be a hyperbolic equilibrium (i.e., $n_0 = 0$, $n_- + n_+ = n$ ). Then the intersections of $W^s(x_0)$ and $W^u(x_0)$ with a sufficiently small neighborhood of $x_0$ contain smooth submanifolds $W^s_{loc}(x_0)$ and $W^u_{loc}(x_0)$ of dimension $n_-$ and $n_+$, respectively.*

*Moreover, $W^s_{loc}(x_0)(W^u_{loc}(x_0))$ is tangent at $x_0$ to $T^s(T^u)$, where $T^s(T^u)$ is the generalized eigenspace corresponding to the union of all eigenvalues of $A$ with* $\text{Re}\,\lambda < 0$ ($\text{Re}\,\lambda > 0$). □

The proof of the theorem, which we are not going to present here, can be carried out along the following lines (Hadamard-Perron). For the unstable manifold, take the linear manifold $T^u$ passing through the equilibrium and apply the map $\varphi^1$ to this manifold, where $\varphi^t$ is the flow corresponding to the system. The image of $T^u$ under $\varphi^1$ is some (nonlinear) manifold of dimension $n_+$ tangent to $T^u$ at $x_0$. Restrict attention to a sufficiently small neighborhood of the equilibrium where the linear part is "dominant" and repeat the procedure. It can be shown that the iterations converge to a smooth invariant submanifold defined in this neighborhood of $x_0$ and tangent to $T^u$ at $x_0$.

The limit is the local unstable manifold $W^u_{\text{loc}}(x_0)$. The local stable manifold $W^s_{\text{loc}}(x_0)$ can be constructed by applying $\varphi^{-1}$ to $T^s$.

**Remark:**
Globally, the invariant sets $W^s$ and $W^u$ are *immersed* manifolds of dimensions $n_-$ and $n_+$, respectively, and have the same smoothness properties as $f$. Having these properties in mind, we will call the sets $W^s$ and $W^u$ the *stable* and *unstable invariant manifolds of* $x_0$, respectively. ◇

**Example 2.2 (Saddles and saddle-foci in $\mathbb{R}^3$) Fig. 2.4** illustrates

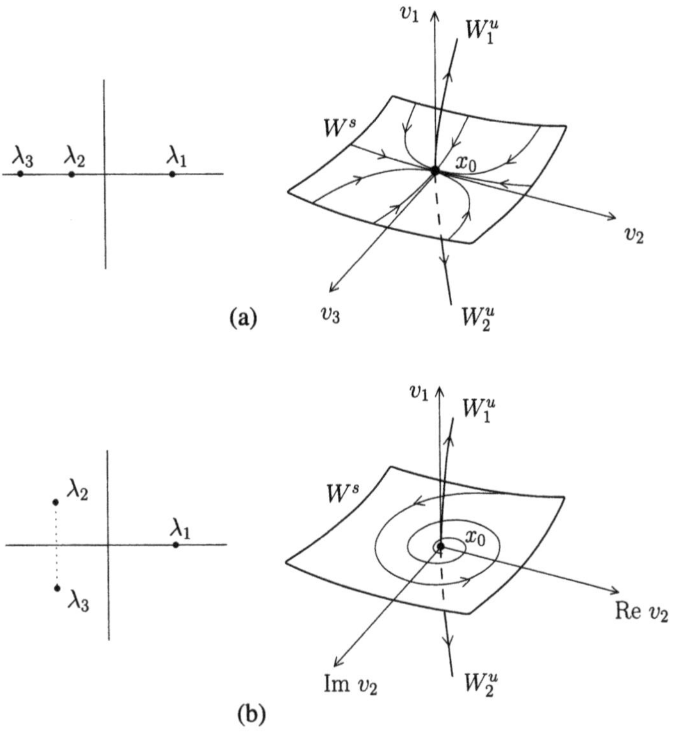


**Fig. 2.4.** (a) Saddle and (b) saddle-focus: The vectors $v_k$ are the eigenvectors corresponding to the eigenvalues $\lambda_k$.

the theorem for the case where $n = 3$, $n_- = 2$, and $n_+ = 1$. In this case, there are two invariant manifolds passing through the equilibrium, namely, the two-dimensional manifold $W^s(x_0)$ formed by all incoming orbits, and the one-dimensional manifold $W^u(x_0)$ formed by two outgoing orbits $W_1^u(x_0)$ and $W_2^u(x_0)$. All orbits not belonging to these manifolds pass near the equilibrium and eventually leave its neighborhood in both time directions.

In case (a) of real simple eigenvalues ($\lambda_3 < \lambda_2 < 0 < \lambda_1$), orbits on $W^s$ form a node, while in case (b) of complex eigenvalues ($\text{Re}\ \lambda_{2,3} < 0 < \lambda_1, \bar{\lambda}_3 = \lambda_2$), $W^s$ carries a focus. Thus, in the first case, the equilibrium is called a *saddle*, while in the second one it is referred to as a *saddle-focus*. The equilibria in these two cases are topologically equivalent. Nevertheless, it is useful to distinguish them, as we shall see in our study of homoclinic orbit bifurcations (Chapter 6). ◇

The following theorem gives the topological classification of hyperbolic equilibria.

**Theorem 2.2** *The phase portraits of system* (2.10) *near two hyperbolic equilibria, $x_0$ and $y_0$, are locally topologically equivalent if and only if these equilibria have the same number $n_-$ and $n_+$ of eigenvalues with* Re $\lambda < 0$ *and with* Re $\lambda > 0$, *respectively.* □

Often, the equilibria $x_0$ and $y_0$ are then also called topologically equivalent. The proof of the theorem is based on two ideas. First, it is possible to show that near a hyperbolic equilibrium the system is locally topologically equivalent to its *linearization:* $\dot{\xi} = A\xi$ (Grobman-Hartman Theorem). This result should be applied both near the equilibrium $x_0$ and near the equilibrium $y_0$. Second, the topological equivalence of two *linear* systems having the same numbers of eigenvalues with Re $\lambda < 0$ and Re $\lambda > 0$ and no eigenvalues on the imaginary axis has to be proved. Example 2.1 is a particular case of such a proof. Nevertheless, the general proof is based on the same idea. See the appendix at the end of this chapter for references.

**Example 2.3 (Generic equilibria of planar systems)** Consider a two-dimensional system

$$\dot{x} = f(x), \quad x = (x_1, x_2)^T \in \mathbb{R}^2,$$

with smooth $f$. Suppose that $x = 0$ is an equilibrium, $f(0) = 0$, and let

$$A = \left. \frac{df(x)}{dx} \right|_{x=0}$$

be its Jacobian matrix. Matrix $A$ has two eigenvalues $\lambda_1, \lambda_2$, which are the roots of the characteristic equation

$$\lambda^2 - \sigma\lambda + \Delta = 0,$$

where $\sigma = \text{tr}\ A$, $\Delta = \det A$.

**Fig. 2.5** displays well-known classical results. There are three topological classes of hyperbolic equilibria on the plane: *stable nodes (foci), saddles, and unstable nodes (foci)*. As we have discussed, nodes and foci (of corresponding stability) are topologically equivalent but can be identified looking at the eigenvalues.

| $(n_+, n_-)$ | Eigenvalues | Phase portrait | Stability |
|---|---|---|---|
| (0, 2) | 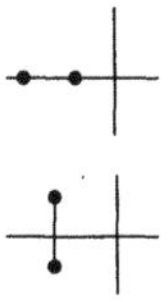 | 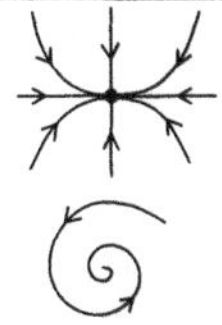 node<br>focus | stable |
| (1, 1) |  | 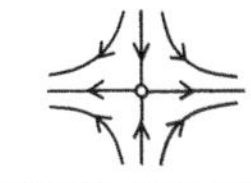 saddle | unstable |
| (2, 0) | 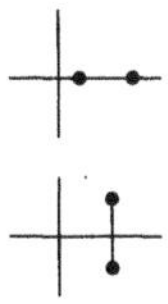 | 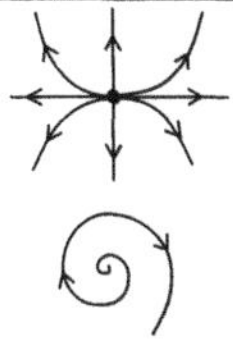 node<br>focus | unstable |

**Fig. 2.5.** Topological classification of hyperbolic equilibria on the plane.

**Definition 2.9** *Nodes and foci are both called* antisaddles.

Stable points have two-dimensional stable manifolds and no unstable manifolds. For unstable equilibria the situation is reversed. Saddles have one-dimensional stable and unstable manifolds, sometimes called *separatrices.* ◊

### 2.2.2 Hyperbolic fixed points in discrete-time systems

Now consider a discrete-time dynamical system

$$x \mapsto f(x), \quad x \in \mathbb{R}^n, \tag{2.11}$$

where the map $f$ is smooth along with its inverse $f^{-1}$ (diffeomorphism). Let $x_0 = 0$ be a fixed point of the system (i.e., $f(x_0) = x_0$) and let $A$ denote the Jacobian matrix $\frac{df}{dx}$ evaluated at $x_0$. The eigenvalues $\mu_1, \mu_2, \dots, \mu_n$ of $A$ are called *multipliers* of the fixed point. Notice that there are no zero multipliers, due to the invertibility of $f$. Let $n_-$, $n_0$, and $n_+$ be the numbers of multipliers of $x_0$ lying inside, on, and outside the unit circle $\{\mu \in \mathbb{C}^1 : |\mu| = 1\}$, respectively.

**Definition 2.10** *A fixed point is called* hyperbolic *if $n_0 = 0$, that is, if there are no multipliers on the unit circle. A hyperbolic fixed point is called a* hyperbolic saddle *if $n_- n_+ \neq 0$.*

Notice that hyperbolicity is a typical property also in discrete time. As in the continuous-time case, we can introduce stable and unstable invariant sets for a fixed point $x_0$ (not necessarily a hyperbolic one):

$$W^s(x_0) = \{x : f^k(x) \to x_0, k \to +\infty\},$$
$$W^u(x_0) = \{x : f^k(x) \to x_0, k \to -\infty\},$$

where $k$ is integer "time" and $f^k(x)$ denotes the $k$th iterate of $x$ under $f$. An analogue of Theorem 2.1 can be formulated.

**Theorem 2.3 (Local Stable Manifold)** *Let $x_0$ be a hyperbolic fixed point, namely, $n_0 = 0$, $n_- + n_+ = n$. Then the intersections of $W^s(x_0)$ and $W^u(x_0)$ with a sufficiently small neighborhood of $x_0$ contain smooth submanifolds $W^s_{loc}(x_0)$ and $W^u_{loc}(x_0)$ of dimension $n_-$ and $n_+$, respectively.*

*Moreover, $W^s_{loc}(x_0)(W^u_{loc}(x_0))$ is tangent at $x_0$ to $T^s(T^u)$, where $T^s(T^u)$ is the generalized eigenspace corresponding to the union of all eigenvalues of $A$ with $|\mu| < 1(|\mu| > 1)$.* □

The proof of the theorem is completely analogous to that in the continuous-time case, if one substitutes $\varphi^1$ by $f$. Globally, the invariant sets $W^s$ and $W^u$ are again *immersed* manifolds of dimension $n_-$ and $n_+$, respectively, and have the same smoothness properties as the map $f$. The manifolds cannot intersect themselves, but their global topology may be very complex, as we shall see later.

The topological classification of hyperbolic fixed points follows from a theorem that is similar to Theorem 2.2 for equilibria in the continuous-time systems.

**Theorem 2.4** *The phase portraits of* (2.11) *near two hyperbolic fixed points, $x_0$ and $y_0$, are locally topologically equivalent if and only if these fixed points have the same number $n_-$ and $n_+$ of multipliers with $|\mu| < 1$ and $|\mu| > 1$, respectively, and the signs of the products of all the multipliers with $|\mu| < 1$ and with $|\mu| > 1$ are the same for both fixed points.* □

As in the continuous-time case, the proof is based upon the fact that near a hyperbolic fixed point the system is locally topologically equivalent to its *linearization:* $x \mapsto Ax$ (discrete-time version of the Grobman-Hartman Theorem). The additional conditions on the products are due to the fact that the dynamical system can define either an *orientation-preserving* or *orientation-reversing* map on the stable or unstable manifold near the fixed point. Recall that a diffeomorphism on $\mathbb{R}^l$ preserves orientation in $\mathbb{R}^l$ if $\det J > 0$, where $J$ is its Jacobian matrix, and reverses it otherwise. Two topologically equivalent maps must have the same orientation properties. The products in Theorem 2.4 are exactly the determinants of the Jacobian matrices of the map (2.11) *restricted* to its stable and unstable local invariant manifolds. It should be clear that one needs only account for *real* multipliers to compute these signs since the product of a complex-conjugate pair of multipliers is always positive.

Let us consider two examples of fixed points.

**Example 2.4 (Stable fixed points in $\mathbb{R}^1$)** Suppose $x_0 = 0$ is a fixed point of a one-dimensional discrete-time system ($n = 1$). Let $n_- = 1$, meaning that the unique multiplier $\mu$ satisfies $|\mu| < 1$. In this case, according to

Theorem 2.3, all orbits starting in some neighborhood of $x_0$ converge to $x_0$. Depending on the sign of the multiplier, we have the two possibilities presented

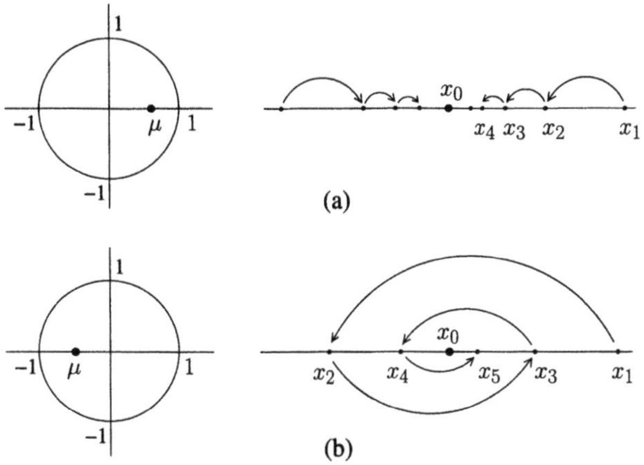


**Fig. 2.6.** Stable fixed points of one-dimensional systems: (a) $0 < \mu < 1$; (b) $-1 < \mu < 0$.

in **Fig. 2.6**. If $0 < \mu < 1$, the iterations converge to $x_0$ *monotonously* (**Fig. 2.6**(a)). If $-1 < \mu < 0$, the convergence is *nonmonotonous* and the phase point "jumps" around $x_0$ while converging to $x_0$ (**Fig. 2.6**(b)). In the first case the map preserves orientation in $\mathbb{R}^1$ while reversing it in the second. It should be clear that "jumping" orbits cannot be transformed into monotonous ones by a continuous map. **Fig. 2.7** presents orbits near the two types of fixed

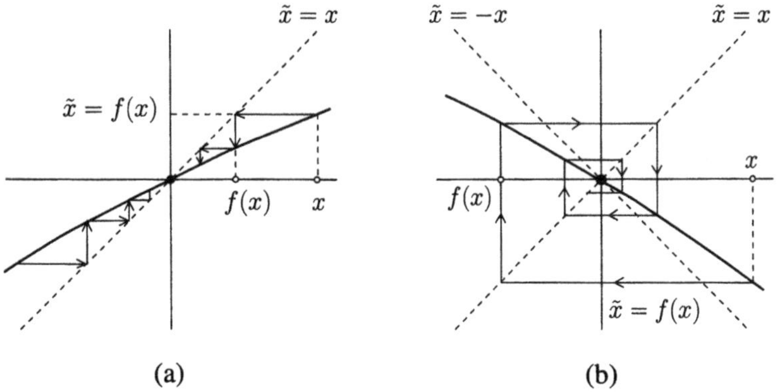


**Fig. 2.7.** Staircase diagrams for stable fixed points.

points using staircase diagrams. ◊

**Example 2.5 (Saddle fixed points in $\mathbb{R}^2$)** Suppose $x_0 = 0$ is a fixed point of a two-dimensional discrete-time system (now $n = 2$). Assume that $n_- = n_+ = 1$, so that there is one (real) multiplier $\mu_1$ outside the unit circle ($|\mu_1| > 1$) and one (real) multiplier $\mu_2$ inside the unit circle ($|\mu_2| < 1$). In our case, there are two invariant manifolds passing through the fixed point, namely the one-dimensional manifold $W^s(x_0)$ formed by orbits converging to $x_0$ under iterations of $f$, and the one-dimensional manifold $W^u(x_0)$ formed by orbits tending to $x_0$ under iterations of $f^{-1}$. Recall that the orbits of a discrete-time system are *sequences* of points. All orbits not belonging to the aforementioned manifolds pass near the fixed point and eventually leave its neighborhood in both "time" directions.

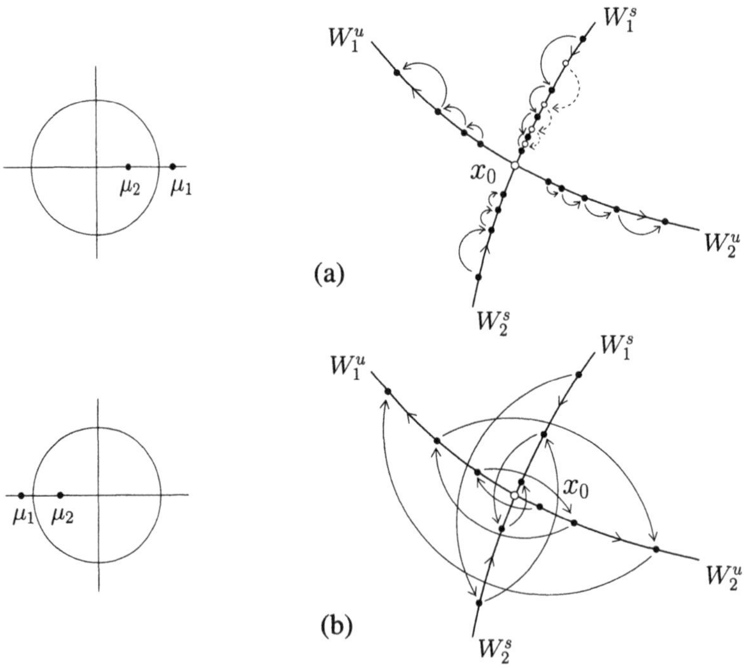


**Fig. 2.8.** Invariant manifolds of saddle fixed points on the plane: (a) positive multipliers; (b) negative multipliers.

**Fig. 2.8** shows two types of saddles in $\mathbb{R}^2$. In the case (a) of positive multipliers, $0 < \mu_2 < 1 < \mu_1$, an orbit starting at a point on $W^s(x_0)$ converges to $x_0$ monotonously. Thus, the stable manifold $W^s(x_0)$ is formed by two invariant branches, $W^s_{1,2}(x_0)$, separated by $x_0$. The same can be said about the unstable manifold $W^u(x_0)$ upon replacing $f$ by its inverse. The restriction of the map onto both manifolds preserves orientation.

If the multipliers are negative (case (b)), $\mu_1 < -1 < \mu_2 < 0$, the orbits on the manifolds "jump" between the two components $W^{s,u}_{1,2}$ separated by

$x_0$. The map reverses orientation in both manifolds. The branches $W_{1,2}^{s,u}$ are invariant with respect to the *second iterate* $f^2$ of the map. ◇

**Remarks:**

(1) The stable and unstable manifolds $W^{s,u}(x_0)$ of a two-dimensional saddle are examples of *invariant curves*: If $x$ belongs to the curve, so does any iterate $f^k(x)$. The invariant curve is not an orbit. Actually, it consists of an *infinite* number of orbits. **Fig. 2.9** shows invariant curves and an orbit near a saddle fixed point with positive multipliers.

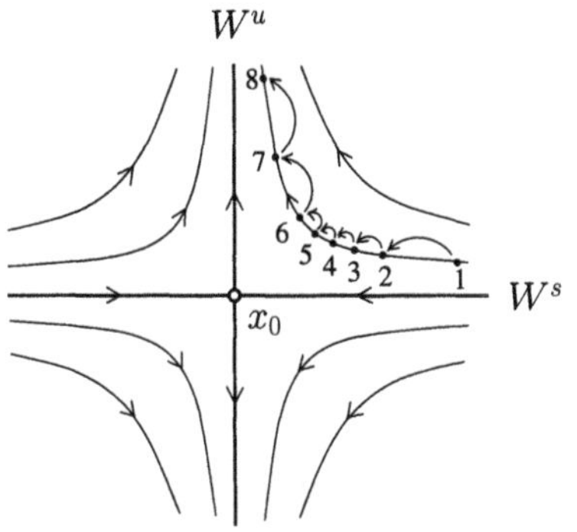


**Fig. 2.9.** Invariant curves and an orbit near a saddle fixed point.

(2) The global behavior of the stable and unstable manifolds $W^{s,u}(x_0)$ of a hyperbolic fixed point can be *very* complex, thus making the word "contain" absolutely necessary in Theorem 2.3.

Return, for example, to the planar case and suppose that $x_0$ is a saddle with positive multipliers. First of all, unlike the stable and unstable sets of an equilibrium in a continuous-time system, the manifolds $W^s(x_0)$ and $W^u(x_0)$ of a generic discrete-time system can *intersect* at nonzero angle (transversally) (see **Fig. 2.10**(a)).

Moreover, one transversal intersection, if it occurs, implies an *infinite* number of such intersections. Indeed, let $x^0$ be a point of the intersection. By definition, it belongs to both invariant manifolds. Therefore, the orbit starting at this point converges to the saddle point $x_0$ under repeated iteration of either $f$ or $f^{-1}$ : $f^k(x^0) \to x_0$ as $k \to \pm\infty$. Each point of this orbit is a point of intersection of $W^s(x_0)$ and $W^u(x_0)$. This infinite number of intersections forces the manifolds to "oscillate" in a complex manner near $x_0$, as sketched in **Fig. 2.10**(b). The resulting "web" is called the *Poincaré homoclinic structure.* The orbit starting at $x^0$ is said to be *homoclinic* to $x_0$. It is the presence of the homoclinic structure that can make the intersection of $W^{s,u}(x_0)$ with any neighborhood of the saddle $x_0$ highly nontrivial.

The dynamical consequences of the existence of the homoclinic structure are also dramatic: It results in the appearance of an *infinite number* of periodic

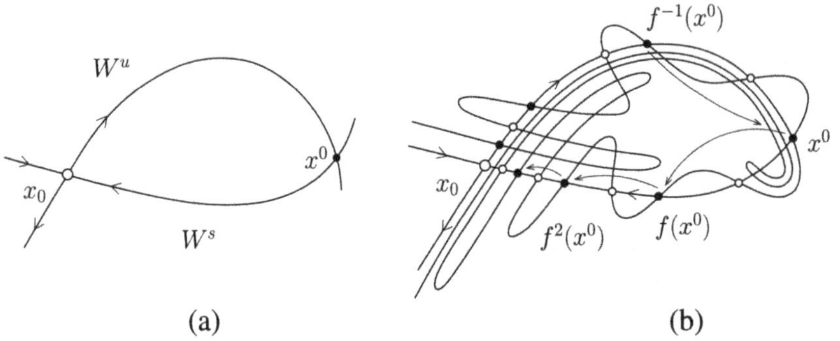


**Fig. 2.10.** Poincaré homoclinic structure.

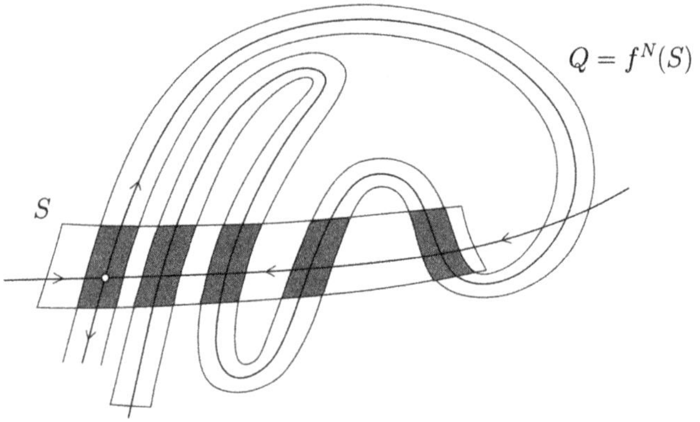


**Fig. 2.11.** Smale horseshoes embedded into the Poincaré homoclinic structure.

points with arbitrary high periods near the homoclinic orbit. This follows from the presence of *Smale horseshoes* (see Chapter 1). **Fig. 2.11** illustrates how the horseshoes are formed. Take a (curvilinear) rectangle $S$ near the stable manifold $W^s(x_0)$ and consider its iterations $f^k S$. If the homoclinic structure is present, for a sufficiently high number of iterations $N$, $f^N S$ will look like the folded and expanded band $Q$ shown in the figure. The intersection of $S$ with $Q$ forms several horseshoes, where each of them implies an infinite number of cycles with arbitrary high periods. ♢

### 2.2.3 Hyperbolic limit cycles

Using the results of the previous section and the Poincaré map construction (see Chapter 1), we can define *hyperbolic limit cycles* in continuous-time systems and describe the topology of phase orbits near such cycles. Consider a continuous-time dynamical system

$$\dot{x} = f(x), \quad x \in \mathbb{R}^n, \tag{2.12}$$

with smooth $f$, and assume that there is an isolated periodic orbit (limit cycle) $L_0$ of (2.12). As in Chapter 1, let $\Sigma$ be a local cross-section to the cycle of dimension $(n-1)$ (codim $\Sigma = 1$) with coordinates $\xi = (\xi_1, \dots, \xi_{n-1})^T$. System (2.12) locally defines a smooth invertible map $P$ (a Poincaré map) from $\Sigma$ to $\Sigma$ along the orbits of (2.12). The point $\xi_0$ of intersection of $L_0$ with $\Sigma$ is a fixed point of the map $P$, $P(\xi_0) = \xi_0$.

Generically, the fixed point $\xi_0$ is hyperbolic, so there exist invariant manifolds

$$W^s(\xi_0) = \{\xi \in \Sigma : P^k(\xi) \to \xi_0, \ k \to +\infty\}$$

and

$$W^u(\xi_0) = \{\xi \in \Sigma : P^{-k}(\xi) \to \xi_0, \ k \to +\infty\},$$

of the dimensions $n_-$ and $n_+$, respectively, where $n_\mp$ are the numbers of eigenvalues of the Jacobian matrix of $P$ at $\xi_0$ located inside and outside the unit circle. Recall that $n_- + n_+ = n-1$ and that the eigenvalues are called *multipliers of the cycle*. The invariant manifolds $W^{s,u}(\xi_0)$ are the intersections with $\Sigma$ of the *stable and unstable manifolds of the cycle*:

$$W^s(L_0) = \{x \ : \ \varphi^t x \to L_0, \ t \to +\infty\},$$
$$W^u(L_0) = \{x \ : \ \varphi^t x \to L_0, \ t \to -\infty\},$$

where $\varphi^t$ is the flow corresponding to (2.12).

We can now use the results on the topological classification of fixed points of discrete-time dynamical systems to classify limit cycles. A limit cycle is called *hyperbolic* if $\xi_0$ is a hyperbolic fixed point of the Poincaré map. Similarly, a hyperbolic cycle is called a *saddle cycle* if it has multipliers both inside and outside the unit circle (i.e., $n_- n_+ \neq 0$). Recall that the product of the multipliers is always *positive* (see Chapter 1); therefore the Poincaré map preserves orientation in $\Sigma$. This imposes some restrictions on the location of the multipliers in the complex plane.

**Example 2.6 (Hyperbolic cycles in planar systems)** Consider a smooth planar system

$$\begin{cases} \dot{x}_1 = f_1(x_1, x_2), \\ \dot{x}_2 = f_2(x_1, x_2), \end{cases}$$

$x = (x_1, x_2)^T \in \mathbb{R}^2$. Let $x_0(t)$ be a solution corresponding to a limit cycle $L_0$ of the system, and let $T_0$ be the (minimal) period of this solution. There is only one multiplier of the cycle, $\mu_1$, which is positive and is given by

$$\mu_1 = \exp\left\{\int_0^{T_0} (\text{div } f)(x_0(t)) \, dt\right\} > 0,$$

where div stands for the *divergence* of the vector field:

$$(\text{div } f)(x) = \frac{\partial f_1(x)}{\partial x_1} + \frac{\partial f_2(x)}{\partial x_2}.$$

If $0 < \mu_1 < 1$, we have a stable hyperbolic cycle and all nearby orbits converge exponentially to it, while for $\mu_1 > 1$ we have an unstable hyperbolic cycle with exponentially diverging neighboring orbits. ◊

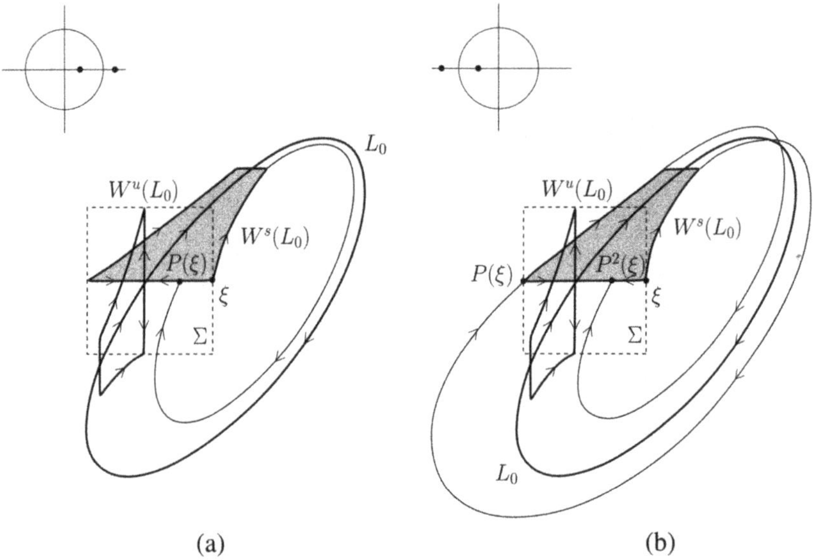


**Fig. 2.12.** Saddle cycles in three-dimensional systems: (a) positive multipliers and (b) negative multipliers.

**Example 2.7 (Saddle cycles in three-dimensional systems)** Example 2.5 provides two types of saddle limit cycles existing in $\mathbb{R}^3$ (see **Fig. 2.12**). If the multipliers of the Poincaré map satisfy

$$0 < \mu_2 < 1 < \mu_1,$$

both invariant manifolds $W^s(L_0)$ and $W^u(L_0)$ of the cycle $L_0$ are *simple* bands (**Fig. 2.12**(a)), while in the case when the multipliers satisfy

$$\mu_1 < -1 < \mu_2 < 0,$$

the manifolds $W^s(L_0)$ and $W^u(L_0)$ are *twisted* bands (called *Möbius strips*) (see **Fig. 2.12**(b)). Other types of saddle cycles in $\mathbb{R}^3$ are impossible since the product of the multipliers of any Poincaré map is positive. Thus, the manifolds $W^s(L_0)$ and $W^u(L_0)$ must both be simple or twisted.

Finally, remark that $W^s(L_0)$ and $W^u(L_0)$ can *intersect* along orbits homoclinic to the cycle $L_0$, giving rise to Poincaré homoclinic structure and Smale horseshoes on the cross-section $\Sigma$. ◊

## 2.3 Bifurcations and bifurcation diagrams

Now consider a dynamical system that depends on parameters. In the continuous-time case we will write it as

$$\dot{x} = f(x, \alpha), \tag{2.13}$$

while in the discrete-time case it is written as

$$x \mapsto f(x, \alpha), \tag{2.14}$$

where $x \in \mathbb{R}^n$ and $\alpha \in \mathbb{R}^m$ represent phase variables and parameters, respectively. Consider the phase portrait of the system.[2] As the parameters vary, the phase portrait also varies. There are two possibilities: either the system remains topologically equivalent to the original one, or its topology changes.

**Definition 2.11** *The appearance of a topologically nonequivalent phase portrait under variation of parameters is called a* bifurcation.

Thus, a bifurcation is a change of the topological type of the system as its parameters pass through a *bifurcation (critical) value.* Actually, the central topic of this book is the classification and analysis of various bifurcations.

**Example 2.8 (Andronov-Hopf bifurcation)** Consider the following planar system that depends on one parameter:

$$\begin{cases} \dot{x}_1 = \alpha x_1 - x_2 - x_1(x_1^2 + x_2^2), \\ \dot{x}_2 = x_1 + \alpha x_2 - x_2(x_1^2 + x_2^2). \end{cases} \tag{2.15}$$

In polar coordinates $(\rho, \theta)$ it takes the form

$$\begin{cases} \dot{\rho} = \rho(\alpha - \rho^2), \\ \dot{\theta} = 1, \end{cases} \tag{2.16}$$

and can be integrated explicitly (see Exercise 6). Since the equations for $\rho$ and $\theta$ are independent in (2.16), we can easily draw phase portraits of the system in a fixed neighborhood of the origin, which is obviously the only equilibrium point (see **Fig. 2.13**). For $\alpha \leq 0$, the equilibrium is a *stable focus* since $\dot{\rho} < 0$ and $\rho(t) \to 0$, if we start from any initial point. On the other hand, if $\alpha > 0$, we have $\dot{\rho} > 0$ for small $\rho > 0$ (the equilibrium becomes an *unstable focus*), and $\dot{\rho} < 0$ for sufficiently large $\rho$. It is easy to see from (2.16) that the system has a *periodic orbit* for any $\alpha > 0$ of radius $\rho_0 = \sqrt{\alpha}$ (at $\rho = \rho_0$ we have $\dot{\rho} = 0$). Moreover, this periodic orbit is stable since $\dot{\rho} > 0$ inside and $\dot{\rho} < 0$ outside the cycle.

Therefore, $\alpha = 0$ is a bifurcation parameter value. Indeed, a phase portrait with a limit cycle cannot be deformed by a one-to-one transformation into

[2] If necessary, one may consider the phase portrait in a parameter-dependent region $U_\alpha \subset \mathbb{R}^n$.

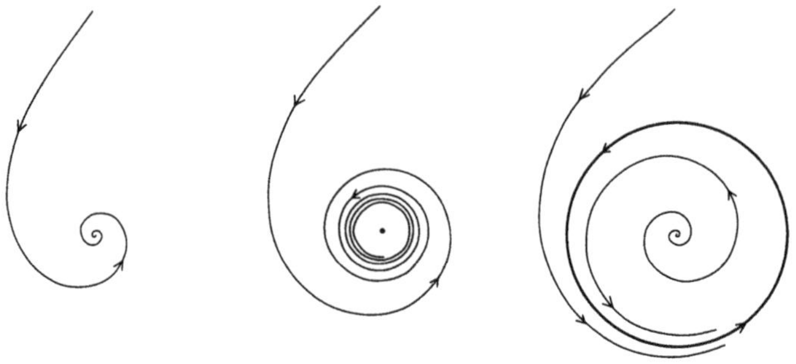


**Fig. 2.13.** Hopf bifurcation.

a phase portrait with only an equilibrium. The presence of a limit cycle is said to be a *topological invariant*. As $\alpha$ increases and crosses zero, we have a bifurcation in system (2.15) called the *Andronov-Hopf bifurcation*. It leads to the appearance, from the equilibrium state, of small-amplitude periodic oscillations. We will use this bifurcation as an example later in this chapter and analyze it in detail in Chapters 3 and 5. ◊

As should be clear, an Andronov-Hopf bifurcation can be detected if we fix *any* small neighborhood of the equilibrium. Such bifurcations are called *local*. One can also define local bifurcations in discrete-time systems as those detectable in any small neighborhood of a fixed point. We will often refer to local bifurcations as *bifurcations of equilibria or fixed points*, although we will analyze not just these points but the whole phase portraits near the equilibria. Those bifurcations of limit cycles which correspond to local bifurcations of associated Poincaré maps are called *local bifurcations of cycles*.

There are also bifurcations that cannot be detected by looking at small neighborhoods of equilibrium (fixed) points or cycles. Such bifurcations are called *global*.

**Example 2.9 (Heteroclinic bifurcation)** Consider the following planar system that depends on one parameter:

$$\begin{cases} \dot{x}_1 = 1 - x_1^2 - \alpha x_1 x_2, \\ \dot{x}_2 = x_1 x_2 + \alpha(1 - x_1^2). \end{cases} \tag{2.17}$$

The system has two saddle equilibria

$$x_{(1)} = (-1, 0), x_{(2)} = (1, 0),$$

for all values of $\alpha$ (see **Fig. 2.14**). At $\alpha = 0$ the horizontal axis is invariant and, therefore, the saddles are connected by an orbit that is asymptotic to one of them for $t \to +\infty$ and to the other for $t \to -\infty$. Such orbits are called *heteroclinic*. Similarly, an orbit that is asymptotic to the same equilibrium

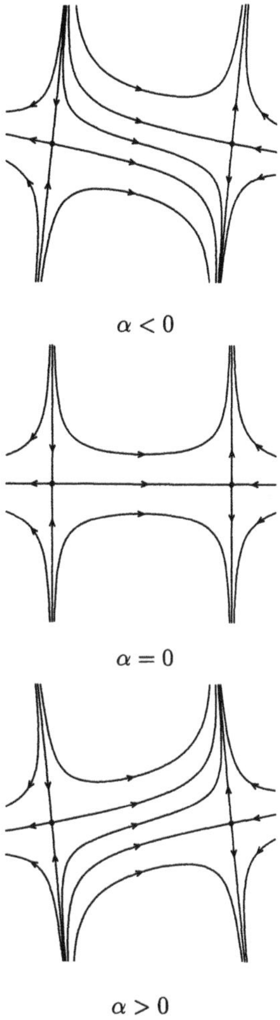


**Fig. 2.14.** Heteroclinic bifurcation.

as $t \to +\infty$ and $t \to -\infty$ is called *homoclinic*. For $\alpha \neq 0$, the $x_1$-axis is no longer invariant, and the connection disappears. This is obviously a global bifurcation. To detect this bifurcation we must fix a region $U$ covering both saddles. We will study hetero- and homoclinic orbit bifurcations in Chapter 6. ◊

There are global bifurcations in which certain local bifurcations are involved. In such cases, looking at the local bifurcation provides only partial information on the behavior of the system. The following example illustrates this possibility.

**Example 2.10 (Saddle-node homoclinic bifurcation)** Let us analyze the following system on the plane:

$$\begin{cases} \dot{x}_1 = x_1(1 - x_1^2 - x_2^2) - x_2(1 + \alpha + x_1), \\ \dot{x}_2 = x_1(1 + \alpha + x_1) + x_2(1 - x_1^2 - x_2^2), \end{cases} \tag{2.18}$$

where $\alpha$ is a parameter. In polar coordinates $(\rho, \theta)$ system (2.18) takes the form

$$\begin{cases} \dot{\rho} = \rho(1 - \rho^2), \\ \dot{\theta} = 1 + \alpha + \rho \cos \theta. \end{cases} \tag{2.19}$$

Fix a thin annulus $U$ around the unit circle $\{(\rho, \theta) : \rho = 1\}$. At $\alpha = 0$, there is a nonhyperbolic equilibrium point of system (2.19) in the annulus:

$$x_0 = (\rho_0, \theta_0) = (1, \pi)$$

(see **Fig. 2.15**). It has eigenvalues $\lambda_1 = 0, \lambda_2 = -2$ (check!). For small positive

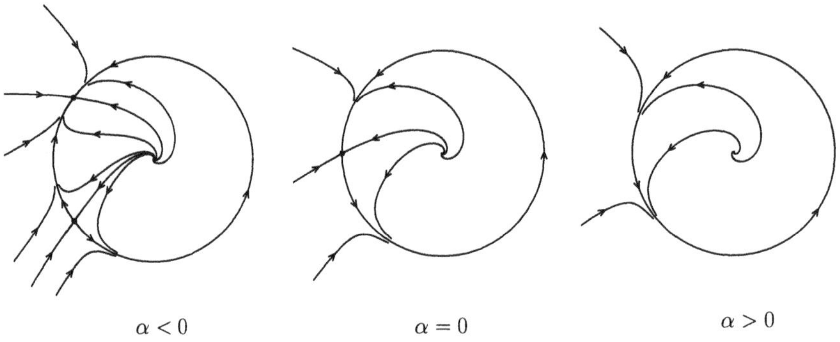


**Fig. 2.15.** Saddle-node homoclinic bifurcation.

values of $\alpha$ the equilibrium disappears, while for small negative $\alpha$ it splits into a saddle and a node (this bifurcation is called a *saddle-node* or *fold* bifurcation; see Chapter 3). This is a local event. However, for $\alpha > 0$ a stable *limit cycle* appears in the system coinciding with the unit circle. This circle is always an invariant set in the system, but for $\alpha \leq 0$ it contains equilibria. Looking at only a small neighborhood of the nonhyperbolic equilibrium, we miss the global appearance of the cycle. Notice that at $\alpha = 0$ there is exactly one orbit that is homoclinic to the nonhyperbolic equilibrium $x_0$. We will discuss such global bifurcations in Chapter 7. ◊

We return now to a general discussion of bifurcations in a parameter-dependent system (2.13) (or (2.14)). Take some value $\alpha = \alpha_0$ and consider a maximal connected parameter set (called a *stratum*) containing $\alpha_0$ and composed by those points for which the system has a phase portrait that is topologically equivalent to that at $\alpha_0$. Taking all such strata in the parameter space $\mathbb{R}^m$, we obtain the *parametric portrait* of the system. For example, system (2.15) exhibiting the Andronov-Hopf bifurcation has a parametric portrait with two strata: $\{\alpha \leq 0\}$ and $\{\alpha > 0\}$. In system (2.17) there are three strata: $\{\alpha < 0\}$, $\{\alpha = 0\}$, and $\{\alpha > 0\}$. Notice, however, that the phase portrait of (2.17) for $\alpha < 0$ is topologically equivalent to that for $\alpha > 0$.

The parametric portrait together with its characteristic phase portraits constitute a *bifurcation diagram.*

**Definition 2.12** *A* bifurcation diagram *of the dynamical system is a stratification of its parameter space induced by the topological equivalence, together with representative phase portraits for each stratum.*

It is desirable to obtain the bifurcation diagram as a result of the qualitative analysis of a given dynamical system. It classifies in a very condensed way all possible modes of behavior of the system and transitions between them (bifurcations) under parameter variations.[3] Note that the bifurcation diagram depends, in general, on the region of phase space considered.

**Remark:**

If a dynamical system has a one- or two-dimensional phase space and depends on only one parameter, its bifurcation diagram can be visualized in the *direct product* of the phase and parameter spaces, $\mathbb{R}^{1,2} \times \mathbb{R}^1$ with the phase portraits represented by one- or two-dimensional *slices* $\alpha = \text{const}$.

Consider, for example, a scalar system

$$\dot{x} = \alpha x - x^3, \ x \in \mathbb{R}^1, \alpha \in \mathbb{R}^1.$$

This system has an equilibrium $x_0 = 0$ for all $\alpha$. This equilibrium is stable for $\alpha < 0$ and unstable for $\alpha > 0$ ($\alpha$ is the eigenvalue of this equilibrium). For $\alpha > 0$, there are two extra equilibria branching from the origin (namely, $x_{1,2} = \pm\sqrt{\alpha}$) which are stable. This bifurcation is often called a *pitchfork bifurcation*, the reason for which becomes immediately clear if one has a look at the bifurcation diagram of the system presented in $(x, \alpha)$-space (see **Fig. 2.16**). Notice that the system demonstrating the pitchfork bifurcation is *invariant* under the transformation $x \mapsto -x$. We will study bifurcations in such *symmetric* systems in Chapter 7. $\diamondsuit$

In the simplest cases, the parametric portrait is composed by a finite number of regions in $\mathbb{R}^m$. Inside each region the phase portrait is topologically equivalent. These regions are separated by *bifurcation boundaries*, which are smooth submanifolds in $\mathbb{R}^m$ (i.e., curves, surfaces). The boundaries can intersect, or meet. These intersections subdivide the boundaries into subregions, and so forth. A bifurcation boundary is defined by specifying a phase object (equilibrium, cycle, etc.) and some *bifurcation conditions* determining the type

[3] Recall that some time-related information on the behavior of the system is lost due to topological equivalence.

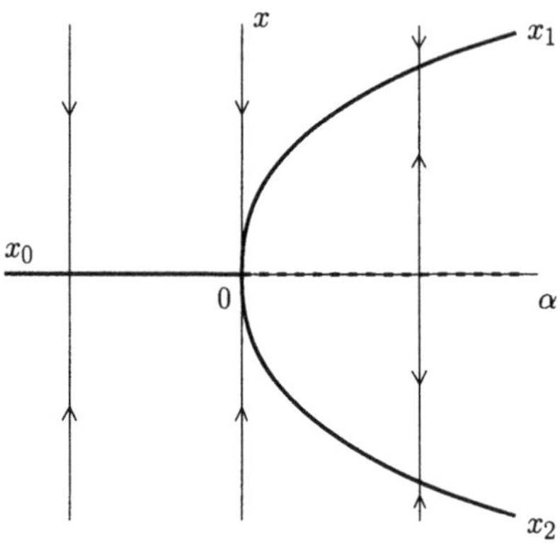


**Fig. 2.16.** Pitchfork bifurcation.

of its bifurcation (Hopf, fold, etc.). For example, the Andronov-Hopf bifurcation of an equilibrium is characterized by one bifurcation condition, namely, the presence of a purely imaginary pair of eigenvalues of the Jacobian matrix evaluated at this equilibrium (cf. Example 2.7):

$$\text{Re } \lambda_{1,2} = 0.$$

When a boundary is crossed, the bifurcation occurs.

**Definition 2.13** *The* codimension *of a bifurcation in system* (2.13) *or* (2.14) *is the difference between the dimension of the parameter space and the dimension of the corresponding bifurcation boundary.*

Equivalently, the codimension (codim for short) is the number of independent conditions determining the bifurcation. This is the most practical definition of the codimension. It makes it clear that the codimension of a certain bifurcation is the same in all generic systems depending on a sufficient number of parameters.

**Remark:**
The bifurcation diagram of even a simple continuous-time system in a bounded region on the plane can be composed by an *infinite* number of strata. The situation becomes more involved for multidimensional continuous-time systems (with $n > 3$). In such systems the bifurcation values can be dense in some parameter regions and the parametric portrait can have a *Cantor* (*fractal*) structure with certain patterns repeated on smaller and smaller scales to infinity. Clearly, the task of fully investigating such a bifurcation diagram is practically impossible. Nevertheless, even partial knowledge of the bifurcation diagram provides important information about the behavior of the system being studied. ◇

## 2.4 Topological normal forms for bifurcations

Fortunately, bifurcation diagrams are not entirely "chaotic." Different strata of bifurcation diagrams in generic systems interact with each other following certain rules. This makes bifurcation diagrams of systems arising in many different applications look similar. To discuss this topic, we have to decide when two dynamical systems have "qualitatively similar" or equivalent bifurcation diagrams. Consider two (for definitiveness, continuous-time) dynamical systems:

$$\dot{x} = f(x, \alpha), \quad x \in \mathbb{R}^n, \ \alpha \in \mathbb{R}^m, \tag{2.20}$$

and

$$\dot{y} = g(y, \beta), \quad y \in \mathbb{R}^n, \ \beta \in \mathbb{R}^m, \tag{2.21}$$

with smooth right-hand sides and the same number of variables and parameters. The following definition is parallel to Definition 2.1, with necessary modifications due to parameter dependence.

**Definition 2.14** *Dynamical system* (2.20) *is called* topologically equivalent *to a dynamical system* (2.21) *if*

(i) *there exists a homeomorphism of the parameter space* $p: \mathbb{R}^m \to \mathbb{R}^m$, $\beta = p(\alpha)$;

(ii) *there is a parameter-dependent homeomorphism of the phase space* $h_\alpha : \mathbb{R}^n \to \mathbb{R}^n$, $y = h_\alpha(x)$, *mapping orbits of the system* (2.20) *at parameter values* $\alpha$ *onto orbits of the system* (2.21) *at parameter values* $\beta = p(\alpha)$, *preserving the direction of time.*

Clearly, the homeomorphism $p$ transforms the parametric portrait of system (2.20) into the parametric portrait of system (2.21), while the homeomorphism $h_\alpha$ maps corresponding phase portraits. By definition, topologically equivalent parameter-dependent systems have (topologically) equivalent bifurcation diagrams.

**Remark:**
Notice that we do *not* require the homeomorphism $h_\alpha$ to depend *continuously* on $\alpha$, which would imply that the map $(x, \alpha) \mapsto (h_{p(\alpha)}(x), p(\alpha))$ be a homeomorphism of the direct product $\mathbb{R}^n \times \mathbb{R}^m$. For this reason, some authors call the above-defined topological equivalence *weak* (or *fiber*) topological equivalence. ◇

As in the constant-parameter case, Definition 2.14 can be modified if one is interested in comparing local behavior of the systems, for example, in a small neighborhood of the origin of the state space, for small parameter values.

**Definition 2.15** *Two systems* (2.20) *and* (2.21) *are called* locally topologically equivalent *near the origin, if there exists a map* $(x, \alpha) \mapsto (h_\alpha(x), p(\alpha))$, *defined in a small neighborhood of* $(x, \alpha) = (0, 0)$ *in the direct product* $\mathbb{R}^n \times \mathbb{R}^m$ *and such that*

(i) $p : \mathbb{R}^m \to \mathbb{R}^m$ *is a homeomorphism defined in a small neighborhood of* $\alpha = 0$, $p(0) = 0$;

(ii) $h_\alpha : \mathbb{R}^n \to \mathbb{R}^n$ *is a parameter-dependent homeomorphism defined in a small neighborhood* $U_\alpha$ *of* $x = 0$, $h_0(0) = 0$, *and mapping orbits of* (2.20) *in* $U_\alpha$ *onto orbits of* (2.21) *in* $h_\alpha(U_\alpha)$, *preserving the direction of time.*

This definition means that one can introduce two small neighborhoods of the origin $U_\alpha$ and $V_\beta$, whose diameters are bounded away from zero uniformly for $\alpha, \beta$ varying in some fixed small neighborhoods of the origin of the corresponding parameter spaces. Then, the homeomorphism $h_\alpha$ maps orbits of (2.20) in $U_\alpha$ onto orbits of (2.21) in $V_{p(\alpha)}$, preserving their orientation.

We now consider the problem of the classification of all possible bifurcation diagrams of generic systems, at least, locally (i.e. near bifurcation boundaries in the parameter space and corresponding critical orbits in the phase space) and up to and including certain codimension. These local diagrams could then serve as "building blocks" to construct the "global" bifurcation diagram of any system. This problem has been solved for equilibrium bifurcations in two-dimensional continuous-time systems up to and including codim 3. In some sense, it has also been solved for bifurcations of equilibria and fixed points in multidimensional continuous- and discrete-time systems up to and including codim 2, although the relevant results are necessarily incomplete (see Chapters 3, 4, 8, and 9). There are also several outstanding results concerning higher-codimension local bifurcations and some global bifurcations of codim 1 and 2.

The classification problem formulated above is simplified due to the following obvious but important observation. The minimal number of free parameters required to meet a codim $k$ bifurcation in a parameter-dependent system is exactly equal to $k$. Indeed, to satisfy a single bifurcation condition, we need, in general, to "tune" a (single) parameter of the system. If there are two conditions to be satisfied, two parameters have to be varied, and so forth. In other words, we have to control $k$ parameters to reach a codim $k$ bifurcation boundary in the parametric portrait of a generic system. On the other hand, it is enough to study a bifurcation of codim $k$ in generic $k$-parameter systems. General $m$-parameter ($m > k$) diagrams near the bifurcation boundary can then be obtained by "shifting" the $k$-parameter diagram in the complementary directions. For example, the Andronov-Hopf bifurcation is a codim 1 (local) bifurcation. Thus, it occurs at isolated parameter values in systems depending on one parameter. In two-parameter systems, it generally occurs on specific curves (one-dimensional manifolds). If we cross this curve at a nonzero angle (transversally), the resulting one-parameter bifurcation diagrams (where the parameter, e.g., is the arclength along a transversal curve) will be topologically equivalent to the original one-parameter diagram. The same will be true if we cross a two-dimensional surface corresponding to the Hopf bifurcation in a system depending on three parameters.

For local bifurcations of equilibria and fixed points, universal bifurcation diagrams are provided by *topological normal forms.*[4] This is one of the central notions in bifurcation theory. Let us discuss it in the continuous-time setting, although it also applies to discrete-time systems. Sometimes it is possible to construct a simple (polynomial in $\xi_i$) system

$$\dot{\xi} = g(\xi, \beta; \sigma), \quad \xi \in \mathbb{R}^n, \ \beta \in \mathbb{R}^k, \ \sigma \in \mathbb{R}^l, \tag{2.22}$$

which has at $\beta = 0$ an equilibrium $\xi = 0$ satisfying $k$ bifurcation conditions determining a codim $k$ bifurcation of this equilibrium. Here $\sigma$ is a vector of the coefficients $\sigma_i, \ i = 1, 2, \dots, l$, of the polynomials involved in (2.22). In all the cases we will consider, there is a *finite* number of regions in the coefficient space corresponding to topologically nonequivalent bifurcation diagrams of (2.22). In the simplest situations, the $\sigma_i$ take only a *finite* number of integer values. For example, all the coefficients $\sigma_i = 1$ except a single $\sigma_{i_0} = \pm 1$. In more complex situations, some components of $\sigma$ may take real values (*modulae*).

Together with system (2.22), let us consider a system

$$\dot{x} = f(x, \alpha), \quad x \in \mathbb{R}^n, \ \alpha \in \mathbb{R}^k, \tag{2.23}$$

having at $\alpha = 0$ an equilibrium $x = 0$.

**Definition 2.16 (Topological normal form)** *System* (2.22) *is called a* topological normal form *for the bifurcation if any generic system* (2.23) *with the equilibrium $x = 0$ satisfying the same bifurcation conditions at $\alpha = 0$ is locally topologically equivalent near the origin to* (2.22) *for some values of the coefficients $\sigma_i$.*

Of course, we have to explain what a *generic* system means. In all the cases we will consider, "generic" means that the system satisfies a finite number of *genericity conditions.* These conditions will have the form of nonequalities:

$$N_i[f] \neq 0, \ i = 1, 2, \dots, s,$$

where each $N_i$ is some (algebraic) function of certain partial derivatives of $f(x, \alpha)$ with respect to $x$ and $\alpha$ evaluated at $(x, \alpha) = (0, 0)$. Thus, a "typical" parameter-dependent system satisfies these conditions. Actually, the value of $\sigma$ is then determined by values of $N_i, \ i = 1, 2, \dots, s$.

It is useful to distinguish those genericity conditions which are determined by the system at the critical parameter values $\alpha = 0$. These conditions can be expressed in terms of partial derivatives of $f(x, 0)$ with respect to $x$ evaluated at $x = 0$, and are called *nondegeneracy conditions.* All the other conditions, in which the derivatives of $f(x, \alpha)$ with respect to the parameters $\alpha$ are involved, are called *transversality conditions.* The role of these two types of conditions

[4] It is possible to construct a kind of topological normal form for certain global bifurcations involving homoclinic orbits.

is different. The nondegeneracy conditions guarantee that the critical equilibrium (*singularity*) is not too degenerate (i.e., typical in a class of equiliubria satisfying given bifurcation conditions), while the transversality conditions assure that the parameters "unfold" this singularity in a generic way.

If a topological normal form is constructed, its bifurcation diagram clearly has a universal meaning since it immanently appears as a part of bifurcation diagrams of generic systems exhibiting the relevant bifurcation. System (2.15) from Example 2.7, by which we have illustrated the Andronov-Hopf bifurcation, corresponds to the case $\sigma = -1$ in the two-dimensional *topological normal form* for this bifurcation:

$$\begin{cases} \dot{\xi}_1 = \beta\xi_1 - \xi_2 + \sigma\xi_1(\xi_1^2 + \xi_2^2), \\ \dot{\xi}_2 = \xi_1 + \beta\xi_2 + \sigma\xi_2(\xi_1^2 + \xi_2^2). \end{cases}$$

The conditions specifying generic systems that demonstrate this bifurcation are the following:

$$\text{(H.1)} \qquad \left.\frac{d}{d\alpha}\text{Re } \lambda_{1,2}(\alpha)\right|_{\alpha=0} \neq 0$$

and

$$\text{(H.2)} \qquad l_1(0) \neq 0.$$

The first condition (transversality) means that the pair of complex-conjugate eigenvalues $\lambda_{1,2}(\alpha)$ crosses the imaginary axis with nonzero speed. The second condition (nondegeneracy) implies that a certain combination of Taylor coefficients of the right-hand sides of the system (up to and including third-order coefficients) does not vanish. An explicit formula for $l_1(0)$ will be derived in Chapter 3, where we also prove that the above system is really a topological normal form for the Hopf bifurcation. We will also show that $\sigma = \text{sign } l_1(0)$.

**Remark:**

There is a closely related notion of *versal deformation* (or *universal unfolding*) for a bifurcation. First, we need to define what we mean by an *induced* system.

**Definition 2.17 (Induced system)** *The system*

$$\dot{y} = g(y, \beta), \quad y \in \mathbb{R}^n, \ \beta \in \mathbb{R}^m,$$

*is said to be* induced *by the system*

$$\dot{x} = f(x, \alpha), \quad x \in \mathbb{R}^n, \ \alpha \in \mathbb{R}^m,$$

*if* $g(y, \beta) = f(y, p(\beta))$, *where* $p : \mathbb{R}^m \to \mathbb{R}^m$ *is a continuous map.*

Notice that the map $p$ is not necessarily a homeomorphism, so it can be noninvertible.

**Definition 2.18 (Versal deformation)** *System* (2.22) *is a* versal deformation *for the corresponding local bifurcation if any system* (2.23)*, with the equilibrium* $x = 0$ *satisfying the same bifurcation conditions and nondegeneracy conditions at* $\alpha = 0$*, is locally topologically equivalent near the origin to a system induced by* (2.22) *for some values of the coefficients* $\sigma_i$.

It can be proved, in many cases, that the topological normal forms we derive are actually versal deformations for the corresponding bifurcations (see also Exercise 7). ◊

## 2.5 Structural stability

There are dynamical systems whose phase portrait (in some domain) does not change qualitatively under all sufficiently small perturbations.

**Example 2.11 (Persistence of a hyperbolic equilibrium)** Suppose that $x_0$ is a hyperbolic equilibrium of a continuous-time system

$$\dot{x} = f(x), \quad x \in \mathbb{R}^n, \tag{2.24}$$

where $f$ is smooth, $f(x_0) = 0$. Consider, together with system (2.24), its one-parameter *perturbation*

$$\dot{x} = f(x) + \varepsilon g(x), \quad x \in \mathbb{R}^n, \tag{2.25}$$

where $g$ is also smooth and $\varepsilon$ is a small parameter; setting $\varepsilon = 0$ brings (2.25) back to (2.24). System (2.25) has an equilibrium $x(\varepsilon)$ for all sufficiently small $|\varepsilon|$ such that $x(0) = x_0$. Indeed, the equation defining equilibria of (2.25) can be written as

$$F(x, \varepsilon) = f(x) + \varepsilon g(x) = 0,$$

with $F(x_0, 0) = 0$. We also have $F_x(x_0, 0) = A_0$, where $A_0$ is the Jacobian matrix of (2.24) at the equilibrium $x_0$. Since $\det A_0 \neq 0$, because $x_0$ is hyperbolic, the Implicit Function Theorem guarantees the existence of a smooth function $x = x(\varepsilon)$, $x(0) = x_0$, satisfying

$$F(x(\varepsilon), \varepsilon) = 0$$

for small values of $|\varepsilon|$. The Jacobian matrix of $x(\varepsilon)$ in (2.25),

$$A_\varepsilon = \left( \frac{df(x)}{dx} + \varepsilon \frac{dg(x)}{dx} \right) \bigg|_{x=x(\varepsilon)},$$

depends smoothly on $\varepsilon$ and coincides with $A_0$ in (2.24) at $\varepsilon = 0$. As already known, the eigenvalues of a matrix that depends smoothly on a parameter change *continuously* with the variation of this parameter.[5] Therefore, $x(\varepsilon)$

[5] The eigenvalues vary smoothly as long as they remain *simple*.

will have no eigenvalues on the imaginary axis for all sufficiently small $|\varepsilon|$ since it has no such eigenvalues at $\varepsilon = 0$. In other words, $x(\varepsilon)$ is a hyperbolic equilibrium of (2.25) for all $|\varepsilon|$ small enough. Moreover, the numbers $n_-$ and $n_+$ of the stable and unstable eigenvalues of $A_\varepsilon$ are fixed for these values of $\varepsilon$. Applying Theorem 2.2, we find that systems (2.24) and (2.25) are locally topologically equivalent near the equilibria. Actually, for every $|\varepsilon|$ small, there is a neighborhood $U_\varepsilon \subset \mathbb{R}^n$ of the equilibrium $x_\varepsilon$ in which system (2.25) is topologically equivalent to (2.24) in $U_0$. In short, all these facts are summarized by saying that "a hyperbolic equilibrium is *structurally stable* under smooth perturbations."

Similar arguments provide the persistence of a hyperbolic equilibrium for all sufficiently small $|\varepsilon|$ in a smooth system

$$\dot{x} = G(x, \varepsilon), \quad x \in \mathbb{R}^n, \ \varepsilon \in \mathbb{R}^1,$$

where $G(x, 0) = f(x)$. $\Diamond$

The parameter $\varepsilon$ from Example 2.11 somehow measures the distance between system (2.24) and its perturbation (2.25); if $\varepsilon = 0$ the systems coincide. There is a general definition of the distance between two smooth dynamical systems. Consider two continuous-time systems

$$\dot{x} = f(x), \quad x \in \mathbb{R}^n, \tag{2.26}$$

and

$$\dot{x} = g(x), \quad x \in \mathbb{R}^n, \tag{2.27}$$

with smooth $f$ and $g$.

**Definition 2.19** *The* distance *between* (2.26) *and* (2.27) *in a closed region* $U \subset \mathbb{R}^n$ *is a positive number* $d_1$ *given by*

$$d_1 = \sup_{x \in U} \left\{ \|f(x) - g(x)\| + \left\| \frac{df(x)}{dx} - \frac{dg(x)}{dx} \right\| \right\}.$$

*The systems are* $\varepsilon$-close *in* $U$ *if* $d_1 \le \varepsilon$.

Here $\|\cdot\|$ means a vector and a matrix norm in $\mathbb{R}^n$, for example:

$$\|x\| = \sqrt{\sum_{i=1,\dots,n} x_i^2}, \quad \|A\| = \sqrt{\sum_{i,j=1,\dots,n} a_{ij}^2}.$$

Thus, two systems are close if their right-hand sides are close to each other, *together* with their first partial derivatives. In this case one usually calls the systems $C^1$-*close*. Clearly, the distance between systems (2.24) and (2.25) is proportional to $|\varepsilon|$: $d_1 = C|\varepsilon|$ for some constant $C > 0$ depending on the upper bounds for $\|g\|$ and $\left\|\frac{dg}{dx}\right\|$ in $U$. Definition 2.19 can be applied verbatim to discrete-time systems.

**Remark:**

The appearance of the first derivatives in the definition of the distance is natural if one wants to ensure that close systems have nearby equilibria of the same topological type (see Example 2.11). It is easy to construct a smooth

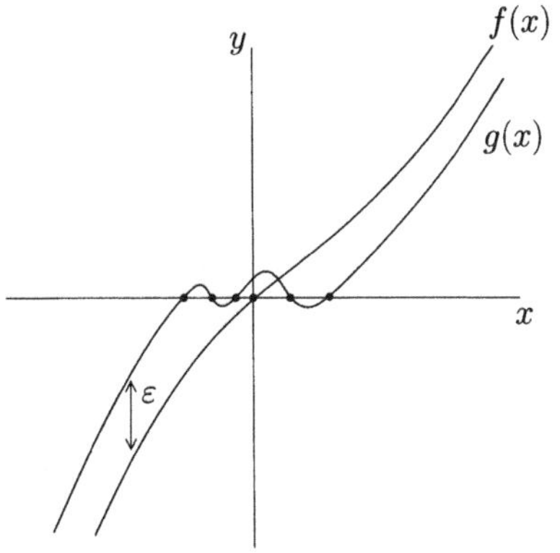


**Fig. 2.17.** Two $C^0$-close functions with different numbers of zeros.

system (2.27) that is $\varepsilon$-close to (2.26) in the $C^0$-*distance*:

$$d_0 = \sup_{x \in U} \{\|f(x) - g(x)\|\},$$

and that has a totally different number of equilibria in any neighborhood of an equilibrium of (2.26) (see **Fig. 2.17** for $n = 1$). ◊

Now we would like to define a structurally stable system, which means that any sufficiently close system is topologically equivalent to the structurally stable one. The following definition seems rather natural.

**Definition 2.20 (Strict structural stability)** *System* (2.26) *is* strictly structurally stable *in the region* $U$ *if any system* (2.27) *that is sufficiently* $C^1$*-close in* $U$ *is topologically equivalent in* $U$ *to* (2.26).

Notice, however, that systems having *hyperbolic* equilibria on the boundary of $U$, or *hyperbolic* cycles touching the boundary (see **Fig. 2.18**), are structurally *unstable* in accordance with this definition since there are small system perturbations moving such equilibria out of $U$, or pushing such cycles to lie (partially) outside of $U$. There are two ways to handle this difficulty.

The first is to consider dynamical systems "in the whole phase space" and to forget about any regions. This way is perfect for dynamical systems defined on a *compact* smooth manifold $X$. In such a case, the "region $U$" in Definition 2.20 (as well as in the definition of the distance) should be substituted by the "compact manifold $X$." Unfortunately, for systems in $\mathbb{R}^n$ this easily leads to complications. For example, the distance between many innocently looking

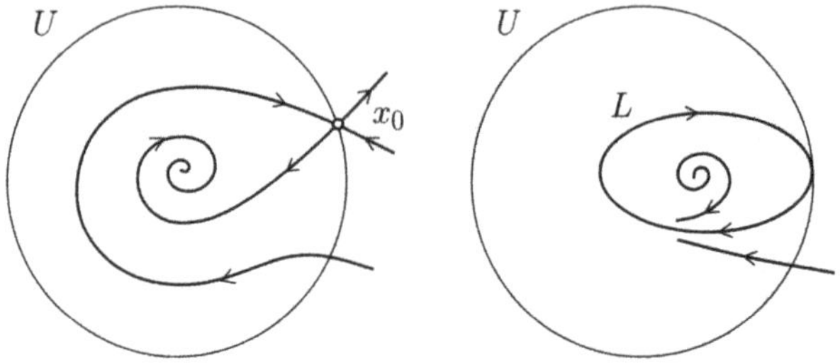


**Fig. 2.18.** Structurally unstable orbits according to Definition 2.20.

systems may be *infinite* if the supremum in $d_1$ is taken over the whole of $\mathbb{R}^n$. Therefore, the second way out is to continue to work with bounded regions but to introduce another definition of structural stability.

**Definition 2.21 (Andronov's structural stability)** *A system* (2.26) *defined in a region* $D \subset \mathbb{R}^n$ *is called* structurally stable *in a region* $D_0 \subset D$ *if for any sufficiently* $C^1$*-close in* $D$ *system* (2.27) *there are regions* $U, V \subset D$, $D_0 \subset U$ *such that* (2.26) *is topologically equivalent in* $U$ *to* (2.27) *in* $V$ (*see* **Fig. 2.19**).

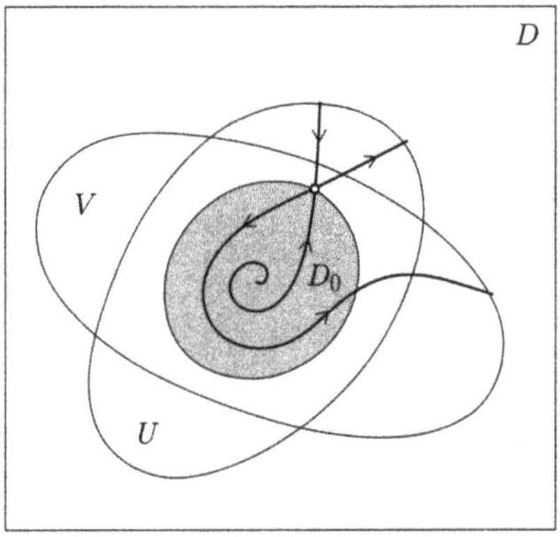


**Fig. 2.19.** Andronov's structural stability.

A parallel definition can be given for discrete-time systems. If (2.26) is structurally stable in $D_0 \subset D$, then it is structurally stable in any region $D_1 \subset D_0$. There are cases when Definitions 2.20 and 2.21 actually coincide.

**Lemma 2.1** *If a system is structurally stable in a region* $D_0$ *with the boundary* $B_0$ *and all its orbits point strictly inside* $B_0$, *then it is strictly structurally stable in* $U = D_0$. □

The following classical theorem gives necessary and sufficient conditions for a continuous-time system in the plane to be structurally stable.

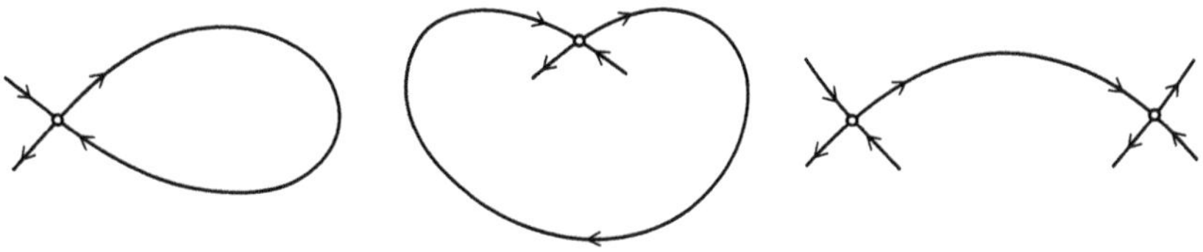

**Fig. 2.20.** Structurally unstable connecting orbits in planar systems.

**Theorem 2.5 (Andronov & Pontryagin [1937])** *A smooth dynamical system*

$$\dot{x} = f(x), \quad x \in \mathbb{R}^2,$$

*is structurally stable in a region $D_0 \subset \mathbb{R}^2$ if and only if*

(i) *it has a finite number of equilibria and limit cycles in $D_0$, and all of them are hyperbolic;*

(ii) *there are no saddle separatrices returning to the same saddle or connecting two different saddles in $D_0$ (see* **Fig. 2.20**). □

**Remark:**

Actually, in their original paper of 1937, Andronov and Pontryagin considered systems with *analytic* right-hand sides in a region $D_0 \subset \mathbb{R}^2$ bounded by a (piecewise) smooth curve. They also assumed that all orbits point strictly inside the region, so they were able to use Definition 2.20. Later, Definition 2.21 was introduced and this restriction on the behavior on the boundary was left out. Moreover, they proved that the homeomorphism $h$ transforming the phase portrait of a perturbed system in $D_0$ into that of the original system can be selected $C^0$-close to the identity map $\text{id}(x) = x$. ◇

This theorem gives the complete description of structurally stable systems on the plane. It is rather obvious, although it has to be proved, that a typical (generic) system on the plane satisfies Andronov-Pontryagin conditions and is, thus, structurally stable. If one considers the bifurcation diagram of a generic planar system depending on $k$ parameters, these are structurally stable systems that occupy $k$-dimensional open *regions* in the parameter space.

One can ask if a similar theorem exists for $n$-dimensional systems. The answer is "no." More precisely, one can establish *sufficient* conditions (called *Morse-Smale conditions*, similar to those in Theorem 2.5) for a continuous-time system to be structurally stable. Nevertheless, there are systems, which do not satisfy these conditions, that are structurally stable. In particular, structurally stable systems can have an *infinite* number of periodic orbits in compact regions. To understand this phenomenon, consider a continuous-time system $\mathbb{R}^3$. Suppose that there is a two-dimensional cross-section $\Sigma$ on which the system defines a Poincaré map generating a *Smale horseshoe* (see Chapter 1 and Example 2.7 in this chapter). Then, the system has an infinite number of saddle cycles in some region of the phase space. A $C^1$-close system will define a $C^1$-close Poincaré map on $\Sigma$. The horseshoe will be slightly deformed, but the geometrical construction we have carried out in Chapter 1 remains valid. Thus, a complex invariant set including an infinite number of saddle cycles will persist under all sufficiently small perturbations. A homeomorphism transforming the corresponding phase portraits can also be constructed.

Moreover, it is possible to construct a system that has no close structurally stable systems. We direct the reader to the literature cited in this chapter's appendix.

## 2.6 Exercises

**(1)** Determine which of the following linear systems has a structurally stable equilibrium at the origin, and sketch its phase portrait:

(a)
$$\begin{cases} \dot{x} = x - 2y, \\ \dot{y} = -2x + 4y; \end{cases}$$

(b)
$$\begin{cases} \dot{x} = 2x + y, \\ \dot{y} = -x; \end{cases}$$

(c)
$$\begin{cases} \dot{x} = x + 2y, \\ \dot{y} = -x - y. \end{cases}$$

**(2)** The following system of partial differential equations is the FitzHugh-Nagumo caricature of the Hodgkin-Huxley equations modeling the nerve impulse propagation along an axon:

$$\begin{cases} \dfrac{\partial u}{\partial t} = \dfrac{\partial^2 u}{\partial x^2} - f_a(u) - v, \\ \frac{\partial v}{\partial t} = bu, \end{cases}$$

where $u = u(x,t)$ represents the membrane potential, $v = v(x,t)$ is a "recovery" variable, $f_a(u) = u(u-a)(u-1), 1 > a > 0, b > 0, -\infty < x < +\infty$, and $t > 0$.

*Traveling waves* are solutions to these equations of the form

$$u(x,t) = U(\xi),\ v(x,t) = V(\xi),\ \xi = x + ct,$$

where $c$ is an a priori unknown wave propagation speed. The functions $U(\xi)$ and $V(\xi)$ are the wave profiles.

(a) Derive a system of three ordinary differential equations for the profiles with "time" $\xi$. (*Hint:* Introduce an extra variable: $W = \dot{U}$.)

(b) Check that for all $c > 0$ the system for the profiles (*the wave system*) has a unique equilibrium with one positive eigenvalue and two eigenvalues with negative real parts. (*Hint:* First, verify this assuming that eigenvalues are real. Then, show that the characteristic equation cannot have roots on the imaginary axis, and finally, use the continuous dependence of the eigenvalues on the parameters.)

(c) Conclude that the equilibrium can be either a saddle or a saddle-focus with a one-dimensional unstable and a two-dimensional stable invariant manifold, and find a condition on the system parameters that defines a boundary between these two cases. Plot several boundaries in the $(a, c)$-plane for different values of $b$ and specify the region corresponding to saddle-foci. (*Hint:* At the boundary the characteristic polynomial $h(\lambda)$ has a double root $\lambda_0 : h(\lambda_0) = h'(\lambda_0) = 0$.)

(d) Sketch possible profiles of traveling *impulses* in both regions. (*Hint:* An impulse corresponds to a solution of the wave system with

$$(U(\xi), V(\xi), W(\xi)) \to (0, 0, 0)$$

as $\xi \to \pm\infty$. See Chapter 6 for further details.)

**(3)** Prove that the system

$$\begin{cases} \dot{x}_1 = -x_1, \\ \dot{x}_2 = -x_2, \end{cases}$$

is locally topologically equivalent near the origin to the system

$$\begin{cases} \dot{x}_1 = -x_1, \\ \dot{x}_2 = -2x_2. \end{cases}$$

(*Hint:* Mimic the proof of Example 2.1 without introducing polar coordinates.) Are the systems diffeomorphic?

**(4) (Diffeomorphic limit cycles)** Show that for diffeomorphic continuous-time systems, corresponding limit cycles have coinciding periods and multipliers. (*Hint:* Use the fact that variational equations around corresponding cycles (considered as autonomous systems with an extra cyclic variable) are diffeomorphic.)

**(5) (Orbital equivalence and global flows)**

(a) Prove that the scalar system

$$\frac{dx}{dt} = x^2, \quad x \in \mathbb{R}^1,$$

having solutions approaching infinity within finite time, and thus defining only *local* flow $\varphi^t : \mathbb{R}^1 \to \mathbb{R}^1$, is orbitally equivalent to the scalar system

$$\frac{dx}{d\tau} = \frac{x^2}{1 + x^2}, \quad x \in \mathbb{R}^1,$$

having no such solutions and therefore defining a *global* flow $\psi^\tau : \mathbb{R}^1 \to \mathbb{R}^1$. How are $t$ and $\tau$ related?

(b) Prove that any smooth system $\dot{x} = f(x)$, $x \in \mathbb{R}^n$, is orbitally equivalent in $\mathbb{R}^n$ to a smooth system defining a global flow $\psi^\tau$ on $\mathbb{R}^n$. (*Hint:* The system

$$\dot{x} = \frac{1}{1 + \|f(x)\|} f(x),$$

where $\|\cdot\|$ is the norm associated with the standard scalar product in $\mathbb{R}^n$, does the job.)

**(6) (One-point parametric portrait)** Construct an autonomous system of differential equations in $\mathbb{R}^3$ depending on two parameters $(\alpha, \beta)$ and having topologically

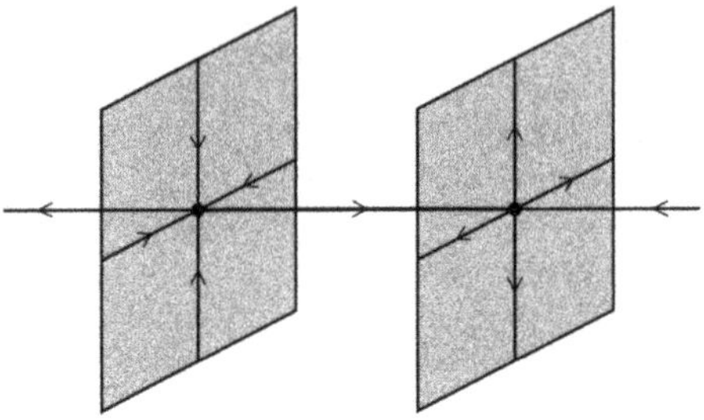

**Fig. 2.21.** Exercise 6.

equivalent phase portraits for all parameter values *except* $(\alpha, \beta) = (0, 0)$. (*Hint:* Use the idea of Example 2.9. At $\alpha = \beta = 0$, the system should have two saddle points with one-dimensional unstable and one-dimensional stable manifolds with coinciding branches (see **Fig. 2.21**).)

**(7) (Induced systems)** Show that the scalar system

$$\dot{y} = \beta y - y^2,$$

which exhibits the *transcritical* bifurcation, is topologically equivalent (in fact, diffeomorphic) to a system induced by the system

$$\dot{x} = \alpha - x^2,$$

which undergoes the fold bifurcation. (*Hint:* See Arrowsmith & Place [1990, p.193].)

**(8) (Proof of Lemma 2.1)**

(a) Prove that a smooth planar system $\dot{x} = f(x)$, $x \in \mathbb{R}^2$, is topologically equivalent (in fact, diffeomorphic) in a region $U$, that is, free of equilibria and periodic orbits and is bounded by two orbits and two smooth curves transversal to orbits, to the system

$$\begin{cases} \dot{y}_1 = 1, \\ \dot{y}_2 = 0, \end{cases}$$

in the unit square $V = \{(y_1, y_2) : |y_1| \leq 1, \ |y_2| \leq 1\}$ (see **Fig. 2.22**).

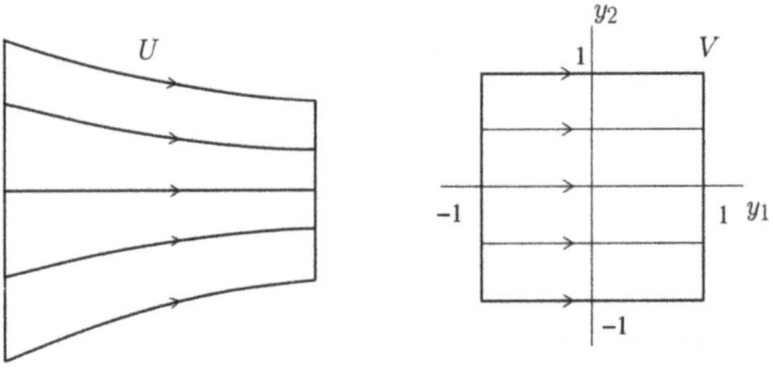


**Fig. 2.22.** Phase portraits in regions $U$ and $V$ are equivalent.

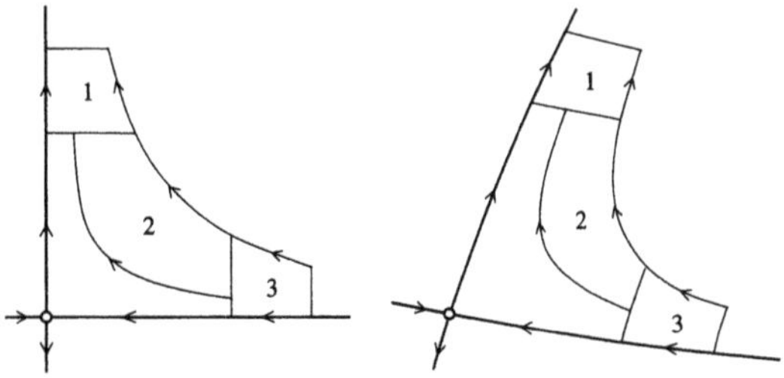


**Fig. 2.23.** Saddles are topologically equivalent.

(b) Generalize this result to $n$-dimensional systems and prove Lemma 2.1.

(c) Prove, using part (a), that two hyperbolic saddle points on the plane have locally topologically equivalent phase portraits. (*Hint:* See **Fig. 2.23**; an explicit map providing the equivalence is constructed in Chapter 6.) Where is the differentiability lost?

## 2.7 Appendix: Bibliographical notes

The notion of topological equivalence of dynamical systems appeared in the paper by Andronov & Pontryagin [1937] devoted to structurally stable systems on the plane. It is extensively used (among other equivalences) in singularity theory to classify singularities of maps and their deformations (Thom [1972], Arnol'd, Varchenko & Guseĭn-Zade [1985], Golubitsky & Schaeffer [1985]).

The local topological equivalence of a nonlinear dynamical system to its linearization at a hyperbolic equilibrium was proved by Grobman [1959] and Hartman [1963]. See Hartman [1964] for details. Local topological equivalence of a map near a hyperbolic fixed point to its linearization has been established by Grobman and Hartman as a by-product of their proofs of the corresponding theorem in the continuous-time case (see also Nitecki [1971]). A constructive proof of the topological equivalence of two linear systems with $n_0 = 0$ and the same $n_-$ and $n_+$ can be found in Arnol'd [1973] and Hale & Koçak [1991].

The Local Stable Manifold Theorem for differential equations originated in works by Hadamard [1901] and Perron [1930]. Complete proofs and generalizations are given by Kelley [1967]; Hirsch, Pugh & Shub [1977] (see also Irwin [1980]). The Local Stable Manifold Theorem for maps is actually the main technical tool used to prove the relevant theorem for differential equations. Therefore, its proof can be found in the cited literature, for example, in Hartman [1964] or Nitecki [1971]. The latter reference also contains a proof that the stable and unstable sets of a hyperbolic fixed point are images of $\mathbb{R}^{n_-}$ and $\mathbb{R}^{n_+}$ under injective immersion. Many results on hyperbolic invariant sets are proved in [Katok & Hasselblatt 1995].

The complex structure generated by the transversal intersection of the stable and unstable manifolds of a hyperbolic fixed point was discovered by Poincaré [1892,1893,1899] while analyzing area-preserving (conservative) maps appearing in celestial mechanics. Further analysis of this phenomenon in the conservative case was undertaken by Birkhoff [1935], with particular emphasis to the statistical properties of corresponding orbits. The nonconservative case was studied by Smale [1963], Neimark [1967], and Shil'nikov [1967*b*]. A nice exposition of this topic is given by Moser [1973].

There are two main approaches to studying bifurcations in dynamical systems. The first one, originating in the works by Poincaré, is to analyze the appearance (*branching*) of new phase objects of a certain type (equilibria or cycles, for example) from some known ones when parameters of the system vary. This approach led to the development of branching theory for equilibrium solutions of finite- and infinite-dimensional nonlinear equations (see, e.g, Vaĭnberg & Trenogin [1974], and Chow & Hale [1982]). The approach also proved to be a powerful tool to study some global bifurcations (see the bibliographical notes in Chapter 6). The second approach, going back to Andronov [1933] and reintroduced by Thom [1972] in order to classify gradient systems $\dot{x} = -\text{grad } V(x, \alpha)$, is to study rearrangements (*bifurcations*) of the whole phase portrait under variations of parameters. In principle, the branching analysis should precede more complete phase portrait study, but there are many cases where complete phase portraits are unavailable and studying certain solutions is the only way to get some information on the bifurcation.

Bifurcations of phase portraits of two-dimensional dynamical systems have been studied in great detail by Andronov and his co-workers in 1930-1950 and summarized in the classical book whose English translation is available as Andronov, Leontovich, Gordon & Maier [1973]. In his famous lectures, Arnol'd [1972] first applied many ideas from singularity theory of differentiable maps to dynamical systems (a similar approach was developed by Takens [1974*a*]). The notions of topological equivalence of parameter-dependent systems (*families*), versal deformations for local bifurcations, as well as many original results, were first presented in Arnold's lectures and then in the book by Arnol'd [1983]. Notice that in the literature in English versal deformations are often called *universal unfoldings* following terminology from singularity theory. A fundamental survey of bifurcation theory, including results on global bifurcations, is given by Arnol'd, Afraimovich, Il'yashenko & Shil'nikov [1994]. The standard graduate-level texts on bifurcation theory include those by Guckenheimer & Holmes [1983], Arrowsmith & Place [1990], Wiggins [1990], and more recent books by Shilnikov, Shilnikov, Turaev & Chua[1998, 2001].

Structurally stable two-dimensional ODE systems were studied by Andronov & Pontryagin [1937] under the name *coarse* (or *rough*) systems. Actually, they included the requirement that the homeomorphism transforming the phase portraits be close to the identity. Peixoto [1962] proved that a typical system on a two-dimensional manifold is structurally stable. To discuss "typicality" one has to specify a space $\mathcal{D}$ of considered dynamical systems. Then, a property is called *typical* (or *generic*) if systems from the intersection of a countable number of open and dense subsets of $\mathcal{D}$ possess this property (see [Wiggins 1990] for an introductory discussion). A class of structurally stable, multidimensional dynamical systems (called *Morse-Smale systems*) has been identified by Smale [1961, 1967]. Such systems have only a finite number of equilibria and cycles, all of which are hyperbolic and have their stable and unstable invariant manifolds intersecting at nonzero angles (*transversally*). There are structurally stable systems that do not satisfy Morse-Smale criteria, in particular, having an *infinite* number of hyperbolic cycles [Smale 1963]. Moreover, structural stability is *not* a typical property for multidimensional dynamical systems, and structurally stable systems are not dense in a space $\mathcal{D}$ of smooth dynamical systems [Smale 1966]. The interested reader is referred to Nitecki [1971], Arnol'd [1983], and Katok & Hasselblatt [1995] for more information.

# 3

# One-Parameter Bifurcations of Equilibria in Continuous-Time Dynamical Systems

In this chapter we formulate conditions defining the simplest bifurcations of equilibria in $n$-dimensional continuous-time systems: the fold and the Hopf bifurcations. Then we study these bifurcations in the lowest possible dimensions: the fold bifurcation for scalar systems and the Hopf bifurcation for planar systems. Appendixes A and B are devoted to technical questions appearing in the analysis of Hopf bifurcation: Effects of higher-order terms and a general theory of Poincaré normal forms, respectively. Chapter 5 shows how to "lift" the results of this chapter to $n$-dimensional situations.

## 3.1 Simplest bifurcation conditions

Consider a continuous-time system depending on a parameter

$$\dot{x} = f(x, \alpha), \quad x \in \mathbb{R}^n, \ \alpha \in \mathbb{R}^1,$$

where $f$ is smooth with respect to both $x$ and $\alpha$. Let $x = x_0$ be a hyperbolic equilibrium in the system for $\alpha = \alpha_0$. As we have seen in Chapter 2, under a small parameter variation the equilibrium moves slightly but remains hyperbolic. Therefore, we can vary the parameter further and monitor the equilibrium. It is clear that there are, generically, only two ways in which the hyperbolicity condition can be violated. Either a simple real eigenvalue approaches zero and we have $\lambda_1 = 0$ (see **Fig. 3.1**(a)), or a pair of simple complex eigenvalues reaches the imaginary axis and we have $\lambda_{1,2} = \pm i\omega_0$, $\omega_0 > 0$ (see **Fig. 3.1**(b)) for some value of the parameter. It is obvious (and can be rigorously formalized) that we need more parameters to allocate extra eigenvalues on the imaginary axis. Notice that this might not be true if the system has some special properties, such as a symmetry (see Chapter 7).

The rest of the chapter will essentially be devoted to the proof that a nonhyperbolic equilibrium satisfying one of the above conditions is structurally *unstable* and to the analysis of the corresponding bifurcations of the local

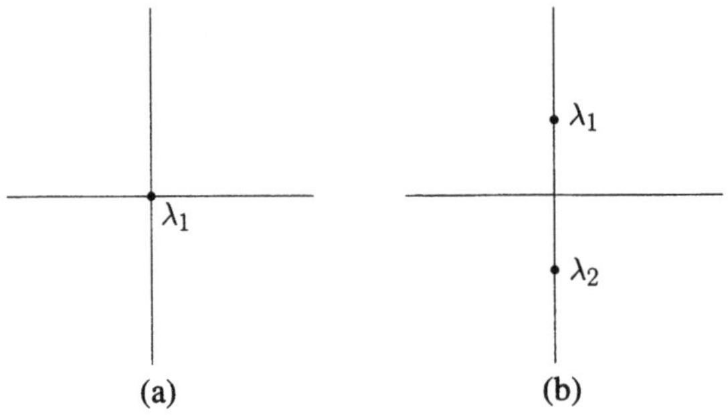


**Fig. 3.1.** Codim 1 critical cases.

phase portrait under variation of the parameter. We have already seen several examples of these bifurcations in Chapter 2. Let us finish this section with the following two definitions.

**Definition 3.1** *The bifurcation associated with the appearance of* $\lambda_1 = 0$ *is called a* fold *(or* tangent*) bifurcation.*

**Remark:**

This bifurcation has a lot of other names, including *limit point, saddle-node bifurcation,* and *turning point.* $\diamondsuit$

**Definition 3.2** *The bifurcation corresponding to the presence of* $\lambda_{1,2} = \pm i\omega_0$, $\omega_0 > 0$, *is called a* Hopf *(or* Andronov-Hopf*) bifurcation.*

Notice that the tangent bifurcation is possible if $n \geq 1$, but for the Hopf bifurcation we need $n \geq 2$.

## 3.2 The normal form of the fold bifurcation

Consider the following one-dimensional dynamical system depending on one parameter:

$$\dot{x} = \alpha + x^2 \equiv f(x, \alpha). \tag{3.1}$$

At $\alpha = 0$ this system has a nonhyperbolic equilibrium $x_0 = 0$ with $\lambda = f_x(0,0) = 0$. The behavior of the system for all the other values of $\alpha$ is also clear (see **Fig. 3.2**). For $\alpha < 0$ there are two equilibria in the system: $x_{1,2}(\alpha) = \pm\sqrt{-\alpha}$, the left one of which is stable, while the right one is unstable. For $\alpha > 0$ there are no equilibria in the system. While $\alpha$ crosses zero from negative to positive values, the two equilibria (stable and unstable) "collide," forming at $\alpha = 0$ an equilibrium with $\lambda = 0$, and disappear. This is a fold bifurcation. The term "collision" is appropriate since the speed of approach $(\frac{d}{d\alpha}x_{1,2}(\alpha))$ of the equilibria tends to infinity as $\alpha \to 0$.

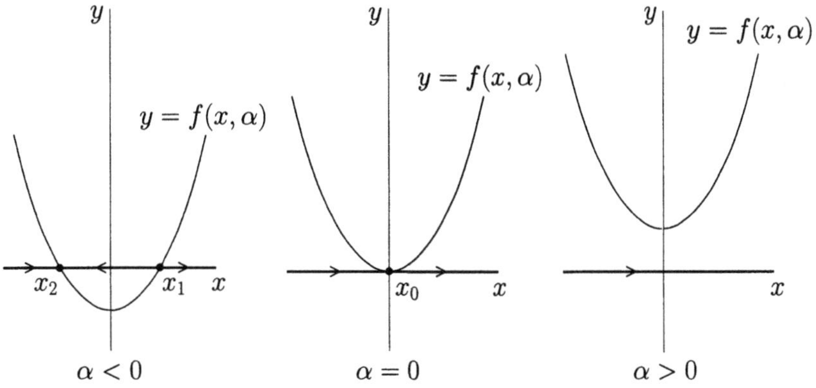


**Fig. 3.2.** Fold bifurcation.

There is another way of presenting this bifurcation: plotting a bifurcation diagram in the direct product of the phase and parameter spaces (simply, the $(x, \alpha)$-plane). The equation

$$f(x, \alpha) = 0$$

defines an *equilibrium manifold*, which is simply the parabola $\alpha = -x^2$ (see **Fig. 3.3**). This presentation displays the bifurcation picture at once. Fixing some $\alpha$, we can easily determine the number of equilibria in the system for this parameter value. The projection of the equilibrium manifold into the parameter axis has a *singularity* of the fold type at $(x, \alpha) = (0, 0)$.

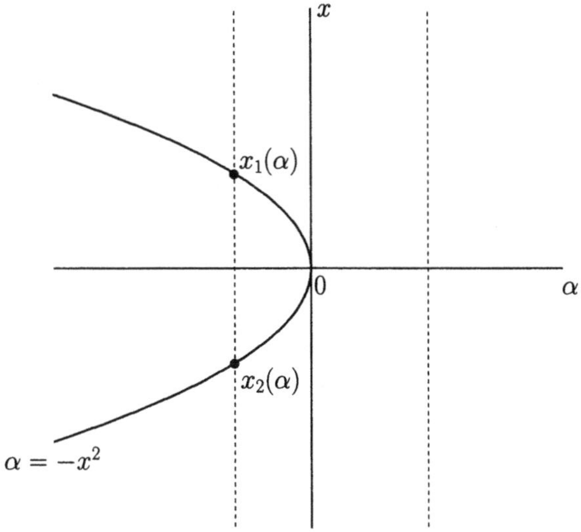


**Fig. 3.3.** Fold bifurcation in the phase-parameter space.

**Remark:**

The system $\dot{x} = \alpha - x^2$ can be considered in the same way. The analysis reveals two equilibria appearing for $\alpha > 0$. $\diamondsuit$

Now add to system (3.1) higher-order terms that can depend smoothly on the parameter. It happens that these terms do not change qualitatively the behavior of the system near the origin $x = 0$ for parameter values close to $\alpha = 0$. Actually, the following lemma holds:

**Lemma 3.1** *The system*

$$\dot{x} = \alpha + x^2 + O(x^3)$$

*is locally topologically equivalent near the origin to the system*

$$\dot{x} = \alpha + x^2.$$

**Proof:**

The proof goes through two steps. It is based on the fact that for scalar systems a homeomorphism mapping equilibria into equilibria will also map their connecting orbits.

*Step 1* (*Analysis of equilibria*). Introduce a scalar variable $y$ and write the first system as

$$\dot{y} = F(y, \alpha) = \alpha + y^2 + \psi(y, \alpha), \tag{3.2}$$

where $\psi = O(y^3)$ is a smooth functions of $(y, \alpha)$ near $(0, 0)$. Consider the equilibrium manifold of (3.2) near the origin $(0, 0)$ of the $(y, \alpha)$-plane:

$$M = \{(y, \alpha) : F(y, \alpha) = \alpha + y^2 + \psi(y, \alpha) = 0\}.$$

The curve $M$ passes through the origin ($F(0, 0) = 0$). By the Implicit Function Theorem (since $F_\alpha(0, 0) = 1$), it can be locally parametrized by $y$:

$$M = \{(y, \alpha) : \alpha = g(y)\},$$

where $g$ is smooth and defined for small $|y|$. Moreover,

$$g(y) = -y^2 + O(y^3)$$

(check!). Thus, for any sufficiently small $\alpha < 0$, there are two equilibria of (3.2) near the origin in (3.2), $y_1(\alpha)$ and $y_2(\alpha)$, which are close to the equilibria of (3.1), i.e., $x_1(\alpha) = +\sqrt{-\alpha}$ and $x_2(\alpha) = -\sqrt{-\alpha}$, for the same parameter value (see **Fig. 3.4**).

*Step 2* (*Homeomorphism construction*). For small $|\alpha|$, construct a parameter-dependent map $y = h_\alpha(x)$ as following. For $\alpha \geq 0$ take the identity map

$$h_\alpha(x) = x.$$

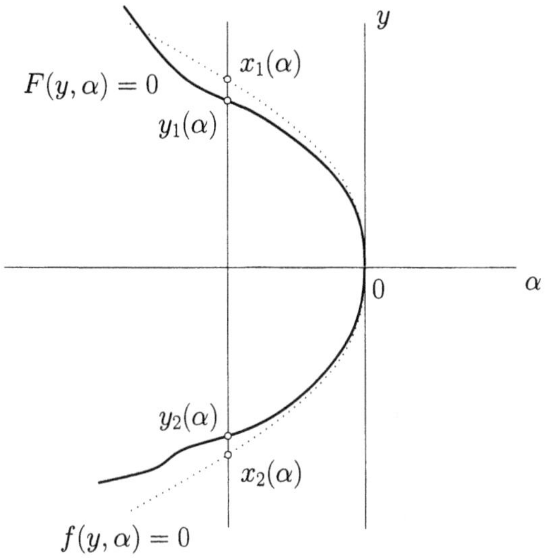


**Fig. 3.4.** Fold bifurcation for the perturbed system.

For $\alpha < 0$ take a linear transformation

$$h_\alpha(x) = a(\alpha) + b(\alpha)x,$$

where the coefficients $a, b$ are uniquely determined by the conditions

$$h_\alpha(x_j(\alpha)) = y_j(\alpha), \quad j = 1, 2,$$

(find them!). The constructed map $h_\alpha : \mathbb{R}^1 \to \mathbb{R}^1$ is a homeomorphism mapping orbits of (3.1) near the origin into the corresponding orbits of (3.2), preserving the direction of time. Chapter 2 identified this property as the *local topological equivalence* of parameter-dependent systems.

Although it is not required in the book for the homeomorphism $h_\alpha$ to depend continuously on $\alpha$ (see Remark after Definition 2.14), this property holds here since $h_\alpha$ tends to the identity map as negative $\alpha \to 0$. □

## 3.3 Generic fold bifurcation

We shall show that system (3.1) (with a possible sign change of the $x^2$-term) is a topological normal form of a generic one-dimensional system having a fold bifurcation. In Chapter 5 we will also see that in some strong sense it describes the fold bifurcation in a generic $n$-dimensional system.

Suppose the system

$$\dot{x} = f(x, \alpha), \quad x \in \mathbb{R}^1, \ \alpha \in \mathbb{R}^1, \tag{3.3}$$

with a smooth $f$ has at $\alpha = 0$ the equilibrium $x = 0$ with $\lambda = f_x(0,0) = 0$. Expand $f(x,\alpha)$ as a Taylor series with respect to $x$ at $x = 0$:

$$f(x,\alpha) = f_0(\alpha) + f_1(\alpha)x + f_2(\alpha)x^2 + O(x^3).$$

Two conditions are satisfied: $f_0(0) = f(0,0) = 0$ (*equilibrium condition*) and $f_1(0) = f_x(0,0) = 0$ (*fold bifurcation condition*).

The main idea of the following simple calculations is this: By smooth invertible changes of the coordinate and the parameter, transform system (3.3) into the form (3.1) up to and including the second-order terms. Then, Lemma 3.1 can be applied, thus making it possible to drop the higher-order terms. While proceeding, we will see that some extra *nondegeneracy* and *transversality conditions* must be imposed to make these transformations possible. These conditions will actually specify which one-parameter system having a fold bifurcation can be considered as *generic*. This idea works for all local bifurcation problems. We will proceed in exactly this way in analyzing the Hopf bifurcation later in this chapter.

*Step 1 (Shift of the coordinate)*. Perform a linear coordinate shift by introducing a new variable $\xi$:

$$\xi = x + \delta, \tag{3.4}$$

where $\delta = \delta(\alpha)$ is an a priori unknown function that will be defined later. The inverse coordinate transformation is

$$x = \xi - \delta.$$

Substituting (3.4) into (3.3) yields

$$\dot{\xi} = \dot{x} = f_0(\alpha) + f_1(\alpha)(\xi - \delta) + f_2(\alpha)(\xi - \delta)^2 + \cdots.$$

Therefore,

$$\begin{aligned}\dot{\xi} = & \left[f_0(\alpha) - f_1(\alpha)\delta + f_2(\alpha)\delta^2 + O(\delta^3)\right] \\ & + \left[f_1(\alpha) - 2f_2(\alpha)\delta + O(\delta^2)\right]\xi \\ & + \left[f_2(\alpha) + O(\delta)\right]\xi^2 \\ & + O(\xi^3).\end{aligned}$$

Assume that

(A.1) $$f_2(0) = \frac{1}{2} f_{xx}(0,0) \neq 0.$$

Then there is a smooth function $\delta(\alpha)$ that annihilates the linear term in the above equation for all sufficiently small $|\alpha|$. This can be justified with the Implicit Function Theorem. Indeed, the condition for the linear term to vanish can be written as

$$F(\alpha, \delta) \equiv f_1(\alpha) - 2f_2(\alpha)\delta + \delta^2\psi(\alpha, \delta) = 0$$

with some smooth function $\psi$. We have

$$F(0,0) = 0, \quad \left.\frac{\partial F}{\partial \delta}\right|_{(0,0)} = -2f_2(0) \neq 0, \quad \left.\frac{\partial F}{\partial \alpha}\right|_{(0,0)} = f_1'(0),$$

which implies (local) existence and uniqueness of a smooth function $\delta = \delta(\alpha)$ such that $\delta(0) = 0$ and $F(\alpha, \delta(\alpha)) \equiv 0$. It also follows that

$$\delta(\alpha) = \frac{f_1'(0)}{2f_2(0)}\alpha + O(\alpha^2).$$

The equation for $\xi$ now contains no linear terms:

$$\dot{\xi} = [f_0'(0)\alpha + O(\alpha^2)] + [f_2(0) + O(\alpha)]\xi^2 + O(\xi^3). \tag{3.5}$$

*Step 2* (*Introduce a new parameter*). Consider as a new parameter $\mu = \mu(\alpha)$ the constant ($\xi$-independent) term of (3.5):

$$\mu = f_0'(0)\alpha + \alpha^2\phi(\alpha),$$

where $\phi$ is some smooth function. We have:

(a) $\mu(0) = 0$;
(b) $\mu'(0) = f_0'(0) = f_\alpha(0,0)$.

If we assume that

(A.2) $$f_\alpha(0,0) \neq 0,$$

then the Inverse Function Theorem implies local existence and uniqueness of a smooth inverse function $\alpha = \alpha(\mu)$ with $\alpha(0) = 0$. Therefore, equation (3.5) now reads

$$\dot{\xi} = \mu + b(\mu)\xi^2 + O(\xi^3),$$

where $b(\mu)$ is a smooth function with $b(0) = f_2(0) \neq 0$ due to the first assumption (A.1).

*Step 3* (*Final scaling*). Let $\eta = |b(\mu)|\xi$ and $\beta = |b(\mu)|\mu$. Then we get

$$\dot{\eta} = \beta + s\eta^2 + O(\eta^3),$$

where $s = \text{sign } b(0) = \pm 1$.

Therefore, the following theorem is proved.

**Theorem 3.1** *Suppose that a one-dimensional system*

$$\dot{x} = f(x, \alpha), \quad x \in \mathbb{R}^1, \ \alpha \in \mathbb{R}^1,$$

*with smooth $f$, has at $\alpha = 0$ the equilibrium $x = 0$, and let $\lambda = f_x(0,0) = 0$. Assume that the following conditions are satisfied:*

(A.1) $f_{xx}(0,0) \neq 0$;
(A.2) $f_\alpha(0,0) \neq 0$.

*Then there are invertible coordinate and parameter changes transforming the system into*

$$\dot{\eta} = \beta \pm \eta^2 + O(\eta^3). \ \square$$

Using Lemma 3.1, we can eliminate $O(\eta^3)$ terms and finally arrive at the following general result.

**Theorem 3.2 (Topological normal form for the fold bifurcation)** *Any generic scalar one-parameter system*

$$\dot{x} = f(x, \alpha),$$

*having at $\alpha = 0$ the equilibrium $x = 0$ with $\lambda = f_x(0,0) = 0$, is locally topologically equivalent near the origin to one of the following normal forms:*

$$\dot{\eta} = \beta \pm \eta^2. \ \square$$

**Remark:**

The genericity conditions in Theorem 3.2 are the nondegeneracy condition (A.1) and the transversality condition (A.2) from Theorem 3.1. $\diamondsuit$

## 3.4 The normal form of the Hopf bifurcation

Consider the following system of two differential equations depending on one parameter:

$$\begin{cases} \dot{x}_1 = \alpha x_1 - x_2 - x_1(x_1^2 + x_2^2), \\ \dot{x}_2 = x_1 + \alpha x_2 - x_2(x_1^2 + x_2^2). \end{cases} \tag{3.6}$$

This system has the equilibrium $x_1 = x_2 = 0$ for all $\alpha$ with the Jacobian matrix

$$A = \begin{pmatrix} \alpha & -1 \\ 1 & \alpha \end{pmatrix}$$

having eigenvalues $\lambda_{1,2} = \alpha \pm i$. Introduce the complex variable $z = x_1 + ix_2, \bar{z} = x_1 - ix_2, |z|^2 = z\bar{z} = x_1^2 + x_2^2$. This variable satisfies the differential equation

$$\dot{z} = \dot{x}_1 + i\dot{x}_2 = \alpha(x_1 + ix_2) + i(x_1 + ix_2) - (x_1 + ix_2)(x_1^2 + x_2^2),$$

and we can therefore rewrite system (3.6) in the *complex* form

$$\dot{z} = (\alpha + i)z - z|z|^2. \tag{3.7}$$

Finally, using the representation $z = \rho e^{i\varphi}$, we obtain

$$\dot{z} = \dot{\rho}e^{i\varphi} + \rho i\dot{\varphi}e^{i\varphi},$$

or

$$\dot{\rho}e^{i\varphi} + i\rho\dot{\varphi}e^{i\varphi} = \rho e^{i\varphi}(\alpha + i - \rho^2),$$

which gives the *polar* form of system (3.6):

$$\begin{cases} \dot{\rho} = \rho(\alpha - \rho^2), \\ \dot{\varphi} = 1. \end{cases} \tag{3.8}$$

Bifurcations of the phase portrait of the system as $\alpha$ passes through zero can easily be analyzed using the polar form since the equations for $\rho$ and $\varphi$ in (3.8) are uncoupled. The first equation (which should obviously be considered only for $\rho \geq 0$) has the equilibrium point $\rho = 0$ for all values of $\alpha$. The equilibrium is linearly stable if $\alpha < 0$; it remains stable at $\alpha = 0$ but *nonlinearly* (so the rate of solution convergence to zero is no longer exponential); for $\alpha > 0$ the equilibrium becomes linearly unstable. Moreover, there is an additional stable equilibrium point $\rho_0(\alpha) = \sqrt{\alpha}$ for $\alpha > 0$. The second equation describes a rotation with constant speed. Thus, by superposition of the motions defined by the two equations of (3.8), we obtain the following bifurcation diagram for the original two-dimensional system (3.6) (see **Fig. 3.5**). The system always has an equilibrium at the origin. This equilibrium is a stable focus for $\alpha < 0$ and an unstable focus for $\alpha > 0$. At the critical parameter value $\alpha = 0$ the equilibrium is nonlinearly stable and topologically equivalent to the focus. Sometimes it is called a *weakly attracting focus.* This equilibrium is surrounded for $\alpha > 0$ by an isolated closed orbit (*limit cycle*) that is unique and stable.

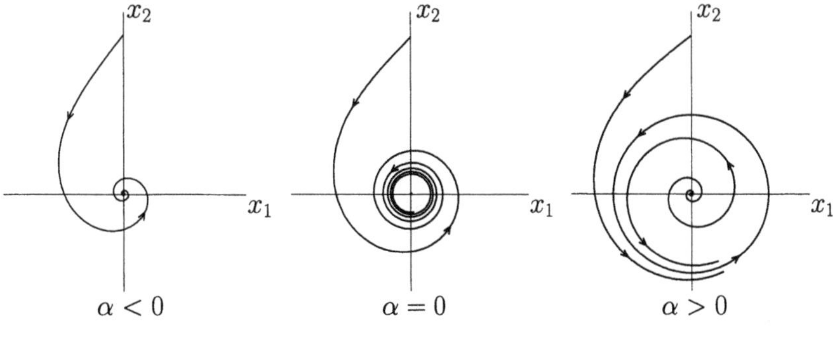


**Fig. 3.5.** Supercritical Hopf bifurcation.

The cycle is a circle of radius $\rho_0(\alpha) = \sqrt{\alpha}$. All orbits starting outside or inside the cycle except at the origin tend to the cycle as $t \to +\infty$. This is an Andronov-Hopf bifurcation.

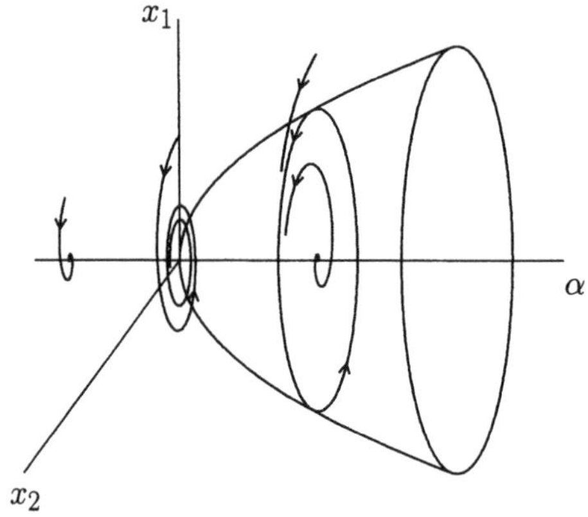


**Fig. 3.6.** Supercritical Hopf bifurcation in the phase-parameter space.

This bifurcation can also be presented in $(x, y, \alpha)$-space (see **Fig. 3.6**). The appearing $\alpha$-family of limit cycles forms a *paraboloid* surface.

A system having nonlinear terms with the opposite sign,

$$\begin{cases} \dot{x}_1 = \alpha x_1 - x_2 + x_1(x_1^2 + x_2^2), \\ \dot{x}_2 = x_1 + \alpha x_2 + x_2(x_1^2 + x_2^2), \end{cases} \tag{3.9}$$

which has the complex form

$$\dot{z} = (\alpha + i)z + z|z|^2,$$

can be analyzed in the same way (see **Figs. 3.7** and **3.8**). The system undergoes the Andronov-Hopf bifurcation at $\alpha = 0$. Contrary to system (3.6), there is an *unstable* limit cycle in (3.9), which disappears when $\alpha$ crosses zero from negative to positive values. The equilibrium at the origin has the same stability for $\alpha \neq 0$ as in system (3.6): It is stable for $\alpha < 0$ and unstable for $\alpha > 0$. Its stability at the critical parameter value is opposite to that in (3.6): It is (nonlinearly) unstable at $\alpha = 0$.

**Remarks:**

(1) We have seen that there are *two* types of Andronov-Hopf bifurcation. The bifurcation in system (3.6) is often called *supercritical* because the cycle exists for positive values of the parameter $\alpha$ ("after" the bifurcation). The bifurcation in system (3.9) is called *subcritical* since the cycle is present "before" the bifurcation. It is clear that this terminology is somehow misleading since "after" and "before" depend on the chosen direction of parameter variation.

(2) In both cases we have a *loss of stability* of the equilibrium at $\alpha = 0$ under increase of the parameter. In the first case (with "$-$" in front of the cubic terms), the stable equilibrium is replaced by a stable limit cycle of small amplitude. Therefore, the system "remains" in a neigborhood of the equilibrium and we have a *soft* or *noncatastrophic* stability loss. In the second case (with "$+$" in front of the cubic terms), the region of attraction of the equilibrium point is bounded by the unstable cycle, which "shrinks" as the parameter approaches its critical value and disappears. Thus, the system is "pushed out" from a neigborhood of the equilibrium, giving us a *sharp* or *catastrophic* loss of stability. If the system loses stability softly, it is well "controllable": If we make the parameter negative again, the system returns to the stable equilibrium. On the contrary, if the system loses its stability sharply, resetting to a negative value of the parameter may not return the system back to the stable equilibrium since it may have left its region of attraction. Notice that the type of Andronov-Hopf bifurcation is determined by the stability of the equilibrium at the critical parameter value.

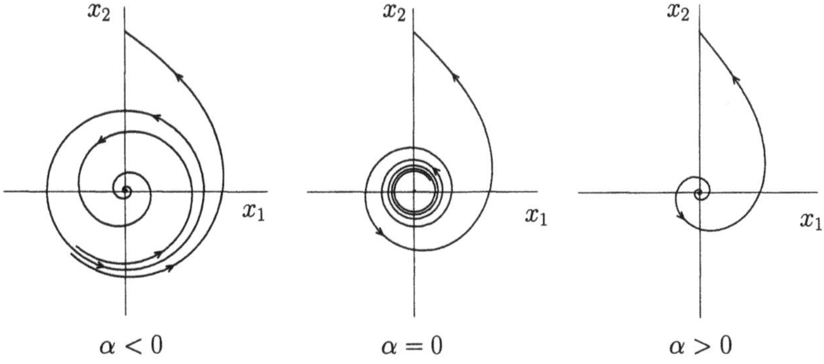


**Fig. 3.7.** Subcritical Hopf bifurcation.

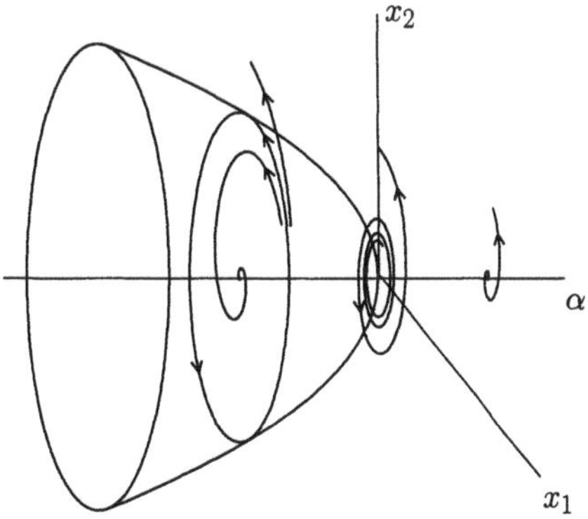


**Fig. 3.8.** Subcritical Hopf bifurcation in the phase-parameter space.

(3) The above interpretation of super- and subcritical Hopf bifurcations should be considered with care. If we consider $\alpha$ as a slow variable and add to system (3.6) the third equation

$$\dot{\alpha} = \varepsilon,$$

with $\varepsilon$ small but positive, then the resulting *time series* $(x(t), y(t), \alpha(t))$ will demonstrate some degree of "sharpness." If the solution starts at some initial point $(x_0, y_0, \alpha_0)$ with $\alpha_0 < 0$, it then converges to the origin and remains very close to it even if $\alpha$ becomes positive, thus demonstrating no oscillations. Only when $\alpha$ reaches some finite positive value will the solution leave the equilibrium "sharply" and start to oscillate with a relatively large amplitude.

(4) Finally, consider a system without nonlinear terms:

$$\dot{z} = (\alpha + i)z.$$

This system also has a family of periodic orbits of increasing amplitude, but all of them are present at $\alpha = 0$ when the system has a *center* at the origin (see **Fig. 3.9**). It can be said that the limit cycle paraboloid "degenerates"

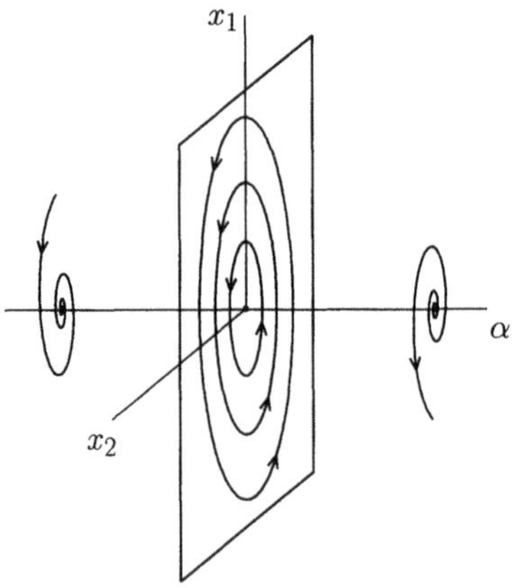


**Fig. 3.9.** "Hopf bifurcation" in a linear system.

into the plane $\alpha = 0$ in $(x, y, \alpha)$-space in this case. This observation makes natural the appearance of small limit cycles in the nonlinear case. $\Diamond$

Let us now add some higher-order terms to system (3.6) and write it in the vector form

$$\begin{pmatrix} \dot{x}_1 \\ \dot{x}_2 \end{pmatrix} = \begin{pmatrix} \alpha & -1 \\ 1 & \alpha \end{pmatrix} \begin{pmatrix} x_1 \\ x_2 \end{pmatrix} - (x_1^2 + x_2^2) \begin{pmatrix} x_1 \\ x_2 \end{pmatrix} + O(\|x\|^4), \tag{3.10}$$

where $x = (x_1, x_2)^T$, $\|x\|^2 = x_1^2 + x_2^2$, and $O(\|x\|^4)$ terms can smoothly depend on $\alpha$. The following lemma will be proved in Appendix A to this chapter.

**Lemma 3.2** *System* (3.10) *is locally topologically equivalent near the origin to system* (3.6). □

Therefore, the higher-order terms do not affect the bifurcation behavior of the system.

## 3.5 Generic Hopf bifurcation

We now shall prove that any generic two-dimensional system undergoing a Hopf bifurcation can be transformed into the form (3.10) with a possible difference in the sign of the cubic terms.

Consider a system

$$\dot{x} = f(x, \alpha), \quad x = (x_1, x_2)^T \in \mathbb{R}^2, \ \alpha \in \mathbb{R}^1,$$

with a smooth function $f$, which has at $\alpha = 0$ the equilibrium $x = 0$ with eigenvalues $\lambda_{1,2} = \pm i\omega_0$, $\omega_0 > 0$. By the Implicit Function Theorem, the system has a unique equilibrium $x_0(\alpha)$ in some neigborhood of the origin for all sufficiently small $|\alpha|$ since $\lambda = 0$ is not an eigenvalue of the Jacobian matrix. We can perform a coordinate shift, placing this equilibrium at the origin. Therefore, we may assume without loss of generality that $x = 0$ is the equilibrium point of the system for $|\alpha|$ sufficiently small. Thus, the system can be written as

$$\dot{x} = A(\alpha)x + F(x, \alpha), \tag{3.11}$$

where $F$ is a smooth vector function whose components $F_{1,2}$ have Taylor expansions in $x$ starting with at least quadratic terms, $F = O(\|x\|^2)$. The Jacobian matrix $A(\alpha)$ can be written as

$$A(\alpha) = \begin{pmatrix} a(\alpha) & b(\alpha) \\ c(\alpha) & d(\alpha) \end{pmatrix}$$

with smooth functions of $\alpha$ as its elements. Its eigenvalues are the roots of the characteristic equation

$$\lambda^2 - \sigma\lambda + \Delta = 0,$$

where $\sigma = \sigma(\alpha) = a(\alpha) + d(\alpha) = \text{tr } A(\alpha)$, and $\Delta = \Delta(\alpha) = a(\alpha)d(\alpha) - b(\alpha)c(\alpha) = \det A(\alpha)$. So,

$$\lambda_{1,2}(\alpha) = \frac{1}{2}\left(\sigma(\alpha) \pm \sqrt{\sigma^2(\alpha) - 4\Delta(\alpha)}\right).$$

The Hopf bifurcation condition implies

$$\sigma(0) = 0, \ \Delta(0) = \omega_0^2 > 0.$$

For small $|\alpha|$ we can introduce

$$\mu(\alpha) = \frac{1}{2}\sigma(\alpha), \; \omega(\alpha) = \frac{1}{2}\sqrt{4\Delta(\alpha) - \sigma^2(\alpha)}$$

and therefore obtain the following representation for the eigenvalues:

$$\lambda_1(\alpha) = \lambda(\alpha), \; \lambda_2(\alpha) = \overline{\lambda(\alpha)},$$

where

$$\lambda(\alpha) = \mu(\alpha) + i\omega(\alpha), \; \mu(0) = 0, \; \omega(0) = \omega_0 > 0.$$

**Lemma 3.3** *By introducing a complex variable $z$, system* (3.11) *can be written for sufficiently small $|\alpha|$ as a single equation:*

$$\dot{z} = \lambda(\alpha)z + g(z, \bar{z}, \alpha), \tag{3.12}$$

*where $g = O(|z|^2)$ is a smooth function of $(z, \bar{z}, \alpha)$.*

**Proof:**

Let $q(\alpha) \in \mathbb{C}^2$ be an eigenvector of $A(\alpha)$ corresponding to the eigenvalue $\lambda(\alpha)$:

$$A(\alpha)q(\alpha) = \lambda(\alpha)q(\alpha),$$

and let $p(\alpha) \in \mathbb{C}^2$ be an eigenvector of the transposed matrix $A^T(\alpha)$ corresponding to its eigenvalue $\overline{\lambda(\alpha)}$:

$$A^T(\alpha)p(\alpha) = \overline{\lambda(\alpha)}p(\alpha).$$

It is always possible to normalize $p$ *with respect to* $q$:

$$\langle p(\alpha), q(\alpha)\rangle = 1,$$

where $\langle \cdot, \cdot \rangle$ means the standard scalar product in $\mathbb{C}^2$: $\langle p, q\rangle = \bar{p}_1 q_1 + \bar{p}_2 q_2$. Any vector $x \in \mathbb{R}^2$ can be uniquely represented for any small $\alpha$ as

$$x = zq(\alpha) + \bar{z}\bar{q}(\alpha) \tag{3.13}$$

for some complex $z$, provided the eigenvectors are specified. Indeed, we have an *explicit* formula to determine $z$:

$$z = \langle p(\alpha), x\rangle.$$

To verify this formula (which results from taking the scalar product with $p$ of both sides of (3.13)), we have to prove that $\langle p(\alpha), \bar{q}(\alpha)\rangle = 0$. This is the case since

$$\langle p, \bar{q}\rangle = \langle p, \frac{1}{\bar{\lambda}}A\bar{q}\rangle = \frac{1}{\bar{\lambda}}\langle A^T p, \bar{q}\rangle = \frac{\lambda}{\bar{\lambda}}\langle p, \bar{q}\rangle$$

and therefore

$$\left(1 - \frac{\lambda}{\bar{\lambda}}\right)\langle p, \bar{q}\rangle = 0.$$

But $\lambda \neq \bar{\lambda}$ because for all sufficiently small $|\alpha|$ we have $\omega(\alpha) > 0$. Thus, the only possibility is $\langle p, \bar{q} \rangle = 0$.

The complex variable $z$ obviously satisfies the equation

$$\dot{z} = \lambda(\alpha)z + \langle p(\alpha), F(zq(\alpha) + \bar{z}\bar{q}(\alpha), \alpha)\rangle,$$

having the required[1] form (3.12) with

$$g(z, \bar{z}, \alpha) = \langle p(\alpha), F(zq(\alpha) + \bar{z}\bar{q}(\alpha), \alpha)\rangle. \square$$

There is no reason to expect $g$ to be an analytic function of $z$ (i.e., $\bar{z}$-independent). Write $g$ as a formal Taylor series in two complex variables ($z$ and $\bar{z}$):

$$g(z, \bar{z}, \alpha) = \sum_{k+l \geq 2} \frac{1}{k!l!} g_{kl}(\alpha) z^k \bar{z}^l,$$

where

$$g_{kl}(\alpha) = \left. \frac{\partial^{k+l}}{\partial z^k \partial \bar{z}^l} \langle p(\alpha), F(zq(\alpha) + \bar{z}\bar{q}(\alpha), \alpha)\rangle \right|_{z=0},$$

for $k + l \geq 2,\ k, l = 0, 1, \ldots$.

**Remarks:**

(1) There are several (equivalent) ways to prove Lemma 3.3. The selected one fits well into the framework of Chapter 5, where we will consider the Hopf bifurcation in $n$-dimensional systems.

(2) Equation (3.13) imposes a linear relation between $(x_1, x_2)$ and the real and imaginary parts of $z$. Thus, the introduction of $z$ can be viewed as a linear invertible change of variables, $y = T(\alpha)x$, and taking $z = y_1 + iy_2$. As it can be seen from (3.13), the components $(y_1, y_2)$ are the coordinates of $x$ in the *real eigenbasis* of $A(\alpha)$ composed by $\{2 \text{ Re } q, -2 \text{ Im } q\}$. In this basis, the matrix $A(\alpha)$ has its *canonical real (Jordan) form:*

$$J(\alpha) = T(\alpha)A(\alpha)T^{-1}(\alpha) = \begin{pmatrix} \mu(\alpha) & -\omega(\alpha) \\ \omega(\alpha) & \mu(\alpha) \end{pmatrix}.$$

(3) Suppose that at $\alpha = 0$ the function $F(x, \alpha)$ from (3.11) is represented as

$$F(x, 0) = \frac{1}{2}B(x, x) + \frac{1}{6}C(x, x, x) + O(\|x\|^4),$$

where $B(x, y)$ and $C(x, y, u)$ are *symmetric multilinear* vector functions of $x, y, u \in \mathbb{R}^2$. In coordinates, we have

---

[1] The vectors $q(\alpha)$ and $p(\alpha)$, corresponding to the simple eigenvalues, can be selected to depend on $\alpha$ as smooth as $A(\alpha)$.

$$B_i(x,y) = \sum_{j,k=1}^{2} \left. \frac{\partial^2 F_i(\xi,0)}{\partial \xi_j \partial \xi_k} \right|_{\xi=0} x_j y_k, \; i = 1,2,$$

and

$$C_i(x,y,u) = \sum_{j,k,l=1}^{2} \left. \frac{\partial^3 F_i(\xi,0)}{\partial \xi_j \partial \xi_k \partial \xi_l} \right|_{\xi=0} x_j y_k u_l, \; i = 1,2.$$

Then,

$$B(zq + \bar{z}\bar{q}, zq + \bar{z}\bar{q}) = z^2 B(q,q) + 2z\bar{z}B(q,\bar{q}) + \bar{z}^2 B(\bar{q},\bar{q}),$$

where $q = q(0), p = p(0)$, so the Taylor coefficients $g_{kl}$, $k + l = 2$, of the quadratic terms in $g(z,\bar{z},0)$ can be expressed by the formulas

$$g_{20} = \langle p, B(q,q) \rangle, \quad g_{11} = \langle p, B(q,\bar{q}) \rangle, \quad g_{02} = \langle p, B(\bar{q},\bar{q}) \rangle,$$

and similar calculations with $C$ give

$$g_{21} = \langle p, C(q,q,\bar{q}) \rangle.$$

(4) The normalization of $q$ is irrelevant in the following. Indeed, suppose that $q$ is normalized by $\langle q,q \rangle = 1$. A vector $\tilde{q} = \gamma q$ is also the eigenvector for any nonzero $\gamma \in \mathbb{C}^1$ but with the normalization $\langle \tilde{q},\tilde{q} \rangle = |\gamma|^2$. Taking $\tilde{p} = \frac{1}{\bar{\gamma}} p$ will keep the relative normalization untouched: $\langle \tilde{p},\tilde{q} \rangle = 1$. It is clear that Taylor coefficients $\tilde{g}_{kl}$ computed using $\tilde{q},\tilde{p}$ will be *different* from the original $g_{kl}$. For example, we can check via the multilinear representation that

$$\tilde{g}_{20} = \gamma g_{20}, \; \tilde{g}_{11} = \bar{\gamma} g_{11}, \; \tilde{g}_{02} = \frac{\bar{\gamma}^2}{\gamma} g_{02}, \; \tilde{g}_{21} = |\gamma|^2 g_{21}.$$

However, this change can easily be neutralized by the linear scaling of the variable: $z = \frac{1}{\gamma} w$, which results in the same equation for $w$ as before.

For example, setting $\langle q,q \rangle = \frac{1}{2}$ corresponds to the standard relation $z = \langle p,x \rangle = x_1 + ix_2$ for a system that already has the real canonical form $\dot{x} = J(\alpha)x$, where $J$ is given above. In this case,

$$q = \frac{1}{2} \begin{pmatrix} 1 \\ -i \end{pmatrix}, \quad p = \begin{pmatrix} 1 \\ -i \end{pmatrix}. \; \Diamond$$

Let us start to make *nonlinear* (complex) coordinate changes that will simplify (3.12). First of all, remove all quadratic terms.

**Lemma 3.4** *The equation*

$$\dot{z} = \lambda z + \frac{g_{20}}{2} z^2 + g_{11} z \bar{z} + \frac{g_{02}}{2} \bar{z}^2 + O(|z|^3), \tag{3.14}$$

*where $\lambda = \lambda(\alpha) = \mu(\alpha) + i\omega(\alpha), \mu(0) = 0, \omega(0) = \omega_0 > 0$, and $g_{ij} = g_{ij}(\alpha)$, can be transformed by an invertible parameter-dependent change of complex coordinate*

$$z = w + \frac{h_{20}}{2}w^2 + h_{11}w\bar{w} + \frac{h_{02}}{2}\bar{w}^2,$$

*for all sufficiently small $|\alpha|$, into an equation without quadratic terms:*

$$\dot{w} = \lambda w + O(|w|^3).$$

**Proof:**

The inverse change of variable is given by the expression

$$w = z - \frac{h_{20}}{2}z^2 - h_{11}z\bar{z} - \frac{h_{02}}{2}\bar{z}^2 + O(|z|^3).$$

Therefore,

$$\dot{w} = \dot{z} - h_{20}z\dot{z} - h_{11}(\dot{z}\bar{z} + z\dot{\bar{z}}) - h_{02}\bar{z}\dot{\bar{z}} + \cdots$$

$$= \lambda z + \left(\frac{g_{20}}{2} - \lambda h_{20}\right) z^2 + \left(g_{11} - \lambda h_{11} - \bar{\lambda} h_{11}\right) z\bar{z} + \left(\frac{g_{02}}{2} - \bar{\lambda} h_{02}\right) \bar{z}^2 + \cdots$$

$$= \lambda w + \frac{1}{2}(g_{20} - \lambda h_{20})w^2 + (g_{11} - \bar{\lambda}h_{11})w\bar{w} + \frac{1}{2}(g_{02} - (2\bar{\lambda} - \lambda)h_{02})\bar{w}^2 + O(|w|^3).$$

Thus, by setting

$$h_{20} = \frac{g_{20}}{\lambda},\ h_{11} = \frac{g_{11}}{\bar{\lambda}},\ h_{02} = \frac{g_{02}}{2\bar{\lambda} - \lambda},$$

we "kill" all the quadratic terms in (3.14). These substitutions are correct because the denominators are nonzero for all sufficiently small $|\alpha|$ since $\lambda(0) = i\omega_0$ with $\omega_0 > 0$. □

**Remarks:**

(1) The resulting coordinate transformation is polynomial with coefficients that are smoothly dependent on $\alpha$. The inverse transformation has the same property but it is not polynomial. Its form can be obtained by the method of unknown coefficients. In some neighborhood of the origin the transformation is *near-identical* because of its linear part.

(2) Notice that the transformation *changes* the coefficients of the cubic (as well as higher-order) terms of (3.14). ◊

Assuming that we have removed all quadratic terms, let us try to eliminate the cubic terms as well. This is "almost" possible: There is only one "resistant" term, as the following lemma shows.

**Lemma 3.5** *The equation*

$$\dot{z} = \lambda z + \frac{g_{30}}{6}z^3 + \frac{g_{21}}{2}z^2\bar{z} + \frac{g_{12}}{2}z\bar{z}^2 + \frac{g_{03}}{6}\bar{z}^3 + O(|z|^4),$$

*where* $\lambda = \lambda(\alpha) = \mu(\alpha) + i\omega(\alpha), \mu(0) = 0, \omega(0) = \omega_0 > 0$, *and* $g_{ij} = g_{ij}(\alpha)$, *can be transformed by an invertible parameter-dependent change of complex coordinate*

$$z = w + \frac{h_{30}}{6}w^3 + \frac{h_{21}}{2}w^2\bar{w} + \frac{h_{12}}{2}w\bar{w}^2 + \frac{h_{03}}{6}\bar{w}^3,$$

*for all sufficiently small* $|\alpha|$, *into an equation with only one cubic term:*

$$\dot{w} = \lambda w + c_1 w^2\bar{w} + O(|w|^4),$$

*where* $c_1 = c_1(\alpha)$.

**Proof:**

The inverse transformation is

$$w = z - \frac{h_{30}}{6}z^3 - \frac{h_{21}}{2}z^2\bar{z} - \frac{h_{12}}{2}z\bar{z}^2 - \frac{h_{03}}{6}\bar{z}^3 + O(|z|^4).$$

Therefore,

$$\begin{aligned}
\dot{w} &= \dot{z} - \frac{h_{30}}{2}z^2\dot{z} - \frac{h_{21}}{2}(2z\bar{z}\dot{z} + z^2\dot{\bar{z}}) - \frac{h_{12}}{2}(\dot{z}\bar{z}^2 + 2z\bar{z}\dot{\bar{z}}) - \frac{h_{03}}{2}\bar{z}^2\dot{\bar{z}} + \cdots \\
&= \lambda z + \left(\frac{g_{30}}{6} - \frac{\lambda h_{30}}{2}\right)z^3 + \left(\frac{g_{21}}{2} - \lambda h_{21} - \frac{\bar{\lambda}h_{21}}{2}\right)z^2\bar{z} \\
&\quad + \left(\frac{g_{12}}{2} - \frac{\lambda h_{12}}{2} - \bar{\lambda}h_{12}\right)z\bar{z}^2 + \left(\frac{g_{03}}{6} - \frac{\bar{\lambda}h_{03}}{2}\right)\bar{z}^3 + \cdots \\
&= \lambda w + \frac{1}{6}(g_{30} - 2\lambda h_{30})w^3 + \frac{1}{2}(g_{21} - (\lambda + \bar{\lambda})h_{21})w^2\bar{w} \\
&\quad + \frac{1}{2}(g_{12} - 2\bar{\lambda}h_{12})w\bar{w}^2 + \frac{1}{6}(g_{03} + (\lambda - 3\bar{\lambda})h_{03})\bar{w}^3 + O(|w|^4).
\end{aligned}$$

Thus, by setting

$$h_{30} = \frac{g_{30}}{2\lambda},\ h_{12} = \frac{g_{12}}{2\bar{\lambda}},\ h_{03} = \frac{g_{03}}{3\bar{\lambda} - \lambda},$$

we can annihilate all cubic terms in the resulting equation except the $w^2\bar{w}$ -term, which we have to treat separately. The substitutions are valid since all the involved denominators are nonzero for all sufficiently small $|\alpha|$.

One can also try to eliminate the $w^2\bar{w}$-term by formally setting

$$h_{21} = \frac{g_{21}}{\lambda + \bar{\lambda}}.$$

This is possible for small $\alpha \neq 0$, but the denominator vanishes at $\alpha = 0$: $\lambda(0) + \bar{\lambda}(0) = i\omega_0 - i\omega_0 = 0$. To obtain a transformation that is smoothly dependent on $\alpha$, set $h_{21} = 0$, which results in

$$c_1 = \frac{g_{21}}{2}. \square$$

**Remark:**

The remaining cubic $w^2\bar{w}$-term is called a *resonant term.* Note that its coefficient is the *same* as the coefficient of the cubic term $z^2\bar{z}$ in the original equation in Lemma 3.5. ◊

We now combine the two previous lemmas.

**Lemma 3.6 (Poincaré normal form for the Hopf bifurcation)**
*The equation*

$$\dot{z} = \lambda z + \sum_{2\le k+l\le 3} \frac{1}{k!l!} g_{kl} z^k \bar{z}^l + O(|z|^4), \tag{3.15}$$

*where* $\lambda = \lambda(\alpha) = \mu(\alpha) + i\omega(\alpha), \mu(0) = 0, \omega(0) = \omega_0 > 0$, *and* $g_{ij} = g_{ij}(\alpha)$, *can be transformed by an invertible parameter-dependent change of complex coordinate, smoothly depending on the parameter,*

$$\begin{aligned} z = w &+ \frac{h_{20}}{2} w^2 + h_{11} w\bar{w} + \frac{h_{02}}{2}\bar{w}^2 \\ &+ \frac{h_{30}}{6} w^3 + \frac{h_{12}}{2} w\bar{w}^2 + \frac{h_{03}}{6}\bar{w}^3, \end{aligned}$$

*for all sufficiently small* $|\alpha|$, *into an equation with only the resonant cubic term:*

$$\dot{w} = \lambda w + c_1 w^2 \bar{w} + O(|w|^4), \tag{3.16}$$

*where* $c_1 = c_1(\alpha)$.

**Proof:**

Obviously, a composition of the transformations defined in Lemmas 3.4 and 3.5 does the job. First, perform the transformation

$$z = w + \frac{h_{20}}{2} w^2 + h_{11} w\bar{w} + \frac{h_{02}}{2}\bar{w}^2, \tag{3.17}$$

with

$$h_{20} = \frac{g_{20}}{\lambda},\ h_{11} = \frac{g_{11}}{\bar{\lambda}},\ h_{02} = \frac{g_{02}}{2\bar{\lambda} - \lambda},$$

defined in Lemma 3.4. This annihilates all the quadratic terms but also changes the coefficients of the cubic terms. The coefficient of $w^2\bar{w}$ will be $\frac{1}{2}\tilde{g}_{21}$, say, instead of $\frac{1}{2}g_{21}$. Then make the transformation from Lemma 3.5 that eliminates all the cubic terms but the resonant one. The coefficient of this term remains $\frac{1}{2}\tilde{g}_{21}$. Since terms of order four and higher appearing in the composition affect only $O(|w|^4)$ terms in (3.16), they can be truncated. □

Thus, all we need to compute to get the coefficient $c_1$ in terms of the given equation (3.15) is a new coefficient $\frac{1}{2}\tilde{g}_{21}$ of the $w^2\bar{w}$-term after the *quadratic* transformation (3.17). We can do this computation in the same manner as in Lemmas 3.4 and 3.5, namely, inverting (3.17). Unfortunately, now we have to know the inverse map up to and including *cubic* terms.[2] However, there is a possibility to avoid explicit inverting of (3.17).

Indeed, we can express $\dot{z}$ in terms of $w, \bar{w}$ in two ways. One way is to substitute (3.17) into the original equation (3.15). Alternatively, since we know the resulting form (3.16) to which (3.15) can be transformed, $\dot{z}$ can be computed by differentiating (3.17),

$$\dot{z} = \dot{w} + h_{20}w\dot{w} + h_{11}(w\dot{\bar{w}} + \bar{w}\dot{w}) + h_{02}\bar{w}\dot{\bar{w}},$$

and then by substituting $\dot{w}$ and its complex conjugate, using (3.16). Comparing the coefficients of the quadratic terms in the obtained expressions for $\dot{z}$ gives the above formulas for $h_{20}, h_{11}$, and $h_{02}$, while equating the coefficients in front of the $w|w|^2$-term leads to

$$c_1 = \frac{g_{20}g_{11}(2\lambda + \bar{\lambda})}{2|\lambda|^2} + \frac{|g_{11}|^2}{\lambda} + \frac{|g_{02}|^2}{2(2\lambda - \bar{\lambda})} + \frac{g_{21}}{2}.$$

This formula gives us the dependence of $c_1$ on $\alpha$ if we recall that $\lambda$ and $g_{ij}$ are smooth functions of the parameter. At the bifurcation parameter value $\alpha = 0$, the previous equation reduces to

$$c_1(0) = \frac{i}{2\omega_0}\left(g_{20}g_{11} - 2|g_{11}|^2 - \frac{1}{3}|g_{02}|^2\right) + \frac{g_{21}}{2}. \tag{3.18}$$

Now we want to transform the Poincaré normal form into the normal form studied in the previous section.

**Lemma 3.7** *Consider the equation*

$$\frac{dw}{dt} = (\mu(\alpha) + i\omega(\alpha))w + c_1(\alpha)w|w|^2 + O(|w|^4),$$

*where $\mu(0) = 0$, and $\omega(0) = \omega_0 > 0$.*

*Suppose $\mu'(0) \neq 0$ and* $\text{Re}\, c_1(0) \neq 0$. *Then, the equation can be transformed by a parameter-dependent linear coordinate transformation, a time rescaling, and a nonlinear time reparametrization into an equation of the form*

$$\frac{du}{d\theta} = (\beta + i)u + su|u|^2 + O(|u|^4),$$

*where $u$ is a new complex coordinate, and $\theta, \beta$ are the new time and parameter, respectively, and* $s = \text{sign}\ \text{Re}\ c_1(0) = \pm 1$.

[2] Actually, only the "resonant" cubic term of the inverse is required:

$$w = z - \frac{h_{20}}{2}z^2 - h_{11}z\bar{z} - \frac{h_{02}}{2}\bar{z}^2 + \frac{1}{2}(3h_{11}h_{20} + 2|h_{11}|^2 + |h_{02}|^2)z^2\bar{z} + \cdots,$$

where the dots now mean all undisplayed terms.

**Proof:**
*Step 1* (*Linear time scaling*). Introduce the new time $\tau = \omega(\alpha)t$. The time direction is preserved since $\omega(\alpha) > 0$ for all sufficiently small $|\alpha|$. Then,

$$\frac{dw}{d\tau} = (\beta + i)w + d_1(\beta)w|w|^2 + O(|w|^4),$$

where

$$\beta = \beta(\alpha) = \frac{\mu(\alpha)}{\omega(\alpha)}, \quad d_1(\beta) = \frac{c_1(\alpha(\beta))}{\omega(\alpha(\beta))}.$$

We can consider $\beta$ as a new parameter because

$$\beta(0) = 0, \quad \beta'(0) = \frac{\mu'(0)}{\omega(0)} \neq 0,$$

and therefore the Inverse Function Theorem guarantees local existence and smoothness of $\alpha$ as a function of $\beta$. Notice that $d_1$ is *complex*.

*Step 2* (*Nonlinear time reparametrization*). Change the time parametrization along the orbits by introducing a new time $\theta = \theta(\tau, \beta)$, where

$$d\theta = (1 + e_1(\beta)|w|^2)\, d\tau$$

with $e_1(\beta) = \text{Im } d_1(\beta)$. The time change is a near-identity transformation in a small neighborhood of the origin. Using the new definition of the time we obtain

$$\frac{dw}{d\theta} = (\beta + i)w + l_1(\beta)w|w|^2 + O(|w|^4),$$

where $l_1(\beta) = \text{Re } d_1(\beta) - \beta e_1(\beta)$ is *real* and

$$l_1(0) = \frac{\text{Re } c_1(0)}{\omega(0)}. \tag{3.19}$$

*Step 3* (*Linear coordinate scaling*). Finally, introduce a new complex variable $u$:

$$w = \frac{u}{\sqrt{|l_1(\beta)|}},$$

which is possible due to Re $c_1(0) \neq 0$ and, thus, $l_1(0) \neq 0$. The equation then takes the required form

$$\frac{du}{d\theta} = (\beta + i)u + \frac{l_1(\beta)}{|l_1(\beta)|}u|u|^2 + O(|u|^4) = (\beta + i)u + su|u|^2 + O(|u|^4),$$

with $s = \text{sign } l_1(0) = \text{sign Re } c_1(0)$. □

**Definition 3.3** *The real function* $l_1(\beta)$ *is called the* first Lyapunov coefficient.

It follows from (3.19) that the first Lyapunov coefficient at $\beta = 0$ can be computed by the formula

$$l_1(0) = \frac{1}{2\omega_0^2} \operatorname{Re}(ig_{20}g_{11} + \omega_0 g_{21}). \tag{3.20}$$

Thus, we need only certain second- and third-order derivatives of the right-hand sides at the bifurcation point to compute $l_1(0)$. Recall that the value of $l_1(0)$ does depend on the normalization of the eigenvectors $q$ and $p$, while its sign (which is what only matters in the bifurcation analysis) is invariant under scaling of $q, p$ obeying the relative normalization $\langle p, q \rangle = 1$. Notice that the equation of $u$ with $s = -1$ written in real form coincides with system (3.10) from the previous section. We now summarize the obtained results in the following theorem.

**Theorem 3.3** *Suppose a two-dimensional system*

$$\frac{dx}{dt} = f(x, \alpha), \quad x \in \mathbb{R}^2, \ \alpha \in \mathbb{R}^1, \tag{3.21}$$

*with smooth $f$, has for all sufficiently small $|\alpha|$ the equilibrium $x = 0$ with eigenvalues*

$$\lambda_{1,2}(\alpha) = \mu(\alpha) \pm i\omega(\alpha),$$

*where $\mu(0) = 0$, $\omega(0) = \omega_0 > 0$.*

*Let the following conditions be satisfied:*

(B.1) $l_1(0) \neq 0$, *where $l_1$ is the first Lyapunov coefficient;*
(B.2) $\mu'(0) \neq 0$.

*Then, there are invertible coordinate and parameter changes and a time reparameterization transforming (3.21) into*

$$\frac{d}{d\tau}\begin{pmatrix} y_1 \\ y_2 \end{pmatrix} = \begin{pmatrix} \beta & -1 \\ 1 & \beta \end{pmatrix}\begin{pmatrix} y_1 \\ y_2 \end{pmatrix} \pm (y_1^2 + y_2^2)\begin{pmatrix} y_1 \\ y_2 \end{pmatrix} + O(\|y\|^4). \ \square$$

Using Lemma 3.2, we can drop the $O(\|y\|^4)$ terms and finally arrive at the following general result.

**Theorem 3.4 (Topological normal form for the Hopf bifurcation)** *Any generic two-dimensional, one-parameter system*

$$\dot{x} = f(x, \alpha),$$

*having at $\alpha = 0$ the equilibrium $x = 0$ with eigenvalues*

$$\lambda_{1,2}(0) = \pm i\omega_0, \ \omega_0 > 0,$$

*is locally topologically equivalent near the origin to one of the following normal forms:*

$$\begin{pmatrix} \dot{y}_1 \\ \dot{y}_2 \end{pmatrix} = \begin{pmatrix} \beta & -1 \\ 1 & \beta \end{pmatrix}\begin{pmatrix} y_1 \\ y_2 \end{pmatrix} \pm (y_1^2 + y_2^2)\begin{pmatrix} y_1 \\ y_2 \end{pmatrix}. \ \square$$

**Remark:**

The genericity conditions assumed in Theorem 3.4 are the nondegeneracy condition (B.1) and the transversality condition (B.2) from Theorem 3.3. ♢

The preceding two theorems together with the normal form analysis of the previous section and formula (3.20) for $l_1(0)$ provide us with all the necessary tools for analysis of the Hopf bifurcation in generic two-dimensional systems. In Chapter 5 we will see how to deal with $n$-dimensional systems where $n > 2$.

**Example 3.1 (Hopf bifurcation in a predator-prey model)** Consider the following system of two differential equations:

$$\begin{cases} \dot{x}_1 = rx_1(1-x_1) - \dfrac{cx_1x_2}{\alpha + x_1}, \\ \dot{x}_2 = -dx_2 + \dfrac{cx_1x_2}{\alpha + x_1}. \end{cases} \tag{3.22}$$

The system describes the dynamics of a simple predator-prey ecosystem (see, e.g., Holling [1965]). Here $x_1$ and $x_2$ are (scaled) population numbers, and $r, c, d,$ and $\alpha$ are parameters characterizing the behavior of isolated populations and their interaction. Let us consider $\alpha$ as a control parameter and assume $c > d$.

To simplify calculations further, let us consider a *polynomial* system that has for $x_1 > -\alpha$ the same orbits as the original one (i.e., orbitally equivalent, see Chapter 2):

$$\begin{cases} \dot{x}_1 = rx_1(\alpha + x_1)(1-x_1) - cx_1x_2, \\ \dot{x}_2 = -\alpha dx_2 + (c-d)x_1x_2 \end{cases} \tag{3.23}$$

(this system is obtained by multiplying both sides of the original system by the function $(\alpha + x_1)$ and introducing a new time variable $\tau$ by $dt = (\alpha + x_1)\, d\tau$).

System (3.23) has a nontrivial equilibrium

$$E_0 = \left( \frac{\alpha d}{c-d}, \frac{r\alpha}{c-d}\left[1 - \frac{\alpha d}{c-d}\right] \right).$$

The Jacobian matrix evaluated at this equilibrium is

$$A(\alpha) = \begin{pmatrix} \dfrac{\alpha r d(c+d)}{(c-d)^2}\left[\dfrac{c-d}{c+d} - \alpha\right] & -\dfrac{\alpha c d}{c-d} \\ \dfrac{\alpha r(c - d(1+\alpha))}{c-d} & 0 \end{pmatrix},$$

and thus

$$\mu(\alpha) = \frac{\sigma(\alpha)}{2} = \frac{\alpha r d(c+d)}{2(c-d)^2}\left[\frac{c-d}{c+d} - \alpha\right].$$

We have $\mu(\alpha_0) = 0$ for

$$\alpha_0 = \frac{c-d}{c+d}.$$

Moreover,

$$\omega^2(\alpha_0) = \frac{rc^2 d(c-d)}{(c+d)^3} > 0. \tag{3.24}$$

Therefore, at $\alpha = \alpha_0$ the equilibrium $E_0$ has eigenvalues $\lambda_{1,2}(\alpha_0) = \pm i\omega(\alpha_0)$ and a Hopf bifurcation takes place.[3] The equilibrium is stable for $\alpha > \alpha_0$ and unstable for $\alpha < \alpha_0$. Notice that the critical value of $\alpha$ corresponds to the passing of the line defined by $\dot{x}_2 = 0$ through the maximum of the curve defined by $\dot{x}_1 = 0$ (see **Fig. 3.10**). Thus, if the line $\dot{x}_2 = 0$ is to the right of

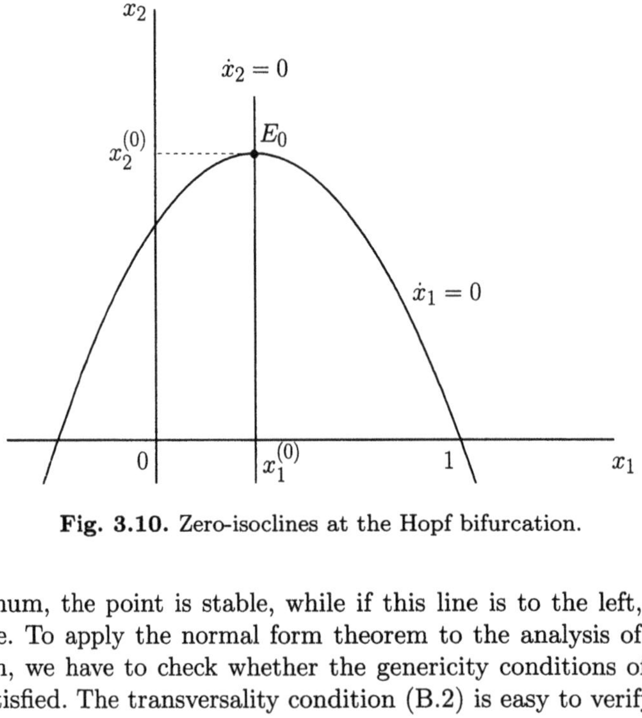


**Fig. 3.10.** Zero-isoclines at the Hopf bifurcation.

the maximum, the point is stable, while if this line is to the left, the point is unstable. To apply the normal form theorem to the analysis of this Hopf bifurcation, we have to check whether the genericity conditions of Theorem 3.3 are satisfied. The transversality condition (B.2) is easy to verify:

$$\mu'(\alpha_0) = -\frac{\alpha_0 rd(c+d)}{2(c-d)^2} = -\frac{rd}{2(c-d)} < 0.$$

To compute the first Lyapunov coefficient, fix the parameter $\alpha$ at its critical value $\alpha_0$. At $\alpha = \alpha_0$, the nontrivial equilibrium $E_0$ has the coordinates

$$x_1^{(0)} = \frac{d}{c+d}, \quad x_2^{(0)} = \frac{rc}{(c+d)^2}.$$

[3] Since (3.23) is only orbitally equivalent to (3.22), the value of $\omega(\alpha_0)$ given by (3.24) *cannot* be used directly to evaluate the period of small oscillations around $E_0$ in the original system.

Translate the origin of the coordinates to this equilibrium by the change of variables

$$\begin{cases} x_1 = x_1^{(0)} + \xi_1, \\ x_2 = x_2^{(0)} + \xi_2. \end{cases}$$

This transforms system (3.23) into

$$\begin{cases} \dot{\xi}_1 = -\dfrac{cd}{c+d}\xi_2 - \dfrac{rd}{c+d}\xi_1^2 - c\xi_1\xi_2 - r\xi_1^3 \equiv F_1(\xi_1, \xi_2), \\ \dot{\xi}_2 = \dfrac{rc(c-d)}{(c+d)^2}\xi_1 + (c-d)\xi_1\xi_2 \equiv F_2(\xi_1, \xi_2). \end{cases}$$

This system can be represented as

$$\dot{\xi} = A\xi + \frac{1}{2}B(\xi, \xi) + \frac{1}{6}C(\xi, \xi, \xi),$$

where $A = A(\alpha_0)$, and the multilinear functions $B$ and $C$ take on the planar vectors $\xi = (\xi_1, \xi_2)^T$, $\eta = (\eta_1, \eta_2)^T$, and $\zeta = (\zeta_1, \zeta_2)^T$ the values

$$B(\xi, \eta) = \begin{pmatrix} -\dfrac{2rd}{c+d}\xi_1\eta_1 - c(\xi_1\eta_2 + \xi_2\eta_1) \\ (c-d)(\xi_1\eta_2 + \xi_2\eta_1) \end{pmatrix}$$

and

$$C(\xi, \eta, \zeta) = \begin{pmatrix} -6r\xi_1\eta_1\zeta_1 \\ 0 \end{pmatrix}.$$

Write the matrix $A(\alpha_0)$ in the form

$$A = \begin{pmatrix} 0 & -\dfrac{cd}{c+d} \\ \dfrac{\omega^2(c+d)}{cd} & 0 \end{pmatrix},$$

where $\omega^2$ is given by formula (3.24).[4] Now it is easy to check that complex vectors

$$q \sim \begin{pmatrix} cd \\ -i\omega(c+d) \end{pmatrix}, \quad p \sim \begin{pmatrix} \omega(c+d) \\ -icd \end{pmatrix},$$

are proper eigenvectors:

$$Aq = i\omega q, \quad A^T p = -i\omega p.$$

To achieve the necessary normalization $\langle p, q \rangle = 1$, we can take, for example,

$$q = \begin{pmatrix} cd \\ -i\omega(c+d) \end{pmatrix}, \quad p = \frac{1}{2\omega cd(c+d)} \begin{pmatrix} \omega(c+d) \\ -icd \end{pmatrix}.$$

[4] It is always useful to express the Jacobian matrix using $\omega$ since this simplifies expressions for the eigenvectors.

The hardest part of the job is done, and now we can simply calculate[5]

$$g_{20} = \langle p, B(q,q)\rangle = \frac{cd(c^2-d^2-rd)+i\omega c(c+d)^2}{(c+d)},$$

$$g_{11} = \langle p, B(q,\bar{q})\rangle = -\frac{rcd^2}{(c+d)}, \quad g_{21} = \langle p, C(q,q,\bar{q})\rangle = -3rc^2d^2,$$

and compute the first Lyapunov coefficient by formula (3.20),

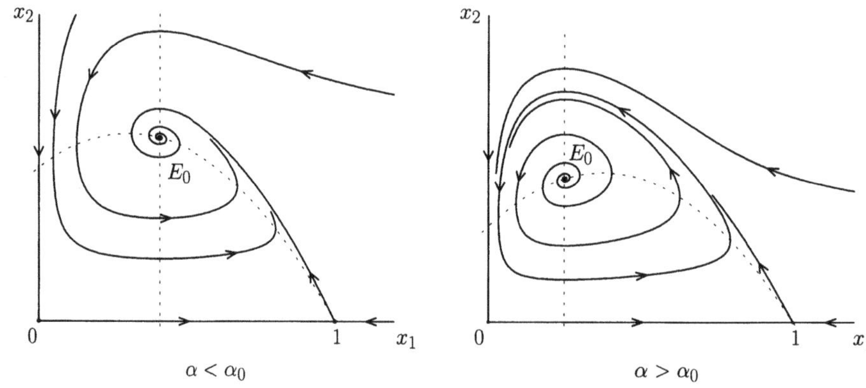


**Fig. 3.11.** Hopf bifurcation in the predator-prey model.

$$l_1(\alpha_0) = \frac{1}{2\omega^2}\,\mathrm{Re}(ig_{20}g_{11}+\omega g_{21}) = -\frac{rc^2d^2}{\omega} < 0.$$

It is clear that $l_1(\alpha_0) < 0$ for all combinations of the fixed parameters. Thus, the nondegeneracy condition (B.1) of Theorem 3.3 holds as well. Therefore, a unique and stable limit cycle bifurcates from the equilibrium via the Hopf bifurcation for $\alpha < \alpha_0$ (see **Fig. 3.11**). ◊

## 3.6 Exercises

**(1) (Fold bifurcation in ecology)** Consider the following differential equation, which models a single population under a constant harvest:

[5] Another way to compute $g_{20}, g_{11}$, and $g_{21}$ (which may be simpler if we use a symbolic manipulation software) is to define the complex-valued function

$$G(z,w) = \bar{p}_1 F_1(zq_1 + w\bar{q}_1, zq_2 + w\bar{q}_2) + \bar{p}_2 F_2(zq_1 + w\bar{q}_1, zq_2 + w\bar{q}_2),$$

where $p, q$ are given above, and to evaluate its formal partial derivatives with respect to $z, w$ at $z = w = 0$, obtaining $g_{20} = G_{zz}$, $g_{11} = G_{zw}$, and $g_{21} = G_{zzw}$. In this way no multilinear functions are necessary. See Exercise 4.

$$\dot{x} = rx\left(1 - \frac{x}{K}\right) - \alpha,$$

where $x$ is the population number; $r$ and $K$ are the *intrinsic growth rate* and the *carrying capacity* of the population, respectively, and $\alpha$ is the *harvest rate*, which is a control parameter. Find a parameter value $\alpha_0$ at which the system has a fold bifurcation, and check the genericity conditions of Theorem 3.1. Based on the analysis, explain what might be a result of overharvesting on the ecosystem dynamics. Is the bifurcation catastrophic in this example?

**(2) (Complex notation)** Verify that

$$\dot{z} = iz + (i+1)z^2 + 2iz\bar{z} + (i-1)\bar{z}^2$$

is a complex form of the system

$$\begin{pmatrix} \dot{x}_1 \\ \dot{x}_2 \end{pmatrix} = \begin{pmatrix} 1 & 2 \\ -1 & -1 \end{pmatrix} \begin{pmatrix} x_1 \\ x_2 \end{pmatrix} + \frac{6}{\sqrt{3}} \begin{pmatrix} 0 \\ x_1 x_2 \end{pmatrix},$$

provided that the eigenvectors are selected in the form

$$q = \frac{1}{2\sqrt{3}} \begin{pmatrix} 2 \\ -1+i \end{pmatrix}, \quad p = \frac{3}{2\sqrt{3}} \begin{pmatrix} 1+i \\ 2i \end{pmatrix}.$$

How will the complex form change if one instead adopts a different setting of $q, p$ satisfying $\langle p, q \rangle = 1$?

**(3) (Nonlinear stability)** Write the system

$$\begin{cases} \dot{x} = -y - xy + 2y^2, \\ \dot{y} = x - x^2 y, \end{cases}$$

in terms of the complex coordinate $z = x + iy$, and compute the normal form coefficient $c_1(0)$ by formula (3.18). Is the origin stable?

**(4) (Hopf bifurcation in the Brusselator)** Consider the Brusselator system (1.9) from Chapter 1:

$$\begin{cases} \dot{x}_1 = A - (B+1)x_1 + x_1^2 x_2 \equiv F_1(x_1, x_2, A, B), \\ \dot{x}_2 = Bx_1 - x_1^2 x_2 \equiv F_2(x_1, x_2, A, B). \end{cases}$$

Fix $A > 0$ and take $B$ as a bifurcation parameter. Using one of the available computer algebra systems, prove that at $B = 1 + A^2$ the system exhibits a supercritical Hopf bifurcation.

(*Hint:* The following sequence of MAPLE commands solves the problem:

```
> with(linalg);
> readlib(mtaylor);
> readlib(coeftayl);
```

The first command above allows us to use the MAPLE linear algebra package. The other two commands load the procedures `mtaylor` and `coeftayl`, which compute the truncated multivariate Taylor series expansion and its individual coefficients, respectively, from the MAPLE library.

```
> F[1]:=A-(B+1)*X[1]+X[1]^2*X[2];
> F[2]:=B*X[1]-X[1]^2*X[2];
> J:=jacobian([F[1],F[2]],[X[1],X[2]]);
> K:=transpose(J);
```

By these commands we input the right-hand sides of the system into MAPLE and compute the Jacobian matrix and its transpose.

```
> sol:=solve({F[1]=0,F[2]=0,trace(J)=0},{X[1],X[2],B});
> assign(sol);
> assume(A>0);
> omega:=sqrt(det(J));
```

These commands solve the following system of equations

$$\begin{cases} F(x_1, x_2, A, B) = 0, \\ \text{tr } F_x(x_1, x_2, A, B) = 0, \end{cases}$$

for $(x_1, x_2, B)$ and allow us to check that $\det F_x = A^2 > 0$ at the found solution. Thus, at $B = 1 + A^2$ the Brusselator has the equilibrium

$$X = \left(A, \frac{1+A^2}{A}\right)^T$$

with purely imaginary eigenvalues $\lambda_{1,2} = \pm i\omega$, $\omega = A > 0$.

```
> ev:=eigenvects(J,'radical');
> q:=ev[1][3][1];
> et:=eigenvects(K,'radical');
> P:=et[2][3][1];
```

These commands show that

$$q = \left(-\frac{iA + A^2}{1 + A^2}, 1\right)^T, \quad p = \left(\frac{-iA + A^2}{A^2}, 1\right)^T,$$

are the critical eigenvectors[6] of the Jacobian matrix $J = F_x$ and its transpose,

$$Jq = i\omega q, \quad J^T p = -i\omega p.$$

```
> s1:=simplify(evalc(conjugate(P[1])*q[1]+conjugate(P[2]*q[2])));
> c:=simplify(evalc(1/conjugate(s1)));
> p[1]:=simplify(evalc(c*P[1]));
> p[2]:=simplify(evalc(c*P[2]));
> simplify(evalc(conjugate(p[1])*q[1]+conjugate(p[2])*q[2]));
```

By the commands above, we achieve the normalization $\langle p, q\rangle = 1$, finally taking

$$q = \left(-\frac{iA + A^2}{1 + A^2}, 1\right)^T, \quad p = \left(-\frac{i(1 + A^2)}{2A}, \frac{1 - iA}{2}\right)^T.$$

[6] Some implementations of MAPLE may produce the eigenvectors in a different form.

```
> F[1]:=A-(B+1)*x[1]+x[1]^2*x[2];
> F[2]:=B*x[1]-x[1]^2*x[2];
> x[1]:=evalc(X[1]+z*q[1]+z1*conjugate(q[1]));
> x[2]:=evalc(X[2]+z*q[2]+z1*conjugate(q[2]));
> H:=simplify(evalc(conjugate(p[1])*F[1]+conjugate(p[2])*F[2]));
```

By means of these commands, we compose $x = X + zq + \bar{z}\bar{q}$ and evaluate the function

$$H(z,\bar{z}) = \langle p, F(X + zq + \bar{z}\bar{q}, A, 1 + A^2)\rangle.$$

(In the MAPLE commands, **z1** stands for $\bar{z}$.)

```
> g[2,0]:=simplify(2*evalc(coeftayl(H,[z,z1]=[0,0],[2,0])));
> g[1,1]:=simplify(evalc(coeftayl(H,[z,z1]=[0,0],[1,1])));
> g[2,1]:=simplify(2*evalc(coeftayl(H,[z,z1]=[0,0],[2,1])));
```

The above commands compute the needed Taylor expansion of $H(z,\bar{z})$ at $(z,\bar{z}) = (0,0)$,

$$H(z,\bar{z}) = i\omega z + \sum_{2\leq j+k\leq 3} \frac{1}{j!k!} g_{jk} z^j \bar{z}^k + O(|z|^4),$$

giving

$$g_{20} = A - i, \quad g_{11} = \frac{(A-i)(A^2-1)}{1+A^2}, \quad g_{21} = -\frac{A(3A-i)}{1+A^2}.$$

```
> l[1]:=factor(1/(2*omega^2)*Re(I*g[2,0]*g[1,1]+omega*g[2,1]));
```

This final command computes the first Lyapunov coefficient

$$l_1 = \frac{1}{2\omega^2}\mathrm{Re}(ig_{20}g_{11} + \omega g_{2,1}) = -\frac{2+A^2}{2A(1+A^2)} < 0,$$

and allows us to check that it is negative.)

**(5)** Check that each of the following systems has an equilibrium that exhibits the Hopf bifurcation at some value of $\alpha$, and compute the first Lyapunov coefficient:

(a) *Rayleigh's equation:*

$$\ddot{x} + \dot{x}^3 - 2\alpha\dot{x} + x = 0;$$

(*Hint:* Introduce $y = \dot{x}$ and rewrite the equation as a system of two differential equations.)

(b) *Van der Pol's oscillator:*

$$\ddot{y} - (\alpha - y^2)\dot{y} + y = 0;$$

(c) *Bautin's example:*

$$\begin{cases} \dot{x} = y, \\ \dot{y} = -x + \alpha y + x^2 + xy + y^2; \end{cases}$$

(d) *Advertising diffusion model* [Feichtinger 1992]:

$$\begin{cases} \dot{x}_1 = \alpha[1 - x_1x_2^2 + A(x_2 - 1)], \\ \dot{x}_2 = x_1x_2^2 - x_2. \end{cases}$$

**(6)** Suppose that a system at the critical parameter values corresponding to the Hopf bifurcation has the form

$$\begin{cases} \dot{x} = -\omega y + \frac{1}{2} f_{xx} x^2 + f_{xy} xy + \frac{1}{2} f_{yy} y^2 \\ \qquad + \frac{1}{6} f_{xxx} x^3 + \frac{1}{2} f_{xxy} x^2 y + \frac{1}{2} f_{xyy} xy^2 + \frac{1}{6} f_{yyy} y^3 + \cdots, \\ \dot{y} = \omega x + \frac{1}{2} g_{xx} x^2 + g_{xy} xy + \frac{1}{2} g_{yy} y^2 \\ \qquad + \frac{1}{6} g_{xxx} x^3 + \frac{1}{2} g_{xxy} x^2 y + \frac{1}{2} g_{xyy} xy^2 + \frac{1}{6} g_{yyy} y^3 + \cdots. \end{cases}$$

Compute Re $c_1(0)$ in terms of the $f$'s and $g$'s. (*Hint:* See Guckenheimer & Holmes [1983, p. 156]. To apply the resulting formula, one needs to transform the system explicitly into its eigenbasis, which can always be avoided by using eigenvectors and complex notation, as described in this chapter.)

## 3.7 Appendix A: Proof of Lemma 3.2

The following statement, which is Lemma 3.2 rewritten in complex form, will be proved in this appendix.

**Lemma 3.8** *The system*

$$\dot{z} = (\alpha + i)z - z|z|^2 + O(|z|^4) \tag{A.1}$$

*is locally topologically equivalent near the origin to the system*

$$\dot{z} = (\alpha + i)z - z|z|^2. \tag{A.2}$$

**Proof:**
*Step 1* (*Existence and uniqueness of the cycle*). Write system (A.1) in polar coordinates $(\rho, \varphi)$:

$$\begin{cases} \dot{\rho} = \rho(\alpha - \rho^2) + \Phi(\rho, \varphi), \\ \dot{\varphi} = 1 + \Psi(\rho, \varphi), \end{cases} \tag{A.3}$$

where $\Phi = O(|\rho|^4), \Psi = O(|\rho|^3)$, and the $\alpha$-dependence of these functions is not indicated to simplify notations. An orbit of $(A.3)$ starting at $(\rho, \varphi) = (\rho_0, 0)$ has the

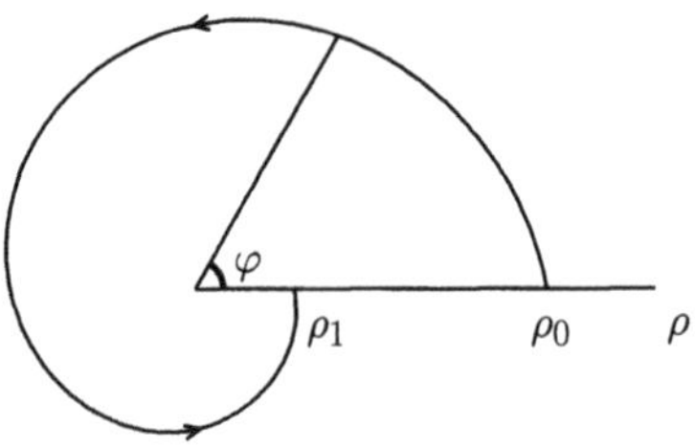


**Fig. 3.12.** Poincaré map for the Hopf bifurcation.

following representation (see **Fig. 3.12**): $\rho = \rho(\varphi; \rho_0), \rho_0 = \rho(0; \rho_0)$ with $\rho$ satisfying the equation

$$\frac{d\rho}{d\varphi} = \frac{\rho(\alpha - \rho^2) + \Phi}{1 + \Psi} = \rho(\alpha - \rho^2) + R(\rho, \varphi), \tag{A.4}$$

where $R = O(|\rho|^4)$. Notice that the transition from (A.3) to (A.4) is equivalent to the introduction of a new time parametrization in which $\dot{\varphi} = 1$, which implies that the return time to the half-axis $\varphi = 0$ is the same for all orbits starting on this axis with $\rho_0 > 0$. Since $\rho(\varphi; 0) \equiv 0$, we can write the Taylor expansion for $\rho(\varphi; \rho_0)$,

$$\rho = u_1(\varphi)\rho_0 + u_2(\varphi)\rho_0^2 + u_3(\varphi)\rho_0^3 + O(|\rho_0|^4). \tag{A.5}$$

Substituting (A.5) into (A.4) and solving the resulting linear differential equations at corresponding powers of $\rho_0$ with initial conditions $u_1(0) = 1, u_2(0) = u_3(0) = 0$, we get

$$u_1(\varphi) = e^{\alpha\varphi},\ u_2(\varphi) \equiv 0,\ u_3(\varphi) = e^{\alpha\varphi}\frac{1 - e^{2\alpha\varphi}}{2\alpha}.$$

Notice that these expressions are *independent* of the term $R(\rho, \varphi)$. Therefore, the return map $\rho_0 \mapsto \rho_1 = \rho(2\pi, \rho_0)$ has the form

$$\rho_1 = e^{2\pi\alpha}\rho_0 - e^{2\pi\alpha}[2\pi + O(\alpha)]\rho_0^3 + O(\rho_0^4) \tag{A.6}$$

for *all* $R = O(\rho^4)$. The map (A.6) can easily be analyzed for sufficiently small $\rho_0$ and $|\alpha|$. There is a neighborhood of the origin in which the map has only a trivial fixed point for small $\alpha < 0$ and an extra fixed point, $\rho_0^{(0)} = \sqrt{\alpha} + \cdots$, for small $\alpha > 0$ (see **Fig. 3.13**). The stability of the fixed points is also easily obtained from (A.6).

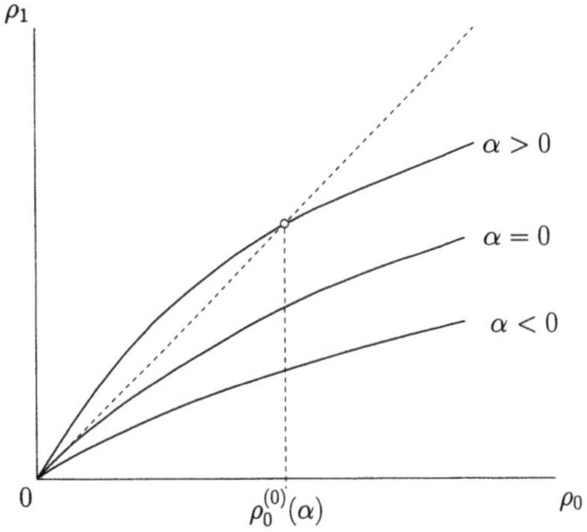


**Fig. 3.13.** Fixed point of the return map.

Taking into account that a positive fixed point of the map corresponds to a limit cycle of the system, we can conclude that system (A.3) (or (A.1)) with any $O(|z|^4)$ terms has a unique (stable) limit cycle bifurcating from the origin and existing for $\alpha > 0$ as in system (A.2). Therefore, in other words, higher-order terms do not affect the limit cycle bifurcation in some neighborhood of $z = 0$ for $|\alpha|$ sufficiently small.

*Step 2* (*Construction of a homeomorphism*). The established existence and uniqueness of the limit cycle is enough for all applications. Nevertheless, extra work must be done to prove the topological equivalence of the phase portraits.

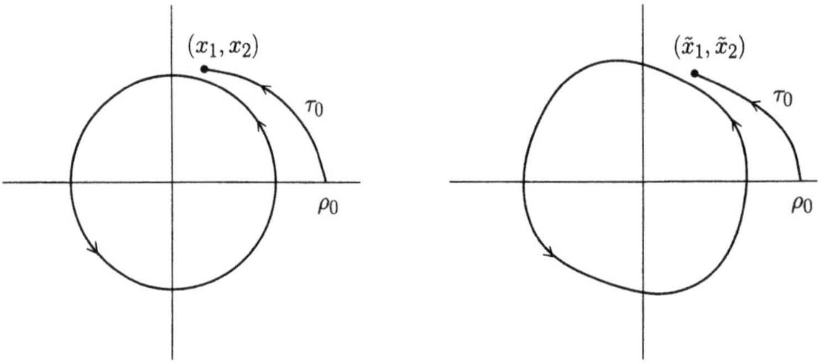


**Fig. 3.14.** Construction of the homeomorphism near the Hopf bifurcation.

Fix $\alpha$ small but positive. Both systems (A.1) and (A.2) have a limit cycle in some neighborhood of the origin. Assume that the time reparametrization resulting in the constant return time $2\pi$ is performed in system (A.1) (see the previous step). Also, apply a linear scaling of the coordinates in system (A.1) such that the point of intersection of the cycle and the horizontal half-axis is at $x_1 = \sqrt{\alpha}$.

Define a map $z \mapsto \tilde{z}$ by the following construction. Take a point $z = x_1 + ix_2$ and find values $(\rho_0, \tau_0)$, where $\tau_0$ is the minimal time required for an orbit of system (A.2) to approach the point $z$ starting from the horizontal half-axis with $\rho = \rho_0$. Now, take the point on this axis with $\rho = \rho_0$ and construct an orbit of system (A.1) on the time interval $[0, \tau_0]$ starting at this point. Denote the resulting point by $\tilde{z} = \tilde{x}_1 + i\tilde{x}_2$ (see **Fig. 3.14**). Set $\tilde{z} = 0$ for $z = 0$.

The map constructed is a homeomorphism that, for $\alpha > 0$, maps orbits of system (A.2) in some neighborhood of the origin into orbits of (A.1) preserving time direction. The case $\alpha < 0$ can be considered in the same way without rescaling the coordinates. $\square$

## 3.8 Appendix B: Poincaré normal forms

Although for the analysis of the Hopf bifurcation only a small portion of the normal form theory is really required, we give in this appendix an elementary introduction to this theory in a rather general setting.

Let $H_k$ be the linear space of vector-functions whose components are homogeneous polynomials of order $k$. Consider a smooth system of ODEs

$$\dot{x} = Ax + f^{(2)}(x) + f^{(3)}(x) + \cdots, \quad x \in \mathbb{R}^n, \tag{B.1}$$

where $f^{(k)} \in H_k$ for $k \geq 2$. Introduce a new variable $y \in \mathbb{R}^n$,

$$x = y + h^{(m)}(y), \tag{B.2}$$

where $h^{(m)} \in H_m$ for some fixed $m \geq 2$. At this moment, $h^{(m)}$ is an arbitrary vector-polynomial from $H_m$. Notice that (B.2) is invertible near the origin and the inverse transformation

$$y = x - h^{(m)}(x) + O(\|x\|^{m+1}) \tag{B.3}$$

is also smooth. Thus

$$\begin{aligned}\dot{y} &= \dot{x} - h_x^{(m)}(x)\dot{x} + O(\|x\|^{m+1}) \\ &= \left[\left(I_n - h_x^{(m)}(x)\right) + O(\|x\|^m)\right]\left(Ax + \sum_{k=2}^{m} f^{(k)}(x) + O(\|x\|^{m+1})\right) \\ &= Ay + \sum_{k=2}^{m-1} f^{(k)}(y) + \left[f^{(m)}(y) - \left(h_y^{(m)}(y)Ay - Ah^{(m)}(y)\right)\right] + O(\|y\|^{m+1}).\end{aligned}$$

Therefore, the new variable $y$ satisfies the equation

$$\dot{y} = Ay + \sum_{k=2}^{m-1} f^{(k)}(y) + \left[f^{(m)}(y) - (L_A h^{(m)})(y)\right] + O(\|y\|^{m+1}), \tag{B.4}$$

where the linear operator $L_A$ is defined by the formula

$$(L_A h)(y) = h_y(y)Ay - Ah(y). \tag{B.5}$$

**Lemma 3.9** *If $h \in H_m$, then $L_A h \in H_m$ for all $m \geq 2$.* □

Notice that all terms of order less than $m$ in equation (B.4) are the same as in equation (B.1), while the terms of order $m$ have changed and differ from $f^{(m)}(y)$ by $-(L_A h^{(m)})(y)$. Introduce now the *homological equation* in $H_m$:

$$L_A h^{(m)} = f^{(m)}. \tag{B.6}$$

If $f^{(m)}$ belongs to the *range* $L_A(H_m)$ of $L_A$, then there is a solution $h^{(m)}$ to (B.6), meaning that there is a transformation (B.2) that eliminates all homogeneous terms of order $m$ in equation (B.1). In general, however, $f^{(m)} = g^{(m)} + r^{(m)}$, where $g^{(m)} \in L_A(H_m)$, while $r^{(m)}$ belongs to a *complement* $\widetilde{H}_m$ to $L_A(H_m)$ in $H_m$. Therefore, only the $g^{(m)}$-part of $f^{(m)}$ can be eliminated from (B.1) by a transformation (B.2). The remaining $r^{(m)}$-terms are called the *resonant terms* of order $m$. Since $\widetilde{H}_m$ is not uniquely defined, the same is true for the resonant terms.

Applying the described elimination procedure recursively for $m = 2, 3, 4, \ldots$, one proves the following theorem.

**Theorem 3.5 (Poincaré, 1879)** *There is a polynomial change of coordinates*

$$x = y + h^{(2)}(y) + h^{(3)}(y) + \cdots + h^{(m)}(y), \ \ h^{(k)} \in H_k,$$

*that transforms a smooth system*

$$\dot{x} = Ax + f(x), \quad x \in \mathbb{R}^n,$$

*with $f(x) = O(\|x\|^2)$ into*

$$\dot{y} = Ay + r^{(2)}(y) + r^{(3)}(y) + \cdots + r^{(m)}(y) + O(\|y\|^{m+1}), \tag{B.7}$$

*where each $r^{(k)}$ contains only resonant terms of order $k$, i.e., $r^{(k)} \in \widetilde{H}_k$ for $k = 2, 3, \ldots, m$.* □

The system (B.7) is called the *Poincaré normal form* of (B.1).

There is still a practical question: How do we select a complement $\widetilde{H}_k \subset H_k$? A natural choice for $\widetilde{H}_k \subset H_k$ is the *orthogonal complement* to $L_A(H_k)$ in $H_k$. To characterize it, one can use the Fredholm Alternative Theorem. Let $\langle \cdot, \cdot \rangle_k$ be a scalar product in $H_k$ and let $L_A^*$ be a linear operator in $H_k$ satisfying

$$\langle u, L_A v \rangle_k = \langle L_A^* u, v \rangle_k$$

for all $u, v \in H_k$. Then the null-space of $L_A^*$,

$$\mathcal{N}_k^* = \{u \in H_k : L_A^* u = 0\}, \tag{B.8}$$

is the orthogonal complement to $L_A(H_k)$. In practice, we can treat the coefficients in front of all monomials in all components of the homogeneous vector-polynomials as coordinates in $H_k$. This allows us to identify $H_k$ for each $k$ with some $\mathbb{R}^{N(k)}$ and $L_A|_{H_k k}$ with an $N(k) \times N(k)$ matrix $M_A^{(k)}$, respectively. With these identifications, we can use the standard scalar product $\langle u, v \rangle = u^T v$ in $\mathbb{R}^{N(k)}$, so that $L_A^*$ will be represented by the transposed matrix $[M_A^{(k)}]^T$. The subspace $\mathcal{N}_k^*$ is then spanned by all eigenvectors of $[M_A^{(k)}]^T$ corresponding to zero eigenvalue.

However, if no generalized eigenvectors are associated with zero eigenvalue of $L_A$ (or, equivalently, $M_A^{(k)}$), the null-space of $L_A$,

$$\mathcal{N}_k = \{v \in H_k : L_A v = 0\}, \tag{B.9}$$

and its range, $L_A(H_k)$, span together the whole $H_k$. Therefore, in this case, we can use $\mathcal{N}_k$ as $\widetilde{H}_k$. With the identification of $H_k$ with $\mathbb{R}^{N(k)}$ described above, the null-space $\mathcal{N}_k$ is spanned by all eigenvectors of $M_A^{(k)}$ corresponding to zero eigenvalue.

**Example B.1 (Poincaré normal form at the Hopf bifurcation)** Consider the system (B.1) with $n = 2$ and

$$A = \begin{pmatrix} 0 & -\omega_0 \\ \omega_0 & 0 \end{pmatrix},$$

where $\omega_0 > 0$. The system has an equilibrium $x = 0$ with two purely imaginary eigenvalues $\lambda_{1,2} = \pm i\omega_0$, $\omega_0 > 0$. Notice that any $2 \times 2$ matrix with two such eigenvalues is similar to $A$. The operator $L_A$ has the form

$$(L_A h)(y) = \omega_0 h_y(y) \begin{pmatrix} -y_2 \\ y_1 \end{pmatrix} - \omega_0 \begin{pmatrix} -h_2(y) \\ h_1(y) \end{pmatrix}. \tag{B.10}$$

First consider the space $H_2$. Any vector-polynomial $f^{(2)} \in H_2$ can be written as

$$f^{(2)}(y) = \begin{pmatrix} a_1 y_1^2 + a_2 y_1 y_2 + a_3 y_2^2 \\ a_4 y_1^2 + a_5 y_1 y_2 + a_6 y_2^2 \end{pmatrix}$$

so that $(a_1, a_2, a_3, a_4, a_5, a_6)$ can be considered as coordinates in $H_2$. Using (B.10), it is easy to check that

$$(L_A f^{(2)})(y) = \omega_0 \begin{pmatrix} (a_1 + a_4)y_1^2 + (-2a_1 + 2a_3 + a_5)y_1 y_2 + (-a_2 + a_6)y_2^2 \\ (-a_1 + a_5)y_1^2 + (-a_2 - 2a_4 + 2a_6)y_1 y_2 + (-a_3 - a_5)y_2^2 \end{pmatrix},$$

so that it acts on $a \in \mathbb{R}^6$ as the multiplication by the matrix

$$M_A^{(2)} = \omega_0 \begin{pmatrix} 0 & 1 & 0 & 1 & 0 & 0 \\ -2 & 0 & 2 & 0 & 1 & 0 \\ 0 & -1 & 0 & 0 & 0 & 1 \\ -1 & 0 & 0 & 0 & 1 & 0 \\ 0 & -1 & 0 & -2 & 0 & 2 \\ 0 & 0 & -1 & 0 & -1 & 0 \end{pmatrix}.$$

Since $\det(M_A^{(2)}) = 9\omega_0^6 \neq 0$, $L_A(H_2) = H_2$, and, therefore, all quadratic terms can be removed by a coordinate transformation

$$x = y + h^{(2)}(y),$$

where $h^{(2)}$ is represented by $[M_A^{(2)}]^{-1}a$. Notice that this transformation *changes* the terms of order $k \geq 3$.

Assume now that all quadratic terms are eliminated, so that (B.1) can be written as

$$\dot{y} = Ay + f^{(3)}(y) + O(\|y\|^4),$$

where

$$f^{(3)}(y) = \begin{pmatrix} b_1 y_1^3 + b_2 y_1^2 y_2 + b_3 y_1 y_2^2 + b_4 y_2^3 \\ b_5 y_1^3 + b_6 y_1^2 y_2 + b_7 y_1 y_2^2 + b_8 y_2^3 \end{pmatrix} \in H_3.$$

We see that $\dim H_3 = 8$ and $(b_1, b_2, \ldots, b_8)$ are the coordinates in $H_3$. One can verify that the action of $L_A$ in $H_3$ is represented by the matrix

$$M_A^{(3)} = \omega_0 \begin{pmatrix} 0 & 1 & 0 & 0 & 1 & 0 & 0 & 0 \\ -3 & 0 & 2 & 0 & 0 & 1 & 0 & 0 \\ 0 & -2 & 0 & 3 & 0 & 0 & 1 & 0 \\ 0 & 0 & -1 & 0 & 0 & 0 & 0 & 1 \\ -1 & 0 & 0 & 0 & 0 & 1 & 0 & 0 \\ 0 & -1 & 0 & 0 & -3 & 0 & 2 & 0 \\ 0 & 0 & -1 & 0 & 0 & -2 & 0 & 3 \\ 0 & 0 & 0 & -1 & 0 & 0 & -1 & 0 \end{pmatrix}.$$

As one might expect, this matrix is singular: $\det M_A^{(3)} = 0$. The transposed matrix $[M_A^{(3)}]^T$ has a two-dimensional null-space $\mathcal{N}_k^*$ spanned by the vectors

$$V^* = (3, 0, 1, 0, 0, 1, 0, 3)^T, \quad W^* = (0, -1, 0, -3, 3, 0, 1, 0)^T$$

(check!). Therefore, a possible choice for $\widetilde{H}_3$ would be

$$\widetilde{H}_3 = \text{Span}\{V^*, W^*\}.$$

For this selection of the complement $\widetilde{H}_3$, the resonant terms in $H_3$ would have the form

$$r_*^{(3)} = \alpha \begin{pmatrix} 3y_1^3 + y_1 y_2^2 \\ y_1^2 y_2 + 3y_2^3 \end{pmatrix} + \beta \begin{pmatrix} -y_1^2 y_2 - 3y_2^3 \\ 3y_1^3 + y_1 y_2^2 \end{pmatrix}$$

with some $\alpha, \beta \in \mathbb{R}^1$.

However, the null-space $\mathcal{N}_k$ of $M_A^{(3)}$ is spanned by two vectors,

$$V = (1, 0, 1, 0, 0, 1, 0, 1)^T, \quad W = (0, -1, 0, -1, 1, 0, 1, 0)^T,$$

while there are no generalized eigenvectors associated to a zero eigenvalue (verify!). Therefore, an alternative choice is

$$\widetilde{H}_3 = \operatorname{span}\{V, W\}.$$

With this selection, the resonant terms in $H_3$ will be

$$r^{(3)} = \alpha \begin{pmatrix} y_1^3 + y_1 y_2^2 \\ y_1^2 y_2 + y_2^3 \end{pmatrix} + \beta \begin{pmatrix} -y_1^2 y_2 - y_2^3 \\ y_1^3 + y_1 y_2^2 \end{pmatrix} = \begin{pmatrix} (\alpha y_1 - \beta y_2)(y_1^2 + y_2^2) \\ (\beta y_1 + \alpha y_2)(y_1^2 + y_2^2) \end{pmatrix}$$

for some $\alpha, \beta \in \mathbb{R}^1$.

Theorem 3.5 now implies that a smooth planar system at the Hopf bifurcation can be transformed into the normal form

$$\begin{pmatrix} \dot{y}_1 \\ \dot{y}_2 \end{pmatrix} = \begin{pmatrix} 0 & -\omega_0 \\ \omega_0 & 0 \end{pmatrix} \begin{pmatrix} y_1 \\ y_2 \end{pmatrix} + (y_1^2 + y_2^2) \begin{pmatrix} \alpha y_1 - \beta y_2 \\ \beta y_1 + \alpha y_2 \end{pmatrix} + O(\|y\|^3).$$

Introducing the complex variable $z = y_1 + iy_2$, we arrive at the familiar equation

$$\dot{z} = i\omega_0 z + c_1 z|z|^2 + O(|z|^4), \quad z \in \mathbb{C}^1,$$

where $c_1 = \alpha + i\beta$. $\Diamond$

When the matrix $A$ in (B.1) is *diagonal*, i.e. has the form

$$A = \begin{pmatrix} \lambda_1 & 0 & \cdots & 0 \\ 0 & \lambda_2 & \cdots & 0 \\ \cdots \\ 0 & 0 & \cdots & \lambda_n \end{pmatrix},$$

one can easily compute the eigenvalues of $L_A$ restricted to $H_m$. Consider a vector-monomial

$$h_j(y) = y_1^{m_1} y_2^{m_2} \cdots y_n^{m_n} \mathrm{e}_j, \tag{B.8}$$

where $m_i \geq 0$ are integer numbers with $m_1 + m_2 + \cdots + m_n = m$, and $\mathrm{e}_j \in \mathbb{R}^n$ is the unit vector along the $y_j$-axis. Clearly, $h \in H_m$, and using (B.5) we see that

$$(L_A h_j)(y) = \left( \sum_{k=1}^{n} \lambda_k m_k - \lambda_j \right) y_1^{m_1} y_2^{m_2} \cdots y_n^{m_n} \mathrm{e}_j.$$

Therefore,

$$L_A h_j = (\langle \lambda, m \rangle - \lambda_j) h_j, \tag{B.9}$$

where

$$\lambda = (\lambda_1, \lambda_2, \ldots, \lambda_n)^T, \quad m = (m_1, m_2, \ldots, m_n)^T,$$

and the usual scalar product $\langle \cdot, \cdot \rangle$ in $\mathbb{R}^n$ is used to simplify notations. The equation (B.9) means that the vector-monomial $h_j$ defined by (B.8) is the eigenvector of $L_A$ corresponding to the eigenvalue

$$\mu_j = \langle \lambda, m \rangle - \lambda_j.$$

Thus, the null-space of $L_A$ is spanned in this case by vector-monomials $h_j$, for which $\mu_j = 0$, i.e.,

$$\lambda_j = \langle \lambda, m \rangle. \tag{B.10}$$

Such monomials are called *resonant monomials* of order $m$, and these are the only monomials present in $r^{(m)}$. For a fixed $m$ and each $j = 1, 2, \ldots, n$, (B.10) implies a condition on the eigenvalues of $A$ (called the *resonance condition* or *resonance*).

If no resonances of order $m$ exist, Theorem B.1 implies that all terms of order $m$ in (B.1) can be eliminated by a polynomial transformation. In the presence of resonances, the resonant monomials satisfying (B.10) cannot be removed from the $j$-component of the right-hand side of (B.1) by all such transformations. This allows us to predict in some cases the Poincaré normal form without any computations.

**Example B.2 (Poincaré normal form for a resonant node)** Consider the system (B.1) with $n = 2$ and

$$A = \begin{pmatrix} 1 & 0 \\ 0 & 2 \end{pmatrix}.$$

The origin $x = 0$ is an unstable node with eigenvalues $\lambda_1 = 1,\ \lambda_2 = 2$ at resonance:

$$\lambda_2 = 2\lambda_1.$$

This is a resonance of order 2 ($m_1 = 0,\ m_2 = 2$) and there are obviously no other resonances of any order. Therefore, there is only one resonant vector-monomial

$$\begin{pmatrix} 0 \\ y_2^2 \end{pmatrix},$$

and (B.1) is locally smoothly equivalent to the system

$$\begin{pmatrix} \dot{y}_1 \\ \dot{y}_2 \end{pmatrix} = \begin{pmatrix} y_1 \\ 2y_2 + \alpha y_2^2 \end{pmatrix} + O(\|y\|^m),$$

where $\alpha \in \mathbb{R}^1$ and $m \geq 3$ can be taken arbitrarily large, provided (B.1) is a $C^\infty$-system. $\Diamond$

Before giving another example, notice that all considerations above are valid verbatim if we replace in (B.1) and in all subsequent formulas $x \in \mathbb{R}^n$ by $z \in \mathbb{C}^n$.

**Example B.3 (Normal form in the Hopf case revisited)** Consider again the system

$$\begin{pmatrix} \dot{x}_1 \\ \dot{x}_2 \end{pmatrix} = \begin{pmatrix} 0 & -\omega_0 \\ \omega_0 & 0 \end{pmatrix} \begin{pmatrix} x_1 \\ x_2 \end{pmatrix} + f(x), \quad x = \begin{pmatrix} x_1 \\ x_2 \end{pmatrix} \in \mathbb{R}^2, \tag{B.11}$$

where $\omega_0 > 0$. If we introduce a complex variable $z = x_1 + ix_2$, then (B.11) can be written as one complex equation

$$\dot{z} = i\omega_0 z + g_1(z, \bar{z}), \quad z \in \mathbb{C}^1,$$

where $g_1$ is a smooth function of $z$ and $\bar{z} = x_1 - ix_2$ (cf. Lemma 3.3). Since $\bar{z}$ satisfies the complex conjugate equation

$$\dot{\bar{z}} = -i\omega_0 \bar{z} + g_2(z, \bar{z}), \quad \bar{z} \in \mathbb{C}^1,$$

where $g_2(z, \bar{z}) = \overline{g_1(z, \bar{z})}$, we can formally introduce a system in $\mathbb{C}^2$,

$$\begin{pmatrix} \dot{z}_1 \\ \dot{z}_2 \end{pmatrix} = \begin{pmatrix} i\omega_0 & 0 \\ 0 & -i\omega_0 \end{pmatrix} \begin{pmatrix} z_1 \\ z_2 \end{pmatrix} + \begin{pmatrix} g_1(z_1, z_2) \\ g_2(z_1, z_2) \end{pmatrix}, \quad \text{(B.12)}$$

in which the first equation is equivalent to (B.11), if we substitute $z_1 = z$ and $z_2 = \bar{z}$. The system (B.12) is called the *complexification* of (B.11). Notice that (B.12) has the diagonal linear part with $\lambda_1 = i\omega_0$ and $\lambda_2 = -i\omega_0$. The sum of these eigenvalues vanishes,

$$\lambda_1 + \lambda_2 = 0,$$

which implies two resonance conditions of order $2k+1$,

$$\begin{aligned} \lambda_1 = \lambda_1 + k(\lambda_1 + \lambda_2) &= (k+1)\lambda_1 + k\lambda_2, \\ \lambda_2 = \lambda_2 + k(\lambda_1 + \lambda_2) &= k\lambda_1 + (k+1)\lambda_2, \end{aligned}$$

for any integer $k \geq 1$. Therefore, the resonant monomials are

$$\begin{pmatrix} \zeta_1(\zeta_1\zeta_2)^k \\ 0 \end{pmatrix} \text{ and } \begin{pmatrix} 0 \\ \zeta_2(\zeta_1\zeta_2)^k \end{pmatrix},$$

where $(\zeta_1, \zeta_2)^T \in \mathbb{C}^2$. Thus, the first equation in the resulting Poincaré normal form is

$$\dot{\zeta}_1 = i\omega_0\zeta_1 + c_1\zeta_1^2\zeta_2 + c_2\zeta_1^3\zeta_2^2 + \cdots,$$

or

$$\dot{\zeta} = i\omega_0\zeta + c_1\zeta|\zeta|^2 + c_2\zeta|\zeta|^4 + O(|\zeta|^6), \quad \zeta \in \mathbb{C}^1,$$

where $c_{1,2} \in \mathbb{C}^1$, while the second equation is the complex-conjugate of the first. This gives the normal form at the Hopf bifurcation up to sixth order. □

It should be stressed that the results outlined in this appendix determine only the *general structure* of the normal forms but give no expressions for their coefficients in terms of the original system. Such expressions can be obtained by the method of unknown coefficients (see this chapter, as well as Chapter 8). There is a version of the normal form theory for maps (see Chapters 4 and 9). In all such computations, symbolic manipulation software is very useful.

## 3.9 Appendix C: Bibliographical notes

The fold bifurcation of equilibria has essentially been known for centuries. Since any scalar system can be written as $\dot{x} = -V_x(x, \alpha)$, for some function $V$, results on the classification of generic parameter-dependent *gradient systems* from catastrophe (or singularity) theory are relevant. Thus, the topological normal form for the fold bifurcation appeared in the list of seven elementary catastrophes by Thom [1972]. As in catastrophe theory, the fold normal form can be derived by using the Malgrange Preparation Theorem (see, for example, Arrowsmith & Place [1990] or Murdock [2003]). We adopted an elementary approach that is more suited for the present book.

Actually, there are many interconnections between bifurcation theory of dynamical systems and singularity theory of smooth functions. The books by Poston & Stewart [1978] and Arnol'd [1984] are recommended as an introduction to this latter subject. It should be noted, however, that most results from singularity theory are directly applicable to the analysis of equilibria but not to the analysis of phase portraits.

The normalization technique used in the analysis of equilibrium bifurcations was developed by Poincaré [1879]. Appendix B gives only a brief exposition of the simplest variant of the normal form theory. This appendix closely follows Guckenheimer & Holmes [1983, Sec. 3.3]. Introductions to normal forms can also be found in Arnol'd [1983] and Vanderbauwhede [1989], where it is explained how to apply this theory to local bifurcation problems. For modern comprehensive presentations, see [Belitskii 2002] and [Murdock 2003]. Theorem 3.4 was stated and briefly proved by Arnol'd [1972, 1983] using Poincaré normal forms.

Phase-portrait bifurcations in a generic one-parameter system on the plane near an equilibrium with purely imaginary eigenvalues were studied first by Andronov & Leontovich [1939]. They used a *succession function* (return map) technique originally due to Lyapunov [1892] without benefiting from the normalization. An explicit expression for the first Lyapunov coefficient in terms of Taylor coefficients of a general planar system was obtained by Bautin [1949]. An exposition of the results by Andronov and his co-workers on this bifurcation can be found in Andronov et al. [1973].

Hopf [1942] proved the appearance of a family of periodic solutions of increasing amplitude in $n$-dimensional systems having an equilibrium with a pair of purely imaginary eigenvalues at some critical parameter value. He did not consider bifurcations of the whole phase portrait. An English-language translation of Hopf's paper is included in Marsden & McCracken [1976]. This very useful book also contains a derivation of the first Lyapunov coefficient and a proof of Hopf's result based on the Implicit Function Theorem.

A much simpler derivation of the Lyapunov coefficient (actually $c_1$) is given by Hassard, Kazarinoff & Wan [1981] using the complex form of the Poincaré normalization. We essentially use their technique, although we do not assume that the Poincaré normal form is known a priori. Formulas to compute Taylor coefficients of the complex equation without an intermediate transformation of the system into its eigenbasis can also be extracted from their book (applying the center manifold reduction technique to the trivial planar case; see Chapter 5). We also extensively use time reparametrization to obtain a simpler normal form, which is then used to prove existence and uniqueness of the cycle and in the analysis of the whole phase-portrait bifurcations (see Appendix A).

There exist other approaches to prove the generation of periodic solutions under the Hopf conditions. An elegant one is to reformulate the problem as that of finding a family of solutions of an abstract equation in a functional space of periodic functions and to apply the *Lyapunov-Schmidt reduction*. This approach, allowing a generalization to infinite-dimensional dynamical systems, is far beyond the scope of this book (see, e.g., Chow & Hale [1982] or Kielhöfer [2004]).

# 4

# One-Parameter Bifurcations of Fixed Points in Discrete-Time Dynamical Systems

In this chapter, which is organized very much like Chapter 3, we present bifurcation conditions defining the simplest bifurcations of fixed points in $n$-dimensional discrete-time dynamical systems: the fold, the flip, and the Neimark-Sacker bifurcations. Then we study these bifurcations in the lowest possible dimension in which they can occur: the fold and flip bifurcations for scalar systems and the Neimark-Sacker bifurcation for planar systems. In Chapter 5 it will be shown how to apply these results to $n$-dimensional systems when $n$ is larger than one or two, respectively.

## 4.1 Simplest bifurcation conditions

Consider a discrete-time dynamical system depending on a parameter

$$x \mapsto f(x, \alpha), \quad x \in \mathbb{R}^n, \ \alpha \in \mathbb{R}^1,$$

where the map $f$ is smooth with respect to both $x$ and $\alpha$. Sometimes we will write this system as

$$\tilde{x} = f(x, \alpha), \quad x, \tilde{x} \in \mathbb{R}^n, \ \alpha \in \mathbb{R}^1,$$

where $\tilde{x}$ denotes the image of $x$ under the action of the map. Let $x = x_0$ be a hyperbolic fixed point of the system for $\alpha = \alpha_0$. Let us monitor this fixed point and its multipliers while the parameter varies. It is clear that there are, generically, only three ways in which the hyperbolicity condition can be violated. Either a simple positive multiplier approaches the unit circle and we have $\mu_1 = 1$ (see **Fig. 4.1**(a)), or a simple negative multiplier approaches the unit circle and we have $\mu_1 = -1$ (**Fig. 4.1**(b)), or a pair of simple complex multipliers reaches the unit circle and we have $\mu_{1,2} = e^{\pm i\theta_0}, 0 < \theta_0 < \pi$ (**Fig. 4.1**(c)), for some value of the parameter. It is obvious that one needs more parameters to allocate extra eigenvalues on the unit circle.

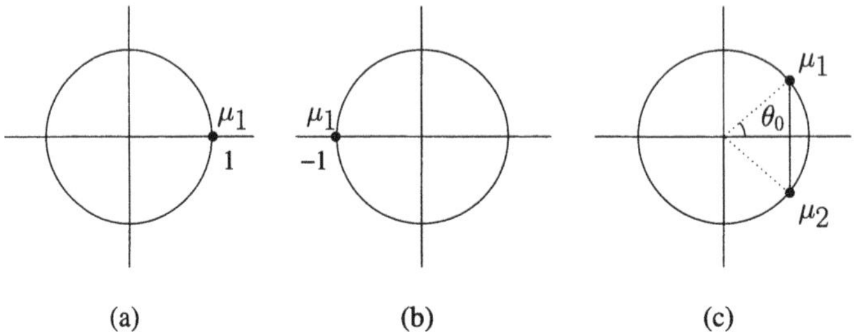


**Fig. 4.1.** Codim 1 critical cases.

The rest of the chapter is devoted to the proof that a nonhyperbolic fixed point satisfying one of the above conditions is *structurally unstable*, and to the analysis of the corresponding bifurcations of the local phase portrait under variation of the parameter. Let us finish this section with the following definitions, the reasoning for which will become clear later.

**Definition 4.1** *The bifurcation associated with the appearance of* $\mu_1 = 1$ *is called a* fold (*or* tangent) *bifurcation.*

**Remark:**

This bifurcation is also referred to as a *limit point, saddle-node bifurcation,* or *turning point,* among other terms. ◇

**Definition 4.2** *The bifurcation associated with the appearance of* $\mu_1 = -1$ *is called a* flip (*or* period-doubling) *bifurcation.*

**Definition 4.3** *The bifurcation corresponding to the presence of* $\mu_{1,2} = e^{\pm i\theta_0}, 0 < \theta_0 < \pi$, *is called a* Neimark-Sacker (*or* torus) *bifurcation.*

Notice that the fold and flip bifurcations are possible if $n \geq 1$, but for the Neimark-Sacker bifurcation we need $n \geq 2$.

**Example 4.1 (Bifurcations of fixed points in the Hénon map)** Consider the Hénon map introduced in Example 1.9 of Section 1.3.2:

$$\begin{pmatrix} x \\ y \end{pmatrix} \mapsto \begin{pmatrix} y \\ \alpha - \beta x - y^2 \end{pmatrix}. \tag{4.1}$$

Let us find curves in the $(\beta, \alpha)$-plane corresponding to the critical cases defined above. More precisely, we look for parameter values for which (4.1) has a fixed point with multipliers satisfying one of the bifurcation conditions visualized in **Fig. 4.1**.

Fixed points of (4.1) can be found from the system

$$\begin{cases} x = y, \\ y = \alpha - \beta x - y^2, \end{cases}$$

implying that all such points are located on the line $y = x$ and their $y$-coordinates satisfy the equation

$$y^2 + (1+\beta)y - \alpha = 0. \tag{4.2}$$

Thus, the map (4.1) can have no more than two fixed points.

Introduce the Jacobian matrix of (4.1):

$$A(x,y,\alpha,\beta) = \begin{pmatrix} 0 & 1 \\ -\beta & -2y \end{pmatrix}.$$

The fold bifurcation condition implies that

$$\det(A(x,y,\alpha,\beta) - I_2) = 0,$$

where $(x,y)$ are the coordinates of a fixed point and $I_2$ is the unit $2\times2$ matrix. This condition is equivalent to

$$1 + \beta + 2y = 0. \tag{4.3}$$

Therefore, the fold bifurcation curve in the $(\alpha,\beta)$-plane can be found by eliminating $y$ from an algebraic system composed of the fixed-point equation (4.2) and the fold bifurcation condition (4.3), i.e.,

$$\begin{cases} y^2 + (1+\beta)y - \alpha = 0, \\ 1 + \beta + 2y = 0. \end{cases}$$

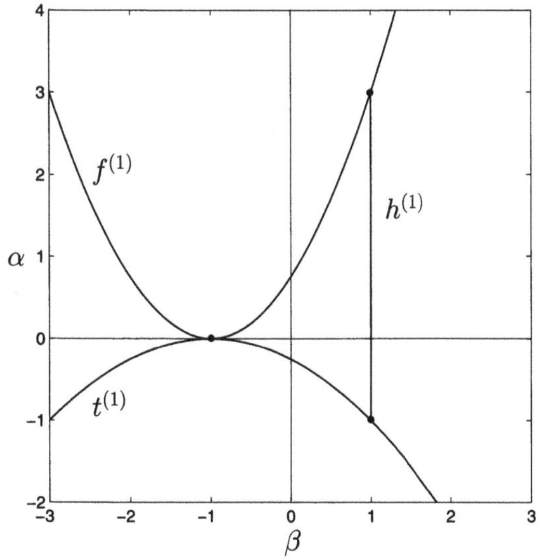


**Fig. 4.2.** Bifurcation curves of the Hénon map (4.1).

The elimination gives

$$t^{(1)} = \left\{ (\beta, \alpha) : \alpha = -\frac{(1+\beta)^2}{4} \right\}.$$

Along this curve, the map (4.1) has a fixed point with multiplier $\mu_1 = 1$.

Similarly, the flip bifurcation condition can be written as

$$\det(A(x, y, \alpha, \beta) + I_2) = 0$$

or

$$1 + \beta + 2y = 0. \tag{4.4}$$

Combining the fixed-point equation (4.2) with the flip bifurcation condition (4.4), we obtain the system

$$\begin{cases} y^2 + (1+\beta)y - \alpha = 0, \\ 1 + \beta - 2y = 0. \end{cases}$$

Eliminating $y$ from this system, we see that the map (4.1) has a fixed point with multiplier $\mu_1 = -1$ when its parameters belong to the curve

$$f^{(1)} = \left\{ (\beta, \alpha) : \alpha = \frac{3(1+\beta)^2}{4} \right\}.$$

Finally, consider the Neimark-Sacker critical case. Notice that the product of the critical multipliers $\mu_{1,2} = e^{\pm i\theta_0}$ is equal to one:

$$\mu_1 \mu_2 = 1.$$

Since the same product is equal to the determinant of $A(x, y, \alpha, \beta)$, we conclude that the Neimark-Sacker condition implies that

$$\det A(x, y, \alpha, \beta) - 1 = 0.$$

Thus, for a fixed point of (4.1) with two multipliers at the unit circle, we have

$$1 - \beta = 0. \tag{4.5}$$

However, here there are two complications. First, the bifurcation condition (4.5) should only be considered when the map (4.1) has fixed points, i.e., for parameter values above the curve $t^{(1)}$ in the $(\beta, \alpha)$-plane. This imposes the restriction $\alpha > -1$. Second, (4.5) is also valid for a fixed point with *two real* multipliers

$$\mu_1 = \nu, \quad \mu_2 = \frac{1}{\nu}, \quad |\nu| > 1.$$

The multiplier $\mu_1$ is outside the unit circle, while the multiplier $\mu_2$ is inside the unit circle. Therefore, we have a saddle fixed point that cannot bifurcate. To exclude this possibility, one has to assume that $\alpha < 3$ (check!). Thus, (4.1) has a fixed point with multipliers $\mu_{1,2} = e^{\pm i\theta_0}, 0 < \theta_0 < \pi$, along the line segment

$$h^{(1)} = \{(\beta, \alpha) : \beta = 1, \quad -1 < \alpha < 3\}$$

that is confined between the curves $t^{(1)}$ and $f^{(1)}$ (see **Fig. 4.2**). ◊

## 4.2 The normal form of the fold bifurcation

Consider the following one-dimensional dynamical system depending on one parameter:

$$x \mapsto \alpha + x + x^2 \equiv f(x, \alpha) \equiv f_\alpha(x). \tag{4.6}$$

The map $f_\alpha$ is invertible for $|\alpha|$ small in a neighborhood of the origin. The system (4.6) has at $\alpha = 0$ a nonhyperbolic fixed point $x_0 = 0$ with $\mu = f_x(0, 0) = 1$. The behavior of the system near $x = 0$ for small $|\alpha|$ is shown in **Fig. 4.3**. For $\alpha < 0$ there are two fixed points in the system: $x_{1,2}(\alpha) = \pm\sqrt{-\alpha}$, the left of which is stable, while the right one is unstable. For $\alpha > 0$ there are no fixed points in the system. While $\alpha$ crosses zero from negative to positive values, the two fixed points (stable and unstable) "collide," forming at $\alpha = 0$ a fixed point with $\mu = 1$, and disappear. This is a fold (tangent) bifurcation in the discrete-time dynamical system.

There is, as usual, another way of presenting this bifurcation: plotting a bifurcation diagram in the direct product of the phase and parameter spaces, namely, in the $(x, \alpha)$-plane. The *fixed-point manifold* $x - f(x, \alpha) = 0$ is simply the parabola $\alpha = -x^2$ (see **Fig. 4.4**). Fixing some $\alpha$, we can easily determine the number of fixed points in the system for this parameter value. At $(x, \alpha) = (0, 0)$ a map projecting the fixed-point manifold onto the $\alpha$-axis has a singularity of the fold type.

**Remark:**
The system $x \mapsto \alpha + x - x^2$ can be considered in the same way. The analysis reveals two fixed points appearing for $\alpha > 0$. ◇

Now add higher-order terms to system (4.6), i.e., consider the system

$$x \mapsto \alpha + x + x^2 + x^3\psi(x, \alpha) \equiv F_\alpha(x), \tag{4.7}$$

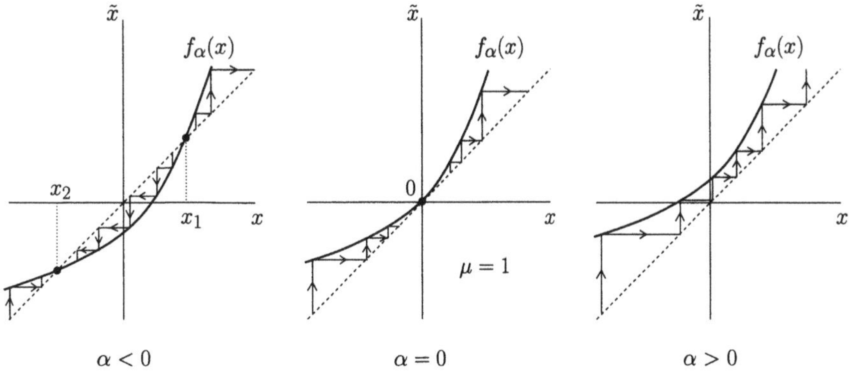


**Fig. 4.3.** Fold bifurcation.

where $\psi = \psi(x, \alpha)$ depends smoothly on $(x, \alpha)$. It is easy to check that in a sufficiently small neighborhood of $x = 0$ the number and the stability of the fixed points are the same for system (4.7) as for system (4.6) at corresponding parameter values, provided $|\alpha|$ is small enough. Moreover, a homeomorphism $h_\alpha$ of a neighborhood of the origin mapping orbits of (4.6) into the corresponding orbits of (4.7) can be constructed for each small $|\alpha|$. This property was called *local topological equivalence* of parameter-dependent systems in Chapter 2. It should be noted that construction of $h_\alpha$ is not as simple as in the continuous-time case (cf. Lemma 3.1). In the present case, a homeomorphism mapping the fixed points of (4.6) into the corresponding fixed points of (4.7) will not necessarily map other orbits of (4.6) into orbits of (4.7). Nevertheless, a homeomorphism $h_\alpha$ satisfying the condition

$$f_\alpha(x) = h_\alpha^{-1}(F_\alpha(h_\alpha(x)))$$

for all $(x, \alpha)$ in a neighbourghood of $(0, 0)$ (cf. Chapter 2) exists. Thus, the following lemma holds.

**Lemma 4.1** *The system*

$$x \mapsto \alpha + x + x^2 + O(x^3)$$

*is locally topologically equivalent near the origin to the system*

$$x \mapsto \alpha + x + x^2. \ \square$$

## 4.3 Generic fold bifurcation

We shall show that system (4.6) (with a possible sign change of the term $x^2$) is a topological normal form of a generic one-dimensional discrete-time system

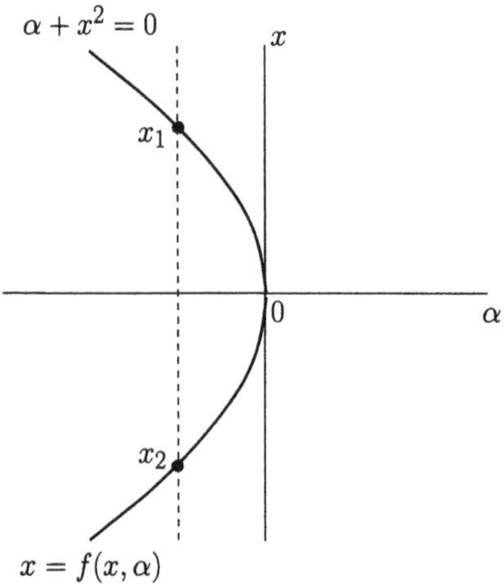


**Fig. 4.4.** Fixed point manifold.

having a fold bifurcation. In Chapter 5 we will also see that in some strong sense it describes the fold bifurcation in a generic $n$-dimensional system.

**Theorem 4.1** *Suppose that a one-dimensional system*

$$x \mapsto f(x, \alpha), \quad x \in \mathbb{R}, \; \alpha \in \mathbb{R}^1, \tag{4.8}$$

*with smooth $f$, has at $\alpha = 0$ the fixed point $x_0 = 0$, and let $\mu = f_x(0,0) = 1$. Assume that the following conditions are satisfied:*

(A.1) $f_{xx}(0,0) \neq 0$;
(A.2) $f_\alpha(0,0) \neq 0$.

*Then there are smooth invertible coordinate and parameter changes transforming the system into*

$$\eta \mapsto \beta + \eta \pm \eta^2 + O(\eta^3).$$

**Proof:**

Expand $f(x, \alpha)$ in a Taylor series with respect to $x$ at $x = 0$:

$$f(x, \alpha) = f_0(\alpha) + f_1(\alpha)x + f_2(\alpha)x^2 + O(x^3).$$

Two conditions are satisfied: $f_0(0) = f(0,0) = 0$ (*fixed-point condition*) and $f_1(0) = f_x(0,0) = 1$ (*fold bifurcation condition*). Since $f_1(0) = 1$, we may write

$$f(x, \alpha) = f_0(\alpha) + [1 + g(\alpha)]x + f_2(\alpha)x^2 + O(x^3),$$

where $g(\alpha)$ is smooth and $g(0) = 0$.

As in the proof of Theorem 3.1 in Chapter 3, perform a coordinate shift by introducing a new variable $\xi$:

$$\xi = x + \delta, \tag{4.9}$$

where $\delta = \delta(\alpha)$ is to be defined suitably. The transformation (4.9) yields

$$\tilde{\xi} = \tilde{x} + \delta = f(x, \alpha) + \delta = f(\xi - \delta, \alpha) + \delta.$$

Therefore,

$$\begin{aligned}\tilde{\xi} = {} & [f_0(\alpha) - g(\alpha)\delta + f_2(\alpha)\delta^2 + O(\delta^3)] \\ & + \xi + [g(\alpha) - 2f_2(\alpha)\delta + O(\delta^2)]\xi \\ & + [f_2(\alpha) + O(\delta)]\xi^2 + O(\xi^3).\end{aligned}$$

Assume that

(A.1) $$f_2(0) = \tfrac{1}{2} f_{xx}(0,0) \neq 0.$$

Then there is a smooth function $\delta(\alpha)$, which annihilates the parameter-dependent linear term in the above map for all sufficiently small $|\alpha|$. Indeed, the condition for that term to vanish can be written as

$$F(\alpha, \delta) \equiv g(\alpha) - 2f_2(\alpha)\delta + \delta^2\varphi(\alpha, \delta) = 0$$

for some smooth function $\varphi$. We have

$$F(0,0) = 0, \quad \left.\frac{\partial F}{\partial \delta}\right|_{(0,0)} = -2f_2(0) \neq 0, \quad \left.\frac{\partial F}{\partial \alpha}\right|_{(0,0)} = g'(0),$$

which implies (local) existence and uniqueness of a smooth function $\delta = \delta(\alpha)$ such that $\delta(0) = 0$ and $F(\alpha, \delta(\alpha)) \equiv 0$. It follows that

$$\delta(\alpha) = \frac{g'(0)}{2f_2(0)}\alpha + O(\alpha^2).$$

The map written in terms of $\xi$ is given by

$$\tilde{\xi} \; = \; [f_0'(0)\alpha + \alpha^2\psi(\alpha)] \; + \; \xi \; + \; [f_2(0) + O(\alpha)]\,\xi^2 \; + \; O(\xi^3), \tag{4.10}$$

where $\psi$ is some smooth function.

Consider as a new parameter $\mu = \mu(\alpha)$ the constant ($\xi$-independent) term of (4.10):

$$\mu = f_0'(0)\alpha + \alpha^2\psi(\alpha).$$

We have

(a) $\mu(0) = 0$;
(b) $\mu'(0) = f_0'(0) = f_\alpha(0,0)$.

If we assume

(A.2) $$f_\alpha(0,0) \neq 0,$$

then the Inverse Function Theorem implies local existence and uniqueness of a smooth inverse function $\alpha = \alpha(\mu)$ with $\alpha(0) = 0$. Therefore, equation (4.10) now reads

$$\tilde{\xi} = \mu + \xi + b(\mu)\xi^2 + O(\xi^3),$$

where $b(\mu)$ is a smooth function with $b(0) = f_2(0) \neq 0$ due to the first assumption (A.1).

Let $\eta = |b(\mu)|\xi$ and $\beta = |b(\mu)|\mu$. Then we get

$$\tilde{\eta} = \beta + \eta + s\eta^2 + O(\eta^3),$$

where $s = \text{sign}\, b(0) = \pm 1$. $\square$

Using Lemma 4.1, we can also eliminate $O(\eta^3)$ terms and finally arrive at the following general result.

**Theorem 4.2 (Topological normal form for the fold bifurcation)**
*Any generic scalar one-parameter system*

$$x \mapsto f(x, \alpha),$$

*having at* $\alpha = 0$ *the fixed point* $x_0 = 0$ *with* $\mu = f_x(0,0) = 1$, *is locally topologically equivalent near the origin to one of the following normal forms:*

$$\eta \mapsto \beta + \eta \pm \eta^2. \ \square$$

**Remark:**
The genericity conditions in Theorem 4.2 are the nondegeneracy condition (A.1) and the transversality condition (A.2) from Theorem 4.1. ◇

## 4.4 The normal form of the flip bifurcation

Consider the following one-dimensional dynamical system depending on one parameter:

$$x \mapsto -(1+\alpha)x + x^3 \equiv f(x, \alpha) \equiv f_\alpha(x). \tag{4.11}$$

The map $f_\alpha$ is invertible for small $|\alpha|$ in a neighborhood of the origin. System (4.11) has the fixed point $x_0 = 0$ for all $\alpha$ with multiplier $\mu = -(1+\alpha)$. The point is linearly stable for small $\alpha < 0$ and is linearly unstable for $\alpha > 0$. At $\alpha = 0$ the point is not hyperbolic since the multiplier $\mu = f_x(0,0) = -1$, but is nevertheless (nonlinearly) stable. There are no other fixed points near the origin for small $|\alpha|$.

Consider now the *second iterate* $f_\alpha^2(x)$ of the map (4.11). If $y = f_\alpha(x)$, then

$$\begin{aligned} f_\alpha^2(x) &= f_\alpha(y) = -(1+\alpha)y + y^3 \\ &= -(1+\alpha)[-(1+\alpha)x + x^3] + [-(1+\alpha)x + x^3]^3 \\ &= (1+\alpha)^2 x - [(1+\alpha)(2 + 2\alpha + \alpha^2)]x^3 + O(x^5). \end{aligned}$$

The map $f_\alpha^2$ obviously has the trivial fixed point $x_0 = 0$. It also has *two* nontrivial fixed points for small $\alpha > 0$:

$$x_{1,2} = f_\alpha^2(x_{1,2}),$$

where $x_{1,2} = \pm\sqrt{\alpha}$ (see **Fig. 4.5**). These two points are stable and constitute a *cycle of period two* for the original map $f_\alpha$. This means that

$$x_2 = f_\alpha(x_1), \ x_1 = f_\alpha(x_2),$$

with $x_1 \neq x_2$. **Fig. 4.6** shows the complete bifurcation diagram of system (4.11) with the help of a staircase diagram. As $\alpha$ approaches zero from above, the period-two cycle "shrinks" and disappears. This is a flip bifurcation.

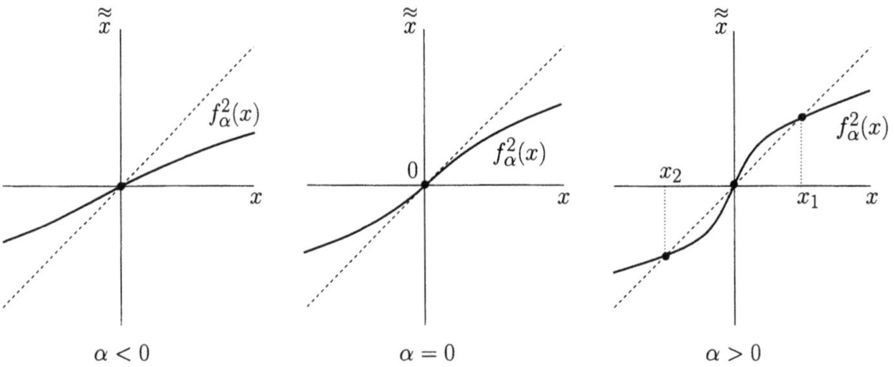


**Fig. 4.5.** Second iterate map near a flip bifurcation.

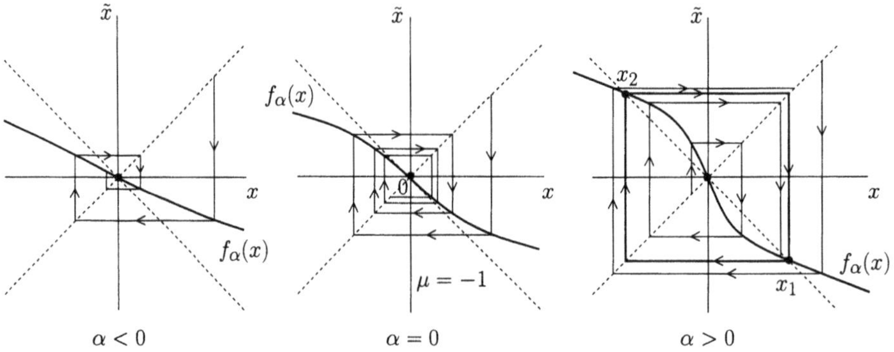


**Fig. 4.6.** Flip bifurcation.

The other way to present this bifurcation is to use the $(x, \alpha)$-plane (see **Fig. 4.7**). In this figure, the horizontal axis corresponds to the fixed point of (4.11) (stable for $\alpha < 0$ and unstable for $\alpha > 0$), while the "parabola" represents the stable cycle of period two $\{x_1, x_2\}$ existing for $\alpha > 0$.

As usual, let us consider the effect of higher-order terms on system (4.11).

**Lemma 4.2** *The system*

$$x \mapsto -(1+\alpha)x + x^3 + O(x^4)$$

*is locally topologically equivalent near the origin to the system*

$$x \mapsto -(1+\alpha)x + x^3. \ \square$$

The analysis of the fixed point and the period-two cycle is a simple exercise. The rest of the proof is not easy and is omitted here.

The case

$$x \mapsto -(1+\alpha)x - x^3 \tag{4.12}$$

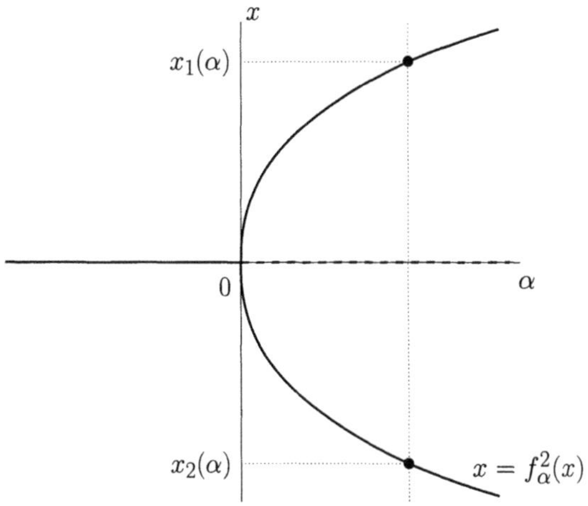


**Fig. 4.7.** A flip corresponds to a pitchfork bifurcation of the second iterate.

can be treated in the same way. For $\alpha \neq 0$, the fixed point $x_0 = 0$ has the same stability as in (4.11). At the critical parameter value $\alpha = 0$ the fixed point is unstable. The analysis of the second iterate of (4.12) reveals an *unstable* cycle of period two for $\alpha < 0$ which disappears at $\alpha = 0$. The higher-order terms do not affect the bifurcation diagram.

**Remark:**
By analogy with the Andronov-Hopf bifurcation, the flip bifurcation in system (4.11) is called *supercritical* or *"soft,"* while the flip bifurcation in system (4.12) is referred to as *subcritical* or *"sharp."* The bifurcation type is determined by the stability of the fixed point at the critical parameter value. ◊

## 4.5 Generic flip bifurcation

**Theorem 4.3** *Suppose that a one-dimensional system*

$$x \mapsto f(x, \alpha), \quad x \in \mathbb{R}^1, \ \alpha \in \mathbb{R}^1,$$

*with smooth $f$, has at $\alpha = 0$ the fixed point $x_0 = 0$, and let $\mu = f_x(0,0) = -1$. Assume that the following nondegeneracy conditions are satisfied:*

(B.1) $\frac{1}{2}(f_{xx}(0,0))^2 + \frac{1}{3}f_{xxx}(0,0) \neq 0$;
(B.2) $f_{x\alpha}(0,0) \neq 0$.

*Then there are smooth invertible coordinate and parameter changes transforming the system into*

$$\eta \mapsto -(1+\beta)\eta \pm \eta^3 + O(\eta^4).$$

**Proof:**

By the Implicit Function Theorem, the system has a unique fixed point $x_0(\alpha)$ in some neighborhood of the origin for all sufficiently small $|\alpha|$ since $f_x(0,0) \neq 1$. We can perform a coordinate shift placing this fixed point at the origin. Therefore, we can assume without loss of generality that $x = 0$ is the fixed point of the system for $|\alpha|$ sufficiently small. Thus, the map $f$ can be written as follows:

$$f(x,\alpha) = f_1(\alpha)x + f_2(\alpha)x^2 + f_3(\alpha)x^3 + O(x^4), \tag{4.13}$$

where $f_1(\alpha) = -[1 + g(\alpha)]$ for some smooth function $g$. Since $g(0) = 0$ and

$$g'(0) = f_{x\alpha}(0,0) \neq 0,$$

according to assumption (B.2), the function $g$ is locally invertible and can be used to introduce a new parameter:

$$\beta = g(\alpha).$$

Our map (4.13) now takes the form

$$\tilde{x} = \mu(\beta)x + a(\beta)x^2 + b(\beta)x^3 + O(x^4),$$

where $\mu(\beta) = -(1+\beta)$, and the functions $a(\beta)$ and $b(\beta)$ are smooth. We have

$$a(0) = f_2(0) = \frac{1}{2}f_{xx}(0,0), \quad b(0) = \frac{1}{6}f_{xxx}(0,0).$$

Let us perform a smooth change of coordinate:

$$x = y + \delta y^2, \tag{4.14}$$

where $\delta = \delta(\beta)$ is a smooth function to be defined. The transformation (4.14) is invertible in some neighborhood of the origin, and its inverse can be found by the method of unknown coefficients:

$$y = x - \delta x^2 + 2\delta^2 x^3 + O(x^4). \tag{4.15}$$

Using (4.14) and (4.15), we get

$$\tilde{y} = \mu y + (a + \delta\mu - \delta\mu^2)y^2 + (b + 2\delta a - 2\delta\mu(\delta\mu + a) + 2\delta^2\mu^3)y^3 + O(y^4).$$

Thus, the quadratic term can be "killed" for all sufficiently small $|\beta|$ by setting

$$\delta(\beta) = \frac{a(\beta)}{\mu^2(\beta) - \mu(\beta)}.$$

This can be done since $\mu^2(0) - \mu(0) = 2 \neq 0$, giving

$$\tilde{y} = \mu y + \left(b + \frac{2a^2}{\mu^2 - \mu}\right) y^3 + O(y^4) = -(1+\beta)y + c(\beta)y^3 + O(y^4)$$

for some smooth function $c(\beta)$, such that

$$c(0) = a^2(0) + b(0) = \frac{1}{4}(f_{xx}(0,0))^2 + \frac{1}{6}f_{xxx}(0,0). \qquad (4.16)$$

Notice that $c(0) \neq 0$ by assumption (B.1).

Apply the rescaling

$$y = \frac{\eta}{\sqrt{|c(\beta)|}}.$$

In the new coordinate $\eta$ the system takes the desired form

$$\tilde{\eta} = -(1+\beta)\eta + s\eta^3 + O(\eta^4),$$

where $s = \text{sign}\ c(0) = \pm 1$. □

Using Lemma 4.2, we arrive at the following general result.

**Theorem 4.4 (Topological normal form for the flip bifurcation)** *Any generic, scalar, one-parameter system*

$$x \mapsto f(x, \alpha),$$

*having at* $\alpha = 0$ *the fixed point* $x_0 = 0$ *with* $\mu = f_x(0,0) = -1$, *is locally topologically equivalent near the origin to one of the following normal forms:*

$$\eta \mapsto -(1+\beta)\eta \pm \eta^3. \ \Box$$

**Remark:**

Of course, the genericity conditions in Theorem 4.4 are the nondegeneracy condition (B.1) and the transversality condition (B.2) from Theorem 4.3. ◊

**Example 4.2 (Ricker's equation)** Consider the following simple population model [Ricker 1954]:

$$x_{k+1} = \alpha x_k e^{-x_k},$$

where $x_k$ is the population density in year $k$, and $\alpha > 0$ is the growth rate. The function on the right-hand side takes into account the negative role of interpopulation competition at high population densities. The above recurrence relation corresponds to the discrete-time dynamical system

$$x \mapsto \alpha x e^{-x} \equiv f(x, \alpha). \qquad (4.17)$$

System (4.17) has a trivial fixed point $x_0 = 0$ for all values of the parameter $\alpha$. At $\alpha_0 = 1$, however, a nontrivial positive fixed point appears:

$$x_1(\alpha) = \ln \alpha.$$

The multiplier of this point is given by the expression

$$\mu(\alpha) = 1 - \ln \alpha.$$

Thus, $x_1$ is stable for $1 < \alpha < \alpha_1$ and unstable for $\alpha > \alpha_1$, where $\alpha_1 = e^2 = 7.38907\ldots$. At the critical parameter value $\alpha = \alpha_1$, the fixed point has multiplier $\mu(\alpha_1) = -1$. Therefore, a flip bifurcation takes place. To apply

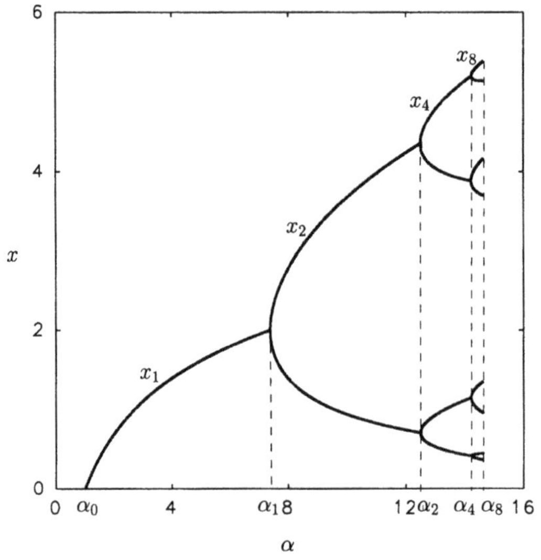


**Fig. 4.8.** Cascade of period-doubling (flip) bifurcations in Ricker's equation.

Theorem 4.4, we need to verify the corresponding nondegeneracy conditions in which all the derivatives must be computed at the fixed point $x_1(\alpha_1) = 2$ and at the critical parameter value $\alpha_1$.

One can check that

$$c(0) = \frac{1}{6} > 0, \quad f_{x\alpha} = -\frac{1}{e^2} \neq 0.$$

Therefore, a unique and stable period-two cycle bifurcates from $x_1$ for $\alpha > \alpha_1$.

The fate of this period-two cycle can be traced further. It can be verified numerically (see Exercise 4) that this cycle loses stability at $\alpha_2 = 12.50925\ldots$ via the flip bifurcation, giving rise to a stable period-four cycle. It bifurcates again at $\alpha_4 = 14.24425\ldots$, generating a stable period-eight cycle that loses its stability at $\alpha_8 = 14.65267\ldots$. The next period doubling takes place at $\alpha_{16} = 14.74212\ldots$ (see **Fig. 4.8**, where several doublings are presented).

It is natural to assume that there is an *infinite* sequence of bifurcation values: $\alpha_{m(k)}$, $m(k) = 2^k$, $k = 1, 2, \ldots$ ($m(k)$ is the period of the cycle before the $k$th doubling). Moreover, one can check that at least the first few elements of this sequence closely resemble a *geometric progression*. In fact, the quotient

$$\frac{\alpha_{m(k)} - \alpha_{m(k-1)}}{\alpha_{m(k+1)} - \alpha_{m(k)}}$$

tends to $\mu_F = 4.6692\ldots$ as $k$ increases. This phenomenon is called *Feigenbaum's cascade* of period doublings, and the constant $\mu_F$ is referred to as the *Feigenbaum constant*. The most surprising fact is that this constant is the same for many different systems exhibiting a cascade of flip bifurcations. This universality has a deep reasoning, which will be discussed in Appendix A to this chapter. $\Diamond$

## 4.6 The "normal form" of the Neimark-Sacker bifurcation

Consider the following two-dimensional discrete-time system depending on one parameter:

$$\begin{aligned}\begin{pmatrix} x_1 \\ x_2 \end{pmatrix} \mapsto (1+\alpha)\begin{pmatrix} \cos\theta & -\sin\theta \\ \sin\theta & \cos\theta \end{pmatrix}\begin{pmatrix} x_1 \\ x_2 \end{pmatrix} \\ + (x_1^2 + x_2^2)\begin{pmatrix} \cos\theta & -\sin\theta \\ \sin\theta & \cos\theta \end{pmatrix}\begin{pmatrix} d & -b \\ b & d \end{pmatrix}\begin{pmatrix} x_1 \\ x_2 \end{pmatrix},\end{aligned} \tag{4.18}$$

where $\alpha$ is the parameter; $\theta = \theta(\alpha), b = b(\alpha)$, and $d = d(\alpha)$ are smooth functions; and $0 < \theta(0) < \pi,\ d(0) \neq 0$.

This system has the fixed point $x_1 = x_2 = 0$ for all $\alpha$ with Jacobian matrix

$$A = (1+\alpha)\begin{pmatrix} \cos\theta & -\sin\theta \\ \sin\theta & \cos\theta \end{pmatrix}.$$

The matrix has eigenvalues $\mu_{1,2} = (1+\alpha)e^{\pm i\theta}$, which makes the map (4.18) invertible near the origin for all small $|\alpha|$. As can be seen, the fixed point at the origin is nonhyperbolic at $\alpha = 0$ due to a complex-conjugate pair of the eigenvalues on the unit circle. To analyze the corresponding bifurcation, introduce the complex variable $z = x_1 + ix_2, \bar{z} = x_1 - ix_2, |z|^2 = z\bar{z} = x_1^2 + x_2^2$, and set $d_1 = d + ib$. The equation for $z$ reads

$$z \mapsto e^{i\theta}z(1 + \alpha + d_1|z|^2) = \mu z + c_1 z|z|^2,$$

where $\mu = \mu(\alpha) = (1+\alpha)e^{i\theta(\alpha)}$ and $c_1 = c_1(\alpha) = e^{i\theta(\alpha)}d_1(\alpha)$ are complex functions of the parameter $\alpha$.

Using the representation $z = \rho e^{i\varphi}$, we obtain for $\rho = |z|$

$$\rho \mapsto \rho|1 + \alpha + d_1(\alpha)\rho^2|.$$

Since

$$|1+\alpha+d_1(\alpha)\rho^2| = (1+\alpha)\left(1+\frac{2d(\alpha)}{1+\alpha}\rho^2+\frac{|d_1(\alpha)|^2}{(1+\alpha)^2}\rho^4\right)^{1/2}$$
$$= 1+\alpha+d(\alpha)\rho^2+O(\rho^4),$$

we obtain the following *polar* form of system (4.18):

$$\begin{cases} \rho \mapsto \rho(1+\alpha+d(\alpha)\rho^2)+\rho^4 R_\alpha(\rho), \\ \varphi \mapsto \varphi+\theta(\alpha)+\rho^2 Q_\alpha(\rho), \end{cases} \tag{4.19}$$

for functions $R$ and $Q$, which are smooth functions of $(\rho, \alpha)$. Bifurcations of the systems's phase portrait as $\alpha$ passes through zero can easily be analyzed using the latter form since the mapping for $\rho$ is *independent* of $\varphi$. The first equation in (4.19) defines a one-dimensional dynamical system that has the fixed point $\rho = 0$ for all values of $\alpha$. The point is linearly stable if $\alpha < 0$; for $\alpha > 0$ the point becomes linearly unstable. The stability of the fixed point at $\alpha = 0$ is determined by the sign of the coefficient $d(0)$. Suppose that $d(0) < 0$; then the origin is (nonlinearly) stable at $\alpha = 0$. Moreover, the $\rho$-map of (4.19) has an additional stable fixed point

$$\rho_0(\alpha) = \sqrt{-\frac{\alpha}{d(\alpha)}} + O(\alpha)$$

for $\alpha > 0$. The $\varphi$-map of (4.19) describes a rotation by an angle depending on $\rho$ and $\alpha$; it is approximately equal to $\theta(\alpha)$. Thus, by superposition of the mappings defined by (4.19), we obtain the bifurcation diagram for the original two-dimensional system (4.18) (see **Fig. 4.9**).

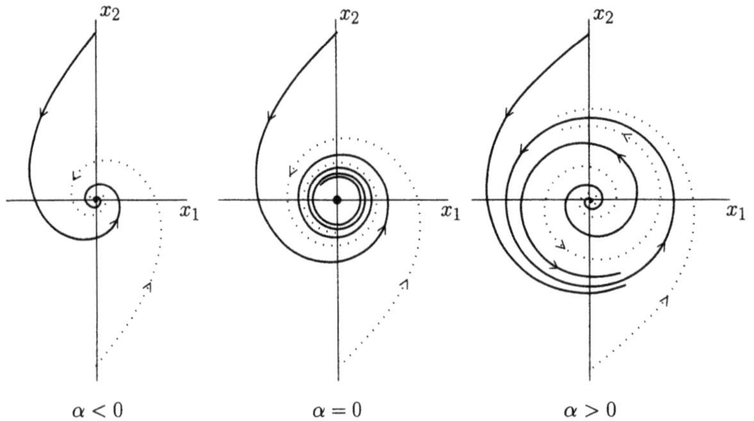


**Fig. 4.9.** Supercritical Neimark-Sacker bifurcation.

The system always has a fixed point at the origin. This point is stable for $\alpha < 0$ and unstable for $\alpha > 0$. The invariant curves of the system near the origin look like the orbits near the stable focus of a continuous-time system for $\alpha < 0$ and like orbits near the unstable focus for $\alpha > 0$. At the critical parameter value $\alpha = 0$ the point is nonlinearly stable. The fixed point is surrounded for $\alpha > 0$ by an isolated *closed invariant curve* that is unique and stable. The curve is a circle of radius $\rho_0(\alpha)$. All orbits starting outside or inside the closed invariant curve, except at the origin, tend to the curve under iterations of (4.19). This is a Neimark-Sacker bifurcation.

This bifurcation can also be presented in $(x_1, x_2, \alpha)$-space. The appearing family of closed invariant curves, parametrized by $\alpha$, forms a *paraboloid* surface.

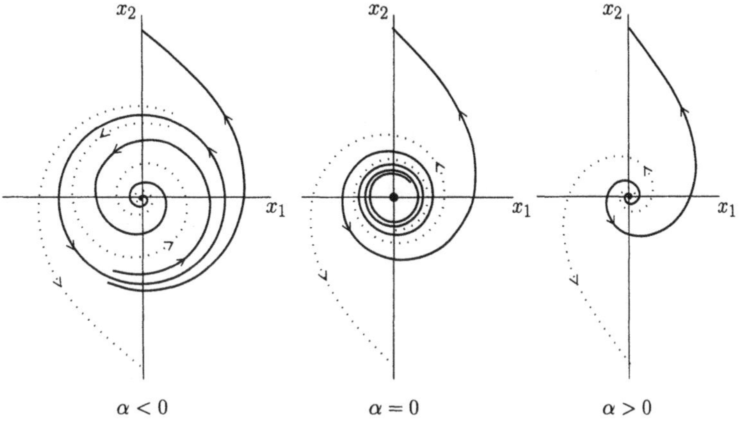


**Fig. 4.10.** Subcritical Neimark-Sacker bifurcation.

The case $d(0) > 0$ can be analyzed in the same way. The system undergoes the Neimark-Sacker bifurcation at $\alpha = 0$. Contrary to the considered case, there is an *unstable* closed invariant curve that disappears when $\alpha$ crosses zero from negative to positive values (see **Fig. 4.10**).

**Remarks:**

(1) As in the cases of the Andronov-Hopf and the flip bifurcations, these two cases are often called *supercritical* and *subcritical* (or, better, *"soft"* and *"sharp"*) Neimark-Sacker bifurcations. As usual, the type of the bifurcation is determined by the stability of the fixed point at the bifurcation parameter value.

(2) The structure of orbits of (4.19) on the invariant circle depends on whether the ratio between the rotation angle $\Delta\varphi = \theta(\alpha) + \rho^2 Q_\alpha(\rho)$ and $2\pi$ is rational or irrational on the circle. If it is rational, all the orbits on the curve are *periodic*. More precisely, if

$$\frac{\Delta\varphi}{2\pi} = \frac{p}{q}$$

with integers $p$ and $q$, all the points on the curve are cycles of period $q$ of the $p$th iterate of the map. If the ratio is irrational, there are no periodic orbits and all the orbits are dense in the circle. ♢

Let us now add higher-order terms to system (4.18); for instance, consider the system

$$\begin{aligned} \begin{pmatrix} x_1 \\ x_2 \end{pmatrix} &\mapsto (1+\alpha)\begin{pmatrix} \cos\theta & -\sin\theta \\ \sin\theta & \cos\theta \end{pmatrix}\begin{pmatrix} x_1 \\ x_2 \end{pmatrix} \\ &+ (x_1^2+x_2^2)\begin{pmatrix} \cos\theta & -\sin\theta \\ \sin\theta & \cos\theta \end{pmatrix}\begin{pmatrix} d & -b \\ b & d \end{pmatrix}\begin{pmatrix} x_1 \\ x_2 \end{pmatrix} + O(\|x\|^4). \end{aligned} \tag{4.20}$$

Here, the $O(\|x\|^4)$ terms can depend smoothly on $\alpha$. Unfortunately, it cannot be said that system (4.20) is locally topologically equivalent to system (4.18). In this case, the higher-order terms do affect the bifurcation behavior of the system. If one writes (4.20) in the polar form, the mapping for $\rho$ will depend on $\varphi$. The system can be represented in a form similar to (4.19) but with $2\pi$-periodic functions $R$ and $Q$. Nevertheless, the phase portraits of systems (4.18) and (4.20) have some important features in common. Namely, the following lemma holds.

**Lemma 4.3** *$O(\|x\|^4)$ terms do not affect the bifurcation of the closed invariant curve in* (4.20)*. That is, a locally unique invariant curve bifurcates from the origin in the same direction and with the same stability as in system* (4.18)*.*
□

The proof of the lemma is rather involved and is given in Appendix B. The geometrical idea behind the proof is simple. We expect that map (4.20) has an invariant curve near the invariant circle of the map (4.18). Fix $\alpha$ and consider the circle

$$S_0 = \left\{(\rho,\varphi) : \rho = \sqrt{-\frac{\alpha}{d(\alpha)}}\right\},$$

which is located near the invariant circle of the "unperturbed" map without $O(\|x\|^4)$ terms. It can be shown that iterations $F^k S_0, k = 1, 2, \ldots$, where $F$ is the map defined by (4.20), converge to a closed invariant curve

$$S_\infty = \{(\rho,\varphi) : \rho = \Psi(\varphi)\},$$

which is not a circle but is close to $S_0$. Here, $\Psi$ is a $2\pi$-periodic function of $\varphi$ describing $S_\infty$ in polar coordinates. To establish the convergence, we have to introduce a new "radial" variable $u$ in a band around $S_0$ (both the band diameter and its width "shrink" as $\alpha \to 0$) and show that the map $F$ defines a *contraction* map $\mathcal{F}$ on a proper function space of $2\pi$-periodic functions $u = u(\varphi)$. Then the Contraction Mapping Principle (see Chapter 1) gives the existence of a fixed point $u^{(\infty)}$ of $\mathcal{F} : \mathcal{F}(u^{(\infty)}) = u^{(\infty)}$. The periodic function $u^{(\infty)}(\varphi)$ represents the closed invariant curve $S_\infty$ we are looking for at $\alpha$

fixed. Uniqueness and stability of $S_\infty$ in the band follow, essentially, from the contraction. It can be verified that outside the band there are no nontrivial invariant sets of (4.20).

**Remarks:**

(1) The orbit structure on the closed invariant curve and the variation of this structure when the parameter changes are generically different in systems (4.18) and (4.20). We will return to the analysis of bifurcations on the invariant curve in Chapter 7. Here we just notice that, generically, there is only a *finite* number of periodic orbits on the closed invariant curve. Let $a(0) < 0$. Then,

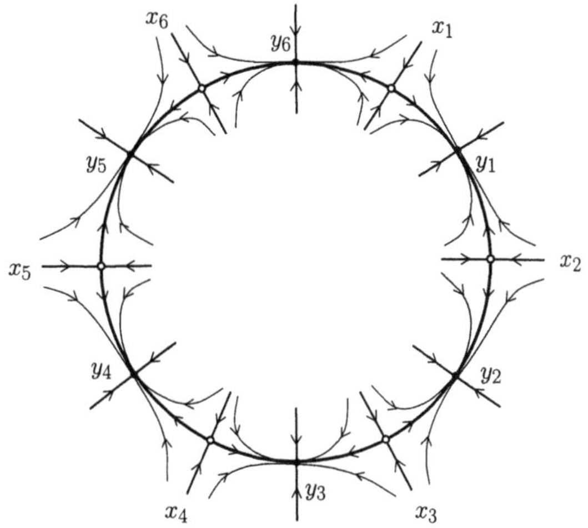


**Fig. 4.11.** Saddle $\{x_1, x_2, \ldots, x_6\}$ and stable $\{y_1, y_2, \ldots, y_6\}$ period-six orbits on the invariant circle.

some iterate $p$ of map (4.20) can have two $q$-periodic orbits: a totally stable "node" cycle of period $q$ and a saddle cycle of period $q$ (see **Fig. 4.11**). The cycles exist in some "parameter window" and disappear on its borders through the fold bifurcation. A generic system exhibits an infinite number of such bifurcations corresponding to different windows.

(2) The bifurcating invariant closed curve in (4.20) has *finite* smoothness: The function $\Psi(\varphi)$ representing it in polar coordinates generically has only a finite number of continuous derivatives with respect to $\varphi$, even if the map (4.20) is differentiable infinitely many times. The number increases as $|\alpha| \to 0$. The nonsmoothness appears when the saddle's unstable (stable) manifolds meet at the "node" points. $\diamondsuit$

## 4.7 Generic Neimark-Sacker bifurcation

We now shall prove that any generic two-dimensional system undergoing a Neimark-Sacker bifurcation can be transformed into the form (4.20).

Consider a system

$$x \mapsto f(x,\alpha), \quad x = (x_1, x_2)^T \in \mathbb{R}^2, \ \alpha \in \mathbb{R}^1,$$

with a smooth function $f$, which has at $\alpha = 0$ the fixed point $x = 0$ with simple eigenvalues $\mu_{1,2} = e^{\pm i\theta_0}$, $0 < \theta_0 < \pi$. By the Implicit Function Theorem, the system has a unique fixed point $x_0(\alpha)$ in some neighborhood of the origin for all sufficiently small $|\alpha|$ since $\mu = 1$ is not an eigenvalue of the Jacobian matrix.[1] We can perform a parameter-dependent coordinate shift, placing this fixed point at the origin. Therefore, we may assume without loss of generality that $x = 0$ is the fixed point of the system for $|\alpha|$ sufficiently small. Thus, the system can be written as

$$x \mapsto A(\alpha)x + F(x,\alpha), \tag{4.21}$$

where $F$ is a smooth vector function whose components $F_{1,2}$ have Taylor expansions in $x$ starting with at least quadratic terms, $F(0,\alpha) = 0$ for all sufficiently small $|\alpha|$. The Jacobian matrix $A(\alpha)$ has two multipliers

$$\mu_{1,2}(\alpha) = r(\alpha)e^{\pm i\varphi(\alpha)},$$

where $r(0) = 1, \varphi(0) = \theta_0$. Thus, $r(\alpha) = 1 + \beta(\alpha)$ for some smooth function $\beta(\alpha), \beta(0) = 0$. Suppose that $\beta'(0) \neq 0$. Then, we can use $\beta$ as a new parameter and express the multipliers in terms of $\beta : \mu_1(\beta) = \mu(\beta), \mu_2(\beta) = \bar{\mu}(\beta)$, where

$$\mu(\beta) = (1+\beta)e^{i\theta(\beta)}$$

with a smooth function $\theta(\beta)$ such that $\theta(0) = \theta_0$.

**Lemma 4.4** *By the introduction of a complex variable and a new parameter, system* (4.21) *can be transformed for all sufficiently small* $|\alpha|$ *into the form*

$$z \mapsto \mu(\beta)z + g(z,\bar{z},\beta), \tag{4.22}$$

*where* $\beta \in \mathbb{R}^1, z \in \mathbb{C}^1, \mu(\beta) = (1+\beta)e^{i\theta(\beta)}$, *and* $g$ *is a complex-valued smooth function of* $z, \bar{z}$, *and* $\beta$ *whose Taylor expansion with respect to* $(z, \bar{z})$ *contains quadratic and higher-order terms:*

$$g(z,\bar{z},\beta) = \sum_{k+l\geq 2} \frac{1}{k!l!} g_{kl}(\beta) z^k \bar{z}^l,$$

*with* $k, l = 0, 1, \ldots$ □

[1] Since $\mu = 0$ is not an eigenvalue, the system is invertible in some neighborhood of the origin for sufficiently small $|\alpha|$.

The proof of the lemma is completely analogous to that from the Andronov-Hopf bifurcation analysis in Chapter 3 and is left as an exercise for the reader.

As in the Andronov-Hopf case, we start by making *nonlinear* (complex) coordinate changes that will simplify the map (4.22). First, we remove all the quadratic terms.

**Lemma 4.5** *The map*

$$z \mapsto \mu z + \frac{g_{20}}{2} z^2 + g_{11} z \bar{z} + \frac{g_{02}}{2} \bar{z}^2 + O(|z|^3), \tag{4.23}$$

*where* $\mu = \mu(\beta) = (1+\beta)e^{i\theta(\beta)}, g_{ij} = g_{ij}(\beta)$, *can be transformed by an invertible parameter-dependent change of complex coordinate*

$$z = w + \frac{h_{20}}{2} w^2 + h_{11} w \bar{w} + \frac{h_{02}}{2} \bar{w}^2,$$

*for all sufficiently small* $|\beta|$, *into a map without quadratic terms:*

$$w \mapsto \mu w + O(|w|^3),$$

*provided that*

$$e^{i\theta_0} \neq 1 \text{ and } e^{3i\theta_0} \neq 1.$$

**Proof:**

The inverse change of variables is given by

$$w = z - \frac{h_{20}}{2} z^2 - h_{11} z \bar{z} - \frac{h_{02}}{2} \bar{z}^2 + O(|z|^3).$$

Therefore, in the new coordinate $w$, the map (4.23) takes the form

$$\begin{aligned}\tilde{w} = \mu w &+ \frac{1}{2}(g_{20} + (\mu - \mu^2)h_{20})w^2 \\ &+ (g_{11} + (\mu - |\mu|^2)h_{11})w\bar{w} \\ &+ \frac{1}{2}(g_{02} + (\mu - \bar{\mu}^2)h_{02})\bar{w}^2 \\ &+ O(|w|^3).\end{aligned}$$

Thus, by setting

$$h_{20} = \frac{g_{20}}{\mu^2 - \mu}, \quad h_{11} = \frac{g_{11}}{|\mu|^2 - \mu}, \quad h_{02} = \frac{g_{02}}{\bar{\mu}^2 - \mu},$$

we "kill" all the quadratic terms in (4.23). These substitutions are valid if the denominators are nonzero for all sufficiently small $|\beta|$ including $\beta = 0$. Indeed, this is the case since

$$\begin{aligned} \mu^2(0) - \mu(0) &= e^{i\theta_0}(e^{i\theta_0} - 1) \neq 0, \\ |\mu(0)|^2 - \mu(0) &= 1 - e^{i\theta_0} \neq 0, \\ \bar{\mu}(0)^2 - \mu(0) &= e^{i\theta_0}(e^{-3i\theta_0} - 1) \neq 0, \end{aligned}$$

due to our restrictions on $\theta_0$. □

**Remarks:**

(1) Let $\mu_0 = \mu(0)$. Then, the conditions on $\theta_0$ used in the lemma can be written as

$$\mu_0 \neq 1, \ \mu_0^3 \neq 1.$$

Notice that the first condition holds automatically due to our initial assumptions on $\theta_0$.

(2) The resulting coordinate transformation is polynomial with coefficients that are smoothly dependent on $\beta$. In some neighborhood of the origin the transformation is *near-identical.*

(3) Notice the transformation *changes* the coefficients of the cubic terms of (4.23). ◇

Assuming that we have removed all quadratic terms, let us try to eliminate the cubic terms as well.

**Lemma 4.6** *The map*

$$z \mapsto \mu z + \frac{g_{30}}{6}z^3 + \frac{g_{21}}{2}z^2\bar{z} + \frac{g_{12}}{2}z\bar{z}^2 + \frac{g_{03}}{6}\bar{z}^3 + O(|z|^4), \tag{4.24}$$

*where $\mu = \mu(\beta) = (1+\beta)e^{i\theta(\beta)}$, $g_{ij} = g_{ij}(\beta)$, can be transformed by an invertible parameter-dependent change of coordinates*

$$z = w + \frac{h_{30}}{6}w^3 + \frac{h_{21}}{2}w^2\bar{w} + \frac{h_{12}}{2}w\bar{w}^2 + \frac{h_{03}}{6}\bar{w}^3,$$

*for all sufficiently small $|\beta|$, into a map with only one cubic term:*

$$w \mapsto \mu w + c_1 w^2\bar{w} + O(|w|^4),$$

*provided that*

$$e^{2i\theta_0} \neq 1 \ \ and \ \ e^{4i\theta_0} \neq 1.$$

**Proof:**

The inverse transformation is

$$w = z - \frac{h_{30}}{6}z^3 - \frac{h_{21}}{2}z^2\bar{z} - \frac{h_{12}}{2}z\bar{z}^2 - \frac{h_{03}}{6}\bar{z}^3 + O(|z|^4).$$

Therefore,

$$\tilde{w} = \lambda w + \frac{1}{6}(g_{30} + (\mu - \mu^3)h_{30})w^3 + \frac{1}{2}(g_{21} + (\mu - \mu|\mu|^2)h_{21})w^2\bar{w}$$
$$+ \frac{1}{2}(g_{12} + (\mu - \bar{\mu}|\mu|^2)h_{12})w\bar{w}^2 + \frac{1}{6}(g_{03} + (\mu - \bar{\mu}^3)h_{03})\bar{w}^3 + O(|w|^4).$$

Thus, by setting

$$h_{30} = \frac{g_{30}}{\mu^3 - \mu}, \quad h_{12} = \frac{g_{12}}{\bar{\mu}|\mu|^2 - \mu}, \quad h_{03} = \frac{g_{03}}{\bar{\mu}^3 - \mu},$$

we can annihilate all cubic terms in the resulting map except the $w^2\bar{w}$-term, which must be treated separately. The substitutions are valid since all the involved denominators are nonzero for all sufficiently small $|\beta|$ due to the assumptions concerning $\theta_0$.

One can also try to eliminate the $w^2\bar{w}$-term by formally setting

$$h_{21} = \frac{g_{21}}{\mu(1 - |\mu|^2)}.$$

This is possible for small $\beta \neq 0$, but the denominator vanishes at $\beta = 0$ *for all* $\theta_0$. Thus, no extra conditions on $\theta_0$ would help. To obtain a transformation that is smoothly dependent on $\beta$, set $h_{21} = 0$, that results in

$$c_1 = \frac{g_{21}}{2}. \ \square$$

**Remarks:**

(1) The conditions imposed on $\theta_0$ in the lemma mean

$$\mu_0^2 \neq 1, \ \mu_0^4 \neq 1,$$

and therefore, in particular, $\mu_0 \neq -1$ and $\mu_0 \neq i$. The first condition holds automatically due to our initial assumptions on $\theta_0$.

(2) The remaining cubic $w^2\bar{w}$-term is called a *resonant term*. Note that its coefficient is the *same* as the coefficient of the cubic term $z^2\bar{z}$ in the original map (4.24). $\diamondsuit$

We now combine the two previous lemmas.

**Lemma 4.7 (Normal form for the Neimark-Sacker bifurcation)**
*The map*

$$z \mapsto \mu z + \frac{g_{20}}{2}z^2 + g_{11}z\bar{z} + \frac{g_{02}}{2}\bar{z}^2$$
$$+ \frac{g_{30}}{6}z^3 + \frac{g_{21}}{2}z^2\bar{z} + \frac{g_{12}}{2}z\bar{z}^2 + \frac{g_{03}}{6}\bar{z}^3$$
$$+ O(|z|^4),$$

*where* $\mu = \mu(\beta) = (1+\beta)e^{i\theta(\beta)}$, $g_{ij} = g_{ij}(\beta)$, *and* $\theta_0 = \theta(0)$ *is such that* $e^{ik\theta_0} \neq 1$ *for* $k = 1,2,3,4$, *can be transformed by an invertible parameter-dependent change of complex coordinate, which is smoothly dependent on the parameter,*

$$z = w + \frac{h_{20}}{2}w^2 + h_{11}w\bar{w} + \frac{h_{02}}{2}\bar{w}^2 + \frac{h_{30}}{6}w^3 + \frac{h_{12}}{2}w\bar{w}^2 + \frac{h_{03}}{6}\bar{w}^3,$$

*for all sufficiently small* $|\beta|$, *into a map with only the resonant cubic term:*

$$w \mapsto \mu w + c_1 w^2\bar{w} + O(|w|^4),$$

*where* $c_1 = c_1(\beta)$. □

The truncated composition of the transformations defined in the two previous lemmas gives the required coordinate change. First, annihilate all the quadratic terms. This will also change the coefficients of the cubic terms. The coefficient of $w^2\bar{w}$ will be $\frac{1}{2}\tilde{g}_{21}$, say, instead of $\frac{1}{2}g_{21}$. Then, eliminate all the cubic terms except the resonant one. The coefficient of this term remains $\frac{1}{2}\tilde{g}_{21}$. Thus, all we need to compute to get the coefficient of $c_1$ in terms of the given equation is a new coefficient $\frac{1}{2}\tilde{g}_{21}$ of the $w^2\bar{w}$-term after the *quadratic* transformation. The computations result in the following expression for $c_1(\alpha)$:

$$c_1 = \frac{g_{20}g_{11}(\bar{\mu} - 3 + 2\mu)}{2(\mu^2-\mu)(\bar{\mu}-1)} + \frac{|g_{11}|^2}{1-\bar{\mu}} + \frac{|g_{02}|^2}{2(\mu^2-\bar{\mu})} + \frac{g_{21}}{2}, \tag{4.25}$$

which gives, for the critical value of $c_1$,

$$c_1(0) = \frac{g_{20}(0)g_{11}(0)(1-2\mu_0)}{2(\mu_0^2-\mu_0)} + \frac{|g_{11}(0)|^2}{1-\bar{\mu}_0} + \frac{|g_{02}(0)|^2}{2(\mu_0^2-\bar{\mu}_0)} + \frac{g_{21}(0)}{2}, \tag{4.26}$$

where $\mu_0 = e^{i\theta_0}$.

We now summarize the obtained results in the following theorem.

**Theorem 4.5** *Suppose a two-dimensional discrete-time system*

$$x \mapsto f(x,\alpha), \quad x \in \mathbb{R}^2, \ \alpha \in \mathbb{R}^1,$$

*with smooth* $f$, *has, for all sufficiently small* $|\alpha|$, *the fixed point* $x = 0$ *with multipliers*

$$\mu_{1,2}(\alpha) = r(\alpha)e^{\pm i\varphi(\alpha)},$$

*where* $r(0) = 1, \varphi(0) = \theta_0$.

*Let the following conditions be satisfied:*

(C.1) $r'(0) \neq 0$;

(C.2) $e^{ik\theta_0} \neq 1$ *for* $k = 1, 2, 3, 4$.

*Then, there are smooth invertible coordinate and parameter changes transforming the system into*

$$\begin{pmatrix} y_1 \\ y_2 \end{pmatrix} \mapsto (1+\beta)\begin{pmatrix} \cos\theta(\beta) & -\sin\theta(\beta) \\ \sin\theta(\beta) & \cos\theta(\beta) \end{pmatrix}\begin{pmatrix} y_1 \\ y_2 \end{pmatrix} + \\ (y_1^2+y_2^2)\begin{pmatrix} \cos\theta(\beta) & -\sin\theta(\beta) \\ \sin\theta(\beta) & \cos\theta(\beta) \end{pmatrix}\begin{pmatrix} d(\beta) & -b(\beta) \\ b(\beta) & d(\beta) \end{pmatrix}\begin{pmatrix} y_1 \\ y_2 \end{pmatrix} + O(\|y\|^4) \tag{4.27}$$

*with* $\theta(0) = \theta_0$ *and* $d(0) = \text{Re}(e^{-i\theta_0}c_1(0))$, *where* $c_1(0)$ *is given by the formula* (4.26).

**Proof:**

The only thing left to verify is the formula for $d(0)$. Indeed, by Lemmas 4.4–4.7, the system can be transformed to the complex Poincaré normal form,

$$w \mapsto \mu(\beta)w + c_1(\beta)w|w|^2 + O(|w|^4),$$

for $\mu(\beta) = (1+\beta)e^{i\theta(\beta)}$. This map can be written as

$$w \mapsto e^{i\theta(\beta)}(1 + \beta + d_1(\beta)|w|^2)w + O(|w|^4),$$

where $d_1(\beta) = d(\beta) + ib(\beta)$ for some real functions $d(\beta)$, $b(\beta)$. A return to the real coordinates $(y_1, y_2), w = y_1 + iy_2$, gives system (4.27). Finally,

$$d(\beta) = \text{Re } d_1(\beta) = \text{Re}(e^{-i\theta(\beta)}c_1(\beta)).$$

Thus,

$$d(0) = \text{Re}(e^{-i\theta_0}c_1(0)). \ \square$$

Using Lemma 4.3, we can state the following general result.

**Theorem 4.6 (Generic Neimark-Sacker bifurcation)** *For any generic two-dimensional one-parameter system*

$$x \mapsto f(x, \alpha),$$

*having at* $\alpha = 0$ *the fixed point* $x_0 = 0$ *with complex multipliers* $\mu_{1,2} = e^{\pm i\theta_0}$, *there is a neighborhood of* $x_0$ *in which a unique closed invariant curve bifurcates from* $x_0$ *as* $\alpha$ *passes through zero.* $\square$

**Remark:**

The genericity conditions assumed in the theorem are the transversality condition (C.1) and the nondegeneracy condition (C.2) from Theorem 4.5 *and* the additional nondegeneracy condition

(C.3) $d(0) \neq 0$.

It should be stressed that the conditions $e^{ik\theta_0} \neq 1$ for $k = 1, 2, 3, 4$ are not merely technical. If they are not satisfied, the closed invariant curve may not appear at all, or there might be several invariant curves bifurcating from the fixed point (see Chapter 9). ◊

The coefficient $d(0)$, which determines the direction of the appearance of the invariant curve in a generic system exhibiting the Neimark-Sacker bifurcation, can be computed via

$$d(0) = \mathrm{Re}\left(\frac{e^{-i\theta_0}g_{21}}{2}\right) - \mathrm{Re}\left(\frac{(1-2e^{i\theta_0})e^{-2i\theta_0}}{2(1-e^{i\theta_0})}g_{20}g_{11}\right) - \frac{1}{2}|g_{11}|^2 - \frac{1}{4}|g_{02}|^2. \tag{4.28}$$

In Chapter 5 we will see how to deal with $n$-dimensional discrete-time systems where $n > 2$ and how to apply the results to limit cycle bifurcations in continuous-time systems.

**Example 4.3 (Neimark-Sacker bifurcation in the delayed logistic equation)** Consider the following recurrence equation:

$$u_{k+1} = ru_k(1 - u_{k-1}).$$

This is a simple population dynamics model, where $u_k$ stands for the density of a population at time $k$, and $r$ is the growth rate. It is assumed that the growth is determined not only by the current population density but also by its density in the past.

If we introduce $v_k = u_{k-1}$, the equation can be rewritten as

$$\begin{cases} u_{k+1} = ru_k(1 - v_k), \\ v_{k+1} = v_k, \end{cases}$$

which, in turn, defines the two-dimensional discrete-time dynamical system,

$$\begin{pmatrix} x_1 \\ x_2 \end{pmatrix} \mapsto \begin{pmatrix} rx_1(1-x_2) \\ x_1 \end{pmatrix} \equiv \begin{pmatrix} F_1(x,r) \\ F_2(x,r) \end{pmatrix}, \tag{4.29}$$

where $x = (x_1, x_2)^T$. The map (4.29) has the fixed point $(0,0)^T$ for all values of $r$. For $r > 1$, a nontrivial positive fixed point $x^0$ appears, with the coordinates

$$x_1^0(r) = x_2^0(r) = 1 - \frac{1}{r}.$$

The Jacobian matrix of the map (4.29) evaluated at the nontrivial fixed point is given by

$$A(r) = \begin{pmatrix} 1 & 1-r \\ 1 & 0 \end{pmatrix}$$

and has eigenvalues

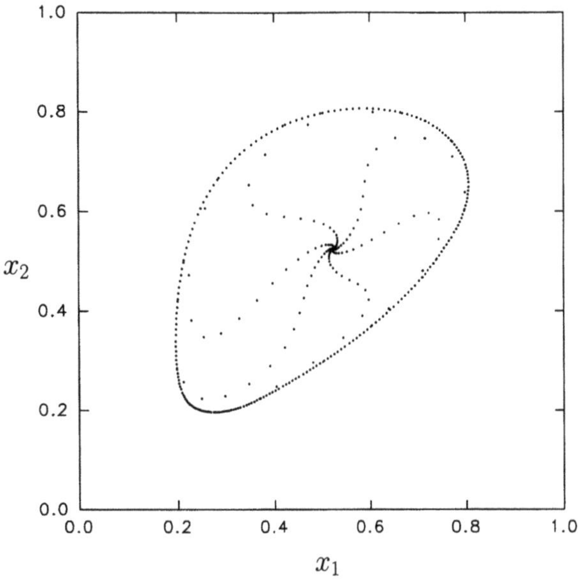


**Fig. 4.12.** Stable invariant curve in the delayed logistic equation.

$$\mu_{1,2}(r) = \frac{1}{2} \pm \sqrt{\frac{5}{4} - r}.$$

If $r > \frac{5}{4}$, the eigenvalues are complex and $|\mu_{1,2}|^2 = \mu_1\mu_2 = r - 1$. Therefore, at $r = r_0 = 2$ the nontrivial fixed point loses stability and we have a Neimark-Sacker bifurcation: The critical multipliers are

$$\mu_{1,2} = e^{\pm i\theta_0}, \ \theta_0 = \frac{\pi}{3} = 60^0.$$

It is clear that conditions (C.1) and (C.2) are satisfied.

To verify the nondegeneracy condition ($C$.3), we have to compute $a(0)$. The critical Jacobian matrix $A_0 = A(r_0)$ have the eigenvectors

$$A_0 q = e^{i\theta_0} q, \ A_0^T p = e^{-i\theta_0} p,$$

where

$$q \sim \left(\frac{1}{2} + i\frac{\sqrt{3}}{2}, \ 1\right)^T, \ p \sim \left(-\frac{1}{2} + i\frac{\sqrt{3}}{2}, \ 1\right)^T.$$

To achieve the normalization $\langle p, q \rangle = 1$, we can take, for example,

$$q = \left(\frac{1}{2} + i\frac{\sqrt{3}}{2}, \ 1\right)^T, \ p = \left(i\frac{\sqrt{3}}{3}, \ \frac{1}{2} - i\frac{\sqrt{3}}{6}\right)^T.$$

Now we compose

$$x = x^0 + zq + \bar{z}\bar{q}$$

and evaluate the function

$$H(z, \bar{z}) = \langle p, F(x^0 + zq + \bar{z}\bar{q}, r_0) - x^0 \rangle.$$

Computing its Taylor expansion at $(z, \bar{z}) = (0, 0)$,

$$H(z, \bar{z}) = e^{i\theta_0} z + \sum_{2 \leq j+k \leq 3} \frac{1}{j!k!} g_{jk} z^j \bar{z}^k + O(|z|^4),$$

gives

$$g_{20} = -2 + i\frac{2\sqrt{3}}{3}, \quad g_{11} = i\frac{2\sqrt{3}}{3}, \quad g_{02} = 2 + i\frac{2\sqrt{3}}{3}, \quad g_{21} = 0,$$

that allows us to find the critical real part

$$\begin{aligned} d(0) &= \text{Re}\left(\frac{e^{-i\theta_0} g_{21}}{2}\right) - \text{Re}\left(\frac{(1 - 2e^{i\theta_0})e^{-2i\theta_0}}{2(1 - e^{i\theta_0})} g_{20} g_{11}\right) - \frac{1}{2}|g_{11}|^2 - \frac{1}{4}|g_{02}|^2 \\ &= -2 < 0. \end{aligned}$$

Therefore, a unique and stable closed invariant curve bifurcates from the nontrivial fixed point for $r > 2$ (see **Fig. 4.12**). $\Diamond$

## 4.8 Exercises

**(1)** Prove that in a small neighborhood of $x = 0$ the number and stability of fixed points and periodic orbits of the maps (4.1) and (4.13) are independent of higher-order terms, provided $|\alpha|$ is sufficiently small. (*Hint*: To prove the absence of long-period cycles, use asymptotic stability arguments.)

**(2)** Show that the normal form coefficient $c$ for the flip bifurcation (4.16) can be computed in terms of the second iterate of the map:

$$c = -\frac{1}{12} \frac{\partial^3}{\partial x^3} f_\alpha^2(x)\bigg|_{(x,\alpha)=(0,0)},$$

where $f_\alpha(x) = f(x, \alpha)$. (*Hint*: Take into account that $f_x(0, 0) = -1$.)

**(3)** **(Logistic map)** Consider the following map [May 1974]:

$$f_\alpha(x) = \alpha x(1 - x),$$

depending on a single parameter $\alpha$.

(a) Show that at $\alpha_1 = 3$ the map exhibits the flip bifurcation, namely, a stable fixed point of $f_\alpha$ becomes unstable, while a stable period-two cycle bifurcates from this point for $\alpha > 3$. (*Hint*: Use the formula from Exercise 2 above.)

(b) Prove that at $\alpha_0 = 1 + \sqrt{8}$ the logistic map has a fold bifurcation generating a stable and an unstable cycle of *period three* as $\alpha$ increases.

**(4) (Second period doubling in Ricker's model)** Verify that the second period doubling takes place in Ricker's map (4.17) at $\alpha_2 = 12.50925\ldots$. (*Hint*: Introduce $y = \alpha x e^{-x}$ and write a system of three equations for the three unknowns $(x, y, \alpha)$ defining a period-two cycle $\{x, y\}$ with multiplier $\mu = -1$. Use one of the standard routines implementing Newton's method (see Chapter 10) to solve the system numerically starting from some suitable initial data.)

**(5) (Hénon map revisited)** Show that the original map introduced by Hénon [1976]

$$\begin{pmatrix} X \\ Y \end{pmatrix} \mapsto \begin{pmatrix} 1 + Y - aX^2 \\ bX \end{pmatrix}$$

can be transformed into map (4.1) by a linear change of the coordinates and parameters.

**(6)** Derive formula (4.26) for $c_1(0)$ for the Neimark-Sacker bifurcation.

**(7) (Discrete-time predator-prey model)**
Consider the following discrete-time system [Maynard Smith 1968]:

$$\begin{cases} x_{k+1} = \alpha x_k(1 - x_k) - x_k y_k, \\ y_{k+1} = \dfrac{1}{\beta} x_k y_k, \end{cases}$$

which is a discrete-time version of the Volterra model. Here $x_k$ and $y_k$ are the prey and predator numbers, respectively, in year (generation) $k$, and it is assumed that in the absence of prey the predators become extinct in one generation.

(a) Prove that a nontrivial fixed point of the map undergoes a Neimark-Sacker bifurcation on a curve in the $(\alpha, \beta)$-plane, and compute the direction of the closed invariant-curve bifurcation.

(b) Guess what happens to the emergent closed invariant curve for parameter values far from the bifurcation curve.

## 4.9 Appendix A: Feigenbaum's universality

As mentioned previously, many one-dimensional, parameter-dependent dynamical systems

$$x \mapsto f_\alpha(x), \quad x \in \mathbb{R}^1, \tag{A.1}$$

exhibit infinite cascades of period doublings. Moreover, the corresponding flip bifurcation parameter values, $\alpha_1, \alpha_2, \ldots, \alpha_i, \ldots$, form (asymptotically) a geometric progression:

$$\frac{\alpha_i - \alpha_{i-1}}{\alpha_{i+1} - \alpha_i} \to \mu_F,$$

as $i \to \infty$, where $\mu_F = 4.6692\ldots$ is a system-independent (universal) constant. The sequence $\{\alpha_i\}$ has a limit $\alpha_\infty$. At $\alpha_\infty$ the dynamics of the system become "chaotic" since its orbits become irregular, nonperiodic sequences.

The phenomenon was first explained for special noninvertible dynamical systems ($A$.1), that belong for all parameter values to some class $\mathcal{Y}$. Namely, a system

$$x \mapsto f(x) \tag{A.2}$$

from this class satisfies the following conditions:

(1) $f(x)$ is an even smooth function, $f : [-1, 1] \to [-1, 1]$;
(2) $f'(0) = 0$, $x = 0$ is the only maximum, $f(0) = 1$;
(3) $f(1) = -a < 0$;
(4) $b = f(a) > a$;
(5) $f(b) = f^2(a) < a$;

where $a$ and $b$ are positive (see **Fig. 4.13**). The function $f_\alpha(x) = 1 - \alpha x^2$ is in this

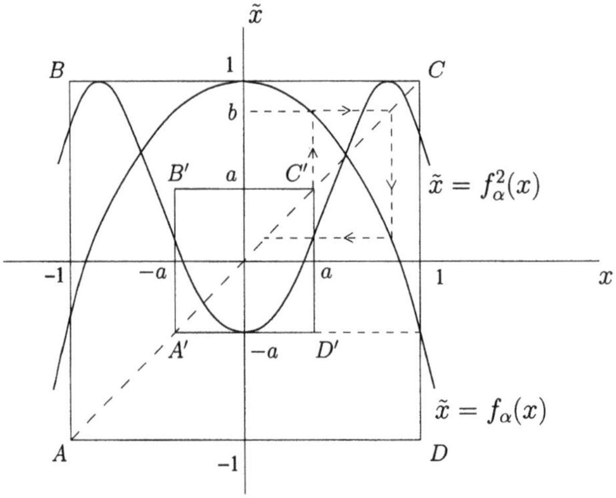


**Fig. 4.13.** A map satisfying conditions (1) through (5) and its second iterate.

class for $\alpha > 1$.

Consider the second iterate $f^2_\alpha$ of a map satisfying conditions (1) through (5). In the square $A'B'C'D'$ (see **Fig. 4.13**), the graph of $f^2_\alpha$, after a coordinate dilatation and a sign change, looks similar to the graph of $f_\alpha$ in the unit square $ABCD$. For example, if $f_\alpha(x) = 1 - \alpha x^2$, then $f^2_\alpha(x) = (1 - \alpha) + 2\alpha^2 x^2 + \cdots$. This observation leads to the introduction of a map defined on functions in $\mathcal{Y}$,

$$(Tf)(x) = -\frac{1}{a} f(f(-ax)), \tag{A.3}$$

where $a = -f(1)$. Notice that $a$ depends on $f$.

**Definition 4.4** *The map $T$ is called the* doubling operator.

It can be checked that map (A.3) transforms a function $f \in \mathcal{Y}$ into some function $Tf \in \mathcal{Y}$. Therefore, we can consider a *discrete-time dynamical system* $\{\mathbb{Z}_+, \mathcal{Y}, T^k\}$. This is a dynamical system with the infinite-dimensional state space $\mathcal{Y}$, which is a *function space.* Moreover, the doubling operator is not invertible in general. Thus, we have to consider only *positive* iterations of $T$.

We shall state the following theorems without proof. They have been proved with the help of a computer and delicate error estimates.

**Theorem 4.7 (Fixed-point existence)** *The map* $T : \mathcal{Y} \to \mathcal{Y}$ *defined by* (A.3) *has a fixed point* $\varphi \in \mathcal{Y}:\ T\varphi = \varphi$. □

It has been found that

$$\varphi(x) = 1 - 1.52763\ldots x^2 + 0.104815\ldots x^4 + 0.0267057\ldots x^6 + \ldots.$$

In Exercise 1 of Chapter 10 we discuss how to obtain some approximations to $\varphi(x)$.

**Theorem 4.8 (Saddle properties of the fixed point)** *The linear part* $L$ *of the doubling operator* $T$ *at its fixed point* $\varphi$ *has only one eigenvalue* $\mu_F = 4.6692\ldots$ *with* $|\mu_F| > 1$. *The rest of the spectrum of* $L$ *is located strictly inside the unit circle.* □

The terms "linear part" and "spectrum" of $L$ are generalizations to the infinite-dimensional case of the notions of the Jacobian matrix and its eigenvalues. An interested reader can find exact definitions in standard textbooks on functional analysis.

Theorems 4.7 and 4.8 mean that the system $\{\mathbb{Z}_+, \mathcal{Y}, T^k\}$ has a saddle fixed point. This fixed point $\varphi$ (a function that is transformed by the doubling operator into itself) has a codim 1 stable invariant manifold $W^s(\varphi)$ and a one-dimensional unstable invariant manifold $W^u(\varphi)$. The stable manifold is composed by functions $f \in \mathcal{Y}$, which become increasingly similar to $\varphi$ under iteration of $T$. The unstable manifold is composed of functions for which *all* their preimages under the action of $T$ remain close to $\varphi$. This is a curve in the function space $\mathcal{Y}$ (**Fig. 4.14** sketches the manifold structure).

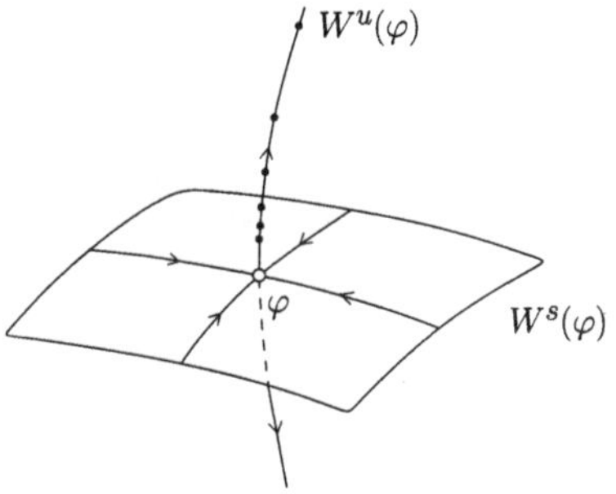


**Fig. 4.14.** Stable and unstable manifolds of the fixed point $\varphi$.

Notice that maps $Tf$ and $f^2$ are topologically equivalent (the relevant homeomorphism is a simple scaling; see (A.3)). Hence, if $Tf$ has a periodic orbit of period $N$, $f^2$ has a periodic orbit of the same period and $f$ therefore has a periodic orbit of period $2N$. This simple observation plays the central role in the following. Consider all maps from $\mathcal{Y}$ having a fixed point with multiplier $\mu = -1$. Such maps form a codim 1 manifold $\Sigma \subset \mathcal{Y}$. The following result has also been established with the help of a computer.

**Theorem 4.9 (Manifold intersection)** *The manifold $\Sigma$ intersects the unstable manifold $W^u(\varphi)$ transversally.* □

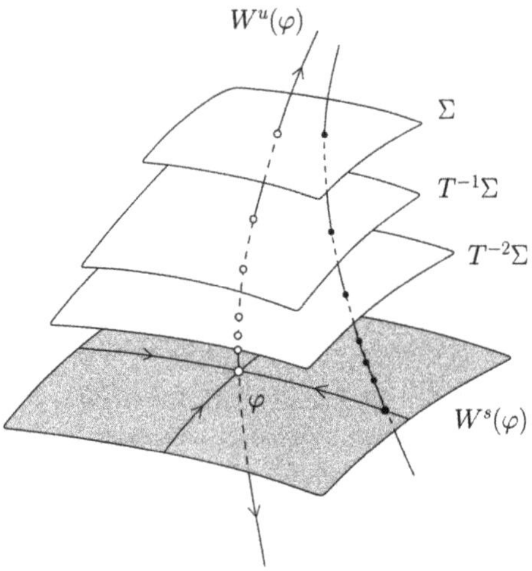


**Fig. 4.15.** Preimages of a surface $\Sigma$ intersecting the unstable manifold $W^u(\varphi)$.

By analogy with a finite-dimensional saddle, it is clear that the preimages $T^{-k}\Sigma$ will accumulate on $W^s(\varphi)$ as $k \to \infty$ (see **Fig. 4.15**). Taking into account the previous observation, we can conclude that $T^{-1}\Sigma$ is composed of maps having a cycle of period two with a multiplier $-1$, that $T^{-2}\Sigma$ is formed by maps having a cycle of period four with a multiplier $-1$, and so forth. Any generic one-parameter dynamical system $f_\alpha$ from the considered class corresponds to a curve $\Lambda$ in $\mathcal{Y}$. If this curve is sufficiently close to $W^u(\varphi)$, it will intersect *all* the preimages $T^{-k}\Sigma$. The points of intersection define a sequence of bifurcation parameter values $\alpha_1, \alpha_2, \ldots$ corresponding to a cascade of period doublings. Asymptotic properties of this sequence are clearly determined by the unstable eigenvalue $\mu_F$. Indeed, let $\xi$ be a coordinate along $W^u(\varphi)$, and let $\xi_k$ denote the coordinate of the intersection of $W^u(\varphi)$ with $T^{-k}\Sigma$. The doubling operator *restricted* to the unstable manifold has the form

$$\xi \mapsto \mu_F \xi + O(\xi^2)$$

and is invertible, with the inverse given by

$$\xi \mapsto \frac{1}{\mu_F}\xi + O(\xi^2).$$

Since

$$\xi_{k+1} = \frac{1}{\mu_F}\xi_k + O(\xi_k^2),$$

we have

$$\frac{\xi_k - \xi_{k-1}}{\xi_{k+1} - \xi_k} \to \mu_F$$

as $k \to \infty$, as does the sequence of the bifurcation parameter values on the curve $\Lambda$.

## 4.10 Appendix B: Proof of Lemma 4.3

In this appendix we prove the following lemma, which is the complex analog of Lemma 4.3.

**Lemma 4.8** *The map*

$$\tilde{z} = e^{i\theta(\alpha)} z(1 + \alpha + d_1(\alpha)|z|^2) + g(z, \bar{z}, \alpha), \tag{B.1}$$

*where $d_1(\alpha) = d(\alpha) + ib(\alpha)$; $b(\alpha)$, $d(\alpha)$, and $\theta(\alpha)$ are smooth real-valued functions; $d(0) < 0, 0 < \theta(0) < \pi, g = O(|z|^4)$ is a smooth complex-valued function of $z, \bar{z}, \alpha$, has a stable closed invariant curve for sufficiently small $\alpha > 0$.*

**Proof:**
*Step 1 (Rescaling and shifting).* First, introduce new variables $(s, \varphi)$ by the formula

$$z = \sqrt{-\frac{\alpha}{d(\alpha)}} e^{i\varphi}(1 + s). \tag{B.2}$$

Substitution of (B.2) into (B.1) gives

$$e^{i\tilde{\varphi}}(1 + \tilde{s}) = e^{i(\varphi + \theta(\alpha))}(1 + s)\left[1 - \alpha(2s + s^2) + i\alpha\nu(\alpha)(1 + s)^2\right] + \alpha^{3/2} h(s, \varphi, \alpha),$$

where

$$\nu(\alpha) = -\frac{b(\alpha)}{d(\alpha)},$$

and $h$ is a smooth complex-valued function of $(s, \varphi, \alpha^{1/2})$. Thus, the map (B.1) in $(s, \varphi)$-coordinates reads

$$\begin{cases} \tilde{s} = (1 - 2\alpha)s - \alpha(3s^2 + s^3) + \alpha^{3/2} p(s, \varphi, \alpha), \\ \tilde{\varphi} = \varphi + \theta(\alpha) + \alpha\nu(\alpha)(1 + s)^2 + \alpha^{3/2} q(s, \varphi, \alpha), \end{cases} \tag{B.3}$$

where $p, q$ are smooth real-valued functions of $(s, \varphi, \alpha^{1/2})$. Now apply the scaling

$$s = \sqrt{\alpha}\xi. \tag{B.4}$$

After rescaling accounting to (B.4), the map (B.3) takes the form

$$\begin{cases} \tilde{\xi} = (1 - 2\alpha)\xi - \alpha^{3/2}(3\xi^2 + \alpha^{1/2}\xi^3) + \alpha p^{(1)}(\xi, \varphi, \alpha), \\ \tilde{\varphi} = \varphi + [\theta(\alpha) + \alpha\nu(\alpha)] + \alpha^{3/2}\nu(\alpha)(2\xi + \alpha^{1/2}\xi^2) + \alpha^{3/2} q^{(1)}(\xi, \varphi, \alpha), \end{cases} \tag{B.5}$$

where

$$p^{(1)}(\xi, \varphi, \alpha) = p(\alpha^{1/2}\xi, \varphi, \alpha), \ q^{(1)}(\xi, \varphi, \alpha) = q(\alpha^{1/2}\xi, \varphi, \alpha),$$

are smooth with respect to $(\xi, \varphi, \alpha^{1/2})$. Denote $\omega(\alpha) = \theta(\alpha) + \alpha\nu(\alpha)$, and notice that $p^{(1)}$ can be written as

$$p^{(1)}(\xi, \varphi, \alpha) = r^{(0)}(\varphi, \alpha) + \alpha^{1/2} r^{(1)}(\xi, \varphi, \alpha).$$

Now (B.5) can be represented by

$$\begin{cases} \tilde{\xi} = (1 - 2\alpha)\xi + \alpha r^{(0)}(\varphi, \alpha) + \alpha^{3/2} r^{(2)}(\xi, \varphi, \alpha), \\ \tilde{\varphi} = \varphi + \omega(\alpha) + \alpha^{3/2} q^{(2)}(\xi, \varphi, \alpha), \end{cases} \tag{B.6}$$

with

$$r^{(2)}(\xi,\varphi,\alpha) = -(3\xi^2 + \alpha^{1/2}\xi^3) + r^{(1)}(\xi,\varphi,\alpha),$$
$$q^{(2)}(\xi,\varphi,\alpha) = \nu(\alpha)(2\xi + \alpha^{1/2}\xi^2) + q^{(1)}(\xi,\varphi,\alpha).$$

The functions $r^{(2)}$ and $q^{(2)}$ have the same smoothness as $p^{(1)}$ and $q^{(1)}$. Finally, perform a coordinate shift, eliminating the term $\alpha r^{(0)}(\varphi,\alpha)$ from the first equation in (B.6):

$$\xi = u + \tfrac{1}{2}r^{(0)}(\varphi,\alpha). \qquad (B.7)$$

This gives a map $F$, which we will work with from now on,

$$F : \begin{cases} \tilde{u} = (1-2\alpha)u + \alpha^{3/2}H_\alpha(u,\varphi), \\ \tilde{\varphi} = \varphi + \omega(\alpha) + \alpha^{3/2}K_\alpha(u,\varphi), \end{cases} \qquad (B.8)$$

where $\omega(\alpha)$ is smooth and

$$H_\alpha(u,\varphi) = r^{(2)}\left(u + \tfrac{1}{2}r^{(0)}(\varphi,\alpha),\varphi,\alpha\right),$$
$$K_\alpha(u,\varphi) = q^{(2)}\left(u + \tfrac{1}{2}r^{(0)}(\varphi,\alpha),\varphi,\alpha\right),$$

are smooth functions of $(u,\varphi,\alpha^{1/2})$ that are $2\pi$-periodic in $\varphi$.

Notice that the band $\{(u,\varphi) : |u| \le 1, \varphi \in [0,2\pi]\}$ corresponds to a band of $O(\alpha)$ width around the circle

$$S_0(\alpha) = \left\{ z : |z|^2 = -\frac{\alpha}{d(\alpha)} \right\}$$

in (B.1), which has an $O(\alpha^{1/2})$ radius in the original coordinate $z$. In what follows, it is conveinient to introduce a number

$$\lambda = \sup_{|u|\le 1,\varphi\in[0,2\pi]} \left\{ |H_\alpha|, |K_\alpha|, \left|\frac{\partial H_\alpha}{\partial u}\right|, \left|\frac{\partial K_\alpha}{\partial u}\right|, \left|\frac{\partial H_\alpha}{\partial \varphi}\right|, \left|\frac{\partial K_\alpha}{\partial \varphi}\right| \right\}. \qquad (B.9)$$

So defined, $\lambda$ depends on $\alpha$ but remains bounded as $\alpha \to 0$.

*Step 2 (Definition of the function space).* We will characterize the closed curves by elements of a function space $U$. By definition, $u \in U$ is a $2\pi$-periodic function $u = u(\varphi)$ satisfying the following two conditions:

(U.1) $|u(\varphi)| \le 1$ for all $\varphi$;
(U.2) $|u(\varphi_1) - u(\varphi_2)| \le |\varphi_1 - \varphi_2|$ for all $\varphi_1, \varphi_2$.

The first property means that $u(\varphi)$ is *absolutely bounded* by unity, while the second means that $u(\varphi)$ is *Lipschitz continuous* with Lipschitz constant equal to one. Space $U$ is a complete metric space with respect to the norm

$$\|u\| = \sup_{\varphi\in[0,2\pi]} |u(\varphi)|.$$

Recall from Chapter 1 that a map $\mathcal{F} : U \to U$ (transforming a function $u(\varphi) \in U$ into some other function $\tilde{u}(\varphi) = (\mathcal{F}u)(\varphi) \in U$) is a *contraction* if there is a number $\varepsilon$, $0 < \varepsilon < 1$, such that

$$\|\mathcal{F}(u_1) - \mathcal{F}(u_2)\| \le \varepsilon \|u_1 - u_2\|$$

for all $u_{1,2} \in U$. A contraction map in a complete normed space has a unique fixed point $u^{(\infty)} \in U$:

$$\mathcal{F}(u^{(\infty)}) = u^{(\infty)}.$$

Moreover, the fixed point $u^{(\infty)}$ is a globally stable equilibrium of the infinite-dimensional dynamical system $\{U, \mathcal{F}\}$, that is,

$$\lim_{k \to +\infty} \|\mathcal{F}^k(u) - u^{(\infty)}\| = 0,$$

for all $u \in U$ (see **Fig. 4.16**). The above two facts are often referred to as the Contraction Mapping Principle.

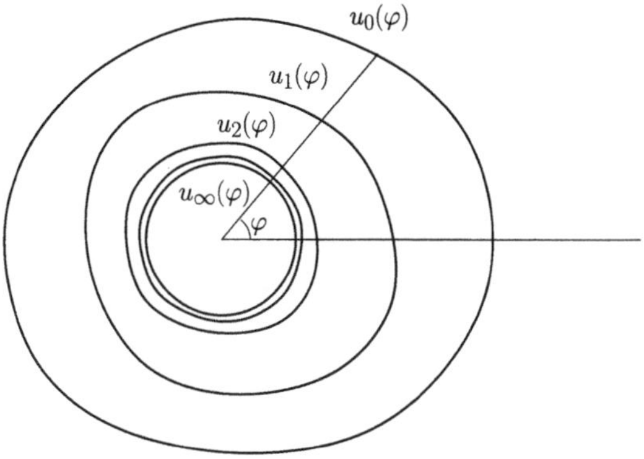


**Fig. 4.16.** Accumulating closed curves.

*Step 3* (*Construction of the map* $\mathcal{F}$). We will consider a map $\mathcal{F}$ *induced* by $F$ on $U$. This means that if $u$ represents a closed curve, then $\tilde{u} = \mathcal{F}(u)$ represents its image under the map $F$ defined by (B.8).

Suppose that a function $u = u(\varphi)$ from $U$ is given. To construct the map $\mathcal{F}$, we have to specify a procedure for each given $\varphi$ that allows us to find the corresponding $\tilde{u}(\varphi) = (\mathcal{F}u)(\varphi)$. Notice, however, that $F$ is nearly a *rotation* by the angle $\omega(\alpha)$ in $\varphi$. Thus, a point $(\tilde{u}(\varphi), \varphi)$ in the resulting curve is the image of a point $(u(\hat{\varphi}), \hat{\varphi})$ in the original curve with a *different* angle coordinate $\hat{\varphi}$ (see **Fig. 4.17**).

To show that $\hat{\varphi}$ is uniquely defined, we have to prove that the equation

$$\varphi = \hat{\varphi} + \omega(\alpha) + \alpha^{3/2} K_\alpha(u(\hat{\varphi}), \hat{\varphi}) \tag{B.10}$$

has a unique solution $\hat{\varphi} = \hat{\varphi}(\varphi)$ for any given $u \in U$. This is the case since the right-hand side of (B.10) is a strictly increasing function of $\hat{\varphi}$. Indeed, let $\varphi_2 > \varphi_1$; then, according to (B.8),

$$\begin{aligned}\tilde{\varphi_2} - \tilde{\varphi_1} &= \varphi_2 - \varphi_1 + \alpha^{3/2}\left[K_\alpha(u(\varphi_2), \varphi_2) - K_\alpha((\varphi_1), \varphi_1)\right] \\ &\ge \varphi_2 - \varphi_1 - \alpha^{3/2}\left|K_\alpha(u(\varphi_2), \varphi_2) - K_\alpha((\varphi_1), \varphi_1)\right|.\end{aligned}$$

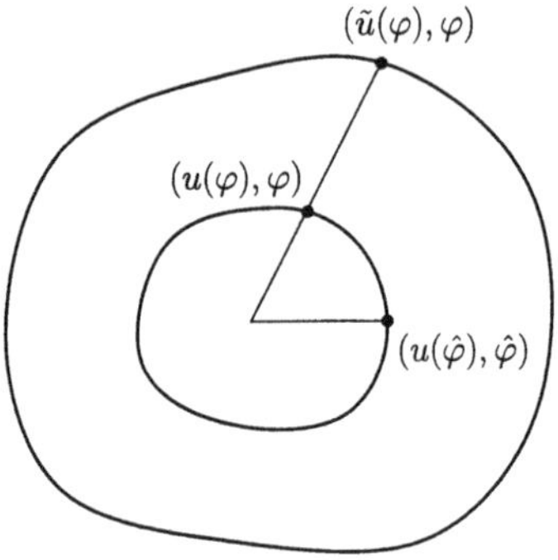


**Fig. 4.17.** Definition of the map.

Taking into account that $K_\alpha$ is a smooth function with (B.9) and (U.2), we get

$$\begin{aligned}|K_\alpha(u(\varphi_2), \varphi_2) - K_\alpha(u(\varphi_1), \varphi_1)| &\leq \lambda[|u(\varphi_2) - u(\varphi_1)| + |\varphi_2 - \varphi_1|] \\ &\leq 2\lambda|\varphi_2 - \varphi_1| \ = \ 2\lambda(\varphi_2 - \varphi_1).\end{aligned}$$

This last estimate can also be written as

$$-|K_\alpha(u(\varphi_2), \varphi_2) - K_\alpha((\varphi_1), \varphi_1)| \geq -2\lambda(\varphi_2 - \varphi_1),$$

which implies

$$\tilde{\varphi_2} - \tilde{\varphi_1} \geq (1 - 2\lambda\alpha^{3/2})(\varphi_2 - \varphi_1).$$

Thus, the right-hand side of (B.10) is a strictly increasing function, provided $\alpha$ is small enough, and its solution $\hat{\varphi}$ is uniquely defined.[2] From the above estimates, it also follows that $\hat{\varphi}(\varphi)$ – that is, the inverse function to the function given by (B.10) – is Lipschitz continuous:

$$|\hat{\varphi}(\varphi_1) - \hat{\varphi}(\varphi_2)| \leq (1 - 2\lambda\alpha^{3/2})^{-1}|\varphi_1 - \varphi_2|. \tag{B.11}$$

Now we can define the map $\tilde{u} = \mathcal{F}(u)$ by the formula

$$\tilde{u}(\varphi) = (1 - 2\alpha)u(\hat{\varphi}) + \alpha^{3/2}K_\alpha(u(\hat{\varphi}), \hat{\varphi}), \tag{B.12}$$

where $\hat{\varphi}$ is the solution of (B.10). The mere definition, of course, is not enough and we have to verify that $\mathcal{F}(u) \in U$, if $u \in U$, namely, to check (U.1) and (U.2) for $\tilde{u} = \mathcal{F}(u)$.

Condition (U.1) for $\tilde{u}$ follows from the estimate

$$|\tilde{u}(\varphi)| \leq (1 - 2\alpha)|u(\hat{\varphi})| + \alpha^{3/2}|H_\alpha(u(\hat{\varphi}), \hat{\varphi})| \leq 1 - 2\alpha + \lambda\alpha^{3/2},$$

where we have used (U.1) for $u$ and the definition (B.9) of $\lambda$. Thus, $|\tilde{u}| \leq 1$ if $\alpha$ is small enough and positive. Condition (U.2) for $\tilde{u}$ is obtained by the sequence of estimates:

[2] Meanwhile, $\hat{\varphi} \approx \varphi - \omega(\alpha)$.

$$\begin{aligned}|\tilde{u}(\varphi_1) - \tilde{u}(\varphi_2)| &\le (1-2\alpha)|u(\hat{\varphi}_1) - u(\hat{\varphi}_2)| \\ &\quad + \alpha^{3/2}|H_\alpha(u(\hat{\varphi}_1), \hat{\varphi}_1) - H_\alpha(u(\hat{\varphi}_2), \hat{\varphi}_2)| \\ &\le (1-2\alpha)|u(\hat{\varphi}_1) - u(\hat{\varphi}_2)| \\ &\quad + \alpha^{3/2}\lambda\Big[|\tilde{u}(\varphi_1) - \tilde{u}(\varphi_2)| + |\hat{\varphi}_1 - \hat{\varphi}_2|\Big] \\ &\le (1 - 2\alpha + 2\lambda\alpha^{3/2})|\hat{\varphi}_1 - \hat{\varphi}_2|,\end{aligned}$$

where the final inequality holds due to the Lipschitz continuity of $u$. Inserting the estimate (B.11), we get

$$|\tilde{u}(\varphi_1) - \tilde{u}(\varphi_2)| \le (1 - 2\alpha + 2\lambda\alpha^{3/2})(1 - 2\lambda\alpha^{3/2})^{-1}|\varphi_1 - \varphi_2|.$$

Thus, (U.2) also holds for $\tilde{u}$ for all sufficiently small positive $\alpha$. Therefore, the map $\tilde{u} = \mathcal{F}(u)$ is well defined.

*Step 4 (Verification of the contraction property).* Now suppose two functions $u_1, u_2 \in U$ are given. What we need to obtain is the estimation of $\|\tilde{u}_1 - \tilde{u}_2\|$ in terms of $\|u_1 - u_2\|$. By the definition (B.12) of $\tilde{u} = \mathcal{F}(u)$,

$$\begin{aligned}\|\tilde{u}_1(\varphi) - \tilde{u}_2(\varphi))\| &\le (1-2\alpha)|u_1(\hat{\varphi}_1) - u_2(\hat{\varphi}_2)| \\ &\quad + \alpha^{3/2}|H_\alpha(u_1(\hat{\varphi}_1), \hat{\varphi}_1) - H_\alpha(u_2(\hat{\varphi}_2), \hat{\varphi}_2)| \\ &\le (1-2\alpha)|u_1(\hat{\varphi}_1) - u_2(\hat{\varphi}_2)| \\ &\quad + \alpha^{3/2}\lambda\Big[|u_1(\hat{\varphi}_1) - u_2(\hat{\varphi}_2)| + |\hat{\varphi}_1 - \hat{\varphi}_2|\Big],\end{aligned} \tag{B.13}$$

where $\hat{\varphi}_1$ and $\hat{\varphi}_2$ are the unique solutions of

$$\varphi = \hat{\varphi}_1 + \omega(\alpha) + \alpha^{3/2}K_\alpha(u_1(\hat{\varphi}_1), \hat{\varphi}_1) \tag{B.14}$$

and

$$\varphi = \hat{\varphi}_2 + \omega(\alpha) + \alpha^{3/2}K_\alpha(u_2(\hat{\varphi}_2), \hat{\varphi}_2), \tag{B.15}$$

respectively. The estimates (B.13) have not solved the problem yet since we have to use only $\|u_1 - u_2\|$ in the right-hand side. First, express $|u_1(\hat{\varphi}_1) - u_2(\hat{\varphi}_2)|$ in terms of $\|u_1 - u_2\|$ *and* $|\hat{\varphi}_1 - \hat{\varphi}_2|$:

$$\begin{aligned}|u_1(\hat{\varphi}_1) - u_2(\hat{\varphi}_2)| &= |u_1(\hat{\varphi}_1) - u_2(\hat{\varphi}_1) + u_2(\hat{\varphi}_1) - u_2(\hat{\varphi}_2)| \\ &\le |u_1(\hat{\varphi}_1) - u_2(\hat{\varphi}_1)| + |u_2(\hat{\varphi}_1) - u_2(\hat{\varphi}_2)| \\ &\le \|u_1 - u_2\| + |\hat{\varphi}_1 - \hat{\varphi}_2|.\end{aligned} \tag{B.16}$$

The last inequality has been obtained using the definition of the norm and the Lipschitz continuity of $u_2$. To complete the estimates, we need to express $|\hat{\varphi}_1 - \hat{\varphi}_2|$ in terms of $\|u_1 - u_2\|$. Subtracting (B.15) from (B.14), transposing, and taking absolute values yield

$$\begin{aligned}|\hat{\varphi}_1 - \hat{\varphi}_2| &\le \alpha^{3/2}|K_\alpha(u_1(\hat{\varphi}_1), \hat{\varphi}_1) - K_\alpha(u_2(\hat{\varphi}_2), \hat{\varphi}_2)| \\ &\le \alpha^{3/2}\lambda[|u_1(\hat{\varphi}_1) - u_2(\hat{\varphi}_2)| + |\hat{\varphi}_1 - \hat{\varphi}_2|].\end{aligned}$$

Inserting (B.16) into this inequality and collecting all the terms involving $|\hat{\varphi}_1 - \hat{\varphi}_2|$ on the left, result in

$$|\hat{\varphi}_1 - \hat{\varphi}_2| \le (1 - 2\alpha^{3/2}\lambda)^{-1}\alpha^{3/2}\lambda\|u_1 - u_2\|. \tag{B.17}$$

Using the estimates (B.16) and (B.17), we can complete (B.13) as follows:

$$\|\tilde{u}_1(\varphi) - \tilde{u}_2(\varphi))\| \leq \epsilon \|u_1 - u_2\|,$$

where

$$\epsilon = (1-2\alpha)\left[1 + \alpha^{3/2}\lambda(1 - 2\alpha^{3/2}\lambda)^{-1}\right] + \alpha^{3/2}\lambda\left[1 + 2\alpha^{3/2}\lambda(1 - 2\alpha^{3/2}\lambda)^{-1}\right].$$

Since

$$\epsilon = 1 - 2\alpha + O(\alpha^{3/2}),$$

the map $\mathcal{F}$ is a contraction in $U$ for small positive $\alpha$. Therfore, it has a unique stable fixed point $u^{(\infty)} \in U$.

*Step 5 (Stability of the invariant curve).* Now take a point $(u_0, \varphi_0)$ within the band $\{(u,\varphi) : |u| \leq 1, \varphi \in [0, 2\pi]\}$. If the point belongs to the curve given by $u^{(\infty)}$, it remains on this curve under iterations of $F$ since the map $\mathcal{F}$ maps this curve into itself. If the point does not lie on the invariant curve, take some (noninvariant) closed curve passing through it represented by $u^{(0)} \in U$, say. Such a curve always exists. Let us apply the iterations of the map $F$ defined by (B.8) to this point. We get a sequence of points

$$\{(u_k, \varphi_k)\}_{k=0}^{\infty}.$$

It is clear that each point from this sequence belongs to the corresponding iterate of the curve $u^{(0)}$ under the map $\mathcal{F}$. We have just shown that the iterations of the curve converge to the invariant curve given by $u^{(\infty)}$. Therefore, the point sequence must also converge to the curve. This proves the stability of the closed invariant curve as the invariant set of the map and completes the proof. □

## 4.11 Appendix C: Bibliographical notes

The dynamics generated by one-dimensional maps is a classical mathematical subject, studied in detail (see Whitley [1983] and van Strien [1991] for surveys). Properties of the fixed points and period-two cycles involved in the fold and flip bifurcations were known long ago. Explicit formulation of the topological normal form theorems for these bifurcations is due to Arnol'd [1983]. A complete proof, that the truncation of the higher-order terms in the normal forms results in locally topologically equivalent systems, happens to be unexpectedly difficult (see Newhouse, Palis & Takens [1983], Arnol'd et al. [1994]).

The appearance of a closed invariant curve surrounding a fixed point while a pair of complex multipliers crosses the unit circle was known to Andronov and studied by Neimark [1959] (without explicit statement of all the genericity conditions). A complete proof was given by Sacker [1965], who discovered the bifurcation independently. It became widely known as "Hopf bifurcation for maps" after Ruelle & Takens [1971] and Marsden & McCracken [1976]. A modern treatment of the Neimark-Sacker bifurcation for planar maps can be found in Iooss [1979], where the normal form coefficient $d(0)$ is computed (see also Wan [1978*b*]). In our Appendix B we follow, essentially, the proof given in Marsden & McCracken [1976].

The normal form theory for maps is presented by Arnol'd [1983]. In our analysis of the codimension-one bifurcations of fixed points we need only a small portion of this theory which we develop "on-line."

A cascade of period-doubling bifurcations in the quadratic map $x \mapsto x^2 - \lambda$ was discovered and investigated by Myrberg [1962]. Independently, similar cascades were observed by mathematical ecologists in one-dimensional discrete-time population models: Shapiro [1974] analyzed a model by Ricker [1954], while May [1974] used the logistic map. Feigenbaum [1978] and Coullet & Eckmann [1980] discovered the universality in such cascades and explained its mechanism based on the properties of the doubling operator. The relevant theorems were proved by Lanford [1980, 1984] with the help of a computer and delicate error estimates (see also Babenko & Petrovich [1983, 1984] and Petrovich [1990]). Analytical proofs have been given much later (see Lyubich [2000] for a survey). Feigenbaum-type universality is also proved for some classes of multidimensional discrete-time dynamical systems.

Both the delayed logistic and discrete-time predator-prey models originate in a book by Maynard Smith [1968]. The fate of the closed invariant curve while a parameter "moves" away from the Neimark-Sacker bifurcation was analyzed for the delayed logistic map by Aronson, Chory, Hall & McGehee [1982].

# 5

# Bifurcations of Equilibria and Periodic Orbits in $n$-Dimensional Dynamical Systems

In the previous two chapters we studied bifurcations of equilibria and fixed points in generic one-parameter dynamical systems having the *minimum possible* phase dimensions. Indeed, the systems we analyzed were either one- or two-dimensional. This chapter shows that these bifurcations occur in "essentially" the same way for generic $n$-dimensional systems. As we shall see, there are certain parameter-dependent one- or two-dimensional *invariant manifolds* on which the system exhibits the corresponding bifurcations, while the behavior off the manifolds is somehow "trivial," for example, the manifolds may be exponentially attractive. Moreover, such manifolds (called *center manifolds*) exist for many dissipative infinite-dimensional dynamical systems. Below we derive quadratic approximations to the center manifolds in finite dimensions and for systems restricted to them at bifurcation parameter values. Using these results, we derive explicit invariant formulas for the critical normal form coefficients at all studied codimension 1 bifurcations of equilibria and fixed points. In Appendix A we consider a reaction-diffusion system on an interval to illustrate the necessary modifications of the technique to handle infinite-dimensional systems.

## 5.1 Center manifold theorems

We are going to formulate without proof the main theorems that allow us to reduce the dimension of a given system near a local bifurcation. Let us start with the *critical* case; we assume in this section that the parameters of the system are fixed at their bifurcation values, which are those values for which there is a nonhyperbolic equilibrium (fixed point). We will treat continuous- and discrete-time cases separately.

### 5.1.1 Center manifolds in continuous-time systems

Consider a continuous-time dynamical system defined by

$$\dot{x} = f(x), \quad x \in \mathbb{R}^n, \tag{5.1}$$

where $f$ is sufficiently smooth, $f(0) = 0$. Let the eigenvalues of the Jacobian matrix $A$ evaluated at the equilibrium point $x_0 = 0$ be $\lambda_1, \lambda_2, \dots, \lambda_n$. Suppose the equilibrium is not hyperbolic and that there are thus eigenvalues with zero real part. Assume that there are $n_+$ eigenvalues (counting multiplicities) with Re $\lambda > 0$, $n_0$ eigenvalues with Re $\lambda = 0$, and $n_-$ eigenvalues with Re $\lambda < 0$ (see **Fig. 5.1**). Let $T^c$ denote the linear (generalized) eigenspace of

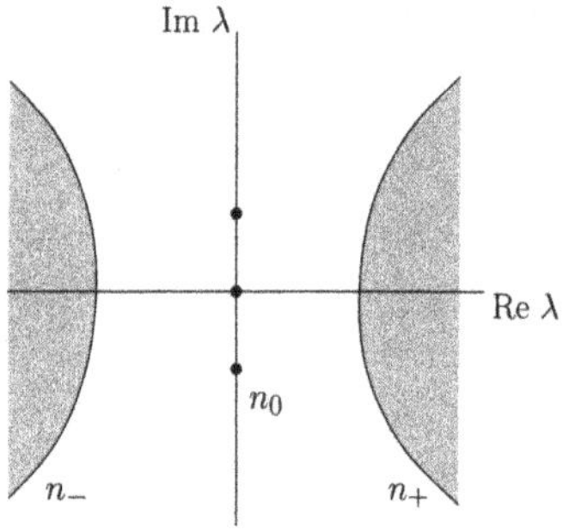


**Fig. 5.1.** Critical eigenvalues of an equilibrium.

$A$ corresponding to the union of the $n_0$ eigenvalues on the imaginary axis. The eigenvalues with Re $\lambda = 0$ are often called *critical*, as is the eigenspace $T^c$. Let $\varphi^t$ denote the flow associated with (5.1). Under the assumptions stated above, the following theorem holds.

**Theorem 5.1 (Center Manifold Theorem)** *There is a locally defined smooth $n_0$-dimensional invariant manifold $W^c_{loc}(0)$ of (5.1) that is tangent to $T^c$ at $x = 0$.*

*Moreover, there is a neighborhood $U$ of $x_0 = 0$, such that if $\varphi^t x \in U$ for all $t \geq 0 (t \leq 0)$, then $\varphi^t x \to W^c_{loc}(0)$ for $t \to +\infty$ $(t \to -\infty)$.* □

**Definition 5.1** *The manifold $W^c_{loc}$ is called the* center manifold.

We are not going to present the proof here. If $n_+ = 0$, the manifold $W^c_{loc}$ can be constructed as a local limit of iterations of $T^c$ under $\varphi^1$. From now on, we drop the subscript "loc" in order to simplify notation. **Figs. 5.2** and **5.3** illustrate the theorem for the fold bifurcation on the plane $(n = 2, n_0 = 1, n_- = 1)$ and for the Hopf bifurcation in $\mathbb{R}^3$ $(n = 3, n_0 = 2, n_- = 1)$. In the first case, the center manifold $W^c$ is tangent to the eigenvector corresponding to $\lambda_1 = 0$, while in the second case, it is tangent to a plane spanned by the real and imaginary parts of the complex eigenvector corresponding to $\lambda_1 = i\omega_0$, $\omega_0 > 0$.

**Remarks:**

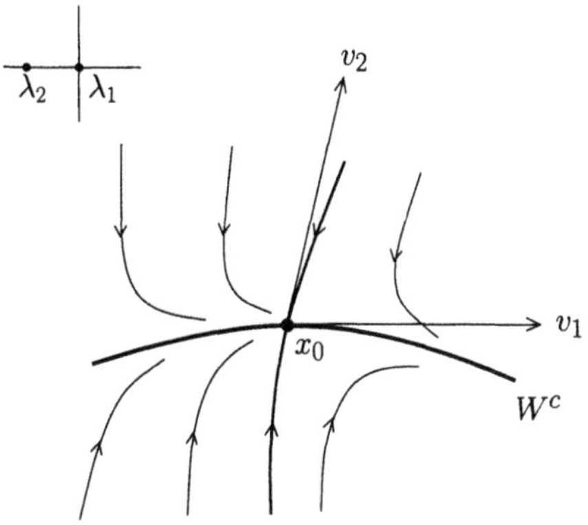


**Fig. 5.2.** A one-dimensional center manifold at the fold bifurcation.

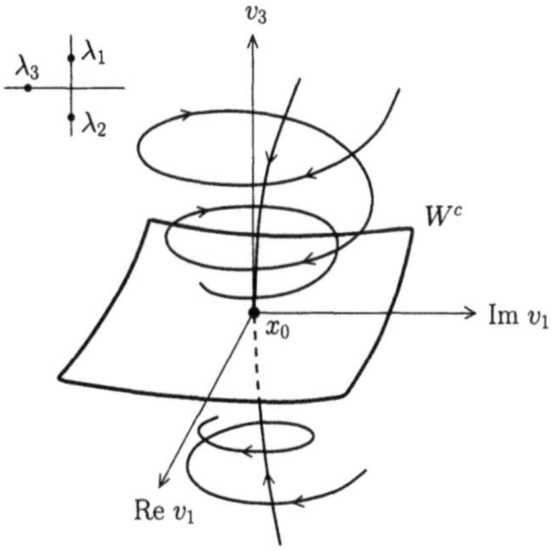


**Fig. 5.3.** A two-dimensional center manifold at the Hopf bifurcation.

(1) The second statement of the theorem means that orbits staying near the equilibrium for $t \geq 0$ or $t \leq 0$ tend to $W^c$ in the corresponding time direction. If we know a priori that *all* orbits starting in $U$ remain in this region forever (a necessary condition for this is $n_+ = 0$), then the theorem implies that these orbits approach $W^c(0)$ as $t \to +\infty$. In this case the manifold is "attracting."

(2) $W^c$ *need not* be unique. The system

$$\begin{cases} \dot{x} = x^2, \\ \dot{y} = -y, \end{cases}$$

has an equilibrium $(x, y) = (0, 0)$ with $\lambda_1 = 0, \lambda_2 = -1$ (a fold case). It possesses a family of one-dimensional center manifolds:

$$W^c_\beta(0) = \{(x, y) \; : \; y = \psi_\beta(x)\},$$

where

$$\psi_\beta(x) = \begin{cases} \beta \exp\left(\frac{1}{x}\right) & \text{for } x < 0, \\ 0 & \text{for } x \geq 0, \end{cases}$$

(see **Fig. 5.4**(a)). The system

$$\begin{cases} \dot{x} = -y - x(x^2 + y^2), \\ \dot{y} = x - y(x^2 + y^2), \\ \dot{z} = -z, \end{cases}$$

has an equilibrium $(x, y, z) = (0, 0, 0)$ with $\lambda_{1,2} = \pm i$, $\lambda_3 = -1$ (Hopf case). There is a family of two-dimensional center manifolds in the system given by

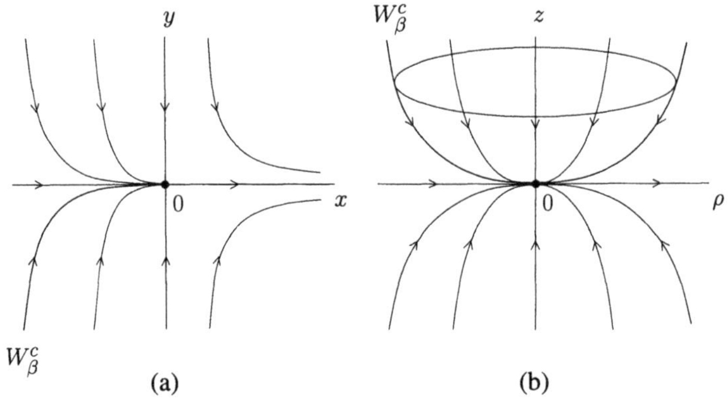


**Fig. 5.4.** Nonuniqueness of the center manifold at (a) fold and (b) Hopf bifurcations.

$$W^c_\beta(0) = \{(x, y, z) \ : \ z = \phi_\beta(x, y)\},$$

where

$$\phi_\beta(x, y) = \begin{cases} \beta \exp\left(-\frac{1}{2(x^2+y^2)}\right) & \text{for } x^2 + y^2 > 0, \\ 0 & \text{for } x = y = 0, \end{cases}$$

(see **Fig. 5.4**(b)). As we shall see, this nonuniqueness is actually irrelevant for applications.

(3) A center manifold $W^c$ has the same *finite* smoothness as $f$ (if $f \in C^k$ with finite $k$, $W^c$ is also a $C^k$ manifold) in some neighborhood $U$ of $x_0$. However, as $k \to \infty$ the neighborhood $U$ may shrink, thus resulting in the nonexistence of a $C^\infty$ manifold $W^c$ for some $C^\infty$ systems (see Exercise 1). ◇

To characterize dynamics near a nonhyperbolic equilibrium $x_0 = 0$ more explicitly, write (5.1) in an *eigenbasis* formed by all (generalized) eigenvectors of $A$ (or their linear combinations if the corresponding eigenvalues are complex). Collecting critical and noncritical components, we can then rewrite (5.1) as

$$\begin{cases} \dot{u} = Bu + g(u,v), \\ \dot{v} = Cv + h(u,v), \end{cases} \tag{5.2}$$

where $u \in \mathbb{R}^{n_0}, v \in \mathbb{R}^{n_+ + n_-}$, $B$ is an $n_0 \times n_0$ matrix with all its $n_0$ eigenvalues on the imaginary axis, while $C$ is an $(n_+ + n_-) \times (n_+ + n_-)$ matrix with no eigenvalue on the imaginary axis[1]. Functions $g$ and $h$ have Taylor expansions starting with at least quadratic terms. A center manifold $W^c$ of system (5.2) can be locally represented as the graph of a smooth function:

$$W^c = \{(u,v) : v = V(u)\}$$

(see **Fig. 5.5**). Here $V : \mathbb{R}^{n_0} \to \mathbb{R}^{n_+ + n_-}$, and due to the tangent property of $W^c$, $V(u) = O(\|u\|^2)$.

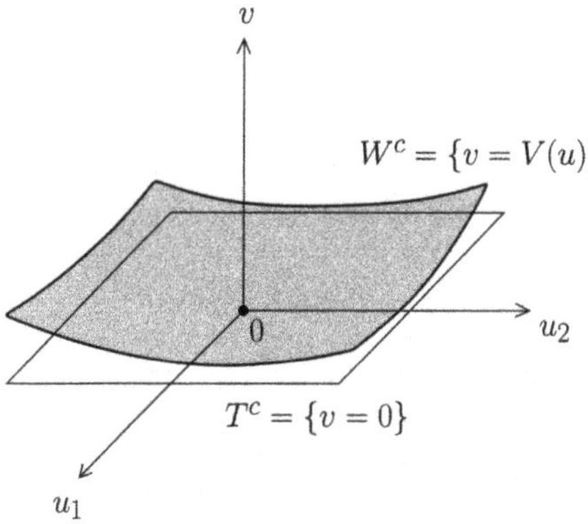


**Fig. 5.5.** Center manifold as the graph of a function $v = V(u)$.

**Theorem 5.2 (Reduction Principle)** *System* (5.2) *is locally topologically equivalent near the origin to the system*

$$\begin{cases} \dot{u} = Bu + g(u, V(u)), \\ \dot{v} = Cv. \end{cases} \tag{5.3}$$

*If there is more than one center manifold, then all the resulting systems* (5.3) *with different $V$ are locally smoothly equivalent.* □

Notice that the equations for $u$ and $v$ are uncoupled in (5.3). The first equation is the *restriction* of (5.2) to its center manifold. Thus, the dynamics of the structurally unstable system (5.2) are essentially determined by this

[1] Actually, any basis in the noncritical eigenspace is allowed. In other words, the matrix $C$ may not be in real canonical (Jordan) form.

restriction since the second equation in (5.3) is linear and has exponentially decaying/growing solutions. For example, if $u = 0$ is the asymptotically stable equilibrium of the restriction and all eigenvalues of $C$ have negative real part, then $(u, v) = (0, 0)$ is the asymptotically stable equilibrium of (5.2). Clearly, the dynamics on the center manifold are determined not only by the linear but also by the *nonlinear* terms of (5.2).

**Example 5.1 (Failure of the tangent approximation)** Consider the planar system

$$\begin{cases} \dot{x} = xy + x^3, \\ \dot{y} = -y - 2x^2. \end{cases} \tag{5.4}$$

There is an equilibrium at $(x, y) = (0, 0)$. Is it stable or unstable? The Jacobian matrix

$$A = \begin{pmatrix} 0 & 0 \\ 0 & -1 \end{pmatrix}$$

has eigenvalues $\lambda_1 = 0,\ \lambda_2 = -1$. Thus, system (5.4) is written in the form (5.2) and has a one-dimensional center manifold $W^c$ represented by a scalar function

$$y = V(x).$$

Let us find the quadratic term in the Taylor expansion of this function:

$$V(x) = \tfrac{1}{2}wx^2 + \cdots.$$

The unknown coefficient $w$ can be found by expressing $\dot{y}$ as

$$\dot{y} = \tfrac{\partial V}{\partial x}\dot{x} = (wx + \cdots)\dot{x} = wx^2y + wx^4 + \cdots = w\left(\tfrac{1}{2}w + 1\right)x^4 + \cdots,$$

or, alternatively, as

$$\dot{y} = -y - 2x^2 = -\left(\tfrac{1}{2}w + 2\right)x^2 + \cdots.$$

Therefore, $w + 4 = 0$ and

$$w = -4.$$

Thus, the center manifold has the following quadratic approximation:

$$V(x) = -2x^2 + O(x^3),$$

and the restriction of (5.4) to its center manifold is given by

$$\dot{x} = xV(x) + x^3 = -2x^3 + x^3 + O(x^4) = -x^3 + O(x^4).$$

Therefore, the origin is *stable* and the phase portrait of the system near the equilibrium is as sketched in **Fig. 5.6**. By restriction of (5.4) onto its critical eigenspace $y = 0$, one gets

$$\dot{x} = x^3.$$

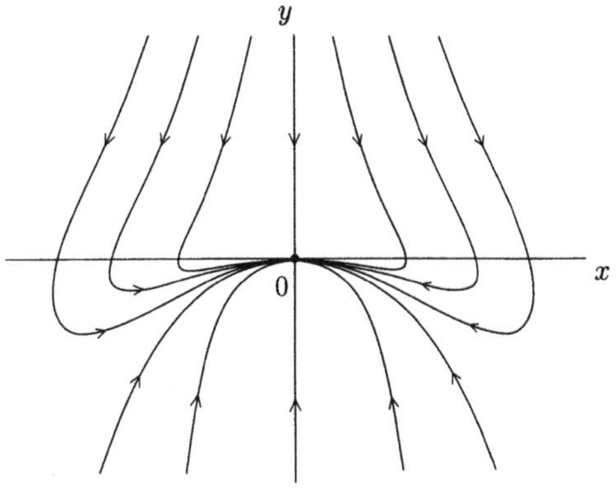


**Fig. 5.6.** Phase portrait of (5.4): The origin is stable.

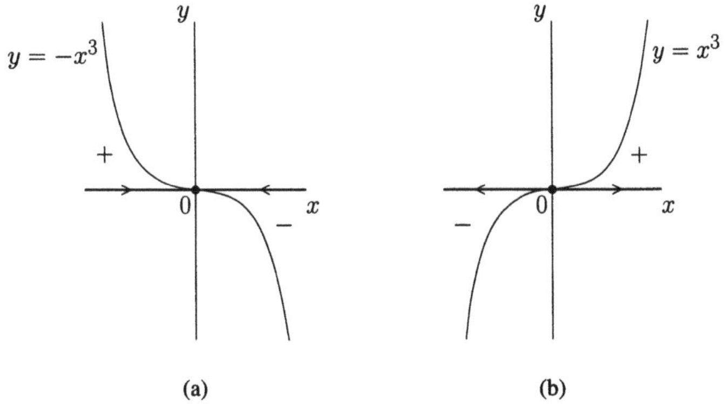


**Fig. 5.7.** Restricted equations: (a) to the center manifold $W^c$; (b) to the tangent line $T^c$.

This equation has an *unstable* point at the origin and thus gives the wrong answer to the stability question. **Fig. 5.7** compares the equations restricted to $W^c$ and $T^c$. ◊

The second equation in (5.3) can be replaced by the equations of a *standard saddle*:

$$\begin{cases} \dot{v} = -v, \\ \dot{w} = \phantom{-}w, \end{cases} \tag{5.5}$$

with $(v, w) \in \mathbb{R}^{n_-} \times \mathbb{R}^{n_+}$. Therefore, the Reduction Principle can be expressed neatly in the following way: *Near a nonhyperbolic equilibrium the system is locally topologically equivalent to the suspension of its restriction to the center manifold by the standard saddle.*

### 5.1.2 Center manifolds in discrete-time systems

Consider now a discrete-time dynamical system defined by

$$x \mapsto f(x), \quad x \in \mathbb{R}^n, \tag{5.6}$$

where $f$ is sufficiently smooth, $f(0) = 0$. Let the eigenvalues of the Jacobian matrix $A$ evaluated at the fixed point $x_0 = 0$ be $\mu_1, \mu_2, \ldots, \mu_n$. Recall, that we call them *multipliers*. Suppose that the equilibrium is not hyperbolic and there are therefore multipliers on the unit circle (with absolute value one). Assume that there are $n_+$ multipliers outside the unit circle, $n_0$ multipliers on the unit circle, and $n_-$ multipliers inside the unit circle (see **Fig. 5.8**). Let $T^c$ denote

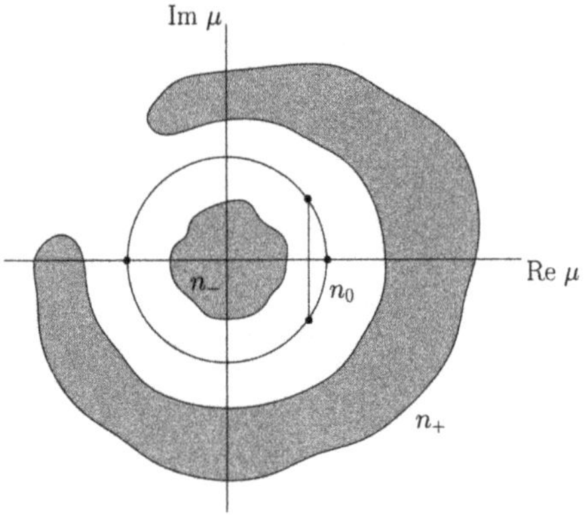


**Fig. 5.8.** Critical multipliers of a fixed point.

the linear invariant (generalized) eigenspace of $A$ corresponding to the union of $n_0$ multipliers on the unit circle. Then, Theorem 5.1 holds verbatim for system (5.6), if we consider only integer time values and set $\varphi^k = f^k$, the $k$th iterate of $f$. Using an eigenbasis of $A$, we can rewrite the system as

$$\begin{pmatrix} u \\ v \end{pmatrix} \mapsto \begin{pmatrix} Bu + g(u,v) \\ Cv + h(u,v) \end{pmatrix} \tag{5.7}$$

with the same notation as before, but the matrix $B$ now has eigenvalues on the unit circle, while all the eigenvalues of $C$ are inside and/or outside it. The center manifold possesses the local representation $v = V(u)$, and the Reduction Principle remains valid.

**Theorem 5.3** *System* (5.7) *is locally topologically equivalent near the origin to the system*

$$\begin{pmatrix} u \\ v \end{pmatrix} \mapsto \begin{pmatrix} Bu + g(u, V(u)) \\ Cv \end{pmatrix}. \tag{5.8}$$

*If there is more than one center manifold, then all the resulting maps* (5.8) *with different* $V$ *are locally smoothly conjugate.* □

The construction of the standard saddle is more involved for the discrete-time case since we have to take into account the *orientation* properties of the map in the expanding and contracting directions. First, suppose for simplicity that there are no multipliers outside the unit circle, (i.e., $n_+ = 0$). Then, if $\det C > 0$, the map $v \mapsto Cv$ in (5.8) can be replaced by

$$v \mapsto \tfrac{1}{2}v,$$

which is a *standard orientation-preserving stable node*. However, if $\det C < 0$, the map $v \mapsto Cv$ in (5.8) must be substituted by

$$\begin{cases} v_1 \mapsto \frac{1}{2}v_1, \\ v_2 \mapsto -\frac{1}{2}v_2, \end{cases}$$

where $v_1 \in \mathbb{R}^{n_- -1}, v_2 \in \mathbb{R}^1$, which is a *standard orientation-reversing stable node*. If there are now $n_+$ multipliers outside the unit circle, the *standard unstable node* $w \mapsto \tilde{w},\ w, \tilde{w} \in \mathbb{R}^{n_+}$, should be added to (5.8). The standard unstable node is defined similarly to the standard stable node but with multiplier 2 instead of $\frac{1}{2}$. Standard stable and unstable nodes together define the *standard saddle* map on $\mathbb{R}^{n_- + n_+}$.

## 5.2 Center manifolds in parameter-dependent systems

Consider now a smooth continuous-time system that depends smoothly on a parameter:

$$\dot{x} = f(x, \alpha), \quad x \in \mathbb{R}^n, \ \alpha \in \mathbb{R}^1. \tag{5.9}$$

Suppose that at $\alpha = 0$ the system has a nonhyperbolic equilibrium $x = 0$ with $n_0$ eigenvalues on the imaginary axis and $(n - n_0)$ eigenvalues with nonzero real parts. Let $n_-$ of them have negative real parts, while $n_+$ have positive real parts. Consider the *extended system*:

$$\begin{cases} \dot{\alpha} = 0, \\ \dot{x} = f(x, \alpha). \end{cases} \tag{5.10}$$

Notice that the extended system (5.10) may be nonlinear, even if the original system (5.9) is linear. The Jacobian matrix of (5.10) at the equilibrium $(\alpha, x) = (0, 0)$ is the $(n+1) \times (n+1)$ matrix

$$J = \begin{pmatrix} 0 & 0 \\ f_\alpha(0,0) & f_x(0,0) \end{pmatrix},$$

having $(n_0 + 1)$ eigenvalues on the imaginary axis and $(n - n_0)$ eigenvalues with nonzero real part. Thus, we can apply the Center Manifold Theorem to system (5.10). The theorem guarantees the existence of a center manifold $\mathcal{W}^c \subset \mathbb{R}^1 \times \mathbb{R}^n, \dim \mathcal{W}^c = n_0 + 1$. This manifold is tangent at the origin to the (generalized) eigenspace of $J$ corresponding to $(n_0+1)$ eigenvalues with zero real part. Since $\dot{\alpha}=0$, the hyperplanes $\Pi_{\alpha_0}=\{(\alpha,x):\alpha=\alpha_0\}$ are also invariant with respect to (5.10). Therefore, the manifold $\mathcal{W}^c$ is foliated by $n_0$-dimensional invariant manifolds

$$W^c_\alpha = \mathcal{W}^c \cap \Pi_\alpha$$

(see **Fig. 5.9**). Thus, we have the following lemma.

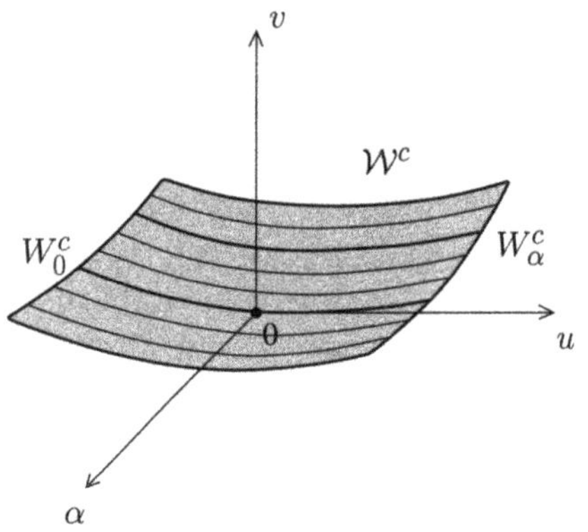


**Fig. 5.9.** Center manifold of the extended system.

**Lemma 5.1** *System* (5.9) *has a parameter-dependent local invariant manifold* $W^c_\alpha$. *If* $n_+=0$, *this manifold is attracting.* □

Notice that $W^c_0$ is a center manifold of (5.10) at $\alpha=0$ as defined in the previous section. Often, the manifold $W^c_\alpha$ is called a *center manifold* for all $\alpha$. For each small $|\alpha|$ we can restrict system (5.9) to $W^c_\alpha$. If we introduce a (parameter-dependent) coordinate system on $W^c_\alpha$ with $u \in \mathbb{R}^{n_0}$ as the coordinate,[2] this restriction will be represented by a smooth system

$$\dot{u} = \Phi(u,\alpha). \tag{5.11}$$

At $\alpha=0$, system (5.11) is equivalent to the restriction of (5.9) to its center manifold $W^c_0$ and will be explicitly computed up to the third-order terms in Section 5.4 for all codim 1 bifurcations.

**Theorem 5.4 (Shoshitaishvili [1972])** *System* (5.9) *is locally topologically equivalent to the suspension of* (5.11) *by the standard saddle* (5.5). *Moreover,* (5.11) *can be replaced by any locally topologically equivalent system.* □

[2] Since $W^c_0$ is tangent to $T^c$, we can parametrize $W^c_\alpha$ for small $|\alpha|$ by coordinates on $T^c$ using a (local) projection from $W^c_\alpha$ onto $T^c$.

This theorem means that all "essential" events near the bifurcation parameter value occur on the invariant manifold $W^c_\alpha$ and are captured by the $n_0$-dimensional system (5.11). A similar theorem can be formulated for discrete-time dynamical systems and for systems with more than one parameter. Let us apply this theorem to the fold and Hopf bifurcations.

**Example 5.2 (Generic fold bifurcation in $\mathbb{R}^2$)** Consider a planar system

$$\dot{x} = f(x, \alpha), \quad x \in \mathbb{R}^2, \ \alpha \in \mathbb{R}^1. \tag{5.12}$$

Assume that at $\alpha = 0$ it has the equilibrium $x_0 = 0$ with one eigenvalue $\lambda_1 = 0$ and one eigenvalue $\lambda_2 < 0$. Lemma 5.1 gives the existence of a smooth, locally defined, one-dimensional attracting invariant manifold $W^c_\alpha$ for (5.12) for small $|\alpha|$. At $\alpha = 0$ the restricted equation (5.11) has the form

$$\dot{u} = bu^2 + O(u^3).$$

If $b \neq 0$ and the restricted equation depends generically on the parameter, then, as proved in Chapter 3, it is locally topologically equivalent to the normal form

$$\dot{u} = \alpha + \sigma u^2,$$

where $\sigma = \text{sign}\, b = \pm 1$. Under these genericity conditions, Theorem 5.4 implies that (5.12) is locally topologically equivalent to the system

$$\begin{cases} \dot{u} = \alpha + \sigma u^2, \\ \dot{v} = -v. \end{cases} \tag{5.13}$$

Equations (5.13) are decoupled. The resulting phase portraits are presented in **Fig. 5.10** for the case $\sigma > 0$. For $\alpha < 0$, there are two hyperbolic equilibria:

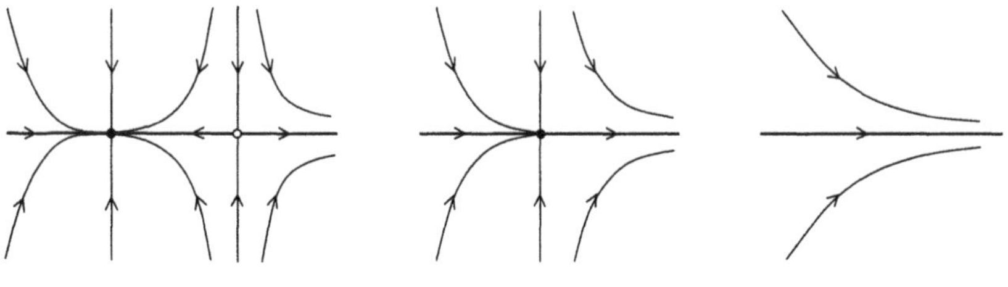

$\alpha < 0$ $\alpha = 0$ $\alpha > 0$

**Fig. 5.10.** Fold bifurcation in the standard system (5.13) for $\sigma = 1$.

a stable node and a saddle. They collide at $\alpha = 0$, forming a nonhyperbolic *saddle-node* point, and disappear. There are no equilibria for $\alpha > 0$. The manifolds $W^c_\alpha$ in (5.13) can be considered as parameter-independent and as given by $v = 0$. Obviously, it is one of the infinite number of choices (see the Remark following Example 5.3). The same events happen in (5.12) on some one-dimensional, parameter-dependent, invariant manifold, that is locally attracting (see **Fig. 5.11**). All the equilibria belong to this manifold. **Figs.**

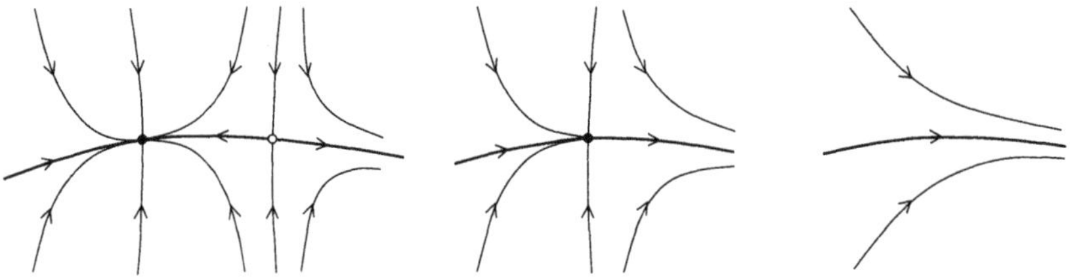


**Fig. 5.11.** Fold bifurcation in a generic planar system.

**5.10** and **5.11** explain why the fold bifurcation is often called the *saddle-node bifurcation*. It should be clear how to generalize these considerations to cover the case $\lambda_2 > 0$, as well as the $n$-dimensional case. $\diamondsuit$

**Example 5.3 (Generic Hopf bifurcation in $\mathbb{R}^3$)** Consider a system

$$\dot{x} = f(x, \alpha), \quad x \in \mathbb{R}^3, \ \alpha \in \mathbb{R}^1. \tag{5.14}$$

Assume that at $\alpha = 0$ it has the equilibrium $x_0 = 0$ with eigenvalues $\lambda_{1,2} = \pm i\omega_0$, $\omega_0 > 0$ and one negative eigenvalue $\lambda_3 < 0$. Lemma 5.1 gives the existence of a parameter-dependent, smooth, local two-dimensional attracting invariant manifold $W^c_\alpha$ of (5.17) for small $|\alpha|$. At $\alpha = 0$ the restricted equation (5.11) can be written in complex form as

$$\dot{z} = i\omega_0 z + g(z, \bar{z}), \quad z \in \mathbb{C}^1.$$

If the Lyapunov coefficient $l_1(0)$ of this equation is nonzero and the restricted equation depends generically on the parameter, then, as proved in Chapter 3, it is locally topologically equivalent to the normal form

$$\dot{z} = (\alpha + i)z + \sigma z^2 \bar{z},$$

where $\sigma = \text{sign } l_1(0) = \pm 1$. Under these genericity conditions, Theorem 5.4 implies that (5.14) is locally topologically equivalent to the system

$$\begin{cases} \dot{z} = (\alpha + i)z + \sigma z^2 \bar{z}, \\ \dot{v} = -v. \end{cases} \tag{5.15}$$

The phase portrait of (5.15) is shown in **Fig. 5.10** for $\sigma = -1$. The supercritical Hopf bifurcation takes place in the invariant plane $v = 0$, which is

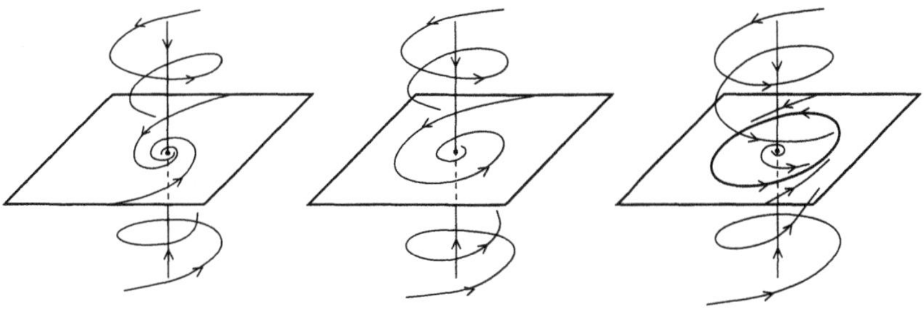


**Fig. 5.12.** Hopf bifurcation in the standard system (5.15) for $\sigma = -1$.

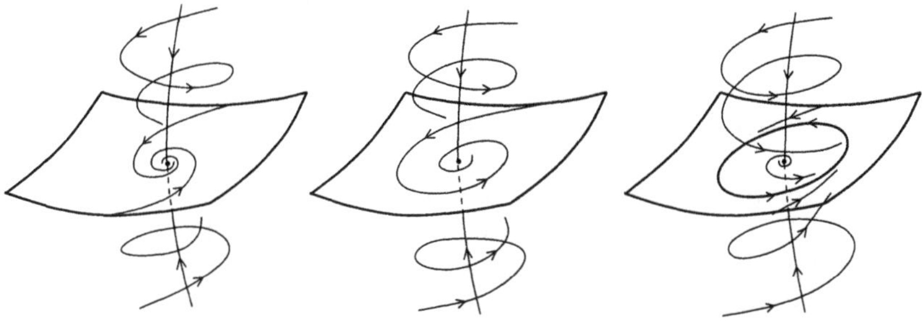


**Fig. 5.13.** Supercritical Hopf bifurcation in a generic three-dimensional system.

attracting. The same events happen for (5.14) on some two-dimensional attracting manifold (see **Fig. 5.13**). The construction allows a generalization to arbitrary dimension $n \geq 3$. ◊

**Remark:**

It should be noted that the manifold $W^c_\alpha$ is *not unique* in either the fold or Hopf cases, but the bifurcating equilibria or cycle belong to any of the center manifolds (cf. Remark (2) after the Center Manifold Theorem in Section 5.1.1). In the fold bifurcation case, the manifold is unique near the saddle and coincides with its unstable manifold as far as it exists. The uniqueness is lost at the stable node. Similarly, in the Hopf bifurcation case, the manifold is unique and coincides with the unstable manifold of the saddle-focus until the stable limit cycle $L_\alpha$, where the uniqueness breaks down. **Fig. 5.14** shows the possible freedom in selecting $W^c_\alpha$ in the Hopf case for $\alpha > 0$ in $(\rho, v)$-coordinates in system (5.15) with $\sigma = -1$. ◊

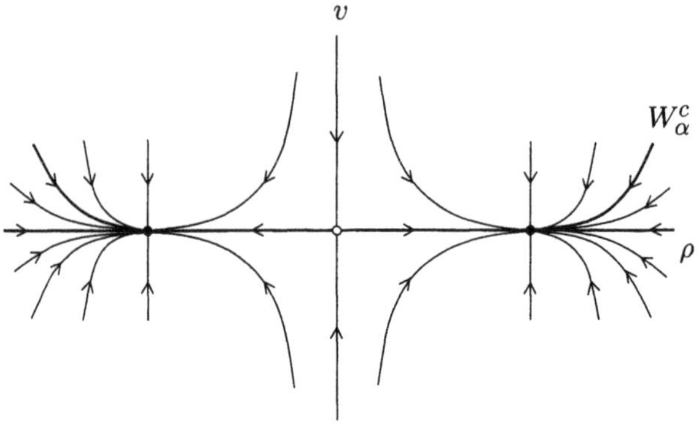


**Fig. 5.14.** Nonuniqueness of the parameter-dependent center manifold near the Hopf bifurcation.

## 5.3 Bifurcations of limit cycles

A combination of the Poincaré map (see Chapter 1) and the center manifold approaches allows us to apply the results of Chapter 4 to limit cycle bifurcations in $n$-dimensional continuous-time systems.

Let $L_0$ be a limit cycle (isolated periodic orbit) of system (5.9) at $\alpha = 0$. Let $P_\alpha$ denote the associated Poincaré map for nearby $\alpha$; $P_\alpha : \Sigma \to \Sigma$, where $\Sigma$ is a local cross-section to $L_0$. If some coordinates $\xi = (\xi_1, \xi_2, \dots, \xi_{n-1})$ are introduced on $\Sigma$, then $\tilde{\xi} = P_\alpha(\xi)$ can be defined to be the point of the next intersection with $\Sigma$ of the orbit of (5.9) having initial point with coordinates $\xi$ on $\Sigma$. The intersection of $\Sigma$ and $L_0$ gives a fixed point $\xi_0$ for $P_0$: $P_0(\xi_0) = \xi_0$. The map $P_\alpha$ is smooth and locally invertible.

Suppose that the cycle $L_0$ is nonhyperbolic, having $n_0$ multipliers on the unit circle. The center manifold theorems then give a parameter-dependent invariant manifold $W^c_\alpha \subset \Sigma$ of $P_\alpha$ on which the "essential" events take place. The Poincaré map $P_\alpha$ is locally topologically equivalent to the suspension of its restriction to this manifold by the standard saddle map. Fix $n = 3$, for simplicity, and consider the implications of this theorem for the limit cycle bifurcations.

### Fold bifurcation of cycles

Assume that at $\alpha = 0$ the cycle has a simple multiplier $\mu_1 = 1$ and its other multiplier satisfies $0 < \mu_2 < 1$. The restriction of $P_\alpha$ to the invariant manifold $W^c_\alpha$ is a one-dimensional map, having a fixed point with $\mu_1 = 1$ at $\alpha = 0$. As has been shown in Chapter 4, this generically implies the collision and disappearance of two fixed points of $P_\alpha$ as $\alpha$ passes through zero. Under our assumption on $\mu_2$, this happens on a one-dimensional attracting invariant manifold of $P_\alpha$; thus, a stable and a saddle fixed point are involved in the bifurcation (see **Fig. 5.15**). Each fixed point of the Poincaré map corresponds

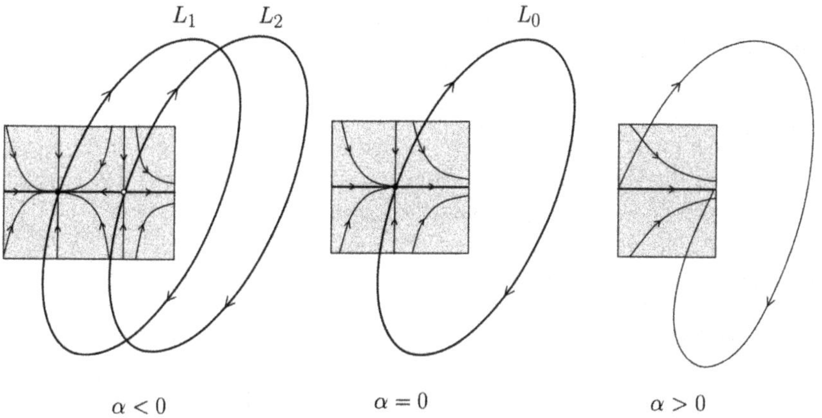


**Fig. 5.15.** Fold bifurcation of limit cycles.

to a limit cycle of the continuous-time system. Therefore, two limit cycles (stable and saddle) collide and disappear in system (5.9) at this bifurcation (see the figure).

## Flip bifurcation of cycles

Suppose that at $\alpha = 0$ the cycle has a simple multiplier $\mu_1 = -1$, while $-1 < \mu_2 < 0$. Then, the restriction of $P_\alpha$ to the invariant manifold will demonstrate

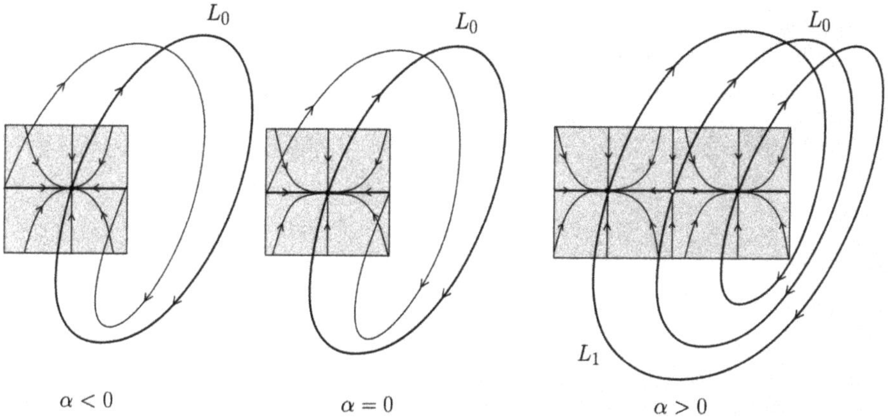


**Fig. 5.16.** Flip bifurcation of limit cycles.

generically the period-doubling (flip) bifurcation: A cycle of period-2 appears for the map, while the fixed point changes its stability (see **Fig. 5.16**). Since the manifold is attracting, the stable fixed point, for example, loses stability and becomes a saddle point, while a stable cycle of period-2 appears. The fixed points correspond to limit cycles of the relevant stability. The cycle of period-two points for the map corresponds to a unique stable limit cycle in (5.9) with *approximately* twice the period of the "basic" cycle $L_0$. The double-period cycle makes two big "excursions" near $L_0$ before the closure. The exact bifurcation scenario is determined by the normal form coefficient of the restricted Poincaré map evaluated at $\alpha = 0$.

### Neimark-Sacker bifurcation of cycles

The last codim 1 bifurcation corresponds to the case when the multipliers are complex and simple and lie on the unit circle: $\mu_{1,2} = e^{\pm i\theta_0}$. The Poincaré map then has a parameter-dependent, two-dimensional, invariant manifold on which a closed invariant curve generically bifurcates from the fixed point (see **Fig. 5.17**). This closed curve corresponds to a two-dimensional *invariant torus* $\mathbb{T}^2$ in (5.9). The bifurcation is determined by the normal form coefficient of the restricted Poincaré map at the critical parameter value. The orbit structure on the torus $\mathbb{T}^2$ is determined by the restriction of the Poincaré map to this closed invariant curve. Thus, generically, there are long-period cycles of different stability types located on the torus (see Chapter 7).

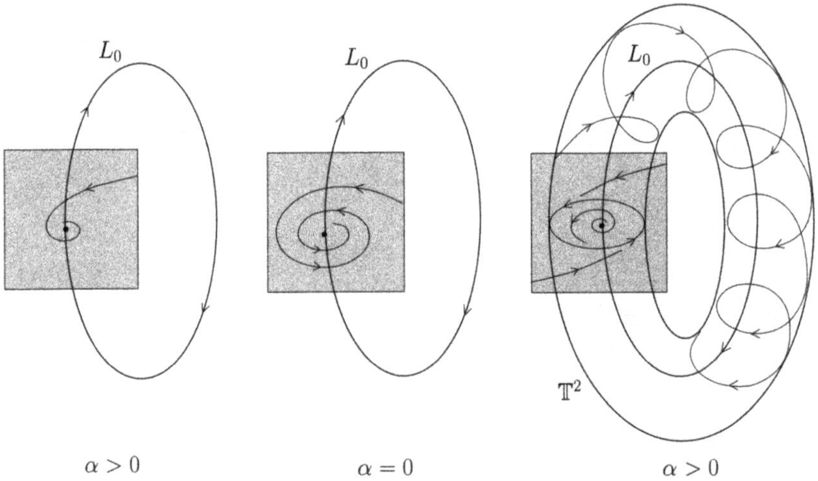


**Fig. 5.17.** Neimark-Sacker bifurcation of a limit cycle.

## 5.4 Computation of center manifolds

As pointed out in the previous sections, the analysis of bifurcations of equilibria and fixed points (and, therefore, limit cycles) in multidimensional systems reduces to that for the differential equations (maps) restricted to the invariant manifold $W^c_\alpha$. Since these bifurcations are determined by the normal form coefficients of the restricted systems at the critical parameter value $\alpha = 0$, we have to be able to compute the center manifold $W^c = W^c_0$ and the ODEs or maps restricted to this manifold up to sufficiently high-order terms.

Unknown coefficients of the Taylor expansion of a function representing the center manifold $W^c$ can be computed via a recursive procedure, each step of which involves solving a linear system of algebraic equations. Even if the center manifold is not unique, the coefficients so obtained are the same for all such manifolds. In the $C^\infty$ case this means that these manifolds can only differ by "flat" functions. Ahead, we derive explicit formulas for the quadratic Taylor coefficients of the center manifolds for all codim 1 bifurcations of equilibria and fixed points. As should now be clear, for these cases $W^c$ is either one- or two-dimensional, $n_0 = 1, 2$. To analyze these bifurcations, we first compute the restricted critical ODEs or maps up to (and including) third-order terms and then find the coefficients of the corresponding normal forms. In Section 8.7 of Chapter 8 and Section 9.7 of Chapter 9 we will use a more advanced technique that combines the computation of the center manifold with the normalization of the restricted equations on this manifold.

In this section we use a *projection method* for center manifold computation, which avoids the transformation of the system into its eigenbasis (i.e., to the form (5.2) or (5.7)). Instead, only eigenvectors corresponding to the critical eigenvalues of $A$ and its transpose $A^T$ are used to "project" the system into the critical eigenspace and its complement. This method is based on the Fredholm Alternative Theorem and can be applied to both continuous- and discrete-time finite-dimensional systems, as well as to some infinite-dimensional systems (see Appendix A), with few modifications.

### 5.4.1 Restricted normalized equations for ODEs

As usual, we start with the continuous-time case. Suppose system (5.1) has the form

$$\dot{x} = Ax + F(x), \quad x \in \mathbb{R}^n, \tag{5.16}$$

where $F(x) = O(\|x\|^2)$ is a smooth function. Write its Taylor expansion near $x = 0$ as

$$F(x) = \frac{1}{2}B(x, x) + \frac{1}{6}C(x, x, x) + O(\|x\|^4), \tag{5.17}$$

where $B(x, y)$ and $C(x, y, z)$ are *multilinear* functions. In coordinates, we have

$$B_i(x, y) = \sum_{j,k=1}^{n} \left.\frac{\partial^2 F_i(\xi)}{\partial \xi_j \partial \xi_k}\right|_{\xi=0} x_j y_k \tag{5.18}$$

and

$$C_i(x, y, z) = \sum_{j,k,l=1}^{n} \left.\frac{\partial^3 F_i(\xi)}{\partial \xi_j \partial \xi_k \partial \xi_l}\right|_{\xi=0} x_j y_k z_l, \tag{5.19}$$

where $i = 1, 2, \dots, n$.

### Fold bifurcation

In this case, $A$ has a simple zero eigenvalue $\lambda_1 = 0$, and the corresponding critical eigenspace $T^c$ is one-dimensional and spanned by an eigenvector $q \in \mathbb{R}^n$ such that $Aq = 0$. Let $p \in \mathbb{R}^n$ be the *adjoint* eigenvector, that is, $A^T p = 0$, where $A^T$ is the transposed matrix.[3] It is possible and convenient to normalize $p$ with respect to $q$ : $\langle p, q \rangle = 1$, where $\langle \cdot, \cdot \rangle$ is the standard scalar product in $\mathbb{R}^n$. The following lemma follows from the Fredholm Alternative Theorem.

**Lemma 5.2** *Let $T^{su}$ denote an $(n-1)$-dimensional linear eigenspace of $A$ corresponding to all eigenvalues other than 0. Then $y \in T^{su}$ if and only if $\langle p, y \rangle = 0$.* $\square$

Using the lemma, we can "decompose" any vector $x \in \mathbb{R}^n$ as

$$x = uq + y,$$

where $uq \in T^c$, $y \in T^{su}$. If $q$ and $p$ are normalized as above, we get explicit expressions for $u$ and $y$:

$$\begin{cases} u = \langle p, x \rangle, \\ y = x - \langle p, x \rangle q. \end{cases} \tag{5.20}$$

Two operators can thus be defined:

$$P_c x = \langle p, x \rangle q, \quad P_{su} x = x - \langle p, x \rangle q.$$

These operators are *projections* onto $T^c$ and $T^{su}$, respectively, and

$$P_c^2 = P_c, \; P_{su}^2 = P_{su}, \; P_c P_{su} = P_{su} P_c = 0.$$

The scalar $u$ and the vector $y$ can be considered as new "coordinates" on $\mathbb{R}^n$. Although $y \in \mathbb{R}^n$, it always satisfies the orthogonality condition $\langle p, y \rangle = 0$. In these new coordinates, system (5.16) can be written as

$$\begin{cases} \dot{u} = \langle p, F(uq + y) \rangle, \\ \dot{y} = Ay + F(uq + y) - \langle p, F(uq + y) \rangle q. \end{cases} \tag{5.21}$$

To obtain these equations, one has to take into account (5.20) and the eigenvector definitions and normalizations. Equivalently, one can apply the above projection operators to system (5.16).

Using the Taylor expansion (5.17) and the equation $Aq = 0$, one can write

$$F(uq + y) = Ay + \frac{1}{2}(u^2 B(q, q) + 2uB(q, y) + B(y, y)) + \frac{1}{6} u^3 C(q, q, q) + \cdots,$$

[3] Recall that $\langle x, Ay \rangle = \langle A^T x, y \rangle$ for any $x, y \in \mathbb{R}^n$.

where the dots represent all undisplayed terms of order three and higher in $(u, y)$. Then (5.21) becomes

$$\begin{cases} \dot{u} = bu^2 + uB(q, y) + ru^3 + \cdots, \\ \dot{y} = Ay + \frac{1}{2}au^2 + \cdots, \end{cases} \tag{5.22}$$

where $u \in \mathbb{R}^1$, $y \in \mathbb{R}^n$ and for $b, r \in \mathbb{R}^1$ and $a \in \mathbb{R}^n$ we have the following invariant expressions:

$$b = \frac{1}{2}\langle p, B(q, q)\rangle, \tag{5.23}$$

$$r = \frac{1}{6}\langle p, C(q, q, q)\rangle, \quad a = B(q, q) - \langle p, B(q, q)\rangle q. \tag{5.24}$$

Actually, in (5.22) only those terms that are essential in the exposition below are displayed.[4]

We seek the second-order term in the Taylor expansion for $y = V(u)$ representing the center manifold:

$$V(u) = \frac{1}{2}w_2 u^2 + O(u^3), \tag{5.25}$$

where $w_2 \in \mathbb{R}^n$ is an unknown vector. Since $V(u) \in T^{su}$ for small $u$, $w_2 \in T^{su}$ implying $\langle p, w_2\rangle = 0$.

Now we proceed as in Example 5.1. The invariance of the center manifold $W^c$ means

$$\dot{y} = V_u(u)\dot{u}, \tag{5.26}$$

where $\dot{y}$ and $\dot{u}$ satisfy (5.22). Substituting the expansion (5.25) into (5.26) and collecting the $u^2$-terms, we see that the vector $w_2$ should satisfy the equation

$$Aw_2 = -a. \tag{5.27}$$

Here, however, we have a problem since $A$ is obviously noninvertible in $\mathbb{R}^n$ ($\lambda = 0$ is its eigenvalue). This difficulty is easy to overcome. Notice that $a \in T^{su}$ since $\langle p, a\rangle = 0$. The restriction of the linear transformation corresponding to $A$ to its invariant subspace $T^{su}$ is invertible. Thus, equation (5.27) has a unique solution $w_2 \in T^{su}$. If we denote this solution by

$$w_2 = -A^{INV}a,$$

the restriction of (5.22) to the center manifold takes the form

$$\dot{u} = bu^2 + cu^3 + O(u^4),$$

where $b$ is defined by (5.23) and

---

[4] For example, $O(\|y\|^2)$ terms in both equations of (5.22) are irrelevant in what follows because they do not affect the quadratic and cubic terms of the restricted equation.

$$c = r - \frac{1}{2}\langle p, B(q, A^{INV}a)\rangle = \frac{1}{6}\langle p, C(q,q,q) - 3B(q, B(q, A^{INV}a))\rangle. \quad (5.28)$$

To check that a fold bifurcation is nondegenerate, we need only to compute $b$ using (5.23). If $b = 0$, the coefficient $c$ must be evaluated (see Section 8.2 in Chapter 8).

In practice, one can compute $w = A^{INV}a$ by solving the $(n+1)$-dimensional *bordered system*

$$\begin{pmatrix} A & q \\ p^T & 0 \end{pmatrix} \begin{pmatrix} w \\ u \end{pmatrix} = \begin{pmatrix} a \\ 0 \end{pmatrix} \quad (5.29)$$

for $w \in \mathbb{R}^n$ and $u \in \mathbb{R}^1$. Here $q$ and $p$ are the above-defined and normalized eigenvectors of $A$ and $A^T$ corresponding to their simple zero eigenvalue, respectively.

**Lemma 5.3** *The $(n+1) \times (n+1)$ matrix of the bordered system* (5.29) *is nonsingular.*

**Proof:** Indeed,

$$\begin{pmatrix} A & q \\ p^T & 0 \end{pmatrix} \begin{pmatrix} \eta \\ \xi \end{pmatrix} = \begin{pmatrix} 0 \\ 0 \end{pmatrix} \quad (5.30)$$

implies $\eta = 0$ and $\xi = 0$, so the null-space of the bordered matrix is trivial. To verify this, suppose that $\eta \in \mathbb{R}^n$ and $\xi \in \mathbb{R}^1$ form a solution to (5.30), i.e.

$$\begin{cases} A\eta + \xi q = 0, \\ \langle p, \eta\rangle = 0. \end{cases}$$

Taking the scalar product of the first equation with $p$, we obtain

$$\langle p, A\eta\rangle + \xi\langle p, q\rangle = 0.$$

However, $\langle p, q\rangle = 1$ and $\langle p, A\eta\rangle = \langle A^T p, \eta\rangle = 0$. Therefore, $\xi = 0$ and the first equation becomes

$$A\eta = 0.$$

This means that $\eta$ is a null-vector of $A$ and, thus, we should have $\eta = \gamma q$ for some $\gamma \in \mathbb{R}^1$. The second equation then gives

$$\gamma\langle p, q\rangle = 0.$$

Since $\langle p, q\rangle = 1$, we infer that $\gamma = 0$, which implies $\eta = 0$. Thus, zero is not an eigenvalue of the bordered matrix and this matrix can be inverted in the ordinary sense. □

Suppose now that $(w, u)^T$ is the unique solution to (5.29) guaranteed by Lemma 5.3. Equivalently,

$$\begin{cases} Aw + uq = a, \\ \langle p, w\rangle = 0. \end{cases}$$

Thus, according to the second equation, $w \in T^{su}$. Taking the scalar product of the first equation with $p$, we now obtain

$$\langle p, Aw\rangle + u\langle p, q\rangle = \langle p, a\rangle.$$

As before, $\langle p, q\rangle = 1$, $\langle p, a\rangle = 0$, and $\langle p, Aw\rangle = \langle A^T p, w\rangle = 0$. Therefore, $u = 0$ and

$$Aw = a.$$

Thus, by definition, $w = A^{INV} a$.

**Remarks:**

(1) We have the following expressions of $b, r, a$ in terms of the directional derivatives of $F$:

$$b = \frac{1}{2}\frac{\partial^2}{\partial u^2}\langle p, F(uq)\rangle\bigg|_{u=0}, \quad r = \frac{1}{6}\frac{\partial^3}{\partial u^3}\langle p, F(uq)\rangle\bigg|_{u=0}, \quad a = \frac{\partial^2}{\partial u^2}F(uq)\bigg|_{u=0} - \sigma q,$$

where, instead of $F$, the right-hand side $f$ of (5.1) can be used.

(2) The choice of normalization for $q$ is irrelevant. Indeed, if the vector $q$ is substituted by $\gamma q$ with some nonzero $\gamma \in \mathbb{R}^1$ but the relative normalization $\langle p, q\rangle = 1$ is preserved, the coefficients of the restricted equation will change, although the equation can easily be scaled back to the original form by the substitution $u \mapsto \frac{1}{\gamma}u$. For the quadratic and cubic terms this can easily be seen from (5.28) and (5.23). ◇

## Hopf bifurcation

In this case, $A$ has a simple pair of complex eigenvalues on the imaginary axis, $\lambda_{1,2} = \pm i\omega_0$, $\omega_0 > 0$, and these eigenvalues are the only eigenvalues with Re $\lambda = 0$. Let $q \in \mathbb{C}^n$ be a *complex* eigenvector corresponding to $\lambda_1$:

$$Aq = i\omega_0 q, \quad A\bar{q} = -i\omega_0 \bar{q}$$

(as in the fold case, its particular normalization is not important). Introduce also the *adjoint* eigenvector $p \in \mathbb{C}^n$ having the properties

$$A^T p = -i\omega_0 p, \quad A^T \bar{p} = i\omega_0 \bar{p},$$

and satisfying the normalization

$$\langle p, q\rangle = 1, \tag{5.31}$$

where $\langle p, q\rangle = \sum_{i=1}^n \bar{p}_i q_i$ is the standard scalar product in $\mathbb{C}^n$ (linear with respect to the second argument). The critical *real* eigenspace $T^c$ corresponding to $\pm i\omega_0$ is now two-dimensional and is spanned by $\{\text{Re } q, \text{Im } q\}$. The real eigenspace $T^{su}$ corresponding to all eigenvalues of $A$ other than $\pm i\omega_0$ is $(n-2)$-dimensional. The following lemma is valid.

**Lemma 5.4** *$y \in T^{su}$ if and only if $\langle p, y\rangle = 0$.* □

Here $y \in \mathbb{R}^n$ is real, while $p \in \mathbb{C}^n$ is complex. Therefore, the condition in the lemma implies two real constraints on $y$ (the real and imaginary parts of $\langle p, y\rangle$ must vanish). As in the previous case, this lemma allows us to decompose any $x \in \mathbb{R}^n$ as

$$x = zq + \bar{z}\bar{q} + y,$$

where $z \in \mathbb{C}^1$, and $zq + \bar{z}\bar{q} \in T^c$, $y \in T^{su}$. The complex variable $z$ is a coordinate on $T^c$. We have

$$\begin{cases} z = \langle p, x\rangle, \\ y = x - \langle p, x\rangle q - \langle \bar{p}, x\rangle \bar{q}. \end{cases} \tag{5.32}$$

(Notice that $\langle p, \bar{q}\rangle = 0$, see Lemma 3.3.) In the coordinates of (5.32), system (5.16) has the form

$$\begin{cases} \dot{z} = i\omega_0 z + \langle p, F(zq + \bar{z}\bar{q} + y)\rangle, \\ \dot{y} = \ \ Ay \ + F(zq + \bar{z}\bar{q} + y) \\ \qquad\qquad - \langle p, F(zq + \bar{z}\bar{q} + y)\rangle q \\ \qquad\qquad - \langle \bar{p}, F(zq + \bar{z}\bar{q} + y)\rangle \bar{q}. \end{cases} \tag{5.33}$$

System (5.33) is $(n+2)$-dimensional, but one has to remember the two real constraints imposed on $y$.

Using (5.17) and the definitions of the eigenvectors, we can rewrite (5.33) in the form

$$\begin{cases} \dot{z} = i\omega_0 z + \frac{1}{2}G_{20}z^2 + G_{11}z\bar{z} + \frac{1}{2}G_{02}\bar{z}^2 + \frac{1}{2}G_{21}z^2\bar{z} \\ \qquad + z\langle p, B(q, y)\rangle + \bar{z}\langle p, B(\bar{q}, y)\rangle + \cdots, \\ \dot{y} = Ay + \frac{1}{2}H_{20}z^2 + H_{11}z\bar{z} + \frac{1}{2}H_{02}\bar{z}^2 + \cdots, \end{cases} \tag{5.34}$$

where $G_{20}, G_{11}, G_{02}, G_{21} \in \mathbb{C}^1$, $H_{ij} \in \mathbb{C}^n$, and the scalar product in $\mathbb{C}^n$ is used. As usual, we display only relevant terms. Complex numbers and vectors involved in (5.34) can be computed by the following formulas:

$$G_{20} = \langle p, B(q, q)\rangle,\ G_{11} = \langle p, B(q, \bar{q})\rangle,\ G_{02} = \langle p, B(\bar{q}, \bar{q})\rangle, \tag{5.35}$$

$$G_{21} = \langle p, C(q, q, \bar{q})\rangle, \tag{5.36}$$

and

$$\begin{cases} H_{20} = B(q, q) - \langle p, B(q, q)\rangle q - \langle \bar{p}, B(q, q)\rangle \bar{q}, \\ H_{11} = B(q, \bar{q}) - \langle p, B(q, \bar{q})\rangle q - \langle \bar{p}, B(q, \bar{q})\rangle \bar{q}. \end{cases} \tag{5.37}$$

Since $y \in \mathbb{R}^n$, we have $\overline{H}_{ij} = H_{ji}$.

The center manifold $W^c$ now has the representation

$$y = V(z, \bar{z}) = \frac{1}{2}w_{20}z^2 + w_{11}z\bar{z} + \frac{1}{2}w_{02}\bar{z}^2 + O(|z|^3),$$

where $\langle p, w_{ij} \rangle = 0$. Collecting the quadratic in $z$ and $\bar{z}$ terms in the invariance condition for $W^c$,

$$\dot{y} = V_z \dot{z} + V_{\bar{z}} \dot{\bar{z}},$$

one can show that the unknown vectors $w_{ij} \in \mathbb{C}^n$ should satisfy the linear equations

$$\begin{cases} (2i\omega_0 I_n - A) w_{20} = H_{20}, \\ \qquad\qquad -A w_{11} = H_{11}, \\ (-2i\omega_0 I_n - A) w_{02} = H_{02}, \end{cases}$$

where $I_n$ is the $n \times n$ unit matrix. These equations have unique solutions $w_{20}, w_{11}$, and $w_{02}$ since the matrices in their left-hand sides are *invertible* in the ordinary sense because $0, \pm 2i\omega_0$ are not eigenvalues of $A$. Thus, this case is even simpler than that of the fold, and the restricted to $W^c$ equation can be written as:

$$\begin{aligned} \dot{z} &= i\omega_0 z + \tfrac{1}{2} G_{20} z^2 + G_{11} z\bar{z} + \tfrac{1}{2} G_{02} \bar{z}^2 \\ &+ \tfrac{1}{2}(G_{21} - 2\langle p, B(q, A^{-1}H_{11})\rangle + \langle p, B(\bar{q}, (2i\omega_0 E - A)^{-1} H_{20})\rangle) z^2 \bar{z} + \cdots, \end{aligned} \tag{5.38}$$

where the scalar product in $\mathbb{C}^n$ is used.

Our next task is to compute the first Lyapunov coeffiecient for the restricted system (5.38) as described in Chapter 3. Substituting of (5.35)–(5.37) into (5.38), taking into account the identities

$$A^{-1} q = \frac{1}{i\omega_0} q, \quad A^{-1}\bar{q} = -\frac{1}{i\omega_0}\bar{q}, \quad (2i\omega_0 I_n - A)^{-1} q = \frac{1}{i\omega_0} q,$$

$$(2i\omega_0 I_n - A)^{-1}\bar{q} = \frac{1}{3i\omega_0}\bar{q},$$

transforms (5.38) into the equation

$$\dot{z} = i\omega_0 z + \frac{1}{2} g_{20} z^2 + g_{11} z\bar{z} + \frac{1}{2} g_{02} \bar{z}^2 + \frac{1}{2} g_{21} z^2 \bar{z} + \cdots,$$

where

$$g_{20} = \langle p, B(q, q)\rangle, \quad g_{11} = \langle p, B(q, \bar{q})\rangle,$$

and

$$\begin{aligned} g_{21} &= \langle p, C(q, q, \bar{q})\rangle \\ &- 2\langle p, B(q, A^{-1}B(q, \bar{q})))\rangle + \langle p, B(\bar{q}, (2i\omega_0 I_n - A)^{-1} B(q, q))\rangle \\ &+ \frac{1}{i\omega_0}\langle p, B(q, q)\rangle \langle p, B(q, \bar{q})\rangle \\ &- \frac{2}{i\omega_0} |\langle p, B(q, \bar{q})\rangle|^2 - \frac{1}{3i\omega_0} |\langle p, B(\bar{q}, \bar{q})\rangle|^2 . \end{aligned}$$

Notice that the terms in the last line are *purely imaginary* while the term in the third line contains the same scalar products as in the product $g_{20}g_{11}$. Thus, the application of formula (3.20) from Chapter 3,

$$l_1(0) = \frac{1}{2\omega_0^2} \operatorname{Re}(ig_{20}g_{11} + \omega_0 g_{21}),$$

gives the following *invariant expression* for the first Lyapunov coefficient:

$$l_1(0) = \frac{1}{2\omega_0} \operatorname{Re}\left[\langle p, C(q,q,\bar{q})\rangle - 2\langle p, B(q, A^{-1}B(q,\bar{q}))\rangle \right.$$
$$\left. + \ \langle p, B(\bar{q}, (2i\omega_0 I_n - A)^{-1}B(q,q))\rangle\right]. \tag{5.39}$$

This formula seems to be the most convenient for analytical treatment of the Hopf bifurcation in $n$-dimensional systems with $n \geq 2$. It does not require a preliminary transformation of the system into its eigenbasis, and it expresses $l_1(0)$ using original linear, quadratic, and cubic terms, assuming that only the critical (ordinary and adjoint) eigenvectors of the Jacobian matrix are known. In Chapter 10 it will be shown how to implement this formula for the numerical evaluation of $l_1(0)$.

**Example 5.4 (Hopf bifurcation in a feedback-control system)** Consider the following nonlinear differential equation depending on positive parameters $(\alpha, \beta)$:

$$\frac{d^3y}{dt^3} + \alpha\frac{d^2y}{dt^2} + \beta\frac{dy}{dt} + y(1-y) = 0,$$

which describes a simple feedback control system of Lur'e type. By introducing $x_1 = y$, $x_2 = \dot{x}_1$, and $x_3 = \dot{x}_2$, we can rewrite the equation as the equivalent third-order system

$$\begin{cases} \dot{x}_1 = x_2, \\ \dot{x}_2 = x_3, \\ \dot{x}_3 = -\alpha x_3 - \beta x_2 - x_1 + x_1^2. \end{cases} \tag{5.40}$$

For all values of $(\alpha, \beta)$, system (5.40) has two equilibria $x^{(0)} = (0,0,0)$ and $x^{(1)} = (1,0,0)$. We will analyze the equilibrium at the origin. The Jacobian matrix of (5.40) evaluated at $x^{(0)}$ has the form

$$\begin{pmatrix} 0 & 1 & 0 \\ 0 & 0 & 1 \\ -1 & -\beta & -\alpha \end{pmatrix}$$

with the characteristic equation

$$\lambda^3 + \alpha\lambda^2 + \beta\lambda + 1 = 0.$$

To find a relation between $\alpha$ and $\beta$ corresponding to the Hopf bifurcation of $x^{(0)}$, substitute $\lambda = i\omega$ into the last equation. This shows that the characteristic polynomial has a pair of purely imaginary roots $\lambda_{1,2} = \pm i\omega,\ \omega > 0$, if

$$\alpha = \alpha_0(\beta) = \frac{1}{\beta},\ \beta > 0. \tag{5.41}$$

It is easy to check that the origin is stable if $\alpha > \alpha_0$ and unstable if $\alpha < \alpha_0$. The transition is caused by a simple pair of complex-conjugate eigenvalues crossing the imaginary axis at $\lambda = \pm i\omega$, where

$$\omega^2 = \beta.$$

The velocity of the crossing is nonzero and the third eigenvalue $\lambda_3$ remains negative for nearby parameter values.[5] Thus, a Hopf bifurcation takes place. In order to analyze the bifurcation (i.e., to determine the direction of the limit cycle bifurcation), we have to compute the first Lyapunov coefficient $l_1(0)$ of the restricted system on the center manifold at the critical parameter values. If $l_1(0) < 0$, the bifurcation is supercritical and a unique stable limit cycle bifurcates from the origin for $\alpha < \alpha_0(\beta)$. As we shall see, this is indeed the case in system (5.40).

Therefore, fix $\alpha$ at its critical value $\alpha_0$ given by (5.41) and leave $\beta$ free to vary. Notice that the elements of the Jacobian matrix are rational functions of $\omega^2$:

$$A = \begin{pmatrix} 0 & 1 & 0 \\ 0 & 0 & 1 \\ -1 & -\omega^2 & -1/\omega^2 \end{pmatrix}.$$

Since the matrix $A$ is not in real canonical form, we will proceed by the projection method.

It is easy to check that the vectors

$$q \sim \begin{pmatrix} 1 \\ i\omega \\ -\omega^2 \end{pmatrix},\quad p \sim \begin{pmatrix} i\omega \\ i\omega^3 - 1 \\ -\omega^2 \end{pmatrix}$$

are eigenvectors of $A$ and $A^T$, respectively, corresponding to the eigenvalues $i\omega$ and $-i\omega$, respectively:

$$Aq = i\omega q,\quad A^T p = -i\omega p.$$

In order to achieve the normalization (5.31) properly, we should scale these vectors. The following scaling, for example, suffices:

$$q = \begin{pmatrix} 1 \\ i\omega \\ -\omega^2 \end{pmatrix},\quad p = \frac{1}{2\omega(\omega^3 + i)} \begin{pmatrix} i\omega \\ i\omega^3 - 1 \\ -\omega^2 \end{pmatrix}.$$

[5] At the critical parameter value (5.41), $\lambda_3 = -\frac{1}{\beta} < 0$.

The linear part of the analysis is now complete.

There is only one nonlinear (quadratic) term in (5.40). Therefore, the bilinear function $B(x,y)$, defined for two vectors $x = (x_1, x_2, x_3)^T \in \mathbb{R}^3$ and $y = (y_1, y_2, y_3)^T \in \mathbb{R}^3$ (see (5.18)), can be expressed as

$$B(x,y) = \begin{pmatrix} 0 \\ 0 \\ 2x_1y_1 \end{pmatrix},$$

while $C(x,y,z) \equiv 0$. Therefore,

$$B(q,q) = B(q,\bar{q}) = \begin{pmatrix} 0 \\ 0 \\ 2q_1^2 \end{pmatrix} = \begin{pmatrix} 0 \\ 0 \\ 2 \end{pmatrix},$$

and solving the corresponding linear systems yields

$$s = A^{-1}B(q,\bar{q}) = \begin{pmatrix} -2 \\ 0 \\ 0 \end{pmatrix}$$

and

$$r = (2i\omega E - A)^{-1}B(q,q) = -\frac{2}{3(1+2i\omega^3)} \begin{pmatrix} 1 \\ 2i\omega \\ -4\omega^2 \end{pmatrix}.$$

Finally, formula (5.39) gives the first Lyapunov coefficient

$$l_1(0) = \frac{1}{2\omega}\,\mathrm{Re}(-4\bar{p}_3q_1s_1 + 2\bar{p}_3\bar{q}_1r_1) = -\frac{\omega^3(1+8\omega^6)}{(1+4\omega^6)(1+\omega^6)}.$$

We can now return to the parameter $\beta$ by making the substitution $\omega^2 = \beta$:

$$l_1(0) = -\frac{(1+8\beta^3)\beta\sqrt{\beta}}{(1+4\beta^3)(1+\beta^3)} < 0.$$

The Lyapunov coefficient is clearly negative for all positive $\beta$. Thus, the Hopf bifurcation is nondegenerate and always supercritical. $\Diamond$

### 5.4.2 Restricted normalized equations for maps

Now we develop the projection technique for the discrete-time case. In this case, we can write system (5.6) as

$$\tilde{x} = Ax + F(x), \quad x \in \mathbb{R}^n, \tag{5.42}$$

where $F(x) = O(\|x\|^2)$ is a smooth function which has the Taylor expansion written in the form (5.17). The following calculations will closely resemble those of the previous sections.

**Fold and flip bifurcations**

Consider the fold and flip cases together. In each case, $A$ has a simple critical eigenvalue (multiplier) $\mu_1 = \pm 1$, and the corresponding critical eigenspace $T^c$ is one-dimensional and spanned by an eigenvector $q \in \mathbb{R}^n$ such that $Aq = \mu_1 q$. Let $p \in \mathbb{R}^n$ be the *adjoint* eigenvector, that is, $A^T p = \mu_1 p$, where $A^T$ is the transposed matrix. Normalize $p$ with respect to $q$ such that $\langle p, q \rangle = 1$. As in the previous section, let $T^{su}$ denote an $(n-1)$-dimensional linear eigenspace of $A$ corresponding to all eigenvalues other than $\mu_1$. Appying Lemma 5.2 to the matrix $(A - \mu_1 E)$ and taking into account that it has common invariant spaces with the matrix $A$, we conclude that $y \in T^{su}$ if and only if $\langle p, y \rangle = 0$.

Now we can "decompose" any vector $x \in \mathbb{R}^n$ as

$$x = uq + y,$$

where $uq \in T^c$, $y \in T^{su}$, and

$$\begin{cases} u = \langle p, x \rangle, \\ y = x - \langle p, x \rangle q. \end{cases}$$

In the coordinates $(u, y)$, the map (5.42) can be written as

$$\begin{cases} \tilde{u} = \mu_1 u + \langle p, F(uq + y) \rangle, \\ \tilde{y} = Ay + F(uq + y) - \langle p, F(uq + y) \rangle q. \end{cases} \tag{5.43}$$

Using Taylor expansion (5.17), we can write (5.43) in the form

$$\begin{cases} \tilde{u} = \mu_1 u + bu^2 + u\langle p, B(q, y) \rangle + ru^3 + \cdots, \\ \tilde{y} = Ay + \frac{1}{2}au^2 + \cdots, \end{cases} \tag{5.44}$$

where $u \in \mathbb{R}^1$, $y \in \mathbb{R}^n$, $b, r \in \mathbb{R}^1$, $a \in \mathbb{R}^n$, and $\langle \cdot, \cdot \rangle$ is the standard scalar product in $\mathbb{R}^n$. Here $b, r$, and $a$ are given by (5.23) and (5.24).

The center manifold of (5.44) has the representation

$$y = V(u) = \frac{1}{2} w_2 u^2 + O(u^3),$$

where $w_2 \in T^{su} \subset \mathbb{R}^n$, so that $\langle p, w_2 \rangle = 0$. The vector $w_2$ satisfies in both the fold and flip cases the equation in $\mathbb{R}^n$

$$(A - I_n) w_2 = -a. \tag{5.45}$$

This equation results from the comparing the coefficients of the $u^2$-terms in the invariance condition for $W^c$,

$$\tilde{y} = V(\tilde{u}),$$

where $\tilde{u}$ and $\tilde{y}$ are given by (5.44).

In the fold case, the matrix $(A - I_n)$ is noninvertible in $\mathbb{R}^n$ since $\mu_1 = 1$ is the eigenvalue of $A$. As in the previous section, $a \in T^{su}$ since $\langle p, a\rangle = 0$. The restriction of the linear transformation corresponding to $(A - I_n)$ to its invariant subspace $T^{su}$ is invertible, so equation (5.45) has a unique solution $w_2 \in T^{su}$. If we denote this solution by

$$w_2 = -(A - I_n)^{INV} a,$$

the restriction of (5.44) to the center manifold takes the form

$$\tilde{u} = u + bu^2 + cu^3 + O(u^4),$$

where the coefficient in front of the $u^2$-terms,

$$b = \frac{1}{2}\langle p, B(q,q)\rangle, \tag{5.46}$$

determines the nondegeneracy of the fold bifurcation. If $b = 0$, the coefficient in front of the $u^3$-term, namely:

$$c = \frac{1}{6}\left(\langle p, C(q,q,q) - 3B(q,(A - I_n)^{INV} a)\rangle\right), \tag{5.47}$$

should be taken into account. As in the continuous-time case, we can compute $w = (A - I_n)^{INV} a$ by solving the following $(n+1)$-dimensional *bordered system*

$$\begin{pmatrix} A - I_n & q \\ p^T & 0 \end{pmatrix} \begin{pmatrix} w \\ u \end{pmatrix} = \begin{pmatrix} a \\ 0 \end{pmatrix}$$

for $w \in \mathbb{R}^n$ and $u \in \mathbb{R}^1$.

In the flip case, the matrix $(A - I_n)$ is invertible in $\mathbb{R}^n$ because $\lambda = 1$ is not an eigenvalue of $A$. Thus, equation (5.45) can be solved directly giving $w_2 = -(A - I_n)^{-1}a$, and the restriction of (5.44) to the center manifold takes the form

$$\tilde{u} = -u + bu^2 + \left(r - \frac{1}{2}\langle p, B(q,(A - I_n)^{-1}a)\rangle\right) u^3 + O(u^4),$$

where $b, r$, and $a$ are again given by (5.23) and (5.24). This restricted map can be simplified further. Using (5.23) and the identity

$$(A - I_n)^{-1} q = -\frac{1}{2} q,$$

we can write the restricted map as

$$\tilde{u} = -u + a_0 u^2 + b_0 u^3 + O(u^4), \tag{5.48}$$

with

$$a_0 = \frac{1}{2}\langle p, B(q,q)\rangle$$

and

$$b_0 = \frac{1}{6}\langle p, C(q,q,q)\rangle - \frac{1}{4}(\langle p, B(q,q)\rangle)^2 - \frac{1}{2}\langle p, B(q,(A-I_n)^{-1}B(q,q))\rangle.$$

It has been shown in Section 4.5 of Chapter 4 that the map (5.48) can be transformed to the normal form

$$\tilde{\xi} = -\xi + c\xi^3 + O(\xi^4),$$

where

$$c = a_0^2 + b_0$$

(see formula (4.16)). Thus, the critical normal form coefficient $c$, that determines the nondegeneracy of the flip bifurcation and allows us to predict the direction of bifurcation of the period-two cycle, is given by the following *invariant formula:*

$$c = \frac{1}{6}\langle p, C(q,q,q)\rangle - \frac{1}{2}\langle p, B(q,(A-I_n)^{-1}B(q,q))\rangle. \tag{5.49}$$

**Neimark-Sacker bifurcation**

In this case, $A$ has a simple pair of complex eigenvalues (multipliers) on the unit circle: $\mu_{1,2} = e^{\pm i\theta_0}, 0 < \theta_0 < \pi$, and these multipliers are the only eigenvalues of $A$ with $|\mu| = 1$. Let $q \in \mathbb{C}^n$ be a *complex* eigenvector corresponding to $\mu_1$:

$$Aq = e^{i\theta_0}q, \quad A\bar{q} = e^{-i\theta_0}\bar{q}.$$

Introduce also the *adjoint* eigenvector $p \in \mathbb{C}^n$ having the properties

$$A^T p = e^{-i\theta_0}p, \quad A^T\bar{p} = e^{i\theta_0}\bar{p},$$

and satisfying the normalization

$$\langle p, q\rangle = 1,$$

where $\langle p, q\rangle = \sum_{i=1}^n \bar{p}_i q_i$ is the standard scalar product in $\mathbb{C}^n$. The critical real eigenspace $T^c$ corresponding to $\mu_{1,2}$ is two-dimensional and is spanned by $\{\mathrm{Re}\ q, \mathrm{Im}\ q\}$. The real eigenspace $T^{su}$ corresponding to all eigenvalues of $A$ other than $\mu_{1,2}$ is $(n-2)$-dimensional. Lemma 5.4 remains valid, i.e., $y \in T^{su}$ if and only if $\langle p, y\rangle = 0$. Notice that $y \in \mathbb{R}^n$ is real, while $p \in \mathbb{C}^n$ is complex. Therefore, the condition $\langle p, y\rangle = 0$ implies two real constraints on $y$. As in the previous sections, decompose $x \in \mathbb{R}^n$ as

$$x = zq + \bar{z}\bar{q} + y,$$

where $z \in \mathbb{C}^1$, and $zq + \bar{z}\bar{q} \in T^c$, $y \in T^{su}$. The complex variable $z$ is a coordinate on $T^c$. We have

$$\begin{cases} z = \langle p, x \rangle, \\ y = x - \langle p, x \rangle q - \langle \bar{p}, x \rangle \bar{q}. \end{cases}$$

In these coordinates, the map (5.42) takes the form

$$\begin{cases} \tilde{z} = e^{i\theta_0} z + \langle p, F(zq + \bar{z}\bar{q} + y) \rangle, \\ \tilde{y} = \ \ Ay \ + F(zq + \bar{z}\bar{q} + y) \\ \qquad\qquad - \langle p, F(zq + \bar{z}\bar{q} + y) \rangle q \\ \qquad\qquad - \langle \bar{p}, F(zq + \bar{z}\bar{q} + y) \rangle \bar{q}. \end{cases} \tag{5.50}$$

System (5.50) is $(n+2)$-dimensional, but we have to remember the two real constraints imposed on $y$. The system can be written in the form

$$\begin{cases} \tilde{z} = e^{i\theta_0} z + \frac{1}{2} G_{20} z^2 + G_{11} z\bar{z} + \frac{1}{2} G_{02} \bar{z}^2 + \frac{1}{2} G_{21} z^2 \bar{z} \\ \qquad + z \langle p, B(q, y) \rangle + \bar{z} \langle p, B(\bar{q}, y) \rangle + \cdots, \\ \tilde{y} = Ay + \frac{1}{2} H_{20} z^2 + H_{11} z\bar{z} + \frac{1}{2} H_{02} \bar{z}^2 + \cdots, \end{cases} \tag{5.51}$$

where $G_{20}, G_{11}, G_{02}, G_{21} \in \mathbb{C}^1$; $H_{ij} = \overline{H}_{ji} \in \mathbb{C}^n$; and the scalar product in $\mathbb{C}^n$ is used. The complex numbers and vectors, involved in (5.51) can be computed by the formulas (5.35)–(5.37).
The center manifold in (5.51) has the representation

$$y = V(z, \bar{z}) = \frac{1}{2} w_{20} z^2 + w_{11} z\bar{z} + \frac{1}{2} w_{02} \bar{z}^2 + O(|z|^3),$$

where $\langle p, w_{ij} \rangle = 0$. The vectors $w_{ij} \in \mathbb{C}^n$ can be found from the linear equations

$$\begin{cases} (e^{2i\theta_0} I_n - A) w_{20} = H_{20}, \\ (I_n - A) w_{11} = H_{11}, \\ (e^{-2i\theta_0} I_n - A) w_{02} = H_{02}. \end{cases}$$

These equations have unique solutions. The matrix $(I_n - A)$ is invertible because 1 is not an eigenvalue of $A$ ($e^{i\theta_0} \neq 1$). If

$$e^{3i\theta_0} \neq 1,$$

the matrices $(e^{\pm 2i\theta_0} I_n - A)$ are also invertible in $\mathbb{C}^n$ because $e^{\pm 2i\theta_0}$ are not eigenvalues of $A$. Thus, generically,[6] the restricted map can be written as

$$\begin{aligned} \tilde{z} = i\omega_0 z &+ \tfrac{1}{2} G_{20} z^2 + G_{11} z\bar{z} + \tfrac{1}{2} G_{02} \bar{z}^2 \\ &+ \tfrac{1}{2}\big(G_{21} + 2\langle p, B(q, (I_n - A)^{-1} H_{11}) \rangle \\ &+ \langle p, B(\bar{q}, (e^{2i\theta_0} I_n - A)^{-1} H_{20}) \rangle\big) z^2 \bar{z} + \cdots . \end{aligned} \tag{5.52}$$

[6] If $e^{3i\theta_0} = 1$, i.e., $e^{2i\theta_0} = e^{-i\theta_0}$, then $w_{20} = (e^{-i\theta_0} I_n - A)^{INV} H_{20}$, $w_{02} = \bar{w}_{20}$, where $INV$ means the inverse in $T^{su}$.

In this generic situation, substituting (5.35)–(5.37) into (5.52), and taking into account the identities

$$(I_n - A)^{-1}q = \frac{1}{1-e^{i\theta_0}}q, \quad (e^{2i\theta_0}I_n - A)^{-1}q = \frac{e^{-i\theta_0}}{e^{i\theta_0}-1}q,$$

and

$$(I_n - A)^{-1}\bar{q} = \frac{1}{1-e^{-i\theta_0}}\bar{q}, \quad (e^{2i\theta_0}I_n - A)^{-1}\bar{q} = \frac{e^{i\theta_0}}{e^{3i\theta_0}-1}\bar{q},$$

transforms (5.52) into the map

$$\tilde{z} = e^{i\theta_0}z + \frac{1}{2}g_{20}z^2 + g_{11}z\bar{z} + \frac{1}{2}g_{02}\bar{z}^2 + \frac{1}{2}g_{21}z^2\bar{z} + \cdots, \tag{5.53}$$

where

$$g_{20} = \langle p, B(q,q)\rangle, \quad g_{11} = \langle p, B(q,\bar{q})\rangle, \quad g_{02} = \langle p, B(\bar{q},\bar{q})\rangle,$$

and

$$\begin{aligned} g_{21} = {} & \langle p, C(q,q,\bar{q})\rangle \\ & + 2\langle p, B(q,(I_n - A)^{-1}B(q,\bar{q}))\rangle + \langle p, B(\bar{q},(e^{2i\theta_0}I_n - A)^{-1}B(q,q))\rangle \\ & + \frac{e^{-i\theta_0}(1-2e^{i\theta_0})}{1-e^{i\theta_0}}\langle p, B(q,q)\rangle\langle p, B(q,\bar{q})\rangle \\ & - \frac{2}{1-e^{-i\theta_0}}\left|\langle p, B(q,\bar{q})\rangle\right|^2 - \frac{e^{i\theta_0}}{e^{3i\theta_0}-1}\left|\langle p, B(\bar{q},\bar{q})\rangle\right|^2. \end{aligned}$$

As shown in Chapter 4, in the absence of strong resonances, i.e.,

$$e^{ik\theta_0} \neq 1, \text{ for } k = 1,2,3,4,$$

the restricted map (5.53) can be transformed into the form

$$\tilde{z} = e^{i\theta_0}z(1 + d_1|z|^2) + O(|z|^4),$$

where the real number $d = \text{Re } d_1$, that determines the direction of bifurcation of a closed invariant curve, can be computed by formula (4.28),

$$d = \text{Re}\left(\frac{e^{-i\theta_0}g_{21}}{2}\right) - \text{Re}\left(\frac{(1-2e^{i\theta_0})e^{-2i\theta_0}}{2(1-e^{i\theta_0})}g_{20}g_{11}\right) - \frac{1}{2}|g_{11}|^2 - \frac{1}{4}|g_{02}|^2.$$

Using this formula with the above-defined coefficients, we obtain the following *invariant expression*

$$\begin{aligned} d = \frac{1}{2}\text{Re}\left\{e^{-i\theta_0}\left[\langle p, C(q,q,\bar{q})\rangle + 2\langle p, B(q,(I_n - A)^{-1}B(q,\bar{q}))\rangle\right.\right. \\ \left.\left. + \ \langle p, B(\bar{q},(e^{2i\theta_0}I_n - A)^{-1}B(q,q))\rangle\right]\right\}. \end{aligned} \tag{5.54}$$

This compact formula allows us to verify the nondegeneracy of the nonlinear terms at a nonresonant Neimark-Sacker bifurcation of $n$-dimensional maps with $n \geq 2$. Note that all the computations can be performed in the original basis.

## 5.5 Exercises

**(1) (Finite smoothness of the center manifold)** Consider the system [Arrowsmith & Place 1990]

$$\begin{cases} \dot{x} = xz - x^3, \\ \dot{y} = y + x^4, \\ \dot{z} = 0. \end{cases}$$

Show that the system has a center manifold given by $y = V(x, z)$, where $V$ is a $C^6$ function in $x$ if $z < \frac{1}{6}$ but only a $C^4$ function in $x$ for $z < \frac{1}{4}$. (*Hint:* Obtain the coefficients $a_j(z)$ of the expansion $V(x, z) = \sum_{j=0} a_j(z)x^j$ and analyze their denominators.)

**(2) (No Neimark-Sacker bifurcation in the Lorenz system)** Prove that the Neimark-Sacker bifurcation of a limit cycle *never occurs* in the Lorenz [1963] system:

$$\begin{cases} \dot{x} = -\sigma x + \sigma y, \\ \dot{y} = -xz + rx - y, \\ \dot{z} = xy - bz, \end{cases} \tag{E.1}$$

where the parameters $(\sigma, r, b)$ are positive. (*Hint:* Use the formula for the multiplier product and the fact that div $f = -(\sigma + b + 1) < 0$, where $f$ is the vector field given by the right-hand side of (E.1).)

**(3)** Prove Lemma 5.2.

**(4)** Verify that

$$P_c^2 = P_c, \; P_{su}^2 = P_{su}, \; P_c P_{su} = P_{su} P_c = 0,$$

where $P_c$ and $P_{su}$ are the projection operators defined after Lemma 5.2 in Section 5.4.

**(5) (Feedback-control model of Moon & Rand [1985])** Show that the origin $(x, y, z) = (0, 0, 0)$ of the system

$$\begin{cases} \dot{x} = y, \\ \dot{y} = -x - xv, \\ \dot{v} = -v + \alpha x^2, \end{cases}$$

is asymptotically stable if $\alpha < 0$ and unstable if $\alpha > 0$.

**(6) (Center manifold in the Lorenz system)**
Compute the second-order approximation to the family of one-dimensional center manifolds of the Lorenz system (E.1) near the origin $(x, y, z) = (0, 0, 0)$ for fixed $(\sigma, b)$ and $r$ close to $r_0 = 1$. Then, calculate the restricted system up to third-order terms in $\xi$. (*Hint:* See Chapter 7.)

**(7) (Hopf bifurcation in the Lorenz system)**
(a) Show that for fixed $b > 0$, $\sigma > b + 1$, and

$$r_1 = \frac{\sigma(\sigma + b + 3)}{\sigma - b - 1}, \tag{E.2}$$

a nontrivial equilibrium of (E.1) exhibits a Hopf bifurcation.

(b) Prove that this Hopf bifurcation is *subcritical* and, therefore, gives rise to a unique saddle limit cycle for $r > r_1$ [Roshchin 1978].

(*Hints:* [Shilnikov, Shilnikov, Turaev & Chua 2001, pp. 877–880]

(i) Write (E.1) as a single third-order equation

$$\dddot{x} + (\sigma + b + 1)\ddot{x} + b(1+\sigma)\dot{x} + b\sigma(1-r)x = \frac{(1+\sigma)\dot{x}^2}{x} + \frac{\dot{x}\ddot{x}}{x} - x^2\dot{x} - \sigma x^3.$$

(ii) Translate the origin to the equilibrium by introducing the new coordinate $\xi = x - x_0$, where $x_0 = \sqrt{b(r-1)}$, thus obtaining the equation

$$\dddot{\xi} + (\sigma + b + 1)\ddot{\xi} + [b(1+\sigma) + x_0^2]\dot{\xi} + [b\sigma(1-r) + 3\sigma x_0^2]\xi = f(\xi, \dot{\xi}, \ddot{\xi}), \tag{E.3}$$

where

$$f(\xi, \dot{\xi}, \ddot{\xi}) = -3\sigma x_0 \xi^2 - 2x_0 \xi\dot{\xi} + \frac{1+\sigma}{x_0}\dot{\xi}^2 + \frac{1}{x_0}\dot{\xi}\ddot{\xi} - \sigma\xi^3 - \xi^2\dot{\xi} - \frac{1+\sigma}{x_0^2}\xi\dot{\xi}^2 - \frac{1}{x_0^2}\xi\dot{\xi}\ddot{\xi} + \cdots$$

and the dots stand for all higher-order terms in $(\xi, \dot{\xi}, \ddot{\xi})$.

(iii) Rewrite (E.3) as a system

$$\dot{U} = AU + F(U), \quad U = (\xi, \dot{\xi}, \ddot{\xi})^T \in \mathbb{R}^3. \tag{E.4}$$

Find the eigenvector and the adjoint eigenvector of $A$ corresponding to its purely imaginary eigenvalues (when (E.2) is satisfied).

(iv) Compute the first Lyapunov coefficient $l_1$ for (E.4) using (5.39). Substitute $\sigma = \sigma^* + b + 1$ and show that $l_1$ is positive for all positive $\sigma^*$ and $b$. Symbolic manipulation software is useful here, but not necessary.)

**(8)** Prove that the origin $(x, y) = (0, 0)$ is a stable fixed point of the planar map

$$\begin{pmatrix} x \\ y \end{pmatrix} \mapsto \begin{pmatrix} 0 & 1 \\ -\frac{1}{2} & \frac{3}{2} \end{pmatrix} \begin{pmatrix} x \\ y \end{pmatrix} - \begin{pmatrix} 0 \\ y^3 \end{pmatrix}$$

using: (a) transformation to its eigenbasis (*Hint:* See Wiggins [1990, pp. 207–209]); (b) the projection technique from Section 5.4 in this chapter (*Hint:* Do not forget that the matrix $(A - E)$ is noninvertible in this case.)

**(9) (Hénon map revisited)** Consider the Hénon [1976] map

$$\begin{pmatrix} x \\ y \end{pmatrix} \mapsto \begin{pmatrix} y \\ \alpha - \beta x - y^2 \end{pmatrix}.$$

Prove that fold and flip bifurcations of its fixed points (see Example 4.1 in Chapter 4) are nondegenerate for $\beta \neq \pm 1$.

**(10) (Adaptive control system of Golden & Ydstie [1988])**

(a) Demonstrate that the fixed point $(x_0, y_0, z_0) = (1, 1, 1 - b - k)$ of the discrete-time dynamical system

$$\begin{pmatrix} x \\ y \\ z \end{pmatrix} \mapsto \begin{pmatrix} y \\ bx + k + yz \\ z - \dfrac{ky}{c+y^2}(bx + k + zy - 1) \end{pmatrix}$$

exhibits the flip bifurcation at

$$b_F = 1 - \left[\frac{1}{2} + \frac{1}{4(c+1)}\right] k,$$

and the Neimark-Sacker bifurcation at

$$b_{NS} = -\frac{c+1}{c+2}.$$

(b) Determine the direction of the period-doubling bifurcation that occurs as $b$ increases and passes through $b_F$.

(c) Compute the normal form coefficient and show that the Neimark-Sacker bifurcation in the system under variation of the parameter $b$ can be either sub- or supercritical depending on the values of $(c, k)$.

**(11) (Hopf bifurcation in Brusselator; read Appendix A first)** The Brusselator on the unit interval is a reaction-diffusion system with two components

$$\begin{cases} \dfrac{\partial X}{\partial t} = d\,\dfrac{\partial^2 X}{\partial r^2} + C - (B+1)X + X^2 Y, \\ \dfrac{\partial Y}{\partial t} = \theta d\,\dfrac{\partial^2 Y}{\partial r^2} + BX - X^2 Y, \end{cases}$$

where $X = X(r,t),\ Y = Y(r,t);\ r \in [0,1];\ t \geq 0;\ A, B, d, \theta > 0$ (see Chapter 1 and Lefever & Prigogine [1968]). Consider the case when $X$ and $Y$ are kept constant at their equilibrium values at the end points:

$$X(0,t) = X(1,t) = C,\ Y(0,t) = Y(1,t) = \frac{B}{C}.$$

Fix

$$C_0 = 1,\ \delta_0 = 2,\ \theta_0 = \frac{1}{2},$$

and show that at

$$B_0 = 1 + C_0^2 + \delta_0(1 + \theta_0) = 5$$

the system exhibits a supercritical Hopf bifurcation giving rise to a stable limit cycle (periodic standing wave). (*Hint:* See Auchmuty & Nicolis [1976] and Hassard et al. [1981].)

**(12) (Hopf transversality condition)** Let $A(\alpha)$ be a parameter-dependent real $(n \times n)$-matrix which has a simple pair of complex eigenvalues $\lambda_{1,2}(\alpha) = \mu(\alpha) \pm i\omega(\alpha),\ \mu(0) = 0,\ \omega(0) > 0$. Prove that

$$\mu'(0) = \text{Re}\ \langle p, A'(0)q \rangle,$$

where $q, p \in \mathbb{C}^n$ satisfy

$$A(0)q = i\omega(0)q,\ A^T(0)p = -i\omega(0)p,\ \langle p, q \rangle = 1.$$

(*Hint:* Differentiate the equation $A(\alpha)q(\alpha) = \lambda_1(\alpha)q(\alpha)$ with respect to $\alpha$ at $\alpha = 0$, and then compute the scalar product of both sides of the resulting equation with $p$.)

## 5.6 Appendix A: Hopf bifurcation in reaction-diffusion systems

Consider a reaction-diffusion system

$$\frac{\partial u}{\partial t} = D\frac{\partial^2 u}{\partial \xi^2} + A(\alpha)u + F(u, \alpha), \tag{A.1}$$

where $u = u(\xi, t)$ is a vector-valued function describing the distribution of $n$ reacting components over a one-dimensional space, $\xi \in [0, \pi]$, at time $t \geq 0$. $D$ is a positive diagonal matrix, $A(\alpha)$ is a parameter-dependent matrix, and $F = O(\|u\|^2)$ is a smooth function depending on a single parameter $\alpha$. Let us assume that $u$ satisfies Dirichlet boundary conditions:

$$u(0, t) = u(\pi, t) = 0 \tag{A.2}$$

for all $t \geq 0$. As we have seen in Chapter 1, system (A.1), (A.2) defines an infinite-dimensional dynamical system $\{\mathbb{R}^1_+, H, \varphi^t_\alpha\}$ on several function spaces $H$. We can take $H$, for example, to be the *completion* of the space $C^2_0([0, \pi], \mathbb{C}^n)$ of twice continuously differentiable, complex-valued, vector functions on the interval $[0, \pi]$ vanishing at $\xi = 0, \pi$, with respect to the norm $\|w\| = \langle w, w\rangle^{1/2}$. Here $\langle \cdot, \cdot \rangle$ is the scalar product defined on functions from $C^2_0([0, \pi], \mathbb{C}^n)$ by

$$\langle w, v\rangle = \frac{1}{\mu_0}\sum_{i=1}^{n}\int_0^{\pi}\left(\bar{w}_i v_i + \frac{d\bar{w}_i}{d\xi}\frac{dv_i}{d\xi} + \frac{d^2\bar{w}_i}{d\xi^2}\frac{d^2 v_i}{d\xi^2}\right) d\xi, \tag{A.3}$$

where $\mu_0 > 0$ is a constant to be specified later. By continuity, this scalar product can be defined for all $u, v \in H$. Thus the introduced space $H$ is a Hilbert space.[7] Of course, (A.1), (A.2) also define a dynamical system on the real subspace of $H$. Actually, any orbit of $\{\mathbb{R}^1_+, H, \varphi^t_\alpha\}$ with initial point at $u_0 \in H$ belongs to $C^2_0([0, \pi], \mathbb{C}^n)$ for $t > 0$, and

$$u(\xi, t) = (\varphi^t_\alpha u)(\xi)$$

is a classical solution to (A.1), (A.2) for $t > 0$.

Obviously, $u_0(\xi) \equiv 0$ is a stationary solution of (A.1). The linear part of (A.1) defines the linearized operator

$$M_\alpha v = D\frac{d^2 v}{d\xi^2} + A(\alpha)v, \tag{A.4}$$

which can be extended to a closed operator $M_\alpha$ in $H$. An operator defined by

$$M^*_\alpha u = D\frac{d^2 u}{dx^2} + A^T(\alpha)u \tag{A.5}$$

can be extended to be a closed *adjoint* operator $M^*_\alpha$ in $H$ with the characteristic property

$$\langle u, M_\alpha v\rangle = \langle M^*_\alpha u, v\rangle$$

whenever both sides are defined. An *eigenvalue* $\lambda_k$ of $M_\alpha$ is a complex number such that $M_\alpha \psi_k = \lambda_k \psi_k$ for some *eigenfunction* $\psi_k$. Equivalently, the eigenvalues and eigenfunctions satisfy the following linear boundary-value *spectral problem*:

[7] Its elements are continuous vector-valued functions defined on the interval.

$$D\frac{d^2\psi_k}{d\xi^2} + A(\alpha)\psi_k = \lambda_k\psi_k,$$
$$\psi_k(0) = \psi_k(\pi) = 0.$$

The *spectrum* of $M_\alpha$ consists entirely of eigenvalues. There is a countable number of eigenvalues. Any eigenfunction in this case has the form

$$\psi_k(\xi) = V_k \sin k\xi$$

for some integer $k \geq 1$. The vector $V_k \in \mathbb{C}^n$ satisfies

$$(-k^2 D + A(\alpha))V_k = \lambda_k V_k.$$

Suppose that at $\alpha = 0$ the operator $M_\alpha$ has a pair of imaginary eigenvalues $\pm i\omega_0$ and all its other eigenvalues lie strictly in the left half-plane of $\mathbb{C}^1$. Assume that the eigenvalues on the imaginary axis correspond to $k = k_0$, and that $i\omega_0$ is a simple eigenvalue of $(-k_0^2 D + A(0))$. The critical eigenspace $T^c \subset H$ of $M_0$ is spanned by the real and imaginary parts of the complex function

$$q(\xi) = V \sin k_0\xi,$$

where $V \in \mathbb{C}^n$ is the eigenvector corresponding to $i\omega_0$.

For systems of the considered class, and, actually, for more general infinite-dimensional systems, the Center Manifold Theorem remains valid (see the bibliographical notes in Appendix B). Under the formulated assumptions, there is a local two-dimensional invariant manifold $W_\alpha^c \subset H$ of the system $\{\mathbb{R}_+^1, H, \varphi_\alpha^t\}$ defined by (A.1), (A.2) which depends on the parameter $\alpha$. The manifold is locally attracting in terms of the norm of $H$ and is tangent to $T^c$ at $\alpha = 0$. Moreover, the manifold is composed of twice continuously differentiable real functions. The restriction of the system onto the manifold $W_\alpha^c$ is given by a smooth system of two ordinary differential equations that depend on $\alpha$. Thus, the restricted system generically exhibits the Hopf bifurcation at $\alpha = 0$, and a unique limit cycle appears for nearby parameter values. The bifurcation is determined by the first Lyapunov coefficient. If the bifurcation is supercritical, the cycle that appears is stable within $W_\alpha^c$ and, therefore, it is stable as a periodic orbit of (A.1), (A.2) in the $H$-norm. This cycle describes a spatially inhomogeneous, time-periodic solution to the reaction-diffusion system. Solutions of this type are sometimes called *spatial-temporal dissipative structures* (*standing waves*).

It is remarkable that formula (5.39) for the first Lyapunov coefficient derived in the finite-dimensional case can be applied almost verbatim in the considered infinite-dimensional case. Fist of all, we need the adjoint eigenfunction $p : M_0^* p = -i\omega_0 p$. It is given by

$$p(\xi) = W \sin k_0\xi,$$

where $(-k_0^2 D + A^T(0))W = -i\omega_0 W,\ W \in \mathbb{C}^n$. We are free to choose $V$ and $W$ such that

$$\langle p, q\rangle = 1. \tag{A.6}$$

There is a useful but simple property of the scalar product (A.3) involving an (adjoint) eigenfunction; namely, such a scalar product in $H$ is proportional to the corresponding scalar product in $L_2$:

$$\langle p, u\rangle = \frac{1}{\mu_0}(1 + k_0^2 + k_0^4)\langle p, u\rangle_{L_2},$$

where

$$\langle p, u\rangle_{L_2} = \sum_{i=1}^{n} \int_0^{\pi} \bar{p}_i(\xi)u_i(\xi)\, d\xi.$$

Therefore, if we assume

$$\mu_0 = 1 + k_0^2 + k_0^4,$$

all the scalar products can be computed in $L_2$. The normalization (A.6) implies

$$\langle W, V\rangle_{\mathbb{C}^n} = \frac{2}{\pi}. \tag{A.7}$$

Now we can decompose any function $u \in H$ (and, in particular, smooth functions corresponding to solutions of (A.1), (A.2)) as in Section 5.4 since an analog of Lemma 5.4 is valid for the operator $M_0$:

$$u = zq + \bar{z}\bar{q} + v,$$

where $z \in \mathbb{C}^1, v \in H, \langle p, v\rangle = 0$. Hence, we can write (A.1), (A.2) in the form (5.33):

$$\begin{cases} \dot{z} = i\omega_0 z + \langle p, F(zq + \bar{z}\bar{q} + v, 0)\rangle, \\ v_t = M_0 v + F(zq + \bar{z}\bar{q} + v, 0) \\ \qquad - \langle p, F(zq + \bar{z}\bar{q} + v, 0)\rangle q \\ \qquad - \langle \bar{p}, F(zq + \bar{z}\bar{q} + v, 0)\rangle \bar{q}. \end{cases} \tag{A.8}$$

Exactly as in Section 5.4, one can show that the behavior of (A.8) on its two-dimensional center manifold $W_0^c$ is determined by the sign of

$$l_1(0) = \frac{1}{2\omega_0}\mathrm{Re}\left[\langle p, C(q, q, \bar{q})\rangle - 2\langle p, B(q, w_{20})\rangle + \langle p, B(\bar{q}, w_{11})\rangle\right], \tag{A.9}$$

where $q$ and $p$ are the above defined eigenfunctions of $M_0$ and $M_0^*$, the multilinear functions $B(\cdot,\cdot)$ and $C(\cdot,\cdot,\cdot)$ are defined by the Taylor expansion of $F$ at $u = 0$ (see (5.17)), while functions $w_{11}$ and $w_{20}$ are the unique solutions to the linear boundary-value problems

$$\begin{cases} (2i\omega_0 E - M_0)w_{20} = B(q, q), \\ -M_0 w_{11} = B(q, \bar{q}). \end{cases} \tag{A.10}$$

Of course, the scalar products $\langle\cdot,\cdot\rangle$ in (A.9) are defined by (A.3).

The boundary-value problems (A.10) can be written in a more "classical" way:

$$D\frac{d^2 w_{20}}{d\xi^2}(\xi) + [A(0) - 2i\omega_0 E]w_{20}(\xi) = -B(q(\xi), q(\xi)),$$

$$D\frac{d^2 w_{11}}{d\xi^2}(\xi) + A(0)w_{11}(\xi) = -B(q(\xi), \bar{q}(\xi)).$$

Here $H_{ij}$ vanish at $\xi = 0, \pi$. Notice that all the functions on the right-hand side have the same spatial dependence $\sin^2(k_0\xi)$. Their exact form is determined by the critical eigenfunctions and the quadratic terms of $F$. The boundary-value problems can be solved, for example, by the Fourier method.

## 5.7 Appendix B: Bibliographical notes

The Center Manifold Theorem in finite dimensions has been proved by Pliss [1964] for the attracting case $n_+ = 0$ and by Kelley [1967] and Hirsch et al. [1977] in general. In the Russian literature the center manifold is often called the *neutral* manifold. Proofs of the Center Manifold Theorem can be found in Carr [1981] and Vanderbauwhede [1989]. A very detail proof based on boundary-value problems is given by Shilnikov, Shilnikov, Turaev & Chua [1998, Ch.5]. Proofs of the Reduction Principles (Theorems 5.2 and 5.3) can be found in [Kirchgraber & Palmer 1990]. Topological normal forms for multidimensional bifurcations of equilibria and limit cycles are based on the works by Shoshitaishvili[1972, 1975], where the topological versality of the suspended system is also established, given that of the restricted system. The first example showing that a $C^\infty$ system may have no $C^\infty$ center manifold was constructed by van Strien [1979].

One-parameter bifurcations of limit cycles and corresponding metamorphoses of the local phase portraits in $n$-dimensional systems were known to mathematicians of Andronov's school since the late 1940s. Their detailed presentation can be found in Neimark [1972] and Butenin, Neimark & Fufaev [1976]. A systematic modern presentation of these and other bifurcations is given in [Shilnikov et al. 2001].

The existence of center manifolds for several classes of partial differential equations and delay differential equations has been established in 1980s. The main technical steps of such proofs are to show that the original system can be formulated as an abstract ordinary differential equation on an appropriate (i.e., Banach or Hilbert) function space $H$, and to use the variation-of-constants formula (Duhamel's integral equation) to prove that this equation defines a smooth dynamical system (*semiflow*) on $H$. For such flows, a general theorem is valid that guarantees existence of a center manifold under certain conditions on the linearized operator. See Marsden & McCracken [1976], Carr [1981], Henry [1981], Hale [1977], and Diekmann et al. [1995], for details and examples.

The projection technique, which avoids putting the linear part in normal form, was originally developed to study bifurcations in some partial differential equations (mainly from hydrodynamics) using the Lyapunov-Schmidt reduction. Our presentation is based on the book by Hassard et al. [1981], where the Hopf bifurcation in continuous-time (finite- and infinite-dimensional) systems is treated. An invariant expression equivalent to (5.39) for the first Lyapunov coefficient was derived by Howard and Kopell in their comments to the translation of the original Hopf paper in Marsden & McCracken [1976] using asymptotic expansions for the bifurcating periodic solution (see also Iooss & Joseph [1980]). Independently, it was obtained by van Gils [1982] and published by Diekmann & van Gils [1984] within an infinite-dimensional context. Computational formulas for the discrete-time flip case were given by Kuznetsov & Rinaldi [1991]. The formula (5.54) to analyze the Neimark-Sacker bifurcation was first derived by Iooss, Arneodo, Coullet & Tresser [1981] using asymptotic expansions for the bifurcating invariant closed curve.

# 6

# Bifurcations of Orbits Homoclinic and Heteroclinic to Hyperbolic Equilibria

In this chapter we will study global bifurcations corresponding to the appearance of *homoclinic* or *heteroclinic* orbits connecting hyperbolic equilibria in continuous-time dynamical systems. First we consider in detail two- and three-dimensional cases where geometrical intuition can be fully exploited. Then we show how to reduce generic $n$-dimensional cases to the considered ones plus a four-dimensional case treated in Appendix A.

## 6.1 Homoclinic and heteroclinic orbits

Consider a continuous-time dynamical system $\{\mathbb{R}^1, \mathbb{R}^n, \varphi^t\}$ defined by a system of ODEs

$$\dot{x} = f(x), \quad x = (x_1, x_2, \dots, x_n)^T \in \mathbb{R}^n, \tag{6.1}$$

where $f$ is smooth. Let $x_0, x_{(1)}$, and $x_{(2)}$ be equilibria of the system.

**Definition 6.1** *An orbit* $\Gamma_0$ *starting at a point* $x \in \mathbb{R}^n$ *is called* homoclinic *to the equilibrium point* $x_0$ *of system* (6.1) *if* $\varphi^t x \to x_0$ *as* $t \to \pm\infty$.

**Definition 6.2** *An orbit* $\Gamma_0$ *starting at a point* $x \in \mathbb{R}^n$ *is called* heteroclinic *to the equilibrium points* $x_{(1)}$ *and* $x_{(2)}$ *of system* (6.1) *if* $\varphi^t x \to x_{(1)}$ *as* $t \to -\infty$ *and* $\varphi^t x \to x_{(2)}$ *as* $t \to +\infty$.

**Fig. 6.1** shows examples of homoclinic and heteroclinic orbits to saddle points if $n = 2$, while **Fig. 6.2** presents relevant examples for $n = 3$.

It is clear that a homoclinic orbit $\Gamma_0$ to the equilibrium $x_0$ belongs to the intersection of its unstable and stable sets: $\Gamma_0 \subset W^u(x_0) \cap W^s(x_0)$. Similarly, a heteroclinic orbit $\Gamma_0$ to the equilibria $x_{(1)}$ and $x_{(2)}$ satisfies $\Gamma_0 \subset W^u(x_{(1)}) \cap W^s(x_{(2)})$. It should be noticed that the Definitions 6.1 and 6.2 do not require the equilibria to be hyperbolic. **Fig. 6.3** shows, for example, a homoclinic orbit to a saddle-node point with an eigenvalue $\lambda_1 = 0$. Actually, orbits homoclinic to hyperbolic equilibria are of particular interest since their presence results in structural instability while the equilibria themselves are structurally stable.

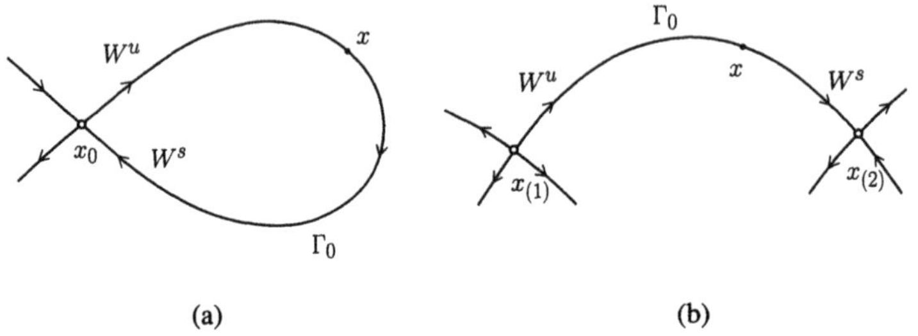


**Fig. 6.1.** (a) Homoclinic and (b) heteroclinic orbits on the plane.

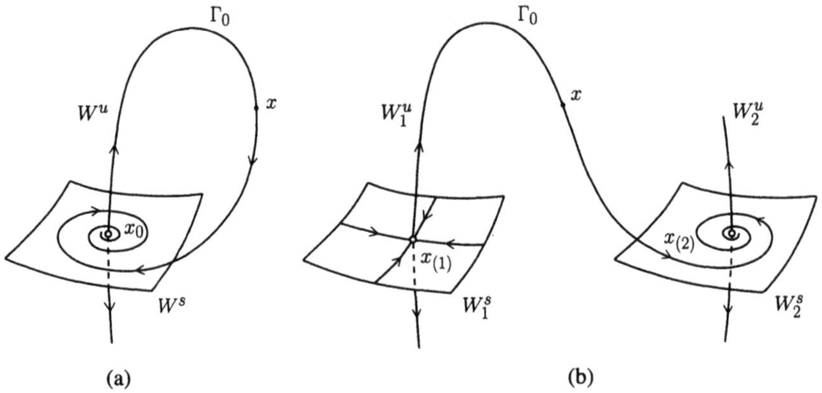


**Fig. 6.2.** (a) Homoclinic and (b) heteroclinic orbits in three-dimensional space.

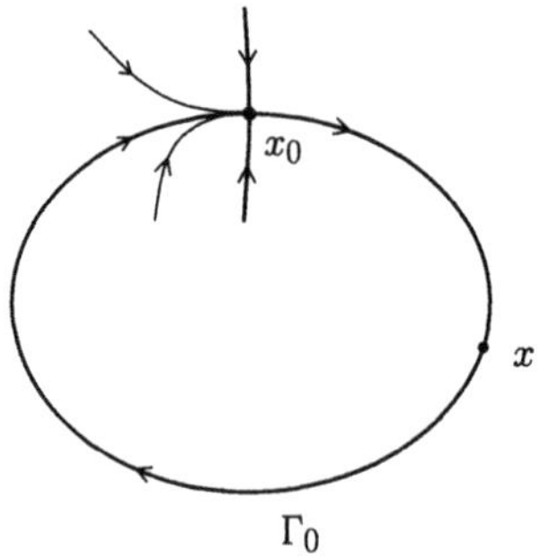


**Fig. 6.3.** Homoclinic orbit $\Gamma_0$ to a saddle-node equilibrium.

**Lemma 6.1** *A homoclinic orbit to a hyperbolic equilibrium of* (6.1) *is structurally unstable.* □

This lemma means that we can perturb a system with an orbit $\Gamma_0$ that is homoclinic to $x_0$ such that the phase portrait in a neighborhood of $\Gamma_0 \cup x_0$ becomes topologically nonequivalent to the original one. As we shall see, the homoclinic orbit simply disappears for generic $C^1$ perturbations of the system. This is a bifurcation of the phase portrait.

To prove the lemma, we need a small portion of transversality theory.

**Definition 6.3** *Two smooth manifolds $M, N \subset \mathbb{R}^n$ intersect* transversally *if there exist $n$ linearly independent vectors that are tangent to at least one of these manifolds at any intersection point.*

For example, a surface and a curve intersecting with a nonzero angle at some point in $\mathbb{R}^3$ are transversal. The main property of transversal intersection is that it persists under small $C^1$ perturbations of the manifolds. In other words, if manifolds $M$ and $N$ intersect transversally, so will all sufficiently $C^1$-close manifolds. Conversely, if the manifolds intersect nontransversally, generic perturbations make them either nonintersecting or transversally intersecting.

Since in this chapter we deal exclusively with saddle (or saddle-focus) hyperbolic equilibria, the sets $W^u$ and $W^s$ are smooth (immersed) invariant manifolds.[1] Any sufficiently $C^1$-close system has a nearby saddle point, and its invariant manifolds $W^{u,s}$ are $C^1$-close to the corresponding original ones in a neighborhood of the saddle.

**Proof of Lemma 6.1:**

Suppose that system (6.1) has a hyperbolic equilibrium $x_0$ with $n_+$ eigenvalues having positive real parts and $n_-$ eigenvalues having negative real parts, $n_\pm > 0$, $n_+ + n_- = n$. Assume that the corresponding stable and unstable manifolds $W^u(x_0)$ and $W^s(x_0)$ intersect along a homoclinic orbit. To prove the lemma, we shall show that the intersection cannot be transversal. Indeed, at any point $x$ of this orbit, the vector $f(x)$ is tangent to both manifolds $W^u(x_0)$ and $W^s(x_0)$. Therefore, we can find no more than $n_+ + n_- - 1 = n - 1$ independent tangent vectors to these manifolds since $\dim W^u = n_+$, $\dim W^s = n_-$. Moreover, any generic perturbation of (6.1) splits the manifolds in that remaining direction and they do not intersect anymore near $\Gamma_0$. $\square$

Let us characterize the behavior of the stable and unstable manifolds near homoclinic bifurcations in two- and three-dimensional systems in more detail.

*Case $n = 2$.* Consider a planar system having a homoclinic orbit to a saddle $x_0$, as shown in the central part of **Fig. 6.4**. Introduce a one-dimensional local cross-section $\Sigma$ to the stable manifold $W^s(x_0)$ near the saddle, as shown in the figure. Select a coordinate $\xi$ along $\Sigma$ such that the point of the intersection with the stable manifold corresponds to $\xi = 0$. This construction can be carried out for all sufficiently close systems. For such systems, however, the unstable manifold $W^u$ generically does not return to the saddle. **Fig. 6.4** illustrates the two possibilities: The manifolds split either "down" or "up." Denote by $\xi^u$ the $\xi$-value of the intersection of $W^u$ with $\Sigma$.

[1] Meanwhile, the manifolds $W^u(x_0)$ and $W^s(x_0)$ intersect transversally at $x_0$.

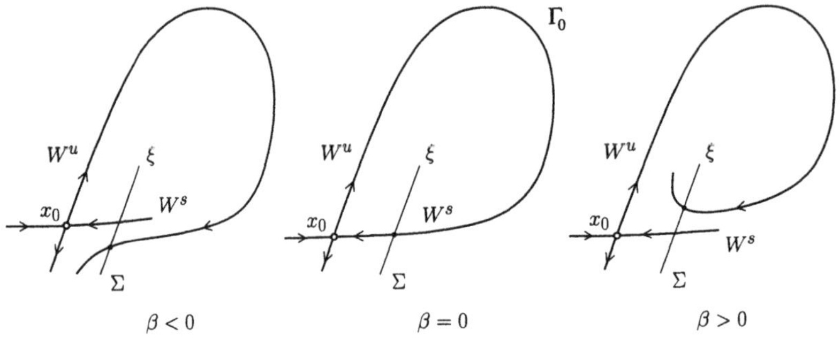


**Fig. 6.4.** Split function in the planar case ($n = 2$).

**Definition 6.4** *The scalar* $\beta = \xi^u$ *is called a* split function.

Actually, the split function is a *functional* defined on the original and perturbed systems. It becomes a smooth function of parameters for a parameter-dependent system. The equation

$$\beta = 0$$

is a bifurcation condition for the homoclinic bifurcation in $\mathbb{R}^2$. Thus, the homoclinic bifurcation in this case has codimension *one*.

**Remark:**

There is a constructive proof of Lemma 6.1 in the planar case due to Andronov. A one-parameter perturbation destroying the homoclinic (heteroclinic) orbit can be constructed explicitly. For example, if a system

$$\begin{cases} \dot{x}_1 = f_1(x_1, x_2), \\ \dot{x}_2 = f_2(x_1, x_2), \end{cases} \tag{6.2}$$

has a homoclinic orbit to a saddle, then the system

$$\begin{cases} \dot{x}_1 = f_1(x_1, x_2) - \alpha f_2(x_1, x_2), \\ \dot{x}_2 = \alpha f_1(x_1, x_2) + f_2(x_1, x_2), \end{cases} \tag{6.3}$$

has no nearby homoclinic orbits to this saddle for any sufficiently small $|\alpha| \neq 0$. System (6.3) is obtained from (6.2) by a *rotation of the vector field.* The proof is left as an exercise to the reader. ◇

*Case* $n = 3$. It is also possible to define a split function in this case. Consider a system in $\mathbb{R}^3$ with a homoclinic orbit $\Gamma_0$ to a saddle $x_0$. Assume that $\dim W^u = 1$ (otherwise, reverse the time direction), and introduce a two-dimensional cross-section $\Sigma$ with coordinates $(\xi, \eta)$ as in **Fig. 6.5**. Suppose that $\xi = 0$ corresponds to the intersection of $\Sigma$ with the stable manifold $W^s$ of $x_0$. As before, this can be done for all sufficiently close systems. Let the point

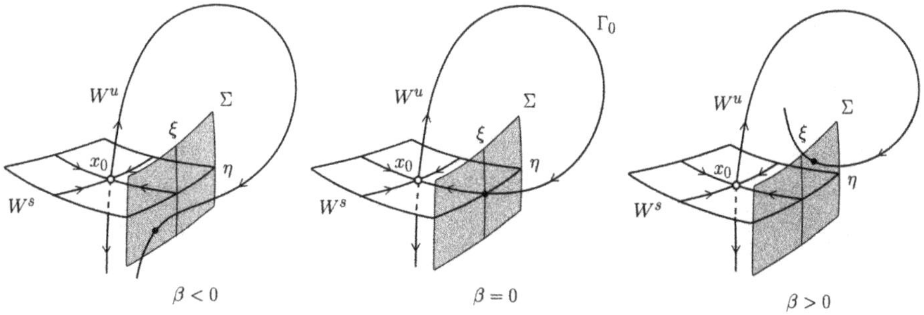


**Fig. 6.5.** Split function in the case $n = 3$.

$(\xi^u, \eta^u)$ correspond to the intersection of $W^u$ with $\Sigma$. Then, a split function can be defined as in the planar case before: $\beta = \xi^u$. Its zero

$$\beta = 0$$

gives a condition for the homoclinic bifurcation in $\mathbb{R}^3$.

**Remarks:**

(1) The preceding cases are examples of nontransversal intersections of the invariant manifolds $W^u$ and $W^s$. One can construct a three-dimensional system with a structurally stable *heteroclinic* orbit connecting two saddles: This orbit must be a transversal intersection of the corresponding *two-dimensional* stable and unstable manifolds.

(2) There are particular classes of dynamical systems (such as Hamiltonian) for which the presence of a nontransversal homoclinic orbit is generic. $\diamondsuit$

Thus, we have found that under certain conditions the presence of a homo-/hetero-clinic orbit $\Gamma_0$ to a saddle/saddles implies a bifurcation. Our goal in the next sections will be to describe the phase portrait bifurcations near such an orbit under small $C^1$ perturbations of the system. "Near" means in a sufficiently small neighborhood $U_0$ of $\Gamma_0 \cup x_0$ or $\Gamma_0 \cup x_{(1)} \cup x_{(2)}$. This task is more complex than for bifurcations of equilibria since it is not easy to construct a continuous-time system that would be a topological normal form for the bifurcation. In some cases ahead, all one-parameter systems satisfying some generic conditions are topologically equivalent in a neighborhood of the corresponding homoclinic bifurcation. In these cases, we will characterize the relevant universal bifurcation diagram by drawing key orbits of the corresponding phase portraits. This will completely describe the diagram up to topological equivalence.

Unfortunately, there are more involved cases in which such an equivalence is absent. In these cases no universal bifurcation diagrams can be presented. Nevertheless, topologically nonequivalent bifurcation diagrams reveal some features in common, and we will give schematic phase portraits describing the bifurcation for such cases as well.

The nontransversal heteroclinic case is somehow trivial since the disappearance of the connecting orbit is the only essential event in $U_0$ (see **Fig. 6.6**). Therefore, in this chapter we will focus on the homoclinic orbit bifurcations and return to nonhyperbolic homoclinic orbits and their associated bifurcations in Chapter 7.

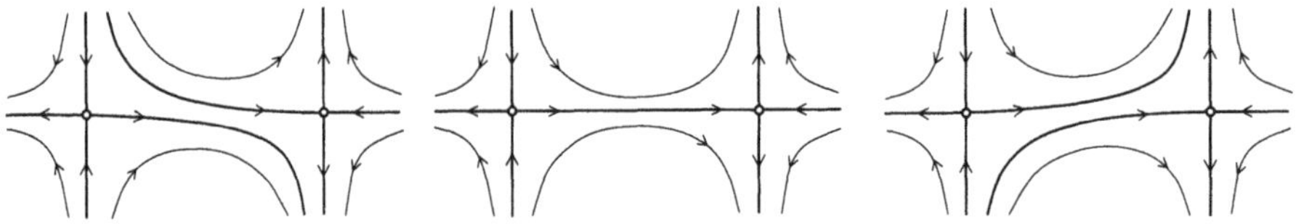


**Fig. 6.6.** Heteroclinic bifurcation on the plane.

## 6.2 Andronov-Leontovich theorem

In the planar case, the homoclinic bifurcation is completely characterized by the following theorem.

**Theorem 6.1 (Andronov & Leontovich [1939])** *Consider a two-dimensional system*

$$\dot{x} = f(x, \alpha), \quad x \in \mathbb{R}^2, \ \alpha \in \mathbb{R}^1, \tag{6.4}$$

*with smooth $f$, having at $\alpha = 0$ a saddle equilibrium point $x_0 = 0$ with eigenvalues $\lambda_1(0) < 0 < \lambda_2(0)$ and a homoclinic orbit $\Gamma_0$. Assume the following genericity conditions hold:*

(H.1) $\sigma_0 = \lambda_1(0) + \lambda_2(0) \neq 0$;
(H.2) $\beta'(0) \neq 0$, *where $\beta(\alpha)$ is the previously defined split function.*

*Then, for all sufficiently small $|\alpha|$, there exists a neighborhood $U_0$ of $\Gamma_0 \cup x_0$ in which a unique limit cycle $L_\beta$ bifurcates from $\Gamma_0$. Moreover, the cycle is stable and exists for $\beta > 0$ if $\sigma_0 < 0$, and is unstable and exists for $\beta < 0$ if $\sigma_0 > 0$.*

The following definition is quite useful.

**Definition 6.5** *The real number $\sigma = \lambda_1 + \lambda_2$ is called the* saddle quantity.

**Figs. 6.7** and **6.8** illustrate the above theorem. If $\alpha = 0$, the system has an orbit homoclinic to the origin. A saddle equilibrium point exists near the origin for all sufficiently small $|\alpha| \neq 0$, while the homoclinic orbit disappears,

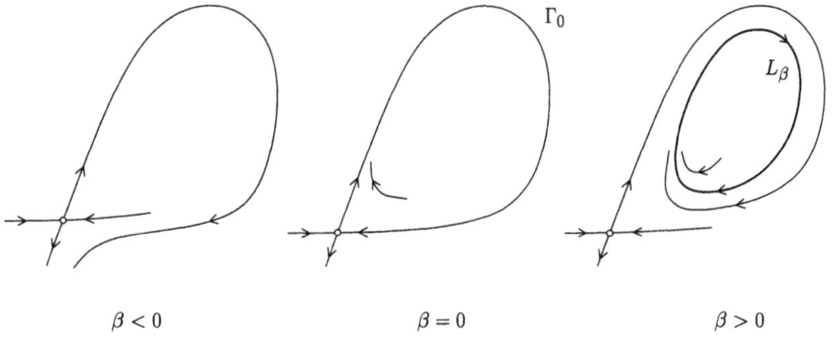


**Fig. 6.7.** Homoclinic bifurcation on the plane ($\sigma_0 < 0$).

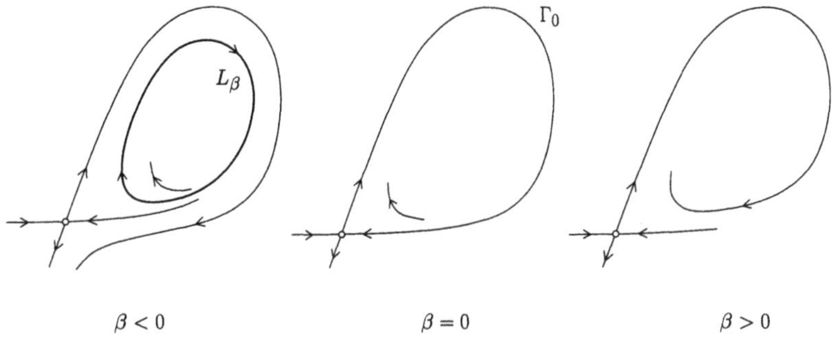


**Fig. 6.8.** Homoclinic bifurcation on the plane ($\sigma_0 > 0$).

splitting "up" or "down." According to condition (H.2), the split function $\beta = \beta(\alpha)$ can be considered as a new parameter.

If the saddle quantity satisfies $\sigma_0 < 0$, the homoclinic orbit at $\beta = 0$ is *stable from the inside*, and the theorem gives the existence of a unique and stable limit cycle $L_\beta \subset U_0$ for $\beta > 0$. For $\beta < 0$ there are no periodic orbits in $U_0$. If the saddle quantity satisfies $\sigma_0 > 0$, the homoclinic orbit at $\beta = 0$ is *unstable from the inside*, and the theorem gives the existence of a unique but unstable limit cycle $L_\beta \subset U_0$ for $\beta < 0$. For $\beta > 0$ there are no periodic orbits in $U_0$. Thus, the sign of $\sigma_0$ determines the direction of bifurcation and the stability of the appearing limit cycle. As usual, the term "direction" has a conventional meaning and is related to our definition of the split function.

As $|\beta| \to 0$, the cycle passes closer and closer to the saddle and becomes increasingly "angled" (see **Fig. 6.9**(a)). Its period $T_\beta$ tends to infinity as $\beta$ approaches zero since a phase point moving along the cycle spends more and more time near the equilibrium (see **Fig. 6.9**(b)). The corresponding time series $(x_1(t), x_2(t))$ demonstrates "peaks" of finite length interspersed by very long "near-equilibrium" intervals.

**Proof of Theorem 6.1:**

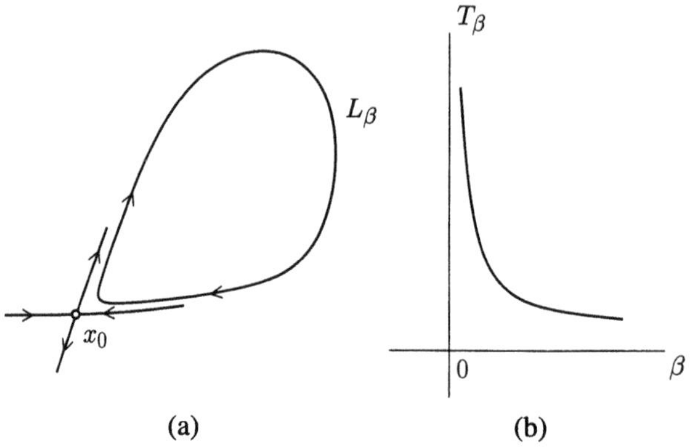


**Fig. 6.9.** A cycle near the homoclinic bifurcation and its period as function of $\beta$.

The main idea of the proof is to introduce two local cross-sections near the saddle, $\Sigma$ and $\Pi$, which are transversal to the stable and the unstable manifolds, respectively (see **Fig. 6.10**). Then it is possible to define a Poincaré

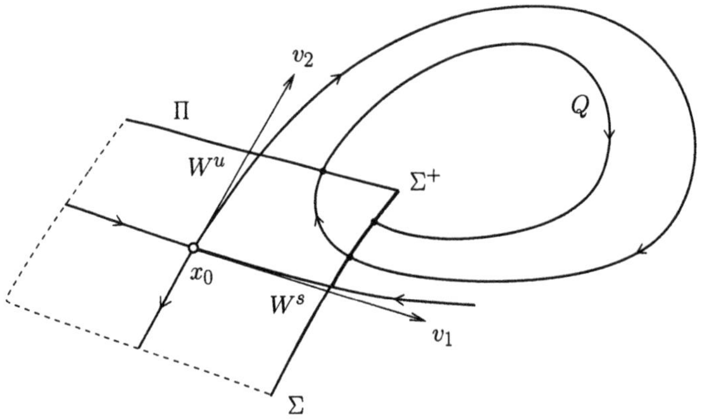


**Fig. 6.10.** Poincaré map for homoclinic bifurcation on the plane.

map $P$ on a half-section $\Sigma^+$,

$$P : \Sigma^+ \to \Sigma,$$

as a composition of a near-to-saddle map $\Delta : \Sigma^+ \to \Pi$ and a map $Q : \Pi \to \Sigma$ near the global part of the homoclinic orbit:

$$P = Q \circ \Delta.$$

Finally, we have to take into account the usual correspondence between limit cycles of (6.4) and fixed points of $P$. The proof proceeds through several steps.

*Step 1* (*Introduction of eigenbasis coordinates*). Without loss of generality, assume that the origin is a saddle equilibrium of (6.4) for all sufficiently small $|\alpha|$. We consider $\beta$ as a new parameter but do not indicate the parameter dependence for a while in order to simplify notation.

There is an invertible linear coordinate transformation that allows us to write (6.4) in the form

$$\begin{cases} \dot{x}_1 = \lambda_1 x_1 + g_1(x_1, x_2), \\ \dot{x}_2 = \lambda_2 x_2 + g_2(x_1, x_2), \end{cases} \tag{6.5}$$

where $x_{1,2}$ denote the new coordinates and $g_{1,2}$ are smooth $O(\|x\|^2)$-functions, $x = (x_1, x_2)^T$, $\|x\|^2 = x_1^2 + x_2^2$.

*Step 2* (*Local linearization of the invariant manifolds*). According to the Local Stable Manifold Theorem (see Chapter 3), the stable and unstable invariant manifolds $W^s$ and $W^u$ of the saddle exist and have the local representations

$$W^s: \; x_2 = S(x_1), \; S(0) = S'(0) = 0;$$
$$W^u: \; x_1 = U(x_2), \; U(0) = U'(0) = 0,$$

with smooth $S, U$ (see **Fig. 6.11**). Introduce new variables $y = (y_1, y_2)^T$ near

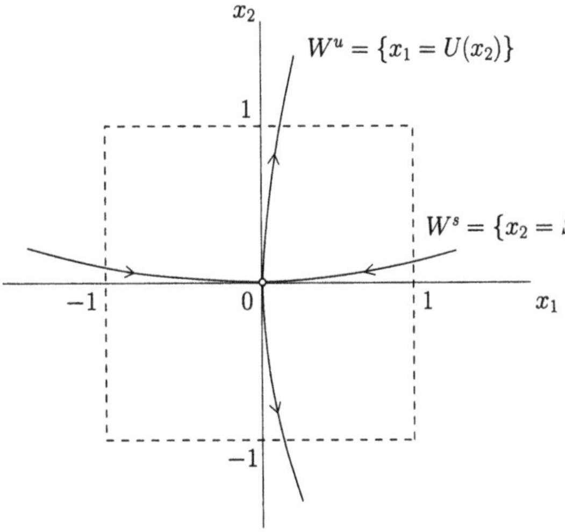


**Fig. 6.11.** Local stable and unstable manifolds in $x$-coordinates.

the saddle:

$$\begin{cases} y_1 = x_1 - U(x_2), \\ y_2 = x_2 - S(x_1). \end{cases}$$

This coordinate change is smooth and invertible in some neighborhood of the origin.[2] We can assume that this neighborhood contains the unit square $\Omega = \{y : -1 < y_{1,2} < 1\}$, which is a matter of an additional linear scaling of system (6.5). Thus, system (6.5) in the new coordinates takes in $\Omega$ the form

$$\begin{cases} \dot{y}_1 = \lambda_1 y_1 + y_1 h_1(y_1, y_2), \\ \dot{y}_2 = \lambda_2 y_2 + y_2 h_2(y_1, y_2), \end{cases} \tag{6.6}$$

where $h_{1,2} = O(\|y\|)$. Notice that (6.6) is a *nonlinear* smooth system with a

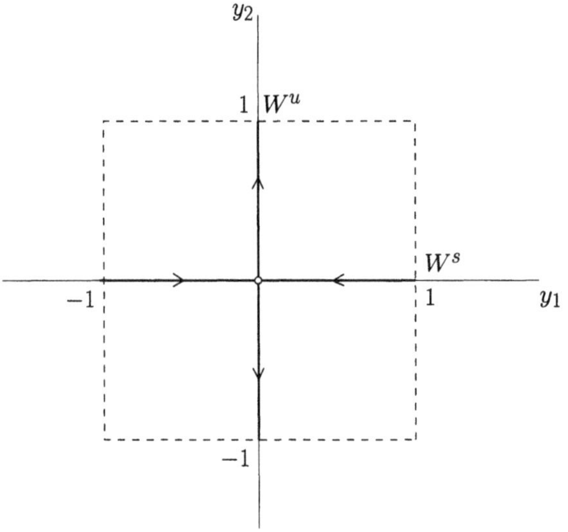


**Fig. 6.12.** Locally linearized stable and unstable manifolds in $y$-coordinates.

saddle at the origin whose invariant manifolds are linear and coincide with the coordinate axes in $\Omega$ (see **Fig. 6.12**).

*Step 3* (*Local $C^1$-linearization of the system*). Now introduce new coordinates $(\xi, \eta)$ in $\Omega$ in which system (6.6) becomes *linear*:

$$\begin{cases} \dot{\xi} = \lambda_1 \xi, \\ \dot{\eta} = \lambda_2 \eta. \end{cases} \tag{6.7}$$

More precisely, we show that the flow corresponding to (6.6) is $C^1$-equivalent in $\Omega$ to the flow generated by the linear system (6.7). To construct the conjugating map

[2] To be more precise, we have to consider a *global* invertible smooth change of the coordinates that coincides with the specified one in a neighborhood of the saddle and is the identity outside some other neighborhood of the saddle. The same should be noticed concerning the map $\Phi$ to be constructed later.

$$\begin{cases} \xi = \varphi(y_1, y_2), \\ \eta = \psi(y_1, y_2), \end{cases}$$

we use the following geometric construction. Take a point $y = (y_1, y_2) \in \Omega$ and the orbit passing through this point (see **Fig. 6.13**(a)). Let $\tau_1$ and $\tau_2$ be the absolute values of the positive and negative times required for such an orbit to reach the boundary of $\Omega$ in system (6.6). It can be checked (Exercise 7(a)) that the pair $(\tau_1, \tau_2)$ is uniquely defined for $y \neq 0$ within each quadrant of $\Omega$.[3] Now find a point $(\xi, \eta)$ in the same quadrant of $\Omega$ with the same "exit" times $\tau_1$ and $\tau_2$ for the corresponding orbit of (6.7) (see **Fig. 6.13**(b)). Let us take $\xi = \eta = 0$ for $y = 0$. Thus, a map $\Phi : (y_1, y_2) \mapsto (\xi, \eta)$ is

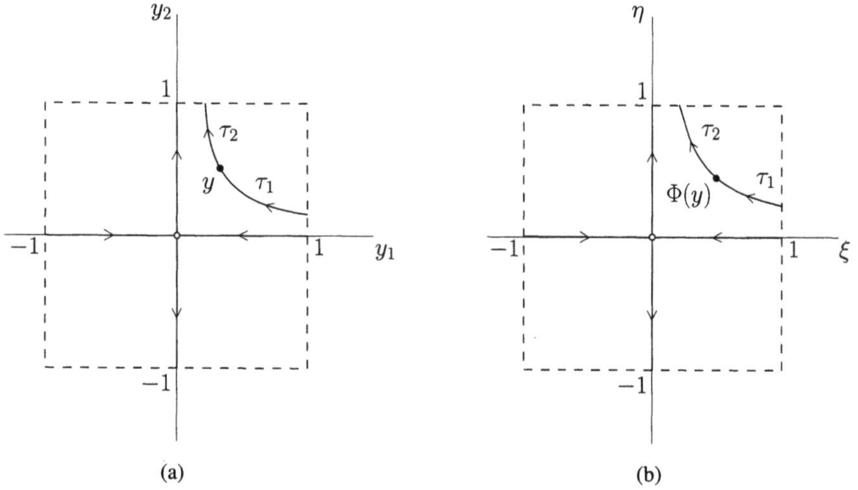


**Fig. 6.13.** Construction of $C^1$-equivalence.

constructed. It clearly maps orbits of the nonlinear system (6.6) into orbits of the linear system (6.7), preserving time parametrization. Map $\Phi : \Omega \to \Omega$ is a homeomorphism transforming each component of the boundary of $\Omega$ into itself; it is identical along the axes. To define a useful coordinate change it must be at least continuously differentiable in $\Omega$. Indeed, the map $\Phi$ is a $C^1$ map. Actually, it is smooth away from the origin but has only first-order continuous partial derivatives at $y = 0$ (the relevant calculations are left to the reader as Exercise 7(b)).

*Step 4* (*Analysis of the composition*). Using the new coordinates $(\xi, \eta)$, we can compute the near-to-saddle map *analytically.* We can assume that the cross-section $\Sigma$ has the representation $\xi = 1, -1 \leq \eta \leq 1$. Then, $\eta$ can be used as a coordinate on it, and $\Sigma^+$ is defined by $\xi = 1, 0 \leq \eta \leq 1$. The map acts from

[3] For points on the coordinate axes we allow one of $\tau_{1,2}$ to be equal to $\pm\infty$.

$\Sigma^+$ into a cross-section $\Pi$, which is defined by $\eta = 1, -1 \le \xi \le 1$, and has $\xi$ as a coordinate (see **Fig. 6.14**). Integrating the linear system (6.7), we obtain

$$\Delta : \ \xi = \eta^{-\frac{\lambda_1}{\lambda_2}}.$$

Notice that the resulting map is *nonlinear* regardless of the linearity of system (6.7). We assumed $\xi = 0$ for $\eta = 0$ by continuity.

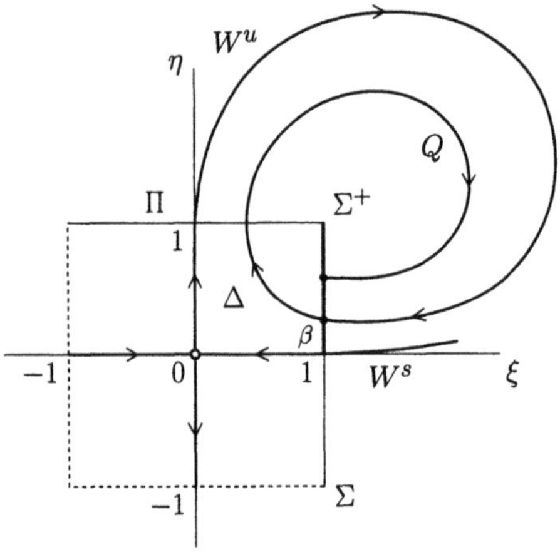


**Fig. 6.14.** Poincaré map in locally linearizing coordinates.

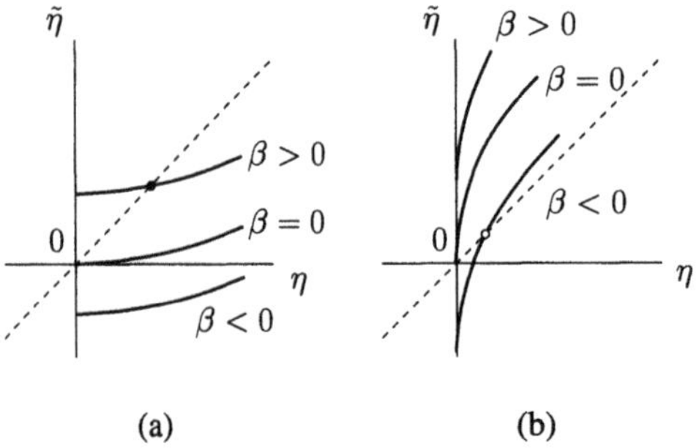


**Fig. 6.15.** Fixed points of the Poincaré map: (a) $\sigma_0 < 0$; (b) $\sigma_0 > 0$.

The global map expressed in $\xi$ and $\eta$ is continuously differentiable and invertible and has the general form

$$Q : \ \eta = \beta + a\xi + O(\xi^2),$$

where $\beta$ is the split function and $a > 0$ since the orbits cannot intersect. Actually, $\lambda_{1,2} = \lambda_{1,2}(\beta)$, $a = a(\beta)$, but as we shall see below only values at

$\beta = 0$ are relevant. Fixed points with small $|\eta|$ of the Poincaré map

$$P : \eta \mapsto \beta + a\eta^{-\frac{\lambda_1}{\lambda_2}} + \cdots$$

can be easily analyzed for small $|\beta|$ (see **Fig. 6.15**). Therefore, we have existence of a positive fixed point (limit cycle) for $\beta > 0$ if $\sigma_0 < 0$ and for $\beta < 0$ if $\sigma_0 > 0$. Stability and uniqueness of the cycle also simply follow from analysis of the map. □

**Remark:**

Until now we have considered only so-called "small" homoclinic orbits in this section. There is another type of homoclinic orbits, namely, "big" homoclinic orbits corresponding to the different return direction. All the results

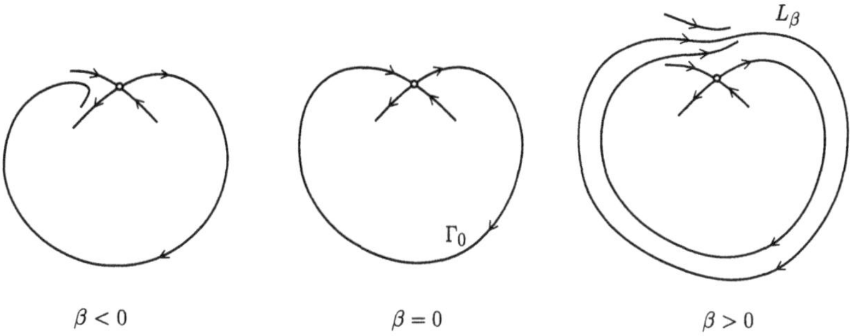


**Fig. 6.16.** Bifurcation of a "big" saddle homoclinic orbit.

obtained are valid for them as well (see **Fig. 6.16**, where a bifurcation diagram for the case $\sigma_0 < 0$ is presented). ◇

**Example 6.1 (Explicit homoclinic bifurcation)** Consider the following system due to Sandstede [1997*a*]:

$$\begin{cases} \dot{x} = -x + 2y + x^2, \\ \dot{y} = (2-\alpha)x - y - 3x^2 + \frac{3}{2}xy, \end{cases} \tag{6.8}$$

where $\alpha$ is a parameter.

The origin $(x, y) = (0, 0)$ is a saddle for all sufficiently small $|\alpha|$. At $\alpha = 0$, this saddle has eigenvalues

$$\lambda_1 = 1, \quad \lambda_2 = -3,$$

with $\sigma_0 = -2 < 0$. Moreover, at this parameter value, there exists a homoclinic orbit to the origin (see **Fig. 6.17**). Indeed, the *Cartesian leaf*,

$$H(x, y) \equiv x^2(1 - x) - y^2 = 0,$$

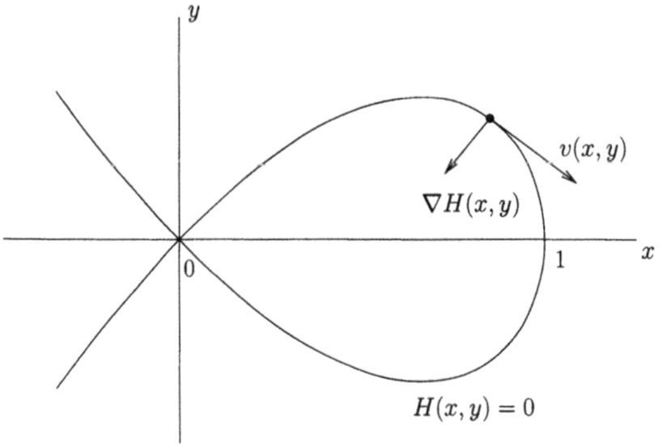


**Fig. 6.17.** The homoclinic orbit of (6.8) at $\alpha = 0$.

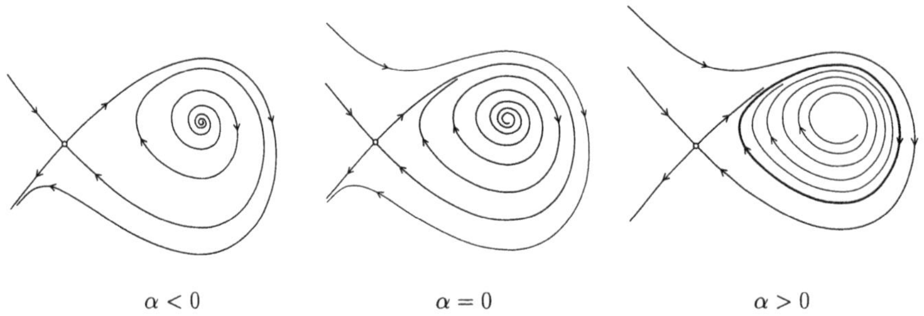


**Fig. 6.18.** Homoclinic bifurcation in (6.8): A stable limit cycle exists for small $\alpha > 0$.

consists of orbits of (6.8) for $\alpha = 0$. One of these orbits is homoclinic to the saddle $0 = (0, 0)$. To verify this fact, we have to prove that the vector field defined by (6.8) with $\alpha = 0$,

$$v(x, y) = \left(-x + 2y + x^2, 2x - y - 3x^2 + \frac{3}{2}xy\right)^T,$$

is tangent to the curve $H(x, y) = 0$ at all nonequilibrium points. Equivalently, it is sufficient to check that $v$ is orthogonal along the curve to the *normal vector* to the curve. A normal vector is given by the gradient of the function $H$:

$$(\nabla H)(x, y) = (2x - 3x^2, -2y)^T.$$

Then, a direct calculation shows that

$$\langle v, \nabla H \rangle = 0$$

along the curve $H = 0$ (check!).

Thus, system (6.8) has an algebraic homoclinic orbit at $\alpha = 0$ with $\sigma_0 < 0$. One can prove that the transversality condition $\beta' \neq 0$ also holds at $\alpha = 0$

(see Exercise 13). Therefore, Theorem 6.1 is applicable, and a unique and stable limit cycle bifurcates from the homoclinic orbit under small variation of $\alpha$ (see **Fig. 6.18**). $\diamondsuit$

**Example 6.2 (Homoclinic bifurcation in a slow-fast system)** Consider the system

$$\begin{cases} \dot{x} = 1 + x - y - x^2 - x^3, \\ \dot{y} = \varepsilon\left[-1 + (1 - 4\alpha)x + 4xy\right], \end{cases} \tag{6.9}$$

where $\alpha$ is a "control" parameter and $0 < \varepsilon \ll 1$. We shall show that the system undergoes a homoclinic bifurcation at some value of $\alpha$ close to zero. More precisely, there is a continuous function $\alpha_0 = \alpha_0(\varepsilon)$ defined for sufficiently small $\varepsilon \geq 0$, $\alpha_0(0) = 0$ such that the system has a homoclinic orbit to a saddle at $\alpha = \alpha_0(\varepsilon)$. Moreover, the genericity conditions of the Andronov-Leontovich theorem are satisfied, and a unique and stable limit cycle bifurcates from the homoclinic orbit under the variation of $\alpha$ for $\alpha < \alpha_0$.

The nontrivial zero-isoclines of (6.9) are graphs of the functions

$$\dot{x} = 0: \; y = (x+1)(1-x^2)$$

and

$$\dot{y} = 0: \; y = \frac{1-x}{4x} + \alpha;$$

their shape at $\alpha = 0$ is presented in **Fig. 6.19**.

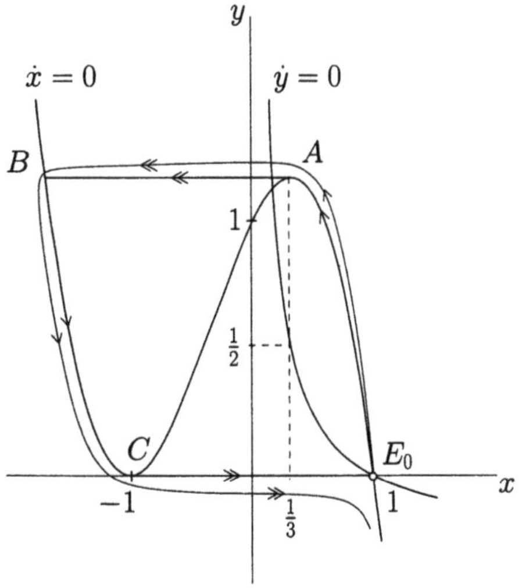


**Fig. 6.19.** Zero-isoclines of (6.9) and the corresponding singular homoclinic orbit.

If $\alpha = 0$, the system has a saddle equilibrium point $E_0 : (x, y) = (1, 0)$ for all $\varepsilon > 0$. It can easily be checked that *near* the saddle $E_0$ the stable invariant manifold $W^s(E_0)$ approaches the $x$-axis while the unstable manifold $W^u(E_0)$ tends to the zero-isocline $\dot{x} = 0$, as $\varepsilon \to 0$. The global behavior of the upper branch $W_1^u$ of the unstable manifold as $\varepsilon \to 0$ is also clear. It approaches a *singular* orbit composed of two *slow* motions along the isocline $\dot{x} = 0$ ($E_0 A$ and $BC$) and two *fast* jumps in the horizontal direction ($AB$ and $CE_0$; the latter happens along the $x$-axis) (see **Fig. 6.19**). This singular orbit returns to $E_0$, thus forming a *singular homoclinic orbit.*

This construction can be carried out for all sufficiently small $\alpha \neq 0$ (see **Fig. 6.20**). The equilibrium point will shift away from the $x$-axis and will have

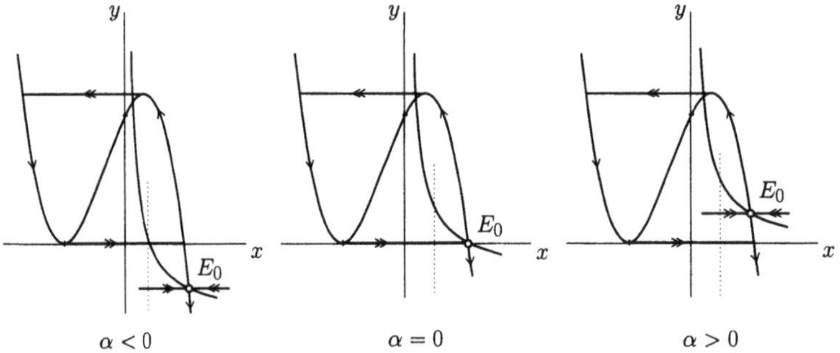


**Fig. 6.20.** Singular homoclinic bifurcation in (6.9).

the $y$-coordinate equal to $\alpha$. Despite this, a singular orbit to which $W_1^u$ tends as $\varepsilon \to 0$ still arrives at a neighborhood of the saddle along the $x$-axis. Therefore, there is a *singular split function* $\beta_0(\alpha)$ measured along a vertical cross-section near the saddle which equals $\alpha$ : $\beta_0(\alpha) = \alpha$. Obviously, $\beta_0'(0) > 0$. Meanwhile, the singular orbit tends to a *singular limit cycle* if $\alpha < 0$.

Thus, we have a generic *singular homoclinic bifurcation* at $\alpha = 0$ in the singular limit $\varepsilon = 0$. This implies the existence of a generic homoclinic bifurcation at $\alpha = \alpha_0(\varepsilon)$ for sufficiently small $\varepsilon > 0$. One can show this using nonstandard analysis. To prove it in a standard way, one has to check that the split function $\beta(\alpha, \varepsilon)$ can be represented for all sufficiently small $\varepsilon > 0$ as

$$\beta(\alpha, \varepsilon) = \beta_0(\alpha) + \varphi(\alpha, \varepsilon),$$

where $\varphi(\alpha, \varepsilon)$ (considered as a function of $\alpha$ for small $|\alpha|$) vanishes uniformly with its first derivative as $\varepsilon \to 0$. Then, elementary arguments[4] show the existence of a unique continuous function $\alpha_0(\varepsilon)$, $\alpha_0(0) = 0$, such that

$$\beta(\alpha_0(\varepsilon), \varepsilon) = 0$$

for all sufficiently small $\varepsilon \geq 0$. Actually, $\alpha_0(\varepsilon)$ is smooth for $\varepsilon > 0$. Therefore, the system has a homoclinic orbit at $\alpha = \alpha_0(\varepsilon)$ for all sufficiently small $\varepsilon$.

[4] The only difficulty that should be overcome is that $\varphi(\alpha, \varepsilon)$ is not differentiable with respect to $\varepsilon$ at $\varepsilon = 0$ (see Exercise 8).

The corresponding saddle quantity $\sigma_0$ is negative and $\beta_\alpha(\alpha_0(\varepsilon), \varepsilon) \neq 0$, thus, Theorem 6.1 is applicable to (6.9). ◊

**More remarks on Theorem 6.1:**

(1) Condition (H.2) of Theorem 6.1 is equivalent to the transversality of the intersection of certain invariant manifolds of the *extended* system

$$\begin{cases} \dot{x} = f(x, \alpha), \\ \dot{\alpha} = 0. \end{cases} \tag{6.10}$$

Let $x_0(\alpha)$ denote a one-parameter family of the saddles in (6.4) for small $|\alpha|$. This family defines an invariant set of (6.10) – a curve of equilibria. This curve has two-dimensional unstable and stable manifolds, $\mathcal{W}^u$ and $\mathcal{W}^s$, whose slices $\alpha =$ const coincide with the corresponding one-dimensional unstable and stable manifolds $W^u$ and $W^s$ of the saddle $x_0(\alpha)$ in (6.4) (see **Fig. 6.21**). Condition (H.2) (meaning that $W^u$ and $W^s$ split with nonzero velocity as $\alpha$

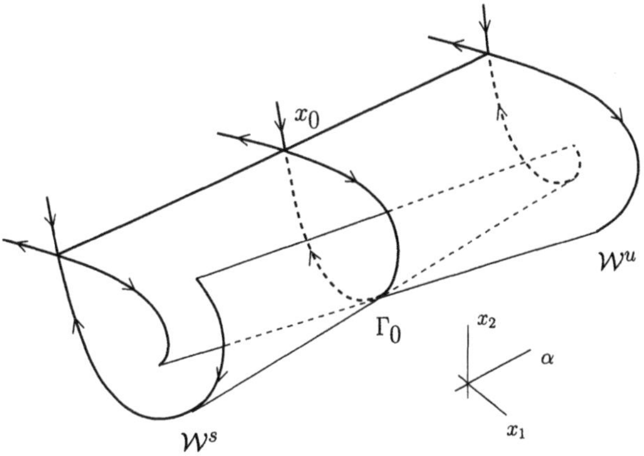


**Fig. 6.21.** Transversal intersection of invariant manifolds $\mathcal{W}^u$ and $\mathcal{W}^s$.

crosses $\alpha = 0$) translates exactly to the transversality of the intersection of $\mathcal{W}^u$ and $\mathcal{W}^s$ along $\Gamma_0$ at $\alpha = 0$ in the three-dimensional state space of (6.10). In Section 6.4.1 we will show that the transversality is equivalent to the *Melnikov condition:*

$$\int_{-\infty}^{+\infty} \exp\left[-\int_0^t \left(\frac{\partial f_1}{\partial x_1} + \frac{\partial f_2}{\partial x_2}\right) d\tau\right] \left(f_1 \frac{\partial f_2}{\partial \alpha} - f_2 \frac{\partial f_1}{\partial \alpha}\right) dt \neq 0,$$

where all expressions involving $f = (f_1, f_2)^T$ are evaluated at $\alpha = 0$ along a solution $x^0(\cdot)$ of (6.4) corresponding to the homoclinic orbit $\Gamma_0$.

(2) One can construct a *topological normal form* for the homoclinic bifurcation on the plane. Consider the $(\xi, \eta)$-plane and introduce two domains:

the unit square

$$\Omega_0 = \{(\xi, \eta) : \ |\xi| \le 1, \ |\eta| \le 1\}$$

and the rectangle

$$\Omega_1 = \{(\xi, \eta) : \ 1 \le \xi \le 2, \ |\eta| \le 1\}$$

(see **Fig. 6.22**). Define a two-dimensional manifold $\Omega$ by glueing $\Omega_0$ and

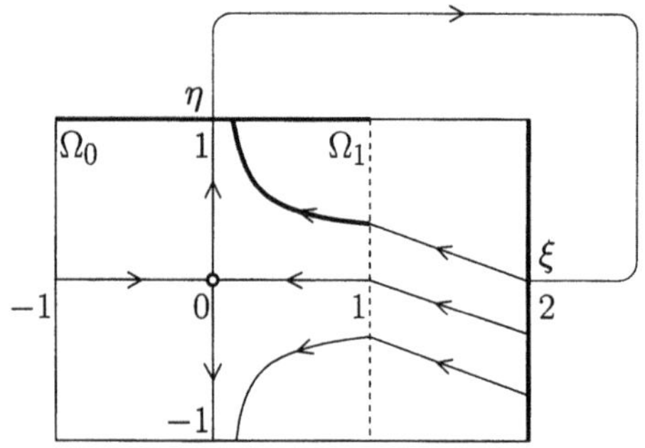


**Fig. 6.22.** Topological normal form for homoclinic bifurcation.

$\Omega_1$ along the vertical segment $\{\xi = 1, \ |\eta| \le 1\}$ and identifying the upper boundary of $\Omega_0$ with the right boundary of $\Omega_1$ (i.e., glueing points $(\xi, 1)$ and $(2, \xi)$ for $|\xi| \le 0$). The resulting manifold is homeomorphic to a simple band.

Consider a system of ODEs in $\Omega$ that is defined by

$$\begin{cases} \dot{\xi} = \lambda_1(\alpha)\xi, \\ \dot{\eta} = \lambda_2(\alpha)\eta, \end{cases} \tag{6.11}$$

in $\Omega_0$ and by

$$\begin{cases} \dot{\xi} = -1, \\ \dot{\eta} = \beta(\alpha)\eta, \end{cases} \tag{6.12}$$

in $\Omega_1$, where $\lambda_{1,2}$ and $\beta$ are smooth functions of a parameter $\alpha$. The behavior of thus defined piecewise-smooth system in $\Omega$ is similar to that of (6.4) near the homoclinic orbit (cf. **Figs. 6.14** and **6.22**). If $\lambda_1 < 0 < \lambda_2$, the constructed system has a saddle at the origin. At $\beta = 0$ this saddle has a homoclinic orbit $\tilde{\Gamma}_0$ composed by two segments of the coordinate axes: $\{\xi = 0, \ 0 < \eta \le 1\}$ and $\{\eta = 0, \ 0 < \xi \le 2\}$. For small $\beta \neq 0$ this homoclinic orbit breaks down, with the parameter $\beta$ playing the role of the split function. Provided the saddle quantity $\sigma_0 = \lambda_1(0) + \lambda_2(0) \neq 0$, a unique limit cycle bifurcates from $\tilde{\Gamma}_0$. This can be seen from the Poincaré map defined by the system (6.11), (6.12) in the cross-section $\{\xi = 1, \ 0 \le \eta \le 1\}$:

$$\eta \mapsto \beta + \eta^{-\frac{\lambda_1}{\lambda_2}}.$$

Actually, the following theorem holds.

**Theorem 6.2** *Under the assumptions of Theorem* 6.1, *the system* (6.4) *is locally topologically equivalent near the homoclinic orbit* $\Gamma_0$ *for nearby parameter values to the system defined by* (6.11) *and* (6.12) *in* $\Omega$ *near* $\tilde{\Gamma}_0$ *for small* $|\beta|$. *Moreover, all such systems with* $\sigma_0 < 0$ ($\sigma_0 > 0$) *are locally topologically equivalent near the respective homoclinic orbits for nearby parameter values.* $\square$

The last statement of the theorem follows from the fact that, for $\sigma_0 < 0$, the constructed system in $\Omega$ is locally topologically equivalent near $\tilde{\Gamma}_0$ for small $|\beta|$ to this system with constant $\lambda_1 = -2$, $\lambda_2 = 1$, while, for $\sigma_0 > 0$, it is equivalent to that with $\lambda_1 = -1, \lambda_2 = 2$. $\diamondsuit$

## 6.3 Homoclinic bifurcations in three-dimensional systems: Shil'nikov theorems

A three-dimensional state space gives rise to a wider variety of homoclinic bifurcations, some of which involve an *infinite number* of periodic orbits. As is known from Chapter 3, the two simplest types of hyperbolic equilibria in $\mathbb{R}^3$ allowing for homoclinic orbits are *saddles* and *saddle-foci*. We assume from now on that these points have a one-dimensional unstable manifold $W^u$ and a two-dimensional stable manifold $W^s$ (otherwise, reverse the time direction). In the saddle case, we assume that the eigenvalues of the equilibrium are simple and satisfy the inequalities $\lambda_1 > 0 > \lambda_2 > \lambda_3$. Then, as we have seen in Chapter 2, all the orbits on $W^s$ approach the equilibrium along a one-dimensional eigenspace of the Jacobian matrix corresponding to $\lambda_2$ except two orbits approaching the saddle along an eigenspace corresponding to $\lambda_3$ (see **Fig. 2.4**(a)).

**Definition 6.6** *The eigenvalues with negative real part that are closest to the imaginary axis are called* leading (*or* principal) *eigenvalues, while the corresponding eigenspace is called a* leading (*or* principal) *eigenspace.*

Thus, almost all orbits on $W^s$ approach a generic saddle along the one-dimensional leading eigenspace. In the saddle-focus case, there are *two* leading eigenvalues $\lambda_2 = \bar{\lambda}_3$, and the leading eigenspace is two-dimensional (see **Fig. 2.4**(b)).

**Definition 6.7** *A* saddle quantity $\sigma$ *of a saddle* (*saddle-focus*) *is the sum of the positive eigenvalue and the real part of a leading eigenvalue.*

Therefore, $\sigma = \lambda_1 + \lambda_2$ for a saddle, and $\sigma = \lambda_1 + \text{Re } \lambda_{2,3}$ for a saddle-focus.

The table below briefly presents some general results Shil'nikov obtained concerning the number and stability of limit cycles generated via homoclinic bifurcations in $\mathbb{R}^3$. The column entries specify the type of the equilibrium having a homoclinic orbit, while the row entries give the possible sign of the corresponding saddle quantity.

| | *Saddle* | *Saddle-focus* |
|---|---|---|
| $\sigma_0 < 0$ | one stable cycle | one stable cycle |
| $\sigma_0 > 0$ | one saddle cycle | $\infty$ saddle cycles |

The following theorems give more precise information.

**Theorem 6.3 (Saddle, $\sigma_0 < 0$)** *Consider a three-dimensional system*

$$\dot{x} = f(x, \alpha), \quad x \in \mathbb{R}^3, \ \alpha \in \mathbb{R}^1, \tag{6.13}$$

*with smooth $f$, having at $\alpha = 0$ a saddle equilibrium point $x_0 = 0$ with real eigenvalues $\lambda_1(0) > 0 > \lambda_2(0) \geq \lambda_3(0)$ and a homoclinic orbit $\Gamma_0$. Assume the following genericity conditions hold:*

(H.1) $\sigma_0 = \lambda_1(0) + \lambda_2(0) < 0$;
(H.2) $\lambda_2(0) \neq \lambda_3(0)$;
(H.3) $\Gamma_0$ *returns to $x_0$ along the leading eigenspace;*
(H.4) $\beta'(0) \neq 0$, *where $\beta(\alpha)$ is the split function defined earlier.*

*Then, the system* (6.13) *has a unique and stable limit cycle $L_\beta$ in a neigborhood $U_0$ of $\Gamma_0 \cup x_0$ for all sufficiently small $\beta > 0$. Moreover, all such systems are locally topologically equivalent near $\Gamma_0 \cup x_0$ for small $|\alpha|$.* □

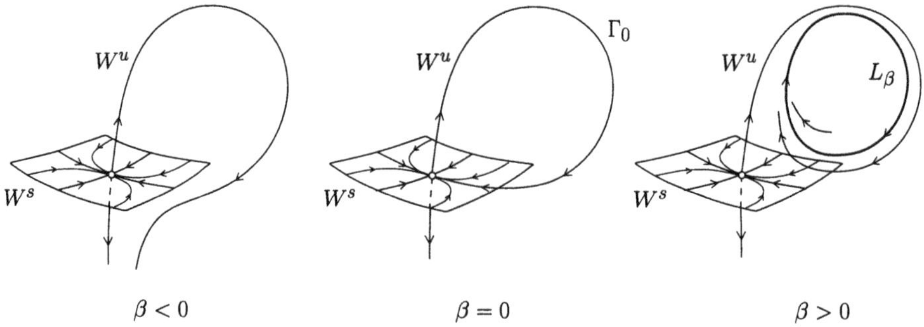


**Fig. 6.23.** Saddle homoclinic bifurcation with $\sigma_0 < 0$.

The theorem is illustrated in **Fig. 6.23**. The unstable manifold $W^u(x_0)$ tends to the cycle $L_\beta$. The period of the cycle tends to infinity as $\beta$ approaches zero. The (nontrivial) multipliers of the cycle are positive and inside the unit circle: $|\mu_{1,2}| < 1$. There are no periodic orbits of (6.13) in $U_0$ for all sufficiently small $\beta \leq 0$. Thus, the bifurcation is completely analogous to that in the planar case. The proof of the theorem will be sketched later (see also Exercise 10).

**Theorem 6.4 (Saddle-focus, $\sigma_0 < 0$)** *Suppose that a three-dimensional system*

$$\dot{x} = f(x, \alpha), \quad x \in \mathbb{R}^3, \ \alpha \in \mathbb{R}^1, \tag{6.14}$$

*with a smooth $f$, has at $\alpha = 0$ a saddle-focus equilibrium point $x_0 = 0$ with eigenvalues $\lambda_1(0) > 0 > \text{Re } \lambda_{2,3}(0)$ and a homoclinic orbit $\Gamma_0$. Assume the following genericity conditions hold:*

(H.1) $\sigma_0 = \lambda_1(0) + \text{Re } \lambda_{2,3}(0) < 0$;
(H.2) $\lambda_2(0) \neq \lambda_3(0)$;
(H.3) $\beta'(0) \neq 0$, *where $\beta(\alpha)$ is the split function.*

*Then, system* (6.14) *has a unique and stable limit cycle $L_\beta$ in a neighborhood $U_0$ of $\Gamma_0 \cup x_0$ for all sufficiently small $\beta > 0$.* □

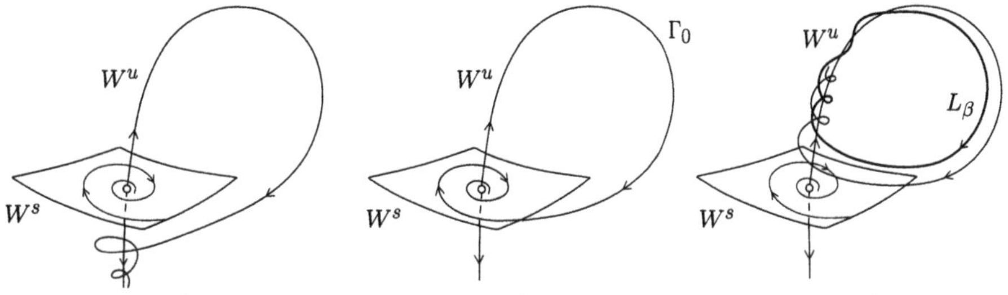


**Fig. 6.24.** Saddle-focus homoclinic bifurcation with $\sigma_0 < 0$.

The theorem is illustrated in **Fig. 6.24**. There are no periodic orbits of (6.14) in $U_0$ for all sufficiently small $\beta \leq 0$. The unstable manifold $W^u(x_0)$ tends to the cycle $L_\beta$. The cycle period tends to infinity as $\beta$ approaches zero. The (nontrivial) multipliers of the cycle are complex, $\mu_2 = \bar{\mu}_1$, and lie inside the unit circle: $|\mu_{1,2}| < 1$.

The analogy with the planar case, however, terminates here. We cannot say that the bifurcation diagrams of all systems (6.10) satisfying (H.1)–(H.3) are topologically equivalent. As a rule, they are *nonequivalent* since the real number

$$\nu_0 = -\frac{\lambda_1(0)}{\text{Re } \lambda_{2,3}(0)} \tag{6.15}$$

is a topological invariant for systems with a homoclinic orbit to a saddle-focus. The nature of this invariant will be clearer later. Thus, although there is a unique limit cycle for $\beta > 0$ in all such systems, the exact topology of their phase portraits can differ. Fortunately, this is not very important in most applications.

Before treating the saddle case with $\sigma_0 > 0$, we have to look at the topology of the invariant manifold $W^s(x_0)$ near $\Gamma_0$ more closely. Suppose we have a three-dimensional system with a saddle equilibrium point $x_0$ having simple eigenvalues and a homoclinic orbit returning along the leading eigenspace to this saddle. Let us fix a small neighborhood $U_0$ of $\Gamma_0 \cap x_0$. The homoclinic orbit $\Gamma_0$ belongs to the stable manifold $W^s(x_0)$ entirely. Therefore, the

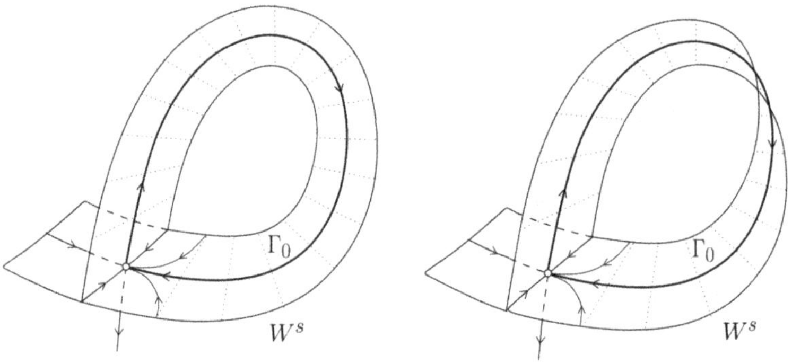


(a) (b)

**Fig. 6.25.** (a) Simple and (b) twisted stable manifolds near a homoclinic orbit to a saddle.

manifold $W^s(x_0)$ can be extended "back in time" along $\Gamma_0$ within the fixed neighborhood. At each point $\varphi^t x \in \Gamma_0$, a *tangent plane* to this manifold is well defined. For $t \to +\infty$, this plane is spanned by the stable eigenvectors $v_2$ and $v_3$. Generically, it approaches the plane spanned by the unstable eigenvector $v_1$ and the nonleading eigenvector $v_3$, as $t \to -\infty$. Thus, generically the manifold $W^s(x_0)$ intersects itself near the saddle along the two exceptional orbits on $W^s(x_0)$ that approach the saddle along the nonleading eigenspace[5] (see **Fig. 6.25**). Therefore, the part of $W^s(x_0)$ in $U_0$ to which belongs the homoclinic orbit $\Gamma_0$ is (generically) a two-dimensional nonsmooth submanifold $\mathcal{M}$. As is well known from elementary topology, such a manifold is topologically equivalent to either a *simple* or a *twisted* band. The latter is known as the *Möbius band.*

**Definition 6.8** *If $\mathcal{M}$ is topologically equivalent to a simple band, the homoclinic orbit $\Gamma_0$ is called* simple (*or* nontwisted). *If $\mathcal{M}$ is topologically equivalent to a Möbius band, the homoclinic orbit is called* twisted.

We are now ready to formulate the relevant theorem.

**Theorem 6.5 (Saddle, $\sigma_0 > 0$)** *Consider a three-dimensional system*

$$\dot{x} = f(x, \alpha), \quad x \in \mathbb{R}^3, \ \alpha \in \mathbb{R}^1, \tag{6.16}$$

[5] This property (often called the *strong inclination property*) was first established by Shil'nikov and is discussed in Exercise 9. See also Chapter 10, where we describe how to verify this property numerically.

*with smooth $f$, having at $\alpha = 0$ a saddle equilibrium point $x_0 = 0$ with real eigenvalues $\lambda_1(0) > 0 > \lambda_2(0) \geq \lambda_3(0)$ and a homoclinic orbit $\Gamma_0$. Assume that the following genericity conditions hold:*

(H.1) $\sigma_0 = \lambda_1(0) + \lambda_2(0) > 0$;
(H.2) $\lambda_2(0) \neq \lambda_3(0)$;
(H.3) $\Gamma_0$ *returns to* $x_0$ *along the leading eigenspace;*
(H.4) $\Gamma_0$ *is simple or twisted;*
(H.5) $\beta'(0) \neq 0$, *where* $\beta(\alpha)$ *is the split function.*

*Then, for all sufficiently small $|\alpha|$, there exists a neighborhood $U_0$ of $\Gamma_0 \cap x_0$ in which a unique saddle limit cycle $L_\beta$ bifurcates from $\Gamma_0$. The cycle exists for $\beta < 0$ if $\Gamma_0$ is nontwisted, and for $\beta > 0$ if $\Gamma_0$ is twisted. Moreover, all such systems (6.16) with simple (twisted) $\Gamma_0$ are locally topologically equivalent in a neighborhood $U_0$ of $\Gamma_0 \cap x_0$ for sufficiently small $|\alpha|$.* □

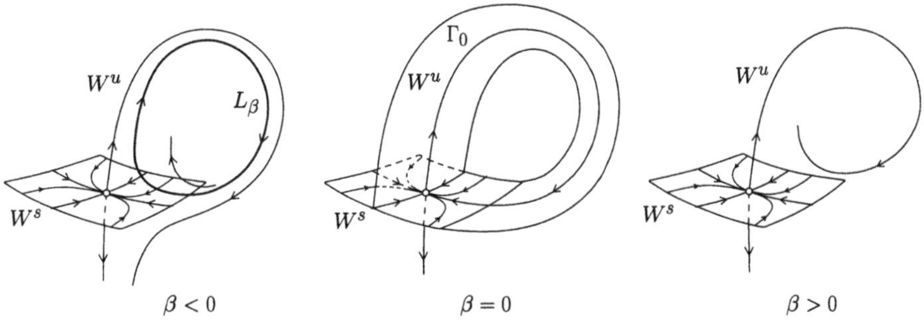


**Fig. 6.26.** Simple saddle homoclinic bifurcation with $\sigma_0 > 0$.

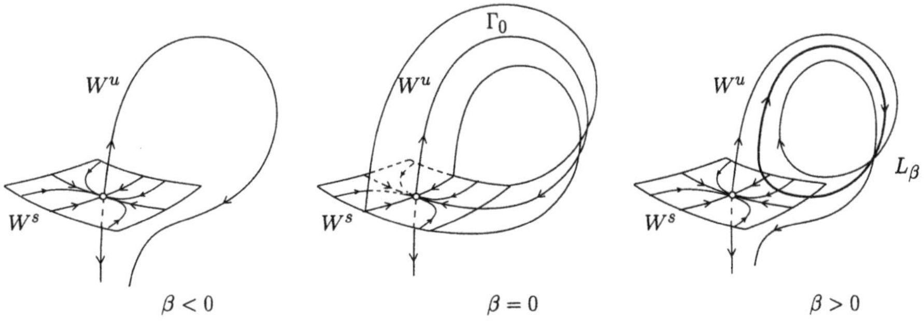


**Fig. 6.27.** Twisted saddle homoclinic bifurcation with $\sigma_0 > 0$.

The bifurcation diagrams to both cases are presented in **Figs. 6.26** and **6.27**, respectively. In both (simple and twisted) cases a unique saddle limit cycle $L_\beta$ bifurcates from the homoclinic orbit. Its period tends to infinity as $\beta$ approaches zero. Remarkably, the direction of the bifurcation is determined by the topology of $\mathcal{M}$.

If the homoclinic orbit is simple, there is a saddle cycle $L_\beta$ for $\beta < 0$. Its multipliers are positive: $\mu_1 > 1 > \mu_2 > 0$. The stable and unstable manifolds $W^{s,u}(L_\beta)$ of the cycle are (locally) simple bands.

If the homoclinic orbit is twisted, there is a saddle cycle $L_\beta$ for $\beta > 0$. Its multipliers are negative: $\mu_1 < -1 < \mu_2 < 0$. The stable and unstable manifolds $W^{s,u}(L_\beta)$ of the cycle are (locally) Möbius bands.

**Sketch of the proof of Theorems 6.3 and 6.5:**

We outline the proof of the theorems in the saddle cases. There are coordinates in $\mathbb{R}^3$ in which the manifolds $W^s(x_0)$ and $W^u(x_0)$ are linear in some neighborhood of $x_0$. Suppose system (6.13) (or (6.16)) is already written in these coordinates and, moreover, locally: $W^s(x_0) \subset \{x_1 = 0\}$, $W^u(x_0) \subset \{x_2 = x_3 = 0\}$. Let the $x_2$-axis be the leading eigenspace and the $x_3$-axis be the nonleading eigenspace. Introduce a rectangular two-dimensional cross-section $\Sigma \subset \{x_2 = \varepsilon_2\}$ and an auxiliary rectangular cross-section $\Pi \subset \{x_1 = \varepsilon_1\}$, where $\varepsilon_{1,2}$ are small enough. Assume that $\Gamma_0$ intersects both local cross-sections (see **Fig. 6.28**). As in the planar case, define a Poincaré map $P : \Sigma^+ \to \Sigma$ along the orbits of (6.13), mapping the upper part $\Sigma^+$ of $\Sigma$ corresponding to $x_1 \geq 0$ into $\Sigma$. Represent $P$ as a composition

$$P = Q \circ \Delta,$$

where $\Delta : \Sigma^+ \to \Pi$ is a near-to-saddle map, and $Q : \Pi \to \Sigma$ is a map along the global part of $\Gamma_0$. The construction can be carried out for all sufficiently small $|\beta|$.

The local map $\Delta$ is "essentially"[6] defined by the linear part of (6.13) near the saddle. It can be seen that the image of $\Sigma^+$ under the action of map $\Delta$, $\Delta\Sigma^+$, looks like a "horn" with a cusp on the $x_1$-axis (on $\Gamma_0$, in other words). Actually, the cross-sections $\Sigma$ and $\Pi$ should be chosen in such a way that $\Delta\Sigma^+ \subset \Pi$. The global map $Q$ maps this "horn" back into the plane $\{x_2 = \varepsilon_2\}$. If $\Gamma_0$ is simple, $P\Sigma^+$ intersects nontrivially with $\Sigma^+$ at $\beta = 0$; otherwise, the intersection with $\Sigma^- \equiv \Sigma \setminus \Sigma^+$ is nontrivial (see **Fig. 6.28**). Note that the transversality of the "horn" to the intersection of $W^s(x_0)$ with $\Sigma$ follows from the orientability or nonorientability of the manifold $\mathcal{M}$.

According to the sign of the saddle quantity $\sigma_0$ and the twisting of the homoclinic orbit, there are several cases of relative position of $P\Sigma^+$ with respect to $\Sigma$ (**Figs. 6.29** and **6.30**). A close look at these figures, actually, completes the proof. If $\sigma_0 < 0$ (Theorem 6.3, **Fig. 6.29**), the map $P$ is a contraction in $\Sigma^+$ for $\beta > 0$ and thus has a unique and stable fixed point in $P\Sigma^+$ corresponding to a stable limit cycle. If $\sigma_0 > 0$ (Theorem 6.5, **Fig.**

[6] Unfortunately, there are obstacles to $C^k$-linearization with $k \geq 1$ in this case. For example, $C^1$-linearization is impossible if $\lambda_2 = \lambda_1 + \lambda_3$ (see Appendix B).

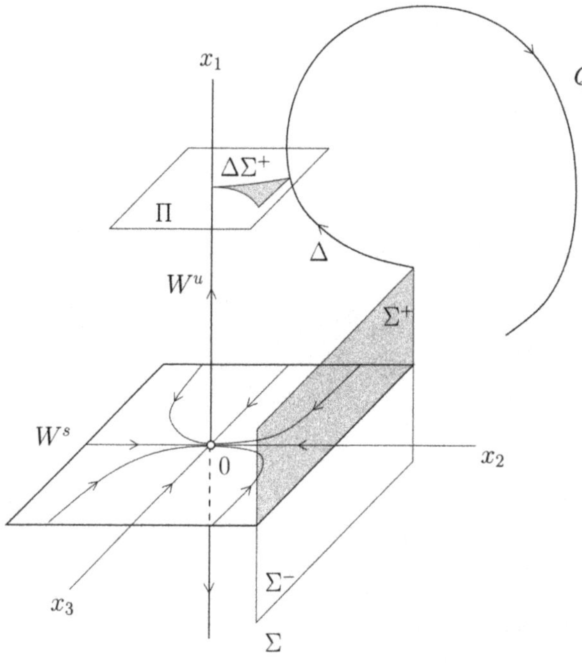


**Fig. 6.28.** Construction of the Poincaré map in the saddle case.

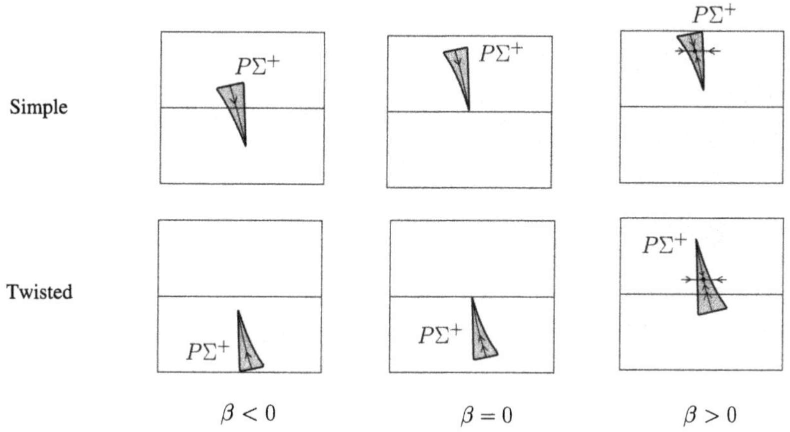


**Fig. 6.29.** The relative position of $P\Sigma^+$ with respect to $\Sigma$ in the case $\sigma_0 < 0$.

**6.30**), the map $P$ contracts along the $x_3$-axis and expands along the "horn." Therefore it has a saddle fixed point in $P\Sigma^+$ for $\beta < 0$ or $\beta > 0$, depending on the twisting of the homoclinic orbit. □

**Remark:**

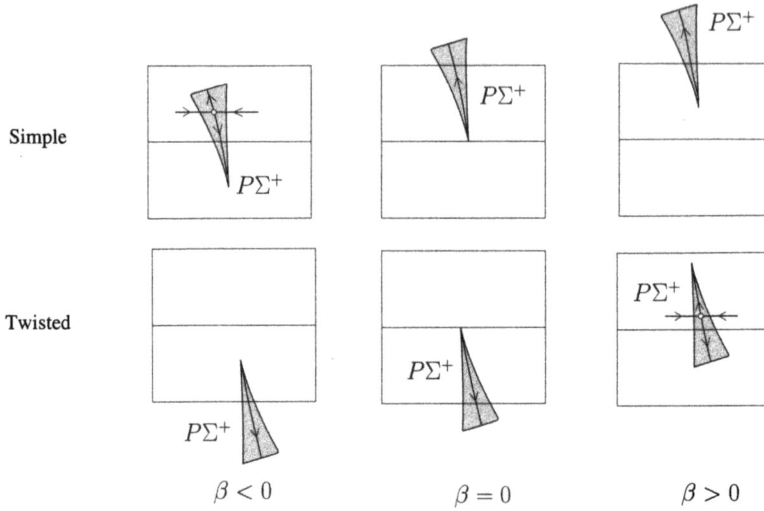


**Fig. 6.30.** The relative position of $P\Sigma^+$ with respect to $\Sigma$ in the case $\sigma_0 > 0$.

Because the map $P$ always acts as a contraction along the $x_3$-axis, the fixed-point analysis reduces (see Exercise 10) to the analysis of a one-dimensional map having the form

$$x_1 \mapsto \beta + Ax_1^{-\frac{\lambda_1}{\lambda_2}} + \cdots$$

that is similar to that in the Andronov-Leontovich theorem but $A$ can be either positive (simple homoclinic orbit) or negative (twisted homoclinic orbit).

Actually, this analogy can be extended further since in this case there is a two-dimensional attracting invariant "center manifold" near the homoclinic orbit (see Section 6.4). ♢

The last case is the most difficult.

**Theorem 6.6 (Saddle-focus, $\sigma_0 > 0$)** *Suppose that a three-dimensional system*

$$\dot{x} = f(x, \alpha), \quad x \in \mathbb{R}^3, \ \alpha \in \mathbb{R}^1, \tag{6.17}$$

*with a smooth $f$, has at $\alpha = 0$ a saddle-focus equilibrium point $x_0 = 0$ with eigenvalues $\lambda_1(0) > 0 > \text{Re}\,\lambda_{2,3}(0)$ and a homoclinic orbit $\Gamma_0$. Assume that the following genericity conditions hold:*

(H.1) $\sigma_0 = \lambda_1(0) + \text{Re}\,\lambda_{2,3}(0) > 0$;
(H.2) $\lambda_2(0) \neq \lambda_3(0)$.

*Then, system (6.17) has an infinite number of saddle limit cycles in a neighborhood $U_0$ of $\Gamma_0 \cup x_0$ for all sufficiently small $|\beta|$.* □

**Sketch of the proof of Theorems 6.4 and 6.6:**

To outline the proof, select a coordinate system in which $W^s(x_0)$ is (locally) the plane $x_1 = 0$, while $W^u(x_0)$ is (also locally) the line $x_2 = x_3 = 0$ (see **Fig. 6.31**). Introduce two-dimensional cross-sections $\Sigma$ and $\Pi$, and represent the Poincaré map $P : \Sigma^+ \to \Sigma$ as a composition $P = Q \circ \Delta$ of two maps: a near-to-saddle $\Delta : \Sigma^+ \to \Pi$ and a global $Q : \Pi \to \Sigma$, as in the proof of Theorems 6.3 and 6.5.[7]

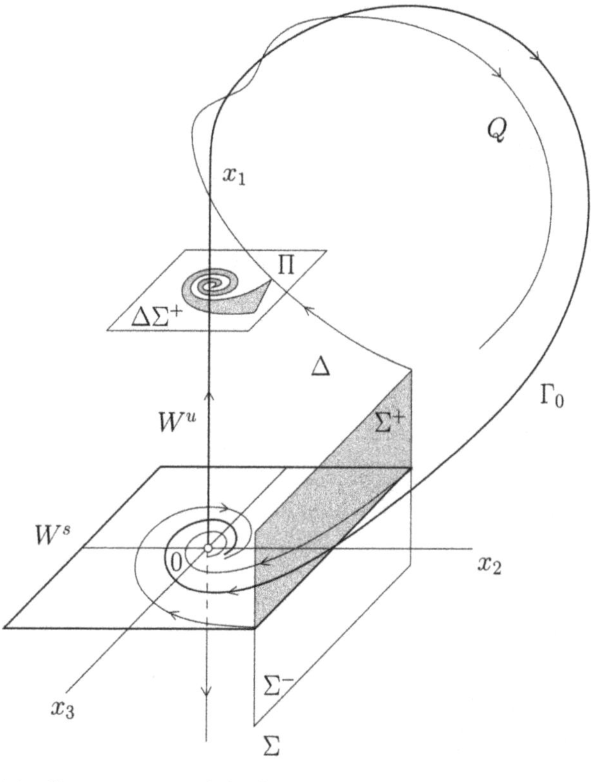


**Fig. 6.31.** Construction of the Poincaré map in the saddle-focus case.

The image $\Delta\Sigma^+$ of $\Sigma^+$ on $\Pi$ is no longer a horn but a "solid spiral" (sometimes called a *Shil'nikov snake*). The global map $Q$ maps the "snake" to the plane containing $\Sigma$.

Assume, first, that $\beta = 0$ and consider the intersection of the "snake" image (i.e., $P\Sigma^+$) with the local cross-section $\Sigma$. The origin of the "snake" is at the intersection of $\Gamma_0$ with $\Sigma$. The intersection of $\Sigma$ with $W^s(x_0)$ splits the

[7] Actually, in the case of the saddle-focus, there is a $C^1$ change of coordinates that locally linearizes the system (see Appendix B). It allows one to compute $\Delta$ analytically.

"snake" into an infinite number of *upper* and *lower* "half-spirals." The preimages $\Sigma_i$ of the upper "half-spirals" $P\Sigma_i$, $i = 1, 2, \dots$, are horizontal strips in $\Sigma^+$ (see **Fig. 6.32**). If the saddle quantity $\sigma_0 > 0$, the intersection $\Sigma_i \cap P\Sigma_i$ is

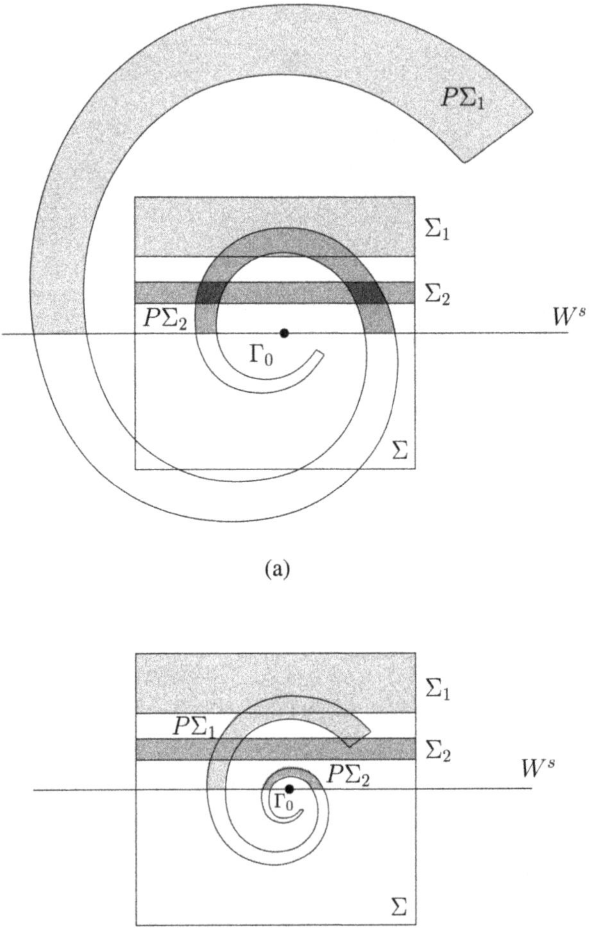


**Fig. 6.32.** Poincaré map structure in the saddle-focus cases: (a) $\sigma_0 > 0$; (b) $\sigma_0 < 0$.

nonempty and consists of two components for $i \geq i_0$, where $i_0$ is some positive number ($i_0 = 2$ in **Fig. 6.32**(a)). Each of these intersections forms a Smale horseshoe (see Chapter 1). It can be checked that the necessary expansion conditions are satisfied. Thus, each horseshoe gives an infinite number of saddle fixed points. These fixed points correspond to saddle limit cycles of (6.17). If $\sigma_0 < 0$, there is some $i_0 > 0$ such that for $i \geq i_0$ the intersection $\Sigma_i \cap P\Sigma_i$

is empty ($i_0 = 2$ in **Fig. 6.32**(b)). Thus, there are no fixed points of $P$ in $\Sigma^+$ close to $\Gamma_0$.

If $\beta \neq 0$, the point corresponding to $\Gamma_0$ is displaced from the horizontal line in $\Sigma$. Therefore, if $\sigma_0 > 0$, there remains only a *finite* number of Smale horseshoes. They still give an infinite number of saddle limit cycles in (6.17) for all sufficiently small $|\beta|$. In the case $\sigma_0 < 0$, the map $P$ is a contraction in $\Sigma^+$ for $\beta > 0$ and thus has a unique attracting fixed point corresponding to a stable limit cycle of (6.14). There are no periodic orbits if $\beta < 0$. □

**Remarks:**

(1) As in the saddle-focus case with $\sigma_0 < 0$, it cannot be said that the bifurcation diagrams of all systems (6.15) satisfying (H.1)–(H.2) are topologically equivalent. The reason is the same: the topological invariance of $\nu_0$ given by (6.15). Actually, the complete topological structure of the phase portrait near the homoclinic orbit is not known, although some substantial information is available due to Shil'nikov. Let $\tilde{\Omega}(\nu)$ be the set of all nonequivalent bi-infinite sequences

$$\omega = \{\dots, \omega_{-2}, \omega_{-1}, \omega_0, \omega_1, \omega_2, \dots\},$$

where $\omega_i$ are nonnegative integers such that

$$\omega_{i+1} < \nu\omega_i$$

for all $i = 0, \pm 1, \pm 2, \dots$, and for some real number $\nu > 0$. Then, at $\beta = 0$ there is a subset of orbits of (6.17) located in a neighborhood $U_0$ of $\Gamma_0 \cup x_0$ for all $t \in \mathbb{R}^1$, whose elements are in one-to-one correspondence with $\tilde{\Omega}(\nu)$, where $\nu$ does not exceed the topological invariant $\nu_0$. The value $\omega_i$ can be viewed as the number of "small" rotations made by the orbit near the saddle after the $i$th "global" turn.

(2) As $\beta$ approaches zero taking positive or negative values, an *infinite* number of bifurcations results. Some of these bifurcations are related to a "basic" limit cycle, which makes one global turn along the homoclinic orbit. It can be shown that this cycle undergoes an infinite number of *tangent* bifurcations as $|\beta|$ tends to zero. To understand the phenomenon, it is useful to compare the dependence on $\beta$ of the period $T_\beta$ of the cycle in the saddle-focus cases with $\sigma_0 < 0$ and $\sigma_0 > 0$. The relevant graphs are presented in **Fig. 6.33**. In the $\sigma_0 < 0$ case, the dependence is monotone, while if $\sigma_0 > 0$ it becomes "wiggly." The presence of wiggles means that the basic cycle disappears and appears via fold bifurcations infinitely many times. Notice that for any sufficiently small $|\beta|$ there is only a finite number of these "basic" cycles (they differ in the number of "small" rotations near the saddle-focus; the higher the period, the more rotations the cycle has). Moreover, the cycle also exhibits an infinite number of *period-doubling* bifurcations. The fold and period-doubling bifurcations are marked by $t$ and $f$, respectively, in the figure. Each of the generated double-period cycles makes two global turns before

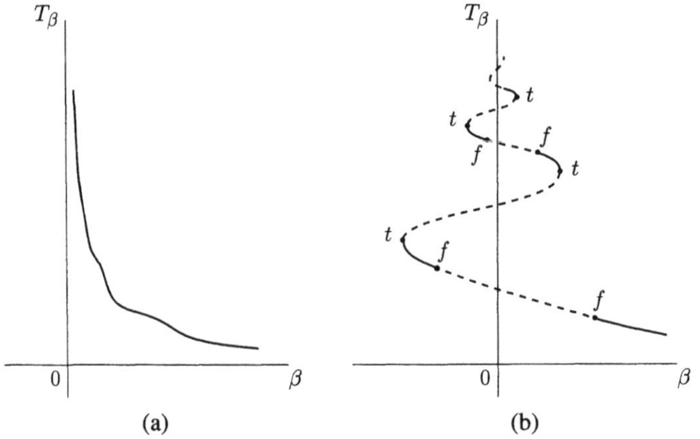


**Fig. 6.33.** Period of the cycle near a saddle-focus homoclinic bifurcation: (a) $\sigma_0 < 0$; (b) $\sigma_0 > 0$.

closure. These cycles themselves bifurcate while approaching the homoclinic orbit, making the picture more involved. Introduce the divergence of (6.17) at the saddle-focus:

$$\sigma_1 = (\text{div } f)(x_0, 0) = \lambda_1 + \lambda_2 + \lambda_3 = \lambda_1 + 2 \text{ Re } \lambda_{2,3}.$$

If $\sigma_1 < 0$ the basic cycle near the bifurcation can be stable (actually, there are only short intervals of $\beta$ within which it is stable). If $\sigma_1 > 0$ there are intervals where the basic cycle is totally unstable (repelling). Similar intervals exist for the secondary cycles generated by period doublinds. Thus, the saddle cycles mentioned in the theorem and coded at $\beta = 0$ by periodic sequences of $\Omega(\nu)$ are not the *only* cycles in $U_0$.

(3) Other bifurcations near the homoclinic orbit are due to *secondary homoclinic orbits.* Under the conditions of Theorem 6.6, there is an infinite sequence of $\beta_i > 0$, $\beta_i \to 0$, for which the system has double homoclinic orbits with different (increasing) number of rotations near the saddle-focus (see **Fig. 6.34**). Other subsidiary homoclinic orbits are also present, like the triple making three global turns before final return.

(4) Recall that in this section we assumed $n_- = \dim W^s = 2$ and $n_+ = \dim W^u = 1$. To apply the results in the opposite case (i.e., $n_- = 1, n_+ = 2$), we have to reverse the direction of time. This boils down to these substitutions: $\lambda_j \mapsto -\lambda_j$, $\sigma_i \mapsto -\sigma_i$, and "stable" $\mapsto$ "repelling." $\diamondsuit$

**Example 6.3 (Complex impulses in the FitzHugh-Nagumo model)** The following system of partial differential equations is the FitzHugh-Nagumo caricature of the Hodgkin-Huxley equations modeling the nerve impulse propagation along an axon (FitzHugh [1961], Nagumo, Arimoto & Yoshizawa [1962]):

$$\begin{cases} \dfrac{\partial u}{\partial t} = \dfrac{\partial^2 u}{\partial x^2} - f_a(u) - v, \\ \dfrac{\partial v}{\partial t} = bu, \end{cases}$$

where $u = u(x,t)$ represents the membrane potential; $v = v(x,t)$ is a phenomenological "recovery" variable; $f_a(u) = u(u-a)(u-1),\ 1 > a > 0,\ b > 0,\ -\infty < x < +\infty,\ t > 0$.

*Traveling waves* are solutions to these equations of the form

$$u(x,t) = U(\xi),\ v(x,t) = V(\xi),\ \xi = x + ct,$$

where $c$ is an a priori unknown wave propagation speed. The functions $U(\xi)$ and $V(\xi)$ define profiles of the waves. They satisfy the system of ordinary differential equations

$$\begin{cases} \dot{U} = W, \\ \dot{W} = cW + f_a(U) + V, \\ \dot{V} = \dfrac{b}{c}U, \end{cases} \tag{6.18}$$

where the dot means differentiation with respect to "time" $\xi$. System (6.18) is called a *wave system*. It depends on three positive parameters $(a, b, c)$. Any bounded orbit of (6.18) corresponds to a traveling wave solution of the FitzHugh-Nagumo system at parameter values $(a, b)$ propagating with velocity $c$.

For all $c > 0$ the wave system has a unique equilibrium $0 = (0,0,0)$ with one positive eigenvalue $\lambda_1$ and two eigenvalues $\lambda_{2,3}$ with negative real parts (see Exercise 2 in Chapter 2). The equilibrium can be either a saddle or a saddle-focus with a one-dimensional unstable and a two-dimensional stable invariant manifold, $W^{u,s}(0)$. The transition between saddle and saddle-focus cases is caused by the presence of a double negative eigenvalue; for fixed $b > 0$ this happens on the curve

$$D_b = \{(a,c) : c^4(4b - a^2) + 2ac^2(9b - 2a^2) + 27b^2 = 0\}.$$

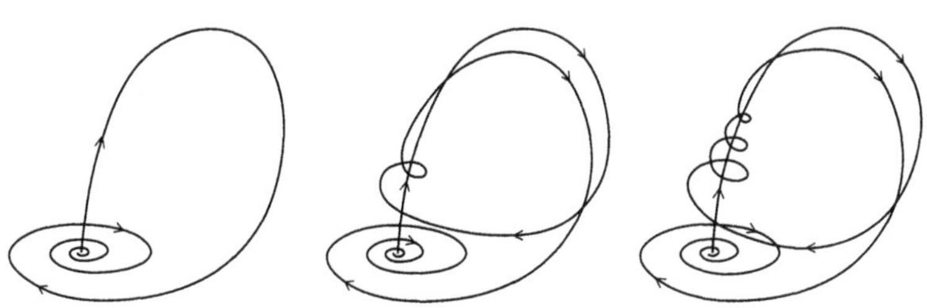

(a) (b) (c)

**Fig. 6.34.** Basic (a) and double (b, c) homoclinic orbits.

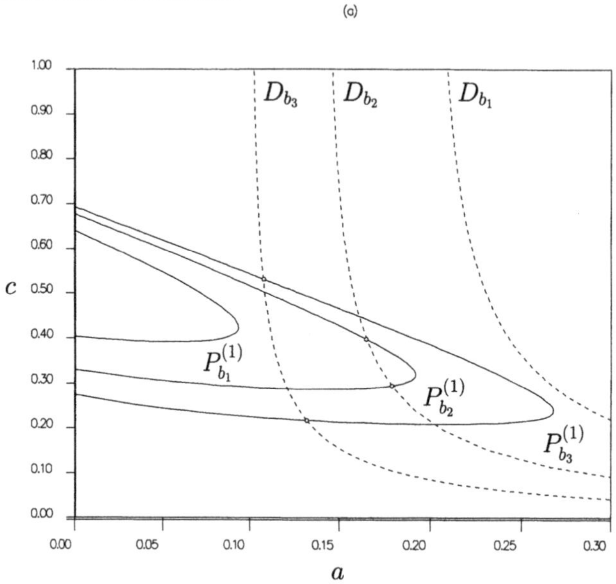


**Fig. 6.35.** Bifurcation curves of the wave system (6.18): $b_1 = 0.01$; $b_2 = 0.005$; $b_3 = 0.0025$.

Several boundaries $D_b$ in the $(a, c)$-plane for different values of $b$ are depicted in **Fig. 6.35** as dashed lines. The saddle-focus region is located below each boundary and disappears as $b \to 0$.

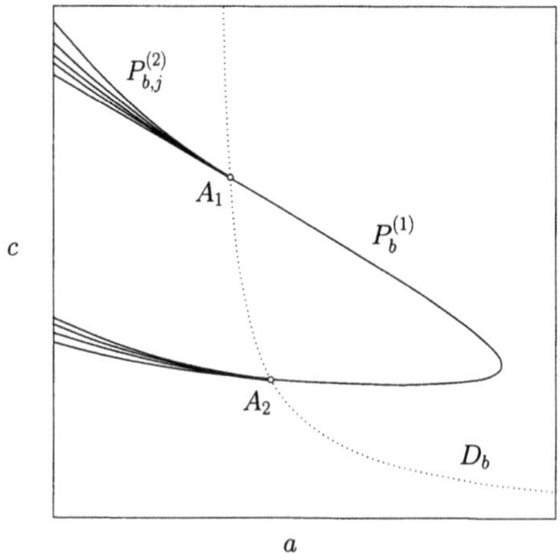


**Fig. 6.36.** Parametric curves $D_b$ and $P_b^{(1)}$ for $b = 0.0025$.

A branch $W_1^u(0)$ of the unstable manifold leaving the origin into the positive octant can return back to the equilibrium, forming a homoclinic orbit

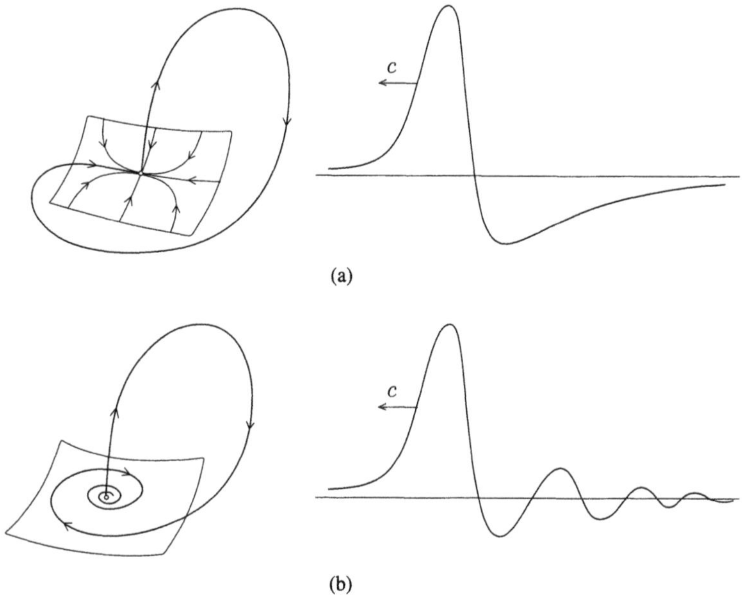


**Fig. 6.37.** Impulses with (a) monotone and (b) oscillating "tails".

$\Gamma_0$ at some parameter values [Hastings 1976]. These parameter values can be found only numerically with the help of the methods described in Chapter 10. **Fig. 6.35** presents several homoclinic bifurcation curves $P_b^{(1)}$ in the $(a, c)$-plane computed by Kuznetsov & Panfilov [1981] for different but fixed values of $b$. Looking at **Fig. 6.35**, we can conclude that for all $b > 0$ the bifurcation curve $P_b^{(1)}$ passes through the saddle-focus region delimited by $D_b$ (see **Fig. 6.36**, where the curves $D_b$ and $P_b^{(1)}$ corresponding to $b = 0.0025$ are superimposed). Actually, for $b > 0.1$, the homoclinic bifurcation curve belongs entirely to the saddle-focus region. Any homoclinic orbit defines a traveling *impulse*. The shape of the impulse depends very much on the type of the corresponding equilibrium: It has a monotone "tail" in the saddle case and an oscillating "tail" in the saddle-focus case (see **Fig. 6.37**).

The saddle quantity $\sigma_0$ is always positive for $c > 0$ (see Exercise 11). Therefore, the phase portraits of (6.18) near the homoclinic curve $P_b^{(1)}$ are described by Theorems 6.5 and 6.6. In particular, near the homoclinic bifurcation curve $P_b^{(1)}$ in the saddle-focus region, system (6.18) has an infinite number of saddle cycles. These cycles correspond to *periodic wave trains* in the FitzHugh-Nagumo model [Feroe 1981]. Secondary homoclinic orbits existing in (6.18) near the primary homoclinic bifurcation correspond to *double traveling impulses* (see **Fig. 6.38**) [Evans, Fenichel & Feroe 1982]. It can be shown using results by Belyakov [1980] (see Kuznetsov & Panfilov [1981]) that

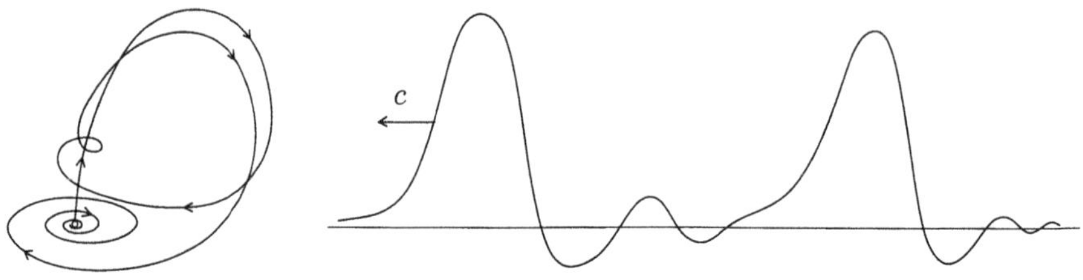


**Fig. 6.38.** A double impulse.

secondary homoclinic bifurcation curves $P_{b,j}^{(2)}$ in (6.18) originate at points $A_{1,2}$ where $P_b^{(1)}$ intersects $D_b$ (see **Fig. 6.36** for a sketch). ◊

## 6.4 Homoclinic bifurcations in $n$-dimensional systems

It has been proved (see references in Appendix B) that there exists a parameter-dependent invariant *center manifold* near homoclinic bifurcations. This allows one to reduce the study of generic bifurcations of orbits homoclinic to hyperbolic equilibria in $n$-dimensional systems with $n > 3$ to that in two-, three-, or four-dimensional systems. In this section, we discuss which homoclinic orbits are generic in $n$-dimensional case and formulate the *Homoclinic Center Manifold Theorem* for such orbits. Then we derive from this theorem some results concerning generic homoclinic bifurcations in $n$-dimensional systems, first obtained by L.P. Shil'nikov without a center-manifold reduction.

### 6.4.1 Regular homoclinic orbits: Melnikov integral

Consider a system

$$\dot{x} = f(x, \alpha), \quad x = (x_1, x_2, \dots, x_n)^T \in \mathbb{R}^n, \ \alpha \in \mathbb{R}^1, \tag{6.19}$$

where $f$ is $C^\infty$ smooth and $n \geq 3$. Suppose, that (6.19) has a hyperbolic equilibrium $x_0$ at $\alpha = 0$, and the Jacobian matrix $A_0 = f_x(x_0, 0)$ has $n_+$ eigenvalues with positive real parts

$$0 < \text{ Re } \lambda_1 \ \leq \text{ Re } \lambda_2 \ \leq \cdots \leq \text{ Re } \lambda_{n_+}$$

and $n_-$ eigenvalues with negative real parts

$$\text{Re } \mu_{n_-} \ \leq \text{ Re } \mu_{n_- - 1} \ \leq \cdots \leq \text{ Re } \mu_1 < 0.$$

For all sufficiently small $|\alpha|$, the equilibrium persists and has unstable and stable local invariant manifolds $W^u$ and $W^s$ that can be globally extended, $\dim W^{u,s} = n_\pm$, $n_+ + n_- = n$. Assume that (6.19) has at $\alpha = 0$ an orbit $\Gamma_0$ homoclinic to $x_0$ and denote by $x^0(t)$ a solution of (6.19) corresponding to $\Gamma_0$.

As we have seen in Section 6.1, the intersection of $W^s(x_0)$ and $W^u(x_0)$ along $\Gamma_0$ cannot be transversal since the vector $\dot{x}^0(t_0) = f(x^0(t_0), 0)$ is tangent to both manifolds at any point $x^0(t_0) \in \Gamma_0$. However, in the generic case, $\dot{x}^0(t_0)$ is the only such vector:

$$T_{x^0(t_0)}W^u(x_0) \cap T_{x^0(t_0)}W^s(x_0) = \text{span}\{\dot{x}^0(t_0)\}.$$

Thus, generically,

$$\text{codim}(T_{x^0(t_0)}W^s(x_0) + T_{x^0(t_0)}W^s(x_0)) = 1.$$

A generic perturbation splits the manifolds $W^s(x_0)$ and $W^u(x_0)$ by an $O(\alpha)$-distance in the remaining direction for $\alpha \neq 0$. Such homoclinic orbits are called *regular*.

As in Section 6.2, introduce the *extended system*:

$$\begin{cases} \dot{x} = f(x, \alpha), \\ \dot{\alpha} = 0, \end{cases} \tag{6.20}$$

with the phase variables $(x, \alpha)^T \in \mathbb{R}^{n+1}$. Let $x_0(\alpha)$ denote a one-parameter family of the saddles in (6.19) for small $|\alpha|$, $x_0(0) = x_0$. This family defines an invariant set of (6.20) – a curve of equilibria. This curve has the unstable and stable manifolds, $\mathcal{W}^u$ and $\mathcal{W}^s$, whose slices $\alpha = \text{const}$ coincide with the corresponding unstable and stable manifolds $W^u$ and $W^s$ of the saddle $x_0(\alpha)$. It is clear that the regularity of the homoclinic orbit translates exactly into the *transversality* of the intersection of $\mathcal{W}^u$ and $\mathcal{W}^s$ along $\Gamma_0$ at $\alpha = 0$ in the $(n+1)$-dimensional phase space of (6.20):

$$T_{(x^0(t_0),0)}\mathcal{W}^u + T_{(x^0(t_0),0)}\mathcal{W}^s = \mathbb{R}^{n+1}.$$

**Fig. 6.21** in Section 6.2 illustrates the case $n = 2$.

The transversality of the intersection of $\mathcal{W}^u$ and $\mathcal{W}^s$ can be expressed analytically. Namely, consider the *linearization* of (6.20) around $x^0(t)$ at $\alpha = 0$:

$$\begin{cases} \dot{u} = f_x(x^0(t), 0)u + f_\alpha(x^0(t), 0)\mu, \\ \dot{\mu} = 0. \end{cases} \tag{6.21}$$

If $(u_0, \mu_0)^T$ is a vector tangent to either $\mathcal{W}^u$ or $\mathcal{W}^s$, then the solution vector $(u(t), \mu(t))^T$ of this system with the initial data $(u_0, \mu_0)^T$ is always tangent to the corresponding invariant manifold. The vector function

$$\begin{pmatrix} \varsigma(t) \\ \mu(t) \end{pmatrix} = \begin{pmatrix} \dot{x}^0(t) \\ 0 \end{pmatrix}$$

is a bounded solution to (6.21) that is tangent to both invariant manifolds $\mathcal{W}^u$ and $\mathcal{W}^s$ along the curve of their intersection.[8] We can multiply this solution

[8] Actually, this solution tends to zero exponentially fast as $t \to \pm\infty$ since $\dot{x}^0(t) = f(x^0(t), 0)$.

by a scalar to get another bounded solution to (6.21). The transversality of the intersection of $\mathcal{W}^u$ and $\mathcal{W}^s$ along $\Gamma_0$ at $\alpha = 0$ means that $(\dot{x}^0(t), 0)^T$ is the unique to within a scalar multiple bounded solution to the extended system (6.21). Taking into account that the equation for $\mu$ in (6.21) is trivial, we can conclude that $\dot{x}^0(t)$ is the unique to within a scalar multiple solution to the *variational equation* around $\Gamma_0$:

$$\dot{u} = A(t)u, \quad u \in \mathbb{R}^n, \tag{6.22}$$

where $A(t) = f_x(x^0(t), 0)$. This implies that the *adjoint variational equation* around $\Gamma_0$:

$$\dot{v} = -A^T(t)v, \quad v \in \mathbb{R}^n, \tag{6.23}$$

has the unique to within a scalar multiple bounded solution $v(t) = \eta(t)$. Indeed, the equations (6.22) and (6.23) have the same number of linearly independent bounded solutions. Actually, the vectors $\zeta(t)$ and $\eta(t)$ are orthogonal for each $t \in \mathbb{R}^1$. Using (6.22) and (6.23), we get

$$\frac{d}{dt}\langle \eta, \zeta \rangle = \langle \dot{\eta}, \zeta \rangle + \langle \eta, \dot{\zeta} \rangle = -\langle A^T\eta, \zeta \rangle + \langle \eta, A\zeta \rangle = -\langle \eta, A\zeta \rangle + \langle \eta, A\zeta \rangle = 0,$$

i.e., $\langle \eta(t), \zeta(t) \rangle = C$. The constant $C$ is zero since both $\eta(t)$ and $\zeta(t)$ tend to zero exponentially fast as $t \to \pm\infty$:

$$\langle \eta(t), \zeta(t) \rangle = 0, \quad t \in \mathbb{R}^1.$$

Meanwhile, similar arguments show that $\eta(t)$ is orthogonal to any vector tangent to either $W^s(x_0)$ or $W^u(x_0)$. Moreover, the transversality is equivalent to the condition

$$M_\alpha(0) = \int_{-\infty}^{+\infty} \langle \eta(t), f_\alpha(x^0(t), 0) \rangle \, dt \neq 0. \tag{6.24}$$

If the intersection of $\mathcal{W}^u$ and $\mathcal{W}^s$ is nontransversal, there exists another bounded solution $(\zeta_0(t), \mu_0)^T$ to (6.21) with $\mu_0 \neq 0$. Taking the scalar product of (6.21) with $\eta$ and integrating over time, we get

$$\begin{aligned}\mu_0 \int_{-\infty}^{+\infty} \langle \eta(t), f_\alpha(x^0(t), 0) \rangle \, dt &= \int_{-\infty}^{+\infty} \langle \eta(t), \dot{\zeta}_0(t) - A(t)\zeta_0(t) \rangle \, dt \\ = \langle \eta(t), \zeta_0(t) \rangle \Big|_{-\infty}^{+\infty} - \int_{-\infty}^{+\infty} &\langle \dot{\eta}(t) + A^T(t)\eta(t), \zeta_0(t) \rangle \, dt = 0.\end{aligned}$$

The integral in (6.24) is called the *Melnikov integral*. This condition allows us to verify the regularity of the manifold splitting in $n$-dimensional systems with $n \geq 2$. Moreover, one can introduce a scalar split function $M(\alpha)$ that measures the displacement of the invariant manifolds $W^s(x_0)$ and $W^u(x_0)$ near the point $x^0(0) \in \Gamma_0$ in the direction defined by the vector $\eta(0)$ and has the property

$$M(\alpha) = M_\alpha(0)\alpha + O(\alpha^2),$$

where $M_\alpha(0)$ is given by (6.24).

In the two-dimensional case the Melnikov integral $M_\alpha(0)$ can be computed more explicitly. Write (6.19) in coordinates:

$$\begin{cases} \dot{x}_1 = f_1(x, \alpha), \\ \dot{x}_2 = f_2(x, \alpha). \end{cases}$$

The solution $\zeta(t)$ to the variational equation (6.22) has the form

$$\zeta(t) = \dot{x}^0(t) = \begin{pmatrix} f_1(x^0(t), 0) \\ f_2(x^0(t), 0) \end{pmatrix}.$$

Since $\eta(t) \perp \zeta(t)$, we have

$$\eta(t) = \varphi(t) \begin{pmatrix} -f_2(x^0(t), 0) \\ f_1(x^0(t), 0) \end{pmatrix}$$

for some scalar function $\varphi(t)$. It is easy to verify that this function satisfies the equation

$$\dot{\varphi}(t) = -(\text{div } f)(x^0(t), 0)\varphi(t),$$

where

$$\text{div } f = \left( \frac{\partial f_1}{\partial x_1} + \frac{\partial f_2}{\partial x_2} \right)$$

is the *divergence* of the vector field $f$. Assuming $\varphi(0) = 1$, we obtain

$$\varphi(t) = e^{-\int_0^t (\text{div } f)(x^0(\tau), 0)\, d\tau}$$

and

$$M_\alpha(0) = \int_{-\infty}^{+\infty} \exp\left[ -\int_0^t \left( \frac{\partial f_1}{\partial x_1} + \frac{\partial f_2}{\partial x_2} \right) d\tau \right] \left( f_1 \frac{\partial f_2}{\partial \alpha} - f_2 \frac{\partial f_1}{\partial \alpha} \right) dt, \quad (6.25)$$

where all expressions with $f = (f_1, f_2)^T$ are evaluated along the homoclinic solution $x^0(\cdot)$ at $\alpha = 0$.

**Remark:**

Suppose that (6.4) is a Hamiltonian system at $\alpha = 0$, and $\alpha$ is a small parameter in front of the perturbation term, i.e.,

$$\dot{x} = J(\nabla H)(x) + \alpha g(x), \quad x \in \mathbb{R}^2, \ \alpha \in \mathbb{R}^1,$$

where

$$J = \begin{pmatrix} 0 & 1 \\ -1 & 0 \end{pmatrix}, \quad \nabla H = \left( \frac{\partial H}{\partial x_1}, \frac{\partial H}{\partial x_2} \right)^T,$$

and $H = H(x)$ is the Hamiltonian function. Then the Melnikov integral (6.25) can be simplifyed further. In such a case, div $f \equiv 0$ and the homoclinic orbit

$\Gamma_0$ belongs to a level curve $\{x : H(x) = H(x_0)\}$. Assume that its interior is a domain $\Omega = \{H(x) \le H(x_0)\}$. Then, applying Green's theorem reduces the Melnikov integral along $\Gamma_0$ to the domain integral

$$M_\alpha(0) = \int_\Omega (\text{div } g)(x^0(t))\, d\omega. \diamondsuit$$

### 6.4.2 Homoclinic center manifolds

To formulate the Homoclinic Center Manifold Theorem, it is useful to distinguish the eigenvalues that are closest to the imaginary axis (see **Fig. 6.39**).

**Definition 6.9** *The eigenvalues with positive (negative) real part that are closest to the imaginary axis are called the unstable (stable)* leading eigenvalues, *while the corresponding eigenspaces are called the unstable (stable)* leading eigenspaces.

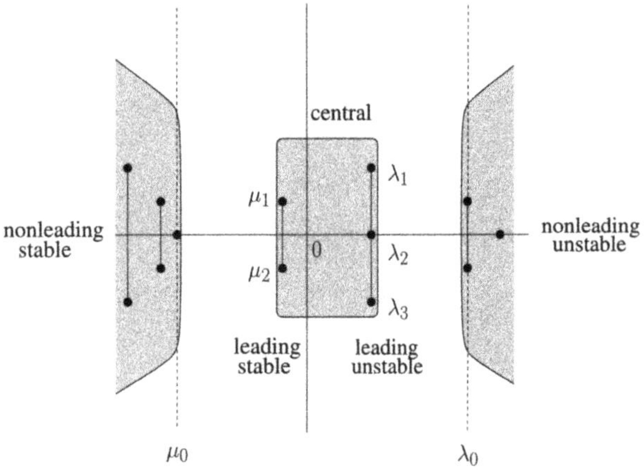


**Fig. 6.39.** Splitting of the eigenvalues.

**Definition 6.10** *The stable and unstable leading eigenvalues together are called* central eigenvalues, *while the corresponding eigenspace is called the* central eigenspace.

Almost all orbits on $W^u(W^s)$ tend to the equilibrium as $t \to -\infty$ $(t \to +\infty)$ along the corresponding leading eigenspace that we denote by $T^u(T^s)$. Exceptional orbits form a *nonleading* manifold $W^{uu}(W^{ss})$ tangent to the eigenspace $T^{uu}(T^{ss})$ corresponding to the nonleading eigenvalues. The central eigenspace $T^c$ is the direct sum of the stable and unstable leading eigenspaces: $T^c = T^u \oplus T^s$. Denote by $\lambda_0$ the minimal Re $\lambda_j$ corresponding to the *nonleading* unstable eigenvalues and by $\mu_0$ the maximal Re $\mu_j$ corresponding to the *nonleading* stable eigenvalues (see **Fig. 6.39**). By the construction,

$$\mu_0 < \text{Re } \mu_1 < 0 < \text{Re } \lambda_1 < \lambda_0,$$

where $\lambda_1$ is a leading unstable eigenvalue and $\mu_1$ is a leading stable eigenvalue. Provided both nonleading eigenspaces are nonempty, introduce two real numbers:

$$g^s = \frac{\mu_0}{\text{Re } \mu_1}, \quad g^u = \frac{\lambda_0}{\text{Re } \lambda_1}.$$

These numbers characterize the *relative gaps* between the corresponding nonleading and leading eigenvalues and satisfy $g^{s,u} > 1$. If one of the nonleading eigenspaces is empty, set formally $g^s = -\infty$ or $g^u = +\infty$.

Now notice that the variational equation (6.22) is a nonautonomous linear system with matrix $A(t)$ that approaches asymptotically a constant matrix, namely

$$\lim_{t \to \pm\infty} A(t) = A_0.$$

Therefore, for $t \to \pm\infty$, solutions of (6.21) behave like solutions of the autonomous linear system

$$\dot{v} = A_0 v$$

and we can introduce four linear subspaces of $\mathbb{R}^n$:

$$E^{uu}(t_0) = \left\{ v_0 : \lim_{t \to -\infty} \frac{v(t)}{\|v(t)\|} \in T^{uu} \right\},$$

$$E^{ss}(t_0) = \left\{ v_0 : \lim_{t \to +\infty} \frac{v(t)}{\|v(t)\|} \in T^{ss} \right\},$$

$$E^{cu}(t_0) = \left\{ v_0 : \lim_{t \to -\infty} \frac{v(t)}{\|v(t)\|} \in T^c \oplus T^{uu} \right\},$$

$$E^{cs}(t_0) = \left\{ v_0 : \lim_{t \to +\infty} \frac{v(t)}{\|v(t)\|} \in T^c \oplus T^{ss} \right\},$$

where $v(t) = \Phi(v_0, t_0, t)$ is the solution to (6.21) with the initial data $v = v_0$ at $t = t_0$, and $\oplus$ stands for the direct sum of the linear subspaces. Finally, define

$$E^c(t_0) = E^{cu}(t_0) \cap E^{cs}(t_0).$$

Now we can formulate without proof the following theorem.

**Theorem 6.7 (Homoclinic Center Manifold)** *Suppose that* (6.19) *has at* $\alpha = 0$ *a hyperbolic equilibrium* $x_0 = 0$ *with a homoclinic orbit*

$$\Gamma_0 = \{x \in \mathbb{R}^n : x = x^0(t), \ t \in \mathbb{R}^1\}.$$

*Assume the following conditions hold:*

(H.1) $\dot{x}^0(0) \in E^c(0)$;
(H.2) $E^{uu}(0) \oplus E^c(0) \oplus E^{ss}(0) = \mathbb{R}^n$.

*Then, for all sufficiently small* $|\alpha|$, (6.19) *has an invariant manifold* $\mathcal{M}_\alpha$ *defined in a small neighborhood* $U_0$ *of* $\Gamma_0 \cup x_0$ *and having the following propertics:*

(i) $x^0(t_0) \in \mathcal{M}_0$ *and the tangent space* $T_{x^0(t_0)}\mathcal{M}_0 = E^c(t_0)$, *for all* $t_0 \in \mathbb{R}^1$;
(ii) *any solution to* (6.19) *that stays inside* $U_0$ *for all* $t \in \mathbb{R}^1$ *belongs to* $\mathcal{M}_\alpha$;
(iii) *each* $\mathcal{M}_\alpha$ *is* $C^k$ *smooth, where* $k \geq 1$ *is the maximal integer number satisfying both*

$$g^s > k \quad \text{and} \quad g^u > k. \quad \square$$

**Definition 6.11** *The manifold* $\mathcal{M}_\alpha$ *is called the* homoclinic center manifold.

**Remarks:**

(1) The conditions (H.1) and (H.2) guarantee that similar conditions hold for all $t_0 \neq 0$. The first condition implies that the homoclinic orbit $\Gamma_0$ approaches the equilibrium $x_0$ along the leading eigenspaces for both $t \to +\infty$ and $t \to -\infty$. The second condition means that the invariant manifolds $W^s(x_0)$ and $W^u(x_0)$ intersect at $\alpha = 0$ along the homoclinic orbit $\Gamma_0$ in the "least possible" nontransversal manner.

(2) The manifold $\mathcal{M}_0$ is exponentially attracting along the $E^{ss}$-directions and exponentially repelling along the $E^{uu}$-directions. The same property holds for $\mathcal{M}_\alpha$ for small $|\alpha| \neq 0$ with $E^{ss,uu}$ replaced by close subspaces.

(3) The homoclinic center manifold has only *finite smoothness* $C^k$ that increases with the relative gaps $g^{s,u}$. The restriction of (6.19) to the invariant manifold $\mathcal{M}_\alpha$ is a $C^k$-system of ODEs, provided proper coordinates on $\mathcal{M}_\alpha$ are choosen. This restricted system has an orbit homoclinic to $x_0$ at $\alpha = 0$.

(4) Actually, under the assumptions of Theorem 6.7, the homoclinic center manifold belongs to the class $C^{k,\beta}$ with some $0 < \beta < 0$, i.e., can locally be represented as the graph of a function whose derivatives of order $k$ are Hölder-continuous with index $\beta$. $\diamondsuit$

The theorem is illustrated for $\mathbb{R}^3$ in **Fig. 6.40**. Only the critical homoclinic center manifold $\mathcal{M}_0$ at $\alpha = 0$ is presented. It is assumed that all eigenvalues of $x_0$ are real and simple: $\mu_2 < \mu_1 < 0 < \lambda_1$. The central eigenspace $T^c$ of the saddle $x_0$ is two-dimensional and is spanned by the (leading) unstable eigenvector $v_1$ ($A_0 v_1 = \lambda_1 v_1$) and the leading stable eigenvector $w_1$ ($A_0 w_1 = \mu_1 w_1$). The manifold $\mathcal{M}_0$ is two-dimensional, contains $\Gamma_0$, and is tangent to $T^c$ at $x_0$. It is exponentially attracting in the $E^{ss}$-direction. The manifold can be either *orientable* (**Fig. 6.40**(a)) or *nonorientable* (**Fig. 6.40**(b)). In this case, condition (H.2) is equivalent to the strong inclination property (see Section 6.3), so the orientability of $\mathcal{M}_0$ depends on whether the closure of $W^u$ near $\Gamma_0$ is orientable or nonorientable (cf. **Fig. 6.25**).

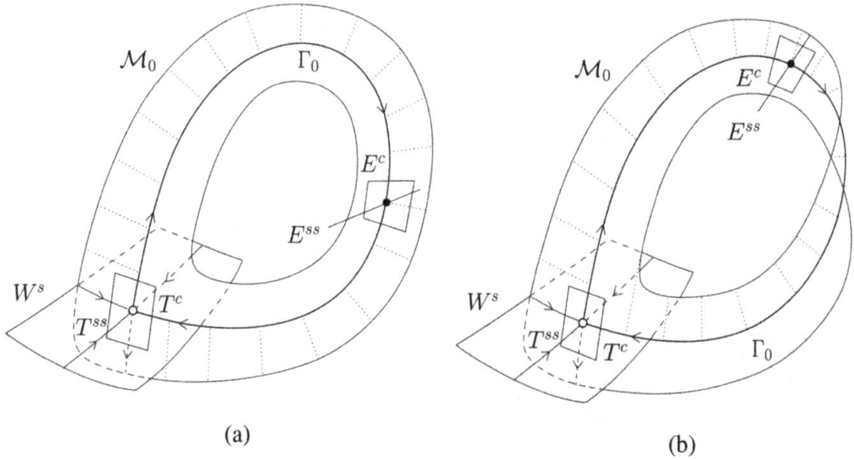


**Fig. 6.40.** Homoclinic center manifold $\mathcal{M}_0$ in $\mathbb{R}^3$.

### 6.4.3 Generic homoclinic bifurcations in $\mathbb{R}^n$

Generically, the leading eigenspaces $T^{s,u}$ are either one- or two-dimensional. In the first case, an eigenspace corresponds to a simple real eigenvalue, while in the second case, it corresponds to a simple pair of complex-conjugate eigenvalues. Reversing the time direction if necessary, we have only three typical configurations of the leading eigenvalues:

(a) (*saddle*) The leading eigenvalues are real and simple: $\mu_1 < 0 < \lambda_1$ (**Fig. 6.41**(a));

(b) (*saddle-focus*) The stable leading eigenvalues are complex and simple: $\mu_1 = \bar{\mu}_2$, while the unstable leading eigenvalue $\lambda_1$ is real and simple (**Fig. 6.41**(b));

(c) (*focus-focus*) The leading eigenvalues are complex and simple: $\lambda_1 = \bar{\lambda}_2$, $\mu_1 = \bar{\mu}_2$ (**Fig. 6.41**(c)).

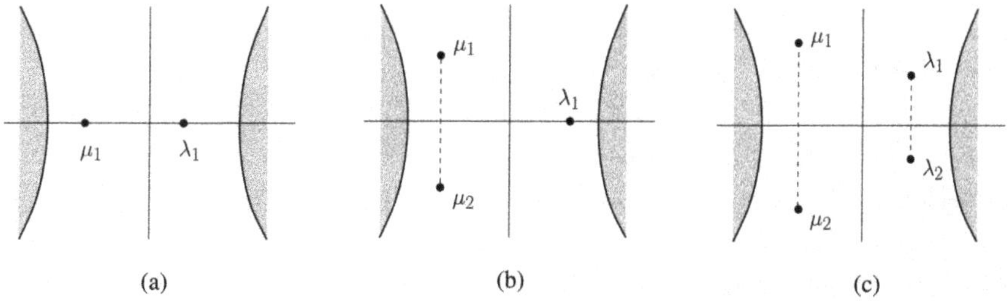


**Fig. 6.41.** Leading eigenvalues in generic Shil'nikov cases.

**Definition 6.12** *The saddle quantity $\sigma$ of a hyperbolic equilibrium is the sum of the real parts of its leading eigenvalues.*

Therefore,

$$\sigma = \text{Re}\ \lambda_1 + \text{Re}\ \mu_1,$$

where $\lambda_1$ is a leading unstable eigenvalue and $\mu_1$ is a leading stable eigenvalue. Assume that the following nondegeneracy condition holds at $\alpha = 0$:

(H.0) *$\sigma_0 \neq 0$ and the leading eigenspaces $T^{s,u}$ are either one- or two-dimensional.*

The following theorems are direct consequences of Theorem 6.7 and the results obtained in Sections 6.2–6.4.

**Theorem 6.8 (Saddle)** *For any generic one-parameter system having a saddle equilibrium point $x_0$ with a homoclinic orbit $\Gamma_0$ at $\alpha = 0$ there exists a neighborhood $U_0$ of $\Gamma_0 \cup x_0$ in which a unique limit cycle $L_\alpha$ bifurcates from $\Gamma_0$ as $\alpha$ passes through zero. Moreover, $\dim W^s(L_\alpha) = n_- + 1$ if $\sigma_0 < 0$, and $\dim W^s(L_\alpha) = n_-$ if $\sigma_0 > 0$.* □

In the saddle case, the homoclinic center manifold $\mathcal{M}_\alpha$ is two-dimensional and is a simple (orientable) or a twisted (nonorientable or Möbius) band. At $\alpha = 0$ the restricted system has a homoclinic orbit. The proof of Theorem 6.1 (see Section 6.2) can be carried out with only a slight modification. Namely, the coefficient $a(\beta)$ of the global map $Q$ can now be either positive (orientable case) or negative (Möbius case). In this case conditions (H.1) and (H.2) imply that the $W^{u,s}$ intersects itself near the saddle along the corresponding non-leading manifold $W^{ss,uu}$. In the three-dimensional case this condition means that the homoclinic orbit $\Gamma_0$ is either simple or twisted (as defined in Section 6.3). Thus, we have an alternative way to prove Theorems 6.3 and 6.5.

**Theorem 6.9 (Saddle-focus)** *For any generic one-parameter system having a saddle-focus equilibrium point $x_0$ with a homoclinic orbit $\Gamma_0$ at $\alpha = 0$ there exists a neighborhood $U_0$ of $\Gamma_0 \cup x_0$ such that one of the following alternatives hold:*

(a) *if $\sigma_0 < 0$, a unique limit cycle $L_\alpha$ bifurcates from $\Gamma_0$ in $U_0$ as $\alpha$ passes through zero, $\dim W^s(L_\alpha) = n_- + 1$;*

(b) *if $\sigma_0 > 0$, the system has an infinite number of saddle limit cycles in $U_0$ for all sufficiently small $|\alpha|$.* □

In this case, the homoclinic center manifold $\mathcal{M}_\alpha$ is three-dimensional. At $\alpha = 0$ the restricted system has a homoclinic orbit to the saddle-focus, so we can repeat the proof of Theorem 6.4 (in case (a)) and that of Theorem 6.6 (in case (b)) on this manifold.

**Theorem 6.10 (Focus-focus)** *For any generic one-parameter system having a focus-focus equilibrium point $x_0$ with a homoclinic orbit $\Gamma_0$ at $\alpha = 0$*

*there exists a neighborhood $U_0$ of $\Gamma_0 \cup x_0$ in which the system has an infinite number of saddle limit cycles in $U_0$ for all sufficiently small $|\alpha|$.* $\square$

Here, the homoclinic center manifold $\mathcal{M}_\alpha$ is four-dimensional and carries a homoclinic orbit to the focus-focus at $\alpha = 0$. Thus, the proof of Theorem 6.11 from Appendix A is valid.

The genericity conditions mentioned in the theorems are the nondegeneracy conditions (H.0), (H.1), and (H.2) listed above, as well as the transversality condition:

(H.3) *the homoclinic orbit $\Gamma_0$ is regular, i.e., the intersection of the tangent spaces $T_{x^0(t)}W^s$ and $T_{x^0(t)}W^u$ at each point $x^0(t) \in \Gamma_0$ is one-dimensional and $W^s$ and $W^u$ split by an $O(\alpha)$ distance as $\alpha$ moves away from zero.*

Recall that $(H.3)$ can be reformulated using the Melnikov integral as

$$\int_{-\infty}^{+\infty} \langle \eta(t), f_\alpha(x^0(t), 0) \rangle \, dt \neq 0,$$

where $\eta(t)$ is the unique to within a scalar multiple bounded solution to the *adjoint variational equation* around $\Gamma_0$:

$$\dot{u} = -A^T(t)u, \quad u \in \mathbb{R}^n.$$

## 6.5 Exercises

**(1)** Construct a one-parameter family of two-dimensional Hamiltonian systems

$$\begin{cases} \dot{x} = H_y, \\ \dot{y} = -H_x, \end{cases}$$

where $H = H(x, y, \alpha)$ is a (polynomial) Hamilton function, having a homoclinic orbit. (*Hint*: Orbits of the system belong to level curves of the Hamiltonian: $H(x, y, \alpha) = \text{const.}$)

**(2) (Homoclinc orbit in a non-Hamiltonian system)** Show that the system

$$\begin{cases} \dot{x} = y, \\ \dot{y} = -x^3 + x + xy, \end{cases}$$

has a saddle at the origin with a "big" homoclinic orbit. (*Hint*: Use the symmetry of the system under a reflection and time reversal: $x \mapsto -x,\ t \mapsto -t$.) Is this orbit nondegenerate?

**(3)** Prove Lemma 6.1 in the planar case using rotation of the vector field. (*Hint*: See Andronov et al. [1973].)

**(4) (Heteroclinic bifurcation)** Prove that the system

$$\begin{cases} \dot{x} = \alpha + 2xy, \\ \dot{y} = 1 + x^2 - y^2, \end{cases}$$

undergoes a heteroclinic bifurcation at $\alpha = 0$.

**(5) (Asymptote of the period)** Find an asymptotic form of the cycle period $T(\beta)$ near the homoclinic bifurcation on the plane. Is this result valid for the $n$-dimensional case? (*Hint*: Use the fact that a point on the cycle spends the most time near the saddle.)

**(6) (Multiplier of the cycle near homoclinic bifurcation)** Show that the (non-trivial) multiplier of the cycle bifurcating from a homoclinic orbit in a planar system approaches zero as $\beta \to 0$. Can this result be generalized to higher dimensions?

**(7) ($C^1$-linearization near the saddle on the plane)**

(a) Draw *isochrones*, $\tau_{1,2} = \text{const}$, of constant "exit" times from the unit square $\Omega$ for the linear system (6.7). Check that these lines are transversal. How will the figure change in the nonlinear case? Prove that the map $\Phi(y)$ constructed in the proof of Theorem 6.1 is a homeomorphism.

(b) Prove that the map $\Phi(y)$ has only first-order continuous partial derivatives at $y = 0$. (*Hint*: $\Phi_y(0) = I_2$, see Deng [1989].)

**(8) (Dependence of orbits upon a singular parameter)** Consider the slow-fast system

$$\begin{cases} \dot{x} = x^2 - y, \\ \dot{y} = -\varepsilon, \end{cases}$$

where $\varepsilon$ is small but positive. Take an orbit of the system starting at

$$\begin{aligned} x_0 &= -(1 + \varepsilon), \\ y_0 &= 1. \end{aligned}$$

Let $y_1 = y_1(\varepsilon)$ be the ordinate of the point of intersection between the orbit and the vertical line $x = 1$.

(a) Show that the derivative of $y_1(\varepsilon)$ with respect to $\varepsilon$ tends to $-\infty$ as $\varepsilon \to 0$. (*Hint*: $y_1'(\varepsilon) = -T(\varepsilon)$, where $T$ is the "flight" time from the initial point $(x_0, y_0)$ to the point $(1, y_1)$.)

(b) Check that the result will not change if we take $x_0 = -(1 + \varphi(\varepsilon))$ with any smooth positive function $\varphi(\varepsilon) \to 0$ for $\varepsilon \to 0$.

(c) Explain the relationship between the above results and nondifferentiability of the split function in the slow-fast planar system used as the example in Section 2.

(d) Prove that, actually, $y_1(\varepsilon) \sim \varepsilon^{2/3}$. (*Hint:* See Mishchenko & Rozov [1980].)

**(9) (Strong inclination property)** Consider a system that is linear,

$$\begin{cases} \dot{x}_1 = \lambda_1 x_1, \\ \dot{x}_2 = \lambda_2 x_2, \\ \dot{x}_3 = \lambda_3 x_3, \end{cases}$$

where $\lambda_1 > 0 > \lambda_2 > \lambda_3$, inside the unit cube $\Omega = \{(x_1, x_2, x_3) : -1 \le x_i \le 1,\ i = 1, 2, 3\}$. Let $\varphi^t$ denote its evolution operator (flow).

(a) Take a line $l_0$ within the plane $x_1 = 1$ passing through the $x_1$-axis and show that its image under the flow (i.e., $l(t) = \varphi^t l$ with $t < 0$) is also a line in some plane $x_1 = \text{const}$ passing through the same axis.

(b) Show that the limit

$$\lim_{t \to -\infty} \varphi^t l$$

is the same for all initial lines except the line $l_0 = \{x_1 = 1\} \cap \{x_3 = 0\}$. What is the limit?

(c) Assume that the system outside the cube $\Omega$ possesses an orbit that is homoclinic to the origin. Using part (b), show that generically the stable manifold $W^s(0)$ intersects itself along the nonleading eigenspace $x_1 = x_2 = 0$. Reformulate the genericity condition as the condition of transversal intersection of $W^s(0)$ with some other invariant manifold near the saddle.

(d) Sketch the shape of the stable manifold in the degenerate case. Guess which phase portraits can appear under perturbations of this degenerate system. (*Hint:* See Yanagida [1987], Deng [1993*a*].)

**(10) (Proofs of Theorems 6.3–6.6 revisited)**

(a) Compute the near-to-saddle map $\Delta$ in the saddle and saddle-focus cases in $\mathbb{R}^3$ assuming that the system is linear inside the unit cube $\Omega = \{(x_1, x_2, x_3) : -1 \leq x_i \leq 1,\ i = 1, 2, 3\}$ and the equilibrium point is located at the origin.

(b) Write a general form of the linear part of the global map $Q : (x_2, x_3) \mapsto (x_1, x_3)$ using the split function as a parameter. How does the formula reflect the twisting of the orbit that is homoclinic to the saddle?

(c) Compose a composition of the maps defined in parts (a) and (b) of the exercise and write the system of equations for its fixed points in the saddle and saddle-focus cases. Analyze the solutions of this system by reducing it to a scalar equation for the $x_1$-coordinate of the fixed points.

**(11)** Show that the saddle quantity $\sigma_0$ of the equilibrium in the wave system for the FitzHugh-Nagumo model is positive. (*Hint*: $\sigma_1 = \lambda_1 + \lambda_2 + \lambda_3 = c > 0$.)

**(12) (Singular homoclinic in $\mathbb{R}^3$)**

(a) Check that the following slow-fast system (cf. Deng [1994]),

$$\begin{cases} \dot{x} = (z+1) + (1-z)[(x-1) - y], \\ \dot{y} = (1-z)[(x-1) + y], \\ \varepsilon\dot{z} = (1-z^2)[z + 1 - m(x+1)] - \varepsilon z, \end{cases}$$

has a homoclinic orbit to the equilibrium $(1, 0, -1)$ in the *singular* limit $\varepsilon = 0$, provided $m = 1$. (*Hint:* First, formally set $\varepsilon = 0$ and analyze the equations on the *slow* manifolds defined by $z = \pm 1$. Then plot the shape of the surface $\dot{z} = 0$ for small but positive $\varepsilon$.)

(b) Could you prove that there is a continuous function $m = m(\varepsilon)$, $m(0) = 1$, defined for $\varepsilon \geq 0$, such that for the corresponding parameter value the system has a saddle-focus with a homoclinic orbit for small $\varepsilon > 0$? What is the sign of the saddle quantity? How many periodic orbits we one expect near the bifurcation?

(c) If part (b) of the exercise is difficult for you, try to find the homoclinic orbit numerically using the boundary-value method described in Chapter 10 and the singular homoclinic orbit from part (a) as the initial guess.

**(13) (Melnikov integral)** Prove that the Melnikov integral (6.24) is nonzero for the homoclinic orbit $\Gamma_0$ in the system (6.8) from Example 6.1. (*Hint:* Find $t_\pm = t_\pm(x)$ along the upper and lower halfs of $\Gamma_0$ by integrating the first equation of (6.8). Then transform the integral (6.24) into the sum of two integrals over $x \in [0,1]$.)

## 6.6 Appendix A: Focus-focus homoclinic bifurcation in four-dimensional systems

In this appendix we study dynamics of four-dimensional systems near an orbit homoclinic to a hyperbolic equilibrium with two complex pairs of eigenvalues (*focus-focus*). This case is similar to the saddle-focus homoclinic case.

Consider a system

$$\dot{x} = f(x, \alpha), \quad x \in \mathbb{R}^4, \ \alpha \in \mathbb{R}^1, \tag{A.1}$$

where $f$ is a smooth function. Assume that at $\alpha = 0$ the system has a hyperbolic equilibrium $x_0 = 0$ with two pairs of complex eigenvalues, namely,

$$\lambda_{1,2}(0) = \rho_1(0) \pm i\omega_1(0), \quad \lambda_{3,4}(0) = \rho_2(0) \pm i\omega_2(0),$$

where

$$\rho_1(0) < 0 < \rho_2(0), \ \omega_{1,2}(0) > 0,$$

(see **Fig. 6.42**). Generically, the *saddle quantity* is nonzero:

(H.1) $\sigma_0 = \rho_1(0) + \rho_2(0) \neq 0.$

Actually, only the case

$$\sigma_0 = \rho_1(0) + \rho_2(0) < 0$$

will be treated, because we can reverse time otherwise. Since $\lambda = 0$ is not an eigen-

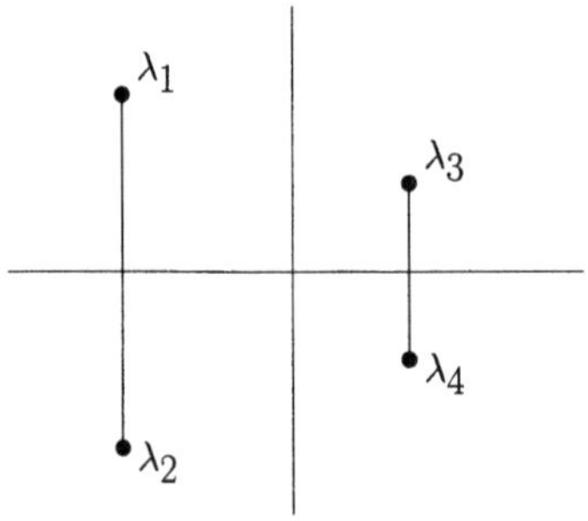


**Fig. 6.42.** Eigenvalues of a focus-focus.

value of the Jacobian matrix $f_x(x_0, 0)$, the Implicit Function Theorem guarantees the persistence of a close hyperbolic equilibrium with two pairs of complex eigenvalues for all sufficiently small $|\alpha|$. Assuming that the origin of coordinates is already shifted to this equlibrium, we can write $(A.1)$ in the form

$$\dot{x} = A(\alpha)x + F(x, \alpha), \tag{A.2}$$

where $F = O(\|x\|^2)$ and the matrix $A(\alpha)$ has the eigenvalues

$$\lambda_{1,2}(\alpha) = \rho_1(\alpha) \pm i\omega_1(\alpha), \quad \lambda_{3,4}(\alpha) = \rho_2(\alpha) \pm i\omega_2(\alpha),$$

with $\rho_i(0)$, $\omega_i(0)$ satisfying the imposed conditions.

The focus-focus equilibrium has the two-dimensional stable and unstable manifolds $W^{s,u}$ that can be globally extended. Suppose that at $\alpha = 0$ the manifolds $W^u$ and $W^s$ intersect along a homoclinic orbit $\Gamma_0$. We assume that the intersection of the tangent spaces to the stable and unstable manifolds is one-dimensional at any point $x \in \Gamma_0$:[9]

(H.2) $\dim\,(T_x W^u \cap T_x W^s) = 1$.

This condition holds generically for systems with a homoclinic orbit to a hyperbolic equilibrium.

The following theorem by Shil'nikov is valid.

**Theorem 6.11** *For any system* (A.2)*, having a focus-focus equilibrium point* $x_0$ *with a homoclinic orbit* $\Gamma_0$ *at* $\alpha = 0$ *and satisfying the nondegeneracy conditions* (H.1) *and* (H.2)*, there exists a neighborhood* $U_0$ *of* $\Gamma_0 \cup x_0$ *in which the system has an infinite number of saddle limit cycles in* $U_0$ *for all sufficiently small* $|\alpha|$. □

**Sketch of the proof:**

First consider the case $\alpha = 0$. Write the system (A.2) in its real eigenbasis. This can be done by applying to (A.2) a nonsingular linear transformation putting $A$ in its real Jordan form. In the eigenbasis, the system at $\alpha = 0$ will take the form

$$\begin{cases} \dot{x}_1 = \rho_1 x_1 - \omega_1 x_2 + G_1(x), \\ \dot{x}_2 = \omega_1 x_1 + \rho_1 x_2 + G_2(x), \\ \dot{x}_3 = \rho_2 x_3 - \omega_2 x_4 + G_3(x), \\ \dot{x}_4 = \omega_2 x_3 + \rho_2 x_4 + G_4(x), \end{cases} \tag{A.3}$$

where old notations for the phase variable are preserved and $G = O(\|x\|^2)$. Now introduce new coordinates $y$ that locally linearize the system (A.3). Due to a theorem by Belitskii (see Appendix B) there exists a nonlinear transformation

$$y = x + g(x),$$

where $g$ is a $C^1$ function ($g_x(0) = 0$), that locally conjugates the flow corresponding to (A.3) with the flow generated by the linear system

$$\begin{cases} \dot{y}_1 = \rho_1 y_1 - \omega_1 y_2, \\ \dot{y}_2 = \omega_1 y_1 + \rho_1 y_2, \\ \dot{y}_3 = \rho_2 y_3 - \omega_2 y_4, \\ \dot{y}_4 = \omega_2 y_3 + \rho_2 y_4. \end{cases} \tag{A.4}$$

In the coordinates $y \in \mathbb{R}^4$ the unstable manifold $W^u$ is locally represented by $y_1 = y_2 = 0$, while the stable manifold $W^s$ is given by $y_3 = y_4 = 0$. Suppose that the linearization (A.4) is valid in the unit 4-cube $\{|y_i| \leq 1,\ i = 1,2,3,4\}$ which can always be achieved by a linear scaling.

Write (A.4) in the polar coordinates

[9] This intersection is spanned by the phase velocity vector $f(x, 0)$ for $x \in \Gamma_0$.

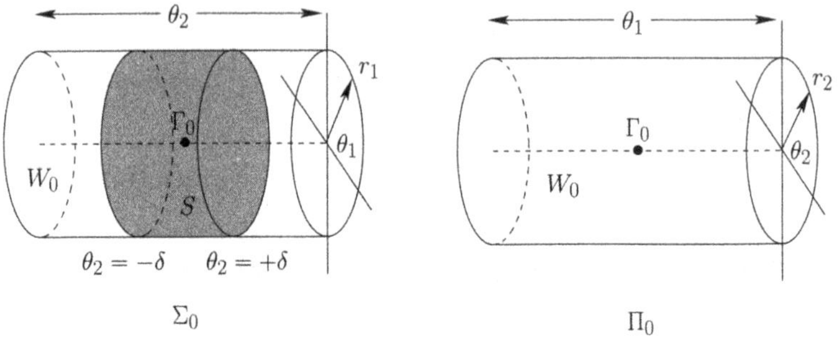


**Fig. 6.43.** Cross-sections $\Sigma_0$ and $\Pi_0$.

$$\begin{cases} \dot{r}_1 = \rho_1 r_1, \\ \dot{\theta}_1 = \omega_1, \\ \dot{r}_2 = \rho_2 r_2, \\ \dot{\theta}_2 = \omega_2, \end{cases} \tag{A.5}$$

by substituting

$$y_1 = r_1 \cos\theta_1,\ y_2 = r_1 \sin\theta_1,\ y_3 = r_2\cos\theta_2,\ y_4 = r_2\sin\theta_2.$$

Introduce two *three-dimensional* cross-sections for (A.5):

$$\Sigma = \{(r_1, \theta_1, r_2, \theta_2) : r_2 = 1\},$$
$$\Pi = \{(r_1, \theta_1, r_2, \theta_2) : r_1 = 1\},$$

and two submanifolds within these cross-sections, namely:

$$\Sigma_0 = \{(r_1, \theta_1, r_2, \theta_2) : r_1 \le 1,\ r_2 = 1\} \subset \Sigma,$$
$$\Pi_0 = \{(r_1, \theta_1, r_2, \theta_2) : r_1 = 1,\ r_2 \le 1\} \subset \Pi.$$

$\Sigma_0$ and $\Pi_0$ are three-dimensional solid tori that can be vizualized as in **Fig. 6.43**, identifying the left and the right faces. The stable manifold $W^s$ intersects $\Sigma_0$ along the center circle $r_1 = 0$, while the unstable manifold $W^u$ intersects $\Pi_0$ along the center circle $r_2 = 0$. Without loss of generality, assume that the homoclinic orbit $\Gamma_0$ crosses $\Sigma_0$ at the point with $\theta_1 = 0$, while its intersection with $\Pi_0$ occurs at $\theta_2 = 0$.

As usual, define a Poincaré map $P : \Sigma \to \Sigma$ along the orbits of the system and represent this map as a composition $P = Q \circ \Delta$ of two maps: a near-to-saddle map $\Delta : \Sigma \to \Pi$ and a map $Q : \Pi \to \Sigma$ near the global part of the homoclinic orbit $\Gamma_0$. Now introduce a three-dimensional solid cylinder $S \subset \Sigma_0$

$$S = \{(r_1, \theta_1, r_2, \theta_2):\ r_1 \le 1,\ r_2 = 1,\ -\delta \le \theta_2 \le \delta\}$$

with some $\delta > 0$ fixed (see **Fig. 6.43**), and trace its image under the Poincaré map $P$.

The map $\Delta : \Sigma \to \Pi$ can be computed explicitly using (A.5). Namely,

$$\Delta : \begin{pmatrix} r_1 \\ \theta_1 \\ 1 \\ \theta_2 \end{pmatrix} \mapsto \begin{pmatrix} 1 \\ \theta_1 + \frac{\omega_1}{\rho_1} \ln \frac{1}{r_1} \\ r_1^{-\frac{\rho_2}{\rho_1}} \\ \theta_2 + \frac{\omega_2}{\rho_1} \ln \frac{1}{r_1} \end{pmatrix}, \tag{A.6}$$

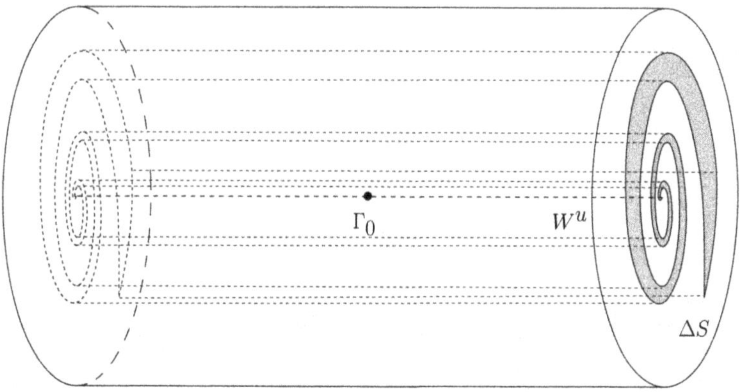


**Fig. 6.44.** The image $\Delta S$ in $\Pi_0$.

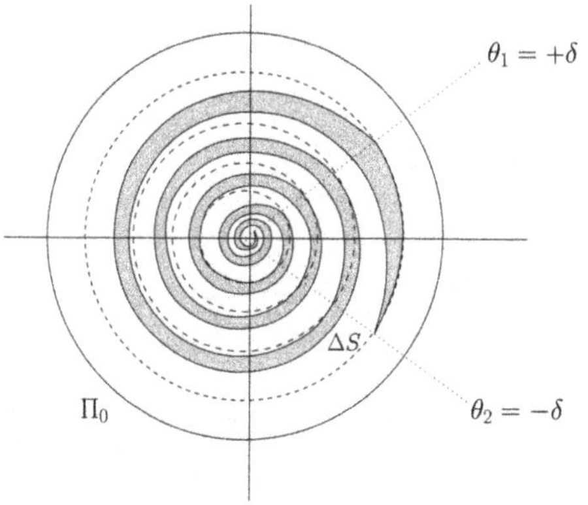


**Fig. 6.45.** The section of $\Delta S$ by the plane $\theta_1 = 0$.

since the flight time from $\Sigma$ to $\Pi$ is equal to

$$T = \frac{1}{\rho_1} \ln \frac{1}{r_1}.$$

According to (A.6), the image $\Delta S \subset \Pi_0$ is a solid *"toroidal scroll"* sketched in **Fig. 6.44**. The section of the image by the plane $\theta_2 = 0$ is presented in **Fig. 6.45**.

The $C^1$ map $Q : \Pi \to \Sigma$ places the scroll back into the cross-section $\Sigma$ by rotating and deforming it such that the image $Q(\Delta S)$ cuts through the cylinder $S$ (see **Fig. 6.46**). The center circle $r_2 = 0$ of $\Pi_0$ is transformed by $Q$ into a curve intersecting the center circle $r_1 = 0$ of $\Sigma_0$ at a nonzero angle due to the condition (H.2).

The geometry of the constructed Poincaré map $P$ implies the presence of three-dimensional analogs of Smale's horseshoe. Indeed, let us partition $S$ into a series of solid annuli: $S = \cup_{k=0}^{\infty} S_k$, where

$$S_k = \left\{ (r_1, \theta_1, r_2, \theta_2) : e^{-\frac{2\pi(k+1)\rho_2}{\omega_1}} < r_1 \le e^{-\frac{2\pi k \rho_2}{\omega_1}},\ r_2 = 1,\ |\theta_2| \le \delta \right\}.$$

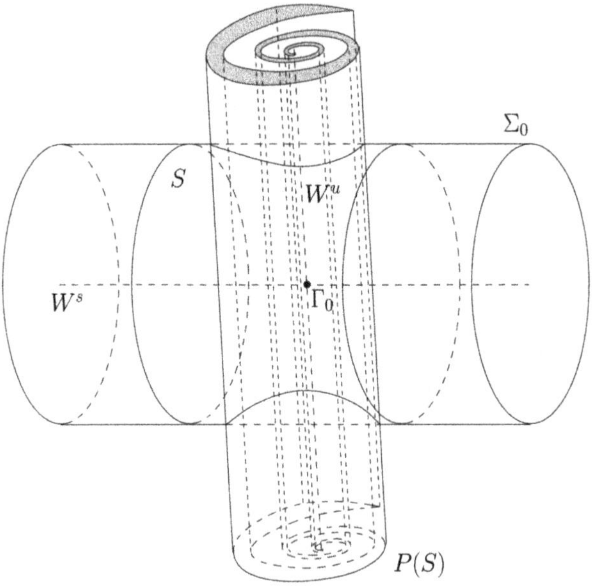


**Fig. 6.46.** The image $P(S)$ and the preimage $S$ in $\Sigma$.

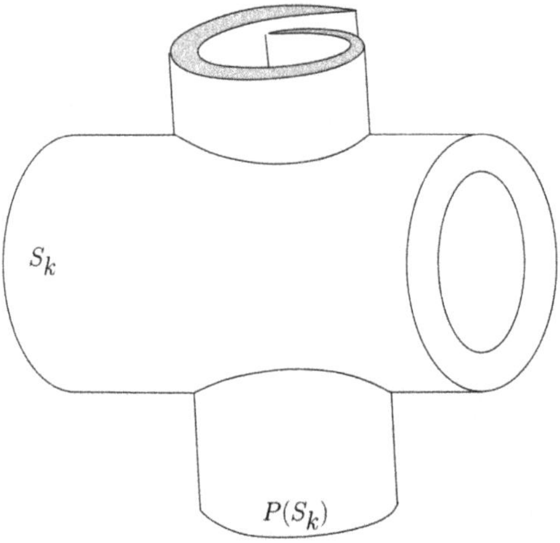


**Fig. 6.47.** The image $P(S_k)$ and the preimage $S_k$ in $\Sigma$.

Provided $k$ is sufficiently large, $S_k$ is mapped by $P$ into a "one-turn scroll" $P(S_k)$ that intersects $S_k$ by two disjoint domains (see **Fig. 6.47**). This is a key feature of Smale's example. Thus, at $\alpha = 0$, the system (A.1) has an infinite number of Smale's horseshoes, each of them implying the existence of a Cantor invariant set containing an infinite number of saddle limit cycles.

If $|\alpha|$ is small but nonzero, the above construction can still be carried out. However, generically, the manifolds $W^s$ and $W^u$ split by $O(\alpha)$ distance, so the image of the center circle of $\Pi_0$ does not intersect that of $\Sigma_0$. Thus, only a finite number of the three-dimensional horseshoes remain. Nevertheless, they still give an infinite number of cycles near $\Gamma_0$.

## 6.7 Appendix B: Bibliographical notes

The homoclinic orbit bifurcation in planar dynamical systems was analyzed by Andronov & Leontovich [1939] (an exposition with much detail can be found in Andronov et al. [1973]). $C^1$-linearization was not known to Andronov; therefore he had to give delicate estimates for the near-to-saddle map (see Wiggins [1988] for such estimates in the $n$-dimensional case). $C^k$-linearization near a hyperbolic equilibrium is studied by Sternberg [1957] and Belitskii [1973, 1979], as well as by many other authors. A theorem by Belitskii provides the $C^1$-equivalence of the flow corresponding to a system in $\mathbb{R}^n$ to the flow generated by its linear part near a hyperbolic equilibrium with eigenvalues $\lambda_1, \lambda_2, \ldots, \lambda_n$ such that

$$\mathrm{Re}\ \lambda_i \neq \mathrm{Re}\ \lambda_j + \mathrm{Re}\ \lambda_k$$

for all combinations of $i, j, k = 1, 2, \ldots, n$. An elementary proof of $C^1$-linearization near a hyperbolic saddle on the plane, which is reproduced in the proof of Theorem 6.1, is due to Deng [1989].

Integrals over homoclinic orbits characterizing splitting of invariant manifolds first appeared in the paper by Melnikov [1963] devoted to *periodic perturbations* of planar autonomous systems. If the unperturbed system has a homoclinic orbit to a saddle equilibrium, the perturbed system (considered as an autonomous system in $\mathbb{R}^2 \times \mathbb{S}^1$) will have a saddle limit cycle with two-dimensional stable and unstable invariant manifolds. These manifolds could intersect along orbits homoclinic to the cycle, giving rise to the Poincaré homoclinic structure and associated chaotic dynamics (see Chapter 2). In a fixed cross-section $t = t_0$, the points corresponding to a homoclinic orbit can be located near the unperturbed homoclinic loop as zeros of the so-called *Melnikov function* (see details in Sanders [1982], Guckenheimer & Holmes [1983], Wiggins [1990]). The generalization of Melnikov's technique to $n$-dimensional situations using the variational and adjoint variational equations is due to Palmer [1984] (see also Lin [1990]). In the papers by Beyn [1990*b*, 1990*a*] the equivalence of the transversality of the intersection of the stable and unstable manifolds in the extended system (6.20) to the nonvanishing of the Melnikov integral (6.24) is proved.

Bifurcations of phase portraits near orbits homoclinic to a hyperbolic equilibrium in $n$-dimensional autonomous systems were first studied by Shil'nikov [1963] and Neimark & Shil'nikov [1965] under simplifying assumptions. The general theory has been developed by Shil'nikov [1968, 1970] (there are also two preceding papers by him in which three- and four-dimensional cases were analyzed: Shil'nikov

[1965, 1967*a*]). The main tool of his analysis is a representation of the near-to-saddle map as the solution to a boundary-value problem (the so-called *parametric representation*); see Deng [1989] for the modern treatment of this technique. This parametrization allowed Shil'nikov to prove one-to-one correspondence between the saddle cycles and periodic sequences of symbols. A particular feature that makes Shil'nikov's main papers difficult to read is the absence of figures. For example, the notion of "orientation" or "twisting" never appeared in his original papers explicitly (it is hidden in the signs of some indirectly defined determinants). A geometrical treatment of the saddle-focus case in $\mathbb{R}^3$ can be found in Guckenheimer & Holmes [1983] and Tresser [1984]. In the latter paper the $C^1$-linearization near the saddle-focus is used. Wiggins [1988, 1990] gives many details concerning homoclinic bifurcations in $\mathbb{R}^3$ and $\mathbb{R}^4$. Appendix A follows his geometrical approach to the focus-focus homoclinic bifurcation.

Arnol'd et al. [1994] provides an excellent survey of codimension 1 homoclinic bifurcations. It includes a proof of topological invariance of $\nu_0$, as well as the construction of topological normal forms for the saddle homoclinic bifurcation. Detailed proofs of many results mentioned in this survey are given by Ilyashenko & Li [1999]. In this text, however, redundunt genericity conditions are often assumed to assure the local $C^k$-linearization of the system near the equilibrium for any finite $k$, whereas $k = 1$ would be sufficient.

The bifurcations of limit cycles near an orbit homoclinic to the saddle-focus were studied by Gaspard [1983], Gaspard, Kapral & Nicolis [1984], and Glendinning & Sparrow [1984]. The existence of the subsidiary homoclinic orbits is proved by Evans et al. [1982] and Gonchenko, Turaev, Gaspard & Nicolis [1997]. Actually, basic results concerning these bifurcations follow from the analysis of a codim 2 bifurcation performed by Belyakov [1980], who studied the homoclinic bifurcation in $\mathbb{R}^3$ near the saddle to saddle-focus transition (see also Belyakov [1974, 1984] for the analysis of other codim 2 saddle-focus cases). Codim 2 homoclinic bifurcations in $\mathbb{R}^3$ have recently attracted much interest (see, e.g., Nozdrachova [1982], Yanagida [1987], Glendinning [1988], Chow, Deng & Fiedler [1990], Kisaka, Kokubu & Oka [1993*a*, 1993*b*], Hirschberg & Knobloch [1993], Deng [1993*a*], Homburg, Kokubu & Krupa [1994], Bykov [1977, 1980, 1993, 1999], Shashkov [1992], and Deng & Sakamoto [1995]). Many results on codim 2 homoclinic and heteroclinic bifurcations are presented in Shilnikov et al. [2001, Chapter 13].

Explicit examples of two- and three-dimensional systems having algebraic homoclinic orbits of codim 1 and 2 have been presented by Sandstede [1997*a*]. Belyakov homoclinic bifurcations play an important role in population dynamics [Kuznetsov, De Feo & Rinaldi 2001].

Homoclinic bifurcations in $n$-dimensional cases with $n > 4$ were treated in the original papers by Shil'nikov and by Ovsyannikov & Shil'nikov [1987] (see also Deng [1993*b*]). The existence of center manifolds near homoclinic bifurcations in $n$-dimensional systems has been established independently by Turaev [1991], Sandstede [1993, 2000], and Homburg [1993, 1996]. Complete proofs can be found in [Shilnikov et al. 1998, Chapter 6].

There is an alternative method to prove the bifurcation of a periodic orbit from the homoclinic orbit: A function space approach based on the Lyapunov-Schmidt method [Lin 1990].

Homoclinic bifurcations in planar slow-fast systems were treated by Diener [1983] in the framework of nonstandard analysis. An elementary treatment of the planar case is given by Kuznetsov, Muratori & Rinaldi [1995] with application to population dynamics. Some higher-dimensional cases have been considered by Szmolyan [1991]. Many examples of three-dimensional slow-fast systems that exhibit homoclinic bifurcations are constructed in Deng [1994, 1995].

# 7

# Other One-Parameter Bifurcations in Continuous-Time Dynamical Systems

The list of possible bifurcations in multidimensional systems is not exhausted by those studied in the previous chapters. Actually, even the complete list of all generic one-parameter bifurcations is unknown. In this chapter we study several unrelated bifurcations that occur in one-parameter continuous-time dynamical systems

$$\dot{x} = f(x, \alpha), \quad x \in \mathbb{R}^n, \ \alpha \in \mathbb{R}^1, \tag{7.1}$$

where $f$ is a smooth function of $(x, \alpha)$. We start by considering global bifurcations of orbits that are homoclinic to nonhyperbolic equilibria. As we shall see, under certain conditions they imply the appearance of complex dynamics. We also briefly touch some other bifurcations generating "strange" behavior, including homoclinic tangency and the "blue-sky" catastrophe. Then we discuss bifurcations occurring on invariant tori. These bifurcations are responsible for such phenomena as frequency and phase locking. Finally, we give a brief introduction to the theory of bifurcations in symmetric systems, which are those systems that are invariant with respect to the representation of a certain symmetry group. After giving some general results on bifurcations in such systems, we restrict our attention to bifurcations of equilibria and cycles in the presence of the simplest symmetry group $\mathbb{Z}_2$, composed of only two elements.

## 7.1 Codim 1 bifurcations of homoclinic orbits to nonhyperbolic equilibria

Let $x^0 = 0$ be a nonhyperbolic equilibrium of system (7.1) at $\alpha = 0$; the Jacobian matrix $A = f_x$ evaluated at $(x^0, 0)$ has eigenvalues with zero real part. As in the hyperbolic case, we introduce two invariant sets:

$$W^s(x^0) = \{x : \varphi^t x \to x^0, t \to +\infty\}, W^u(x^0) = \{x : \varphi^t x \to x^0, t \to -\infty\},$$

where $\varphi^t$ is the flow associated with (7.1); recall that $W^s(x^0)$ is called the *stable set* of $x^0$, while $W^u(x^0)$ is called the *unstable set* of $x^0$. If these sets are both nonempty, they could intersect; in other words, there may exist homoclinic orbits approaching $x^0$ in both time directions. Since the presence of a nonhyperbolic equilibrium is already a degeneracy, the codimension of such a singularity is greater than or equal to one. As we can easily see, if the equilibrium has a pair of complex-conjugate eigenvalues on the imaginary axis, we need more than one parameter to tune in order to get a homoclinic orbit to this equilibrium. Consider, for example, a system depending on several parameters in $\mathbb{R}^3$, having an equilibrium $x^0$ with one positive eigenvalue $\lambda_3 > 0$ and a pair of complex-conjugate eigenvalues that can cross the imaginary axis. To obtain a *Shil'nikov-Hopf bifurcation*, we have to spend one parameter to satisfy the Hopf bifurcation condition $\lambda_{1,2} = \pm i\omega_0$, and another parameter to place the unstable one-dimensional manifold $W^u(x^0)$ of the equilibrium on its stable set $W^s(x^0)$ (in fact, the center manifold $W^c(x^0)$). Thus, the Shil'nikov-Hopf bifurcation has codim 2. Therefore, since we are interested here in codim 1 bifurcations, let us instead consider the case when a simple zero eigenvalue is the only eigenvalue of the Jacobian matrix on the imaginary axis. We will start with the two-dimensional case.

### 7.1.1 Saddle-node homoclinic bifurcation on the plane

Suppose that for $\alpha = 0$, system (7.1) with $n = 2$ has the equilibrium $x^0 = 0$ with a simple zero eigenvalue $\lambda_1 = 0$. According to the Center Manifold Theorem (see Chapter 5), for $\alpha = 0$ there is a one-dimensional manifold $W_0^c(x^0)$ tangent to the eigenvector of $A$ corresponding to $\lambda_1 = 0$. This manifold is locally attracting or repelling, depending on the sign of the second eigenvalue $\lambda_2 \neq 0$. The restriction of (7.1) to $W_0^c$ at $\alpha = 0$ has the form

$$\dot{\xi} = b\xi^2 + O(\xi^3), \tag{7.2}$$

where, generically, $b \neq 0$. Under this nondegeneracy condition, the system is locally topologically equivalent at $\alpha = 0$ near the origin to the normal form

$$\begin{cases} \dot{\xi}_1 = b\xi_1^2, \\ \dot{\xi}_2 = \sigma\xi_2, \end{cases}$$

where $\sigma = \text{sign}\ \lambda_2$ (see **Fig. 7.1**, where the two cases with $a > 0$ are shown). These equilibria are called *saddle-nodes*. Notice that in **Fig. 7.1**(a) the stable set $W^s(x^0)$ is the left half-plane $\xi_1 \leq 0$, while the unstable set $W^u(x^0)$ is the right half-axis $\{\xi_1 \geq 0, \xi_2 = 0\}$. In **Fig. 7.1**(b) the unstable set $W^u(x^0)$ is given by $\{\xi_1 \geq 0\}$, and the stable set $W^s(x^0)$ by $\{\xi_1 \leq 0, \xi_2 = 0\}$. There are infinitely many center manifolds passing through the saddle-node (see Section 5.1.1 in Chapter 5); a part of each center manifold $W_0^c$ belongs to the stable set of the saddle-node, while the other part belongs to the unstable set of the equilibrium.

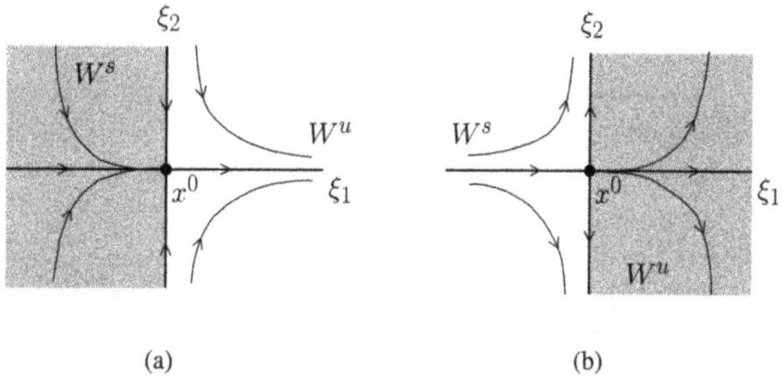


**Fig. 7.1.** Planar saddle-nodes: (a) $\lambda_2 < 0$; (b) $\lambda_2 > 0$.

If the restriction of (7.1) to its parameter-dependent center manifold $W^c_\alpha$ written in a proper coordinate $\xi$,

$$\dot{\xi} = \beta(\alpha) + b(\alpha)\xi^2 + O(\xi^3), \tag{7.3}$$

depends generically on the parameter, a fold bifurcation occurs: The saddle-node equilibrium either disappears or bifurcates into a saddle $x^1$ and a node $x^2$.

Consider the case $b > 0, \lambda_2 < 0$, and assume that there is an orbit $\Gamma_0$ homoclinic to the saddle-node $x^0$. Clearly, there may be at most one homoclinic orbit to such an equilibrium, and this orbit must locally coincide with the one-dimensional unstable set $W^u(x^0)$. Thus, there is only one way the homoclinic orbit can leave the saddle-node. However, it can return back to the saddle-node along any of the infinitely many orbits composing the stable set $W^s(x^0)$. This "freedom" implies that the presence of a homoclinic orbit to the saddle-node is not an extra bifurcation condition imposed on the system, and therefore the codimension of the singularity is still one, which is that of the fold bifurcation. Any of the orbits tending to the saddle-node, apart from the two exceptional orbits that bound the stable set (the vertical axis in **Fig. 7.1**(a) or (b)), can be considered as a part of the center manifold $W^c(x^0)$. Thus, generically, the closure of the homoclinic orbit is smooth and coincides with one of the center manifolds near the saddle-node.

If the parameter is varied such that the equilibrium disappears ($\beta > 0$), a stable limit cycle $L_\beta$ is born near the former smooth homoclinic orbit $\Gamma_0$. This fact is almost obvious if we consider a cross-section transversal to the center manifold. The Poincaré map defined on this section for $\beta > 0$ is a contraction due to $\lambda_2 < 0$. Let us summarize the discussion by formulating the following theorem.

**Theorem 7.1 (Andronov & Leontovich [1939])** *Suppose the system*

$$\dot{x} = f(x, \alpha), \quad x \in \mathbb{R}^2, \ \alpha \in \mathbb{R}^1,$$

*with smooth $f$, has at $\alpha = 0$ the equilibrium $x^0 = 0$ with $\lambda_1 = 0, \lambda_2 < 0$, and there exists an orbit $\Gamma_0$ that is homoclinic to this equilibrium.*

*Assume that the following genericity conditions are satisfied:*

(SNH.1) *the system exhibits a generic fold bifurcation at* $\alpha = 0$, *so that its restriction to the center manifold can be transformed to the form*

$$\dot{\xi} = \beta(\alpha) + b(\alpha)\xi^2 + O(\xi^3),$$

*where* $b(0) > 0$ *and* $\beta'(0) \neq 0$;

(SNH.2) *the homoclinic orbit* $\Gamma_0$ *departs from and returns to the saddle-node along one of its center manifolds, meaning that the closure of* $\Gamma_0$ *is smooth.*

*Then there is a neighborhood* $U_0$ *of* $\Gamma_0 \cup x^0$ *in which the system has a bifurcation diagram topologically equivalent to the one presented in* **Fig. 7.2.** □

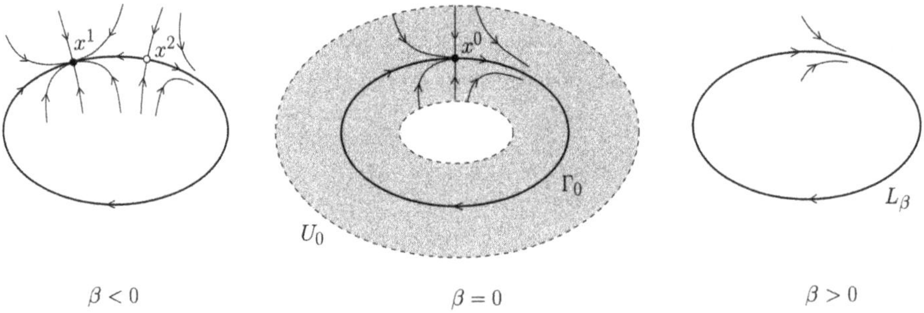


**Fig. 7.2.** Saddle-node homoclinic bifurcation.

**Remarks:**

(1) Example 2.10 from Chapter 2 provides an explicit planar system undergoing a generic saddle-node homoclinic bifurcation. It also happens in Bazykin's predator-prey system, which will be considered in Chapter 8 (Example 8.3).

(2) The saddle-node homoclinic bifurcation is a global bifurcation in which a local bifurcation is also involved. Looking at only a small neighborhood of the saddle-node equilibrium, we miss the appearance of the cycle.

(3) If we approach the saddle-node homoclinic bifurcation from parameter values for which the cycle is present ($\beta > 0$ in our consideration), the cycle period $T_\beta$ tends to infinity (see Exercise 7.1). A phase point moving along the cycle spends more and more time near the place where the saddle-node will appear: It "feels" the approaching fold bifurcation.

(4) The case $\lambda_2 > 0$ brings nothing new. The only difference from the considered one is that the appearing cycle is *unstable*. Actually, this case can be reduced to that in Theorem 7.1 by reversing time. ◇

### 7.1.2 Saddle-node and saddle-saddle homoclinic bifurcations in $\mathbb{R}^3$

Now consider a system (7.1) in $\mathbb{R}^3$ having at $\alpha = 0$ the equilibrium $x^0 = 0$ with a simple zero eigenvalue and no other eigenvalues on the imaginary axis. There are more possibilities for such equilibria to allow for different kinds of homoclinic orbit.

#### Saddle-nodes and saddle-saddles

As in the planar case, at $\alpha = 0$ there is a one-dimensional manifold $W_0^c(x^0)$ tangent to the eigenvector of $A$ corresponding to the zero eigenvalue. The restriction of (7.1) to $W_0^c$, at $\alpha = 0$, in this case has the same form (7.2),

$$\dot{\xi} = b\xi^2 + O(\xi^3),$$

where, generically, $b \neq 0$. Under this condition, the system is locally topologically equivalent at $\alpha = 0$ near the origin to the system

$$\begin{cases} \dot{\xi}_1 = b\xi_1^2, \\ \dot{\xi}_2 = \sigma_1\xi_2, \\ \dot{\xi}_3 = \sigma_2\xi_3, \end{cases} \tag{7.4}$$

where $(\sigma_1, \sigma_2)$ are the signs of the real parts of the nonzero eigenvalues. Thus,

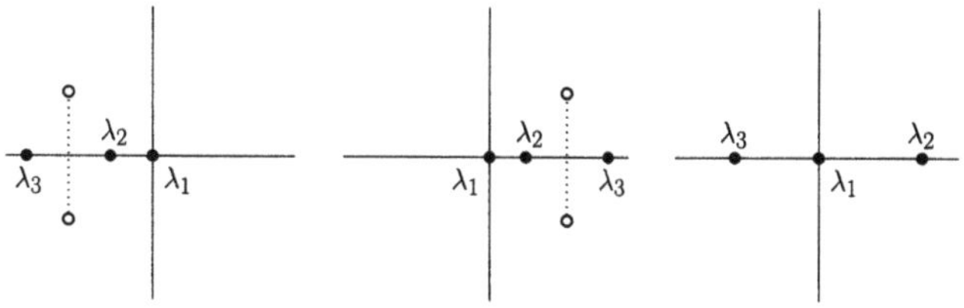


**Fig. 7.3.** Three types of equilibria with $\lambda_1 = 0$.

we have three obvious possibilities (see **Fig. 7.3**): (a) two nonzero eigenvalues are located in the left half-plane, $\sigma_1 = \sigma_2 = -1$; (b) two nonzero eigenvalues are located in the right half-plane, $\sigma_1 = \sigma_2 = 1$; (c) one of the nonzero eigenvalues is to the right of the imaginary axis, while the other is to the left, $\sigma_1 = 1, \sigma_2 = -1$. Suppose that $b > 0$; otherwise, reverse time.

In case (a), system (7.4) reads

$$\begin{cases} \dot{\xi}_1 = b\xi_1^2, \\ \dot{\xi}_2 = -\xi_2, \\ \dot{\xi}_3 = -\xi_3, \end{cases} \tag{7.5}$$

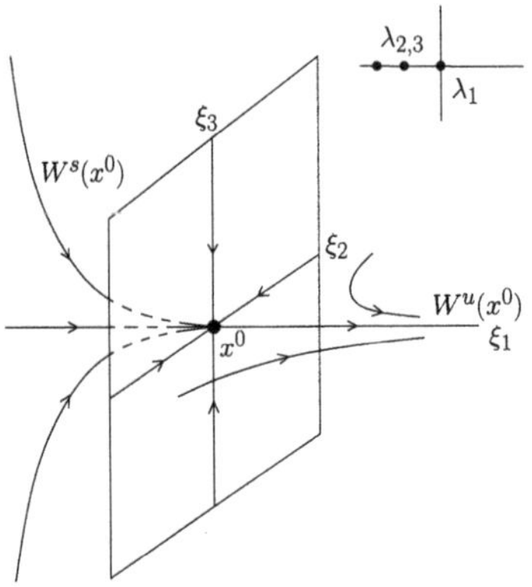


**Fig. 7.4.** Saddle-node equilibrium with two stable eigenvalues.

and has the phase portrait presented in **Fig. 7.4**. The stable set $W^s(x^0)$ is the half-space $\{\xi_1 \leq 0\}$, within which the majority of the orbits tend to the equilibrium tangent to the $\xi_1$-axis. The unstable set $W^u(x^0)$ is the half-axis $\{\xi_1 \geq 0, \xi_2 = \xi_3 = 0\}$.

In case (b), system (7.4) has the form

$$\begin{cases} \dot{\xi}_1 = b\xi_1^2, \\ \dot{\xi}_2 = \ \ \xi_2, \\ \dot{\xi}_3 = \ \ \xi_3. \end{cases} \tag{7.6}$$

Its phase portrait is presented in **Fig. 7.5**. The stable set is now one-

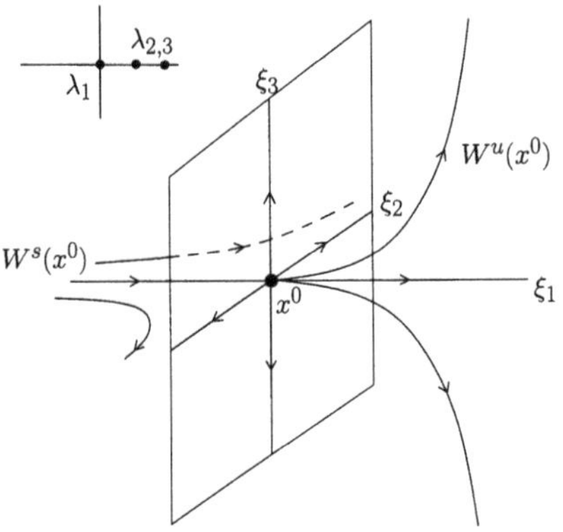


**Fig. 7.5.** Saddle-node equilibrium with two unstable eigenvalues.

dimensional, while the unstable set is three-dimensional. As in the planar case, the equilibrium for either (a) or (b) is called a *saddle-node*. Note that the nonzero eigenvalues can constitute a complex-conjugate pair. In such a case, orbits of the original system within the corresponding half-space tend to the equilibrium by spiraling.

In case (c), the system (7.4) turns out to be

$$\begin{cases} \dot{\xi}_1 = b\xi_1^2, \\ \dot{\xi}_2 = \xi_2, \\ \dot{\xi}_3 = -\xi_3. \end{cases} \tag{7.7}$$

Its phase portrait is presented in **Fig. 7.6**. It can be constructed by taking

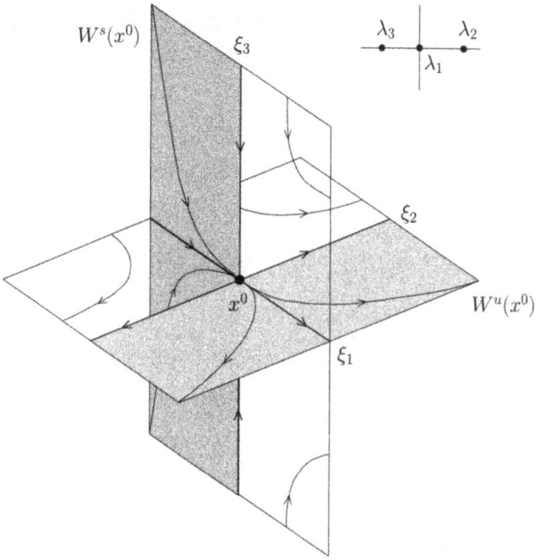


**Fig. 7.6.** Saddle-saddle equilibrium.

into account that all the coordinate planes $\xi_k = 0, k = 1, 2, 3$, are invariant with respect to (7.7). Notice that both the stable and the unstable sets are *two-dimensional* half-planes approaching each other transversally:

$$W^s = \{\xi : \xi_1 \leq 0, \xi_2 = 0\}, \ W^u = \{\xi : \xi_1 \geq 0, \xi_3 = 0\}.$$

In this case, the equilibrium $x^0$ is called a *saddle-saddle.*

Since the restriction of (7.1) to the center manifold has the form

$$\dot{\xi} = \beta(\alpha) + b(\alpha)\xi^2 + O(\xi^3),$$

a generic fold bifurcation takes place if $b(0) \neq 0$ and $\beta'(0) \neq 0$, leading either to the disappearance of the equilibrium (for $b(0)\beta > 0$) or to the appearance of two hyperbolic ones (for $b(0)\beta < 0$). In the case of a saddle-node, one of the bifurcating equilibria is saddle, while the other is (stable or unstable) three-dimensional node. On the contrary, in the saddle-saddle case, both appearing equilibria are (topologically different) saddles.

### Saddle-node homoclinic orbit

If there is an orbit $\Gamma_0$ homoclinic to a saddle-node, then, generically, a unique limit cycle appears when the equilibria disappear. The bifurcation is similar to the planar one. The stability of the cycle is determined by the sign of $\sigma_{1,2}$. More precisely, the following theorem holds.

**Theorem 7.2 (Shil'nikov [1966])** *Suppose the system*

$$\dot{x} = f(x, \alpha), \quad x \in \mathbb{R}^3, \; \alpha \in \mathbb{R}^1,$$

*with smooth $f$, has at $\alpha = 0$ the equilibrium $x = 0$ with $\lambda_1 = 0, \text{Re}\,\lambda_{2,3} < 0$ (or $\text{Re}\,\lambda_{2,3} > 0$), and there exists an orbit $\Gamma_0$ that is homoclinic to this equilibrium.*

*Assume that the following genericity conditions are satisfied:*

(SNH.1) *the system exhibits a generic fold bifurcation at $\alpha = 0$, so that its restriction to the center manifold can be transformed to the form*

$$\dot{\xi} = \beta(\alpha) + b(\alpha)\xi^2 + O(\xi^3),$$

*where $b(0) \neq 0$ and $\beta'(0) \neq 0$;*

(SNH.2) *the homoclinic orbit $\Gamma_0$ departs from and returns to $x^0$ along one of its center manifolds, meaning that the closure of $\Gamma_0$ is smooth.*

*Then there is a neighborhood $U_0$ of $\Gamma_0 \cup x^0$ in which the system has a unique stable (or repelling) limit cycle $L_\beta$ for small $|\beta|$ corresponding to the disappearance of the equilibria, and no limit cycles for small $|\beta|$ when two hyperbolic equilibria exist.* □

The theorem is illustrated in **Fig. 7.7**, where $a > 0$ and the two stable

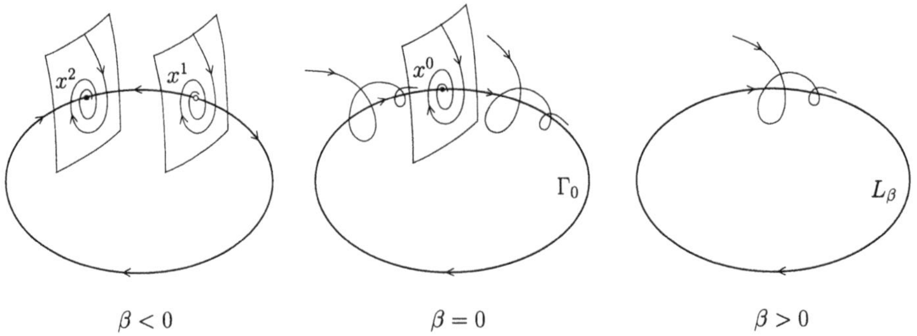


**Fig. 7.7.** Saddle-node homoclinic bifurcation in $\mathbb{R}^3$.

eigenvalues $\lambda_{2,3}$ are complex. In this case, the multipliers of the appearing cycle $L_\beta$ are also complex. The period of the cycle $T_\beta \to \infty$, as $\beta \to 0$.

## Saddle-saddle with one homoclinic orbit

If there is a saddle-saddle equilibrium $x^0$ with a single homoclinic orbit $\Gamma_0$, then, generically, a unique limit cycle appears when the equilibria disappear. The cycle is saddle since the Poincaré map defined on a transversal cross-section to any center manifold is contracting in the direction of the stable eigenvector of the saddle-saddle and expanding in the direction of its unstable eigenvector. We also have to require that the stable set $W^s(x^0)$ intersects the unstable set $W^u(x^0)$ along the homoclinic orbit $\Gamma_0$ transversally. These heuristic arguments can be formalized by the following theorem.

**Theorem 7.3 (Shil'nikov [1966])** *Suppose the system*

$$\dot{x} = f(x, \alpha), \quad x \in \mathbb{R}^3, \ \alpha \in \mathbb{R}^1,$$

*with smooth $f$, has at $\alpha = 0$ the equilibrium $x = 0$ with $\lambda_1 = 0, \lambda_2 > 0, \lambda_3 < 0$, and there exists a single orbit $\Gamma_0$ homoclinic to this equilibrium.*

*Assume that the following genericity conditions are satisfied:*

(SNH.1) *the system exhibits a generic fold bifurcation at $\alpha = 0$, such that its restriction to the center manifold can be transformed to the form*

$$\dot{\xi} = \beta(\alpha) + b(\alpha)\xi^2 + O(\xi^3),$$

*where $b(0) \neq 0$ and $\beta'(0) \neq 0$;*

(SNH.2) *the homoclinic orbit $\Gamma_0$ departs from and returns to $x^0$ along one of its center manifolds, meaning that the closure of $\Gamma_0$ is smooth.*

(SNH.3) *the stable set $W^s(x^0)$ transversally intersects the unstable set $W^u(x^0)$ along the homoclinic orbit $\Gamma_0$.*

*Then there is a neighborhood $U_0$ of $\Gamma_0 \cup x^0$ in which the system has a unique saddle limit cycle $L_\beta$ for small $|\beta|$ corresponding to the disappearance of the equilibria, and no limit cycles for small $|\beta|$ when two saddle equilibria exist.* □

The theorem is illustrated in **Fig. 7.8**. Actually, all systems exhibiting the bifurcation described by this and the previous theorem are topologically equivalent in $U_0$ for small $|\alpha|$.

**Remark:**

The topology of the stable and the unstable invariant manifolds of the appearing cycle $L_\beta$ is determined by the global behavior of the stable and the unstable set of the saddle-saddle around $\Gamma_0$ at $\alpha = 0$. The transversality of the intersection $W^u(x^0)$ with $W^s(x^0)$ implies that the closure of each of these sets in a tubular neighborhood $U_0$ of $\Gamma_0$ is either a nontwisted or twisted two-dimensional band. In **Fig. 7.8**, the manifold $W^u(x^0)$ is shown as orientable at $\beta = 0$. If these manifolds are nontwisted (orientable), then the appearing limit cycle has positive multipliers: $0 < \mu_1 < 1 < \mu_2$. On the contrary, if

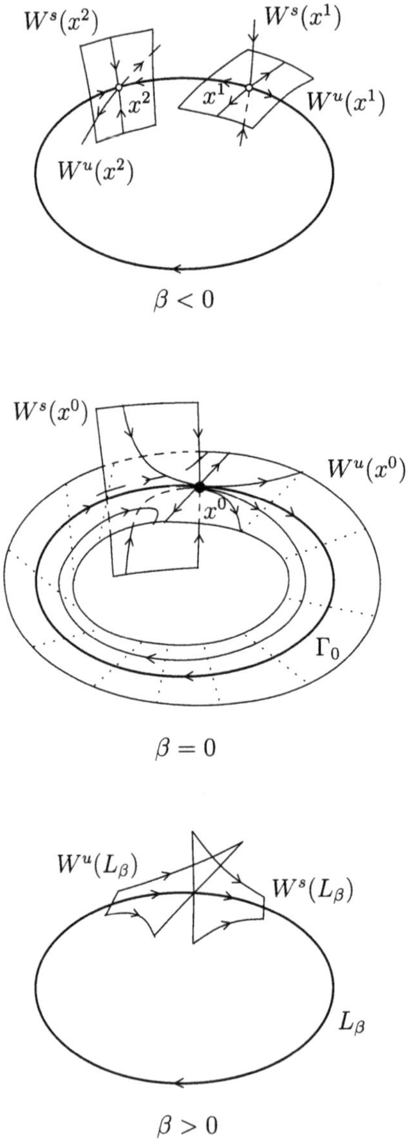


**Fig. 7.8.** Saddle-saddle bifurcation with one homoclinic orbit.

the manifolds are twisted (nonorientable), then the cycle multipliers are both negative: $\mu_1 < -1 < \mu_2 < 0$. In the former case, the stable and the unstable invariant manifolds of the cycle are also nontwisted, while in the latter case, they are both twisted. ◊

### Saddle-saddle with more than one homoclinic orbit

Since the stable and the unstable sets of a saddle-saddle are two-dimensional, they can intersect along *more than one* homoclinic orbit. Such an intersection leads to a bifurcation that has no analog in the planar systems: It gives rise to an *infinite number* of saddle limit cycles when the equilibria disappear. Let us formulate the corresponding theorem due to Shil'nikov in the case where there are *two* homoclinic orbits, $\Gamma_1$ and $\Gamma_2$, present at $\alpha = 0$.

**Theorem 7.4 (Shil'nikov [1969])** *Suppose the system*

$$\dot{x} = f(x, \alpha), \quad x \in \mathbb{R}^3, \ \alpha \in \mathbb{R}^1,$$

*with smooth $f$, has at $\alpha = 0$ the equilibrium $x = 0$ with $\lambda_1 = 0, \lambda_2 > 0, \lambda_3 < 0$, and there exist two orbits, $\Gamma_1$ and $\Gamma_2$, homoclinic to this equilibrium.*

*Assume that the following genericity conditions are satisfied:*

(SNH.1) *the system exhibits a generic fold bifurcation at $\alpha = 0$, such that its restriction to the center manifold can be transformed to the form*

$$\dot{\xi} = \beta(\alpha) + b(\alpha)\xi^2 + O(\xi^3),$$

*where $b(0) \neq 0$ and $\beta'(0) \neq 0$;*

(SNH.2) *both homoclinic orbits $\Gamma_{1,2}$ depart from and return to $x^0$ along its center manifolds, meaning that the closure of each $\Gamma_{1,2}$ is smooth;*

(SNH.3) *the stable set $W^s(x^0)$ intersects the unstable set $W^u(x^0)$ transversally along two homoclinic orbits $\Gamma_{1,2}$.*

*Then, there is a neighborhood $U_0$ of $\Gamma_1 \cup \Gamma_2 \cup x^0$ in which an infinite number of saddle limit cycles exists for small positive or negative $\beta$, depending on the parameter direction corresponding to the disappearance of the equilibria.*

*Moreover, there is a one-to-one correspondence between the orbits located entirely inside $U_0$ for such values of $\beta$, and all nonequivalent sequences $\{\omega_i\}_{i=-\infty}^{+\infty}$ of two symbols, $\omega_i \in \{1, 2\}$.*

**Outline of the proof:**

Introduce coordinates $(\xi_1, \xi_2, \xi_3)$ in such a way that, for $\alpha = 0$, the saddle-saddle is located at the origin and its unstable set $W^u(0)$ is given, within the unit cube

$$\{\xi : |\xi_k| \leq 1, k = 1, 2, 3\},$$

by $\{\xi_1 \geq 0, \xi_3 = 0\}$; the stable set $W^s(x^0)$ is defined in the same cube by $\{\xi_1 \leq 0, \xi_2 = 0\}$ (see **Fig. 7.9** and compare it with the phase portrait of (7.7)). Consider two faces of the cube:

$$\Pi_1 = \{\xi : \xi_1 = -1, |\xi_{2,3}| \leq 1\}, \ \Pi_2 = \{\xi : \xi_1 = 1, |\xi_{2,3}| \leq 1\}.$$

Let $A_1, B_1$ be the points where the orbits $\Gamma_1, \Gamma_2$ intersect with the plane $\Pi_1$ as they enter the unit cube while returning to the saddle-saddle. Similarly,

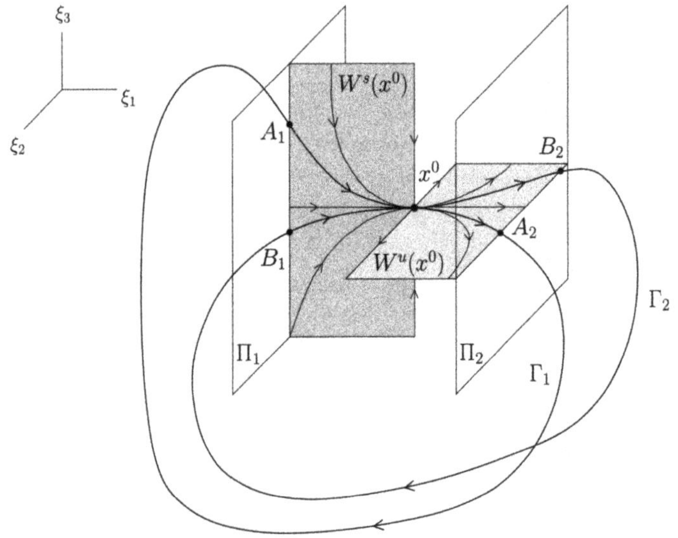


**Fig. 7.9.** A saddle-saddle with two homoclinic orbits.

denote by $A_2, B_2$ the intersection points of $\Gamma_1, \Gamma_2$ with the plane $\Pi_2$ as these orbits leave the cube.

Take a small value of $|\beta|$ with its sign corresponding to the disappearance of the equilibrium. Then, the Poincaré map along orbits of the system defined on $\Pi_1$,

$$P_\beta : \Pi_1 \to \Pi_1,$$

can be represented as the composition of a "local" map $\Delta_\beta : \Pi_1 \to \Pi_2$ along orbits passing through the cube, and a "global" map $Q_\beta : \Pi_2 \to \Pi_1$:

$$P_\beta = Q_\beta \circ \Delta_\beta.$$

Notice that $\Delta_\beta$ is undefined for $\beta = 0$, as well as when there are hyperbolic equilibria inside the cube.

By solving the linearized system inside the cube, one can show that the square $\Pi_1$ is contracted by the map $\Delta_\beta$ in the $\xi_3$-direction and expanded in the $\xi_2$-direction. Thus, the intersection of its image $\Delta_\beta \Pi_1$ with the square $\Pi_2$ would be a horizontal strip $\Sigma = \Delta_\beta \Pi_1 \cap \Pi_2$ (see **Fig. 7.10**(b)), which gets thinner and thinner as $\beta \to 0$ (explain why).

The strip $\Sigma$ contains the points $A_2$ and $B_2$. Since for $\beta = 0$ the map $Q_\beta$ sends the point $A_2$ into the point $A_1$ and the point $B_2$ into the point $B_1$,

$$Q_\beta(A_2) = A_1, \;\; Q_\beta(B_2) = B_1,$$

there would be some neighborhoods of $A_2, B_2$ in $\Pi_2$, that $Q_\beta$ maps into neighborhoods of $A_1$ and $B_1$ in $\Pi_1$, respectively, for $|\beta| > 0$. Therefore, the image $Q_\beta(\Sigma)$ will intersect the square $\Pi_1$ in two strips, $\Sigma_1$ and $\Sigma_2, \Sigma_1 \cup \Sigma_2 =$

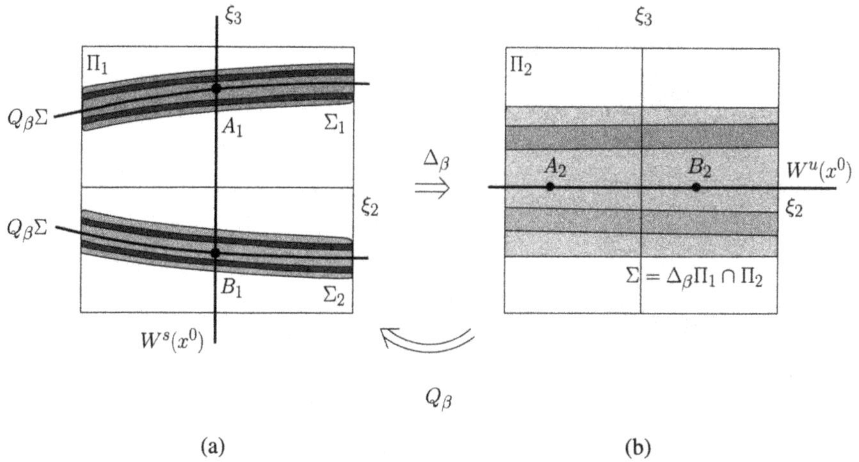


**Fig. 7.10.** Cross-section near a saddle-saddle.

$Q_\beta \Sigma_0 \cap \Pi_1$ (see **Fig. 7.10**(a)), containing $A_1$ and $B_1$, respectively. Due to the transversality assumption, $\Sigma_{1,2}$ intersect the vertical axis at a nonzero angle near the points $A_1$ and $B_1$, respectively.

Thus, the intersection of the image of $\Pi_1$ under the Poincaré map $P_\beta = Q_\beta \circ \Delta_\beta$ with $\Pi_1$ has the standard features of the Smale horseshoe (see Chapter 1). For example, applying the construction once more, we first obtain two strips inside $\Sigma$, and then two narrow strips inside each $\Sigma_{1,2}$, and so forth (see **Fig. 7.10**). Inverting the procedure, we get vertical strips with a Cantor structure. The presence of the Smale horseshoe implies the possibility to code orbits near $\Gamma_1 \cup \Gamma_2$ by sequences of two symbols, say $\{1, 2\}$. Equivalent sequences code the same orbit. □

In the present context, this coding has a clear geometrical interpretation. Indeed, let $\gamma$ be an orbit located in a neighborhood $U_0$ of the homoclinic orbits $\Gamma_{1,2}$ for all $t \in (-\infty, +\infty)$. Then it passes outside the cube near either $\Gamma_1$ or $\Gamma_2$. The elements of the corresponding sequence $\omega = \{\ldots, \omega_{-2}, \omega_{-1}, \omega_0, \omega_1, \omega_2, \ldots\}$ specify whether the orbit $\gamma$ makes its $i$th passage near $\Gamma_1$ or $\Gamma_2$; in the former case, $\omega_i = 1$, while in the latter, $\omega_i = 2$. For example, the sequence

$$\{\ldots, 1, 1, 1, 1, 1, \ldots\}$$

corresponds to a unique saddle cycle orbiting around $\Gamma_1$. The sequence

$$\{\ldots, 2, 2, 2, 2, 2, \ldots\}$$

describes a saddle cycle located near $\Gamma_2$, while the periodic sequence

$$\{\ldots, 1, 2, 1, 2, 1, 2, \ldots\}$$

corresponds to a cycle making its first trip near $\Gamma_1$, its second near $\Gamma_2$, and so on (see **Fig. 7.11**).

The case when there are more than two homoclinic orbits, $\Gamma_1, \Gamma_2, \ldots, \Gamma_N$, say, to the saddle-saddle at the critical parameter value can be treated similarly. In such a case, orbits located entirely inside a neighborhood of $\Gamma_1 \cup \Gamma_2 \cup \cdots \cup \Gamma_N \cup x^0$ are coded by sequences of $N$ symbols, for example, $\omega_i \in \{1, 2, \ldots, N\}$.

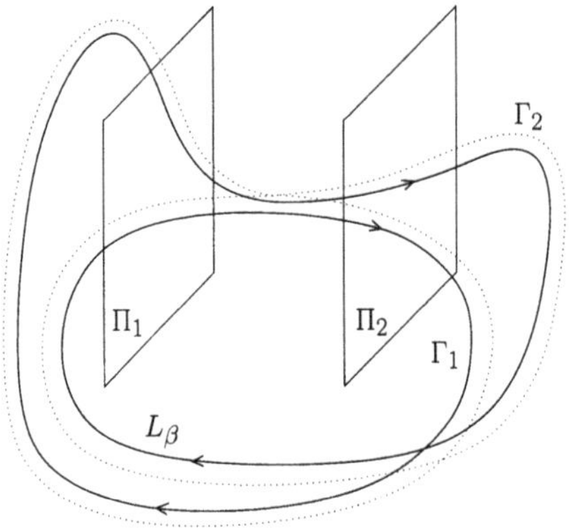


**Fig. 7.11.** A cycle corresponding to the sequence $\{\ldots, 1, 2, 1, 2, \ldots\}$.

We finish the consideration of nonhyperbolic homoclinic bifurcations by pointing out that the results presented in this section for three-dimensional systems can be generalized into arbitrary finite dimensions of the phase space (see the appropriate references in the bibliographical notes).

## 7.2 Bifurcations of orbits homoclinic to limit cycles

Several other codim 1 bifurcations in generic one-parameter systems have been analyzed theoretically. Let us briefly discuss some of them without pretending to give a complete picture.

### 7.2.1 Nontransversal homoclinic orbit to a hyperbolic cycle

Consider a three-dimensional system (7.1) with a hyperbolic limit cycle $L_\alpha$. Its stable and unstable two-dimensional invariant manifolds, $W^s(L_\alpha)$ and $W^u(L_\alpha)$, can intersect along homoclinic orbits, tending to $L_\alpha$ as $t \to \pm\infty$. Generically, such an intersection is transversal. As we have seen in Chapter 2, it implies the presence of an infinite number of saddle limit cycles near the homoclinic orbits. However, at a certain parameter value, say $\alpha = 0$, the manifolds can become *tangent* to each other and then no longer intersect (see **Fig.**

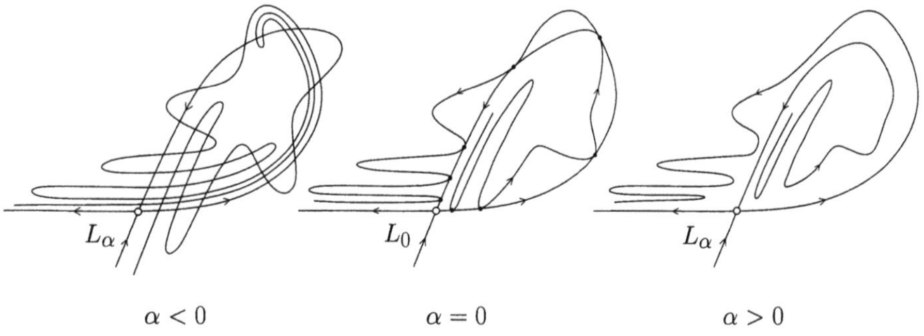


**Fig. 7.12.** Homoclinic tangency.

**7.12**, where a cross-section to the homoclinic structure is sketched). At $\alpha = 0$ there is a homoclinic orbit $\Gamma_0$ to $L_0$ along which the manifolds $W^s(L_0)$ and $W^u(L_0)$ generically have a quadratic tangency. As we shall demonstrate below, passing the critical parameter value is accompanied by an infinite number of period-doubling and fold bifurcations of limit cycles located near $\Gamma_0$.

First, we establish a relationship between properly defined *return maps* near the homoclinic tangency and the Hénon map (1.4) introduced in Example 1.9 of Chapter 1. Denote by $P_\alpha$ the Poincaré map defined on a "global" two-dimensional cross-section to $L_\alpha$ in the considered three-dimensional system (7.1). The intersection of $L_\alpha$ with the cross-section is a saddle fixed point $O$ of $P_\alpha$. Let the multipliers of $O$ be denoted by $\gamma$ and $\lambda$, so that

$$0 < |\lambda| < 1 < |\gamma|.$$

For simplicity, we shall consider only the case when $\lambda$ and $\gamma$ are both positive and their product satisfies

$$\lambda\gamma < 1.$$

Since both multipliers depend smoothly on $\alpha$, it is sufficient to check the above inequalities at $\alpha = 0$. As in **Fig. 7.12**, the stable and unstable manifolds of $L_\alpha$ will intersect the cross-section along the stable and unstable manifolds of the saddle fixed point $O$. An orbit of (7.1) homoclinic to $L_\alpha$ will be represented by a discrete orbit of $P_\alpha$ homoclinic to $O$. The stable and unstable manifolds of the fixed point $O$ intersect transversally or touch each other tangentially along the homoclinic orbits.

We are interested in orbits located entirely in a small neighborhood of the critical homoclinic orbit $\Gamma_0$. This neighborhood consists of a small neighborhood $D_0$ of $O$ and a finite number of small neighborhoods of points on the homoclinic orbit outside $D_0$ (see **Fig. 7.13**). The return map is constructed as the composition of two maps: "local," defined in $D_0$, and "global," defined along a part of the homoclinic orbit located outside $D_0$.

Let us fix for a while $\alpha = 0$ and consider two points of the critical homoclinic orbit $\Gamma_0$ in $D_0$: $M^+$ on the stable manifold of $O$ and $M^-$ on the unstable

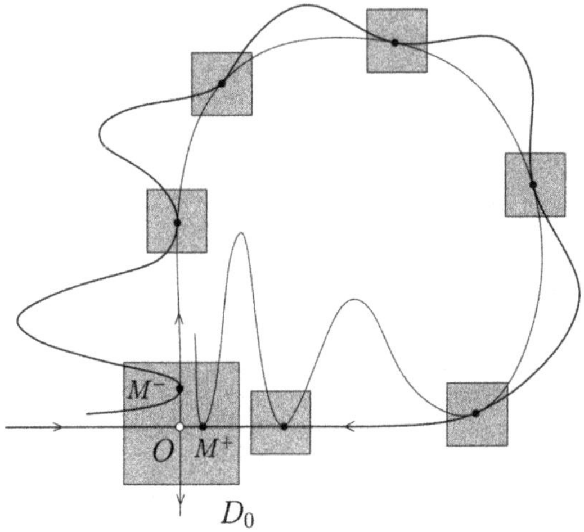


**Fig. 7.13.** A neighborhood of the critical homoclinic orbit in the cross-section.

manifold of $O$. There is an integer number $n_0$ such that $M^+ = P_0^{n_0}(M^-)$. Next we can choose two small neighborhoods in $D_0$: $\Pi^+$ (of point $M^+$) and $\Pi^-$ (of point $M^-$). Consider the forward images of $\Pi^+$ under $P_0$ (see **Fig. 7.14**(a)). For all sufficiently large $k$, the difference $\Pi^- \setminus P_0^k(\Pi^+)$ consists of two disjoint components. Denote $P_0^k(\Pi^+) \cap \Pi^-$ by $\sigma_k^1$. One can also iterate $\Pi^-$ under $P_0^{-1}$ to obtain domains $\sigma_k^0 = P_0^{-k}(\Pi^-) \cap \Pi^+$ (see **Fig. 7.14**(b)), which are the preimages of $\sigma_k^1$, i.e., $\sigma_k^1 = P_0^k(\sigma_k^0)$. Clearly, only orbits starting in $\sigma_k^0$ may end

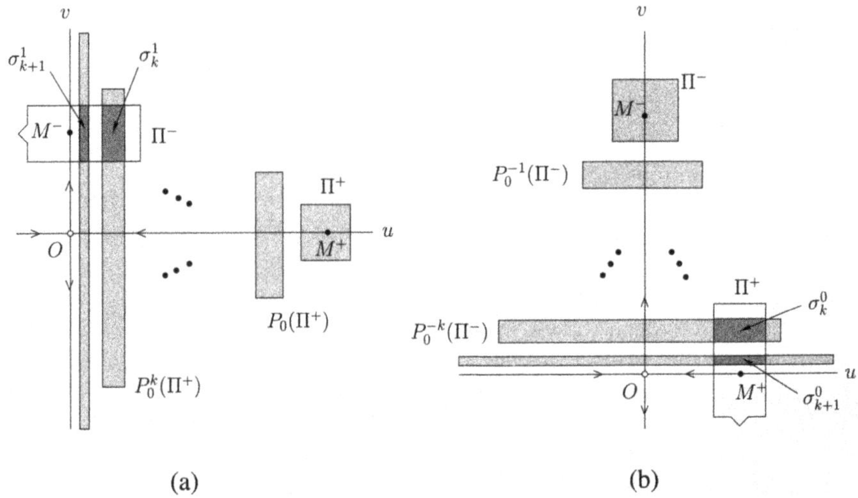


**Fig. 7.14.** Domains of definition for the "local" map.

up in $\Pi^-$ after $k$ iterates of $P_0$ and then return to $\sigma_k^0$ after $n_0$ iterates of $P_0$. Therefore, the return map, whose fixed points are the $(k+n_0)$-periodic cycles of $P_0$, has to be defined in $\sigma_k^0$. This construction can be carried out for all $\alpha$ with sufficiently small $|\alpha|$.

Introduce a special coordinate system $(u, v)$ in the cross-section, so that

$$O = (0,0), \quad M^- = (0, v^-), \quad M^+ = (u^+, 0),$$

and the local stable and unstable manifolds of $O$ coincide in $D_0$ with the horizontal and vertical coordinate axes, respectively, for all sufficiently small $|\alpha|$. Let $(u_0, v_0) \in \Pi^+$ and $(u_1, v_1) \in \Pi^-$. In these coordinates, the "local" map

$$P_\alpha^k : \Pi^+ \to \Pi^-$$

(defined on $\sigma_k^0$) is close to the linearization of the $k$th iterate of $P_\alpha$, namely

$$\Delta_\alpha^k : \begin{cases} u_1 = \lambda^k u_0, \\ v_1 = \gamma^k v_0. \end{cases} \tag{7.8}$$

Since the homoclinic tangency is quadratic , the "global" map $P_\alpha^{n_0} : \Pi^- \to \Pi^+$ must be close to the map

$$Q_\alpha : \begin{cases} \tilde{u}_0 = U^+ + a u_1 + b(v_1 - V^-), \\ \tilde{v}_0 = \mu + c u_1 + d(v_1 - V^-)^2, \end{cases} \tag{7.9}$$

where $\mu, a, b, c, d, U^+$, and $V^-$ depend smoothly on $\alpha$, $\mu(0) = 0$, and $U^+(0) = u^+$, $V^-(0) = v^-$. The "split function" $\mu = \mu(\alpha)$ measures the $v$-distance between the stable and the unstable manifolds near $M^+$. Assuming $\mu'(0) \neq 0$, we can use $\mu$ as a new parameter and consider $\lambda, \gamma, a, b, c, d, U^+$, and $V^-$ as smooth functions of $\mu$. Moreover, $d \neq 0$ due to the quadratic nature of the tangency, and $bc < 0$ since the Poincaré map $P_\alpha$ is an orientation-preserving diffeomorphism.

The composition $Q_\alpha \circ \Delta_\alpha^k$ of the maps defined by (7.8) and (7.9), which approximates the return map $P_\alpha^{n_0+k}$ on $\sigma_k^0$, can, therefore, be written as

$$\begin{cases} \tilde{u}_0 = U^+ + a\lambda^k u_0 + b(\gamma^k v_0 - V^-), \\ \tilde{v}_0 = \mu + c\lambda^k u_0 + d(\gamma^k v_0 - V^-)^2. \end{cases} \tag{7.10}$$

Introduce

$$\begin{cases} \xi = \phantom{\gamma^k} u_0 - U^+, \\ \eta = \gamma^k v_0 - V^-. \end{cases}$$

The map (7.10) written in the $(\xi, \eta)$-coordinates will take the form

$$\begin{cases} \tilde{\xi} = a\lambda^k U^+ + a\lambda^k \xi + b\eta, \\ \tilde{\eta} = -V^- + \gamma^k \mu + c(\lambda\gamma)^k U^+ + c(\lambda\gamma)^k \xi + d\gamma^k \eta^2. \end{cases} \tag{7.11}$$

Finally, introduce new coordinates by

$$\begin{cases} X = -\dfrac{d\gamma^k}{b}(\xi - a\lambda^k U^+), \\ Y = -\dfrac{d\gamma^k}{b}(a\lambda^k \xi + b\eta). \end{cases}$$

This brings (7.11) into the form

$$\begin{pmatrix} X \\ Y \end{pmatrix} \mapsto \begin{pmatrix} Y \\ M_k + bc(\lambda\gamma)^k X - Y^2 \end{pmatrix} + O(\lambda^k), \tag{7.12}$$

where

$$M_k = -d\gamma^{2k}(\mu - \gamma^{-k}V^- + O(\lambda^k)) + O(\lambda^k). \tag{7.13}$$

The same result can be obtained without truncating the "local" and "global" maps (see references in the appendix to this chapter).

Recalling that $|\lambda| < 1$ and $|\lambda\gamma| < 1$, one can conclude that the return map near the critical homoclinic orbit is well-approximated by the Hénon map with Jacobian $-bc|\lambda\gamma|^k \to 0$ as $k \to \infty$. When $\mu$ varies (together with $\alpha$) in a small but fixed interval around $\mu = 0$, $M_k$ takes arbitrary big real values, provided that $k$ is big enough. Thus, the Hénon map approximates the return maps defined on infinitely many domains $\sigma_k^0 \subset \Pi^+$ of the cross-section near the critical homoclinic orbit $\Gamma_0$. Since the Hénon map exhibits the fold and flip bifurcations (see Example 4.1 in Chapter 4), infinite sequences of such bifurcations can be expected for the original map. Indeed, for (7.12) fold and flip bifurcations happen at

$$M_k^{(t)} = -\frac{1}{4} + O(\lambda^k), \quad M_k^{(f)} = \frac{3}{4} + O(\lambda^k),$$

respectively (consider the limit $\beta \to 0$ in Example 4.1). Relation (7.13) implies the existence of two infinite sequences of bifurcation parameter values,

$$\mu_k^{(t)} = \gamma^{-k}V^- - \frac{1}{4d(0)}\gamma^{-2k} + o(\gamma^{-2k}),$$
$$\mu_k^{(f)} = \gamma^{-k}V^- + \frac{3}{4d(0)}\gamma^{-2k} + o(\gamma^{-2k}),$$

accumulating on $\mu = 0$ as $k \to \infty$. Note that all these bifurcations occur when $\mu > 0$, i.e., when the invariant manifolds of $L_\alpha$ have no intersection (see case $\alpha > 0$ in **Fig. 7.12**).

### 7.2.2 Homoclinic orbits to a nonhyperbolic limit cycle

Suppose a three-dimensional system (7.1) has at $\alpha = 0$ a nonhyperbolic limit cycle $N_0$ with a simple multiplier $\mu_1 = 1$, while the second multiplier satisfies $|\mu_2| < 1$. Under generic perturbations, this cycle will either disappear or split into two hyperbolic cycles, $N_\alpha^{(1)}$ and $N_\alpha^{(2)}$ (the fold bifurcation for cycles, see

Section 5.3 in Chapter 5). However, the locally unstable manifold $W^u(N_0)$ of the cycle can "return" to the cycle $N_0$ at the critical parameter value $\alpha = 0$ forming a set composed of homoclinic orbits which approach $N_0$ as $t \to \pm\infty$. There are two cases, depending on whether the closure of $W^u(N_0)$ is a manifold or not.

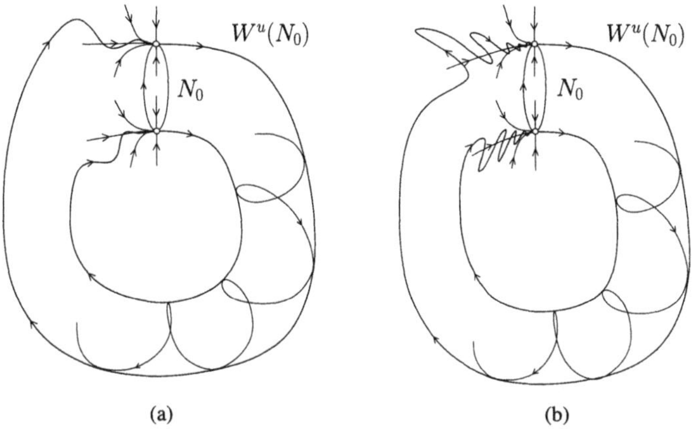


**Fig. 7.15.** Homoclinic structure of a saddle-node cycle depicted via a global Poincaré map: (a) smooth and (b) nonsmooth cases.

(1) *Torus case.* If $W^u(N_0)$ forms a *torus*, two subcases are still possible (see **Fig. 7.15**, where a global Poincaré section to $N_0$ is used, so the torus appears as two concentric curves). Depending on whether the torus is smooth (**Fig. 7.15**(a)) or not (**Fig. 7.15**(b)), the disappearance of the cycle $N_0$ under parameter variation leads either to the creation of a smooth invariant torus or a "strange" attracting invariant set that contains an infinite number of saddle and stable limit cycles. For systems in more than three dimensions, there may exist several tori (or Klein bottles) at the critical parameter value, leading to more diverse and complicated bifurcation pictures.

(2) *"Blue-sky" case.* A bifurcation known as a "blue-sky" catastrophe appears in the case when $W^u(N_0)$ is not a manifold. More precisely, at $\alpha = 0$ the unstable set $W^u(N_0)$ of the cycle $N_0$ can become a tube that returns to $N_0$ developing a "French horn" (**Fig. 7.16**). In a local cross-section to the cycle $N_0$, the spiraling end of this horn appears as an infinite sequence of "circles" accumulating at the point corresponding to $N_0$.

When the cycle $N_0$ splits into two hyperbolic cycles, no other periodic orbits exist near the horn. On the other side of the bifurcation, when the cycle $N_0$ disappears, there emerges a unique and stable hyperbolic limit cycle $L_\alpha$ that makes one global turn and several local turns following the horn. As $\alpha$ approaches $\alpha = 0$, the cycle $L_\alpha$ makes more and more turns near the would-be critical cycle $N_0$. At the bifurcation parameter value, $L_\alpha$ becomes an

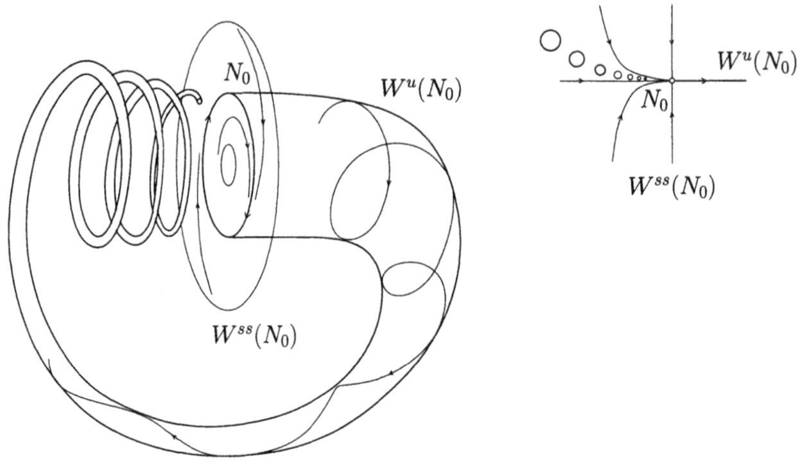


**Fig. 7.16.** The invariant "French horn" near a "blue-sky" bifurcation.

orbit homoclinic to $N_0$, and its length and period become infinite. Thus, the stable limit cycle $L_\alpha$ disappears as its length $l_\alpha$ and period $T_\alpha$ tend to infinity, while it remains bounded and located at a finite distance from all equilibrium points. This bifurcation is called the *"blue-sky" catastrophe* of $L_\alpha$. In other words, the cycle $L_\alpha$ is "broken" at the critical parameter value by another cycle $N_0$ that appears in a "transverse" direction to $L_\alpha$ and then splits into two hyperbolic cycles.

**Example 7.1 ("Blue-sky" bifurcation model)** Consider the following system due to Gavrilov & Shilnikov [2000]:

$$\begin{cases} \dot{x} = x[2 + \mu - b(x^2 + y^2)] + z^2 + y^2 + 2y, \\ \dot{y} = -z^3 - (y+1)(z^2 + y^2 + 2y) - 4x + \mu y, \\ \dot{z} = z^2(y+1) + x^2 - \varepsilon, \end{cases} \tag{7.14}$$

where $\mu, \varepsilon$ are positive parameters and $b = 10$ is fixed. If $\mu = \varepsilon = 0$, the circle

$$C_0 = \{(x, y, z) : x = 0,\ z^2 + (y+1)^2 = 1\}$$

is an invariant curve of (7.14). It consists of two equilibria: $E_0 = (0, 0, 0)$ and $E_1 = (0, -2, 0)$, and two connecting orbits: from $E_0$ to $E_1$ and from $E_1$ to $E_0$. The equilibrium $E_0$ has one zero eigenvalue $\lambda_1 = 0$ and two purely imaginary eigenvalues $\lambda_{2,3} = \pm i\omega_0$ with $\omega_0 = 2$, while the equilibrium $E_1$ has one zero eigenvalue $\lambda_1 = 0$ and two real eigenvalues $\lambda_{2,3} < 0$ (check!). Thus, both equilibria are nonhyperbolic and should bifurcate when the parameters change. One can prove that there is a curve $\mathcal{B}$ in the $(\mu, \varepsilon)$-plane along which the system (7.14) has a limit cycle $N_0$ with a simple unit multiplier. This cycle shrinks to the equilibrium $E_0$ when we approach the origin of the parameter plane along $\mathcal{B}$. Crossing the curve $\mathcal{B}$ near the origin results in a generic fold

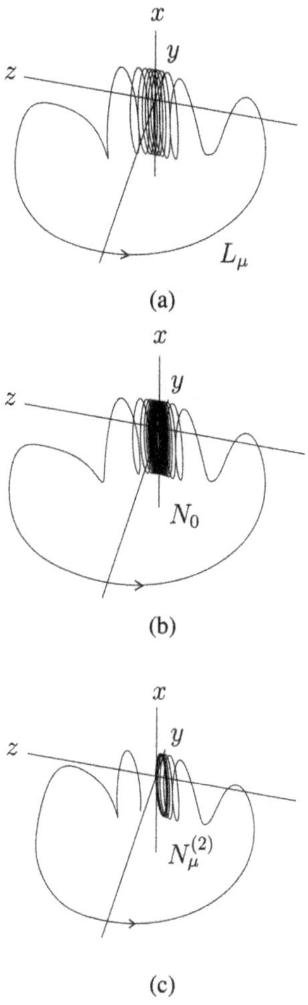


**Fig. 7.17.** "Blue-sky" bifurcation in (7.14): (a) $\mu = 0.4$; (b) $\mu = 0.3$; and (c) $\mu = 0.25$.

bifurcation of this cycle: Two small hyperbolic cycles (one stable and one saddle) collide, forming the cycle $N_0$ at the critical parameter values, and disappear. Moreover, for parameter values corresponding to the curve $\mathcal{B}$, the equilibrium $E_1$ does not exist and the two-dimensional unstable set $W^u(N_0)$ returns to $N_0$ near $C_0$ forming a "Fernch horn" configuration as in **Fig. 7.16.** Therefore, there is another limit cycle in (7.14) that stays near the circle $C_0$ and undergoes the "blue-sky" bifurcation if we cross $\mathcal{B}$ sufficiently close to $(\mu, \varepsilon) = (0, 0)$. **Fig. 7.17** shows phase orbits of (7.14) corresponding to $\varepsilon = 0.02$ at three different values of $\mu$. For $\mu = 0.4$, there is a stable limit cycle $L_\mu$ that makes a number of transversal turns near the origin. The blue-sky bifurcation happens at $\mu \approx 0.3$. For $\mu = 0.25$, an orbit starting near the (invisible) saddle cycle $N_\mu^{(1)}$ approaches the stable cycle $N_\mu^{(2)}$ and stays there forever. $\Diamond$

For systems with more than three phase variables, the "blue-sky" bifurcation may generate an infinite number of saddle limit cycles which belong to a *Smale-Williams solenoid attractor.* When the parameter approaches its critical value, the attractor does not bifurcate but the period and length of any cycle in it tend to infinity.

## 7.3 Bifurcations on invariant tori

Continuous-time dynamical systems with phase-space dimension $n > 2$ can have invariant tori. As we have seen in Chapters 4 and 5, an invariant two-dimensional torus $\mathbb{T}^2$ appears through a generic Neimark-Sacker bifurcation. For example, a stable cycle in $\mathbb{R}^3$ can lose stability when a pair of complex-conjugate multipliers crosses the unit circle. Then, provided there are no strong resonances and the cubic normal form coefficient has the proper sign, a smooth,[1] stable, invariant torus bifurcates from the cycle. In this section, we discuss changes of the orbit structure on an invariant two-torus under variation of the parameters of the system.

### 7.3.1 Reduction to a Poincaré map

Let $\mathbb{T}^2$ be a smooth, invariant two-torus of (7.1) at $\alpha = 0$. For simplicity, we can think of a three-dimensional system. Introduce a cross-section $\Sigma$, codim $\Sigma = 1$, to the torus (see **Fig. 7.18**). The intersection $\mathbb{T}^2 \cap \Sigma$ is a closed curve $S$, topologically (or even (finite-)smoothly) equivalent to the unit circle $\mathbb{S}^1$. Let us consider only the case when any orbit starting at a point $x \in S$ returns to $S$. Then, a Poincaré map

$$P : S \to S$$

is defined. Alternatively, we can consider a Poincaré map defined by (7.1) on the cross-section $\Sigma$. The closed curve $S$ is obviously an invariant curve of this map; its restriction to $S$ is the map $P$ introduced above. The map $P$ and its inverse are both differentiable. The standard relationship between fixed points of $P$ and limit cycles of (7.1) exists. All such cycles belong to the torus.

Assume that the invariant torus $\mathbb{T}^2$ persists under small parameter variations, meaning that there is a close invariant torus of (7.1) for all $\alpha$ with sufficiently small $|\alpha|$. Then, the above constriction can be carried out for all

[1] Finitely differentiable.

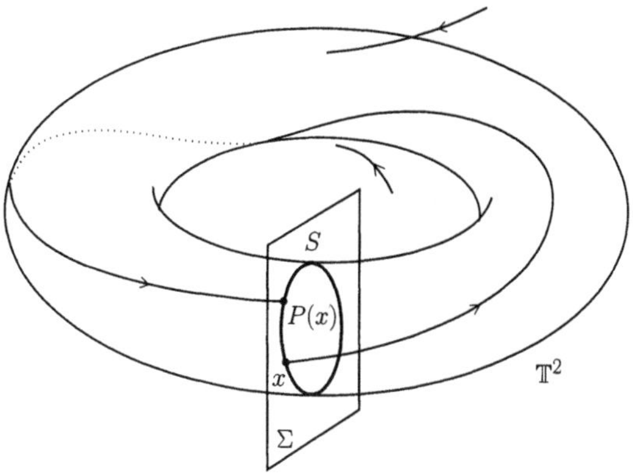


**Fig. 7.18.** Poincaré map on the torus.

nearby $\alpha$, resulting in a Poincaré map $P_\alpha : S \to S$, smoothly depending on the parameter.

**Remark:**
It can be proved (see the bibliographical notes) that a stable invariant torus $\mathbb{T}^2$ persists as a manifold under small parameter variations if it is *normally hyperbolic*, i.e., the convergence of nearby orbits to $\mathbb{T}^2$ is stronger than orbit convergence on the torus, provided proper measures of convergence are introduced. ◇

The problem now is to classify possible orbit structures of $P_\alpha : S \to S$ and to analyze their metamorphoses under variation of the parameter. To proceed, let us introduce canonical coordinates on $\mathbb{T}^2$. Namely, parametrize the torus by two angular coordinates $\psi, \varphi$ (mod $2\pi$). Using these coordinates, we can map the torus onto the square,

$$U = \{(\psi, \varphi) : 0 \leq \psi, \varphi \leq 2\pi\},$$

with opposite sides identified (see **Fig. 7.19**). Assume that the intersection $S = \mathbb{T}^2 \cap \Sigma$ is given by $\psi = 0$. Consider an orbit $\gamma$ on $\mathbb{T}^2$ starting at a point $(0, \varphi_0)$ on $S$. By our assumption, $\gamma$ returns to $S$ at some point $(2\pi, P(\varphi_0)) = (0, P(\varphi_0))$, where $P : \mathbb{S}^1 \to \mathbb{S}^1$ is a smooth function.[2] Since orbits on the torus do not intersect,

$$P'(\varphi) > 0,$$

so the map $P$ preserves the orientation of $S$.

### 7.3.2 Rotation number and orbit structure

A fixed point $\varphi_0$ of the map $P, P(\varphi_0) = \varphi_0$, corresponds to a cycle on $\mathbb{T}^2$ making one revolution along the parallel and some $p$ revolutions along the

[2] We use the same notation for both the map of the curve $S$ and the function it defines on the unit circle $\mathbb{S}^1$.

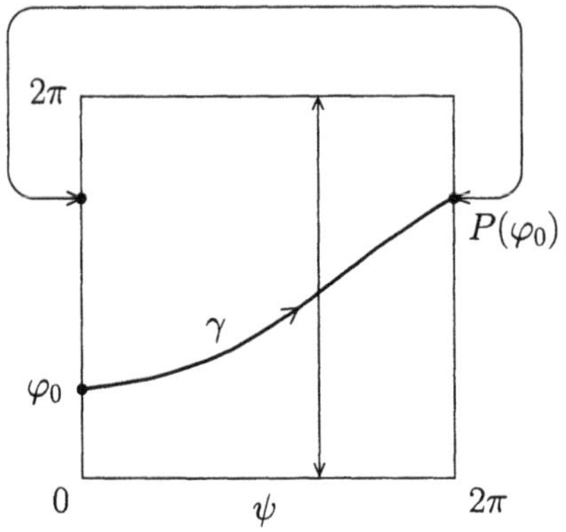


**Fig. 7.19.** An orbit $\gamma$ on the torus.

meridian before closure (see **Fig. 7.20**(a)). A cycle of period $q$,

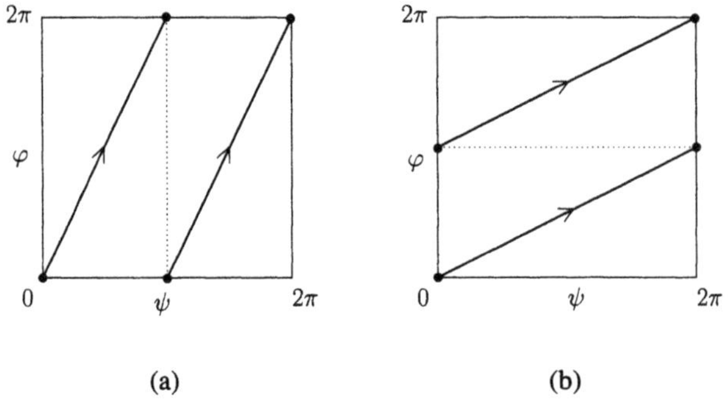


**Fig. 7.20.** Cycles on the torus: (a) $p = 2,\ q = 1$; (b) $p = 1,\ q = 2$.

$$\{\varphi_0, P(\varphi_0), P^2(\varphi_0), \ldots, P^q(\varphi_0) = \varphi_0\},$$

corresponds to a cycle on $\mathbb{T}^2$ that makes $q$ revolutions along the parallel and some $p$ revolutions along the meridian (**Fig. 7.20**(b)). Such a cycle is called a *$(p,q)$-cycle.* The local theory of fixed points and periodic orbits of $P$ on $S$ is the same as that for scalar maps. In particular, a fixed point $\varphi_0$ is stable (unstable) if $P'(\varphi_0) < 1$ ($P'(\varphi_0) > 1$). Points with $P'(\varphi_0) \neq 1$ are called *hyperbolic.* As usual, these notions can be extended to $q$-periodic orbits by considering $P^q$, the $q$th iterate of $P$. Clearly, if a stable $(p,q)$-cycle exists, an unstable $(p,q)$-cycle must also exist since stable and unstable fixed points of $P^q$ have to alternate (see **Fig. 7.21**).

The difference $a(\varphi) = P(\varphi) - \varphi$ is called the *angular function.*

**Definition 7.1** *The* rotation number *of $P : S \to S$ is defined by*

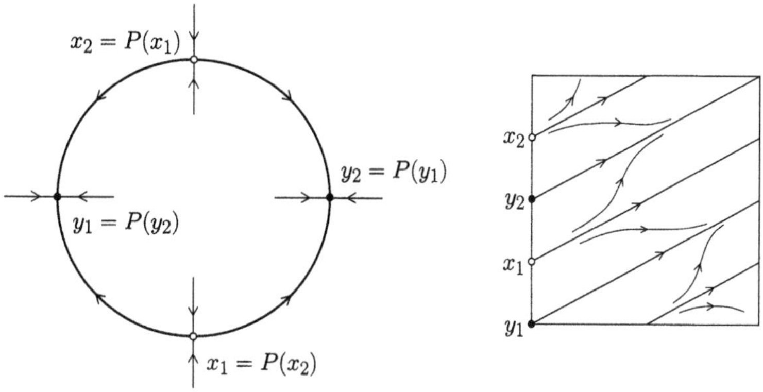


**Fig. 7.21.** Stable and unstable $(1, 2)$-cycles on the torus.

$$\rho = \frac{1}{2\pi} \lim_{k \to \infty} \frac{a(\varphi) + a(P(\varphi)) + \cdots + a(P^{k-1}(\varphi))}{k}.$$

One can prove that the limit in the definition exists and is independent of the point $\varphi \in S$. Thus, $\rho$ is well defined. It characterizes the average angle by which $P$ "rotates" $S$. For a rigid rotation through angle $2\pi\nu$,

$$P(\varphi) = \varphi + 2\pi\nu \pmod{2\pi}, \tag{7.15}$$

we have $\rho = \nu$. The role of the rotation number is clarified by the following two statements, which we give without proof.

**Lemma 7.1** *The rotation number of the map $P : S \to S$ is rational, $\rho = \frac{p}{q}$, if and only if $P$ has a $(p, q)$-periodic orbit.* □

Note that Lemma 7.1 does not state that the periodic orbit is unique.

**Lemma 7.2 (Denjoy [1932])** *If the map $P : S \to S$ is at least twice differentiable and its rotation number is irrational, then $P$ is topologically equivalent to rigid rotation through the angle $2\pi\rho$.* □

Under the conditions of the lemma, any orbit of $P$ on $S$ is dense, as is true for the rigid rotation (7.15) with irrational $\nu$. There are examples of $C^1$ diffeomorphisms that do not satisfy Denjoy's lemma. However, if the right-hand side $f$ of (7.1) is sufficiently smooth and its invariant torus $\mathbb{T}^2$ is also smooth enough, $P$ must satisfy Denjoy's differentiability condition.

### 7.3.3 Structural stability and bifurcations

Let us now address the problem of structural stability of systems on tori. We can use Chapter 2's definitions of the distance between dynamical systems and their structural stability simply by making the substitution $\mathbb{T}^2$ for $U$. This problem is equivalent to that for discrete-time dynamical systems on the circle $S$. The following theorem provides the complete characterization of structurally stable systems on $S$.

**Theorem 7.5** *A smooth dynamical system $P : S \to S$ is structurally stable if and only if its rotation number is rational and all periodic orbits are hyperbolic.* □

If the rotation number is irrational, we can always introduce an arbitrary small perturbation, resulting in a topologically nonequivalent system. Actually, such a perturbation will generate long-period cycles instead of dense orbits, resulting in a rational rotation number. The phase portrait of a structurally stable system on $\mathbb{T}^2$ is therefore rather simple: There is an even number of hyperbolic limit cycles of $(p, q)$ type; all other orbits tend to one of these cycles in the correct time direction. The $q$th iterate of the Poincaré map reveals an even number of fixed points of $P^q$ on $S$ having alternating stability.

Consider a one-parameter map $P_\alpha : S \to S$ corresponding to system (7.1) with an invariant torus. Let $\alpha = \alpha^0$ provide a structurally stable system. By Theorem 7.5 there is an open interval $(\alpha^0 - \varepsilon, \alpha^0 + \varepsilon)$ with $\varepsilon > 0$ within which the system has topologically equivalent phase portraits. What are the boundaries of this interval? Or, in other words, how does the rotation number change?

First of all, let us point out that bifurcations *can* take place even if the rotation number is *constant*. Indeed, the system may have an even number of hyperbolic $(p, q)$-cycles on the torus. While these cycles collide and disappear pairwise under parameter variation, the rotation number remains constant ($\rho = \frac{p}{q}$), provided that there remain at least two such cycles on the torus. However, when the last two cycles (a stable and an unstable one) collide and disappear, the rotation number becomes irrational until another "structurally stable window" opens. Inside the windows, the asymptotic behavior of the system is *periodic*, while it is *quasiperiodic* outside. In the former case, there are at least two limit cycles (possibly with a very high period) on $\mathbb{T}^2$, while in the latter case, the torus is filled by dense nonperiodic orbits.

The bifurcation from quasiperiodic behavior to periodic oscillations is called a *phase locking*. In periodically forced systems this phenomenon appears as a *frequency locking*. Suppose, for simplicity, that we have a two-dimensional, periodically forced system of ODEs that depends on a parameter. Assume that the associated period-return (Poincaré) map has an attracting closed invariant curve. If the rotation number of the map restricted to this curve is rational, the system exhibits periodic oscillations with a period that is an integer multiple of the forcing period. The frequency of the oscillations is "locked" at the external forcing frequency.

**Remark:**

Theorem 7.5 establishes a delicate relationship between "genericity" and structural stability. Consider a map $P_\alpha : S \to S$, depending on a single parameter. The set of parameter values for which $P_\alpha$ is structurally stable is open and dense. However, the measure of this set might be small compared with that of the parameter set corresponding to irrational rotation numbers. Thus, a map chosen "randomly" from this family would have an irrational rotation number with a high probability. ◇

### 7.3.4 Phase locking near a Neimark-Sacker bifurcation: Arnold tongues

We can apply the developed theory to an invariant torus (curve) appearing via a Neimark-Sacker bifurcation. Consider a two-dimensional discrete-time system near such a bifurcation. This system can be viewed as generated by the Poincaré map (restricted to a center manifold, if necessary) associated with a limit cycle of a continuous-time system (7.1). This map can be transformed, for nearby parameter values, by means of smooth, invertible, and smoothly parameter-dependent transformations to the form

$$z \mapsto \lambda z + c_1 z|z|^2 + O(|z|^4), \quad z \in \mathbb{C}^1, \tag{7.16}$$

where $\lambda$ and $c_1$ are smooth complex-valued functions of $\alpha$ (see Chapter 4). Consider, for a while, Re $\lambda$ and Im $\lambda$ as two independent parameters. On the plane of these parameters, the unit circle $|\lambda| = 1$ corresponds to the Neimark-Sacker bifurcation locus (see **Fig. 7.22**). Assume that the bifurcation occurs

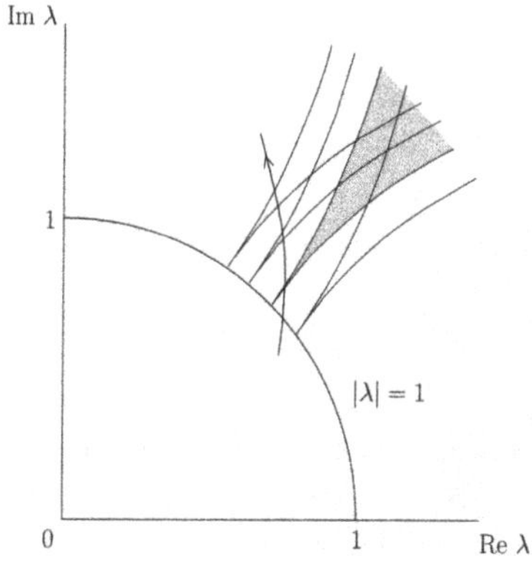


**Fig. 7.22.** Arnold tongues near the Neimark-Sacker bifurcation.

away from strong resonances and is supercritical. Then, a stable closed invariant curve exists for nearby parameter values outside the circle. Parameter regions in the (Re $\lambda$, Im $\lambda$)-plane corresponding to rational rotation numbers approach the circle $|\lambda| = 1$ at all rational points,

$$\lambda = e^{i\theta}, \ \theta = \frac{2\pi p}{q},$$

as narrow tongues.[3] These regions are called *Arnold tongues.* Recall that our system (7.1) depends on a single parameter $\alpha$. Therefore, it defines a curve in the $\lambda$-plane traced by $\lambda(\alpha)$. Near the circle $|\lambda| = 1$, this curve crosses an infinite number of Arnold tongues corresponding to various rational rotation numbers. Therefore, near a generic Neimark-Sacker bifurcation, an infinite number of long-periodic cycles are born and die as the parameter varies. Far from the Neimark-Sacker bifurcation curve, the tongues can intersect. At such parameter values, the invariant torus does not exist[4] and two independent fold bifurcations merely happen with unrelated remote cycles.

**Example 7.2 (Arnold tongue in the perturbed delayed logistic map)** Consider the recurrence equation

$$x_{k+1} = rx_k(1 - x_{k-1}) + \varepsilon, \tag{7.17}$$

where $x_k$ is the density of a population at year $k$, $r$ is the growth rate, and $\varepsilon$ is the migration rate. For $\varepsilon = 0$, this model was studied in Chapter 4 (see Example 4.2).

As in Chapter 4, introduce $y_k = x_{k-1}$ and rewrite (7.17) as a planar dynamical system

$$\begin{pmatrix} x \\ y \end{pmatrix} \mapsto \begin{pmatrix} rx(1-y) + \varepsilon \\ x \end{pmatrix}. \tag{7.18}$$

The analysis in Chapter 4 revealed a supercritical Neimark-Sacker bifurcation of (7.18) at $r = 2$ for $\varepsilon = 0$. There is a curve $h^{(1)}$ in the $(r, \varepsilon)$-plane passing through the point $(r, \varepsilon) = (2, 0)$ on which the fixed point of (7.18),

$$x^0 = y^0 = \frac{1+\varepsilon}{2},$$

undergoes a Neimark-Sacker bifurcation. The curve $h^{(1)}$ is given by the expression

$$h^{(1)} = \left\{ (r, \varepsilon) : r = \frac{2}{1+\varepsilon} \right\}.$$

Iterating the map (7.18) with $\varepsilon = 0$ for $r$ slightly greater than 2 (e.g., at $r = 2.1$ or $r = 2.15$) yields a closed invariant curve apparently filled by quasiperiodic orbits. An example of such a curve was shown in **Fig. 4.12** in Chapter 4. However, taking $r = 2.177$ results in a stable cycle of *period seven.* Thus, these parameter values belong to a phase-locking window. There is also an unstable (saddle) cycle of period seven, which is hard to detect by numerical simulations. These two cycles are located on the closed invariant curve. Actually, the curve is composed of the unstable manifolds of the saddle cycle.

---

[3] Their width $w$ at a distance $d$ from the circle satisfies $w \sim d^{(q-2)/2}$, as $d \to 0$.

[4] Otherwise, orbits on such a torus define two different rotation numbers, which is impossible.

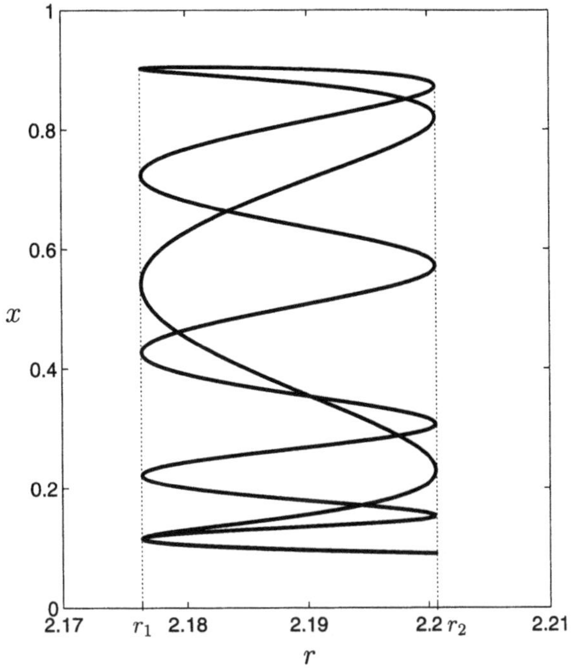


**Fig. 7.23.** Bifurcation diagram of period-seven cycles.

The seventh-iterate map (7.18), therefore, has seven *stable fixed points* and the same number of *unstable fixed points.* While we increase or decrease the parameter $r$, keeping $\varepsilon = 0$, the stable and unstable fixed points collide and disappear at fold bifurcations. Plotting the coordinates of all the fixed points against $r$ reveals the peculiar *closed curve* shown in **Fig. 7.23.**[5] All seven stable fixed points collide with their respective saddle points simultaneously at the fold points $r_{1,2}$ since, actually, there are only two period-seven cycles of the opposite stability, that collide at these parameter values $r_{1,2}$. Note that each stable fixed point of the seventh-iterate map collides with one immediately neighboring unstable fixed point at the fold bifurcation at $r = r_1$ and the other one at $r = r_2$. Thus, each stable fixed point can be thought to migrate between its two neighboring unstable fixed points as $r$ varies from $r_1$ to $r_2$.

The continuation of the boundary points of the phase-locking interval gives two fold bifurcation curves, $t_1^{(7)}$ and $t_2^{(7)}$, for cycles of period seven (see **Fig. 7.24**). They form a typical Arnold tongue, approaching a point on the Neimark-Sacker curve $h^{(1)}$ where the multipliers of the original fixed point have the representation

[5] In Chapter 10 we present a continuation technique by means of which the fixed-point curve in **Fig. 7.23** is computed.

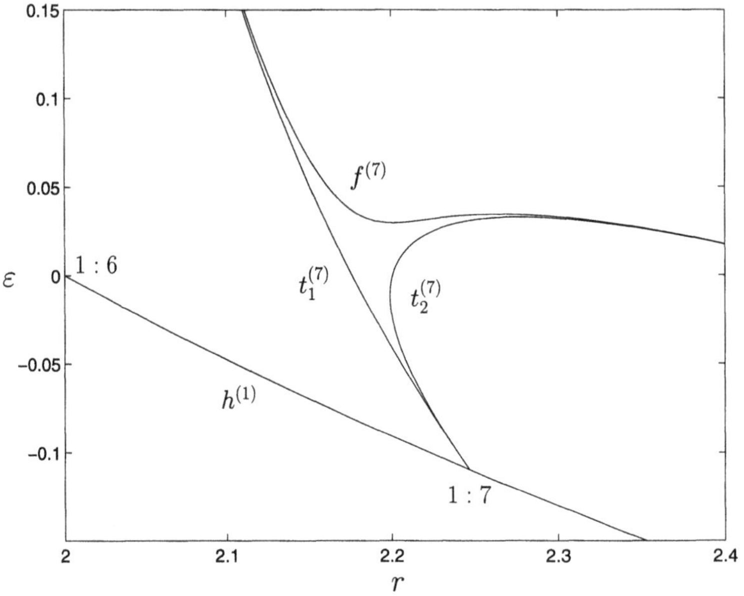


**Fig. 7.24.** 1:7 Arnold tongue in (7.17).

$$\mu_{1,2} = e^{\pm i\theta_0}, \ \theta_0 = \frac{2\pi}{7},$$

in accordance with the theory. Note that the point $(r, \varepsilon) = (2, 0)$ is the origin of another Arnold tongue, corresponding to cycles of period six (cf. Example 4.2). This 1:6 tongue is not shown in **Fig. 7.24**.

It is worthwhile mentioning that the stable period-seven cycle exhibits a *period doubling* if $\varepsilon$ increases and passes a certain critical value, while $r$ is fixed. The critical parameter values form a curve $f^{(7)}$, also presented in **Fig. 7.24**. Above (and near) this curve, a stable cycle of period 14 is present, while the closed invariant curve no longer exists. This is one of the possible ways in which an invariant curve can loose its smoothness and disappear. ◇

## 7.4 Bifurcations in symmetric systems

In this section we touch on an important topic: bifurcations in systems with symmetry. First we summarize some general results on symmetric systems, including a symmetric version of the Center Manifold Theorem. Then, we analyze bifurcations of equilibria and limit cycles in the presence of the simplest discrete symmetry.

Symmetric systems appear naturally in many applications. Often the symmetry reflects certain spatial invariance of the dynamical system or its finite-dimensional approximation. Normal forms for many bifurcations also have certain symmetries. As we shall see, some bifurcations can have a smaller codimension in the class of systems with a specified symmetry, and the corresponding bifurcations usually have some unique features. In contrast, some bifurcations become impossible in the presence of certain symmetries.

### 7.4.1 General properties of symmetric systems

Suppose we have a (compact) group $G$ that can be *represented* in $\mathbb{R}^n$ by matrices $\{T_g\}$:

$$T_e = I_n, \quad T_{g_1 g_2} = T_{g_1} T_{g_2},$$

for any $g_{1,2} \in G$. Here $e \in G$ is the *group unit* ($eg = ge = g$), while $I_n$ is the $n \times n$ unit matrix.

**Definition 7.2** *A continuous-time system*

$$\dot{x} = f(x), \quad x \in \mathbb{R}^n, \tag{7.19}$$

*is called* invariant with respect to the representation $\{T_g\}$ of the group $G$ (*or, simply,* $G$-equivariant) *if*

$$T_g f(x) = f(T_g x) \tag{7.20}$$

*for all* $g \in G$ *and all* $x \in \mathbb{R}^n$.

**Example 7.3.** The famous Lorenz system

$$\begin{cases} \dot{x} = -\sigma x + \sigma y, \\ \dot{y} = rx - y - xz, \\ \dot{z} = -bz + xy, \end{cases} \tag{7.21}$$

is invariant with respect to the transformation

$$T : \begin{pmatrix} x \\ y \\ z \end{pmatrix} \mapsto \begin{pmatrix} -x \\ -y \\ z \end{pmatrix}.$$

A matrix $R$ corresponding to this transformation ($R^2 = I$), together with the unit matrix $I$, forms a three-dimensional representation of the group $\mathbb{Z}_2$ (see Section 7.4.2). $\Diamond$

Equation (7.20) implies that the linear transformation

$$y = T_g x, \quad g \in G,$$

does not change system (7.19). Indeed,

$$\dot{y} = T_g \dot{x} = T_g f(x) = f(T_g x) = f(y).$$

Therefore, if $x(t)$ is a solution to (7.19), then $y(t) = T_g x(t)$ is also a solution. For example, if $(x_0(t), y_0(t), z_0(t))$, where $z_0 \not\equiv 0$, is a homoclinic solution to the origin in the Lorenz system (7.21), then there is another homoclinic orbit to the same equilibrium, given by $(-x_0(t), -y_0(t), z_0(t))$.

**Definition 7.3** *The* fixed-point subspace $X^G \subset \mathbb{R}^n$ *is the set*

$$X^G = \{x \in \mathbb{R}^n : T_g x = x, \text{ for all } g \in G\}.$$

The set $X^G$ is a linear subspace of $\mathbb{R}^n$. This subspace is an *invariant set* of (7.19), because $x \in X^G$ implies $\dot{x} \in X^G$. Indeed,

$$T_g \dot{x} = T_g f(x) = f(T_g x) = f(x) = \dot{x}$$

for all $g \in G$. For system (7.21), we have only one symmetry transformation $T$, so the fixed-point subspace is the $z$-axis, $X^G = \{(x, y, z) : x = y = 0\}$. This axis is obviously invariant under the flow associated with the system.

Let us explain the modifications that symmetry brings to the Center Manifold Theorem. Suppose we have a smooth $G$-equivariant system (7.19). Let $x^0$ be an equilibrium point that belongs to the fixed-point subspace $X^G$, and assume the Jacobian matrix $A = f_x$ evaluated at $x^0$ has $n_0$ eigenvalues (counting multiplicity) on the imaginary axis. Let $X^c$ denote the corresponding critical eigenspace of $A$. The following lemma is a direct consequence of the identity $T_g A = A T_g$, for all $g \in G$, that can be obtained by differentiating (7.20) with respect to $x$ at $x = x^0$.

**Lemma 7.3** *$X^c$ is $G$-invariant; in other words, if $v \in X^c$, then $T_g v \in X^c$ for all $g \in G$.* □

Therefore, it is possible to consider the *restriction* $\{T_g^c\}$ of $\{T_g\}$ on $X^c$. If we fix some coordinates $\xi = (\xi_1, \xi_2, \dots, \xi_{n_0})$ on $X^c$, $\{T_g^c\}$ will be given by certain $n_0 \times n_0$ matrices.

**Theorem 7.6 (Ruelle [1973])** *Any center manifold $W^c$ of the equilibrium $x^0 \in X^G$ of* (7.19) *is locally $G$-invariant.*

*Moreover, there are local coordinates $\xi \in \mathbb{R}^{n_0}$ on $W^c$ in which the restriction of* (7.19) *to $W^c$,*

$$\dot{\xi} = \psi(\xi), \quad \xi \in \mathbb{R}^{n_0}, \tag{7.22}$$

*is invariant with respect to the restriction of $\{T_g\}$ to $X^c$,*

$$T_g^c \psi(\xi) = \psi(T_g^c \xi). \ \square$$

Attraction properties of $W^c$ are determined, in the usual way, by the eigenvalues of $A$ with Re $\lambda \neq 0$. If (7.19) depends on parameters, one can construct a parameter-dependent center manifold $W_\alpha^c$ by appending the equation $\dot{\alpha} = 0$ to the system, as in Chapter 5. The restricted system (7.22) depends on $\alpha$ and is invariant with respect to $T_g^c$ at each fixed $\alpha$. Similar results are valid for $G$-equavaliant discrete-time systems.

### 7.4.2 $\mathbb{Z}_2$-equivariant systems

The simplest possible nontrivial group $G$ consists of two distinct elements $\{e, r\}$ such that

$$r^2 = e, \quad re = er = r, \quad e^2 = e.$$

This group is usually denoted by $\mathbb{Z}_2$. Let $\{I_n, R\}$ be a linear representation of $\mathbb{Z}_2$ in $\mathbb{R}^n$, where $I_n$ is the $n \times n$ unit matrix and the $n \times n$ matrix $R$ satisfies

$$R^2 = I_n.$$

The matrix $R$ defines the *symmetry transformation*: $x \mapsto Rx$.[6] It is easy to verify (Exercise 3) that the space $\mathbb{R}^n$ can be decomposed into a direct sum

$$\mathbb{R}^n = X^+ \oplus X^-,$$

where $Rx = x$ for $x \in X^+$, and $Rx = -x$ for $x \in X^-$. Therefore, $R$ is the identity on $X^+$ and a *central reflection* on $X^-$. According to Definition 7.4, $X^+$ is the fixed-point subspace associated with $G$. Let $n^{\pm} = \dim X^{\pm}$, $n^+ \geq 0, n^- \geq 1$. Clearly, there is a basis in $\mathbb{R}^n$ in which the matrix $R$ has the form

$$R = \begin{pmatrix} I_{n^+} & 0 \\ 0 & -I_{n^-} \end{pmatrix},$$

where $I_m$ is the $m \times m$ unit matrix. From now on, we can assume that such a basis is fixed, and we can thus consider a smooth parameter-dependent system

$$\dot{x} = f(x, \alpha), \quad x \in \mathbb{R}^n, \ \alpha \in \mathbb{R}^1, \tag{7.23}$$

that satisfies

$$Rf(x, \alpha) = f(Rx, \alpha),$$

for all $(x, \alpha) \in \mathbb{R}^n \times \mathbb{R}^1$, where $R$ is the matrix defined above. The fixed-point subspace $X^+$ is an invariant set of (7.23).

There is a simple classification of equilibria and periodic solutions of the $\mathbb{Z}_2$-invariant system (7.23).

**Definition 7.4** *An equilibrium $x^0$ of* (7.23) *is called* fixed *if $Rx^0 = x^0$.*

Thus, the symmetry transformation maps a fixed equilibrium into itself. If an equilibrium is not fixed, $Rx^0 = x^1 \neq x^0$, then $x^1$ is also an equilibrium of (7.23) (check!) and $Rx^1 = x^0$.

**Definition 7.5** *Two equilibria $x^0$ and $x^1$ of* (7.23) *are called* $R$-conjugate *if $x^1 = Rx^0$.*

Similar terminology can be introduced for periodic solutions.

**Definition 7.6** *A periodic solution $x_f(t)$ of* (7.23) *is called* fixed *if $Rx_f(t) = x_f(t)$ for all $t \in \mathbb{R}^1$.*

Obviously, the closed orbit corresponding to a fixed periodic solution belongs to $X^+$ and is invariant under the symmetry transformation $R$ (see **Fig. 7.25**(a)). It can exist if $n^+ \geq 2$. However, there is another type of periodic solution that defines a closed orbit that is $R$-invariant but not fixed.

[6] Sometimes, we will also use the symbol $R$ to denote this transformation. In such cases, $I_n$ will mean the identity map.

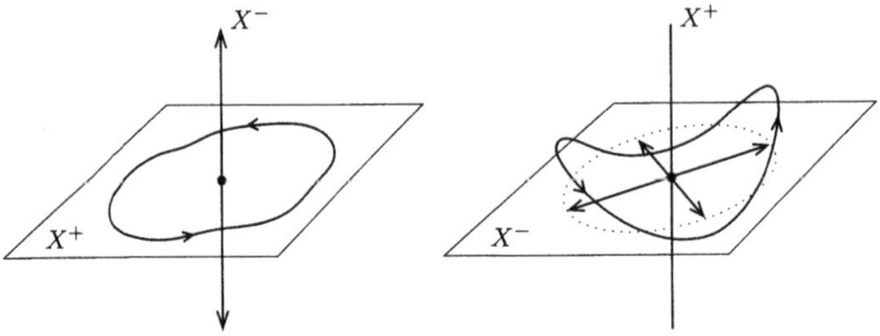


**Fig. 7.25.** Invariant cycles: (a) $F$-cycle; (b) $S$-cycle.

**Definition 7.7** *A periodic solution $x_s(t)$ of (7.23) with (minimal) period $T_s$ is called* symmetric *if*

$$Rx_s(t) = x_s\left(t + \frac{T_s}{2}\right)$$

*for all $t \in \mathbb{R}^1$.*

Thus, a symmetric periodic solution is transformed into itself by applying $R$ and shifting the time by half of the period. The orbit corresponding to a symmetric solution cannot intersect $X^+$ and requires $n^- \geq 2$ to exist. Its projection to $X^-$ is symmetric with respect to the central reflection (see **Fig. 7.25**(b)). Notice that the $X^+$-components of the symmetric periodic solution oscillate with the double frequency.

We call a limit cycle $L$ of (7.23) fixed (symmetric) if the corresponding periodic solution is fixed (symmetric), and we denote them by $F$- and $S$-cycle, respectively. Both $F$- and $S$-cycles are $R$-invariant as curves in $\mathbb{R}^n$: $R(L) = L$. We leave the proof of the following lemma as an exercise to the reader.

**Lemma 7.4** *Any $R$-invariant cycle of (7.23) is either an $F$- or an $S$-cycle.* $\square$

Of course, there may exist noninvariant limit cycles, $R(L) \neq L$. If $x^0(t)$ is a periodic solution corresponding to such a cycle, then $x^1(t) = Rx^0(t)$ is another periodic solution of (7.23).

**Definition 7.8** *Two noninvariant limit cycles are called* $R$-conjugate *if two of their corresponding periodic solutions satisfy $x^1(t) = Rx^0(t)$ for all $t \in \mathbb{R}^1$.*

### 7.4.3 Codim 1 bifurcations of equilibria in $\mathbb{Z}_2$-equivariant systems

Our aim now is to analyze generic bifurcations of equilibria in $\mathbb{Z}_2$-equivariant systems. Clearly, the bifurcations of $R$-conjugate equilibria happen in the same way as in generic systems, being merely "doubled" by the symmetry transformation $R$. For example, two pairs of $R$-conjugate equilibria of opposite stability can collide and disappear via the fold bifurcation. Thus, one can expect new phenomena only if the bifurcating equilibrium is of the fixed type. Let us analyze the following simple example.

**Example 7.4 (Symmetric pitchfork bifurcation)** Consider the scalar system

$$\dot{x} = \alpha x - x^3, \quad x \in \mathbb{R}^1, \ \alpha \in \mathbb{R}^1. \tag{7.24}$$

The system is obviously $\mathbb{Z}_2$-equivariant. Indeed, in this case, $Rx = -x$ (reflection) and $n^+ = 0, n^- = 1$. At $\alpha = 0$, system (7.24) has the fixed equilibrium $x^0 = 0$ with eigenvalue zero. The bifurcation diagram of (7.24) is simple (see **Fig. 7.26**). There is always a trivial equilibrium $x^0 = 0$, which

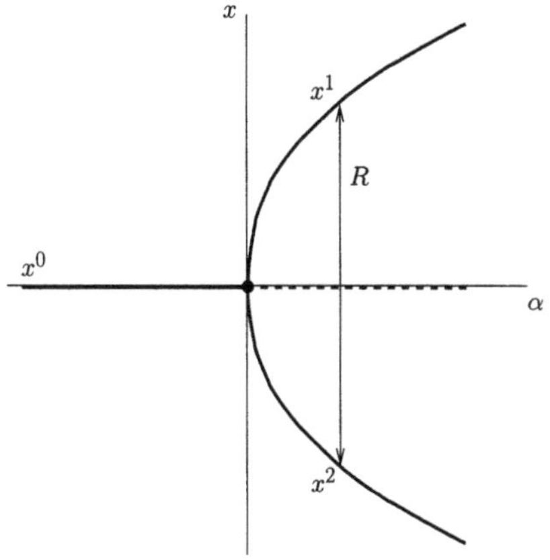


**Fig. 7.26.** Symmetric pitchfork bifurcation.

is linearly stable for $\alpha < 0$ and unstable for $\alpha > 0$. It is *fixed* according to Definition 7.5. There are also two stable nontrivial equilibria, $x^{1,2}(\alpha) = \pm\sqrt{\alpha}$, existing for $\alpha > 0$ and $R$-conjugate, $Rx^1(\alpha) = x^2(\alpha)$.

Any $\mathbb{Z}_2$-equivariant system

$$\dot{x} = \alpha x - x^3 + O(x^5)$$

is locally topologically equivalent near the origin to (7.24). Indeed, such a system has the form

$$\dot{x} = \alpha x - x^3 + x\psi_\alpha(x^2), \tag{7.25}$$

with some $\psi_\alpha(x^2) = O(x^4)$ since any odd function vanishing at $x = 0$ can be represented as $x\varphi(x)$, where $\varphi(x)$ is even and, thus, $\varphi(x) = \psi(x^2)$. It is clear that $x = 0$ is always an equilibrium of (7.25). The nontrivial equilibria satisfy the equation

$$\alpha - x^2 + \psi_\alpha(x^2) = 0,$$

which can easily be analyzed near $(x, \alpha) = (0, 0)$ by means of the Implicit Function Theorem. This proves that the number and stability of the equilibria in (7.24) and (7.25) are the same for corresponding small parameter values with small $|\alpha|$. The homeomorphism $h_\alpha : \mathbb{R}^1 \to \mathbb{R}^1$ that identifies the phase portraits of the systems can be constructed to satisfy

$$h_\alpha(-x) = -h_\alpha(x),$$

for all $(x, \alpha)$. In other words, the homeomorphism can be defined by an odd function of $x$. $\Diamond$

This example has a fundamental meaning, due to the following theorem.

**Theorem 7.7 (Bifurcations at a zero eigenvalue)** *Suppose that a $\mathbb{Z}_2$-equivariant system*

$$\dot{x} = f(x, \alpha), \quad x \in \mathbb{R}^n, \ \alpha \in \mathbb{R}^1,$$

*with smooth $f$, $Rf(x, \alpha) = f(Rx, \alpha)$, $R^2 = I_n$, has at $\alpha = 0$ the fixed equilibrium $x^0 = 0$ with simple zero eigenvalue $\lambda_1 = 0$, and let $v \in \mathbb{R}^n$ be the corresponding eigenvector.*

*Then the system has a one-dimensional $R$-invariant center manifold $W^c_\alpha$, and one of the following alternatives generically takes place:*

(i) (fold) *If $v \in X^+$, then $W^c_\alpha \subset X^+$ for all sufficiently small $|\alpha|$, and the restriction of the system to $W^c_\alpha$ is locally topologically equivalent near the origin to the normal form*

$$\dot{\xi} = \beta \pm \xi^2;$$

(ii) (pitchfork) *If $v \in X^-$, then $W^c_\alpha \cap X^+ = x^0$ for all sufficiently small $|\alpha|$, and the restriction of the system to $W^c_\alpha$ is locally topologically equivalent near the origin to the normal form*

$$\dot{\xi} = \beta\xi \pm \xi^3. \ \Box$$

In case (i), the standard fold bifurcation happens within the invariant subspace $X^+$, giving rise to two fixed-type equilibria. The genericity conditions for this case are those formulated in Chapter 3 for the nonsymmetric fold.

In case (ii), the pitchfork bifurcation studied in Example 7.3 happens, resulting in the appearance of two $R$-conjugate equilibria, while the fixed equilibrium changes its stability. The genericity conditions include nonvanishing of the cubic term of the restriction of the system to the center manifold at $\alpha = 0$. We leave the reader to work out the details.

The presence of $\mathbb{Z}_2$-symmetry in a system having a purely imaginary pair of eigenvalues brings nothing new to nonsymmetric Hopf bifurcation theory. Namely, one can prove the following.

**Theorem 7.8 (Bifurcation at purely imaginary eigenvalues)** *Suppose that a $\mathbb{Z}_2$-invariant system*

$$\dot{x} = f(x, \alpha), \quad x \in \mathbb{R}^n, \ \alpha \in \mathbb{R}^1,$$

*with smooth $f$, $Rf(x, \alpha) = f(Rx, \alpha)$, $R^2 = I_n$, has at $\alpha = 0$ the fixed equilibrium $x^0 = 0$ with a simple pair of imaginary eigenvalues $\lambda_{1,2} = \pm i\omega_0$, $\omega_0 > 0$.*

*Then, generically, the system has a two-dimensional $R$-invariant center manifold $W^c_\alpha$, and the restriction of the system to $W^c_\alpha$ is locally topologically equivalent near the origin to the normal form:*

$$\begin{pmatrix} \dot{\xi}_1 \\ \dot{\xi}_2 \end{pmatrix} = \begin{pmatrix} \beta & -1 \\ 1 & \beta \end{pmatrix} \begin{pmatrix} \xi_1 \\ \xi_2 \end{pmatrix} \pm (\xi_1^2 + \xi_2^2) \begin{pmatrix} \xi_1 \\ \xi_2 \end{pmatrix}. \ \square$$

This normal form is the standard normal form for a generic Hopf bifurcation. It describes the appearance (or disappearance) of a unique limit cycle having amplitude $\sqrt{\beta}$. There is a subtle difference, depending on whether the critical eigenspace $X^c$ belongs to $X^+$ or $X^-$. If $X^c \subset X^+$, then $W^c_\alpha \subset X^+$ and the standard Hopf bifurcation happens within the invariant subspace $X^+$. The bifurcating cycle is of type $F$. In contrast, if $X^c \subset X^-$, then $W^c_\alpha \cap X^+ = x^0$ and the system restricted to the center manifold is $Z_2$-invariant with respect to the transformation

$$R \begin{pmatrix} \xi_1 \\ \xi_2 \end{pmatrix} = - \begin{pmatrix} \xi_1 \\ \xi_2 \end{pmatrix}.$$

The bifurcating small-amplitude limit cycle is of type $S$.

### 7.4.4 Codim 1 bifurcations of cycles in $\mathbb{Z}_2$-equivariant systems

As was the case for equilibria, bifurcations of noninvariant limit cycles happen in the same manner as those in generic systems. Bifurcations of $F$- and $S$-cycles are very different and have to be treated separately.

#### Codim 1 bifurcations of *F*-cycles

Consider a fixed limit cycle $L_F$ of (7.23), and select a codim 1 hyperplane $\Sigma$ that is transversal to the cycle and $R$-invariant, $R(\Sigma) = \Sigma$. Let $P_\alpha$ be the Poincaré map defined on $\Sigma$ near its intersection with $L_F$ (see **Fig. 7.27**).

**Lemma 7.5** *The Poincaré map $P_\alpha : \Sigma \to \Sigma$ is $G$-equivariant,*

$$R_\Sigma \circ P_\alpha = P_\alpha \circ R_\Sigma,$$

*where $R_\Sigma$ is the restriction of the map $R$ to $\Sigma$.*

**Proof:**

Let an orbit $\gamma$ of (7.23) start at a point $u \in \Sigma$ and return to a point $v \in \Sigma$ close to $u$: $v = P_\alpha(u)$. The $R$-conjugate orbit $\tilde{\gamma} = R(\gamma)$ starts at the point $\tilde{u} = R_\Sigma(u)$ and returns to $\Sigma$ at the point $\tilde{v} = R_\Sigma(v)$. Since $\tilde{v} = P_\alpha(\tilde{u})$, we have

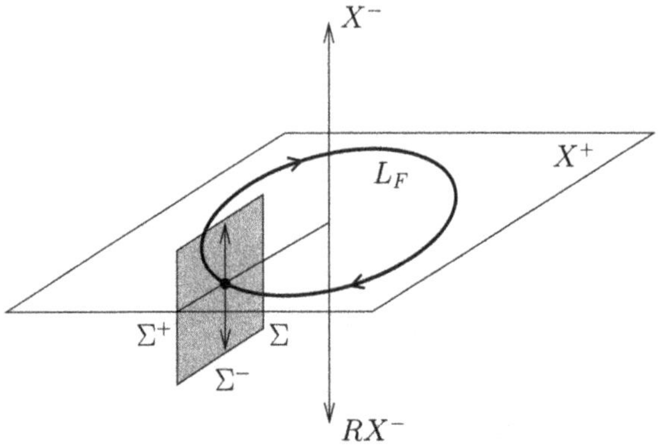


**Fig. 7.27.** Poincaré map for an $F$-cycle.

$$R_\Sigma(P_\alpha(u)) = P_\alpha(R_\Sigma(u))$$

for all $u \in \Sigma$ such that both sides of the equation are defined. □

Therefore, the analysis of bifurcations of $F$-cycles is reduced to that of fixed points in a discrete-time dynamical system (map) having $\mathbb{Z}_2$-symmetry. Introduce local coordinates $\xi \in \mathbb{R}^{n-1}$ on $\Sigma$ so that $\xi = 0$ corresponds to the cycle $L_F$. Let us use the same symbol $R$ instead of $R_\Sigma$, and decompose $\Sigma$ by

$$\Sigma = \Sigma^+ \oplus \Sigma^-,$$

where $\xi \in \Sigma^\pm$ if $R\xi = \pm\xi$. Consider each codimension one case separately. We give the following theorems, which are obvious enough, without proofs.

**Theorem 7.9 (Bifurcations at $\mu = 1$)** *Suppose that a $\mathbb{Z}_2$-equivariant system*

$$\dot{x} = f(x, \alpha), \quad x \in \mathbb{R}^n, \quad \alpha \in \mathbb{R}^1,$$

*with smooth $f$, $Rf(x,\alpha) = f(Rx,\alpha)$, $R^2 = I_n$, has at $\alpha = 0$ an $F$-cycle $L_0$ with a simple multiplier $\mu_1 = 1$, which is the only multiplier with $|\mu| = 1$. Let $v$ be the corresponding eigenvector of the Jacobian matrix of the Poincaré map $P_0$ associated with the cycle.*

*Then the map $P_\alpha$ has a one-dimensional $R$-invariant center manifold $W^c_\alpha$, and the restriction of $P_\alpha$ to this manifold is, generically, locally topologically equivalent near the cycle to one of the following normal forms:*

(i) (fold) *If $v \in \Sigma^+$, then*

$$\eta \mapsto \beta + \eta \pm \eta^2;$$

(ii) (pitchfork) *If $v \in \Sigma^-$, then*

$$\eta \mapsto (1 + \beta)\eta \pm \eta^3. \ \square$$

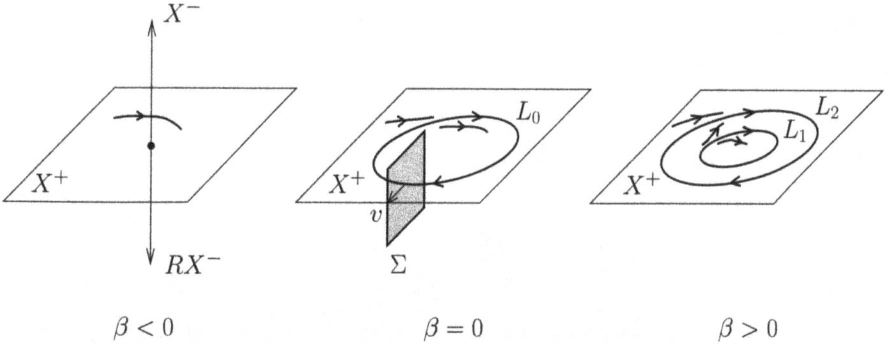


**Fig. 7.28.** Fold bifurcation of an $F$-cycle.

In case (i), $W^c_\alpha \subset \Sigma^+$, and we have the standard fold bifurcation giving rise to two $F$-cycles $L_1, L_2 \in X^+$, with different stability (see **Fig. 7.28**).

In case (ii), $W^c_\alpha \cap \Sigma^+ = 0$, and we have the appearance (or disappearance) of two $R$-conjugate limit cycles $L_{1,2}, L_2 = R(L_1)$, as the original $F$-cycle changes its stability (see **Fig. 7.29**).

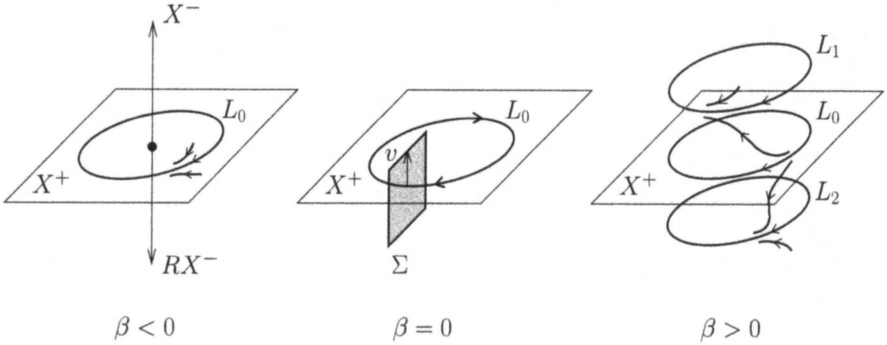


**Fig. 7.29.** Pitchfork bifurcation of an $F$-cycle.

**Theorem 7.10 (Bifurcation at $\mu = -1$)** *Suppose that a $\mathbb{Z}_2$-equivariant system*

$$\dot{x} = f(x, \alpha), \quad x \in \mathbb{R}^n, \ \alpha \in \mathbb{R}^1,$$

*with smooth $f$, $Rf(x,\alpha) = f(Rx,\alpha)$, $R^2 = I_n$, has at $\alpha = 0$ an $F$-cycle $L_0$ with a simple multiplier $\mu_1 = -1$, which is the only multiplier with $|\mu| = 1$. Let $v$ be the corresponding eigenvector of the Jacobian matrix of the associated Poincaré map $P_0$.*

*Then the map $P_\alpha$ has a one-dimensional $R$-invariant center manifold $W^c_\alpha$, and the restriction of $P_\alpha$ to this manifold is, generically, locally topologically equivalent near the cycle to the normal form:*

$$\eta \mapsto -(1+\beta)\eta \pm \eta^3.$$

*Moreover, the double-period limit cycle corresponding to the fixed points of* $P_\alpha^2$ *has* $F$*-type if* $v \in \Sigma^+$ *and* $S$*-type if* $v \in \Sigma^-$. □

**Theorem 7.11 (Bifurcation at complex multipliers $|\mu_{1,2}| = 1$)** *Suppose that a* $\mathbb{Z}_2$*-equivariant system*

$$\dot{x} = f(x, \alpha), \quad x \in \mathbb{R}^n, \quad \alpha \in \mathbb{R}^1,$$

*with smooth* $f$, $Rf(x,\alpha) = f(Rx,\alpha)$, $R^2 = I_n$, *has at* $\alpha = 0$ *an* $F$*-cycle* $L_0$ *with simple multipliers* $\mu_{1,2} = e^{\pm i\theta_0}$, *which are the only multipliers with* $|\mu| = 1$.

*Then the map* $P_\alpha$ *has a two-dimensional* $R$*-invariant center manifold* $W_\alpha^c$ *on which a unique invariant closed curve generically bifurcates from the fixed point corresponding to* $L_0$. *This curve corresponds to an invariant two-torus* $\mathbb{T}^2$ *of the system,* $R(\mathbb{T}^2) = \mathbb{T}^2$. □

**Remark:**

The fold and pitchfork bifurcations are the only possible codim 1 bifurcations in generic, one-parameter, $\mathbb{Z}_2$-equivariant systems in $\mathbb{R}^3$. ◊

### Codim 1 bifurcations of $S$-cycles

As one can see, in this case the cross-section $\Sigma$ *cannot* be selected to be $R$-invariant. Instead, one can choose two secant hyperplanes to the cycle $L_0$, $\Sigma_1$ and $\Sigma_2$, such that

$$R(\Sigma_1) = \Sigma_2,$$

and the Poincaré map $P_\alpha$ can be represented for all sufficiently small $|\alpha|$ as the composition of two maps $Q_\alpha^{(1)} : \Sigma_1 \to \Sigma_2$ and $Q_\alpha^{(2)} : \Sigma_2 \to \Sigma_1$ defined near the cycle

$$P_\alpha = Q_\alpha^{(2)} \circ Q_\alpha^{(1)}$$

(see **Fig. 7.30**).

**Lemma 7.6** *There is a smooth map* $Q_\alpha : \Sigma_1 \to \Sigma_1$ *such that*

$$P_\alpha = Q_\alpha^2. \tag{7.26}$$

**Proof:**

In the proper coordinates the map $Q_\alpha^{(2)}$ coincides with $Q_\alpha^{(1)}$. More precisely, due to the symmetry of the system,

$$Q_\alpha^{(2)} \circ R = R \circ Q_\alpha^{(1)},$$

or, equivalently, $Q_\alpha^{(2)} = R \circ Q_\alpha^{(1)} \circ R^{-1}$. Now introduce a map

$$Q_\alpha = R^{-1} \circ Q_\alpha^{(1)}$$

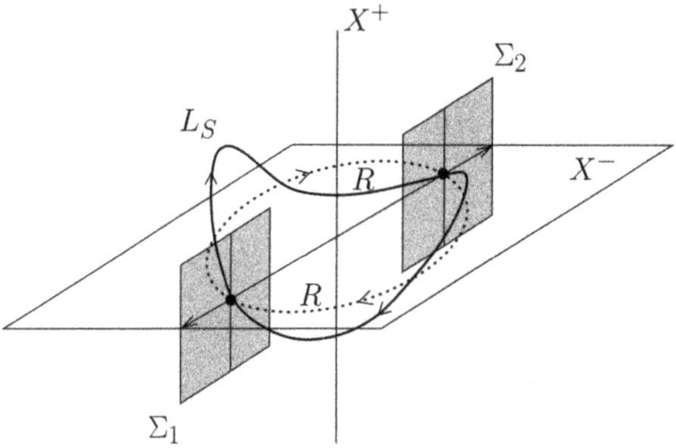


**Fig. 7.30.** Poincaré map for an $S$-cycle.

transforming $\Sigma_1$ into itself: First we allow a point to fly along the orbit of the system from $\Sigma_1$ to $\Sigma_2$ and then apply the inverse symmetry transformation $R^{-1}$ placing it back to $\Sigma_1$. We have

$$P_\alpha = R \circ Q_\alpha^{(1)} \circ R^{-1} \circ Q_\alpha^{(1)} = R \circ (R \circ R^{-1}) \circ Q_\alpha^{(1)} \circ R^{-1} \circ Q_\alpha^{(1)} = R^2 \circ Q_\alpha^2,$$

that gives (7.26) since $R^2 = I$. □

Consequently, the analysis of bifurcations of $S$-cycles is reduced to that for fixed points of the map $Q_\alpha$ that has no special symmetry. However, equation (7.26) imposes strong restrictions on possible bifurcations.

**Proposition 7.1** *An $S$-cycle cannot have the simple multiplier $\mu = -1$.*

**Proof:**

Let $A$ and $B$ be the Jacobian matrices of $P_0$ and $Q_0$ evaluated at their common fixed point. Then (7.26) implies

$$A = B^2.$$

If $\mu$ is a simple real eigenvalue of $A$, then there exists a simple real eigenvalue $\lambda$ of $B$. Therefore, $\mu = \lambda^2 > 0$. □

Thus, only the cases $\mu_1 = 1$ and $\mu_{1,2} = e^{\pm i\theta_0}$ have to be considered.

**Theorem 7.12 (Bifurcation at $\mu = 1$)** *Suppose that a $\mathbb{Z}_2$-equivariant system*

$$\dot{x} = f(x, \alpha), \quad x \in \mathbb{R}^n, \quad \alpha \in \mathbb{R}^1,$$

*with smooth $f$, $Rf(x,\alpha) = f(Rx,\alpha)$, $R^2 = I_n$, has at $\alpha = 0$ an $S$-cycle $L_0$ with a simple multiplier $\mu_1 = 1$, which is the only multiplier with $|\mu| = 1$. Let $v$ be the corresponding eigenvector of the Jacobian matrix $A = B^2$ of the associated Poincaré map $P_0 = Q_0^2$.*

*Then the map $P_\alpha$ has a one-dimensional R-invariant center manifold $W^c_\alpha$, and the restriction of $P_\alpha$ to this manifold is, generically, locally topologically equivalent near the cycle to one of the following normal forms:*

(i) (fold) *If $Bv = v$, then*

$$\eta \mapsto \beta + \eta \pm \eta^2;$$

(ii) (pitchfork) *If $Bv = -v$, then*

$$\eta \mapsto (1+\beta)\eta \pm \eta^3.$$

**Outline of the proof:**

In case (i), we have a standard fold bifurcation of $Q_\alpha$ ($\lambda_1 = 1$). Therefore, on its center manifold (which is also a center manifold for $P_\alpha$) the map $Q_\alpha$ is generically equivalent to

$$\xi \mapsto \gamma + \xi \pm \xi^2.$$

The fixed points of this map correspond to $S$-cycles of the system. Its second iterate

$$\xi \mapsto (2\gamma + \cdots) + (1 + \cdots)\xi \pm (2 + \cdots)\xi^2 + \cdots$$

is topologically equivalent to the normal form (i).

In case (ii), we have a standard flip bifurcation for $Q_\alpha$ ($\lambda_1 = -1, \lambda_1^2 = 1$). On its center manifold, the map $Q_\alpha$ is equivalent to

$$\xi \mapsto -(1+\gamma)\xi \pm \xi^3.$$

The cycle of period two of this map corresponds to a *pair* of $R$-conjugate cycles of the original system. The second iterate of this map,

$$\xi \mapsto (1 + 2\gamma + \cdots)\xi \mp (2 + \cdots)\xi^3 + \cdots,$$

is topologically equivalent to the normal form (ii). □

**Theorem 7.13 (Bifurcation at complex multipliers $|\mu_{1,2}| = 1$)** *Suppose that a $\mathbb{Z}_2$-equivariant system*

$$\dot{x} = f(x, \alpha), \quad x \in \mathbb{R}^n, \ \alpha \in \mathbb{R}^1,$$

*with smooth $f$, $Rf(x,\alpha) = f(Rx, \alpha)$, $R^2 = I_n$, has at $\alpha = 0$ an $S$-cycle $L_0$ with simple multipliers $\mu_1 = e^{\pm i\theta_0}$, which are the only multipliers with $|\mu| = 1$.*

*Then the map $P_\alpha$ has a two-dimensional R-invariant center manifold $W^c_\alpha$ on which a unique invariant closed curve generically bifurcates from the fixed point corresponding to $L_0$. This curve corresponds to an invariant two-torus $\mathbb{T}^2$ of the system, $R(\mathbb{T}^2) = \mathbb{T}^2$.* □

**Remark:**

A system in $\mathbb{R}^3$ that is invariant under the transformation $Rx = -x, x \in \mathbb{R}^3$ cannot exhibit the Neimark-Sacker bifurcation of an $S$-cycle. ◇

## 7.5 Exercises

**(1) (Asymptotics of the cycle period near saddle-node homoclinic bifurcation)** Find an asymptotic expression for the period $T_\beta$ of the cycle as a function of $\beta$, when it approaches the homoclinic orbit at a saddle-node bifurcation. (*Hint:* The leading term of the expansion is given by

$$T_\beta \sim \int_{-1}^{1} \frac{d\xi}{\beta + \xi^2},$$

where $\xi$ is a coordinate on a center manifold $W^c_\beta$ near the saddle-node.)

**(2) (Arnold's circle map)** Consider the following two-parameter smooth map $P_{\alpha,\varepsilon} : \mathbb{S}^1 \to \mathbb{S}^1$,

$$P_{\alpha,\varepsilon}(\varphi) = \varphi + \alpha + \varepsilon \sin\theta,$$

where $0 \leq \varepsilon < 1$, $\varphi, \alpha \pmod{2\pi}$. Compute asymptotic expressions for the curves that bound a region in the $(\alpha, \varepsilon)$-plane corresponding to the map having rotation number $\rho = \frac{1}{2}$. (*Hint*: A rotation number $\rho = \frac{1}{2}$ implies the presence of cycles of period two, and the boundaries are defined by the fold bifurcation of these cycles.)

**(3) (Symmetry decomposition)** Let $R$ be a real $n \times n$ matrix such that $R^2 = I_n$. Prove that the space $\mathbb{R}^n$ can be decomposed as $\mathbb{R}^n = X^+ \oplus X^-$, where $Rx = x$ for $x \in X^+$, and $Rx = -x$ for $x \in X^-$. (*Hint*: Any eigenvalue $\lambda$ of $R$ satisfies $\lambda^2 = 1$.)

**(4) (Hopf bifurcation in $\mathbb{Z}_2$-equivariant planar systems)** Prove that the Hopf bifurcation never happens in the planar systems invariant under the transformation

$$R\begin{pmatrix} x \\ y \end{pmatrix} = \begin{pmatrix} x \\ -y \end{pmatrix}.$$

(*Hint*: Any real matrix $A$ satisfying $AR = RA$, where $R$ is defined above, is diagonal.)

**(5) (Pitchfork bifurcation in the Lorenz system)** Prove that the equilibrium $(x, y, z) = (0, 0, 0)$ of the Lorenz system (7.21) exhibits a nondegenerate pitchfork bifurcation at $r_0 = 1$, for any fixed positive $(\sigma, b)$. (*Hints*:

(a) Verify that at $r_0 = 1$ the equilibrium at the origin has a simple zero eigenvalue, and compute the corresponding eigenvector $v$. Check that $Av = -v$, where $A$ is the Jacobian matrix of (7.21) evaluated at $x = y = z = 0$ for $r = 1$, so that case (ii) of Theorem 7.6 is applicable.

(b) Compute the second-order approximation to the center manifold $W^c$ at $r_0 = 1$ and prove that it is $R$-invariant.

(c) Check that the restriction of the system to the center manifold has no quadratic term. Could one expect this a priori?

(d) Compute the coefficient of the cubic term as a function of $(\sigma, b)$ and verify that it is nonzero for positive parameter values.)

**(6) (Normal form for $O(2)$-symmetric Hopf bifurcation)** Consider the following smooth, four-dimensional system written as two complex equations:

$$\begin{cases} \dot{z}_1 = z_1(\beta + i\omega(\beta) + A(\beta)|z_1|^2 + B(\beta)|z_2|^2), \\ \dot{z}_2 = z_2(\beta + i\omega(\beta) + B(\beta)|z_1|^2 + A(\beta)|z_2|^2), \end{cases} \tag{7.27}$$

where $\beta$ is the bifurcation parameter, $\omega(0) > 0$, $A(\beta)$ and $B(\beta)$ are complex-valued functions, and for $a(\beta) = \text{Re } A(\beta), b(\beta) = \text{Re } B(\beta)$,

$$a(0)b(0)(a^2(0) - b^2(0)) \neq 0.$$

This is a (truncated) normal form for the Hopf bifurcation with a four-dimensional center manifold of $O(2)$-equivariant systems (see van Gils & Mallet-Paret [1986], Kuznetsov [1984, 1985]. Notice that the critical pair of eigenvalues $\pm i\omega(0)$ is double.

(a) Verify that system (7.27) is invariant with respect to the representation of the orthogonal group $O(2)$ in $\mathbb{C}^2$ by the transformations

$$R(z_1, z_2) = (z_2, z_1), \; T_\theta(z_1, z_2) = (e^{i\theta} z_1, e^{-i\theta} z_2) \;\; (\theta \bmod 2\pi). \tag{7.28}$$

(b) Write system (7.27) in polar coordinates $z_k = \rho_k e^{i\varphi}, k = 1, 2$, and check that equations for $\rho_k$ are independent of those for $\varphi_k$.

(c) Introduce $r_k = \rho_k^2, k = 1, 2$, and derive a quadratic planar system for $r_k$. Assume $a(0) < 0$ and obtain the bifurcation diagrams of the resulting system as $\beta$ varies. (*Hint*: There are three subcases: (i) $b(0) < a(0)$; (ii) $b(0) > a(0), a(0) + b(0) < 0$; (iii) $b(0) > 0, a(0) + b(0) > 0$. In all cases, the amplitude system cannot have limit cycles.)

(d) Interpret the results of part (c) in terms of the four-dimensional system (7.27). Prove that, besides the trivial equilibrium at the origin, the system can have a pair of $R$-conjugate limit cycles, and/or a two-dimensional $R-$ and $T_\theta$-invariant torus foliated by closed orbits. Explain why this structurally unstable orbit configuration on the torus persists under parameter variations. Prove that when they exist simultaneously, the cycles and the torus have opposite stability.

(e) Show that any smooth system

$$\begin{cases} \dot{z}_1 = \lambda(\alpha) z_1 + f_1(z_1, \bar{z}_1, z_2, \bar{z}_2, \alpha), \\ \dot{z}_1 = \lambda(\alpha) z_2 + f_2(z_1, \bar{z}_1, z_2, \bar{z}_2, \alpha), \end{cases} \tag{7.29}$$

which is invariant with respect to the transformations (7.28) and has $\lambda(0) = i\omega(0)$, can be reduced to within cubic terms by smooth and smoothly parameter-dependent invertible transformations to the form (7.27), where $\beta = \beta(\alpha)$. Verify that the resulting transformation preserves the symmetry (i.e., is invariant under (7.28)).

(f) Prove that the limit cycles and the torus survive under adding any $O(2)$-equivariant higher-order terms to the truncated normal form (7.27).

(g) Assume that (7.27) is a truncated normal form of the equations on a center manifold of a reaction-diffusion system on a two-dimensional domain $\Omega$, having the spatial symmetry group $O(2)$, composed of rotations and a reflection. Convince yourself that the cycles in the system on the center manifold correspond to *rotating waves* in the reaction-diffusion system, while the torus describes *standing waves* in the system.

## 7.6 Appendix: Bibliographical notes

The saddle-node homoclinic bifurcation was described, among other planar codim 1 bifurcations, by Andronov in the 1940s (see Andronov et al. [1973]). Multidimensional theorems on saddle-node and saddle-saddle homoclinic bifurcations are due to Shil'nikov [1963, 1966, 1969]. Our presentation of the saddle-saddle multiple-homoclinic case follows a lecture given by Yu.S. Il'yashenko at Moscow State University in 1987 (for details, see [Ilyashenko & Li 1999]). Two-parameter unfolding of a nontransversal homoclinic orbit to a saddle-saddle equilibrium has been analyzed by Champneys, Härterich & Sandstede [1996], who also constructed a polynomial system having two orbits homoclinic to a saddle-saddle. Another codim 2 case, when the homoclinic orbit returns to a saddle-node along the noncentral direction, was analyzed by Lukyanov [1982] for planar systems and by Chow & Lin [1990] and Deng [1990] in general.

Nontransversal intersections of the invariant manifolds of a saddle cycle were first studied by Gavrilov & Shilnikov [1972, 1973] who analyzed how Smale horseshoes are created and destroyed near the critical parameter value; see also [Gruzdev & Neimark 1975]. Details of this complicated phenomenon were studied by Newhouse et al. [1983] and Palis & Takens [1993]. The latter book contains a proof that the Poincaré map defined near the nontransversal homoclinic orbit to a saddle cycle can be approximated by the Hénon map. The corresponding codim 2 bifurcation, when the saddle cycle in neutral, was studied by Gonchenko & Gonchenko [2000]. This leads to the *generalized Hénon map* investigated by Gonchenko, Kuznetsov & Meijer [2004]. The analysis of bifurcations of homoclinic orbits to a nonhyperbolic cycle was initiated by Afraimovich & Shil'nikov [1972, 1974, 1982]; see [Ilyashenko & Li 1999] for a modern treatment of these phenomena.

The "blue-sky" problem was first formulated by Palis & Pugh [1975]. Medvedev [1980] has constructed the first explicit example of this bifurcation on the Klein bottle. However, the constructed limit cycle exhibited infinitely many fold bifurcations while approaching the "blue-sky" parameter value. The "French horn" mechanism of the "blue-sky" bifurcation of a stable cycle is proposed and analyzed by Turaev & Shil'nikov [1995] and Gavrilov & Shilnikov [2000] (for details, see Shilnikov et al. [2001, Section 12.4]). Some authors naively tend to consider any homoclinic bifurcation as a "blue-sky" catastrophe.

Bifurcations of continuous-time systems on tori and the associated bifurcations of maps of a circle is a classical topic dating back to Poincaré. A good introduction to the theory of differential equations on the torus can be found in Arnol'd [1983], including the proof of Denjoy's theorem (see Denjoy [1932] and also Nitecki [1971]). The persistence of normally hyperbolic invariant manifolds (including tori) under perturbations is proved by Fenichel [1971]. When the normal hyperbolicity is lost, the torus can break-up. Possible break-up scenaria are classified by Arnol'd, Afraimovich, Il'yashenko & Shil'nikov [1994] and further analyzed by Broer, Simó & Tatjer [1998]. For an example showing the complexity in the bifurcation sequence leading to the break-up of an invariant torus, see also Aronson et al. [1982].

Bifurcation with symmetry is a huge and rapidly developing field. The standard references here are the books by Golubitsky & Schaeffer [1985] and by Golubitsky, Stewart & Schaeffer [1988]. Their reading, however, requires a rather high mathematical sophistication. The Center Manifold Theorem in the presence of a compact symmetry group was formulated by Ruelle [1973]. Its generalization to the noncompact case has been proved by Sandstede, Scheel & Wulff [1997]. The main results on limit cycles and their bifurcations in the presence of discrete symmetries were obtained by Fiedler [1988] and Nikolaev [1994]. Our presentation of cycle bifurcations in $\mathbb{Z}_2$-equivariant systems closely follows Nikolaev [1992, 1995] (see also Nikolaev & Shnol [1998*a*, 1998*b*]).

# 8

# Two-Parameter Bifurcations of Equilibria in Continuous-Time Dynamical Systems

This chapter is devoted to bifurcations of equilibria in generic two-parameter systems of differential equations. First, we make a complete list of such bifurcations. Then, we derive a *parameter-dependent normal form* for each bifurcation in the minimal possible phase dimension and specify relevant genericity conditions. Next, we truncate higher-order terms and present the bifurcation diagrams of the resulting system. The analysis is completed by a discussion of the effect of the higher-order terms. In those cases where the higher-order terms do not qualitatively alter the bifurcation diagram, the truncated systems provide topological normal forms for the relevant bifurcations. The results of this chapter can be applied to $n$-dimensional systems by means of the parameter-dependent version of the Center Manifold Theorem and Theorem 5.4 (see Chapter 5). We close this chapter with the derivation of the critical normal form coefficients for all codim 2 bifurcations using a combined reduction/normalization technique.

The reader is warned that the parameter and coordinate transformations required to put a system into the normal form can lead to lengthy intermediate calculations and expressions that can make the theory seem unnecessarily complicated. While many such expressions are included here, the reader is advised against trying to follow the calculations "by hand." Instead, we strongly urge you to use one of the symbolic manipulation packages, which are well suited for such problems (see Exercise 15 at the end of the chapter and the bibliographical notes to Chapter 10).

## 8.1 List of codim 2 bifurcations of equilibria

Consider a two-parameter system

$$\dot{x} = f(x, \alpha), \tag{8.1}$$

where $x = (x_1, x_2, \ldots, x_n)^T \in \mathbb{R}^n$, $\alpha = (\alpha_1, \alpha_2)^T \in \mathbb{R}^2$, and $f$ is a sufficiently smooth function of $(x, \alpha)$.

### 8.1.1 Bifurcation curves

Suppose that at $\alpha = \alpha^0$, system (8.1) has an equilibrium $x = x^0$ for which either the fold or Hopf bifurcation conditions are satisfied. Then, generically, there is a *bifurcation curve* $\mathcal{B}$ in the $(\alpha_1, \alpha_2)$-plane along which the system has an equilibrium exhibiting the same bifurcation. Let us consider two simple examples.

**Example 8.1 (Fold bifurcation curve in a scalar system)** Assume that at $\alpha = \alpha^0 = (\alpha_1^0, \alpha_2^0)^T$ the system

$$\dot{x} = f(x, \alpha), \quad x \in \mathbb{R}^1, \ \alpha = (\alpha_1, \alpha_2)^T \in \mathbb{R}^2, \tag{8.2}$$

has an equilibrium $x = x^0$ with eigenvalue $\lambda = f_x(x^0, \alpha^0) = 0$. Consider the system of scalar nonlinear equations

$$\begin{cases} f(x, \alpha) = 0, \\ f_x(x, \alpha) = 0. \end{cases} \tag{8.3}$$

This is a system of two equations in $\mathbb{R}^3$ with coordinates $(x, \alpha_1, \alpha_2)$. Generically, it defines a smooth one-dimensional manifold (curve) $\Gamma \subset \mathbb{R}^3$ passing through the point $(x^0, \alpha_1^0, \alpha_2^0)$ (see **Fig. 8.1**). Here "generically" means that

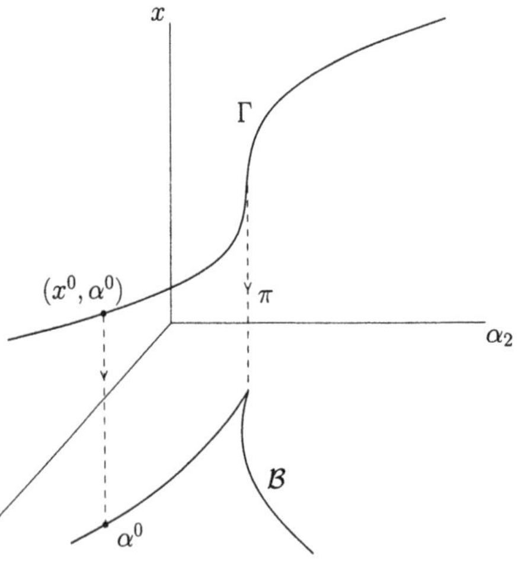


**Fig. 8.1.** A bifurcation curve $\Gamma$ and its corresponding bifurcation boundary $\mathcal{B}$.

the rank of the Jacobian matrix of (8.3),

$$J = \begin{pmatrix} f_x & f_{\alpha_1} & f_{\alpha_2} \\ f_{xx} & f_{x\alpha_1} & f_{x\alpha_2} \end{pmatrix},$$

is maximal, i.e., equal to 2. For example, if the conditions of Theorem 3.1 (see Chapter 3) for the fold bifurcation are satisfied with respect to $\alpha_1$ at $\alpha^0$:

(A.1) $f_{xx}(x^0, \alpha^0) \neq 0$;
(A.2) $f_{\alpha_1}(x^0, \alpha^0) \neq 0$;

then rank $J = 2$ at $(x^0, \alpha^0)$ since

$$\det \begin{pmatrix} f_x & f_{\alpha_1} \\ f_{xx} & f_{x\alpha_1} \end{pmatrix} = \det \begin{pmatrix} 0 & f_{\alpha_1} \\ f_{xx} & f_{x\alpha_1} \end{pmatrix} = -f_{xx} f_{\alpha_1} \neq 0.$$

In this case, the Implicit Function Theorem provides the (local) existence of two smooth functions:

$$\begin{cases} x = X(\alpha_2), \\ \alpha_1 = A(\alpha_2), \end{cases}$$

satisfying (8.3) and such that

$$X(\alpha_2^0) = x^0, \quad A(\alpha_2^0) = \alpha_1^0.$$

These functions define the curve $\Gamma$ parametrized by $\alpha_2$ near the point $(x^0, \alpha^0)$. By continuity, the genericity conditions (A.1) and (A.2) will be satisfied at nearby points on $\Gamma$. Therefore, the construction can be repeated to extend the curve farther.

If $f_{\alpha_1} = 0$ but $f_{\alpha_2} \neq 0$ at a certain point where $f_{xx} \neq 0$, similar arguments give the local existence of $\Gamma$ parametrized by $\alpha_1$. Even if $f_{xx} = 0$ at some point, which would mean that the nondegeneracy condition (A.1) is violated, system (8.3) can still define a curve, provided that

$$\det \begin{pmatrix} f_{\alpha_1} & f_{\alpha_2} \\ f_{x\alpha_1} & f_{x\alpha_2} \end{pmatrix} \neq 0.$$

At such a point, rank $J = 2$ as before, and the curve $\Gamma$ is locally parametrized by $x$.

Each point $(x, \alpha) \in \Gamma$ defines an equilibrium point $x$ of system (8.2) with zero eigenvalue at the parameter value $\alpha$ (see system (8.3)). The *standard projection*

$$\pi : (x, \alpha) \mapsto \alpha$$

maps $\Gamma$ onto a curve $\mathcal{B} = \pi\Gamma$ in the parameter plane (see **Fig. 8.1**). A fold bifurcation takes place on this curve. ◊

**Example 8.2 (Hopf bifurcation curve in a planar system)** Consider a planar system

$$\dot{x} = f(x, \alpha), \quad x = (x_1, x_2)^T \in \mathbb{R}^2, \quad \alpha = (\alpha_1, \alpha_2)^T \in \mathbb{R}^2, \tag{8.4}$$

having, at $\alpha = \alpha^0 = (\alpha_1^0, \alpha_2^0)^T$, an equilibrium $x^0 = (x_1^0, x_2^0)^T$ with a pair of eigenvalues on the imaginary axis: $\lambda_{1,2} = \pm i\omega_0$. Consider now the following system of three scalar equations in $\mathbb{R}^4$ with coordinates $(x_1, x_2, \alpha_1, \alpha_2)$:

$$\begin{cases} f(x,\alpha) = 0, \\ \operatorname{tr}(f_x(x,\alpha)) = 0, \end{cases} \tag{8.5}$$

where tr stands for the sum of the diagonal matrix elements (trace). Clearly, $(x^0, \alpha^0)$ satisfies (8.5) since the trace equals the sum of the eigenvalues of $f_x$. We leave the reader to show that the Jacobian matrix of (8.5) has maximal rank (equal to 3) at $(x^0, \alpha^0)$ if the equilibrium $x^0$ exhibits a generic Hopf bifurcation at $\alpha^0$. Actually, the rank remains equal to 3 under less restrictive assumptions. Therefore, system (8.5) defines a curve $\Gamma$ in $\mathbb{R}^4$ passing through $(x^0, \alpha^0)$. Each point on the curve specifies an equilibrium of (8.4) with $\lambda_{1,2} = \pm i\omega_0$, $\omega_0 > 0$, as long as $\det(f_x(x,\alpha)) > 0$. The standard projection of $\Gamma$ onto the $(\alpha_1, \alpha_2)$-plane yields the Hopf bifurcation boundary $\mathcal{B} = \pi\Gamma$.

Notice that the second equation in (8.5) is also satisfied by an equilibrium with *real* eigenvalues

$$\lambda_1 = \tau, \ \lambda_2 = -\tau,$$

where $\tau > 0$. In this case, $\det(f_x(x,\alpha)) < 0$ and the equilibrium is called a *neutral saddle*. For a neutral saddle, the saddle quantity $\sigma = \lambda_1 + \lambda_2 = 0$ (see Chapter 6). $\diamondsuit$

The constructions of Examples 8.1 and 8.2 can be generalized to an arbitrarily high phase-space dimension $n$. Suppose, as before, that at $\alpha = \alpha^0$, system (8.1) has an equilibrium $x = x^0$ satisfying either the fold or Hopf bifurcation conditions. In each case, a smooth scalar function $\psi = \psi(x,\alpha)$ can be constructed in terms of the elements of the Jacobian matrix $f_x$. Adding this function to the equilibrium equation yields the system

$$\begin{cases} f(x,\alpha) = 0, \\ \psi(x,\alpha) = 0, \end{cases} \tag{8.6}$$

which, generically, defines a curve $\Gamma$ passing through the point $(x^0, \alpha^0)$ in $\mathbb{R}^{n+2}$ with coordinates $(x, \alpha)$. $\Gamma$ consists of equilibria satisfying the defining bifurcation condition. The standard projection of $\Gamma$ onto the $\alpha$-plane results in the corresponding bifurcation boundary $\mathcal{B}$.

The function $\psi$ is most easily constructed in the case of the fold bifurcation. System (8.6), with

$$\psi = \psi_t(x,\alpha) = \det(f_x(x,\alpha)), \tag{8.7}$$

defines a curve of equilibria having at least one zero eigenvalue. Indeed, $\psi_t$ is the product of all the eigenvalues of $f_x$ and thus vanishes at an equilibrium with a zero eigenvalue. One can check that rank $J = n+1$ at a generic fold point $(x^0, \alpha^0)$, where $J$ is the Jacobian matrix of (8.6) with respect to $(x, \alpha)$.

A function $\psi = \psi_H(x,\alpha)$ can also be constructed for the Hopf bifurcation. Namely,

$$\psi_H(x,\alpha) = \det\left(2f_x(x,\alpha) \odot I_n\right), \tag{8.8}$$

where $\odot$ denotes the *bialternate product* of two matrices. This product is a certain square matrix of order $\frac{1}{2}n(n-1)$. The function $\psi_H$ is equal to the product of all formally distinct sums of the eigenvalues of $f_x$:

$$\psi_H = \prod_{i>j}(\lambda_i + \lambda_j);$$

and it therefore vanishes at an equilibrium having a pair of eigenvalues with zero sum. It can also be shown that rank $J = n + 1$ at a generic Hopf bifurcation. We will return to the precise definition and practical computation of the bialternate product in Chapter 10.

### 8.1.2 Codimension two bifurcation points

Let the parameters $(\alpha_1, \alpha_2)$ be varied simultaneously to track a bifurcation curve $\Gamma$ (or $\mathcal{B}$). Then, the following events might happen to the monitored nonhyperbolic equilibrium at some parameter values:

(i) extra eigenvalues can approach the imaginary axis, thus changing the dimension of the center manifold $W^c$;

(ii) some of the genericity conditions for the codim 1 bifurcation can be violated.

For nearby parameter values we can expect the appearance of new phase portraits of the system, implying that a codim 2 bifurcation has occurred. It is worthwhile to recall that the different genericity conditions for either the fold or Hopf bifurcation have differing natures. As we saw in Chapter 3, some conditions (called "nondegeneracy conditions") imply that a certain coefficient in the normal form of the equation on the center manifold is nonzero at the critical point. These coefficients can be computed in terms of the Taylor coefficients of $f(x, 0)$ at the equilibrium. In contrast, there are conditions (called "transversality conditions") in which certain derivatives of $f(x, \alpha)$ with respect to some parameter $\alpha_i$ are involved. These two types of conditions play differing roles in the bifurcation analysis. The nondegeneracy conditions essentially determine the number and stability of the equilibria and cycles appearing under parameter perturbations, while the transversality conditions merely suggest the introduction of a new parameter to "unfold" the bifurcation (see Chapter 3). Thus, only violating a nondegeneracy condition can produce new phase portraits. For example, if

$$\frac{\partial}{\partial \alpha_i} \text{Re } \lambda_{1,2}(\alpha) = 0$$

at the Hopf bifurcation point, then, generically, the eigenvalues do not cross the imaginary axis as $\alpha_i$ passes the critical value. This results in the same local phase portrait for both sub- and supercritical parameter values.

Let us first follow the fold bifurcation curve $\mathcal{B}_t$. A typical point in this curve defines an equilibrium with a simple zero eigenvalue $\lambda_1 = 0$ and no other eigenvalues on the imaginary axis. The restriction of (8.1) to a center manifold $W^c$ has the form

$$\dot{\xi} = b\xi^2 + O(\xi^3). \tag{8.9}$$

The formula for the coefficient $b$ was derived in Chapter 5. By definition, the coefficient $a$ is nonzero at a nondegenerate fold bifurcation point. While the curve is being tracked, the following singularities can be met:

(1) An additional real eigenvalue $\lambda_2$ approaches the imaginary axis, and $W^c$ becomes two-dimensional:

$$\lambda_{1,2} = 0$$

(see **Fig. 8.2**(a)). These are the conditions for the *Bogdanov-Takens* (or *double-zero*) bifurcation. To have this bifurcation, we need $n \geq 2$.

(2) Two extra complex eigenvalues $\lambda_{2,3}$ arrive at the imaginary axis, and $W^c$ becomes three-dimensional:

$$\lambda_1 = 0, \ \lambda_{2,3} = \pm i\omega_0,$$

for $\omega_0 > 0$ (see **Fig. 8.2**(b)). These conditions correspond to the *fold-Hopf* bifurcation, sometimes called a *Gavrilov-Guckenheimer* or a *zero-pair* bifurcation. We obviously need $n \geq 3$ for this bifurcation to occur.

(3) The eigenvalue $\lambda_1 = 0$ remains simple and the only one on the imaginary axis ($\dim W^c = 1$), but the normal form coefficient $a$ in (8.9) vanishes:

$$\lambda_1 = 0, \ b = 0.$$

These are the conditions for a *cusp* bifurcation, which is possible in systems with $n \geq 1$. Notice that this bifurcation is undetectable by looking at only the eigenvalues of the equilibrium since quadratic terms of $f(x, 0)$ are involved in the computation of $b$. Bifurcations of this type are sometimes referred to as "degeneracy of nonlinear terms."

Let us now follow a Hopf bifurcation curve $\mathcal{B}_H$ in system (8.1). At a typical point on this curve, the system has an equilibrium with a simple pair of purely imaginary eigenvalues $\lambda_{1,2} = \pm i\omega_0, \ \omega_0 > 0$, and no other eigenvalues with Re $\lambda = 0$. The center manifold $W^c$ is two-dimensional in this case, and there

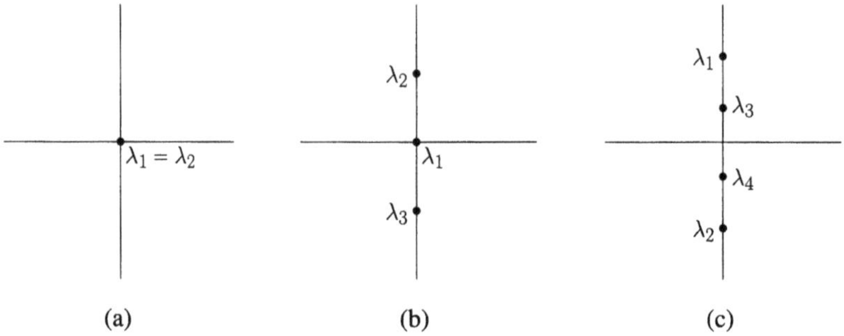


**Fig. 8.2.** Linear singularities of codim 2.

are (polar) coordinates $(\rho, \varphi)$ for which the restriction of (8.1) to this manifold is orbitally equivalent to

$$\begin{cases} \dot{\rho} = l_1\rho^3 + O(\rho^4), \\ \dot{\varphi} = 1 + O(\rho^3). \end{cases} \tag{8.10}$$

The formula for the coefficient $l_1$ was derived in Chapters 3 and 5. By definition, $l_1 \neq 0$ at a nondegenerate Hopf point.

While moving along the curve, we can encounter the following new possibilities:

(4) Two extra complex-conjugate eigenvalues $\lambda_{3,4}$ approach the imaginary axis, and $W^c$ becomes four-dimensional:

$$\lambda_{1,2} = \pm i\omega_0, \ \lambda_{3,4} = \pm i\omega_1,$$

with $\omega_{0,1} > 0$ (**Fig. 8.2**(c)). These conditions define the *Hopf-Hopf* or *two-pair* bifurcation. It is possible only if $n \geq 4$.

(5) Finally, the first Lyapunov coefficient $l_1$ might vanish while $\lambda_{1,2} = \pm i\omega_0$ remain simple and, therefore, $\dim W^c = 2$:

$$\lambda_{1,2} = \pm i\omega_0, \ l_1 = 0.$$

At this point, a "soft" Andronov-Hopf bifurcation turns into a "sharp" one (or vice versa). We call this event a *Bautin* bifurcation (see the bibliographical notes); it is often called a *generalized* (or *degenerate*) *Hopf* bifurcation. It is possible if $n \geq 2$. As with the cusp bifurcation, the Bautin bifurcation cannot be detected by merely monitoring the eigenvalues. We have to take into account the quadratic and cubic Taylor series coefficients of the right-hand side of (8.1) at the equilibrium.

Clearly, the Bogdanov-Takens bifurcation can also be located along a Hopf bifurcation curve, as $\omega_0$ approaches zero. At this point, two purely imaginary eigenvalues collide and we have a double zero eigenvalue. If we continue to trace the curve defined by (8.6) with $\psi = \psi_H$ given by (8.8), we will follow a *neutral saddle* equilibrium with real eigenvalues $\lambda_1 = -\lambda_2$. Obviously, a fold-Hopf bifurcation can also be found while tracing a Hopf bifurcation curve.

Thus, we have identified five bifurcation points that one can meet in generic two-parameter systems while moving along codim 1 curves. Each of these bifurcations is characterized by two independent conditions (and is therefore of codim 2). There are no other codim 2 bifurcations in generic continuous-time systems.

Sections 8.2–8.6 of this chapter are devoted to a systematic study of these bifurcations in the least possible phase-space dimensions. The analysis of each codim 2 bifurcation will be organized in a similar manner to the study of codim 1 bifurcations:

(i) First, we derive the simplest parameter-dependent form to which any generic two-parameter system exhibiting the bifurcation can be transformed

by smooth invertible changes of coordinates and parameters and (if necessary) time reparametrizations. In the course of this derivation, certain *nondegeneracy and transversality conditions* will be imposed on the system to make the transformation possible. These conditions explicitly specify which systems are "generic."
(ii) Then we truncate higher-order terms and present bifurcation diagrams of the resulting system, sometimes called the *"approximate normal form"* or *"model system."* For this system to have a nondegenerate bifurcation diagram, some extra genericity conditions might have to be imposed at this stage.
(iii) Finally, we discuss the influence of the higher-order terms.

It turns out that for the cusp, Bautin, and Bogdanov-Takens bifurcations the higher-order terms do not qualitatively affect the bifurcation diagrams, and the model systems provide *topological normal forms* for the corresponding bifurcations (see Chapter 2 for a definition). Notice that these are exactly those codim 2 bifurcations possible in generic *scalar* or *planar* systems. For the remaining codim 2 bifurcations (fold-Hopf and Hopf-Hopf cases, with minimal phase dimensions $n = 3$ and 4, respectively), the situation is more involved since higher-order terms do change bifurcation diagrams. We discuss which features of the behavior of the system will persist, if one takes these terms into account, and which will not. In any case, the study of the approximate normal form provides important information on the behavior of the system near the bifurcation point.

The obtained results can be applied to $n$-dimensional systems using Theorem 5.4 due to Shoshitaishvili. The theorem implies that all "essential" events near the critical parameter values occur on an invariant center manifold $W^c_\alpha$ that is exponentially attracting or repelling in the transverse directions (*normally hyperbolic*). The obtained diagrams describe bifurcations of the systems restricted to $W^c_\alpha$. To determine the bifurcation scenario for a given system, the corresponding critical normal form coefficients have to be computed.[1] In Section 8.7 we derive the critical normal form coefficients for all codim 2 bifurcations using a combined reduction/normalization technique.

## 8.2 Cusp bifurcation

### 8.2.1 Normal form derivation

Suppose the system

$$\dot{x} = f(x, \alpha), \quad x \in \mathbb{R}^1, \ \alpha \in \mathbb{R}^2, \tag{8.11}$$

[1] We assume that the system depends generically on the parameters, so that the transversality conditions are satisfied. In principle, this has to be verified separately.

with a smooth function $f$, has at $\alpha = 0$ the equilibrium $x = 0$ for which the cusp bifurcation conditions are satisfied, namely $\lambda = f_x(0,0) = 0$ and $b = \frac{1}{2} f_{xx}(0,0) = 0$. As in the analysis of the fold bifurcation in Chapter 3, expansion of $f(x, \alpha)$ as a Taylor series with respect to $x$ at $x = 0$ yields

$$f(x, \alpha) = f_0(\alpha) + f_1(\alpha)x + f_2(\alpha)x^2 + f_3(\alpha)x^3 + O(x^4).$$

Since $x = 0$ is an equilibrium, we have $f_0(0) = f(0,0) = 0$. The cusp bifurcation conditions yield $f_1(0) = f_x(0,0) = 0$ and $f_2(0) = \frac{1}{2} f_{xx}(0,0) = 0$.

As in Chapter 3, let us analyze the simplification of the right-hand side of (8.11) that can be achieved by a parameter-dependent *shift* of the coordinate

$$\xi = x + \delta(\alpha). \tag{8.12}$$

Substituting (8.12) into (8.11), taking into account the expansion of $f(x, \alpha)$, yields

$$\begin{aligned} \dot{\xi} = & \left[f_0(\alpha) - f_1(\alpha)\delta + \delta^2 \varphi(\alpha, \delta)\right] + \left[f_1(\alpha) - 2f_2(\alpha)\delta + \delta^2 \phi(\alpha, \delta)\right] \xi \\ & + \left[f_2(\alpha) - 3f_3(\alpha)\delta + \delta^2 \psi(\alpha, \delta)\right] \xi^2 + \left[f_3(\alpha) + \delta\theta(\alpha, \delta)\right] \xi^3 + O(\xi^4) \end{aligned}$$

for some smooth functions $\varphi, \phi, \psi$, and $\theta$. Since $f_2(0) = 0$, we cannot use the Implicit Function Theorem to select a function $\delta(\alpha)$ to eliminate the linear term in $\xi$ in the above equation (as we did in Chapter 3). However, there is a smooth shift function $\delta(\alpha), \delta(0) = 0$, which annihilates the *quadratic* term in the equation for all sufficiently small $\|\alpha\|$, provided that

(C.1) $$f_3(0) = \frac{1}{6} f_{xxx}(0,0) \neq 0.$$

To see this, denote the coefficient in front of $\xi^2$ by $F(\alpha, \delta)$:

$$F(\alpha, \delta) = f_2(\alpha) - 3f_3(\alpha)\delta + \delta^2 \psi(\alpha, \delta).$$

We have

$$F(0,0) = 0, \quad \left.\frac{\partial F}{\partial \delta}\right|_{(0,0)} = -3f_3(0) \neq 0.$$

Therefore, the Implicit Function Theorem gives the (local) existence and uniqueness of a smooth scalar function $\delta = \delta(\alpha)$, such that $\delta(0) = 0$ and

$$F(\alpha, \delta(\alpha)) \equiv 0,$$

for $\|\alpha\|$ small enough. The equation for $\xi$, with $\delta(\alpha)$ constructed as above, contains no quadratic terms. Now we can introduce *new parameters* $\mu = (\mu_1, \mu_2)$ by setting

$$\begin{cases} \mu_1(\alpha) = f_0(\alpha) - f_1(\alpha)\delta(\alpha) + \delta^2(\alpha)\varphi(\alpha, \delta(\alpha)), \\ \mu_2(\alpha) = f_1(\alpha) - 2f_2(\alpha)\delta(\alpha) + \delta^2(\alpha)\phi(\alpha, \delta(\alpha)). \end{cases} \tag{8.13}$$

Here $\mu_1$ is the $\xi$-independent term in the equation, while $\mu_2$ is the coefficient in front of $\xi$. Clearly, $\mu(0) = 0$. The parameters (8.13) are well defined if the Jacobian matrix of the map $\mu = \mu(\alpha)$ is nonsingular at $\alpha_1 = \alpha_2 = 0$:

$$\text{(C.2)} \qquad \det\left(\frac{\partial \mu}{\partial \alpha}\right)\Bigg|_{\alpha=0} = \det\begin{pmatrix} f_{\alpha_1} & f_{\alpha_2} \\ f_{x\alpha_1} & f_{x\alpha_2} \end{pmatrix}\Bigg|_{\alpha=0} \neq 0.$$

Then the Inverse Function Theorem implies the local existence and uniqueness of a smooth inverse function $\alpha = \alpha(\mu)$ with $\alpha(0) = 0$. Therefore, the equation for $\xi$ now reads

$$\dot{\xi} = \mu_1 + \mu_2\xi + c(\mu)\xi^3 + O(\xi^4),$$

where $c(\mu) = f_3(\alpha(\mu)) + \delta(\alpha(\mu))\theta(\alpha(\mu), \delta(\alpha(\mu)))$ is a smooth function of $\mu$ and

$$c(0) = f_3(0) = \frac{1}{6} f_{xxx}(0,0) \neq 0$$

due to (C.1).

Finally, perform a *linear scaling*

$$\eta = \frac{\xi}{|c(\mu)|},$$

and introduce new parameters:

$$\beta_1 = \frac{\mu_1}{|c(\mu)|},$$
$$\beta_2 = \mu_2.$$

This gives

$$\dot{\eta} = \beta_1 + \beta_2\eta + s\eta^3 + O(\eta^4), \tag{8.14}$$

where $s = \text{sign } c(0) = \pm 1$, and the $O(\eta^4)$ terms can depend smoothly on $\beta$.

Thus, the following lemma is proved.

**Lemma 8.1** *Suppose that a one-dimensional system*

$$\dot{x} = f(x, \alpha), \quad x \in \mathbb{R}^1, \ \alpha \in \mathbb{R}^2,$$

*with smooth $f$, has at $\alpha = 0$ the equilibrium $x = 0$, and let the cusp bifurcation conditions hold:*

$$\lambda = f_x(0,0) = 0, \ b = \frac{1}{2} f_{xx}(0,0) = 0.$$

*Assume that the following genericity conditions are satisfied:*

(C.1) $f_{xxx}(0,0) \neq 0$;
(C.2) $(f_{\alpha_1} f_{x\alpha_2} - f_{\alpha_2} f_{x\alpha_1})(0,0) \neq 0$.

*Then there are smooth invertible coordinate and parameter changes transforming the system into*

$$\dot{\eta} = \beta_1 + \beta_2\eta \pm \eta^3 + O(\eta^4). \ \square$$

**Remarks:**

(1) Notice that (C.2) implies that a unique and smooth fold bifurcation curve $\Gamma$, defined by

$$\begin{cases} f(x,\alpha) = 0, \\ f_x(x,\alpha) = 0, \end{cases}$$

passes through $(x, \alpha) = (0, 0)$ in $\mathbb{R}^3$-space with coordinates $(x, \alpha)$ and can be locally parametrized by $x$ (see Section 8.1.1).

(2) Given $f_x(0,0) = f_{xx}(0,0) = 0$, the nondegeneracy condition (C.1) and the transversality condition (C.2) together are equivalent to the regularity (nonsingularity of the Jacobian matrix) of a map $T : \mathbb{R}^3 \to \mathbb{R}^3$ defined by

$$(x, \alpha) \mapsto (f(x,\alpha), f_x(x,\alpha), f_{xx}(x,\alpha))$$

at the point $(x, \alpha) = (0, 0)$. ♢

System (8.14) with the $O(\eta^4)$ terms truncated is called the *approximate normal form* for the cusp bifurcation. In the following subsections we study its bifurcation diagrams and see that higher-order terms do not actually change them. This justifies calling

$$\dot{\eta} = \beta_1 + \beta_2\eta \pm \eta^3$$

the *topological normal form* for the cusp bifurcation.

### 8.2.2 Bifurcation diagram of the normal form

Consider the normal form corresponding to $s = -1$:

$$\dot{\eta} = \beta_1 + \beta_2\eta - \eta^3. \qquad (8.15)$$

Its bifurcation diagram is easy to analyze. System (8.15) can have from one to three equilibria. A fold bifurcation occurs at a bifurcation curve $T$ on the $(\beta_1, \beta_2)$-plane that is given by the projection of the curve

$$\Gamma : \begin{cases} \beta_1 + \beta_2\eta - \eta^3 = 0, \\ \beta_2 - 3\eta^2 = 0, \end{cases}$$

onto the parameter plane. Elimination of $\eta$ from these equations gives the projection

$$T = \{(\beta_1, \beta_2) : \ 4\beta_2^3 - 27\beta_1^2 = 0\}.$$

It is called a *semicubic parabola* (see **Fig. 8.3**). The curve $T$ has two branches, $T_1$ and $T_2$, which meet tangentially at the cusp point $(0, 0)$. The resulting wedge divides the parameter plane into two regions. In region **1**, inside the wedge, there are three equilibria of (8.15), two stable and one unstable; in region **2**, outside the wedge, there is a single equilibrium, which is stable

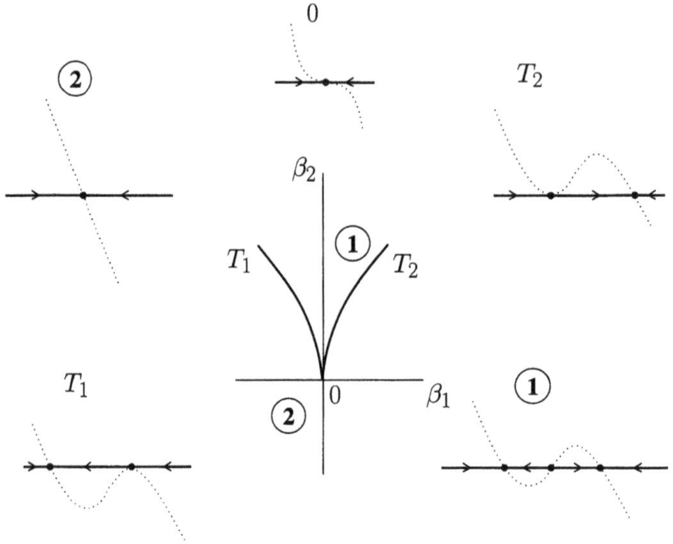


**Fig. 8.3.** One-dimensional cusp bifurcation.

(**Fig. 8.3**). As we can easily check, a nondegenerate fold bifurcation (with respect to the parameter $\beta_1$) takes place if we cross either $T_1$ or $T_2$ at any point other than the origin. If the curve $T_1$ is crossed from region **1** to region **2**, the right stable equilibrium collides with the unstable one and both disappear. The same happens to the left stable equilibrium and the unstable equilibrium at $T_2$. If we approach the cusp point from inside region **1**, all three equilibria merge together into a *triple* root of the right-hand side of (8.15).

A useful way to present this bifurcation is to plot the *equilibrium manifold* of (8.15),

$$\mathcal{M} = \{(\eta, \beta_1, \beta_2) : \ \beta_1 + \beta_2\eta - \eta^3 = 0\},$$

in $\mathbb{R}^3$ (see **Fig. 8.4**). The standard projection of $\mathcal{M}$ onto the $(\beta_1, \beta_2)$-plane has singularities of the fold type along $\Gamma$ except the origin, where a *cusp singularity* shows up. Notice that the curve $\Gamma$ is smooth everywhere and has no geometrical singularity at the cusp point. It is the projection that makes the fold parametric boundary nonsmooth.

The cusp bifurcation implies the presence of the phenomenon known as *hysteresis*. More precisely, a catastrophic "jump" to a different stable equilibrium (caused by the disappearance of a traced stable equilibrium via a fold bifurcation as the parameters vary) happens at branch $T_1$ or $T_2$ depending on whether the equilibrium being traced belongs initially to the upper or lower sheet of $\mathcal{M}$ (see **Fig. 8.5**). If we make a roundtrip in the parameter plane, crossing the wedge twice, a jump occurs on *each* branch of $T$.

The case $s = 1$ can be treated similarly or reduced to the considered case using the substitutions $t \to -t, \beta_1 \to -\beta_1, \beta_3 \to -\beta_2$. In this case, the

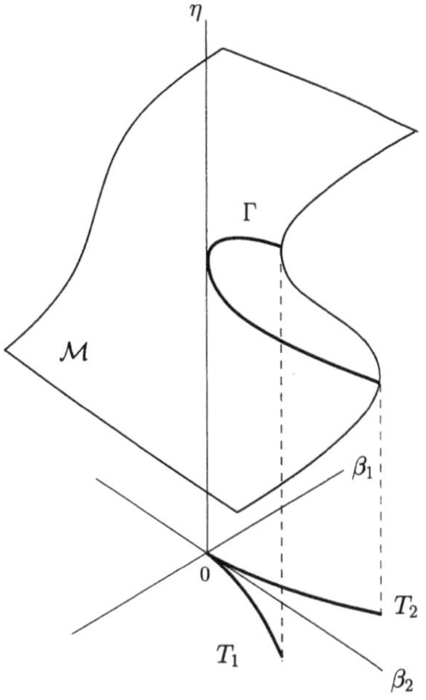


**Fig. 8.4.** Equilibrium manifold near a cusp bifurcation.

truncated system typically has either one unstable equilibrium or one stable and two unstable equilibria that can pairwise collide and disappear through fold bifurcations.

### 8.2.3 Effect of higher-order terms

The following lemma actually indicates that the higher-order terms in (8.14) are irrelevant.

**Lemma 8.2** *The system*

$$\dot{\eta} = \beta_1 + \beta_2\eta \pm \eta^3 + O(\eta^4)$$

*is locally topologically equivalent near the origin to the system*

$$\dot{\eta} = \beta_1 + \beta_2\eta \pm \eta^3. \ \Box$$

An elementary proof of the lemma is sketched in the hints to Exercise 2 for this chapter. We can now complete the analysis of the cusp bifurcation by formulating a general theorem.

**Theorem 8.1 (Topological normal form for the cusp bifurcation)** *Any generic scalar two-parameter system*

$$\dot{x} = f(x, \alpha)$$

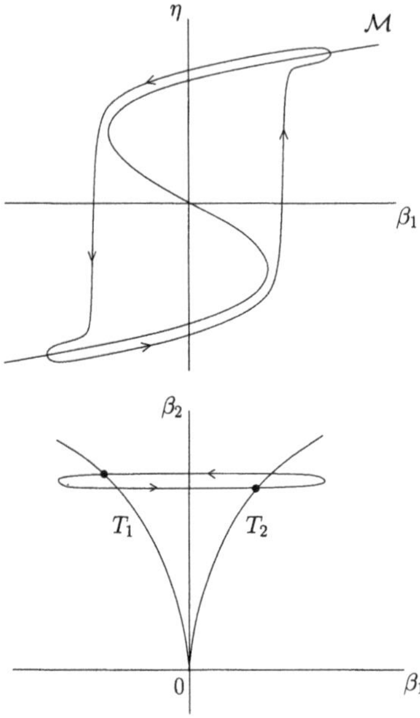


**Fig. 8.5.** Hysteresis near a cusp bifurcation.

*having at* $\alpha = 0$ *an equilibrium* $x = 0$ *exhibiting the cusp bifurcation is locally topologically equivalent near the origin to one of the normal forms*

$$\dot{\eta} = \beta_1 + \beta_2\eta \pm \eta^3. \ \square$$

If an $n$-dimensional system has a cusp bifurcation, the above theorem should be applied to the equation on the center manifold (see Chapter 5). Recall that $c(0)$ (and thus the sign in the normal form) can be computed by formulas (5.28) and (5.23) from Chapter 5. We give an alternative derivation of $c(0)$ in Section 8.7. Shoshitaishvili's Theorem gives the following topological normal forms for this case:

$$\begin{cases} \dot{\eta} &= \beta_1 + \beta_2\eta \pm \eta^3, \\ \dot{\zeta}_- &= -\zeta_-, \\ \dot{\zeta}_+ &= \zeta_+, \end{cases}$$

where $\eta \in \mathbb{R}^1$, $\zeta_\pm \in \mathbb{R}^{n_\pm}$, and $n_-$ and $n_+$ are the numbers of eigenvalues of the critical equilibrium with Re $\lambda > 0$ and Re $\lambda < 0$. **Fig. 8.6** presents the bifurcation diagram for $n = 2$ in the case where $c(0) < 0$ and the second eigenvalue at the cusp point is negative ($n_- = 1, n_+ = 0$).

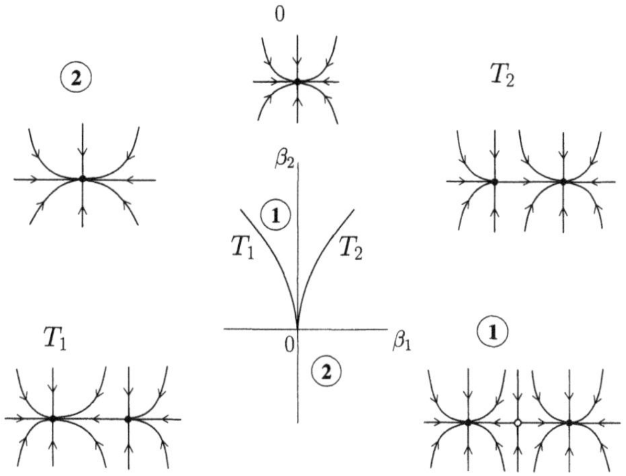


**Fig. 8.6.** Cusp bifurcation on the plane.

## 8.3 Bautin (generalized Hopf) bifurcation

### 8.3.1 Normal form derivation

Assume that the system

$$\dot{x} = f(x, \alpha), \quad x \in \mathbb{R}^2, \ \alpha \in \mathbb{R}^2, \tag{8.16}$$

with $f$ smooth has at $\alpha = 0$ the equilibrium $x = 0$, which satisfies the Bautin bifurcation conditions. More precisely, the equilibrium has purely imaginary eigenvalues $\lambda_{1,2} = \pm i\omega_0$, $\omega_0 > 0$, and the first Lyapunov coefficient vanishes: $l_1 = 0$. Since $\lambda = 0$ is not an eigenvalue, the equilibrium in general moves as $\alpha$ varies but remains isolated and close to the origin for all sufficiently small $\|\alpha\|$. As in the analysis of the Andronov-Hopf bifurcation, we can always perform a (parameter-dependent) shift of coordinates that puts this equilibrium at $x = 0$ for all $\alpha$ with $\|\alpha\|$ small enough, and assume from now on that $f(0, \alpha) \equiv 0$.

Lemma 3.3 from Chapter 3 allows us to write (8.16) in the complex form

$$\dot{z} = (\mu(\alpha) + i\omega(\alpha))z + g(z, \bar{z}, \alpha), \quad z \in \mathbb{C}^1, \tag{8.17}$$

where $\mu, \omega$, and $g$ are smooth functions of their arguments, $\mu(0) = 0, \omega(0) = \omega_0$, and formally

$$g(z, \bar{z}, \alpha) = \sum_{k+l \geq 2} \frac{1}{k!l!} g_{kl}(\alpha) z^k \bar{z}^l$$

for smooth functions $g_{kl}(\alpha)$.

**Lemma 8.3 (Poincaré normal form for the Bautin bifurcation)**
*The equation*

$$\dot{z} = \lambda(\alpha)z + \sum_{2\le k+l\le 5} \frac{1}{k!l!} g_{kl}(\alpha)z^k\bar{z}^l + O(|z|^6), \tag{8.18}$$

*where $\lambda(\alpha) = \mu(\alpha) + i\omega(\alpha)$, $\mu(0) = 0$, $\omega(0) = \omega_0 > 0$, can be transformed by an invertible parameter-dependent change of the complex coordinate, smoothly depending on the parameters:*

$$z = w + \sum_{2\le k+l\le 5} \frac{1}{k!l!} h_{kl}(\alpha)w^k\bar{w}^l, \quad h_{21}(\alpha) = h_{32}(\alpha) = 0,$$

*for all sufficiently small $\|\alpha\|$, into the equation*

$$\dot{w} = \lambda(\alpha)w + c_1(\alpha)w|w|^2 + c_2(\alpha)w|w|^4 + O(|w|^6). \ \square \tag{8.19}$$

The lemma can be proved using the same method as for Lemma 3.6 in Chapter 3. By Lemma 3.6, we can assume that all the quadratic and nonresonant cubic terms in (8.18) are already eliminated: $g_{20} = g_{11} = g_{02} = g_{30} = g_{12} = g_{21} = 0$, and $\frac{1}{2}g_{21} = c_1$. Then, by a proper selection of $h_{ij}$ with $i+j = 4$, we can annihilate all the order-four terms in (8.18), having the coefficient of the resonant cubic term $c_1(\alpha)$ untouched while changing the coefficients of the fifth- and higher-order terms. Finally, we can "remove" all the fifth-order terms except the *resonant* one shown in (8.19). These calculations make a good exercise in symbolic manipulations.

The coefficients $c_1(\alpha)$ and $c_2(\alpha)$ are complex. They can be made simultaneously real by a time reparametrization.

**Lemma 8.4** *System* (8.19) *is locally orbitally equivalent to the system*

$$\dot{w} = (\nu(\alpha) + i)w + l_1(\alpha)w|w|^2 + l_2(\alpha)w|w|^4 + O(|w|^6), \tag{8.20}$$

*where $\nu(\alpha)$, $l_1(\alpha)$, and $l_2(\alpha)$ are real functions, $\nu(0) = 0$.*

**Proof:**

First, introduce the new time $\tau = \omega(\alpha)t$. The direction of time is preserved for all sufficiently small $\|\alpha\|$ since $\omega(0) = \omega_0 > 0$. This gives

$$\frac{dw}{d\tau} = (\nu(\alpha) + i)w + d_1(\alpha)w|w|^2 + d_2(\alpha)w|w|^4 + O(|w|^5), \tag{8.21}$$

with

$$\nu(\alpha) = \frac{\mu(\alpha)}{\omega(\alpha)}, \ d_1(\alpha) = \frac{c_1(\alpha)}{\omega(\alpha)}, \ d_2(\alpha) = \frac{c_2(\alpha)}{\omega(\alpha)}.$$

Notice that $\nu$, $d_1$, and $d_2$ are smooth and that $d_{1,2}$ are still complex-valued.

Next, change the time parametrization along the orbits of (8.21) by introducing a new time $\theta$ such that

$$d\tau = (1 + e_1(\alpha)|w|^2 + e_2(\alpha)|w|^4)\ d\theta,$$

where the real functions $e_{1,2}$ have yet to be defined. In terms of $\theta$, (8.21) can be written as

$$\frac{dw}{d\theta} = (\nu+i)w+((\nu+i)e_1+d_1)w|w|^2+((\nu+i)e_2+e_1d_1+d_2)w|w|^4+O(|w|^5).$$

Therefore, setting

$$e_1(\alpha) = -\text{Im } d_1(\alpha), \; e_2(\alpha) = -\text{Im } d_2(\alpha) + [\text{Im } d_1(\alpha)]^2$$

yields

$$\frac{dw}{d\theta} = (\nu(\alpha) + i)w + l_1(\alpha)w|w|^2 + l_2(\alpha)w|w|^4 + O(|w|^5),$$

where

$$l_1(\alpha) = \text{Re } d_1(\alpha) - \nu(\alpha) \text{ Im } d_1(\alpha) = \frac{\text{Re } c_1(\alpha)}{\omega(\alpha)} - \mu(\alpha)\frac{\text{Im } c_1(\alpha)}{\omega^2(\alpha)} \tag{8.22}$$

is the first Lyapunov coefficient introduced in Chapter 3, and

$$l_2(\alpha) = \text{Re } d_2(\alpha) - \text{Re } d_1(\alpha) \text{ Im } d_1(\alpha) + \nu(\alpha)\left([\text{Im } d_1(\alpha)]^2 - \text{Im } d_2(\alpha)\right).$$

The functions $\nu(\alpha)$, $l_1(\alpha)$, and $l_2(\alpha)$ are smooth and real-valued. □

**Definition 8.1** *The real function $l_2(\alpha)$ is called the* second Lyapunov coefficient.

Recall that $c_1 = c_1(\alpha)$ used in (8.22) can be computed by the formula (see Chapter 3)

$$c_1 = \frac{g_{21}}{2} + \frac{g_{20}g_{11}(2\lambda + \bar{\lambda})}{2|\lambda|^2} + \frac{|g_{11}|^2}{\lambda} + \frac{|g_{02}|^2}{2(2\lambda - \bar{\lambda})},$$

where $\lambda = \lambda(\alpha)$ and $g_{kl} = g_{kl}(\alpha)$.

At the Bautin bifurcation point, where

$$\mu(0) = 0, \; l_1(0) = \frac{\text{Re } c_1(0)}{\omega_0} = \frac{1}{2\omega_0}\left(\text{Re } g_{21}(0) - \frac{1}{\omega_0}\text{ Im}(g_{20}(0)g_{11}(0))\right) = 0,$$

we obtain

$$l_2(0) = \frac{\text{Re } c_2(0)}{\omega_0}.$$

The following formula gives a rather compact expression for $l_2(0)$ *at the Bautin point*

$$\begin{aligned} 12l_2(0) &= \frac{1}{\omega_0}\text{Re } g_{32} \\ &+ \frac{1}{\omega_0^2}\text{Im } \left[g_{20}\bar{g}_{31} - g_{11}(4g_{31} + 3\bar{g}_{22}) - \frac{1}{3}g_{02}(g_{40} + \bar{g}_{13}) - g_{30}g_{12}\right] \end{aligned}$$

$$+\frac{1}{\omega_0^3}\{\text{Re}\left[g_{20}(\bar{g}_{11}(3g_{12}-\bar{g}_{30})+g_{02}\left(\bar{g}_{12}-\frac{1}{3}g_{30}\right)+\frac{1}{3}\bar{g}_{02}g_{03})\right.$$
$$\left.+\,g_{11}(\bar{g}_{02}\left(\frac{5}{3}\bar{g}_{30}+3g_{12}\right)+\frac{1}{3}g_{02}\bar{g}_{03}-4g_{11}g_{30})\right]$$
$$+\,3\,\text{Im}(g_{20}g_{11})\,\text{Im}\,g_{21}\}$$
$$+\frac{1}{\omega_0^4}\left\{\text{Im}\left[g_{11}\bar{g}_{02}\left(\bar{g}_{20}^2-3\bar{g}_{20}g_{11}-4g_{11}^2\right)\right]\right.$$
$$\left.+\;\text{Im}(g_{20}g_{11})\left[3\,\text{Re}(g_{20}g_{11})-2|g_{02}|^2\right]\right\}, \qquad (8.23)$$

where all the $g_{kl}$ are evaluated at $\alpha = 0$. In deriving this formula, we have taken the equation $l_1(0) = 0$ (or, equivalently, $\text{Re}\ g_{21} = \frac{1}{\omega_0}\text{Im}(g_{20}g_{11})$) into account.

Suppose that at a Bautin point

(B.1) $$l_2(0) \neq 0.$$

A neighborhood of the point $\alpha = 0$ can be parametrized by two new parameters, the zero locus of the first one corresponding to the Hopf bifurcation condition, while the simultaneous vanishing of both specifies the Bautin point. Clearly, we might consider $\nu(\alpha)$ as the first parameter and $l_1(\alpha)$ as the second one. Notice that both are defined for all sufficiently small $\|\alpha\|$ and vanish at $\alpha = 0$. Thus, let us introduce new parameters $(\mu_1, \mu_2)$ by the map

$$\begin{cases} \mu_1 = \nu(\alpha), \\ \mu_2 = l_1(\alpha), \end{cases} \qquad (8.24)$$

assuming its regularity at $\alpha = 0$:

(B.2) $$\det\left.\begin{pmatrix} \frac{\partial \nu}{\partial \alpha_1} & \frac{\partial \nu}{\partial \alpha_2} \\ \frac{\partial l_1}{\partial \alpha_1} & \frac{\partial l_1}{\partial \alpha_2} \end{pmatrix}\right|_{\alpha=0} = \frac{1}{\omega_0}\det\left.\begin{pmatrix} \frac{\partial \mu}{\partial \alpha_1} & \frac{\partial \mu}{\partial \alpha_2} \\ \frac{\partial l_1}{\partial \alpha_1} & \frac{\partial l_1}{\partial \alpha_2} \end{pmatrix}\right|_{\alpha=0} \neq 0.$$

This condition can easily be expressed in terms of $\mu(\alpha)$, $\text{Re}\ c_1(\alpha)$, and $\text{Im}\ c_1(\alpha)$ since $\omega_0 \neq 0$. It is equivalent to local smooth invertibility of the map (8.24), so we can write $\alpha$ in terms of $\mu$, thus obtaining the equation

$$\dot{w} = (\mu_1 + i)w + \mu_2 w|w|^2 + L_2(\mu)w|w|^4 + O(|w|^6),$$

where $L_2(\mu) = l_2(\alpha(\mu))$ is a smooth function of $\mu$, such that $L_2(0) = l_2(0) \neq 0$ due to (B.1). Then, rescaling

$$w = \sqrt[4]{|L_2(\mu)|}\ u, \quad u \in \mathbb{C}^1,$$

and defining the parameters

$$\begin{cases} \beta_1 = \mu_1, \\ \beta_2 = \sqrt{|L_2(\mu)|}\ \mu_2, \end{cases}$$

yield the normal form

$$\dot{u} = (\beta_1 + i)u + \beta_2 u|u|^2 + su|u|^4 + O(|u|^6).$$

Here $s = \text{sign}\ l_2(0) = \pm 1$, where $l_2(0)$ is given by (8.23).

Summarizing the results obtained, we can formulate the following theorem.

**Theorem 8.2** *Suppose that a planar system*

$$\dot{x} = f(x, \alpha), \quad x \in \mathbb{R}^2, \ \alpha \in \mathbb{R}^2,$$

*with smooth $f$, has the equilibrium $x = 0$ with eigenvalues*

$$\lambda_{1,2}(\alpha) = \mu(\alpha) \pm i\omega(\alpha),$$

*for all $\|\alpha\|$ sufficiently small, where $\omega(0) = \omega_0 > 0$. For $\alpha = 0$, let the Bautin bifurcation conditions hold:*

$$\mu(0) = 0, \ l_1(0) = 0,$$

*where $l_1(\alpha)$ is the first Lyapunov coefficient. Assume that the following genericity conditions are satisfied:*

(B.1) *$l_2(0) \neq 0$, where $l_2(0)$ is the second Lyapunov coefficient given by* (8.23)*;*

(B.2) *the map $\alpha \mapsto (\mu(\alpha), l_1(\alpha))^T$ is regular at $\alpha = 0$.*

*Then, by the introduction of a complex variable, applying smooth invertible coordinate transformations that depend smoothly on the parameters, and performing smooth parameter and time changes, the system can be reduced to the complex form*

$$\dot{z} = (\beta_1 + i)z + \beta_2 z|z|^2 + sz|z|^4 + O(|z|^6), \tag{8.25}$$

*where $s = \text{sign}\ l_2(0) = \pm 1$.* □

We will proceed in the same way as in analyzing the cusp bifurcation. First, we will study the approximate normal form resulting from (8.25) by dropping the $O(|z|^6)$ terms. As we shall then see, this approximate normal form is also the topological normal form for the Bautin bifurcation.

### 8.3.2 Bifurcation diagram of the normal form

Set $s = -1$ and write system (8.25) without the $O(|z|^6)$ terms in polar coordinates $(\rho, \varphi)$, where $z = \rho e^{i\varphi}$:

$$\begin{cases} \dot{\rho} = \rho(\beta_1 + \beta_2 \rho^2 - \rho^4), \\ \dot{\varphi} = 1. \end{cases} \tag{8.26}$$

The equations in (8.26) are independent. The second equation describes a rotation with unit angular velocity. The trivial equilibrium $\rho = 0$ of the first equation corresponds to the only equilibrium, $z = 0$, of the truncated system. Positive equilibria of the first equation in (8.26) satisfy

$$\beta_1 + \beta_2 \rho^2 - \rho^4 = 0 \tag{8.27}$$

and describe circular limit cycles. Equation (8.27) can have zero, one, or two positive solutions (cycles). These solutions branch from the trivial one along the line

$$H = \{(\beta_1, \beta_2) : \beta_1 = 0\}$$

and collide and disappear at the half-parabola

$$T = \{(\beta_1, \beta_2) : \beta_2^2 + 4\beta_1 = 0, \ \beta_2 > 0\}.$$

The stability of the cycles is also clearly detectable from the first equation in (8.26). The bifurcation diagram of (8.26) is depicted in **Fig. 8.7**. The line

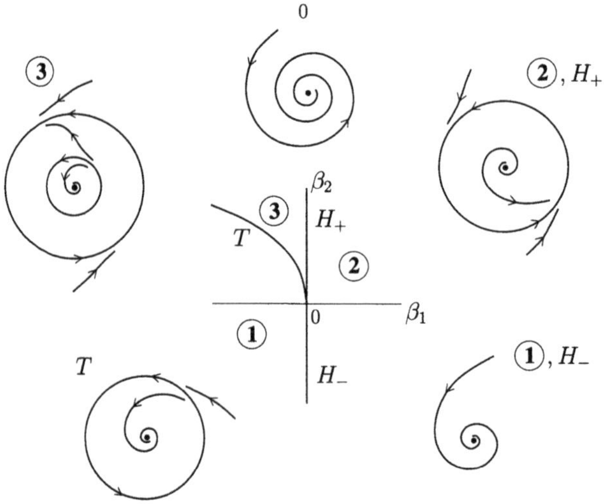


**Fig. 8.7.** Bautin bifurcation with $s = -1$.

$H$ corresponds to the Hopf bifurcation: Along this line the equilibrium has eigenvalues $\lambda_{1,2} = \pm i$. The equilibrium is stable for $\beta_1 < 0$ and unstable for $\beta_1 > 0$. The first Lyapunov coefficient $l_1(\beta) = \beta_2$. Therefore, the Bautin bifurcation point $\beta_1 = \beta_2 = 0$ separates two branches, $H_-$ and $H_+$, corresponding to a Hopf bifurcation with negative and with positive Lyapunov coefficient, respectively (i.e., "soft" and "sharp"). A stable limit cycle bifurcates from the equilibrium if we cross $H_-$ from left to right, while an unstable cycle appears if we cross $H_+$ in the opposite direction. The cycles collide and disappear on the curve $T$, corresponding to a nondegenerate fold bifurcation of the cycles (studied in Chapters 4 and 5). Along this curve the system has a critical limit cycle with multiplier $\mu = 1$ and a nonzero normal form coefficient $a$ of the Poincaré map. The curves divide the parameter plane into three regions (see **Fig. 8.7**).

To fully understand the bifurcation diagram, let us make an excursion on the parameter plane around the Bautin point counterclockwise, starting at a point in region **1**, where the system has a single stable equilibrium and no cycles. Crossing the Hopf bifurcation boundary $H_-$ from region **1** to region **2** implies the appearance of a unique and stable limit cycle, which survives when we enter region **3**. Crossing the Hopf boundary $H_+$ creates an extra unstable cycle inside the first one, while the equilibrium regains its stability. Two cycles of opposite stability exist inside region **3** and disappear at the curve $T$ through a fold bifurcation that leaves a single stable equilibrium, thus completing the circle.

The case $s = 1$ in (8.25) can be treated similarly or can be reduced to the one studied by the transformation $(z, \beta, t) \mapsto (\bar{z}, -\beta, -t)$.

### 8.3.3 Effect of higher-order terms

**Lemma 8.5** *The system*

$$\dot{z} = (\beta_1 + i)z + \beta_2 z|z|^2 \pm z|z|^4 + O(|z|^6)$$

*is locally topologically equivalent near the origin to the system*

$$\dot{z} = (\beta_1 + i)z + \beta_2 z|z|^2 \pm z|z|^4. \ \square$$

The proof of the lemma can be obtained by deriving the Taylor expansion of the Poincaré map for the first system and analyzing its fixed points. It turns out that the terms of order less than six are independent of $O(|z|^6)$ terms and thus coincide with those for the second system. This means that the two maps have the same number of fixed points for corresponding parameter values and that these points undergo similar bifurcations as the parameters vary near the origin. Then, one can construct a homeomorphism (actually, a diffeomorphism) identifying the parametric portraits of the systems near the origin and a homeomorphism that maps the phase portrait of the first system near the origin into that of the second system for all parameter values (as in Appendix A to Chapter 3). Therefore, we can complete the analysis of the Bautin bifurcation by stating the following theorem.

**Theorem 8.3 (Topological normal form for Bautin bifurcation)** *Any generic planar two-parameter system*

$$\dot{x} = f(x, \alpha),$$

*having at* $\alpha = 0$ *an equilibrium* $x = 0$ *that exhibits the Bautin bifurcation, is locally topologically equivalent near the origin to one of the following complex normal forms:*

$$\dot{z} = (\beta_1 + i)z + \beta_2 z|z|^2 \pm z|z|^4. \ \square$$

This theorem means that the described normal form captures the topology of any two-dimensional system having a Bautin bifurcation and satisfying the genericity conditions (B.1) and (B.2). In particular, although the limit cycles in such a system would not be perfect circles, we can expect the existence of two of them for nearby parameter values. Moreover, they will collide and disappear along a curve emanating from the codim 2 point.

The Bautin bifurcation is the first example that demonstrates the appearance of *limit cycle bifurcations* near codim 2 bifurcations of equilibria. In this case, by purely local analysis (computing the Lyapunov coefficients $l_1$ and $l_2$ at a Hopf point), we can *prove* the existence of a fold bifurcation of limit cycles for nearby parameter values.

The multidimensional case of the Bautin bifurcation can be treated by the center manifold reduction to the studied planar case. Then, (8.16) should be considered as the equations on the center manifold. Shoshitaishvili's Theorem gives the topological normal form for the Bautin bifurcation in $\mathbb{R}^n$:

$$\begin{cases} \dot{z} &= (\beta_1 + i)z + \beta_2 z|z|^2 + sz|z|^4, \\ \dot{\zeta}_- &= -\zeta_-, \\ \dot{\zeta}_+ &= \zeta_+, \end{cases}$$

where $s = \operatorname{sign} l_2(0) = \pm 1$, $z \in \mathbb{C}^1$, $\zeta_\pm \in \mathbb{R}^{n_\pm}$, and $n_-$ and $n_+$ are the numbers of eigenvalues of the critical equilibrium with $\operatorname{Re} \lambda > 0$ and $\operatorname{Re} \lambda < 0$, respectively, so that $n_- + n_+ + 2 = n$.

We return to the computation of $l_2(0)$ in the $n$-dimensional case in Section 8.7.

## 8.4 Bogdanov-Takens (double-zero) bifurcation

### 8.4.1 Normal form derivation

Consider a planar system

$$\dot{x} = f(x, \alpha), \quad x \in \mathbb{R}^2, \ \alpha \in \mathbb{R}^2, \tag{8.28}$$

where $f$ is smooth. Suppose that (8.28) has, at $\alpha = 0$, the equilibrium $x = 0$ with two zero eigenvalues (the Bogdanov-Takens condition), $\lambda_{1,2}(0) = 0$.

*Step 0 (Preliminary transformation).* We can write (8.28) at $\alpha = 0$ in the form

$$\dot{x} = A_0 x + F(x), \tag{8.29}$$

where $A_0 = f_x(0,0)$ and $F(x) = f(x,0) - A_0 x$ is a smooth function, $F(x) = O(\|x\|^2)$. The bifurcation conditions imply that

$$\Delta(0) = \det A_0 = 0, \;\; \sigma(0) = \operatorname{tr} A_0 = 0.$$

Assume that

(BT.0) $$A_0 \neq 0,$$

that is, $A_0$ has at least one nonzero element. Then there exist two real linearly independent vectors, $q_{0,1} \in \mathbb{R}^2$, such that

$$A_0 q_0 = 0, \quad A_0 q_1 = q_0. \tag{8.30}$$

The vector $q_0$ is the eigenvector of $A_0$ corresponding to the eigenvalue 0, while $q_1$ is the *generalized eigenvector* of $A_0$ corresponding to this eigenvalue. Moreover, there exist similar *adjoint* eigenvectors $p_{1,2} \in \mathbb{R}^2$ of the transposed matrix $A_0^T$:

$$A_0^T p_1 = 0, \quad A_0^T p_0 = p_1. \tag{8.31}$$

The vectors $q_1$ and $p_0$ are not uniquely defined even if $q_0$ and $p_1$ are fixed.[2] Nevertheless, we can always select four vectors satisfying (8.30) and (8.31) such that

$$\langle q_0, p_0 \rangle = \langle q_1, p_1 \rangle = 1, \tag{8.32}$$

where $\langle \cdot, \cdot \rangle$ stands for the standard scalar product in $\mathbb{R}^2 : \langle x, y \rangle = x_1 y_1 + x_2 y_2$, and

$$\langle q_1, p_0 \rangle = \langle q_0, p_1 \rangle = 0. \tag{8.33}$$

If $q_0$ and $q_1$ are selected as basis, then any vector $x \in \mathbb{R}^2$ can be uniquely represented as

$$x = y_1 q_0 + y_2 q_1,$$

for some real $y_{1,2} \in \mathbb{R}^1$. Taking into account (8.32) and (8.33), we find that these new coordinates $(y_1, y_2)$ are given by

$$\begin{cases} y_1 = \langle p_0, x \rangle, \\ y_2 = \langle p_1, x \rangle. \end{cases} \tag{8.34}$$

In the coordinates $(y_1, y_2)$, system (8.29) takes the form

$$\begin{pmatrix} \dot{y}_1 \\ \dot{y}_2 \end{pmatrix} = \begin{pmatrix} 0 & 1 \\ 0 & 0 \end{pmatrix} \begin{pmatrix} y_1 \\ y_2 \end{pmatrix} + \begin{pmatrix} \langle p_0, F(y_1 q_0 + y_2 q_1) \rangle \\ \langle p_1, F(y_1 q_0 + y_2 q_1) \rangle \end{pmatrix}. \tag{8.35}$$

Notice the particular form of the Jacobian matrix, which is the zero Jordan block of order 2.

[2] For example, if $q_1$ satisfies the second equation of (8.30), then $q_1' = q_1 + \gamma q_0$ with any $\gamma \in \mathbb{R}^1$ also satisfies this equation.

Let us use the same coordinates $(y_1, y_2)$ for all $\alpha$ with $\|\alpha\|$ small. In these coordinates, system (8.28) reads:

$$\begin{pmatrix} \dot{y}_1 \\ \dot{y}_2 \end{pmatrix} = \begin{pmatrix} \langle p_0, f(y_1 q_0 + y_2 q_1, \alpha)\rangle \\ \langle p_1, f(y_1 q_0 + y_2 q_1, \alpha)\rangle \end{pmatrix} \tag{8.36}$$

and for $\alpha = 0$ reduces to (8.35). Expand the right-hand side of (8.36) as a Taylor series with respect to $y$ at $y = 0$:

$$\begin{cases} \dot{y}_1 = y_2 + a_{00}(\alpha) + a_{10}(\alpha)y_1 + a_{01}(\alpha)y_2 \\ \qquad + \frac{1}{2}a_{20}(\alpha)y_1^2 + a_{11}(\alpha)y_1y_2 + \frac{1}{2}a_{02}(\alpha)y_2^2 + P_1(y,\alpha), \\ \dot{y}_2 = b_{00}(\alpha) + b_{10}(\alpha)y_1 + b_{01}(\alpha)y_2 \\ \qquad + \frac{1}{2}b_{20}(\alpha)y_1^2 + b_{11}(\alpha)y_1y_2 + \frac{1}{2}b_{02}(\alpha)y_2^2 + P_2(y,\alpha), \end{cases} \tag{8.37}$$

where $a_{kl}(\alpha), b_{kl}(\alpha)$ and $P_{1,2}(y,\alpha) = O(\|y\|^3)$ are smooth functions of their arguments. We have

$$a_{00}(0) = a_{10}(0) = a_{01}(0) = b_{00}(0) = b_{10}(0) = b_{01}(0) = 0.$$

The functions $a_{kl}(\alpha)$ and $b_{kl}(\alpha)$ can be expressed in terms of the right-hand side $f(x,\alpha)$ of (8.28) and the vectors $v_{0,1}, w_{0,1}$. For example,

$$a_{20}(\alpha) = \left.\frac{\partial^2}{\partial y_1^2}\langle p_0, f(y_1q_0 + y_2q_1, \alpha)\rangle\right|_{y=0},$$

$$b_{20}(\alpha) = \left.\frac{\partial^2}{\partial y_1^2}\langle p_1, f(y_1q_0 + y_2q_1, \alpha)\rangle\right|_{y=0},$$

$$b_{11}(\alpha) = \left.\frac{\partial^2}{\partial y_1 \partial y_2}\langle p_1, f(y_1q_0 + y_2q_1, \alpha)\rangle\right|_{y=0}.$$

Now we start transforming (8.37) into a simpler form by smooth invertible transformations (smoothly depending upon parameters) and time reparametrization. At a certain point, we will introduce new parameters.

*Step 1* (*Reduction to a nonlinear oscillator*). Introduce new variables $(u_1, u_2)$ by denoting the right-hand side of the first equation in (8.37) by $u_2$ and renaming $y_1$ to be $u_1$:

$$\begin{cases} u_1 = y_1, \\ u_2 = y_2 + a_{00} + a_{10}y_1 + a_{01}y_2 + \frac{1}{2}a_{20}y_1^2 + a_{11}y_1y_2 + \frac{1}{2}a_{02}y_2^2 + P_1(y,\cdot). \end{cases}$$

This transformation is invertible in some neighborhood of $y = 0$ for small $\|\alpha\|$ and depends smoothly on the parameters. If $\alpha = 0$, the origin $y = 0$ is a fixed point of this map. The transformation brings (8.37) into

$$\begin{cases} \dot{u}_1 = u_2, \\ \dot{u}_2 = g_{00}(\alpha) + g_{10}(\alpha)u_1 + g_{01}(\alpha)u_2 \\ \qquad + \frac{1}{2}g_{20}(\alpha)u_1^2 + g_{11}(\alpha)u_1u_2 + \frac{1}{2}g_{02}(\alpha)u_2^2 + Q(u,\alpha), \end{cases} \tag{8.38}$$

for certain smooth functions $g_{kl}(\alpha)$, $g_{00}(0) = g_{10}(0) = g_{01}(0) = 0$, and a smooth function $Q(u, \alpha) = O(\|u\|^3)$. We can verify that

$$\begin{aligned} g_{20}(0) &= b_{20}(0), \\ g_{11}(0) &= a_{20}(0) + b_{11}(0), \\ g_{02}(0) &= b_{02}(0) + 2a_{11}(0). \end{aligned} \tag{8.39}$$

Furthermore, we have

$$\begin{aligned} g_{00}(\alpha) &= b_{00}(\alpha) + \cdots, \\ g_{10}(\alpha) &= b_{10}(\alpha) + a_{11}(\alpha)b_{00}(\alpha) - b_{11}(\alpha)a_{00}(\alpha) + \cdots, \\ g_{01}(\alpha) &= b_{01}(\alpha) + a_{10}(\alpha) + a_{02}(\alpha)b_{00}(\alpha) \\ &\quad - (a_{11}(\alpha) + b_{02}(\alpha))a_{00}(\alpha) + \cdots, \end{aligned} \tag{8.40}$$

where dots represent all terms containing at least one product of some $a_{kl}, b_{ij}$ with $k + l \leq 1$ ($i + j \leq 1$). Since $a_{kl}(\alpha)$ and $b_{kl}(\alpha)$ vanish at $\alpha = 0$, for all $k + l \leq 1$, the displayed terms are sufficient to compute the first partial derivatives of $g_{00}(\alpha)$, $g_{10}(\alpha)$, and $g_{01}(\alpha)$ with respect to $(\alpha_1, \alpha_2)$ at $\alpha = 0$.

Note that system (8.38) can be written as a single second-order differential equation for $w = u_1$:

$$\ddot{w} = G(w, \alpha) + \dot{w}H(w, \alpha) + \dot{w}^2 Z(w, \dot{w}, \alpha),$$

which provides the general form for the equation of motion of a nonlinear oscillator.

*Step 2* (*Parameter-dependent shift*). A parameter-dependent shift of coordinates in the $u_1$-direction

$$\begin{cases} u_1 = v_1 + \delta(\alpha), \\ u_2 = v_2, \end{cases}$$

transforms (8.38) into

$$\begin{cases} \dot{v}_1 = v_2, \\ \dot{v}_2 = g_{00} + g_{10}\delta + O(\delta^2) \\ \qquad + \left(g_{10} + g_{20}\delta + O(\delta^2)\right) v_1 + \left(g_{01} + g_{11}\delta + O(\delta^2)\right) v_2 \\ \qquad + \frac{1}{2}(g_{20} + O(\delta))v_1^2 + (g_{11} + O(\delta))v_1v_2 + \frac{1}{2}(g_{02} + O(\delta))v_2^2 \\ \qquad + O(\|v\|^3). \end{cases}$$

Assume that

(BT.1) $$g_{11}(0) = a_{20}(0) + b_{11}(0) \neq 0.$$

Then, standard arguments based on the Implicit Function Theorem provide local existence of a smooth function

$$\delta = \delta(\alpha) \approx -\frac{g_{01}(\alpha)}{g_{11}(0)},$$

annihilating the term proportional to $v_2$ in the equation for $v_2$, which leads to the system

$$\begin{cases} \dot{v}_1 = v_2, \\ \dot{v}_2 = h_{00}(\alpha) + h_{10}(\alpha)v_1 \\ \qquad + \frac{1}{2}h_{20}(\alpha)v_1^2 + h_{11}(\alpha)v_1v_2 + \frac{1}{2}h_{02}(\alpha)v_2^2 + R(v,\alpha), \end{cases} \tag{8.41}$$

where $h_{kl}(\alpha)$ and $R(v,\alpha) = O(\|v\|^3)$ are smooth. We find

$$h_{00}(\alpha) = g_{00}(\alpha) + \cdots, \quad h_{10}(\alpha) = g_{10}(\alpha) - \frac{g_{20}(0)}{g_{11}(0)}g_{01}(\alpha) + \cdots, \tag{8.42}$$

where again only the terms needed to compute the first partial derivatives with respect to $(\alpha_1, \alpha_2)$ at $\alpha = 0$ are kept (see (8.40)). Clearly, $h_{00}(0) = h_{10}(0) = 0$. The only relevant values of $h_{kl}(\alpha)$, $k + l = 2$, are, as we shall see, at $\alpha = 0$. These terms are given by

$$h_{20}(0) = g_{20}(0), \quad h_{11}(0) = g_{11}(0), \quad h_{02}(0) = g_{02}(0), \tag{8.43}$$

where $g_{kl}(0)$, $k + l = 2$, are determined by (8.39).

*Step 3* (*Time reparametrization and second reduction to a nonlinear oscillator*). Introduce the new time $\tau$ via the equation

$$dt = (1 + \theta v_1)\, d\tau,$$

where $\theta = \theta(\alpha)$ is a smooth function to be defined later. The direction of time is preserved near the origin for small $\|\alpha\|$. Assuming that a dot over a variable now means differentiation with respect to $\tau$, we obtain

$$\begin{cases} \dot{v}_1 = v_2 + \theta v_1 v_2, \\ \dot{v}_2 = h_{00} + (h_{10} + h_{00}\theta)v_1 + \frac{1}{2}(h_{20} + 2h_{10}\theta)v_1^2 + h_{11}v_1v_2 + \frac{1}{2}h_{02}v_2^2 \\ \qquad + O(\|v\|^3). \end{cases}$$

The above system has a similar form to (8.37), which is a bit discouraging. However, we can reduce it once more to a nonlinear oscillator by a coordinate transformation similar to that in the first step:

$$\begin{cases} \xi_1 = v_1, \\ \xi_2 = v_2 + \theta v_1 v_2, \end{cases}$$

mapping the origin into itself for all $\theta$. The system in $(\xi_1, \xi_2)$-coordinates takes the form

$$\begin{cases} \dot{\xi}_1 = \xi_2, \\ \dot{\xi}_2 = f_{00}(\alpha) + f_{10}(\alpha)\xi_1 + \frac{1}{2}f_{20}(\alpha)\xi_1^2 + f_{11}(\alpha)\xi_1\xi_2 + \frac{1}{2}f_{02}(\alpha)\xi_2^2 \\ \qquad + O(\|\xi\|^3), \end{cases} \tag{8.44}$$

where

$$f_{00}(\alpha) = h_{00}(\alpha), \quad f_{10}(\alpha) = h_{10}(\alpha) + h_{00}(\alpha)\theta(\alpha),$$

and

$$\begin{aligned} f_{20}(\alpha) &= h_{20}(\alpha) + 2h_{10}(\alpha)\theta(\alpha), \\ f_{11}(\alpha) &= h_{11}(\alpha), \\ f_{02}(\alpha) &= h_{02}(\alpha) + 2\theta(\alpha). \end{aligned}$$

Now we can take

$$\theta(\alpha) = -\frac{h_{02}(\alpha)}{2}$$

to eliminate the $\xi_2^2$-term, thus specifying the time reparametrization. Consequently, we have

$$\begin{cases} \dot{\xi}_1 = \xi_2, \\ \dot{\xi}_2 = \mu_1(\alpha) + \mu_2(\alpha)\xi_1 + A(\alpha)\xi_1^2 + B(\alpha)\xi_1\xi_2 + O(\|\xi\|^3), \end{cases} \tag{8.45}$$

where

$$\mu_1(\alpha) = h_{00}(\alpha), \quad \mu_2(\alpha) = h_{10}(\alpha) - \frac{1}{2}h_{00}(\alpha)h_{02}(\alpha), \tag{8.46}$$

and

$$A(\alpha) = \frac{1}{2}\left(h_{20}(\alpha) - h_{10}(\alpha)h_{02}(\alpha)\right), \quad B(\alpha) = h_{11}(\alpha). \tag{8.47}$$

*Step 4* (*Final scaling and setting of new parameters*). Introduce a new time (and denote it by $t$ again)

$$t = \left|\frac{B(\alpha)}{A(\alpha)}\right| \tau.$$

Since $B(0) = h_{11}(0) = g_{11}(0) = a_{20}(0) + b_{11}(0) \neq 0$ due to $(BT.1)$, the time scaling above will be well defined if we further assume

(BT.2) $$2A(0) = h_{20}(0) = g_{20}(0) = b_{20}(0) \neq 0.$$

Simultaneously, perform a scaling by introducing the new variables

$$\eta_1 = \frac{A(\alpha)}{B^2(\alpha)}\xi_1, \quad \eta_2 = \operatorname{sign}\left(\frac{B(\alpha)}{A(\alpha)}\right)\frac{A^2(\alpha)}{B^3(\alpha)}\xi_2.$$

Notice that the denominators are nonzero at $\alpha = 0$ because $A(0) \neq 0$ and $B(0) \neq 0$. In the coordinates $(\eta_1, \eta_2)$, system (8.45) takes the form

$$\begin{cases} \dot{\eta}_1 = \eta_2, \\ \dot{\eta}_2 = \beta_1 + \beta_2\eta_1 + \eta_1^2 + s\eta_1\eta_2 + O(\|\eta\|^3), \end{cases} \tag{8.48}$$

with

$$s = \operatorname{sign}\left(\frac{B(0)}{A(0)}\right) = \operatorname{sign}\left(\frac{a_{20}(0) + b_{11}(0)}{b_{20}(0)}\right) = \pm 1,$$

and

$$\beta_1(\alpha) = \frac{B^4(\alpha)}{A^3(\alpha)}\mu_1(\alpha),$$
$$\beta_2(\alpha) = \frac{B^2(\alpha)}{A^2(\alpha)}\mu_2(\alpha).$$

Obviously, $\beta_1(0) = \beta_2(0) = 0$. In order to define an invertible smooth change of parameters near the origin, we have to assume the regularity of the map $\alpha \mapsto \beta$ at $\alpha = 0$:

(BT.3) $$\det \left(\frac{\partial \beta}{\partial \alpha}\right)\bigg|_{\alpha=0} \neq 0.$$

This condition is equivalent to the regularity of the map $\alpha \mapsto \mu$ at $\alpha = 0$ and can be expressed more explicitly if we take into account formulas (8.46), (8.42), and (8.40). Indeed, the following lemma can be proved by straightforward calculations.

**Lemma 8.6** *Let system* (8.37) *be written as*

$$\dot{y} = P(y, \alpha), \quad y \in \mathbb{R}^2, \ \alpha \in \mathbb{R}^2,$$

*and the nondegeneracy conditions* (BT.1) *and* (BT.2) *are satisfied. Then the transversality condition* (BT.3) *is equivalent to the regularity of the map*

$$(y,\alpha) \mapsto \left(P(y,\alpha), \operatorname{tr}\left(\frac{\partial P(y,\alpha)}{\partial y}\right), \det\left(\frac{\partial P(y,\alpha)}{\partial y}\right)\right)$$

*at the point* $(y, \alpha) = (0, 0)$. $\square$

The map in the lemma is a map from $\mathbb{R}^4$ to $\mathbb{R}^4$, so its regularity means the nonvanishing of the determinant of its Jacobian matrix. Since the linear change of coordinates $x \mapsto y$ defined by (8.34) is regular, we can merely check the regularity of the map

$$(x,\alpha) \mapsto \left(f(x,\alpha), \operatorname{tr}\left(\frac{\partial f(x,\alpha)}{\partial x}\right), \det\left(\frac{\partial f(x,\alpha)}{\partial x}\right)\right)$$

at the point $(x, \alpha) = (0, 0)$.

Therefore, in this subsection we have proved the following theorem.

**Theorem 8.4** *Suppose that a planar system*

$$\dot{x} = f(x, \alpha), \quad x \in \mathbb{R}^2, \ \alpha \in \mathbb{R}^2,$$

*with smooth $f$, has at $\alpha = 0$ the equilibrium $x = 0$ with a double zero eigenvalue:*

$$\lambda_{1,2}(0) = 0.$$

*Assume that the following genericity conditions are satisfied:*

(BT.0) *the Jacobian matrix* $A(0) = f_x(0,0) \neq 0$;
(BT.1) $a_{20}(0) + b_{11}(0) \neq 0$;
(BT.2) $b_{20}(0) \neq 0$;
(BT.3) *the map*

$$(x,\alpha) \mapsto \left( f(x,\alpha), \operatorname{tr}\left(\frac{\partial f(x,\alpha)}{\partial x}\right), \det\left(\frac{\partial f(x,\alpha)}{\partial x}\right)\right)$$

*is regular at point* $(x,\alpha) = (0,0)$.

*Then there exist smooth invertible variable transformations smoothly depending on the parameters, a direction-preserving time reparametrization, and smooth invertible parameter changes, which together reduce the system to*

$$\begin{cases} \dot{\eta}_1 = \eta_2, \\ \dot{\eta}_2 = \beta_1 + \beta_2\eta_1 + \eta_1^2 + s\eta_1\eta_2 + O(\|\eta\|^3), \end{cases}$$

*where* $s = \operatorname{sign}[b_{20}(a_{20}(0) + b_{11}(0))] = \pm 1$. □

The coefficients $a_{20}(0)$, $b_{20}(0)$, and $b_{11}(0)$ can be computed in terms of $f(x,0)$ by the formulas given after system (8.37).

**Remarks:**

(1) While time reparametrization was essential in the derivation of the intermediate parameter-dependent normal form (8.45), it is not necessary at the critical parameter values. One can prove that (8.37) is smoothly equivalent at $\alpha = 0$ to

$$\begin{cases} \dot{\xi}_1 = \xi_2, \\ \dot{\xi}_2 = A(0)\xi_1^2 + B(0)\xi_1\xi_2 + O(\|\xi\|^3), \end{cases} \tag{8.49}$$

where $A(0) = \frac{1}{2}b_{20}(0)$ and $B(0) = a_{20}(0) + b_{11}(0)$ as before.

(2) There are several (equivalent) normal forms for the Bogdanov-Takens bifurcation. The normal form (8.48) was introduced by Bogdanov, while Takens derived the normal form

$$\begin{cases} \dot{\eta}_1 = \eta_2 + \beta_2\eta_1 + \eta_1^2 + O(\|\eta\|^3), \\ \dot{\eta}_2 = \beta_1 + s\eta_1^2 + O(\|\eta\|^3), \end{cases} \tag{8.50}$$

where $s = \pm 1$. The proof of the equivalence of these two normal forms is left to the reader as an exercise. ◇

### 8.4.2 Bifurcation diagram of the normal form

Take $s = -1$ and consider system (8.48) without $O(\|\eta\|^3)$ terms:

$$\begin{cases} \dot{\eta}_1 = \eta_2, \\ \dot{\eta}_2 = \beta_1 + \beta_2\eta_1 + \eta_1^2 - \eta_1\eta_2. \end{cases} \tag{8.51}$$

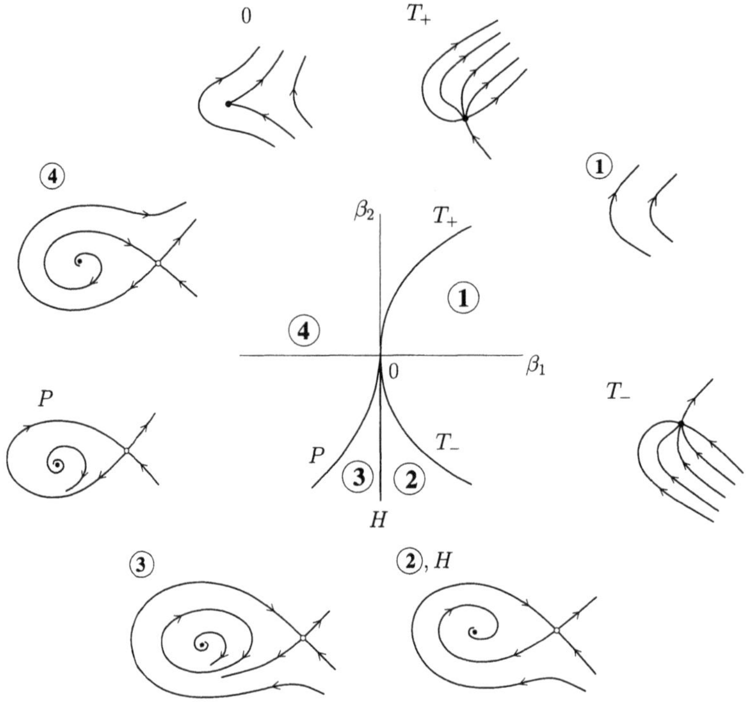


**Fig. 8.8.** Bogdanov-Takens bifurcation with $s = -1$.

This is the first case where the analysis of a truncated system is nontrivial. More precisely, bifurcations of equilibria are easy to analyze, while the study of limit cycles (actually, the uniqueness of the cycle) is rather involved.

The bifurcation diagram of system (8.51) is presented in **Fig. 8.8**. Any equilibria of the system are located on the horizontal axis, $\eta_2 = 0$, and satisfy the equation

$$\beta_1 + \beta_2\eta_1 + \eta_1^2 = 0. \tag{8.52}$$

Equation (8.52) can have between zero and two real roots. The discriminant parabola

$$T = \{(\beta_1, \beta_2) : 4\beta_1 - \beta_2^2 = 0\} \tag{8.53}$$

corresponds to a fold bifurcation: Along this curve system (8.51) has an equilibrium with a zero eigenvalue. If $\beta_2 \neq 0$, then the fold bifurcation is nondegenerate and crossing $T$ from right to left implies the appearance of two equilibria. Let us denote the left one by $E_1$ and the right one by $E_2$:

$$E_{1,2} = (\eta_{1,2}^0, 0) = \left( \frac{-\beta_2 \mp \sqrt{\beta_2^2 - 4\beta_1}}{2}, 0 \right).$$

The point $\beta = 0$ separates two branches $T_-$ and $T_+$ of the fold curve corresponding to $\beta_2 < 0$ and $\beta_2 > 0$, respectively. We can check that passage through $T_-$ implies the coalescence of a stable node $E_1$ and a saddle point $E_2$, while crossing $T_+$ generates an unstable node $E_1$ and a saddle $E_2$. There is a nonbifurcation curve (not shown in the figure) located at $\beta_1 > 0$ and passing through the origin at which the equilibrium $E_1$ undergoes a node to focus transition.

The vertical axis $\beta_1 = 0$ is a line on which the equilibrium $E_1$ has a pair of eigenvalues with zero sum: $\lambda_1 + \lambda_2 = 0$. The lower part,

$$H = \{(\beta_1, \beta_2) : \ \beta_1 = 0, \ \beta_2 < 0\}, \tag{8.54}$$

corresponds to a nondegenerate Andronov-Hopf bifurcation ($\lambda_{1,2} = \pm i\omega$), while the upper half-axis is a nonbifurcation line corresponding to a neutral saddle. The Hopf bifurcation gives rise to a stable limit cycle since $l_1 < 0$ (Exercise 10(b)). The cycle exists near $H$ for $\beta_1 < 0$. The equilibrium $E_2$ remains a saddle for all parameter values to the left of the curve $T$ and does not bifurcate. There are no other local bifurcations in the dynamics of (8.51).

Make a roundtrip near the Bogdanov-Takens point $\beta = 0$, starting from region **1** where there are no equilibria (and thus no limit cycles are possible). Entering from region **1** into region **2** through the component $T_-$ of the fold curve yields two equilibria: a saddle and a stable node. Then the node turns into a focus and loses stability as we cross the Hopf bifurcation boundary $H$. A stable limit cycle is present for close parameter values to the left of $H$. If we continue the journey clockwise and finally return to region **1**, no limit cycles must remain. Therefore, there must be global bifurcations "destroying" the cycle somewhere between $H$ and $T_+$. We know of only two such bifurcations of codim 1 in planar systems: a saddle homoclinic bifurcation (Chapter 6) and a saddle-node homoclinic bifurcation (Chapter 7). Since the saddle-node equilibrium at the fold bifurcation cannot have a homoclinic orbit, the only possible candidate for the global bifurcation is the appearance of an orbit homoclinic to the saddle $E_2$. Thus, there should exist at least one bifurcation curve originating at $\beta = 0$ along which system (8.51) has a saddle homoclinic bifurcation. As we trace the homoclinic orbit along the curve $P$ toward the Bogdanov-Takens point, the looplike orbit shrinks and disappears.

**Lemma 8.7** *There is a unique smooth curve $P$ corresponding to a saddle homoclinic bifurcation in system* (8.51) *that originates at $\beta = 0$ and has the local representation*

$$P = \left\{(\beta_1, \beta_2) : \ \beta_1 = -\frac{6}{25}\beta_2^2 + o(\beta_2^2), \ \beta_2 < 0\right\}. \tag{8.55}$$

*Moreover, for $\|\beta\|$ small, system* (8.51) *has a unique and hyperbolic stable cycle for parameter values inside the region bounded by the Hopf bifurcation curve $H$ and the homoclinic bifurcation curve $P$, and no cycles outside this region.* □

In Appendix A we outline a "standard" proof of this lemma based on a "blowing-up" by a *singular scaling* and Pontryagin's technique of *perturbation of Hamiltonian systems.* The proof gives expression (8.55) (see also Exercise 14(a)) and can be applied almost verbatim to the complete system (8.48), with $O(\|\eta\|^3)$ terms kept.

Due to the lemma, the stable cycle born through the Hopf bifurcation does not bifurcate in region **3**. As we move clockwise, it "grows" and approaches the saddle, turning into a homoclinic orbit at $P$. Notice that the hyperbolicity of the cycle near the homoclinic bifurcation follows from the fact that the saddle quantity $\sigma_0 < 0$ along $P$. To complete our roundtrip, note that there are no cycles in region 4 located between the curve $P$ and the branch $T_+$ of the fold curve. An unstable node and a saddle, existing for the parameter values in this region, collide and disappear at the fold curve $T_+$. Let us also point out that at $\beta = 0$ the critical equilibrium with a double zero eigenvalue has exactly two asymptotic orbits (one tending to the equilibrium for $t \to +\infty$ and one approaching it as $t \to -\infty$). These orbits form a peculiar "cuspoidal edge" (see **Fig. 8.8**).

The case $s = +1$ can be treated similarly. Since it can be reduced to the one studied by the substitution $t \mapsto -t$, $\eta_2 \mapsto -\eta_2$, the parametric portrait remains as it was but the cycle becomes *unstable* near the Bogdanov-Takens point.

### 8.4.3 Effect of higher-order terms

**Lemma 8.8** *The system*

$$\begin{cases} \dot{\eta}_1 = \eta_2, \\ \dot{\eta}_2 = \beta_1 + \beta_2\eta_1 + \eta_1^2 \pm \eta_1\eta_2 + O(\|\eta\|^3), \end{cases}$$

*is locally topologically equivalent near the origin to the system*

$$\begin{cases} \dot{\eta}_1 = \eta_2, \\ \dot{\eta}_2 = \beta_1 + \beta_2\eta_1 + \eta_1^2 \pm \eta_1\eta_2. \ \square \end{cases}$$

We give only an outline of the proof. Take $s = -1$ and develop the $O(\|\eta\|^3)$ term in system (8.48) into a Taylor series in $\eta_1$. This results in

$$\begin{cases} \dot{\eta}_1 = \eta_2, \\ \dot{\eta}_2 = \beta_1 + \beta_2\eta_1 + \eta_1^2(1 + \eta_1 P(\eta_1, \beta)) - \eta_1\eta_2(1 + \eta_1 Q(\eta_1, \beta)) \\ \qquad + \eta_2^2(\eta_1 R(\eta, \beta) + \eta_2 S(\eta, \beta)), \end{cases} \tag{8.56}$$

where $P, Q, R,$ and $S$ are some smooth functions.

It is an easy exercise in the Implicit Function Theorem to prove the existence of both a fold bifurcation curve and a Hopf bifurcation curve in (8.56) that are close to the corresponding curves $T$ and $H$ in (8.51). The nondegeneracy conditions for these bifurcations can also be verified rather straightforwardly.

An analog of Lemma 8.7 can be proved for system (8.56) practically by repeating step by step the proof outlined in Appendix A.[3] Then, a homeomorphism (actually, diffeomorphism) mapping the parameter portrait of (8.56) into that of (8.51) can be constructed, as well as a (parameter-dependent) homeomorphism identifying the corresponding phase portraits.

As usual, let us formulate a general theorem.

**Theorem 8.5 (Topological normal form for BT bifurcation)** *Any generic planar two-parameter system*

$$\dot{x} = f(x, \alpha),$$

*having, at $\alpha = 0$, an equilibrium that exhibits the Bogdanov-Takens bifurcation, is locally topologically equivalent near the equilibrium to one of the following normal forms:*

$$\begin{cases} \dot{\eta}_1 = \eta_2, \\ \dot{\eta}_2 = \beta_1 + \beta_2\eta_1 + \eta_1^2 \pm \eta_1\eta_2. \;\square \end{cases}$$

As for the Bautin bifurcation, the Bogdanov-Takens bifurcation gives rise to a *limit cycle* bifurcation, namely, the appearance of the homoclinic orbit, for nearby parameter values. Thus, we can prove analytically (by verifying the bifurcation conditions and the genericity conditions (BT.1)-(BT.3)) the existence of this *global* bifurcation in the system. Again, this is one of few regular methods to detect homoclinic bifurcations analytically.

The multidimensional case of the Bogdanov-Takens bifurcation brings nothing new since it can be reduced to the planar case using the Center Manifold Theorem. Shoshitaishvili's Theorem gives the following topological normal form for the Bogdanov-Takens bifurcation in $\mathbb{R}^n$:

$$\begin{cases} \dot{\eta}_1 &= \eta_2, \\ \dot{\eta}_2 &= \beta_1 + \beta_2\eta_1 + \eta_1^2 + s\eta_1\eta_2, \\ \dot{\zeta}_- &= -\zeta_-, \\ \dot{\zeta}_+ &= \zeta_+, \end{cases}$$

where $s = \text{sign } A(0)B(0) = \pm 1$, $(\eta_1, \eta_2)^T \in \mathbb{R}^2$, $\zeta_\pm \in \mathbb{R}^{n_\pm}$, and $n_-$ and $n_+$ are the numbers of eigenvalues of the critical equilibrium with Re $\lambda > 0$ and Re $\lambda < 0$ so that $n_- + n_+ + 2 = n$.

We will explain in Section 8.7 how to find $A(0)$ and $B(0)$. Here we only note that *linear approximation* of the center manifold at the critical parameter values is sufficient to compute them.

**Example 8.3 (Bazykin [1985])** Consider the system of two differential equations

---

[3] To proceed in *exactly* the same way as in Appendix A, one has to eliminate the $P(\eta_1, \beta)$-term from (8.56) by proper variable and time transformations. Then, the equilibria of (8.56) will coincide with those of (8.51).

$$\begin{cases} \dot{x}_1 = x_1 - \dfrac{x_1 x_2}{1 + \alpha x_1} - \varepsilon x_1^2, \\ \dot{x}_2 = -\gamma x_2 + \dfrac{x_1 x_2}{1 + \alpha x_1} - \delta x_2^2 . \end{cases}$$

The equations model the dynamics of a predator-prey ecosystem. The variables $x_1$ and $x_2$ are (scaled) population numbers of prey and predator, respectively, while $\alpha, \gamma, \varepsilon$, and $\delta$ are nonnegative parameters describing the behavior of isolated populations and their interaction (see Bazykin [1985]). If $\delta = 0$, the system reduces to a model that differs only by scaling from the system considered in Chapter 3 (see Example 3.1). Assume that $\varepsilon \ll 1$ and $\gamma = 1$ are fixed. The bifurcation diagram of the system with respect to the two remaining parameters $(\alpha, \delta)$ exhibits all codim 2 bifurcations possible in planar systems.

The system has two trivial equilibria,

$$O = (0,0), \quad E_0 = \left(\frac{1}{\varepsilon}, 0\right).$$

The equilibrium $O$ is always a saddle, while $E_0$ is a stable node for $\alpha > 1 - \varepsilon$ and is the only attractor for these parameter values. At the line

$$k = \{(\alpha, \delta) : \ \alpha = 1 - \varepsilon\}$$

a nontrivial equilibrium bifurcates from $E_0$ into the positive quadrant while $E_0$ turns into a saddle (transcritical bifurcation). Actually, the system can have between one and three positive equilibria, $E_1, E_2, E_3$, to the left of the line $k$. These equilibria collide pairwise and disappear via the fold (tangent) bifurcation at a curve $t$,

$$\begin{aligned} t = \big\{(\alpha, \delta) : \ & 4\varepsilon(\alpha - 1)^3 + \left[(\alpha^2 - 20\alpha - 8)\varepsilon^2 + 2\alpha\varepsilon(\alpha^2 - 11\alpha + 10) \right. \\ & + \left. \alpha^2(\alpha - 1)^2\right]\delta - 4(\alpha + \varepsilon)^3\delta^2 = 0\big\} . \end{aligned}$$

This curve delimits a region resembling "lips" which is sketched in **Fig. 8.9**(a).[4] Inside the region, the system has three equilibrium points (two antisaddles $E_1, E_3$, and a saddle $E_2$). Outside the region, at most one nontrivial equilibrium might exist. There are two cusp singular points, $C_1$ and $C_2$, on the curve $t$. One can check that the corresponding cusp bifurcations are generic. While approaching any of these points from inside the "lips," three positive equilibria simultaneously collide, and only one of them survives outside.

Parameter values for which the system has an equilibrium with $\lambda_1 + \lambda_2 = 0$ belong to the curve

$$\begin{aligned} h = \big\{(\alpha, \delta) : \ & 4\varepsilon[\alpha(\alpha - 1) + \varepsilon(\alpha + 1)] + \left[2(\varepsilon + 1)\alpha^2 + (3\varepsilon^2 - 2\varepsilon - 1)\alpha \right. \\ & + \left. \varepsilon(\varepsilon^2 - 2\varepsilon + 5)\right]\delta + (\alpha + \varepsilon - 1)^2\delta^2 = 0\big\} . \end{aligned}$$

---

[4] In reality, the region is much more narrow than that in **Fig. 8.9**.

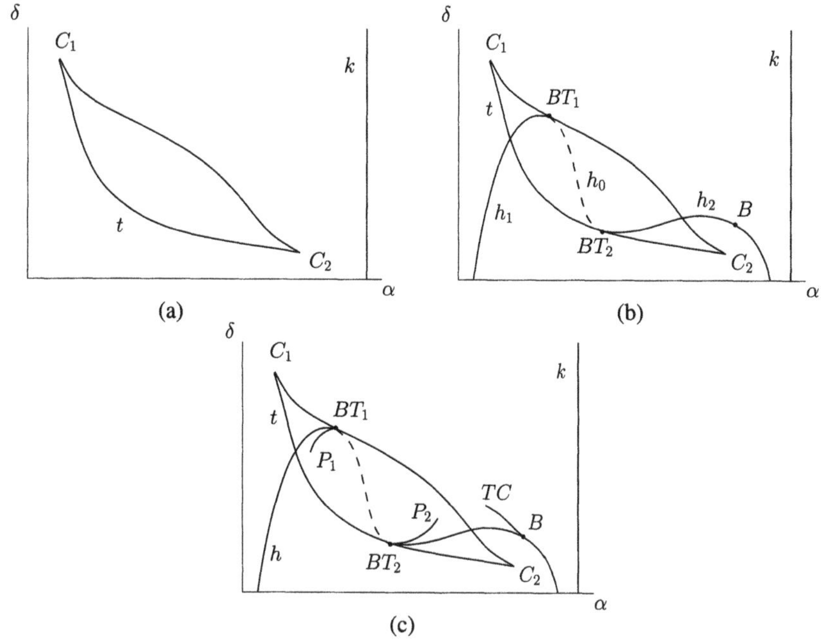


**Fig. 8.9.** Construction of the parametric portrait.

The curve $h$ has two tangencies with the curve $t$ at points $BT_1$ and $BT_2$ (see **Fig. 8.9**(b)). For the corresponding parameter values there is an equilibrium with a double zero eigenvalue. One can check that the system exhibits generic Bogdanov-Takens bifurcations at these points. A part $h_0$ between the points $BT_1$ and $BT_2$ corresponds to a neutral saddle, while the outer branches define two Hopf bifurcation curves, $h_1$ and $h_2$. The last local codim 2 bifurcation appears at a point $B$ on the Hopf curve $h_2$. This is a Bautin point at which the first Lyapunov coefficient $l_1$ vanishes but the second Lyapunov coefficient $l_2 < 0$. We have $l_1 < 0$ along $h_2$ to the right of the point $B$, $l_1 > 0$ between points $BT_2$ and $B$, and $l_1 < 0$ to the left of $BT_1$. Negative values of $l_1$ ensure the appearance of a stable limit cycle via a Hopf bifurcation, while positive values lead to an unstable cycle nearby.

Based on the theory we have developed for the Bautin and the Bogdanov-Takens bifurcations in the previous sections, we can make some conclusions about *limit cycle bifurcations* in the system. There is a bifurcation curve $TC$ originating at the Bautin point $B$ at which the fold bifurcation of limit cycles takes place: A stable and an unstable limit cycle collide, forming a nonhyperbolic cycle with multiplier $\mu = 1$, and disappear (see **Fig. 8.9**(c)). Two other limit cycle bifurcation curves, $P_1$ and $P_2$, emanate from the Bogdanov-Takens points $BT_{1,2}$. These are, of course, the homoclinic bifurcation curves, along which the "central" saddle $E_2$ has a homoclinic orbit around one of the "peripheral" antisaddles $E_{1,3}$.

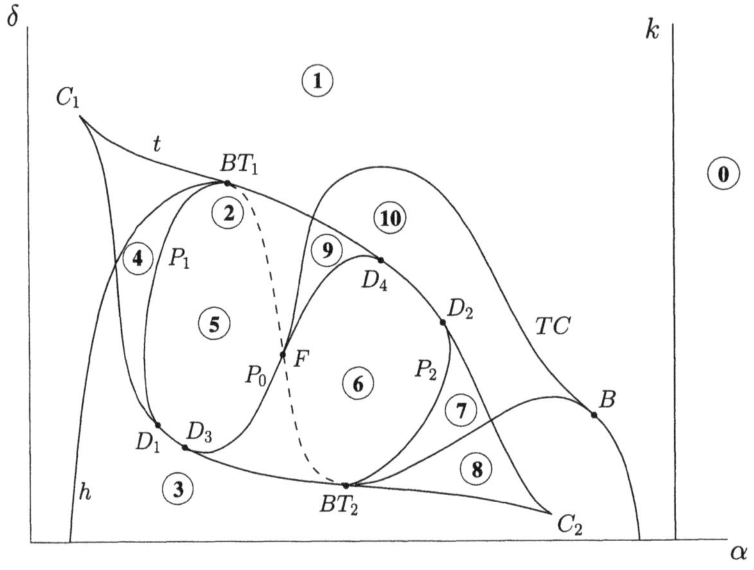


**Fig. 8.10.** Parametric portrait of Bazykin's predator-prey system.

The completion of the bifurcation diagram requires numerical methods, as described in Chapter 10. The resulting parametric portrait is sketched in **Fig. 8.10**, while relevant phase portraits are presented in **Fig. 8.11**.[5] The homoclinic curves $P_{1,2}$ terminate at points $D_{1,2}$ on the fold curve $t$. These points are codim 2 saddle-node homoclinic points at which the homoclinic orbit returns to the saddle-node along a *noncentral* manifold (see **Fig. 8.12**(a)). The homoclinic curves $P_{1,2}$ are tangent to $t$ at $D_{1,2}$. A bit more surprising is the end point of the cycle fold curve $TC$: It terminates at a point in the branch $h_0$ corresponding to a neutral saddle. This is another codim 2 global bifurcation that we have not studied yet. Namely, the curve $TC$ terminates at the point $F$, where $h_0$ intersects with another saddle homoclinic curve $P_0$. This curve $P_0$ corresponds to the appearance of a *"big homoclinic loop"* (see **Fig. 8.12**(b)). It terminates at points $D_{3,4}$ on the fold curve $t$, similar to the points $D_{1,2}$. At the point $F$, there is a "big" homoclinic orbit to the saddle $E_2$ with zero saddle quantity $\sigma_0$ (see Chapter 6). The stability of the cycle generated via destruction of the big homoclinic orbit is opposite along the two branches of $P_0$ separated by $F$. This leads to the existence of a fold curve for cycles nearby. Numerical continuation techniques show that it is the fold curve $TC$ originating at the point $B$ that terminates at $F$. The curves $TC$ and $P_0$ have an *infinite-order* tangency at the point $F$. The intervals $D_1D_3$ and $D_2D_4$ of the tangent curve $t$ correspond to the homoclinic saddle-node bifurcation

[5] The *central projection* of the plane $(x_1, x_2)$ onto the lower hemisphere of the unit sphere $x_1^2 + x_2^2 + (x_3 - 1)^2 = 1$ is used to draw the phase portraits.

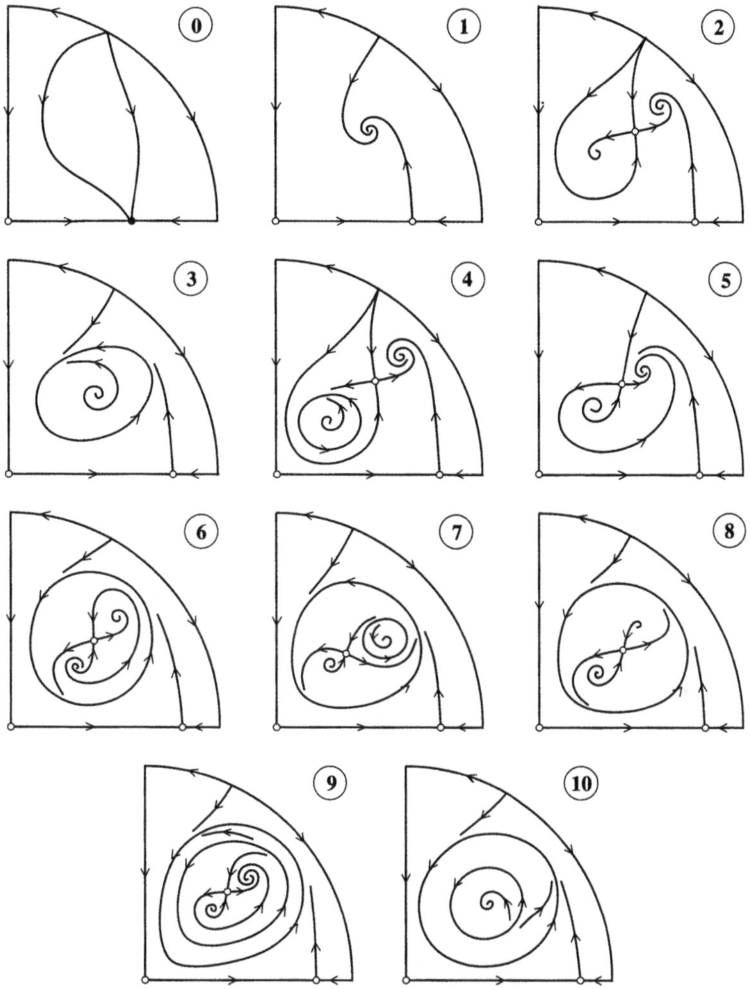


**Fig. 8.11.** Generic phase portraits of Bazykin's predator-prey system.

of codim 1 (studied in Chapter 7, Theorem 7.1). Leaving the "lips" through any of these intervals generates a limit cycle, stable near $D_1D_3$ and unstable near $D_2D_4$; the unstable cycle is located *inside* the "big" stable cycle. The parametric portrait is completed, and we recommend that the reader "walk" around it, tracing various metamorphoses of the phase portrait.

No theorem guarantees that the studied system cannot have more than two limit cycles, even if $\varepsilon \ll 1$ and $\gamma = 1$. Nevertheless, numerous simulations confirm that the phase portraits presented in **Fig. 8.11** are indeed the only possible ones in the system for generic parameter values $(\alpha, \delta)$. The ecological interpretation of the described bifurcation diagram can be found in

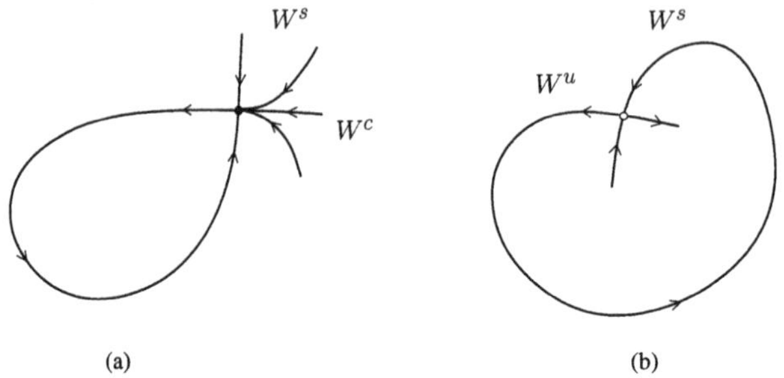


**Fig. 8.12.** (a) Noncentral saddle-node homoclinic orbit; (b) "big" homoclinic orbit.

Bazykin [1985]. Let us just point out here that the system exhibits nontrivial coexistence of equilibrium and oscillatory behavior. ◊

## 8.5 Fold-Hopf bifurcation

Now we have a smooth three-dimensional system depending on two parameters:

$$\dot{x} = f(x, \alpha), \quad x \in \mathbb{R}^3, \ \alpha \in \mathbb{R}^2. \tag{8.57}$$

Suppose that at $\alpha = 0$ the system has the equilibrium $x = 0$ with one zero eigenvalue $\lambda_1 = 0$ and a pair of purely imaginary eigenvalues $\lambda_{2,3} = \pm i\omega_0$ with $\omega_0 > 0$.

### 8.5.1 Derivation of the normal form

Expand the right-hand side of (8.57) with respect to $x$ at $x = 0$:

$$\dot{x} = a(\alpha) + A(\alpha)x + F(x, \alpha), \tag{8.58}$$

where $a(0) = 0$, $F(x, \alpha) = O(\|x\|^2)$. Since the eigenvalues $\lambda_1 = 0$ and $\lambda_{2,3} = \pm i\omega_0$ of the matrix $A(0)$ are *simple*, the matrix $A(\alpha)$ has simple eigenvalues

$$\lambda_1(\alpha) = \nu(\alpha), \ \lambda_{2,3}(\alpha) = \mu(\alpha) \pm i\omega(\alpha), \tag{8.59}$$

for all sufficiently small $\|\alpha\|$, where $\nu$, $\mu$, and $\omega$ are smooth functions of $\alpha$ such that

$$\nu(0) = \mu(0) = 0, \ \omega(0) = \omega_0 > 0.$$

Notice that these eigenvalues are the eigenvalues of the equilibrium $x = 0$ at $\alpha = 0$ but typically $a(\alpha) \neq 0$ for nearby parameter values and the matrix $A(\alpha)$ is not the Jacobian matrix of any equilibrium point of (8.58). Nevertheless, the matrix $A(\alpha)$ is well defined and has two smoothly parameter-dependent eigenvectors $q_0(\alpha) \in \mathbb{R}^3$ and $q_1(\alpha) \in \mathbb{C}^3$ corresponding to the eigenvalues $\nu(\alpha)$ and $\lambda(\alpha) = \mu(\alpha) + i\omega(\alpha)$, respectively:

$$A(\alpha)q_0(\alpha) = \nu(\alpha)q_0(\alpha), \quad A(\alpha)q_1(\alpha) = \lambda(\alpha)q_1(\alpha).$$

Moreover, the adjoint eigenvectors $p_0(\alpha) \in \mathbb{R}^3$ and $p_1(\alpha) \in \mathbb{C}^3$ can be defined by

$$A^T(\alpha)p_0(\alpha) = \nu(\alpha)p_0(\alpha), \quad A^T(\alpha)p_1(\alpha) = \bar{\lambda}(\alpha)p_1(\alpha).$$

Normalize the eigenvectors such that

$$\langle p_0, q_0 \rangle = \langle p_1, q_1 \rangle = 1,$$

for all $\|\alpha\|$ small.[6] The following orthogonality properties follow from the Fredholm Alternative Theorem:

$$\langle p_1, q_0 \rangle = \langle p_0, q_1 \rangle = 0.$$

Now any real vector $x$ can be represented as

$$x = uq_0(\alpha) + zq_1(\alpha) + \bar{z}\bar{q}_1(\alpha),$$

with

$$\begin{cases} u = \langle p_0(\alpha), x \rangle, \\ z = \langle p_1(\alpha), x \rangle. \end{cases}$$

In the coordinates $u \in \mathbb{R}^1$ and $z \in \mathbb{C}^1$ system (8.58) reads

$$\begin{cases} \dot{u} = \Gamma(\alpha) + \nu(\alpha)u + g(u, z, \bar{z}, \alpha), \\ \dot{z} = \Omega(\alpha) + \lambda(\alpha)z + h(u, z, \bar{z}, \alpha). \end{cases} \tag{8.60}$$

Here

$$\Gamma(\alpha) = \langle p_0(\alpha), a(\alpha) \rangle, \quad \Omega(\alpha) = \langle p_1(\alpha), a(\alpha) \rangle, \tag{8.61}$$

are smooth functions of $\alpha$, $\Gamma(0) = 0$, $\Omega(0) = 0$, and

$$\begin{aligned} g(u, z, \bar{z}, \alpha) &= \langle p_0(\alpha), F(uq_0(\alpha) + zq_1(\alpha) + \bar{z}\bar{q}_1(\alpha), \alpha) \rangle, \\ h(u, z, \bar{z}, \alpha) &= \langle p_1(\alpha), F(uq_0(\alpha) + zq_1(\alpha) + \bar{z}\bar{q}_1(\alpha), \alpha) \rangle, \end{aligned} \tag{8.62}$$

are smooth functions of $u, z, \bar{z}, \alpha$ whose Taylor expansions in the first three arguments start with quadratic terms:

$$g(u, z, \bar{z}, \alpha) = \sum_{j+k+l \geq 2} \frac{1}{j!k!l!} g_{jkl}(\alpha) u^j z^k \bar{z}^l,$$

and

$$h(u, z, \bar{z}, \alpha) = \sum_{j+k+l \geq 2} \frac{1}{j!k!l!} h_{jkl}(\alpha) u^j z^k \bar{z}^l.$$

Clearly, $\gamma(\alpha)$ is real, and since $g$ must be real, we have $g_{jkl}(\alpha) = \bar{g}_{jlk}(\alpha)$. Therefore, $g_{jkl}$ is real for $k = l$. Obviously, $g_{jkl}$ and $h_{jkl}$ can be computed by formal differentiation of the expressions given by (8.62) with respect to $u, z$, and $\bar{z}$.

Using the standard technique, we can simplify the linear part and eliminate *nonresonant terms* in (8.60) by a change of variables.

[6] As usual, $\langle v, w \rangle = \bar{v}_1 w_1 + \bar{v}_2 w_2 + \bar{v}_3 w_3$ for two complex vectors $v, w \in \mathbb{C}^3$.

**Lemma 8.9 (Poincaré normal form)** *Assume*

(ZH.1) $g_{200}(0) \neq 0$.

*Then there is a locally defined smooth, invertible variable transformation, smoothly depending on the parameters, that for all sufficiently small* $\|\alpha\|$ *reduces* (8.60) *into the form*

$$\left\{\begin{aligned} \dot{v} &= \gamma(\alpha) + G_{200}(\alpha)v^2 + G_{011}(\alpha)|w|^2 + G_{300}(\alpha)v^3 \\ &\quad + G_{111}(\alpha)v|w|^2 + O(\|(v,w,\bar{w})\|^4), \\ \dot{w} &= \Lambda(\alpha)w + H_{110}(\alpha)vw + H_{210}(\alpha)v^2w + H_{021}(\alpha)w|w|^2 \\ &\quad + O(\|(v,w,\bar{w})\|^4), \end{aligned}\right. \tag{8.63}$$

*where* $v \in \mathbb{R}^1$, $w \in \mathbb{C}^1$, *and* $\|(v,w,\bar{w})\|^2 = v^2 + |w|^2$. *In* (8.63), $\gamma(\alpha)$ *and* $G_{jkl}(\alpha)$ *are real-valued smooth functions, while* $\Lambda(\alpha)$ *and* $H_{jkl}(\alpha)$ *are complex-valued smooth functions. Moreover,* $\gamma(0) = 0$, $\Lambda(0) = i\omega_0$,

$$G_{200}(0) = \frac{1}{2}g_{200}(0),\ G_{011}(0) = g_{011}(0),\ H_{110}(0) = h_{110}(0), \tag{8.64}$$

*and*

$$G_{300}(0) = \frac{1}{6}g_{300}(0) - \frac{1}{\omega_0}\ \mathrm{Im}(g_{110}(0)h_{200}(0)), \tag{8.65}$$

$$G_{111}(0) = g_{111}(0) - \frac{1}{\omega_0}\left[2\ \mathrm{Im}(g_{110}(0)h_{011}(0)) + \mathrm{Im}(g_{020}(0)h_{101}(0))\right], \tag{8.66}$$

$$\begin{aligned} H_{210}(0) &= \frac{1}{2}h_{210}(0) \\ &\quad + \frac{i}{2\omega_0}\left[h_{200}(0)(h_{020}(0) - 2g_{110}(0)) - |h_{101}(0)|^2 - h_{011}(0)\bar{h}_{200}(0)\right], \end{aligned} \tag{8.67}$$

$$\begin{aligned} H_{021}(0) &= \frac{1}{2}h_{021}(0) \\ &\quad + \frac{i}{2\omega_0}\left(h_{011}(0)h_{020}(0) - \frac{1}{2}g_{020}(0)h_{101}(0) - 2|h_{011}(0)|^2 - \frac{1}{3}|h_{002}(0)|^2\right). \end{aligned} \tag{8.68}$$

**Sketch of the proof:**

Let us first prove the lemma for $\alpha = 0$, which is that for $\Gamma = \nu = \Omega = 0$. Perform a nonlinear change of variables in (8.60)

$$\left\{\begin{aligned} v &= u + \tfrac{1}{2}V_{020}z^2 + \tfrac{1}{2}V_{002}\bar{z}^2 + V_{110}uz + V_{101}u\bar{z}, \\ w &= z + \tfrac{1}{2}W_{200}u^2 + \tfrac{1}{2}W_{020}z^2 + \tfrac{1}{2}W_{002}\bar{z}^2 + W_{101}u\bar{z} \\ &\quad + W_{011}z\bar{z}, \end{aligned}\right. \tag{8.69}$$

where $V_{jlk}$ and $W_{jkl}$ are unknown coefficients to be defined later. The transformation (8.69) is invertible near $(u, z) = (0, 0)$ and reduces (8.60) into the form (8.63) up to third-order terms if we take

$$V_{020} = -\frac{g_{020}}{2i\omega_0}, \quad V_{002} = \frac{g_{002}}{2i\omega_0}, \quad V_{110} = -\frac{g_{110}}{i\omega_0}, \quad V_{101} = \frac{g_{101}}{i\omega_0},$$

and

$$W_{200} = \frac{h_{200}}{i\omega_0}, \quad W_{020} = -\frac{h_{020}}{i\omega_0}, \quad W_{101} = \frac{h_{101}}{2i\omega_0}, \quad W_{002} = \frac{h_{002}}{3i\omega_0}, \quad W_{011} = \frac{h_{011}}{i\omega_0},$$

where all the $g_{jkl}$ and $h_{jkl}$ have to be evaluated at $\alpha = 0$. These coefficients are selected exactly in order to annihilate all the quadratic terms in the resulting system except those present in (8.63). Then, one can eliminate all nonresonant order-three terms without changing the coefficients in front of the resonant ones displayed in (8.63). To verify the expressions for $G_{jkl}(0)$ and $H_{jkl}(0)$, one has to invert (8.69) up to and including *third-order* terms.[7]

To prove the lemma for $\alpha \neq 0$ with small $\|\alpha\|$, we have to perform a parameter-dependent transformation that coincides with (8.69) at $\alpha = 0$ but contains a small *affine part* for $\alpha \neq 0$ to counterbalance the appearance of "undesired" linear terms in (8.63). For example, we can take

$$\left\{\begin{array}{l} v = u + \delta_0(\alpha) + \delta_1(\alpha)u + \delta_2(\alpha)z + \delta_3(\alpha)\bar{z} \\ \qquad + \frac{1}{2}V_{020}(\alpha)z^2 + \frac{1}{2}V_{002}(\alpha)\bar{z}^2 + V_{110}(\alpha)uz + V_{101}(\alpha)u\bar{z}, \\ w = z + \Delta_0(\alpha) + \Delta_1(\alpha)u + \Delta_2(\alpha)z + \Delta_3(\alpha)\bar{z} \\ \qquad + \frac{1}{2}W_{200}(\alpha)u^2 + \frac{1}{2}W_{020}(\alpha)z^2 + \frac{1}{2}W_{002}(\alpha)\bar{z}^2 + W_{101}(\alpha)u\bar{z} \\ \qquad + W_{011}(\alpha)z\bar{z}, \end{array}\right. \tag{8.70}$$

with $\delta_k(0) = 0$ and $\Delta_k(0) = 0$. To prove that it is possible to select the parameter-dependent coefficients of (8.70) to eliminate all constant, linear, and quadratic terms except those shown in (8.63), for all small $\|\alpha\|$, one has to apply the Implicit Function Theorem using the assumptions on the eigenvalues of $A(0)$ and condition (ZH.1). We leave the details to the reader. □

Making a nonlinear time reparametrization in (8.63) and performing an extra variable transformation allows one to simplify the system further. As the following lemma shows, all but one resonant cubic term can be "removed" under certain nondegeneracy conditions.

**Lemma 8.10 (Gavrilov normal form)** *Assume that:*

(ZH.1) $G_{200}(0) \neq 0$;
(ZH.2) $G_{011}(0) \neq 0$.

[7] A way to avoid explicitly inverting (8.69) is to compare the equations for $\dot{v}$ and $\dot{w}$ expressed in terms of $(u, z)$ obtained by differentiating (8.69) and substituting $(\dot{u}, \dot{z})$ using (8.60) with those obtained by substitution of (8.69) into (8.63).

*Then, system (8.63) is locally smoothly orbitally equivalent near the origin to the system*

$$\begin{cases} \dot{u} = \delta(\alpha) + B(\alpha)u^2 + C(\alpha)|z|^2 + O(\|(u,z,\bar{z})\|^4), \\ \dot{z} = \Sigma(\alpha)z + D(\alpha)uz + E(\alpha)u^2 z + O(\|(u,z,\bar{z})\|^4), \end{cases} \quad (8.71)$$

*where* $\delta(\alpha)$, $B(\alpha)$, $C(\alpha)$, *and* $E(\alpha)$ *are smooth real-valued functions, while* $\Sigma(\alpha)$ *and* $D(\alpha)$ *are smooth complex-valued functions. Moreover,* $\delta(0) = 0$, $\Sigma(0) = \Lambda(0) = i\omega_0$, *and*

$$B(0) = G_{200}(0), \quad C(0) = G_{011}(0), \quad (8.72)$$

$$D(0) = H_{110}(0) - i\omega_0 \frac{G_{300}(0)}{G_{200}(0)}, \quad (8.73)$$

$$E(0) = \text{Re}\left[H_{210}(0) + H_{110}(0)\left(\frac{\text{Re } H_{021}(0)}{G_{011}(0)} - \frac{3G_{300}(0)}{2G_{200}(0)} + \frac{G_{111}(0)}{2G_{011}(0)}\right) - \frac{H_{021}(0)G_{200}(0)}{G_{011}(0)}\right]. \quad (8.74)$$

**Proof:**

As in Lemma 8.9, start with $\alpha = 0$. Make the following time reparametrization in (8.63):

$$dt = (1 + e_1 v + e_2|w|^2)\, d\tau, \quad (8.75)$$

with the constants $e_{1,2} \in \mathbb{R}^1$ to be determined. Simultaneously, introduce new variables, again denoted by $u$ and $z$, via

$$\begin{cases} u = v + \frac{1}{2}e_3 v^2, \\ z = w + Kvw, \end{cases} \quad (8.76)$$

where $e_3 \in \mathbb{R}^1$ and $K \in \mathbb{C}^1$ are "unknown coefficients." The reparametrization (8.75) preserves the direction of time near the origin, and the transformation (8.76) is locally invertible. In the new variables and time, system (8.63) takes the form (8.71) if we set

$$e_1 = -\frac{G_{300}(0)}{G_{200}(0)}, \quad K = -\frac{i\omega_0 e_2 + H_{021}(0)}{G_{011}(0)},$$

$$e_3 = 2 \text{ Re } K - e_1 - \frac{G_{111}(0)}{G_{011}(0)},$$

and then tune the remaining free parameter $e_2$ to annihilate the imaginary part of the coefficient of the $u^2z$-term. This is always possible since this coefficient has the form

$$\Psi - i\omega_0 \frac{G_{200}(0)}{G_{011}(0)} e_2,$$

with a purely imaginary factor in front of $e_2$. Direct calculations show that $\text{Re}\ \Psi = E(0)$, where $E(0)$ is given by (8.74) in the lemma statement.

We leave the reader to verify that a similar construction can be carried out for small $\alpha \neq 0$ with the help of the Implicit Function Theorem if one considers $e_{1,2}$ in (8.75) as functions of $\alpha$ and replaces (8.76) by

$$\begin{cases} u = v + e_4(\alpha)v + \frac{1}{2}e_3(\alpha)v^2, \\ z = w + K(\alpha)vw, \end{cases}$$

for smooth functions $e_{3,4}(\alpha)$, $e_4(0) = 0$, and $K(\alpha)$. □

Finally, by a linear scaling of the variables and time in (8.71),

$$u = \frac{B(\alpha)}{E(\alpha)}\xi,\ z = \frac{B^3(\alpha)}{C(\alpha)E^2(\alpha)}\zeta,\ t = \frac{E(\alpha)}{B^2(\alpha)}\tau,$$

we obtain

$$\begin{cases} \dot{\xi} = \beta_1(\alpha) + \xi^2 + s|\zeta|^2 + O(\|(\xi,\zeta,\bar{\zeta})\|^4), \\ \dot{\zeta} = (\beta_2(\alpha) + i\omega_1(\alpha))\zeta + (\theta(\alpha) + i\vartheta(\alpha))\xi\zeta + \xi^2\zeta + O(\|(\xi,\zeta,\bar{\zeta})\|^4), \end{cases} \tag{8.77}$$

for $s = \text{sign}[B(0)C(0)] = \pm 1$ and

$$\beta_1(\alpha) = \frac{E^2(\alpha)}{B^3(\alpha)}\delta(\alpha),$$

$$\beta_2(\alpha) = \frac{E(\alpha)}{B^2(\alpha)}\text{Re}\ \Sigma(\alpha),$$

$$\theta(\alpha) + i\vartheta(\alpha) = \frac{D(\alpha)}{B(\alpha)},$$

$$\omega_1(\alpha) = \frac{E(\alpha)}{B^2(\alpha)}\ \text{Im}\ \Sigma(\alpha).$$

Since $B(0)C(0) \neq 0$ due to (ZH.1) and (ZH.2), we have only to assume that

(ZH.3) $E(0) \neq 0$,

for the scaling to be valid. Notice that $\tau$ has the same direction as $t$ only if $E(0) > 0$. We should have this in mind when interpreting stability results. We took the liberty of introducing this possible time reverse to reduce the number of distinct cases. We can also assume that $\omega_1(\alpha) > 0$ since the transformation $\zeta \mapsto \bar{\zeta}$ changes the sign of $\omega_1$.

If we impose an additional transversality condition, namely,

(ZH.4) *the map* $\alpha \mapsto \beta$ *is regular at* $\alpha = 0$,

then $(\beta_1, \beta_2)$ can be considered as new parameters and $\theta$ as a function of $\beta$ (to save symbols). Condition (ZH.4) is equivalent to the regularity at $\alpha = 0$ of the map $\alpha \mapsto (\gamma(\alpha), \mu(\alpha))^T$ (see (8.59) and (8.61)).

We summarize the results obtained in this section by formulating a theorem.

**Theorem 8.6** *Suppose that a three-dimensional system*

$$\dot{x} = f(x, \alpha), \quad x \in \mathbb{R}^3, \; \alpha \in \mathbb{R}^2, \tag{8.78}$$

*with smooth $f$, has at $\alpha = 0$ the equilibrium $x = 0$ with eigenvalues*

$$\lambda_1(0) = 0, \; \lambda_{2,3}(0) = \pm i\omega_0, \; \omega_0 > 0.$$

*Let*

(ZH.1) $g_{200}(0) \neq 0$;
(ZH.2) $g_{011}(0) \neq 0$;
(ZH.3) $E(0) \neq 0$, *where $E(0)$ can be computed using* (8.64)–(8.68) *and* (8.74);
(ZH.4) *the map $\alpha \mapsto (\gamma(\alpha), \mu(\alpha))^T$ is regular at $\alpha = 0$.*

*Then, by introducing a complex variable, making smooth and smoothly parameter-dependent transformations, reparametrizing time (reversing it if $E(0) < 0$), and introducing new parameters, we can bring system* (8.78) *into the form*

$$\begin{cases} \dot{\xi} = \beta_1 + \xi^2 + s|\zeta|^2 + O(\|(\xi, \zeta, \bar{\zeta})\|^4), \\ \dot{\zeta} = (\beta_2 + i\omega_1)\zeta + (\theta + i\vartheta)\xi\zeta + \xi^2\zeta + O(\|(\xi, \zeta, \bar{\zeta})\|^4), \end{cases} \tag{8.79}$$

*where $\xi \in \mathbb{R}^1$ and $\zeta \in \mathbb{C}^1$ are new variables; $\beta_1$ and $\beta_2$ are new parameters; $\theta = \theta(\beta)$, $\vartheta = \vartheta(\beta)$, $\omega_1 = \omega_1(\beta)$ are smooth real-valued functions; $\omega_1(0) \neq 0$; and*

$$s = \text{sign}[g_{200}(0) g_{011}(0)] = \pm 1,$$

$$\theta(0) = \frac{\text{Re } h_{110}(0)}{g_{200}(0)}. \; \square$$

Only $s$ and $\theta(0)$ are important in what follows. Assume that

(ZH.5) $\theta(0) \neq 0$.

**Remark:**
It is a matter of taste which cubic term to keep in the normal form. An alternative to (8.79) used by Gavrilov is the following normal form due to Guckenheimer:

$$\begin{cases} \dot{\xi} = \beta_1 + \xi^2 + s|\zeta|^2 + O(\|(\xi, \zeta, \bar{\zeta})\|^4), \\ \dot{\zeta} = (\beta_2 + i\omega_1)\zeta + (\theta + i\vartheta)\xi\zeta + \zeta|\zeta|^2 + O(\|(\xi, \zeta, \bar{\zeta})\|^4), \end{cases} \tag{8.80}$$

with the $\zeta|\zeta|^2$-term kept in the second equation instead of the $\xi^2\zeta$-term present in (8.79). Of course, the alternative choice leads to equivalent bifurcation diagrams. $\diamondsuit$

### 8.5.2 Bifurcation diagram of the truncated normal form

In coordinates $(\xi, \rho, \varphi)$ with $\zeta = \rho e^{i\varphi}$, system (8.77) *without* $O(\| \cdot \|^4)$-terms can be written as

$$\begin{cases} \dot{\xi} = \beta_1 + \xi^2 + s\rho^2, \\ \dot{\rho} = \rho(\beta_2 + \theta\xi + \xi^2), \\ \dot{\varphi} = \omega_1 + \vartheta\xi, \end{cases} \tag{8.81}$$

the first two equations of which are *independent* of the third one. The equation for $\varphi$ describes a rotation around the $\xi$-axis with almost constant angular velocity $\dot{\varphi} \approx \omega_1$, for $|\xi|$ small. Thus, to understand the bifurcations in (8.81), we need to study only the planar system for $(\xi, \rho)$ with $\rho \geq 0$:

$$\begin{cases} \dot{\xi} = \beta_1 + \xi^2 + s\rho^2, \\ \dot{\rho} = \rho(\beta_2 + \theta\xi + \xi^2). \end{cases} \tag{8.82}$$

This system is often called a (*truncated*) *amplitude system.* If considered in the whole $(\xi, \rho)$-plane, system (8.82) is $\mathbb{Z}_2$-symmetric since the reflection $\rho \mapsto -\rho$ leaves it invariant. The bifurcation diagrams of (8.82) corresponding to different possible cases are depicted in **Figs. 8.13**, **8.14**, and **8.16**, **8.17**. In all these cases, system (8.82) can have between zero and three equilibria in a small neighborhood of the origin for $\|\beta\|$ small. Two equilibria with $\rho = 0$ exist for $\beta_1 < 0$ and are given by

$$E_{1,2} = (\xi_{1,2}^{(0)}, 0) = \left(\mp\sqrt{-\beta_1}, 0\right).$$

These equilibria appear via a generic fold bifurcation on the line

$$S = \{(\beta_1, \beta_2) : \ \beta_1 = 0\}.$$

The bifurcation line $S$ has two branches, $S_+$ and $S_-$, separated by the point $\beta = 0$ and corresponding to $\beta_2 > 0$ and $\beta_2 < 0$, respectively (see **Fig. 8.13**, for example). Crossing the branch $S_+$ gives rise to an unstable node and a saddle, while passing through $S_-$ implies a stable node and a saddle.

The equilibria $E_{1,2}$ can bifurcate further; namely, a nontrivial equilibrium with $\rho > 0$,

$$E_3 = (\xi_3^{(0)}, \rho_3^{(0)}) = \left(-\frac{\beta_2}{\theta} + o(\beta_2), \sqrt{-\frac{1}{s}\left(\beta_1 + \frac{\beta_2^2}{\theta^2} + o(\beta_2^2)\right)}\right),$$

can branch from either $E_1$ or $E_2$ (here we use the assumption (ZH.5) for the first time). Clearly, (8.82) might have another nontrivial equilibrium with

$$\xi = -\theta + \cdots, \quad \rho^2 = -s\theta^2 + \cdots,$$

where dots represent terms that vanish as $\beta \to 0$. We do not worry about this equilibrium since it is located outside any sufficiently small neighborhood of the origin in the phase plane and does not interact with any of our $E_k$, $k = 1, 2, 3$. The nontrivial equilibrium $E_3$ appears at the bifurcation curve

$$H = \left\{ (\beta_1, \beta_2) : \ \beta_1 = -\frac{\beta_2^2}{\theta^2} + o(\beta_2^2) \right\}.$$

If $s\theta > 0$, the appearing equilibrium $E_3$ is a saddle, while it is a node for $s\theta < 0$. The node is stable if it exists for $\theta\beta_2 > 0$ and unstable if the opposite inequality holds.[8]

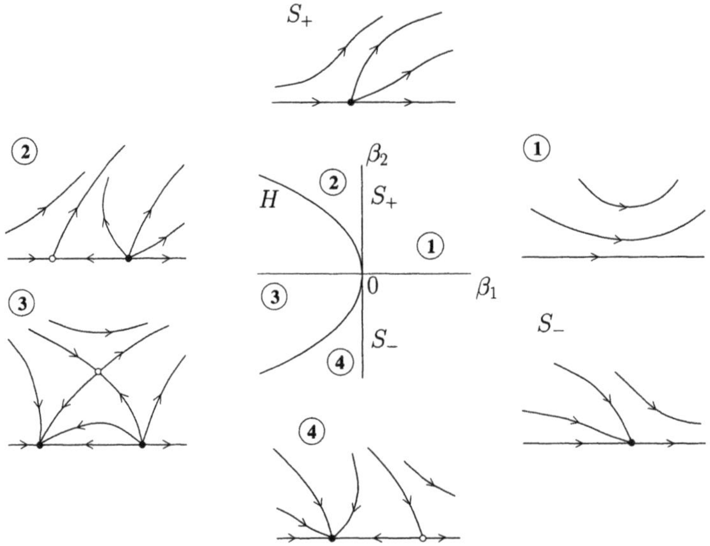


**Fig. 8.13.** Bifurcation diagram of the amplitude system (8.81) ($s = 1$, $\theta > 0$).

If $s\theta > 0$, the nontrivial equilibrium $E_3$ does not bifurcate, and the bifurcation diagrams of (8.82) are those presented in **Fig. 8.13** (for $s = 1$, $\theta > 0$) and **Fig. 8.14** (for $s = -1$, $\theta < 0$).

**Remark:**

For $s = 1$, there is a subtle difference between the cases $\theta(0) > 1$ and $0 < \theta(0) \leq 1$ that appears only at the critical parameter value $\beta = 0$ (see **Fig. 8.15**). ♢

If $s\theta < 0$, the equilibrium $E_3$ has two purely imaginary eigenvalues for parameter values belonging to the line

$$T = \{(\beta_1, \beta_2) : \beta_2 = 0, \ \theta\beta_1 > 0\}.$$

---

[8] A careful reader will have recognized that a *pitchfork bifurcation*, as studied in Chapter 7, takes place at $H$ due to the $\mathbb{Z}_2$-symmetry of (8.82).

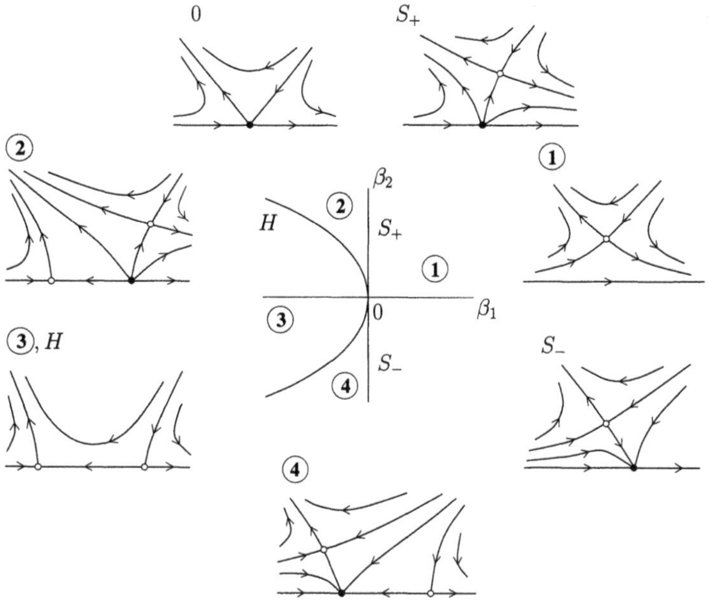


**Fig. 8.14.** Bifurcation diagram of the amplitude system (8.81) ($s = -1,\ \theta < 0$).

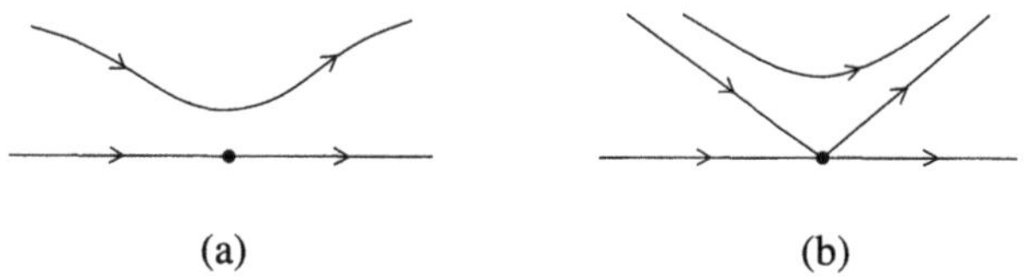


**Fig. 8.15.** Critical phase portraits for $s = 1$: (a) $0 < \theta(0) \leq 1$; (b) $\theta(0) > 1$.

We can check that the corresponding first Lyapunov coefficient is given by

$$l_1 = -\frac{C_+}{\theta\sqrt{\theta\beta_1}}$$

for some constant $C_+ > 0$. Thus, the Lyapunov coefficient $l_1$ is nonzero along $T$ for sufficiently small $\|\beta\| > 0$. Therefore, a nondegenerate Hopf bifurcation takes place if we cross the line $T$ in a neighborhood of $\beta = 0$, and a unique limit cycle exists for nearby parameter values. Its stability depends on the sign of $l_1$.

It can be proved that (8.82) can have at most *one* limit cycle in a sufficiently small neighborhood of the origin in the $(\xi, \rho)$-plane for small $\|\beta\|$. The proof is difficult and is omitted here.[9] The fate of this limit cycle is rather different depending on whether $s = 1$ or $s = -1$ (see **Figs. 8.16** and **8.17**).

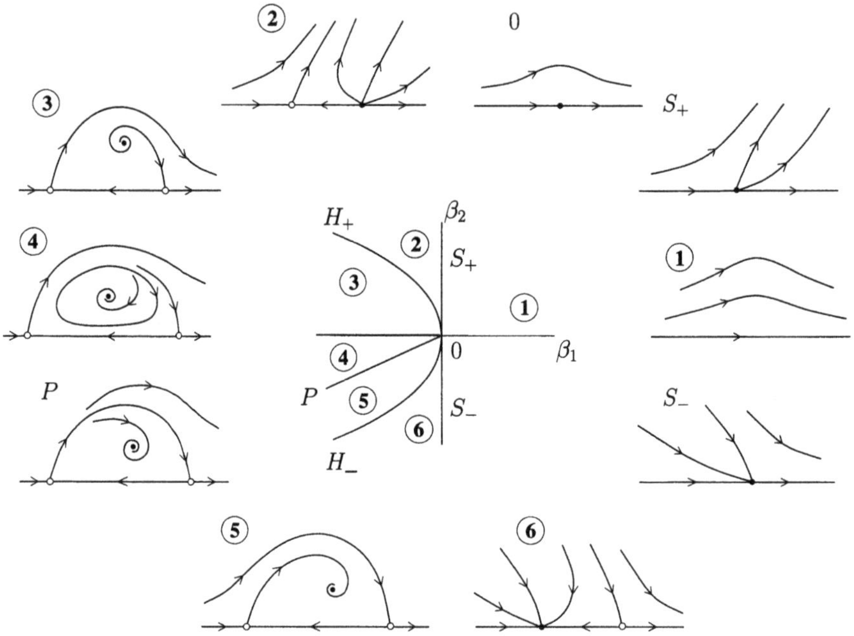


**Fig. 8.16.** Bifurcation diagram of the amplitude system (8.81) ($s = 1$, $\theta < 0$).

If $s = 1$ (and $\theta < 0$), the limit cycle is *unstable* and coexists with the two trivial equilibria $E_{1,2}$, which are saddles. Under parameter variation, the cycle can approach a *heteroclinic cycle* formed by the separatrices of these saddles: Its period tends to infinity and the cycle disappears. Notice that due to the symmetry the $\xi$-axis is always invariant, so one orbit that connects the saddles $E_{1,2}$ is always present. The second connection appears along a curve originating at $\beta = 0$ and having the representation

$$P = \left\{ (\beta_1, \beta_2) : \ \beta_2 = \frac{\theta}{3\theta - 2}\beta_1 + o(\beta_1), \ \beta_1 < 0 \right\}$$

(see Exercise 14(b)). The resulting bifurcation diagram is presented in **Fig. 8.16**.

If $s = -1$ (and $\theta > 0$), a *stable* limit cycle appears through the Hopf bifurcation when there are no equilibria with $\rho = 0$. When we move clockwise

[9] It is based on a singular rescaling and Pontryagin's techniques of perturbation of Hamiltonian systems (as in the Bogdanov-Takens case but more involved, see Appendix A and Exercise 14(b)).

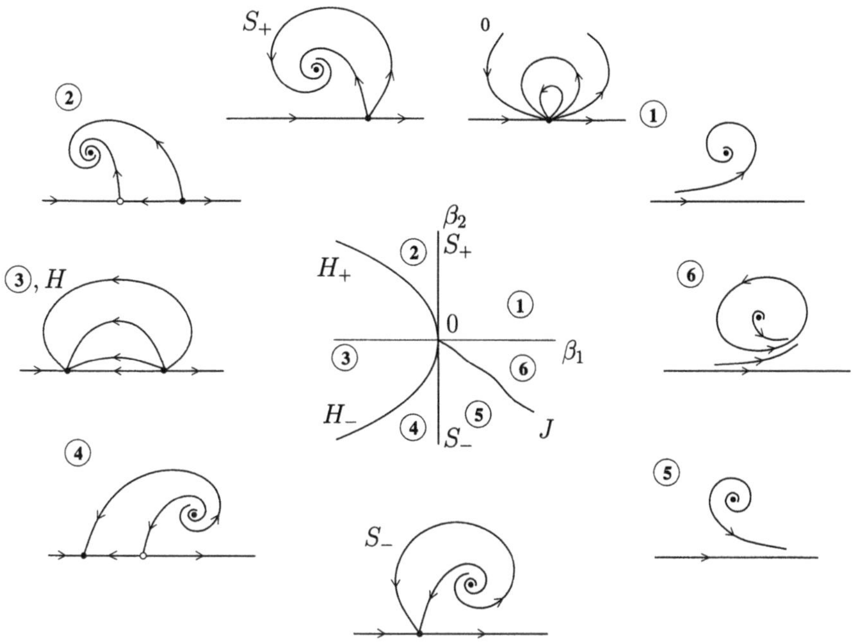


**Fig. 8.17.** Bifurcation diagram of the amplitude system (8.81) ($s = -1$, $\theta > 0$).

around the origin in the parameter plane, the cycle must disappear somehow before entering region **3**, where no cycle can exist since no nontrivial equilibrium is left. A little thinking reveals that the cycle cannot "die" via a homoclinic or heteroclinic bifurcation. To understand what happens with the cycle born via the Hopf bifurcation, fix a small neighborhood $U_0$ of the origin in the phase plane. Then, as we move clockwise around $\beta = 0$ on the parameter plane, the cycle "grows" and touches the boundary of $U_0$. Afterward, the cycle becomes *invisible* for anyone looking only at the interior of $U_0$. We cannot get rid of this phenomenon, called *cycle blow-up*, by decreasing the neighborhood $U_0$. We also cannot make the neighborhood very big since all of the previous analysis is valid only in a sufficiently small neighborhood of the origin. Thus, for $s = -1$, there is a "bifurcation" curve $J$, originating at $\beta = 0$ and depending on the region in which we consider the system (8.82), at which the cycle reaches the boundary of this region (see **Fig. 8.17**). The strangest feature of this phenomenon is that the cycle appears through the Hopf bifurcation and then approaches the boundary of any small but *fixed* region if we make a roundtrip along an *arbitrary small* circle on the parameter plane centered at the origin $\beta = 0$ (the curve $J$ terminates at the origin). It means that the diameter of the cycle increases under parameter variation arbitrarily fast when the radius of the circle shrinks. That is why the cycle is said to exhibit a blow-up.

Now we can use the obtained bifurcation diagrams for (8.82) to reconstruct bifurcations in the three-dimensional *truncated* normal form (8.81) by "suspension" of the rotation in $\varphi$. The equilibria $E_{1,2}$ with $\rho = 0$ in (8.82) correspond to *equilibrium points* of (8.81). Thus, the curve $S$ is a fold bifurcation curve for (8.81) at which two equilibria appear, a node and a saddle-focus. The nontrivial equilibrium $E_3$ in (8.82) corresponds to a *limit cycle* in (8.81)

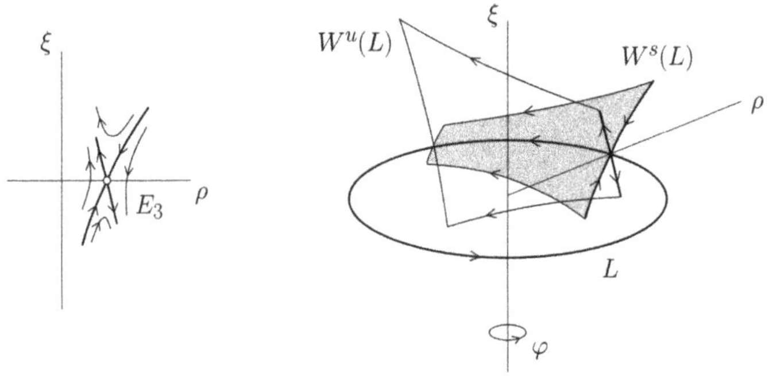


**Fig. 8.18.** A nontrivial equilibrium corresponds to a cycle.

of the same stability as $E_3$ (see **Fig. 8.18**). The pitchfork curve $H$, at which a small cycle bifurcates from an equilibrium, clearly corresponds to a Hopf bifurcation in (8.81). One could naturally expect the presence of these two local bifurcation curves near the fold-Hopf bifurcation. The limit cycle in (8.82) corresponds to an *invariant torus* in (8.81) (see **Fig. 8.19**). Therefore, the

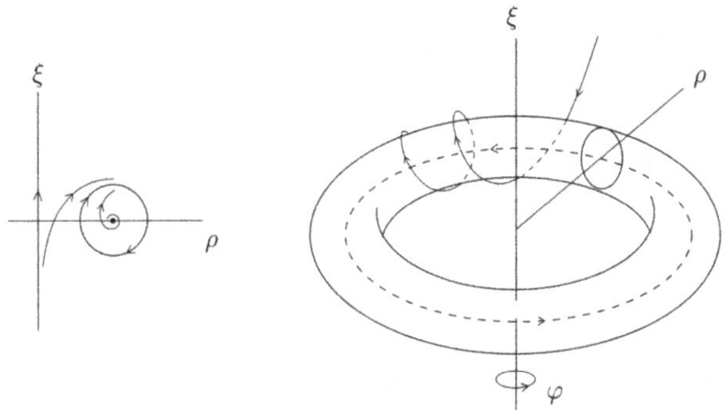


**Fig. 8.19.** A cycle corresponds to a torus.

Hopf bifurcation curve $T$ describes the Neimark-Sacker bifurcation of the cycle, at which it loses stability and a stable torus appears "around" it. This torus then either approaches a heteroclinic set composed of a spherelike surface and the $\xi$-axis (see **Fig. 8.20**) or reaches the boundary of the considered region and "disappears" by blow-up.

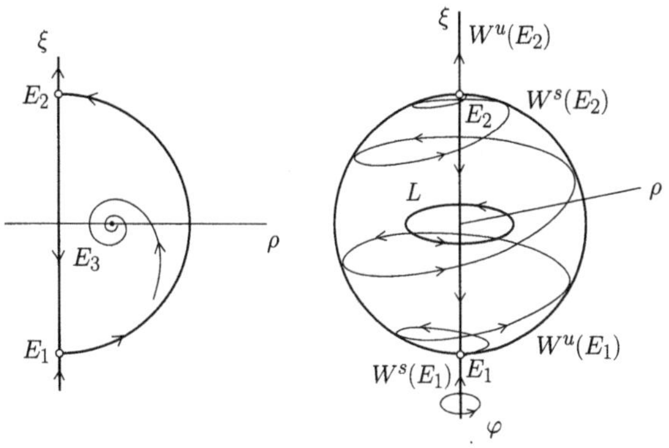


**Fig. 8.20.** A heteroclinic orbit corresponds to a "sphere".

### 8.5.3 Effect of higher-order terms

Recall that the previous results concern the *truncated* normal form (8.81). As we shall see, some of these results "survive" if we consider the whole system (8.77), while others do not. In the majority of the cases, (8.81) is *not* a normal form for (8.77).

Let us start with positive information. Writing system (8.77) in the same coordinates $(\xi, \rho, \varphi)$ as (8.81), we get

$$\begin{cases} \dot{\xi} = \beta_1 + \xi^2 + s\rho^2 + \Theta_\beta(\xi, \rho, \varphi), \\ \dot{\rho} = \rho(\beta_2 + \theta\xi + \xi^2) + \Psi_\beta(\xi, \rho, \varphi), \\ \dot{\varphi} = \omega_1 + \vartheta\xi + \Phi_\beta(\xi, \rho, \varphi), \end{cases} \tag{8.83}$$

where $\Theta_\beta, \Psi_\beta = O((\xi^2+\rho^2)^2)$, and $\Phi_\beta = O(\xi^2+\rho^2)$ are smooth functions that are $2\pi$-periodic in $\varphi$. Using the Implicit Function Theorem, one can show that, for sufficiently small $\|\beta\|$, system (8.83) exhibits the same *local* bifurcations in a small neighborhood of the origin in the phase space as (8.81). More precisely, it has at most two equilibria, which appear via the fold bifurcation on a curve that is close to $S$. The equilibria undergo the Hopf bifurcation at a curve close to $H$, thus giving rise to a unique limit cycle. If $s\theta < 0$, this cycle loses stability and generates a torus via the Neimark-Sacker bifurcation taking place at some curve close to the curve $T$. The nondegeneracy conditions for these bifurcations can be verified rather simply. Actually, leading-order terms of the Taylor expansions for functions representing these bifurcation curves in (8.83) coincide with those for (8.81). Therefore, we can say that, generically, "the interaction between fold and Hopf bifurcations leads to tori."

If $s = 1$ and $\theta > 0$, one can establish more – namely that the accounting for higher-order terms does not qualitatively change the whole bifurcation diagram of (8.81).

**Lemma 8.11** *If $s = 1$ and $\theta > 0$, then system* (8.83) *is locally topologically equivalent near the origin to system* (8.81). □

In this case, we have only the fold and the Hopf bifurcation curves on the parameter plane. One of the equilibria in the $\xi$-axis is always a node. The cycle born through the Hopf bifurcation is of the saddle type. No tori are possible.

Moreover, in this case we do not need to consider the cubic terms at all. Taking into account only the quadratic terms is sufficient.

**Lemma 8.12** *If $s = 1$ and $\theta > 0$, then system* (8.83) *is locally topologically equivalent near the origin to the system*

$$\begin{cases} \dot{\xi} = \beta_1 + \xi^2 + \rho^2, \\ \dot{\rho} = \beta_2 \rho + \theta \xi \rho, \\ \dot{\varphi} = \omega_1 + \vartheta \xi. \ \square \end{cases} \tag{8.84}$$

Moreover, the bifurcation diagram remains equivalent if we take $\omega_1 = 1$ and substitute the functions $\theta(\beta)$ and $\vartheta(\beta)$ by constant values $\theta = \theta(0)$ and $\vartheta = 0$. Therefore, the following theorem can be formulated.

**Theorem 8.7 (Simple fold-Hopf bifurcation)** *Suppose that a system*

$$\dot{x} = f(x, \alpha), \quad x \in \mathbb{R}^3, \ \alpha \in \mathbb{R}^2,$$

*with smooth $f$, has at $\alpha = 0$ the equilibrium $x = 0$ with eigenvalues*

$$\lambda_1(0) = 0, \ \lambda_{2,3} = \pm i\omega_0, \ \omega_0 > 0.$$

*Let the following genericity conditions hold:*

(ZH0.1) $g_{200}(0) g_{011}(0) > 0$;

(ZH0.2)

$$\theta_0 = \frac{\text{Re } h_{110}(0)}{g_{200}(0)} > 0;$$

(ZH0.3) *the map $\alpha \mapsto (\gamma(\alpha), \mu(\alpha))^T$ is regular at $\alpha = 0$.*

*Then, the system is locally topologically equivalent near the origin to the system*

$$\begin{cases} \dot{\rho} = \beta_2 \rho + \theta_0 \zeta \rho, \\ \dot{\varphi} = 1, \\ \dot{\zeta} = \beta_1 + \zeta^2 + \rho^2, \end{cases}$$

*where $(\rho, \varphi, \zeta)$ are cylindrical polar coordinates.* □

In all other cases, adding generic higher-order terms results in topologically nonequivalent bifurcation diagrams. The reason for this is that phase portraits of system (8.81) have some *degenerate* features that disappear under "perturbation" by generic higher-order terms.

Let us first explain why even another "simple" case $s = -1$, $\theta < 0$, is sensitive to adding higher-order terms. In system (8.82) the $\xi$-axis is *invariant* since $\rho = 0$ implies $\dot{\rho} = 0$. Then, in region **3** in **Fig. 8.14**, system (8.82) has two saddle-focus points with $\rho = 0$, one of them with two-dimensional stable and one-dimensional unstable manifolds, while the other has one-dimensional stable and two-dimensional unstable invariant manifolds. The invariant axis *connects* these saddle-foci: We have a *heteroclinic orbit* for all the parameter values in region **3**.

On the contrary, the term $\Psi_\beta(\xi, \rho, \varphi) = O((\xi^2 + \rho^2)^2)$ in (8.83) does not necessarily vanish for $\rho = 0$. Then, the $\xi$-axis is no longer always invariant, and the heteroclinic connection normally disappears.[10] Thus, generically, the bifurcation diagrams of (8.81) and (8.83) are not equivalent.

**Remark:**

The situation with the invariance of the $\xi$-axis is more delicate than one might conclude from the above explanation. By a suitable change of variables, one can make the terms of the Taylor expansion of $\Psi_\beta$ at $(\xi, \rho) = (0,0)$ proportional to $\rho$ up to an *arbitrary high* order, retaining the invariance of the axis in the truncated form. The "tail," however, is still not proportional to $\rho$ in the generic situation. Such properties are called *flat*, because we can decompose the function $\Psi$ by

$$\Psi_\beta(\xi, \rho, \varphi) = \rho^4 \Psi_\beta^{(0)}(\xi, \rho, \varphi) + \Psi_\beta^{(1)}(\xi, \rho, \varphi),$$

with all the partial derivatives of $\Psi^{(1)}$ with respect to $\rho$ equal to zero at $\rho = 0$ (*flat function* of $\rho$). However, generically, $\Psi_\beta^{(1)}(\xi, 0, \varphi) \neq 0$. $\Diamond$

If $s\theta < 0$, a torus is present in the truncated normal form (8.81) and the situation becomes much more complex. The torus created by the Neimark-Sacker bifurcation exists in (8.83) only for parameter values *near* the corresponding bifurcation curve. If we move away from the curve, the torus loses its smoothness and is destroyed. The complete sequence of events is unknown and is likely to involve an infinite number of bifurcations since any weak resonance point on the Neimark-Sacker curve is the root of an Arnold phase-locking tongue (see Chapter 7).

More detailed information is available for the case $s = 1$, $\theta < 0$. Recall that in this case the truncated normal form (8.81) has the curve $P$ at which there exists a spherelike surface formed by the coincidence of the two-dimensional invariant manifolds of the saddle-foci. This is an extremely degenerate structure

[10] Since we need to tune two parameters to restore the connection, there might be only isolated points on the $(\beta_1, \beta_2)$-plane where this connection is still present in (8.83).

that disappears when generic higher-order terms are added. Instead, these invariant manifolds intersect transversally, forming a *heteroclinic structure* (see

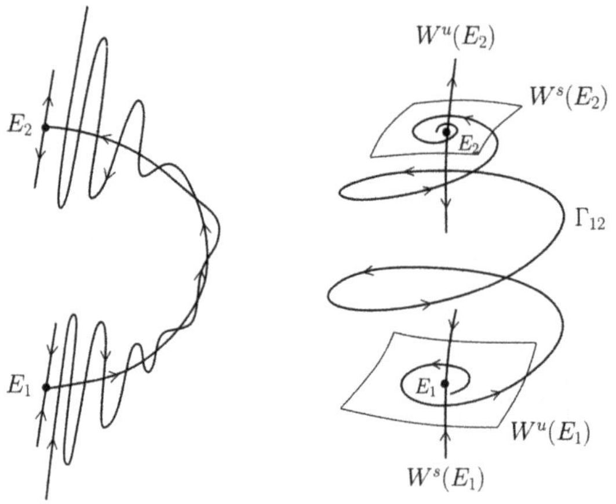


**Fig. 8.21.** (a) A cross-section of the intersecting stable and unstable manifolds; (b) a heteroclinic orbit connecting the saddle-foci.

**Fig. 8.21**, where a cross-section of this structure is sketched together with a heteroclinic orbit $\Gamma_{12}$). Therefore, the torus cannot approach the "sphere," since it simply does not exist, and thus must disappear before. It is also clear from continuity arguments that the region of existence of the transversal heteroclinic structure should be bounded by some curves corresponding to a *tangency* of the invariant manifolds along a heteroclinic orbit connecting the saddle-focus equilibria.

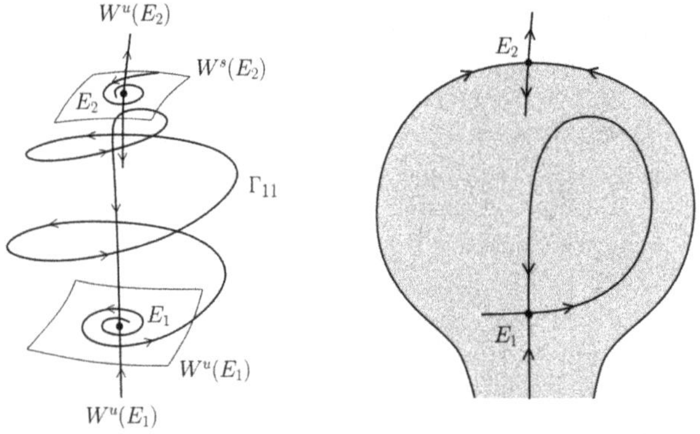


(a) (b)

**Fig. 8.22.** (a) An orbit homoclinic to the saddle $E_1$; (b) boundary of attraction.

Finally, let us point out that *homoclinic* orbits to a saddle-focus are also possible and have actually been proved to be present in the generic case. Such an orbit $\Gamma_{11}$ can begin by spiraling along an unstable two-dimensional manifold of one of the saddle-foci, pass near the second one, and return along the stable one-dimensional invariant manifold back to the first saddle-focus (see **Fig. 8.22**(a)). A homoclinic orbit to the opposite saddle-focus is also possible. Actually, there are two curves that intersect each other infinitely many times, emanating from the origin of the parameter plane, which correspond to these two homoclinic bifurcations. One can check that if a homoclinic orbit exists *and*

$$-2 < \theta < 0,$$

then the corresponding saddle quantity $\sigma_0$ (see Chapter 6) satisfies the Shil'nikov "chaotic condition" implying the presence of Smale horseshoes. Moreover, one of these homoclinic orbits is located inside an attracting region (see **Fig. 8.22**(b)) bounded by the two-dimensional stable manifold of the second saddle-focus.[11] Therefore, a stable "strange" dynamics exists near the fold-Hopf bifurcation in this case.

In summary, if $s = 1$, $\theta < 0$, system (8.83) may have, in addition to local bifurcation curves, a bifurcation set corresponding to global bifurcations (heteroclinic tangencies, homoclinic orbits) and bifurcations of long-periodic limit cycles (folds and period-doubling cascades), which is located near the heteroclinic cycle curve $P$ of the truncated normal form (8.81).

**Remarks:**

(1) Actually, the truncated planar normal form (8.82) has a more fundamental meaning if considered in the class of systems on the $(\xi, \rho)$-plane, which are *invariant* under the two-dimensional representation of the group $\mathbb{Z}_2$: $(\xi, \rho) \mapsto (\xi, -\rho)$. A perturbation of (8.82) by higher-order terms that leaves it in this class of symmetric systems can be written (cf. Chapter 7) as

$$\begin{cases} \dot{\xi} = \beta_1 + \xi^2 + s\rho^2 + \tilde{\Theta}_\beta(\xi, \rho^2), \\ \dot{\rho} = \rho(\beta_2 + \theta\xi + \xi^2) + \rho^4 \tilde{\Psi}_\beta(\xi, \rho^2), \end{cases} \tag{8.85}$$

with $\Theta_\beta$, $\rho^4\Psi_\beta(\xi, \rho^2) = O((\xi^2 + \rho^2)^2)$. System (8.85) always has the invariant axis $\rho = 0$. It has been proved that (8.85) with $s\theta \neq 0$ is locally topologically equivalent to (8.82). Moreover, the homeomorphism identifying the phase portraits can be selected to commute with the transformation $(\xi, \rho) \mapsto (\xi, -\rho)$ for all parameter values.

Therefore, system (8.82) is a topological normal form for a generic $\mathbb{Z}_2$-symmetric planar system with the invariant axis $x_2 = 0$,

$$\begin{cases} \dot{x}_1 = G(x_1, x_2^2, \alpha), \\ \dot{x}_2 = x_2 H(x_1, x_2^2, \alpha), \end{cases}$$

[11] Recall that all the objects we speak about will actually be attracting in the original system only if $E(0) > 0$.

having at $\alpha = 0$ the equilibrium $x = 0$ with *two zero* eigenvalues.

(2) A careful reader might ask why the quadratic terms are not enough for the analysis of the truncated system (8.82) if $s\theta < 0$ and why we have also to keep the cubic term. The reason is that in the system

$$\begin{cases} \dot{\xi} = \beta_1 + \xi^2 + s\rho^2, \\ \dot{\rho} = \rho(\beta_2 + \theta\xi), \end{cases}$$

with $s\theta < 0$, the Hopf bifurcation is *degenerate*. Moreover, if $s = 1,\ \theta < 0$, this bifurcation occurs simultaneously with the creation of the heteroclinic cycle formed by the separatrices of the saddles $E_{1,2}$. More precisely, the system is

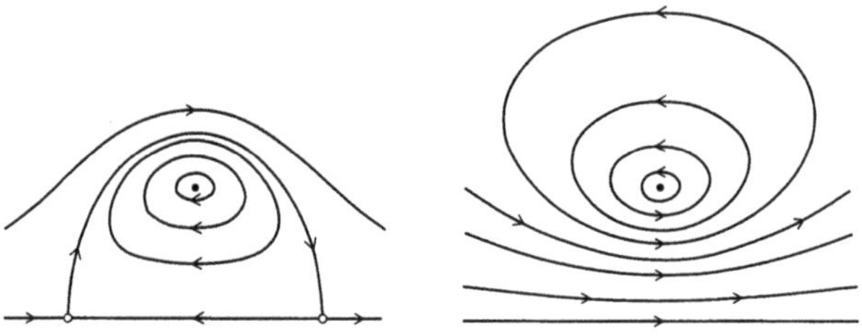

(a) (b)

**Fig. 8.23.** (a) $s = 1, \theta < 0$; (b) $s = -1, \theta > 0$.

*integrable* along the "Hopf line" $T$, and for corresponding parameter values the nontrivial equilibrium is a *center* surrounded by a family of closed orbits. This family is bounded by the heteroclinic cycle for $s = 1,\ \theta < 0$ (see **Fig. 8.23**(a)), but remains unbounded for $s = -1,\ \theta > 0$ (**Fig. 8.23**(b)). Actually, the system is *orbitally equivalent* to a Hamiltonian system for $\rho > 0$. Therefore, system (8.82) can be considered near the line $T$ as a "perturbation" of a Hamiltonian system by the (only relevant) cubic term. This term "stabilizes" the bifurcation diagram by making the Hopf bifurcation nondegenerate and splits the heteroclinic curve off the vertical axis. These properties allow one to prove the uniqueness of the limit cycle in the system and derive an asymptotic formula for the heteroclinic curve $P$. The interested reader is directed to Appendix A and Exercise 14(b). $\diamondsuit$

The multidimensional case of the fold-Hopf bifurcation reduces to the considered one by the Center Manifold Theorem. Notice that we need only the *linear* approximation of the center manifold to distinguish between the "simple" and "difficult" cases. However, the second-order approximation is required if we want to find $E(0)$. We return to the computation of the coefficients of (8.63) in the $n$-dimensional case in Section 8.7.

## 8.6 Hopf-Hopf bifurcation

Consider a four-dimensional smooth system depending on two parameters:

$$\dot{x} = f(x, \alpha), \quad x \in \mathbb{R}^4, \ \alpha \in \mathbb{R}^2. \tag{8.86}$$

Let (8.86) have at $\alpha = 0$ the equilibrium $x = 0$ with two distinct pairs of purely imaginary eigenvalues:

$$\lambda_{1,4} = \pm i\omega_1, \ \lambda_{2,3} = \pm i\omega_2,$$

with $\omega_1 > \omega_2 > 0$. Since there is no zero eigenvalue, system (8.86) has an equilibrium point close to $x = 0$ for all $\alpha$ with $\|\alpha\|$ small. Suppose that a parameter-dependent shift of the coordinates that places the origin at this equilibrium point has been performed, so that we can assume without loss of generality that $x = 0$ is the equilibrium of (8.86) for all small $\|\alpha\|$ : $f(0, \alpha) \equiv 0$.

### 8.6.1 Derivation of the normal form

Write system (8.86) in the form

$$\dot{x} = A(\alpha)x + F(x, \alpha), \tag{8.87}$$

where $F(x, \alpha) = O(\|x\|^2)$ is a smooth function. The matrix $A(\alpha)$ has two pairs of simple complex-conjugate eigenvalues

$$\lambda_{1,4}(\alpha) = \mu_1(\alpha) \pm i\omega_1(\alpha), \ \lambda_{2,3}(\alpha) = \mu_2(\alpha) \pm i\omega_2(\alpha),$$

for all sufficiently small $\|\alpha\|$, where $\mu_{1,2}$ and $\omega_{1,2}$ are smooth functions of $\alpha$ and

$$\mu_1(0) = \mu_2(0) = 0, \ \omega_1(0) > \omega_2(0) > 0.$$

Since the eigenvalues are simple, there are two complex eigenvectors, $q_{1,2}(\alpha) \in \mathbb{C}^4$, corresponding to the eigenvalues $\lambda_{1,2}(\alpha) = \mu_{1,2}(\alpha) + i\omega_{1,2}(\alpha)$:

$$A(\alpha)q_1(\alpha) = \lambda_1(\alpha)q_1(\alpha), \ A(\alpha)q_2(\alpha) = \lambda_2(\alpha)q_2(\alpha).$$

As usual, introduce the adjoint eigenvectors $p_{1,2}(\alpha) \in \mathbb{C}^4$ by

$$A^T(\alpha)p_1(\alpha) = \bar{\lambda}_1(\alpha)p_1(\alpha), \ A^T(\alpha)p_2(\alpha) = \bar{\lambda}_2(\alpha)p_2(\alpha),$$

where $T$ denotes transposition. These eigenvectors can be selected to depend smoothly on $\alpha$, normalized using the standard scalar product in $\mathbb{C}^4$,

$$\langle p_1, q_1 \rangle = \langle p_2, q_2 \rangle = 1,$$

and made to satisfy the orthogonality conditions

$$\langle p_2, q_1 \rangle = \langle p_1, q_2 \rangle = 0,$$

thus forming a *biorthogonal set* of vectors in $\mathbb{C}^4$. Note that we always use the scalar product $\langle v, w \rangle = \sum_{k=1}^{4} \bar{v}_k w_k$, which is linear in its *second* argument. Any real vector $x \in \mathbb{R}^4$ can be represented for each small $\|\alpha\|$ as

$$x = z_1 q_1 + \bar{z}_1 \bar{q}_1 + z_2 q_2 + \bar{z}_2 \bar{q}_2,$$

where

$$z_1 = \langle p_1, x \rangle, \quad z_2 = \langle p_2, x \rangle,$$

are new complex coordinates, $z_{1,2} \in \mathbb{C}^1$ (cf. Section 3.5 in Chapter 3). In these coordinates, system (8.87) takes the form

$$\begin{cases} \dot{z}_1 = \lambda_1(\alpha) z_1 + g(z_1, \bar{z}_1, z_2, \bar{z}_2, \alpha), \\ \dot{z}_2 = \lambda_2(\alpha) z_2 + h(z_1, \bar{z}_1, z_2, \bar{z}_2, \alpha), \end{cases} \tag{8.88}$$

where

$$\begin{aligned} g(z_1, \bar{z}_1, z_2, \bar{z}_2, \alpha) &= \langle p_1, F(z_1 q_1 + \bar{z}_1 \bar{q}_1 + z_2 q_2 + \bar{z}_2 \bar{q}_2, \alpha) \rangle, \\ h(z_1, \bar{z}_1, z_2, \bar{z}_2, \alpha) &= \langle p_2, F(z_1 q_1 + \bar{z}_1 \bar{q}_1 + z_2 q_2 + \bar{z}_2 \bar{q}_2, \alpha) \rangle \end{aligned}$$

(for simplicity, the dependence of $p_k, q_l$ on the parameters is not indicated). The functions $g$ and $h$ are complex-valued smooth functions of their arguments and have formal Taylor expansions with respect to the first four arguments that start with quadratic terms:

$$g(z_1, \bar{z}_1, z_2, \bar{z}_2, \alpha) = \sum_{j+k+l+m \geq 2} g_{jklm}(\alpha) z_1^k \bar{z}_1^l z_2^l \bar{z}_2^m,$$

and

$$h(z_1, \bar{z}_1, z_2, \bar{z}_2, \alpha) = \sum_{j+k+l+m \geq 2} h_{jklm}(\alpha) z_1^k \bar{z}_1^l z_2^l \bar{z}_2^m.$$

**Lemma 8.13 (Poincaré normal form)** *Assume that*

$$\text{(HH.0)} \qquad k\omega_1(0) \neq l\omega_2(0), \;\; k, l > 0, \;\; k + l \leq 5.$$

*Then, there exists a locally defined, smooth and smoothly parameter-dependent, invertible transformation of the complex variables that reduces* (8.88) *for all sufficiently small* $\|\alpha\|$ *into the form*

$$\begin{cases} \dot{w}_1 = \lambda_1(\alpha) w_1 + G_{2100}(\alpha) w_1 |w_1|^2 + G_{1011}(\alpha) w_1 |w_2|^2 \\ \qquad + G_{3200}(\alpha) w_1 |w_1|^4 + G_{2111}(\alpha) w_1 |w_1|^2 |w_2|^2 + G_{1022}(\alpha) w_1 |w_2|^4 \\ \qquad + O(\|(w_1, \bar{w}_1, w_2, \bar{w}_2)\|^6), \\ \dot{w}_2 = \lambda_2(\alpha) w_2 + H_{1110}(\alpha) w_2 |w_1|^2 + H_{0021}(\alpha) w_2 |w_2|^2 \\ \qquad + H_{2210}(\alpha) w_2 |w_1|^4 + H_{1121}(\alpha) w_2 |w_1|^2 |w_2|^2 + H_{0032}(\alpha) w_2 |w_2|^4 \\ \qquad + O(\|(w_1, \bar{w}_1, w_2, \bar{w}_2)\|^6), \end{cases} \tag{8.89}$$

*where $w_{1,2} \in \mathbb{C}^1$ and $\|(w_1, \bar{w}_1, w_2, \bar{w}_2)\|^2 = |w_1|^2 + |w_2|^2$. The complex-valued functions $G_{jklm}(\alpha)$ and $H_{jklm}(\alpha)$ are smooth; moreover,*

$$\begin{aligned} G_{2100}(0) = {} & g_{2100} + \frac{i}{\omega_1} g_{1100} g_{2000} + \frac{i}{\omega_2}(g_{1010} h_{1100} - g_{1001} \bar{h}_{1100}) \\ & - \frac{i}{2\omega_1 + \omega_2} g_{0101} \bar{h}_{0200} - \frac{i}{2\omega_1 - \omega_2} g_{0110} h_{2000} \\ & - \frac{i}{\omega_1} |g_{1100}|^2 - \frac{2i}{3\omega_1} |g_{0200}|^2, \end{aligned} \tag{8.90}$$

$$\begin{aligned} G_{1011}(0) = {} & g_{1011} + \frac{i}{\omega_2}(g_{1010} h_{0011} - g_{1001} \bar{h}_{0011}) \\ & + \frac{i}{\omega_1}(2 g_{2000} g_{0011} - g_{1100} \bar{g}_{0011} - g_{0011} \bar{h}_{0110} - g_{0011} h_{1010}) \\ & - \frac{2i}{\omega_1 + 2\omega_2} g_{0002} \bar{h}_{0101} - \frac{2i}{\omega_1 - 2\omega_2} g_{0020} h_{1001} \\ & - \frac{i}{2\omega_1 - \omega_2} |g_{0110}|^2 - \frac{i}{2\omega_1 + \omega_2} |g_{0101}|^2, \end{aligned} \tag{8.91}$$

$$\begin{aligned} H_{1110}(0) = {} & h_{1110} + \frac{i}{\omega_1}(g_{1100} h_{1010} - \bar{g}_{1100} h_{0110}) \\ & + \frac{i}{\omega_2}(2 h_{0020} h_{1100} - h_{0011} \bar{h}_{1100} - g_{1010} h_{1100} - \bar{g}_{1001} h_{1100}) \\ & + \frac{2i}{2\omega_1 - \omega_2} g_{0110} h_{2000} - \frac{2i}{2\omega_1 + \omega_2} \bar{g}_{0101} h_{0200} \\ & - \frac{i}{2\omega_2 - \omega_1} |h_{1001}|^2 - \frac{i}{\omega_1 + 2\omega_2} |h_{0101}|^2, \end{aligned} \tag{8.92}$$

$$\begin{aligned} H_{0021}(0) = {} & h_{0021} + \frac{i}{\omega_1}(g_{0011} h_{1010} - \bar{g}_{0011} h_{0110}) + \frac{i}{\omega_2} h_{0011} h_{0020} \\ & - \frac{i}{2\omega_2 - \omega_1} g_{0020} h_{1001} - \frac{i}{2\omega_2 + \omega_1} \bar{g}_{0002} h_{0101} \\ & - \frac{i}{\omega_2} |h_{0011}|^2 - \frac{2i}{3\omega_2} |h_{0002}|^2, \end{aligned} \tag{8.93}$$

*where all the $g_{jklm}$ and $h_{jklm}$ have to be evaluated at $\alpha = 0$.* □

Notice that the expressions in the last line of each of the preceding formulas are purely imaginary. The lemma can be proved by the standard normalization technique. The hint to Exercise 15 in this chapter explains how to perform the necessary calculations using one of the computer algebra systems. We do not give here the explicit formulas for the coefficients of the fifth-order resonant terms due to their length (see references in Appendix B). As we shall see, the given formulas are enough to distinguish between "simple" and "difficult" Hopf-Hopf bifurcations.

By the introduction of a new time and a variable transformation in which cubic "resonant terms" are involved, one can simplify normal form (8.89) further.

**Lemma 8.14** *Assume that:*

(HH.1) Re $G_{2100}(0) \neq 0$;
(HH.2) Re $G_{1011}(0) \neq 0$;
(HH.3) Re $H_{1110}(0) \neq 0$;
(HH.4) Re $H_{0021}(0) \neq 0$.

*Then, the system*

$$\begin{cases} \dot{w}_1 = \lambda_1(\alpha)w_1 + G_{2100}(\alpha)w_1|w_1|^2 + G_{1011}(\alpha)w_1|w_2|^2 \\ \qquad + G_{3200}(\alpha)w_1|w_1|^4 + G_{2111}(\alpha)w_1|w_1|^2|w_2|^2 + G_{1022}(\alpha)w_1|w_2|^4 \\ \qquad + O(\|(w_1, \bar{w}_1, w_2, \bar{w}_2)\|^6), \\ \dot{w}_2 = \lambda_2(\alpha)w_2 + H_{1110}(\alpha)w_2|w_1|^2 + H_{0021}(\alpha)w_2|w_2|^2 \\ \qquad + H_{2210}(\alpha)w_2|w_1|^4 + H_{1121}(\alpha)w_2|w_1|^2|w_2|^2 + H_{0032}(\alpha)w_2|w_2|^4 \\ \qquad + O(\|(w_1, \bar{w}_1, w_2, \bar{w}_2)\|^6), \end{cases} \tag{8.94}$$

*is locally smoothly orbitally equivalent near the origin to the system*

$$\begin{cases} \dot{v}_1 = \lambda_1(\alpha)v_1 + P_{11}(\alpha)v_1|v_1|^2 + P_{12}(\alpha)v_1|v_2|^2 \\ \qquad + iR_1(\alpha)v_1|v_1|^4 + S_1(\alpha)v_1|v_2|^4 \\ \qquad + O(\|(v_1, \bar{v}_1, v_2, \bar{v}_2)\|^6), \\ \dot{v}_2 = \lambda_2(\alpha)v_2 + P_{21}(\alpha)v_2|v_1|^2 + P_{22}(\alpha)v_2|v_2|^2 \\ \qquad + S_2(\alpha)v_2|v_1|^4 + iR_2(\alpha)v_2|v_2|^4 \\ \qquad + O(\|(v_1, \bar{v}_1, v_2, \bar{v}_2)\|^6), \end{cases} \tag{8.95}$$

*where $v_{1,2} \in \mathbb{C}^1$ are new complex variables, $P_{jk}(\alpha)$ and $S_k(\alpha)$ are complex-valued smooth functions, and $R_k(\alpha)$ are real-valued smooth functions.*

**Proof:**

Introduce a new time $\tau$ in (8.94) via

$$dt = (1 + e_1|w_1|^2 + e_2|w_2|^2)\, d\tau,$$

where the real functions $e_1 = e_1(\alpha)$ and $e_2 = e_2(\alpha)$ will be defined later. The resulting system has the same resonant terms as (8.94) but with modified coefficients.

Then, perform a smooth invertible transformation involving "resonant" cubic terms:

$$\begin{cases} v_1 = w_1 + K_1 w_1|w_1|^2, \\ v_2 = w_2 + K_2 w_2|w_2|^2, \end{cases} \tag{8.96}$$

where $K_j = K_j(\alpha)$ are complex-valued functions to be determined. In the new variables $(v_1, v_2)$ the system takes the form

$$\begin{cases} \dot{v}_1 = \lambda_1 v_1 + \sum_{j+k+l+m\geq 3} \hat{G}_{jklm} v_1^k \bar{v}_1^l v_2^l \bar{v}_2^m, \\ \dot{v}_2 = \lambda_2 v_2 + \sum_{j+k+l+m\geq 3} \hat{H}_{jklm} v_1^k \bar{v}_1^l v_2^l \bar{v}_2^m, \end{cases} \tag{8.97}$$

where the dot now means a derivative with respect to $\tau$, and

$$\hat{G}_{2100} = G_{2100} + \lambda_1 e_1 + (\lambda_1 + \bar{\lambda}_1)K_1, \quad \hat{G}_{1011} = G_{1011} + \lambda_1 e_2,$$

$$\hat{H}_{1110} = H_{1110} + \lambda_2 e_1, \quad \hat{H}_{0021} = H_{0021} + \lambda_2 e_2 + (\lambda_2 + \bar{\lambda}_2)K_2,$$

and

$$\begin{aligned} \hat{G}_{3200} &= G_{3200} + G_{2100}e_1 + (K_1\bar{\lambda}_1 - \overline{K}_1\lambda_1)e_1 \\ &\quad + K_1\overline{G}_{2100} - \overline{K}_1 G_{2100} - 2(\lambda_1 + \bar{\lambda}_1)K_1^2 - |K_1|^2\bar{\lambda}_1, && (8.98) \\ \hat{G}_{2111} &= G_{2111} + G_{1011}e_1 + G_{2100}e_2 + (\lambda_1 + \bar{\lambda}_1)e_2 K_1 \\ &\quad + 2K_1 \operatorname{Re} G_{1011}, && (8.99) \\ \hat{G}_{1022} &= G_{1022} + G_{1011}e_2 - 2\lambda_1 e_2 \operatorname{Re} K_2 - 2G_{1011} \operatorname{Re} K_2, \\ \hat{H}_{2210} &= H_{2210} + H_{1110}e_1 - 2\lambda_2 e_1 \operatorname{Re} K_1 - 2H_{1110} \operatorname{Re} K_1, \\ \hat{H}_{1121} &= H_{1121} + H_{0021}e_1 + H_{1110}e_2 + (\lambda_2 + \bar{\lambda}_2)e_1 K_2 \\ &\quad + 2K_2 \operatorname{Re} H_{1110}, && (8.100) \\ \hat{H}_{0032} &= H_{0032} + H_{0021}e_2 + (K_2\bar{\lambda}_2 - \overline{K}_2\lambda_2)e_2 \\ &\quad + K_2\overline{H}_{0021} - \overline{K}_2 H_{0021} - 2(\lambda_2 + \bar{\lambda}_2)K_2^2 - |K_2|^2\bar{\lambda}_2. && (8.101) \end{aligned}$$

Notice that the transformation (8.96) *brings in* fourth-order terms, which were absent in the Poincaré normal form. Elimination of these terms alters the fifth-order terms in (8.97). However, due to the particular form of (8.96), the annihilation of the fourth-order terms that appear in (8.97) does not alter the coefficients of the fifth-order *resonant* terms given above (check!). Then, these coefficients will not be changed by the elimination of the nonresonant fifth-order terms. So, assume that such eliminations have already been made from (8.97) so that it contains only the resonant terms up to fifth order.

Taking into account (8.99) and (8.100), we can then make $\hat{G}_{2111} = 0$ and $\hat{H}_{1121} = 0$ by setting

$$K_1 = -\frac{G_{2111} + G_{1011}e_1 + G_{2100}e_2}{(\lambda_1 + \bar{\lambda}_1)e_2 + 2 \operatorname{Re} G_{1011}}$$

and

$$K_2 = -\frac{H_{1121} + H_{0021}e_1 + H_{1110}e_2}{(\lambda_2 + \bar{\lambda}_2)e_1 + 2 \operatorname{Re} H_{1110}}.$$

Recall that $K_{1,2}$ are functions of $\alpha$. This setting is valid for all sufficiently small $\|\alpha\|$, due to assumptions (HH.2) and (HH.3).

We still have two free coefficients, namely $e_1$ and $e_2$. We fix them by requiring

$$\begin{cases} \operatorname{Re} \hat{G}_{3200} = 0, \\ \operatorname{Re} \hat{H}_{0032} = 0, \end{cases} \tag{8.102}$$

for all sufficiently small $\|\alpha\|$. We claim that there are smooth functions $e_{1,2}(\alpha)$ satisfying (8.102) for all sufficiently small $\|\alpha\|$. To see this, note that for $\alpha = 0$ we have $2 \text{ Re } \lambda_{1,2} = \lambda_{1,2} + \bar{\lambda}_{1,2} = 0$, and the system (8.102) reduces to a linear system for $(e_1(0), e_2(0))$ according to (8.98) and (8.101), namely:

$$\begin{cases} \text{Re } G_{2100}(0)e_1(0) = -\text{Re } G_{3200}(0), \\ \text{Re } H_{0021}(0)e_2(0) = -\text{Re } H_{0032}(0), \end{cases} \tag{8.103}$$

with a nonzero determinant

$$\begin{vmatrix} G_{2100}(0) & 0 \\ 0 & H_{0021}(0) \end{vmatrix} = G_{2100}(0)H_{0021}(0) \neq 0,$$

due to (HH.1) and (HH.4). System (8.103) obviously has the unique solution

$$e_1(0) = -\frac{\text{Re } G_{3200}(0)}{\text{Re } G_{2100}(0)}, \quad e_2(0) = -\frac{\text{Re } H_{0032}(0)}{\text{Re } H_{0021}(0)}.$$

Therefore, by the Implicit Function Theorem, (8.102) has a unique solution $(e_1(\alpha), e_2(\alpha))$ for all sufficiently small $\|\alpha\|$ with $e_{1,2}(\alpha)$ depending smoothly on $\alpha$.

Thus, $\hat{G}_{1022} = 0$, $\hat{H}_{2210} = 0$, Re $\hat{G}_{3200} = 0$, and Re $\hat{H}_{0032} = 0$ for all small $\alpha$, so system (8.97) has the form (8.95) with $P_{11} = \hat{G}_{2100}$, $P_{12} = \hat{G}_{1011}$, $P_{21} = \hat{H}_{1110}$, $P_{22} = \hat{H}_{0021}$, $S_1 = \hat{G}_{1022}$, $S_2 = \hat{H}_{2210}$, $R_1 = \text{Im } \hat{G}_{3200}$, and $R_2 = \text{Im } \hat{H}_{0032}$. We easily check that

$$\text{Re } P_{11}(0) = \text{Re } G_{2100}(0), \quad \text{Re } P_{12}(0) = \text{Re } G_{1011}(0), \tag{8.104}$$

$$\text{Re } P_{21}(0) = \text{Re } H_{1110}(0), \quad \text{Re } P_{22}(0) = \text{Re } H_{0021}(0), \tag{8.105}$$

and

$$\begin{aligned} \text{Re } S_1(0) = \text{Re } G_{1022}(0) & \\ + \text{Re } G_{1011}(0) & \left[ \frac{\text{Re } H_{1121}(0)}{\text{Re } H_{1110}(0)} - 2\frac{\text{Re } H_{0032}(0)}{\text{Re } H_{0021}(0)} \right. \\ & \left. - \frac{\text{Re } G_{3200}(0) \text{ Re } H_{0021}(0)}{\text{Re } G_{2100}(0) \text{ Re } H_{1110}(0)} \right], \end{aligned} \tag{8.106}$$

$$\begin{aligned} \text{Re } S_2(0) = \text{Re } H_{2210}(0) & \\ + \text{Re } H_{1110}(0) & \left[ \frac{\text{Re } G_{2111}(0)}{\text{Re } G_{1011}(0)} - 2\frac{\text{Re } G_{3200}(0)}{\text{Re } G_{2100}(0)} \right. \\ & \left. - \frac{\text{Re } G_{2100}(0) \text{ Re } H_{0032}(0)}{\text{Re } G_{1011}(0) \text{ Re } H_{0021}(0)} \right]. \end{aligned} \tag{8.107}$$

This proves the lemma. □

Let

$$v_1 = r_1 e^{i\varphi_1}, \quad v_2 = r_2 e^{i\varphi_2}.$$

In polar coordinates $(r_1, r_2, \varphi_1, \varphi_2)$, system (8.95) can be written as

$$\begin{cases} \dot{r}_1 = r_1 \left(\mu_1(\alpha) + p_{11}(\alpha) r_1^2 + p_{12}(\alpha) r_2^2 + s_1(\alpha) r_2^4\right) \\ \qquad + \Phi_1(r_1, r_2, \varphi_1, \varphi_2, \alpha), \\ \dot{r}_2 = r_2 \left(\mu_2(\alpha) + p_{21}(\alpha) r_1^2 + p_{22}(\alpha) r_2^2 + s_2(\alpha) r_1^4\right) \\ \qquad + \Phi_2(r_1, r_2, \varphi_1, \varphi_2, \alpha), \\ \dot{\varphi}_1 = \omega_1(\alpha) + \Psi_1(r_1, r_2, \varphi_1, \varphi_2, \alpha), \\ \dot{\varphi}_2 = \omega_2(\alpha) + \Psi_2(r_1, r_2, \varphi_1, \varphi_2, \alpha). \end{cases}$$

Here

$$p_{jk} = \text{Re } P_{jk}, \quad s_j = \text{Re } S_j, \quad j, k = 1, 2,$$

are smooth functions of $\alpha$; the real functions $\Phi_k$ and $\Psi_k$ are smooth functions of their arguments and are $2\pi$-periodic in $\varphi_j$, $\Phi_k = O((r_1^2 + r_2^2)^3)$, $\Psi_k(0, 0, \varphi_1, \varphi_2) = 0$, $k = 1, 2$.

If the map $(\alpha_1, \alpha_2) \mapsto (\mu_1(\alpha), \mu_2(\alpha))$ is regular at $\alpha = 0$, that is,

$$\det \left( \frac{\partial \mu}{\partial \alpha} \right) \bigg|_{\alpha=0} \neq 0,$$

one can use $(\mu_1, \mu_2)$ to parametrize a small neighborhood of the origin of the parameter plane and consider $\omega_k, p_{jk}, s_k, \Phi_k$, and $\Psi_k$ as functions of $\mu$.

We conclude this section by formulating the following theorem.

**Theorem 8.8** *Consider a smooth system*

$$\dot{x} = f(x, \alpha), \quad x \in \mathbb{R}^4, \; \alpha \in \mathbb{R}^2,$$

*which has, for $\alpha = 0$, the equilibrium $x = 0$ with eigenvalues*

$$\lambda_k(\alpha) = \mu_k(\alpha) \pm i\omega_k(\alpha), \quad k = 1, 2,$$

*such that*

$$\mu_1(0) = \mu_2(0) = 0, \quad \omega_{1,2}(0) > 0.$$

*Let the following nondegeneracy conditions be satisfied:*

(HH.0) $k\omega_1(0) \neq l\omega_2(0), \; k, l > 0, \; k + l \leq 5$;
(HH.1) $p_{11}(0) = \text{Re } G_{2100}(0) \neq 0$;
(HH.2) $p_{12}(0) = \text{Re } G_{1011}(0) \neq 0$;
(HH.3) $p_{21}(0) = \text{Re } H_{1110}(0) \neq 0$;
(HH.4) $p_{22}(0) = \text{Re } H_{0021}(0) \neq 0$;

*where $G_{2100}(0), G_{1011}(0), H_{1110}(0)$, and $H_{0021}(0)$ are given by (8.90)–(8.93), and*

(HH.5) *the map $\alpha \mapsto \mu(\alpha)$ is regular at $\alpha = 0$.*

*Then, the system is locally smoothly orbitally equivalent near the origin to the system*

$$\begin{cases} \dot{r}_1 = r_1(\mu_1 + p_{11}(\mu)r_1^2 + p_{12}(\mu)r_2^2 + s_1(\mu)r_2^4) + \Phi_1(r_1, r_2, \varphi_1, \varphi_2, \mu), \\ \dot{r}_2 = r_2(\mu_2 + p_{21}(\mu)r_1^2 + p_{22}(\mu)r_2^2 + s_2(\mu)r_1^4) + \Phi_2(r_1, r_2, \varphi_1, \varphi_2, \mu), \\ \dot{\varphi}_1 = \omega_1(\mu) + \Psi_1(r_1, r_2, \varphi_1, \varphi_2, \mu), \\ \dot{\varphi}_2 = \omega_2(\mu) + \Psi_2(r_1, r_2, \varphi_1, \varphi_2, \mu), \end{cases} \tag{8.108}$$

*where* $\Phi_k = O((r_1^2 + r_2^2)^3)$ *and* $\Psi_k = o(1)$ *are* $2\pi$*-periodic in* $\varphi_k$, *and the coefficients* $p_{jk}(0)$, *and* $s_k(0)$, $j, k = 1, 2$, *can be computed using the formulas* (8.104)–(8.107), *provided that the resonant coefficients* $G_{jklm}(0)$ *and* $H_{jklm}(0)$ *are known for* $j + k + l + m = 3$ *and* 5. □

### 8.6.2 Bifurcation diagram of the truncated normal form

We now truncate higher-order terms in (8.108) and consider the system

$$\begin{cases} \dot{r}_1 = r_1(\mu_1 + p_{11}r_1^2 + p_{12}r_2^2 + s_1r_2^4), \\ \dot{r}_2 = r_2(\mu_2 + p_{21}r_1^2 + p_{22}r_2^2 + s_2r_1^4), \\ \dot{\varphi}_1 = \omega_1, \\ \dot{\varphi}_2 = \omega_2, \end{cases} \tag{8.109}$$

where, for simplicity, the dependence of $p_{jk}, s_k$, and $\omega_k$ on $\mu$ is not indicated. The first pair of equations in (8.109) is *independent* of the second pair. The last two equations describe rotations in the planes $r_2 = 0$ and $r_1 = 0$ with angular velocities $\omega_1$ and $\omega_2$, respectively. Therefore, the bifurcation diagram of (8.109) is determined by that of the planar system

$$\begin{cases} \dot{r}_1 = r_1(\mu_1 + p_{11}r_1^2 + p_{12}r_2^2 + s_1r_2^4), \\ \dot{r}_2 = r_2(\mu_2 + p_{21}r_1^2 + p_{22}r_2^2 + s_2r_1^4). \end{cases} \tag{8.110}$$

This system is often called a (*truncated*) *amplitude system.* It is enough to study it only for $r_1 \geq 0,\ r_2 \geq 0$. Formally, the system is invariant under the transformations $r_1 \mapsto -r_1$ and $r_2 \mapsto -r_2$. Notice that an equilibrium $E_0$ with $r_1 = r_2 = 0$ of (8.110) corresponds to the *equilibrium point* at the origin of the four-dimensional system (8.109). Possible equilibria in the invariant coordinate axes of (8.110) correspond to *cycles* of (8.109) with $r_1 = 0$ or $r_2 = 0$, while a nontrivial equilibrium with $r_{1,2} > 0$ of (8.110) generates a *two-dimensional torus* of (8.109). Finally, if a limit cycle is present in the amplitude system (8.110), then (8.109) has a *three-dimensional torus.* The stability of all these invariant sets in (8.109) is clearly detectable from that of the corresponding objects in (8.110).

The study of the amplitude system simplifies if we use squares $\rho_{1,2}$ of the amplitudes:

$$\rho_k = r_k^2,\ k = 1, 2.$$

The equations for $\rho_{1,2}$ read

$$\begin{cases} \dot{\rho}_1 = 2\rho_1(\mu_1 + p_{11}\rho_1 + p_{12}\rho_2 + s_1\rho_2^2), \\ \dot{\rho}_2 = 2\rho_2(\mu_2 + p_{21}\rho_1 + p_{22}\rho_2 + s_2\rho_1^2), \end{cases} \tag{8.111}$$

and are also referred to as the *amplitude equations.* Now the polynomials in the right-hand sides are of order three.

There are two essentially different types of bifurcation diagrams of (8.111), depending on whether $p_{11}$ and $p_{22}$ have the same or opposite signs. For each of these cases, we also have different subcases.

### "Simple" case: $p_{11}p_{22} > 0$

Consider the case

$$p_{11} < 0, \ p_{22} < 0.$$

The case where $p_{11}$ and $p_{22}$ are positive can be reduced to this one by reversing time. Introducing new phase variables and rescaling time in (8.111) according to

$$\xi_1 = -p_{11}\rho_1, \quad \xi_2 = -p_{22}\rho_2, \quad \tau = 2t,$$

yield

$$\begin{cases} \dot{\xi}_1 = \xi_1(\mu_1 - \xi_1 - \theta\xi_2 + \Theta\xi_2^2), \\ \dot{\xi}_2 = \xi_2(\mu_2 - \delta\xi_1 - \xi_2 + \Delta\xi_1^2), \end{cases} \tag{8.112}$$

where

$$\theta = \frac{p_{12}}{p_{22}}, \quad \delta = \frac{p_{21}}{p_{11}}, \quad \Theta = \frac{s_1}{p_{22}^2}, \quad \Delta = \frac{s_2}{p_{11}^2}. \tag{8.113}$$

The scaling is nonsingular since $p_{11}p_{22} \neq 0$ by (HH.1) and (HH.4). Only the values $\theta(0), \delta(0), \Theta(0)$, and $\Delta(0)$ matter in what follows. Notice that $\theta \neq 0$ and $\delta \neq 0$, due to (HH.2) and (HH.3), respectively.

System (8.112) has an equilibrium $E_0 = (0,0)$ for all $\mu_{1,2}$. Two trivial equilibria

$$E_1 = (\mu_1, 0), \ E_2 = (0, \mu_2),$$

bifurcate from the origin at the bifurcation lines

$$H_1 = \{(\mu_1, \mu_2) : \ \mu_1 = 0\}$$

and

$$H_2 = \{(\mu_1, \mu_2) : \ \mu_2 = 0\},$$

respectively. There may also exist a nontrivial equilibrium in a small phase-space neighborhood of the origin for sufficiently small $\|\mu\|$, namely

$$E_3 = \left( -\frac{\mu_1 - \theta\mu_2}{\theta\delta - 1} + O(\|\mu\|^2), \frac{\delta\mu_1 - \mu_2}{\theta\delta - 1} + O(\|\mu\|^2) \right).$$

For this expression to be valid, we need to assume that $\theta\delta - 1 \neq 0$, which is equivalent to the condition

(HH.6) $$\det \begin{pmatrix} p_{11}(0) & p_{12}(0) \\ p_{21}(0) & p_{22}(0) \end{pmatrix} \neq 0.$$

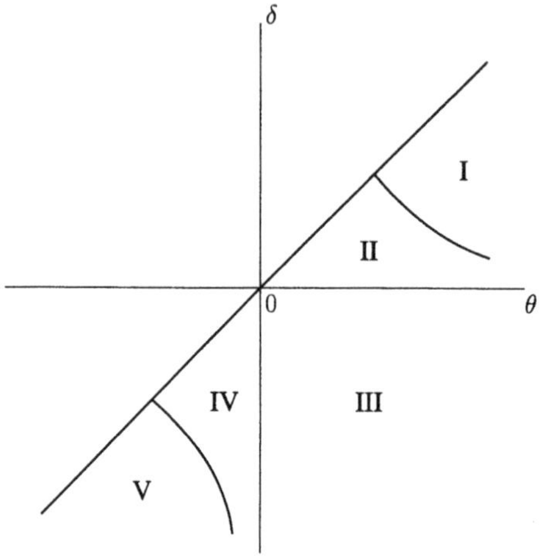


**Fig. 8.24.** Five subregions on the $(\theta, \delta)$-plane in the "simple" case.

The nontrivial equilibrium $E_3$ collides with a trivial one and disappears from the positive quadrant on the bifurcation curves

$$T_1 = \left\{(\mu_1, \mu_2) : \ \mu_1 = \theta\mu_2 + O(\mu_2^2), \ \mu_2 > 0\right\}$$

and

$$T_2 = \left\{(\mu_1, \mu_2) : \ \mu_2 = \delta\mu_1 + O(\mu_1^2), \ \mu_1 > 0\right\}.$$

These are the only bifurcations that the nontrivial equilibrium $E_3$ can exhibit in the "simple" case. Moreover, one can easily check that in this case the planar system (8.112) can have no periodic orbits.

We can assume without loss of generality that

$$\theta \geq \delta$$

(otherwise, reverse time and exchange the subscripts in (8.112)). Under all these assumptions, there are *five* topologically different bifurcation diagrams of (8.112), corresponding to the following cases:

I. $\theta > 0, \ \delta > 0, \ \theta\delta > 1$;
II. $\theta > 0, \ \delta > 0, \ \theta\delta < 1$;
III. $\theta > 0, \ \delta < 0$;
IV. $\theta < 0, \ \delta < 0, \ \theta\delta < 1$;
V. $\theta < 0, \ \delta < 0, \ \theta\delta > 1$.

Each case specifies a region in the $(\theta, \delta)$-half-plane $\theta \geq \delta$ (see **Fig. 8.24**). The $(\mu_1, \mu_2)$-parametric portraits corresponding to regions I–V are shown in **Fig. 8.25**, while the only possible *fifteen* generic phase portraits occupy **Fig. 8.26**. Notice that phase portraits **11**, **12**, **13**, **14**, and **15** can be obtained from those in regions **2**, **3**, **6**, **8**, and **9** by the reflection $(\xi_1, \xi_2) \mapsto (\xi_2, \xi_1)$.

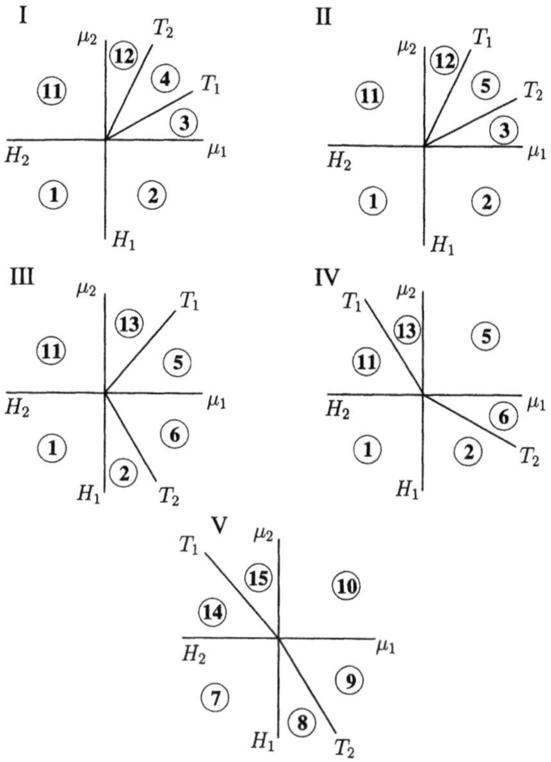


**Fig. 8.25.** Parametric portraits of (8.112) (the "simple" case).

Actually, in all these cases, the topology of the bifurcation diagram is *independent* of the cubic terms, so we can merely set $\Theta = \Delta = 0$ in (8.112).

**Lemma 8.15** *System* (8.112) *is locally topologically equivalent near the origin to the system*

$$\begin{cases} \dot{\xi}_1 = \xi_1(\mu_1 - \xi_1 - \theta\xi_2), \\ \dot{\xi}_2 = \xi_2(\mu_2 - \delta\xi_1 - \xi_2). \quad \square \end{cases} \tag{8.114}$$

### "Difficult" case: $p_{11}p_{22} < 0$

Similarly to the previous case, assume that the conditions (HH.1)–(HH.4), and (HH.6) hold, and consider only

$$p_{11} > 0, \ p_{22} < 0.$$

Introducing new phase variables and rescaling time in (8.111) by setting

$$\xi_1 = p_{11}\rho_1, \quad \xi_2 = -p_{22}\rho_2, \quad \tau = 2t,$$

we obtain the system

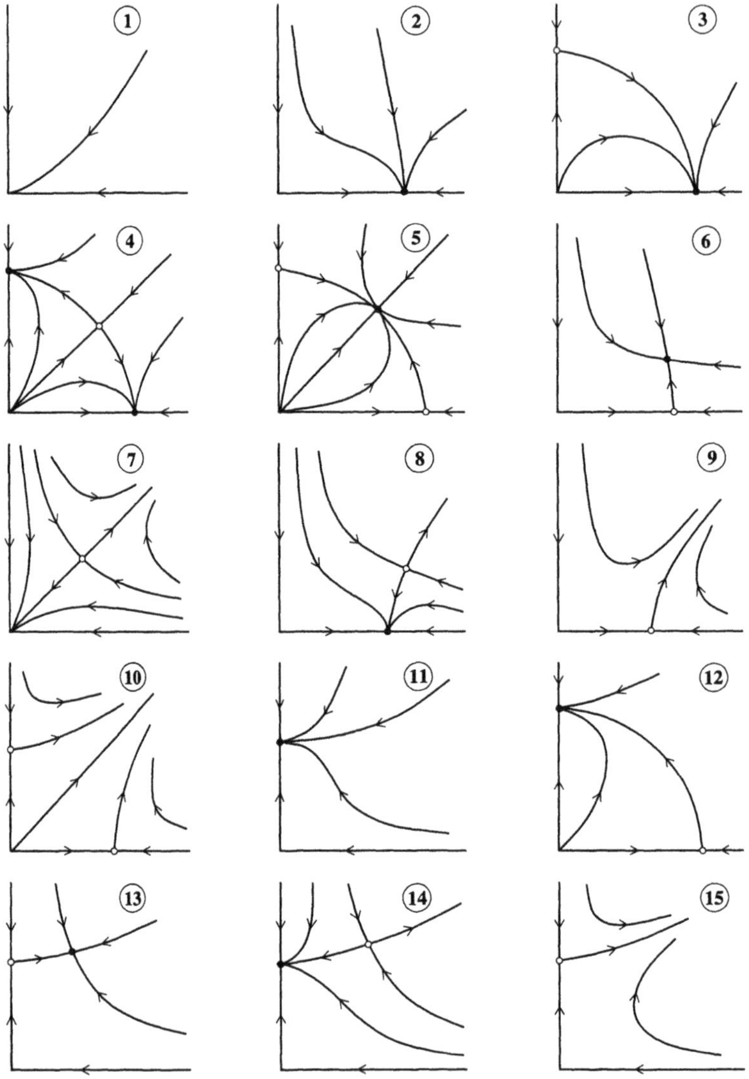


**Fig. 8.26.** Generic phase portraits of (8.112).

$$\begin{cases} \dot{\xi}_1 = \xi_1(\mu_1 + \xi_1 - \theta\xi_2 + \Theta\xi_2^2), \\ \dot{\xi}_2 = \xi_2(\mu_2 + \delta\xi_1 - \xi_2 + \Delta\xi_1^2), \end{cases} \tag{8.115}$$

where $\theta, \delta, \Theta$, and $\Delta$ are given by (8.113), as before. The trivial and the nontrivial equilibria of (8.115) have the representations

$$E_1 = (-\mu_1, 0), \quad E_2 = (0, \mu_2),$$

and

$$E_3 = \left(\frac{\mu_1 - \theta\mu_2}{\theta\delta - 1} + O(\|\mu\|^2), \frac{\delta\mu_1 - \mu_2}{\theta\delta - 1} + O(\|\mu\|^2)\right)$$

whenever their coordinates are nonnegative. Bifurcation lines corresponding to the appearance of the equilibria $E_{1,2}$ are formally the same as in the "simple" case and coincide with the coordinate axes. The nontrivial equilibrium $E_3$ collides with the trivial ones at the curves

$$T_1 = \{(\mu_1, \mu_2) : \ \mu_1 = \theta\mu_2 + O(\mu_2^2), \ \mu_2 > 0\}$$

and

$$T_2 = \{(\mu_1, \mu_2) : \ \mu_2 = \delta\mu_1 + O(\mu_1^2), \ \mu_1 < 0\}.$$

However, the nontrivial equilibrium $E_3$ can bifurcate, and system (8.115) may have *limit cycles.* The Hopf bifurcation of the equilibrium $E_3$ happens at a curve $C$ with the following characterization:

$$C = \left\{(\mu_1, \mu_2) : \ \mu_2 = -\frac{\delta - 1}{\theta - 1}\mu_1 + O(\mu_1^2), \ \mu_1 - \theta\mu_2 > 0, \ \delta\mu_1 - \mu_2 > 0\right\}.$$

Clearly, we have to assume in this case that $\delta \neq 1$ and $\theta \neq 1$ to avoid the tangency of $C$ with the $\mu_1$- or $\mu_2$-axes, that is,

(HH.7) $p_{11}(0) \neq p_{12}(0)$;
(HH.8) $p_{21}(0) \neq p_{22}(0)$.

A final nondegeneracy condition is that the first Lyapunov coefficient $l_1$ be *nonzero* along the Hopf curve $C$ near the origin. One can check that for sufficiently small $\|\mu\|$

$$\text{sign}\ l_1 = \text{sign}\left\{\frac{\delta - 1}{\theta - 1}\left[\theta(\theta - 1)\Delta + \delta(\delta - 1)\Theta\right]\right\}.$$

Thus, we assume that $\theta(\theta - 1)\Delta + \delta(\delta - 1)\Theta \neq 0$, or equivalently,

(HH.9) $$(p_{21}(p_{21} - p_{11})s_1 + p_{12}(p_{12} - p_{22})s_2)(0) \neq 0.$$

Suppose that $l_1 < 0$. The opposite case can be treated similarly. There are *six* essentially different bifurcation diagrams of (8.115), if we consider only the case $p_{11} > 0, \ p_{22} < 0, \ l_1 < 0$, and restrict attention to the half-plane $\theta \geq \delta$. The subcases are as follows:

I. $\theta > 1, \ \delta > 1$;
II. $\theta > 1, \ \delta < 1, \ \theta\delta > 1$;
III. $\theta > 0, \ \delta > 0, \ \theta\delta < 1$;
IV. $\theta > 0, \ \delta < 0$;
V. $\theta < 0, \ \delta < 0, \ \theta\delta < 1$;
VI. $\theta < 0, \ \delta < 0, \ \theta\delta > 1$.

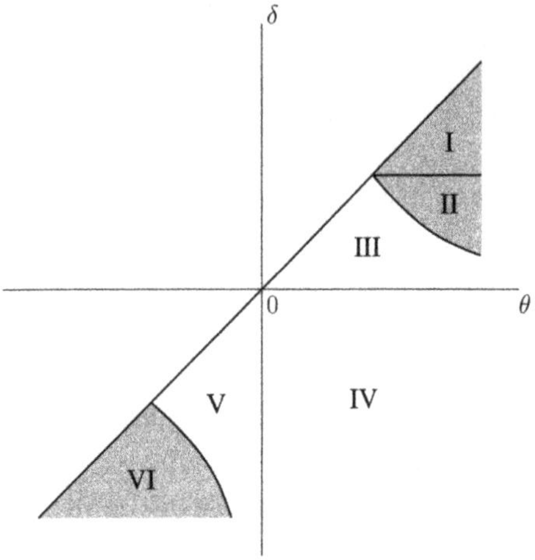


**Fig. 8.27.** Six subregions on the $(\theta, \delta)$-plane in the "difficult" case. A cycle exists in the three shaded subregions.

The parametric portraits corresponding to the regions I–VI shown in **Fig. 8.27** are depicted in **Fig. 8.28**, while the *twenty-one* distinct generic phase portraits that appear are shown in **Fig. 8.29**.

The Hopf bifurcation and consequent existence of cycles are only possible in cases I, II, and VI. A careful analysis based on Pontryagin's technique and nontrivial estimates of Abelian integrals shows that system (8.115) can have *no more than one* limit cycle (see the bibliographical notes). This cycle is born via the Hopf bifurcation at the curve $C$. Its ultimate fate depends on whether we fall into case I, II, or VI.

In cases I and II we have a *cycle blow-up* similar to that in one case of the fold-Hopf bifurcation (see Section 8.5.2). More precisely, there is a bifurcation curve $J$, depending on the considered neighborhood $U_0$ of the origin in the phase space, on which the cycle generated by the Hopf bifurcation touches the boundary of $U_0$ and "disappears" for the observer. In case VI, the cycle disappears in a more "visible" way, namely via a *heteroclinic bifurcation*. In this case, the cycle coexists with three *saddles* ($E_0, E_1$, and $E_2$). If they exist, the saddles, $E_0, E_1$, and $E_0, E_2$ are connected by orbits belonging to the invariant coordinate axis. For parameter values along the curve

$$Y = \left\{ (\mu_1, \mu_2) : \ \mu_2 = -\frac{\delta - 1}{\theta - 1}\mu_1 - \frac{(\theta - 1)\Delta + (\delta - 1)\Theta}{(\theta - 1)^3}\mu_1^2 + O(\mu_1^3) \right\},$$

which is tangent to the Hopf bifurcation curve $C$ (see Exercise 14(c)), the two separatrices of the saddles $E_1$ and $E_2$ that belong to the positive quadrant coincide. A heteroclinic cycle is formed by these orbits (see **Fig. 8.30**); it is stable from the inside due to (HH.9) and our assumption $l_1 < 0$.

**Remark:**

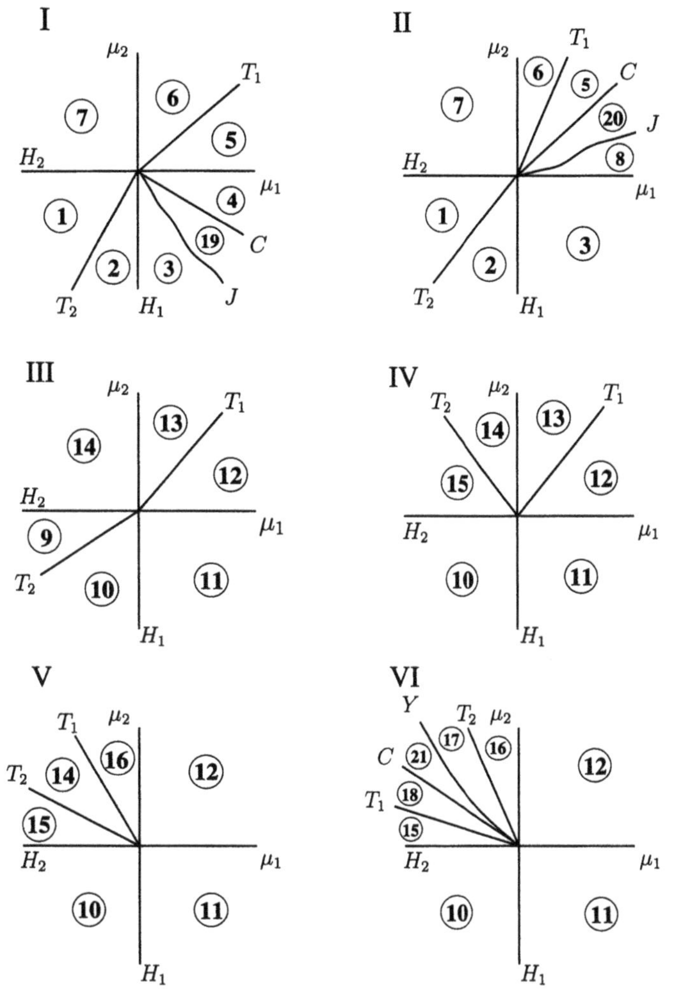


**Fig. 8.28.** Parametric portraits of (8.115) (the "difficult" case).

There is a subtle difference in the bifurcation diagrams within the cases III and IV, depending on whether $\theta < 1$ or $\theta > 1$. This difference appears only at $\mu = 0$, giving rise to topologically different critical phase portraits. All critical phase portraits are given in **Fig. 8.31**, where IIIa and IVa correspond to $\theta > 1$. $\diamondsuit$

Recalling the interpretation of equilibria and cycles of the amplitude system (8.111) in the four-dimensional *truncated* normal form (8.110), we can establish a relationship between bifurcations in these two systems. The curves $H_{1,2}$ at which the trivial equilibria appear in (8.111) obviously correspond to Hopf bifurcation curves in (8.110). These are the two "independent" Hopf

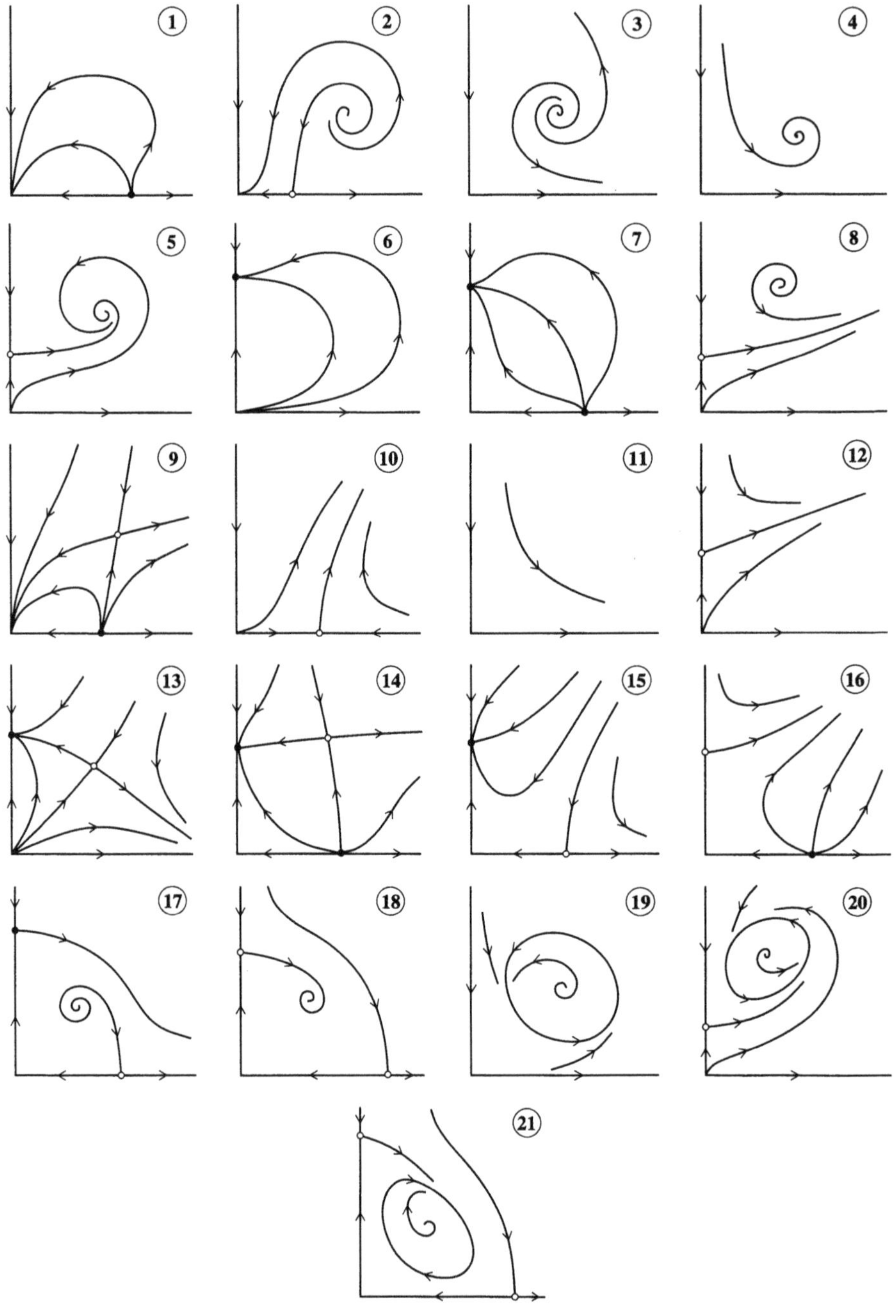


**Fig. 8.29.** Generic phase portraits of (8.115).

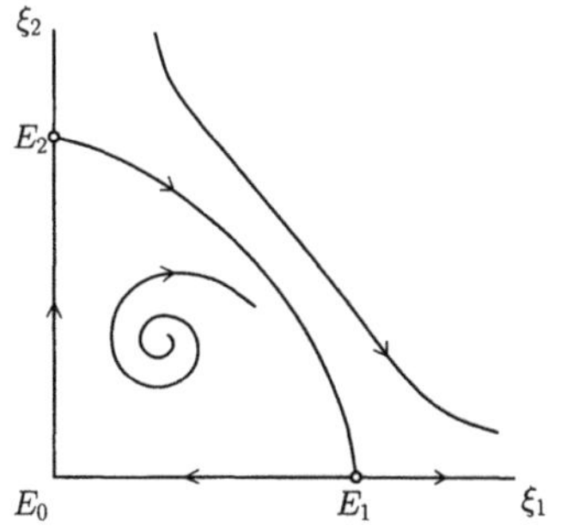


**Fig. 8.30.** A heteroclinic "triangle".

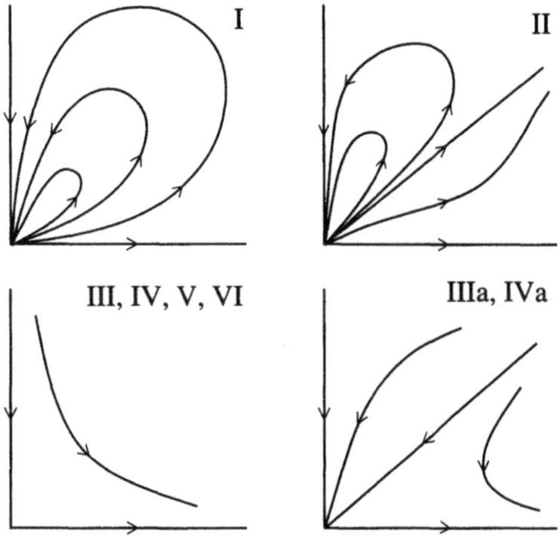


**Fig. 8.31.** Critical phase portraits of (8.115) at $\mu_1 = \mu_2 = 0$.

bifurcations caused by the two distinct pairs of eigenvalues passing through the imaginary axis. Crossing a bifurcation curve $T_1$ (or $T_2$) results in the branching of a two-dimensional torus from a cycle. Therefore, the curves $T_{1,2}$ correspond to Neimark-Sacker bifurcations in (8.110). On the curve $C$, system (8.110) exhibits a bifurcation that we have not yet encountered, namely, branching of a three-dimensional torus from the two-dimensional torus. The curves $J$ describe blow-ups of three-dimensional tori, while the curve $Y$ implies the presence of a heteroclinic coincidence of the three-dimensional stable and unstable invariant manifolds of a cycle and a three-torus.

Our next task is to discuss what will remain from the obtained bifurcation picture if we "turn on" the higher-order terms in the four-dimensional normal form (8.109).

### 8.6.3 Effect of higher-order terms

The effect of adding higher-order terms is even more dramatic for this bifurcation than for the fold-Hopf bifurcation. Actually, a generic system (8.108) is *never* topologically equivalent to the truncated normal form (8.109).

However, the truncated normal form does capture some information on the behavior of the whole system. Namely, the following lemma holds.

**Lemma 8.16** *If conditions* (HH.1)–(HH.4),

$$p_{jk}(0) \neq 0, \ j,k = 1,2,$$

*and* (HH.6),

$$(p_{11}p_{22} - p_{12}p_{21})(0) \neq 0,$$

*hold for system* (8.108), *then it has, for sufficiently small* $\|\mu\|$, *bifurcation curves* $\tilde{H}_k$ *and* $\tilde{T}_k$, $k = 1,2$, *at which nondegenerate Hopf bifurcations of the equilibrium and nondegenerate Neimark-Sacker bifurcations of limit cycles take place, which are tangent to the corresponding bifurcation lines* $H_k$ *and* $T_k$ *of the truncated system* (8.109). □

From Lemmas 8.15 and 8.16 it follows that a generic four-dimensional system exhibiting a Hopf-Hopf bifurcation also has the corresponding bifurcation curves in its parametric portrait near this codim 2 point. Crossing these curves results in the appearance of *limit cycles* and invariant *two-dimensional tori* nearby. Thus, we may say that "Hopf-Hopf interaction leads to tori."

However, the orbit structure on a torus in the full system (8.108) is generically *different* from that in the truncated system (8.109) due to phase locking. Indeed, any two-torus of (8.109) is either filled in by a dense quasiperiodic orbit (if the ratio between $\omega_1(\mu)$ and $\omega_2(\mu)$ is irrational) or filled with periodic orbits (if this ratio is rational), while in (8.108) the higher-order terms "select" only a finite (even) number of hyperbolic limit cycles on the two-torus for generic parameter values (see Chapter 7). Moreover, the tori exist and remain smooth only near the bifurcation curves $T_{1,2}$. Away from these curves the tori lose smoothness and are destroyed. Notice that Lemma 8.16 does not guarantee the presence of a bifurcation curve corresponding to the curve $C$ in the truncated system at which a three-torus bifurcates from a two-torus.

The truncated system demonstrates other degeneracies that do not survive the addition of generic higher-order terms. There are regions in the parameter plane in which the equilibrium at the origin is a saddle with two-dimensional stable and unstable manifolds, while simultaneously there is a saddle limit cycle within one of the coordinate planes $r_k = 0$ with a two-dimensional stable and a three-dimensional unstable invariant manifold. The situation is degenerate since the stable manifold of the cycle coincides with the unstable manifold of the equilibrium for all parameter values in such a region. Such a coincidence is nontransversal and will disappear under the addition of higher-order terms. Such terms only slightly displace the saddle and the cycle but destroy the invariance of the coordinate plane. The phase portrait of system (8.109) at the bifurcation curve $Y$ is also degenerate and does not persist under higher-order perturbations. Along this curve the truncated system has two saddle cycles within the coordinate planes $r_{1,2} = 0$ that have a common three-dimensional invariant manifold corresponding to the orbit connecting the trivial saddles in system (8.111). This intersection is also nontransversal and disappears if generic higher-order terms are added, forming instead a more complex *heteroclinic structure*. Thus, "strange" dynamics involving Smale horseshoes exist near a generic Hopf-Hopf bifurcation. The corresponding parametric portrait has, in addition to local bifurcation curves $H_k$ and $T_k$, a bifurcation set corresponding to global bifurcations (heteroclinic tangencies of equilibrium and cycle invariant manifolds, homoclinic orbits) and associated bifurcations of long-periodic limit cycles.

**Remarks:**

(1) Similar to the fold-Hopf case, the planar system (8.110) is a topological normal form for two-dimensional systems that are invariant under the representation of the group $\mathbb{Z}_2$ by the transformation $(x_1, x_2) \mapsto (-x_1, -x_2)$, have invariant coordinate axis $x_{1,2} = 0$, and at $\alpha_1 = \alpha_2 = 0$ possess an equilibrium $x = 0$ with a double zero eigenvalue. Indeed, any such system has the form

$$\begin{cases} \dot{x}_1 = x_1 G(x_1^2, x_2^2, \alpha), \\ \dot{x}_2 = x_2 H(x_1^2, x_2^2, \alpha), \end{cases}$$

and can be transformed under the nondegeneracy conditions (HH.1)–(HH.4) and the transversality condition (HH.5) into a system that is orbitally equivalent near the origin to

$$\begin{cases} \dot{r}_1 = r_1(\mu_1 + p_{11}r_1^2 + p_{12}r_2^2 + s_1 r_2^4 + \Phi(r_1^2, r_2^2, \mu)), \\ \dot{r}_2 = r_2(\mu_2 + p_{21}r_1^2 + p_{22}r_2^2 + s_2 r_1^4 + \Psi(r_1^2, r_2^2, \mu)), \end{cases} \tag{8.116}$$

where $\Phi, \Psi = O((r_1^2 + r_2^2)^3)$. Then, one can prove that system (8.116) is locally topologically equivalent to (8.110), provided that conditions (HH.1)–(HH.6) (and, if necessary, (HH.7)–(HH.9)) hold. The homeomorphism identifying the phase portraits, as well as the transformation into (8.116), can be selected to commute with the symmetry.

(2) Keeping only quadratic terms in (8.110) is not enough for studying the bifurcations in the "difficult" case $p_{11}p_{22} < 0$. Indeed, the system

$$\begin{cases} \dot{\xi}_1 = \xi_1(\mu_1 + \xi_1 - \theta\xi_2), \\ \dot{\xi}_2 = \xi_2(\mu_2 + \delta\xi_1 - \xi_2), \end{cases} \tag{8.117}$$

which is obtained from (8.115) by setting $\Theta = \Delta = 0$, thus violating (HH.9), is degenerate along the "Hopf curve"

$$C_0 = \left\{ (\mu_1, \mu_2) : \ \mu_2 = -\frac{\delta - 1}{\theta - 1}\mu_1, \ \mu_1 - \theta\mu_2 > 0, \ \delta\mu_1 - \mu_2 > 0 \right\}.$$

Actually, it is *orbitally equivalent* to a Hamiltonian system for $\xi_{1,2} > 0$, and the nontrivial equilibrium $E_3$ is surrounded by a family of periodic orbits.

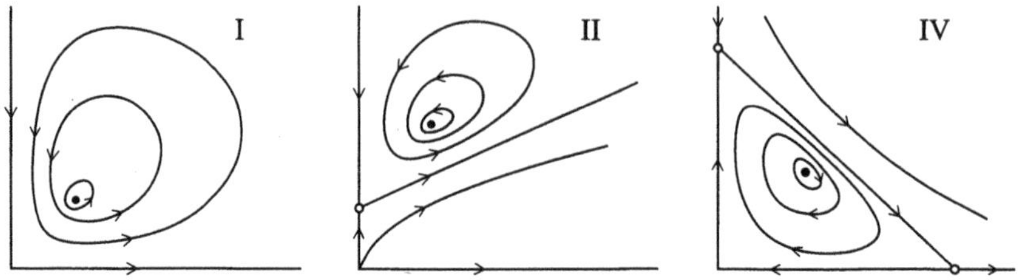


**Fig. 8.32.** Phase portraits of (8.117) on the curve $C_0$.

Three possible phase portraits corresponding to the cases I, II, and VI are depicted in **Fig. 8.32**. The cubic terms in (8.115) make the Hopf bifurcation nondegenerate whenever it exists and split a heteroclinic curve $Y$ in the case VI. The reader can obtain more details while solving this chapter's Exercise 14(c). See also the bibliographical notes. ◇

Finally, let us mention that the multidimensional case of the Hopf-Hopf bifurcation reduces as usual to the considered four-dimensional one by means of the Center Manifold Theorem (see the next section).

## 8.7 Critical normal forms for $n$-dimensional systems

To apply the theory of the codim 2 bifurcations developed in this chapter to particular models, one needs to verify the nondegeneracy conditions at the bifurcation point – in other words, to compute the coefficients of the normal form on the critical center manifold up to certain order. In principle, these coefficients could be found by first computing the Taylor expansion of the center manifold and then evaluating the corresponding normal form coefficients. This is how the explicit formulas (5.23) and (5.39) for such coefficients at the codim 1 bifurcations of equilibria have been obtained in Chapter 5.

In this section we derive the critical coefficients for all codim 2 bifurcations using another reduction/normalization technique. In this approach, the center-manifold reduction and normalization are performed simultaneously. The resulting formulas involve only critical eigenvectors of the Jacobian matrix and its transpose, as well as the Taylor expansion of the system at the critical equilibrium in the original basis. This makes them equally suitable for both symbolic and numerical evaluation.

### 8.7.1 The method

Suppose (8.1) has an equilibrium $x = 0$ at $\alpha = 0$ and represent $F(x) = f(x, 0)$ as

$$F(x) = Ax + \frac{1}{2}B(x,x) + \frac{1}{6}C(x,x,x) + \frac{1}{24}D(x,x,x,x) + \frac{1}{120}E(x,x,x,x,x) + \ldots, \tag{8.118}$$

where $A = f_x(0,0)$, the multilinear functions $B$ and $C$ are given by (5.18) and (5.19) in Chapter 5, and

$$D_i(x,y,z,v) = \sum_{j,k,l,m=1}^{n} \left.\frac{\partial^4 F_i(\xi)}{\partial\xi_j \partial\xi_k \partial\xi_l \partial\xi_m}\right|_{\xi=0} x_j y_k z_l v_m,$$

$$E_i(x,y,z,v,w) = \sum_{j,k,l,m,s=1}^{n} \left.\frac{\partial^5 F_i(\xi)}{\partial\xi_j \partial\xi_k \partial\xi_l \partial\xi_m \partial\xi_s}\right|_{\xi=0} x_j y_k z_l v_m w_s,$$

for $i = 1, 2, \ldots, n$.

Further suppose that the Jacobian matrix $A = f_x(0,0)$ of (8.1) has $n_c$ eigenvalues (counting multiplicities) with zero real part and denote by $T^c$ the corresponding generalized critical eigenspace of $A$. Write (8.1) at $\alpha = 0$ as

$$\dot{x} = F(x), \quad x \in \mathbb{R}^n, \tag{8.119}$$

and restrict it to its $n_c$-dimensional invariant center manifold parametrized by $w \in \mathbb{R}^{n_c}$:

$$x = H(w), \quad H : \mathbb{R}^{n_c} \to \mathbb{R}^n. \tag{8.120}$$

The restricted equation can be written as

$$\dot{w} = G(w), \quad G : \mathbb{R}^{n_c} \to \mathbb{R}^{n_c}. \tag{8.121}$$

Substitution of (8.120) and (8.121) into (8.119) gives the *homological equation*

$$H_w(w)G(w) = F(H(w)), \tag{8.122}$$

which is merely the condition for the center manifold $W^c$ to be invariant. Now expand the functions $G, H$ in (8.122) into multivariant Taylor series,

$$G(w) = \sum_{|\nu|\geq 1} \frac{1}{\nu!} g_\nu w^\nu, \quad H(w) = \sum_{|\nu|\geq 1} \frac{1}{\nu!} h_\nu w^\nu,$$

and assume that the restricted equation (8.121) is put into the normal form up to a certain order. This is the crucial assumption! We assume, therefore, that such a smooth parametrization of the center manifold is looked at, where the restriction has the simplest (i.e., normal) form.[12]

The coefficients $g_\nu$ of the normal form (8.121) and the coefficients $h_\nu$ of the Taylor expansion for $H(w)$ are unknown but can be found from (8.122) by a recursive procedure from lower- to higher-order terms.[13] Collecting the

[12] A normal form should be used to which the restricted equation can be transformed by means of smooth coordinate transformations but not time reparametrization since the derivative in (8.121) has the same meaning as in (8.119).

[13] Obviously, one has $\sum_{|\nu|=1} h_\nu w^\nu \in T^c$.

coefficients of the $w^\nu$-terms in (8.122) gives a linear system for the coefficient $h_\nu$,

$$Lh_\nu = R_\nu. \tag{8.123}$$

Here the matrix $L$ is determined by the Jacobian matrix $A$ and its critical eigenvalues, while the right-hand side $R_\nu$ depends on the coefficients of $G$ and $H$ of order less than or equal to $|\nu|$, as well as on the terms of order less than or equal to $|\nu|$ of the Taylor expansion (8.118) for $F$. As we shall see, when $R_\nu$ involves only known quantities, the system (8.123) has a solution because either $L$ is nonsingular or $R_\nu$ satisfies Fredholm's solvability condition

$$\langle p, R_\nu \rangle = 0,$$

where $p$ is a null-vector of the adjoint matrix $\overline{L}^T$. When $R_\nu$ depends on the unknown coefficient $g_\nu$ of the normal form, $L$ is singular and the above solvability condition gives the expression for $g_\nu$.

For all codim 2 bifurcations except Bogdanov-Takens, the invariant subspace of $L(\overline{L}^T)$ corresponding to zero eigenvalue is one-dimensional in $\mathbb{C}^n$, i.e., there are unique (up to scaling) null-vectors $q$ and $p$;

$$Lq = 0, \quad \overline{L}^T p = 0, \quad \langle p, q \rangle = 1,$$

and no generalized null-vectors. Then, the unique solution $h_\nu$ to (8.123) satisfying $\langle p, h_\nu \rangle = 0$ can be obtained by solving the nonsingular $(n+1)$-dimensional *bordered system*

$$\begin{pmatrix} L & q \\ \bar{p}^T & 0 \end{pmatrix} \begin{pmatrix} h_\nu \\ s \end{pmatrix} = \begin{pmatrix} R_\nu \\ 0 \end{pmatrix}. \tag{8.124}$$

We write $h_\nu = L^{INV} R_\nu$ as in Section 5.4.1 of Chapter 5.

The Taylor expansion of $H(w)$ simultaneously defines the expansions of the center manifold, the normalizing transformation on it, and the normal form itself. Since we know from the previous sections which terms are present in the normal form, the described procedure is a powerful tool to compute their coefficients at the bifurcation parameter values. In the following sections, this method will be applied to all codim 2 cases.

### 8.7.2 Cusp bifurcation

At the cusp bifurcation, the system (8.119) has the equilibrium $x = 0$ with a simple zero eigenvalue $\lambda_1 = 0$ and no other critical eigenvalues. Let $q, p \in \mathbb{R}^n$ satisfy

$$Aq = 0, \quad A^T p = 0, \quad \langle p, q \rangle = 1.$$

Any point $y \in T^c$ can be represented as

$$y = wq, \quad w \in \mathbb{R}^1,$$

where $w = \langle p, y \rangle$. The homological equation (8.122) has the form

$$H_w \dot{w} = F(H(w)),$$

where

$$F(H) = AH + \frac{1}{2}B(H,H) + \frac{1}{6}C(H,H,H) + O(\|H\|^4)$$

(see (8.118)),

$$H(w) = wq + \frac{1}{2}h_2 w^2 + \frac{1}{6}h_3 w^3 + O(w^4)$$

with unknown vectors $h_i \in \mathbb{R}^n$, and

$$\dot{w} = bw^2 + cw^3 + O(w^4) \tag{8.125}$$

with unknown coefficients $b$ and $c$. Substituting these expressions into the homological equation gives

$$\begin{aligned} &bw^2 q + (cq + bh_2)w^3 = \\ &\quad \frac{1}{2}w^2[Ah_2 + B(q,q)] + \frac{1}{6}w^3[Ah_3 + 3B(q,h_2) + C(q,q,q)] + O(w^4). \end{aligned} \tag{8.126}$$

The $w^2$-terms in (8.126) give the equation for $h_2$

$$Ah_2 = -B(q,q) + 2bq, \tag{8.127}$$

where the matrix $A$ is obviously singular. The solvability of this system implies

$$\langle p, -B(q,q) + 2b_0 q \rangle = -\langle p, B(q,q) \rangle + 2b\langle p, q \rangle = 0$$

and allows one to find $b$, namely

$$b = \frac{1}{2}\langle p, B(q,q) \rangle$$

in accordance with the formula (5.23) for the fold bifurcation in Chapter 5. With this value of $b$, the linear system (8.127) becomes

$$Ah_2 = -B(q,q) + \langle p, B(q,q) \rangle q$$

and its unique solution $h_2 = -A^{INV}[B(q,q) - \langle p, B(q,q) \rangle q]$ satisfying $\langle p, h_2 \rangle = 0$ can be computed by solving the nonsingular $(n+1)$-dimensional bordered system

$$\begin{pmatrix} A & q \\ p^T & 0 \end{pmatrix} \begin{pmatrix} h_2 \\ s \end{pmatrix} = \begin{pmatrix} -B(q,q) + \langle p, B(q,q) \rangle q \\ 0 \end{pmatrix}.$$

Collecting the $w^3$-terms in (8.126) yields

$$cq + bh_2 = \frac{1}{6}Ah_3 + \frac{1}{2}B(q,h_2) + \frac{1}{6}C(q,q,q),$$

which is equivalent to another singular system,

$$Ah_3 = cq + bh_2 - \frac{1}{6}[C(q,q,q) + 3B(q,h_2)].$$

Its solvability implies

$$c\langle p,q\rangle + b\langle p,h_2\rangle - \frac{1}{6}\langle p, C(q,q,q) + 3B(q,h_2)\rangle = 0.$$

Since $\langle p,h_2\rangle = 0$, we obtain the following expression for the coefficient $c$:

$$c = \frac{1}{6}\langle p, C(q,q,q) + 3B(q,h_2)\rangle.$$

Now recall that $b = 0$ at the cusp bifurcation. Under this condition, the coefficient $c$ in the normal form (8.125) can be expressed shortly as

$$c = \frac{1}{6}\langle p, C(q,q,q) - 3B(q, A^{INV}B(q,q))\rangle. \tag{8.128}$$

This expression has been already derived in Chapter 5; see equation (5.28).

### 8.7.3 Bautin bifurcation

At the Bautin bifurcation, the system (8.119) has an equilibrium with a simple pair of purely imaginary eigenvalues $\lambda_{1,2} = \pm i\omega_0,\ \omega_0 > 0$, and no other critical eigenvalues. As in the Hopf case, introduce two complex eigenvectors

$$Aq = i\omega_0 q, \quad A^T p = -i\omega_0 p,$$

and normalize them according to

$$\langle p,q\rangle = 1.$$

Any vector $y \in T^c \subset \mathbb{R}^n$ can be represented as

$$y = wq + \overline{w}\,\overline{q},$$

where $w = \langle p,y\rangle \in \mathbb{C}^1$. The homological equation (8.122) now takes the form

$$H_w\dot{w} + H_{\overline{w}}\dot{\overline{w}} = F(H(w,\overline{w})), \tag{8.129}$$

where $F$ is given by (8.118),

$$H(w,\overline{w}) = wq + \overline{w}\,\overline{q} + \sum_{1\le j+k\le 5} \frac{1}{j!k!} h_{jk} w^j \overline{w}^k + O(|w|^6),$$

with $h_{jk} \in \mathbb{C}^n,\ h_{kj} = \overline{h}_{jk}$, and

$$\dot{w} = i\omega_0 w + c_1 w|w|^2 + c_2 w|w|^4 + O(|w|^6), \tag{8.130}$$

where $c_{1,2} \in \mathbb{C}^1$. This is equation (8.19) at $\alpha = 0$.

Collecting the coefficients of the quadratic terms in (8.129) and solving the nonsingular linear systems that appear, we get

$$\begin{aligned} h_{20} &= (2i\omega_0 I_n - A)^{-1} B(q,q), \\ h_{11} &= -A^{-1} B(q,\bar{q}). \end{aligned}$$

The coefficient in front of the $w^3$-term in (8.129) leads to the expression for $h_{30}$,

$$h_{30} = (3i\omega_0 I_n - A)^{-1}[C(q,q,q) + 3B(q,h_{20})],$$

while the $w^2\bar{w}$-terms give the singular system for $h_{21}$:

$$(i\omega_0 I_n - A)h_{21} = C(q,q,\bar{q}) + B(\bar{q},h_{20}) + 2B(q,h_{11}) - 2c_1 q. \tag{8.131}$$

The solvability of this system is equivalent to

$$\langle p, C(q,q,\bar{q}) + B(\bar{q},h_{20}) + 2B(q,h_{11}) - 2c_1 q\rangle = 0,$$

so the cubic coefficient in the normal form (8.130) can be expressed as

$$c_1 = \frac{1}{2}\langle p, C(q,q,\bar{q}) + B(\bar{q},(2i\omega_0 I_n - A)^{-1}B(q,q)) - 2B(q,A^{-1}B(q,\bar{q}))\rangle$$

and

$$l_1(0) = \frac{1}{\omega_0}\mathrm{Re}\ c_1$$

coincides with the expression (5.39) for the first Lyapunov coefficient derived in Chapter 5. Then from (8.131) follows

$$h_{21} = (i\omega_0 I_n - A)^{INV}[C(q,q,\bar{q}) + B(\bar{q},h_{20}) + 2B(q,h_{11}) - 2c_1 q].$$

Here the complex vector $h_{21}$ satisfying $\langle p, h_{21}\rangle = 0$ can be found by solving the nonsingular $(n+1)$-dimensional complex bordered system

$$\begin{pmatrix} i\omega_0 I_n - A & q \\ \bar{p}^T & 0 \end{pmatrix}\begin{pmatrix} h_{21} \\ s \end{pmatrix} = \begin{pmatrix} C(q,q,\bar{q}) + B(\bar{q},h_{20}) + 2B(q,h_{11}) - 2c_1 q \\ 0 \end{pmatrix}.$$

For the fourth-order coefficients, we get

$$\begin{aligned} h_{40} &= (4i\omega_0 I_n - A)^{-1}[D(q,q,q,q) + 6C(q,q,h_{20}) + 4B(q,h_{30}) + 3B(h_{20},h_{20})], \\ h_{31} &= (2i\omega_0 I_n - A)^{-1}[D(q,q,q,\bar{q}) + 3C(q,q,h_{11}) + 3C(q,\bar{q},h_{20}) \\ &\qquad + 3B(h_{20},h_{11}) + B(\bar{q},h_{30}) + 3B(q,h_{21}) - 6c_1 h_{20}], \\ h_{22} &= -A^{-1}[D(q,q,\bar{q},\bar{q}) + 4C(q,\bar{q},h_{11}) + C(\bar{q},\bar{q},h_{20}) + C(q,q,\bar{h}_{20}) \\ &\qquad + 2B(h_{11},h_{11}) + 2B(q,\bar{h}_{21}) + 2B(\bar{q},h_{21}) + B(\bar{h}_{20},h_{20}) \\ &\qquad - 4h_{11}(c_1 + \bar{c}_1)]. \end{aligned}$$

Taking into account the equality $\langle p, h_{21}\rangle = 0$, one can check that the solvability condition of a linear system for $h_{32}$ provides the formula

$$l_2(0) = \frac{1}{\omega_0}\text{Re } c_2$$

for the second Lyapunov coefficient (see Lemma 8.4), where

$$\begin{aligned} c_2 = \frac{1}{12}\langle p, &E(q,q,q,\bar{q},\bar{q}) \\ &+ D(q,q,q,\bar{h}_{20}) + 3D(q,\bar{q},\bar{q},h_{20}) + 6D(q,q,\bar{q},h_{11}) \\ &+ C(\bar{q},\bar{q},h_{30}) + 3C(q,q,\bar{h}_{21}) + 6C(q,\bar{q},h_{21}) + 3C(q,\bar{h}_{20},h_{20}) \\ &+ 6C(q,h_{11},h_{11}) + 6C(\bar{q},h_{20},h_{11}) \\ &+ 2B(\bar{q},h_{31}) + 3B(q,h_{22}) + B(\bar{h}_{20},h_{30}) + 3B(\bar{h}_{21},h_{20}) + 6B(h_{11},h_{21})\rangle, \end{aligned}$$

with all $h_{jk}$ defined earlier. Notice that $h_{40}$ does not enter the expression for $c_2$. Also recall that at the Bautin bifurcation $l_1(0) = 0$ or $c_1 + \bar{c}_1 = 0$, so the last term in $h_{22}$ vanishes.

### 8.7.4 Bogdanov-Takens bifurcation

At the Bogdanov-Takens bifurcation, the system (8.119) has two eigenvalues, $\lambda_{1,2} = 0$, and there exist two real linearly independent (generalized) eigenvectors, $q_{0,1} \in \mathbb{R}^n$, such that

$$Aq_0 = 0, \quad Aq_1 = q_0.$$

Moreover, there exist similar vectors $p_{1,0} \in \mathbb{R}^n$ of the transposed matrix $A^T$:

$$A^T p_1 = 0, \quad A^T p_0 = p_1.$$

As in Section 8.4, we can select these vectors to satisfy

$$\langle p_0, q_0\rangle = \langle p_1, q_1\rangle = 1, \quad \langle p_0, q_1\rangle = \langle p_1, q_0\rangle = 0.$$

Any vector $y \in T^c$ can be uniquely represented as

$$y = w_0 q_0 + w_1 q_1,$$

where $w_0 = \langle p_0, y\rangle$, $w_1 = \langle p_1, y\rangle$. The homological equation (8.122) has the form

$$H_{w_0}\dot{w}_0 + H_{w_1}\dot{w}_1 = F(H(w_0, w_1)), \tag{8.132}$$

where

$$F(H) = AH + \frac{1}{2}B(H,H) + O(\|H\|^3),$$

$$H(w_0, w_1) = w_0 q_0 + w_1 q_1 + \frac{1}{2}h_{20}w_0^2 + h_{11}w_0 w_1 + \frac{1}{2}h_{02}w_1^2 + O(\|(w_0, w_1)\|^3)$$

with unknown $h_{jk} \in \mathbb{R}^n$, and $\dot{w}_0, \dot{w}_1$ are defined by the critical normal form

$$\begin{cases} \dot{w}_0 = w_1 \\ \dot{w}_1 = a w_0^2 + b w_0 w_1 + O(\|w\|^3), \end{cases} \tag{8.133}$$

with unknown coefficients $a$ and $b$. This system is equivalent to (8.49) with $A(0) = a$ and $B(0) = b$.

Substituting the above expressions into (8.132) and collecting the $w_0^2$-terms gives the singular linear system for $h_{20}$:

$$Ah_{20} = 2aq_1 - B(q_0, q_0). \tag{8.134}$$

The solvability condition for this system is

$$\langle p_1, 2aq_1 - B(q_0, q_0)\rangle = 2a\langle p_1, q_1\rangle - \langle p_1, B(q_0, q_0)\rangle = 0,$$

which gives

$$a = \frac{1}{2}\langle p_1, B(q_0, q_0)\rangle. \tag{8.135}$$

Taking the scalar product of both sides of (8.134) with $p_0$ yields $\langle p_0, Ah_{20}\rangle = 2a\langle p_0, q_1\rangle - \langle p_0, B(q_0, q_0)\rangle$, which implies

$$\langle p_1, h_{20}\rangle = -\langle p_0, B(q_0, q_0)\rangle. \tag{8.136}$$

The $w_0 w_1$-terms in (8.132) give the linear system

$$Ah_{11} = bq_1 + h_{20} - B(q_0, q_1). \tag{8.137}$$

Its solvability means

$$\langle p_1, bq_1 + h_{20} - B(q_0, q_1)\rangle = b\langle p_1, q_1\rangle + \langle p_1, h_{20}\rangle - \langle p_1, B(q_0, q_1)\rangle = 0.$$

Taking into account (8.136), we get

$$b = \langle p_0, B(q_0, q_0)\rangle + \langle p_1, B(q_0, q_1)\rangle. \tag{8.138}$$

Thus, the coefficients $a$ and $b$ of the normal form (8.133) are computed.

Notice that solutions to the singular linear systems (8.134) and (8.137) with $a$ and $b$ given by (8.135) and (8.138) are *not* unique. For example, a scalar multiple of $q_0$ can be added to $h_{20}$. This freedom can be used to assure that the right-hand side of the linear system for $h_{02}$ appearing from (8.132),

$$Ah_{02} = 2h_{11} - B(q_1, q_1), \tag{8.139}$$

is orthogonal to $p_1$. Indeed, from (8.137) it follows that

$$\langle p_1, h_{11}\rangle = \langle p_0, h_{20}\rangle - \langle p_0, B(q_0, q_1)\rangle.$$

Using this identity, we get

$$\langle p_1, 2h_{11} - B(q_1, q_1)\rangle = 2\langle p_0, h_{20}\rangle - 2\langle p_0, B(q_0, q_1)\rangle - \langle p_1, B(q_1, q_1)\rangle.$$

The substitution $h_{20} \mapsto h_{20} + \gamma q_0$ with a properly selected $\gamma$ makes the right-hand side of this equation equal to zero. This implies that (8.139) is solvable for $h_{02}$.

### 8.7.5 Fold-Hopf bifurcation

At the fold-Hopf bifurcation, the system (8.119) has an equilibrium with a simple zero eigenvalue and a pair of purely imaginary simple eigenvalues of the Jacobian matrix $A = f_x(0,0)$,

$$\lambda_1 = 0, \ \lambda_{2,3} = \pm i\omega_0,$$

with $\omega_0 > 0$, and no other critical eigenvalues. Introduce two eigenvectors, $q_0 \in \mathbb{R}^n$ and $q_1 \in \mathbb{C}^n$,

$$Aq_0 = 0, \quad Aq_1 = i\omega_0 q_1,$$

and two adjoint eigenvectors, $p_0 \in \mathbb{R}^n$ and $p_1 \in \mathbb{C}^n$,

$$A^T p_0 = 0, \quad A^T p_1 = -i\omega_0 p_1.$$

Normalize them such that

$$\langle p_0, q_0 \rangle = \langle p_1, q_1 \rangle = 1.$$

The following orthogonality properties hold:

$$\langle p_1, q_0 \rangle = \langle p_0, q_1 \rangle = 0.$$

Now any vector $y \in T^c \in \mathbb{R}^n$ can be represented as

$$y = w_0 q_0 + w_1 q_1 + \overline{w}_1 \overline{q}_1,$$

where $w_0 = \langle p_0, y \rangle \in \mathbb{R}^1$ and $w_1 = \langle p_1, y \rangle \in \mathbb{C}^1$. The homological equation (8.122) can be written as

$$H_{w_0}\dot{w}_0 + H_{w_1}\dot{w}_1 + H_{\overline{w}_1}\dot{\overline{w}}_1 = F(H(w_0, w_1, \overline{w}_1)), \tag{8.140}$$

where

$$F(H) = AH + \frac{1}{2}B(H,H) + \frac{1}{6}C(H,H,H) + O(\|H\|^4),$$

$$\begin{aligned} H(w_0, w_1, \overline{w}_1) &= w_0 q_0 + w_1 q_1 + \overline{w}_1 \overline{q}_1 \\ &+ \sum_{2 \le j+k+l \le 3} \frac{1}{j!k!l!} h_{jkl} w_0^j w_1^k \overline{w}_1^l + O(\|(w_0, w_1, \overline{w}_1)\|^4), \end{aligned}$$

$h_{jkl} \in \mathbb{C}^n, h_{jlk} = \overline{h}_{jkl}$, and $(\dot{w}_0, \dot{w}_1)$ are defined by the normal form (8.63) at $\alpha = 0$,

$$\begin{cases} \dot{w}_0 = G_{200} w_0^2 + G_{011}|w_1|^2 + G_{300} w_0^3 + G_{111} w_0 |w_1|^2 \\ \qquad + O(\|(w_0, w_1, \overline{w}_1)\|^4) \\ \dot{w}_1 = i\omega_0 w_1 + H_{110} w_0 w_1 + H_{210} w_0^2 w_1 + H_{021} w_1 |w_1|^2 \\ \qquad + O(\|(w_0, w_1, \overline{w}_1)\|^4), \end{cases} \tag{8.141}$$

where we use $(w_0, w_1)$ instead of $(v, w)$ and all coefficients are taken at their critical values.

Collecting the $w_0^j w_1^k \bar{w}_1^l$-terms in (8.140) with $j+k+l = 2$, one gets from the solvability conditions the expressions for the quadratic coefficients in (8.141),

$$G_{200} = \frac{1}{2}\langle p_0, B(q_0, q_0)\rangle, \ H_{110} = \langle p_1, B(q_0, q_1)\rangle, \ G_{011} = \langle p_0, B(q_1, \bar{q}_1)\rangle,$$

and the following formulas for the coefficients $h_{jkl}$ with $j + k + l = 2$:

$$h_{200} = -A^{INV}[B(q_0, q_0) - \langle p_0, B(q_0, q_0)\rangle q_0], \tag{8.142}$$
$$h_{020} = (2i\omega_0 I_n - A)^{-1} B(q_1, q_1), \tag{8.143}$$
$$h_{110} = (i\omega_0 I_n - A)^{INV}[B(q_0, q_1) - \langle p_1, B(q_0, q_1)\rangle q_1], \tag{8.144}$$
$$h_{011} = -A^{INV}[B(q_1, \bar{q}_1) - \langle p_0, B(q_1, \bar{q}_1)\rangle q_0]. \tag{8.145}$$

Here the vectors $h_{200}$ and $h_{011}$ can be computed by solving the nonsingular $(n + 1)$-dimensional real bordered systems

$$\begin{pmatrix} A & q_0 \\ p_0^T & 0 \end{pmatrix} \begin{pmatrix} h_{200} \\ s \end{pmatrix} = \begin{pmatrix} -B(q_0, q_0) + \langle p_0, B(q_0, q_0)\rangle q_0 \\ 0 \end{pmatrix}$$

and

$$\begin{pmatrix} A & q_0 \\ p_0^T & 0 \end{pmatrix} \begin{pmatrix} h_{011} \\ s \end{pmatrix} = \begin{pmatrix} -B(q_1, \bar{q}_1) + \langle p_0, B(q_1, \bar{q}_1)\rangle q_0 \\ 0 \end{pmatrix},$$

while the vector $h_{110}$ can be found by solving the nonsingular $(n + 1)$-dimensional complex bordered system

$$\begin{pmatrix} i\omega_0 I_n - A & q_1 \\ \bar{p}_1^T & 0 \end{pmatrix} \begin{pmatrix} h_{110} \\ s \end{pmatrix} = \begin{pmatrix} B(q_0, q_1) - \langle p_1, B(q_0, q_1)\rangle q_1 \\ 0 \end{pmatrix}.$$

Finally, the solvability conditions applied to the systems coming out from the resonant $w_0^j w_1^k \bar{w}_1^l$-terms in (8.140) with $j + k + l = 3$ yield

$$\begin{aligned}
G_{300} &= \frac{1}{6}\langle p_0, C(q_0, q_0, q_0) + 3B(q_0, h_{200})\rangle, \\
G_{111} &= \langle p_0, C(q_0, q_1, \bar{q}_1) + B(q_1, \bar{h}_{110}) + B(\bar{q}_1, h_{110}) + B(q_0, h_{011})\rangle, \\
H_{210} &= \frac{1}{2}\langle p_1, C(q_0, q_0, q_1) + 2B(q_0, h_{110}) + B(q_1, h_{200})\rangle, \\
H_{021} &= \frac{1}{2}\langle p_1, C(q_1, q_1, \bar{q}_1) + 2B(q_1, h_{011}) + B(\bar{q}_1, h_{020})\rangle,
\end{aligned}$$

where $h_{jkl}$ are defined by (8.142)–(8.145). Thus, all the coefficients in (8.141) are computed.

**Example 8.4 (Lorenz-84 model).** Consider the simplified model of atmospheric circulation [Lorenz 1984]

$$\begin{cases} \dot{x} = -y^2 - z^2 - ax + aF, \\ \dot{y} = xy - bxz - y + G, \\ \dot{z} = bxy + xz - z, \end{cases} \tag{8.146}$$

where $(F, G)$ are parameters and $a = \frac{1}{4}, b = 4$. One can show (see Exercise 12) that at

$$F_0 = \frac{3907}{2320} = 1.684051724\ldots, \quad G_0 = \frac{1297}{9280}\sqrt{145} = 1.682968552\ldots,$$

the system (8.146) has the equilibrium

$$(x_0, y_0, z_0) = \left(\frac{9}{8}, \; -\frac{1}{1160}\sqrt{145}, \; \frac{9}{290}\sqrt{145}\right)$$

exhibiting a *fold-Hopf bifurcation*. Indeed, the Jacobian matrix of (8.146) evaluated at the critical equilibrium,

$$A = \begin{pmatrix} -\frac{1}{4} & \frac{\sqrt{145}}{580} & -\frac{9\sqrt{145}}{145} \\ -\frac{\sqrt{145}}{8} & \frac{1}{8} & -\frac{9}{2} \\ \frac{4\sqrt{145}}{145} & \frac{9}{2} & \frac{1}{8} \end{pmatrix},$$

has the eigenvalues

$$\lambda_1 = 0, \quad \lambda_{2,3} = \pm i\omega_0, \quad \omega_0 = \frac{1}{1160}\sqrt{27561455} = 4.525776271\ldots > 0.$$

The vectors in $\mathbf{C}^3$

$$q_0 = \left(1, \; -\frac{1007}{188065}\sqrt{145}, -\frac{5252}{188065}\sqrt{145}\right)^T,$$

$$q_1 = \left(\frac{2}{145}\sqrt{145}, \; 1, \; -\frac{1}{36} - \frac{i}{5220}\sqrt{27561455}\right)^T,$$

$$p_0 = \left(\frac{188065}{190079}, \; -\frac{2594}{190079}\sqrt{145}, 0\right)^T,$$

$$p_1 = \left(\frac{1007}{380158}\sqrt{145} - \frac{i}{380158}\sqrt{145}\sqrt{27561455}, \right.$$
$$\left. \frac{188065}{380158} + \frac{i}{380158}\sqrt{27561455}, \; -\frac{18\,i}{190079}\sqrt{27561455}\right)^T,$$

satisfy

$$Aq_0 = A^T p_0 = 0, \quad Aq_1 = i\omega_0 q_1, \quad A^T p_1 = -i\omega_0 p_1,$$

with the normalization conditions

$$\langle p_0, q_0\rangle = \langle p_1, q_1\rangle = 1.$$

There are no cubic terms in (8.146), while the components of the bilinear function $B(p,q)$ are given at $b = 4$ by the formulas

$$B_i(p,q) = p^T H^i q, \quad i = 1,2,3,$$

with the Hessian matrices

$$H^1 = \begin{pmatrix} 0 & 0 & 0 \\ 0 & -2 & 0 \\ 0 & 0 & -2 \end{pmatrix}, \quad H^2 = \begin{pmatrix} 0 & 1 & -4 \\ 1 & 0 & 0 \\ -4 & 0 & 0 \end{pmatrix}, \quad H^3 = \begin{pmatrix} 0 & 4 & 1 \\ 4 & 0 & 0 \\ 1 & 0 & 0 \end{pmatrix}.$$

Following the procedure described above, one gets the values for the normal form coefficients:

$$\begin{aligned}
G_{200} &= -\frac{62051}{190079}, \\
H_{110} &= \frac{252130}{190079} + \frac{141\,i}{190079}\sqrt{145}\sqrt{190079}, \\
G_{011} &= -\frac{6915604}{1710711}, \\
G_{300} &= \frac{12801407360}{36130026241}, \\
G_{111} &= \frac{9729482240}{325170236169}, \\
H_{210} &= -\frac{18320144480}{36130026241} + \frac{479799208640\,i}{6867559257863039}\sqrt{145}\sqrt{190079}, \\
H_{021} &= -\frac{2213399552}{325170236169} + \frac{25579201429264\,i}{26886494494533797685}\sqrt{145}\sqrt{190079}.
\end{aligned}$$

This gives the following numerical values for the coefficients of the Gavrilov normal form (8.71) in Section 8.5:

$$B(0) = -\frac{62051}{190079} = -0.32644848\ldots, \quad C(0) = -\frac{6915604}{1710711} = -4.04253202\ldots,$$

and

$$E(0) = \frac{33652980958948512}{20391681953530129} = 1.65032885\ldots > 0,$$

while, by Theorem 8.6,

$$\theta(0) \equiv \frac{\text{Re}\, H_{110}(0)}{2G_{200}} = -\frac{126065}{62051} = -2.03163527\ldots < 0.$$

Thus, the case $s = \text{sign}\,(b(0)c(0)) = 1$, $\theta < 0$ occurs without time reversing (see Section 8.5). One can also check transversality with respect to the parameters $(F, G)$. Therefore, a nontrivial invariant set bifurcates from the critical equilibrium under small parameter variations. $\Diamond$

### 8.7.6 Hopf-Hopf bifurcation

At the Hopf-Hopf bifurcation, the system (8.119) has an equilibrium with two pairs of purely imaginary simple eigenvalues of the Jacobian matrix $A = f_x(0,0)$,

$$\lambda_{1,4} = \pm i\omega_1, \ \lambda_{2,3} = \pm i\omega_2,$$

with $\omega_1 > \omega_2 > 0$, and no other critical eigenvalues. Assume that the conditions (HH.0) from Lemma 8.13 hold:

$$k\omega_1 \neq l\omega_2, \ k,l > 0, k+l \leq 5. \tag{8.147}$$

Since the eigenvalues are simple, there are two complex eigenvectors, $q_{1,2} \in \mathbb{C}^n$, corresponding to these eigenvalues:

$$Aq_1 = i\omega_1 q_1, \ Aq_2 = i\omega_2 q_2.$$

Introduce the adjoint eigenvectors $p_{1,2} \in \mathbb{C}^n$ by

$$A^T p_1 = -i\omega_1 p_1, \ A^T p_2 = -i\omega_2 p_2,$$

where $T$ denotes transposition. These eigenvectors can be normalized using the standard scalar product in $\mathbb{C}^n$,

$$\langle p_1, q_1\rangle = \langle p_2, q_2\rangle = 1,$$

and satisfy the orthogonality conditions

$$\langle p_2, q_1\rangle = \langle p_1, q_2\rangle = 0.$$

Any vector $y \in T^c \subset \mathbb{R}^n$ can be represented as

$$y = w_1 q_1 + \overline{w}_1 \overline{q}_1 + w_2 q_2 + \overline{w}_2 \overline{q}_2, \quad w_i \in \mathbb{C}^1,$$

where $w_1 = \langle p_1, y\rangle$, $w_2 = \langle p_2, y\rangle$. Therefore, the homological equation (8.122) can be written as

$$H_{w_1}\dot{w}_1 + H_{\overline{w}_1}\dot{\overline{w}}_1 + H_{w_2}\dot{w}_2 + H_{\overline{w}_2}\dot{\overline{w}}_2 = F(H(w_1, \overline{w}_1, w_2, \overline{w}_2)), \tag{8.148}$$

where $F$ is defined by (8.118),

$$\begin{aligned} H(w_1, \overline{w}_1, w_2, \overline{w}_2) &= w_1 q_1 + \overline{w}_1 \overline{q}_1 + w_2 q_2 + \overline{w}_2 \overline{q}_2 \\ &+ \sum_{j+k+l+m \geq 2} \frac{1}{j!k!l!m!} h_{jklm} w_1^j \overline{w}_1^k w_2^l \overline{w}_2^m, \end{aligned}$$

where $h_{jklm} \in \mathbb{C}^n$, $h_{kjml} = \overline{h}_{jklm}$, and $(\dot{w}_1, \dot{\overline{w}}_2)$ are specified by the normal form (8.89) at $\alpha = 0$:

$$\begin{cases} \dot{w}_1 = i\omega_1 w_1 + G_{2100} w_1 |w_1|^2 + G_{1011} w_1 |w_2|^2 \\ \quad + G_{3200} w_1 |w_1|^4 + G_{2111} w_1 |w_1|^2 |w_2|^2 + G_{1022} w_1 |w_2|^4 \\ \quad + O(\|(w_1, \bar{w}_1, w_2, \bar{w}_2)\|^6) \\ \dot{w}_2 = i\omega_2 w_2 + H_{1110} w_2 |w_1|^2 + H_{0021} w_2 |w_2|^2 \\ \quad + H_{2210} w_2 |w_1|^4 + H_{1121} w_2 |w_1|^2 |w_2|^2 + H_{0032} w_2 |w_2|^4 \\ \quad + O(\|(w_1, \bar{w}_1, w_2, \bar{w}_2)\|^6), \end{cases} \tag{8.149}$$

where all coefficients are taken at their critical values.

Collecting the coefficients of the $w_1^j \bar{w}_1^k w_2^l \bar{w}_2^m$-terms with $j+k+l+m=2$ in (8.148) gives the following expressions for $h_{jklm}$:

$$h_{1100} = -A^{-1} B(q_1, \bar{q}_1), \tag{8.150}$$
$$h_{2000} = (2i\omega_1 I_n - A)^{-1} B(q_1, q_1), \tag{8.151}$$
$$h_{1010} = [i(\omega_1 + \omega_2) I_n - A]^{-1} B(q_1, q_2), \tag{8.152}$$
$$h_{1001} = [i(\omega_1 - \omega_2) I_n - A]^{-1} B(q_1, \bar{q}_2), \tag{8.153}$$
$$h_{0020} = (2i\omega_2 I_n - A)^{-1} B(q_2, q_2), \tag{8.154}$$
$$h_{0011} = -A^{-1} B(q_2, \bar{q}_2). \tag{8.155}$$

All matrices involved in (8.150)–(8.155) are invertible in the ordinary sense due to the assumptions (8.147) on the critical eigenvalues.

Collecting the coefficients in front of the nonresonant $w_1^j \bar{w}_1^k w_2^l \bar{w}_2^m$-terms with $j+k+l+m=3$ in (8.148), one obtains the following expressions for $h_{jklm}$:

$$\begin{aligned} h_{3000} &= (3i\omega_1 I_n - A)^{-1} [C(q_1, q_1, q_1) + 3B(h_{2000}, q_1)], \\ h_{2010} &= [i(2\omega_1 + \omega_2) I_n - A]^{-1} [C(q_1, q_1, q_2) + B(h_{2000}, q_2) + 2B(h_{1010}, q_1)], \\ h_{2001} &= [i(2\omega_1 - \omega_2) I_n - A]^{-1} [C(q_1, q_1, \bar{q}_2) + B(h_{2000}, \bar{q}_2) + 2B(h_{1001}, q_1)], \\ h_{1020} &= [i(\omega_1 + 2\omega_2) I_n - A]^{-1} [C(q_1, q_2, q_2) + B(h_{0020}, q_1) + 2B(h_{1010}, q_2)], \\ h_{1002} &= [i(\omega_1 - 2\omega_2) I_n - A]^{-1} [C(q_1, \bar{q}_2, \bar{q}_2) + B(\bar{h}_{0020}, q_1) + 2B(h_{1001}, \bar{q}_2)], \\ h_{0030} &= (3i\omega_2 I_n - A)^{-1} [C(q_2, q_2, q_2) + 3B(h_{0020}, q_2)]. \end{aligned}$$

All matrices in these expressions are invertible. Collecting the coefficients of the resonant cubic terms in (8.148), one obtains the resonant cubic coefficients in the normal form (8.149):

$$\begin{aligned} G_{2100} &= \frac{1}{2} \langle p_1, C(q_1, q_1, \bar{q}_1) + B(h_{2000}, \bar{q}_1) + 2B(h_{1100}, q_1) \rangle, \\ G_{1011} &= \langle p_1, C(q_1, q_2, \bar{q}_2) + B(h_{1010}, \bar{q}_2) + B(h_{1001}, q_2) + B(h_{0011}, q_1) \rangle, \\ H_{1110} &= \langle p_2, C(q_1, \bar{q}_1, q_2) + B(h_{1100}, q_2) + B(h_{1010}, \bar{q}_1) + B(\bar{h}_{1001}, q_1) \rangle, \\ H_{0021} &= \frac{1}{2} \langle p_2, C(q_2, q_2, \bar{q}_2) + B(h_{0020}, \bar{q}_2) + 2B(h_{0011}, q_2) \rangle. \end{aligned}$$

Similarly collecting the fourth- and fifth-order terms, one can compute all remaining coefficients in (8.149). The resulting formulas are lengthy and can be found elsewhere (see references in Appendix 2).

## 8.8 Exercises

**(1) (Cusp points in Bazykin system)** Compute the $(\alpha, \delta)$-coordinates of the cusp bifurcation points $C_{1,2}$ in the system

$$\begin{cases} \dot{x}_1 = x_1 - \dfrac{x_1 x_2}{1 + \alpha x_1} - \varepsilon x_1^2, \\ \dot{x}_2 = -\gamma x_2 + \dfrac{x_1 x_2}{1 + \alpha x_1} - \delta x_2^2, \end{cases}$$

for $\gamma = 1$. (*Hint:* The cusp points are *triple* roots of the polynomial

$$P(x_1) = \delta(1 + \alpha x_1)^2(1 - \varepsilon x_1) - (1 - \alpha)x_1 + 1,$$

with $x_1$ as the first coordinate of nontrivial equilibrium points.)

**(2) (Lemma 8.2)** Prove that a smooth system

$$\dot{x} = \alpha_1 + \alpha_2 x - x^3 + F(x, \alpha), \tag{E.1}$$

where $x \in \mathbb{R}^1$, $\alpha \in \mathbb{R}^2$, and $F(x, \alpha) = O(x^4)$, is locally topologically equivalent near $(x, \alpha) = (0, 0)$ to the system

$$\dot{x} = \alpha_1 + \alpha_2 x - x^3. \tag{E.2}$$

(*Hints:*

(a) Derive a system of two equations for a curve $\tilde{\Gamma}$ in the $(x, \alpha)$-space corresponding to the fold bifurcation in (E.1). Show that this curve is well-defined near the origin and can be locally parametrized by $\eta$. (*Hint to hint:* See Example 8.1.)

(b) Compute the leading terms of the Taylor expansions of the functions $\alpha_1 = A_1(x)$, $\alpha_2 = A_2(x)$, representing the curve $\tilde{\Gamma}$ near $x = 0$. Show that the projection $\tilde{T}$ of $\tilde{\Gamma}$ onto the $(\alpha_1, \alpha_2)$-plane has two branches near the origin, $\tilde{T}_{1,2}$, located in opposite (with respect to the axis $\alpha_1 = 0$) half-planes of the parameter plane and terminating at the point $\alpha = 0$, being tangent to this axis.

(c) Explain why there are no other bifurcation curves near the origin in (E.1). Construct a local homeomorphism of the parameter plane that maps the curves $\tilde{T}_{1,2}$ into the corresponding curves $T_{1,2}$ of the truncated system (E.2) (see Section 8.2.2). Show that the resulting map is differentiable, and compute several of its Taylor coefficients at $\alpha = 0$.

(d) Show that in a region containing the upper half-axis $\{\alpha : \alpha_1 = 0, \alpha_2 > 0\}$ system (E.1) has three equilibria near the origin, while in the parameter region containing the lower half-axis $\{\alpha : \alpha_1 = 0, \alpha_2 < 0\}$ there is only one equilibrium. Compare the number and stability of the equilibria of (E.1) and (E.2) in the corresponding regions, along the fold branches, and at the origin of the parameter plane.

(e) Construct a parameter-dependent local homeomorphism mapping equilibria of (E.1) into those of (E.2) for all sufficiently small $\|\alpha\|$. Show that this map provides local topological equivalence of the studied systems.)

**(3) (Bautin bifurcation in a predator-prey system)** The following system is a generalization of Volterra equations by Bazykin & Khibnik [1981]:

$$\begin{cases} \dot{x} = \dfrac{x^2(1-x)}{n+x} - xy, \\ \dot{y} = -\gamma y(m-x), \end{cases}$$

(see also Bazykin [1985]). Here $n, m$, and $\gamma$ are positive parameters.

(a) Derive the equation for the Hopf bifurcation curve in the system and show that it is independent of $\gamma$.

(b) Using the algorithm from Chapter 3 (Section 3.5), compute the expression for the first Lyapunov coefficient along the Hopf curve and show that it vanishes at a Bautin point when

$$(m, n) = \left(\frac{1}{4}, \frac{1}{8}\right).$$

(c) Compute the second Lyapunov coefficient using (8.23) and prove that the Bautin bifurcation is nondegenerate for all $\gamma > 0$.

(d) Sketch the bifurcation diagram (the parametric portrait on the $(m, n)$-plane and all possible phase portraits) of the system. (*Hint:* See Bazykin [1985, p. 42].)

**(4) (Bautin bifurcation in a laser model)** Show that the following model of a laser with controllable resonator [Bautin & Leontovich 1976, pp. 320-329]:

$$\begin{cases} \dot{m} = Gm\left(n - \dfrac{\beta}{\rho m + 1} - 1\right), \\ \dot{n} = \alpha - (m+1)n, \end{cases}$$

where $m \geq 0,\ \rho > 0,\ G > 1$, and $\alpha, \beta > 0$, has a Bautin bifurcation point on the Hopf curve on the $(\alpha, \beta)$-plane if

$$\rho < 1$$

and

$$\rho - 1 + G\rho > 0.$$

(*Hint:* Parametrize the Hopf bifurcation curve by the $m$-coordinate of the nontrivial equilibrium. See Bautin & Leontovich [1976, pp. 320 – 329].)

**(5) (Regularity of Hopf bifurcation curve)** Show that the Jacobian matrix of (8.5) has maximal rank (equal to 3) at $(x^0, \alpha^0)$ if the equilibrium $x^0$ exhibits a generic Hopf bifurcation in the sense of Chapter 3 at $\alpha^0$.

**(6) (Bialternate product for $n = 3$)** Given a $3 \times 3$ matrix $A$, construct another $3 \times 3$ matrix $B \equiv 2A \odot I$, whose determinant equals the product of all formally distinct sums of the eigenvalues of $A$:

$$\det B = (\lambda_1 + \lambda_2)(\lambda_2 + \lambda_3)(\lambda_1 + \lambda_3).$$

(*Hint:* The elements of $B$ are certain *linear combinations* of those of $A$ (see Appendix B to Chapter 10 for the answer).)

**(7) (Bogdanov-Takens points)**

(a) By calculation of the normal form coefficients, prove that the averaged forced Van der Pol oscillator [Holmes & Rand 1978]

$$\begin{cases} \dot{x}_1 = \alpha_1 x_2 + x_1(1 - x_1^2 - x_2^2), \\ \dot{x}_2 = \alpha_1 x_1 + x_2(1 - x_1^2 - x_2^2) - \alpha_2, \end{cases}$$

exhibits a nondegenerate Bogdanov-Takens bifurcation at $(\alpha_1, \alpha_2) = \left(\frac{1}{2}, \frac{1}{2}\right)$. (*Hint:* The critical equilibrium with a double zero eigenvalue has coordinates $(x_1, x_2) = \left(\frac{1}{2}, \frac{1}{2}\right)$.)

(b) Show that a prototype reference adaptive control system with the so-called $\sigma$-modification adaptation law (see Salam & Bai [1988]),

$$\begin{cases} \dot{x}_1 = x_1 - x_1 x_2 + 1, \\ \dot{x}_2 = \alpha_1 x_2 + \alpha_2 x_1^2, \end{cases}$$

has a nondegenerate Bogdanov-Takens bifurcation at $\alpha = \left(-\frac{2}{3}, \frac{8}{81}\right)$. Sketch the phase portraits of the system near the BT-point.

**(8) (Bogdanov-Takens bifurcations in predator-prey systems)**

(a) Show that the following predator-prey model by Bazykin [1974] (cf. Example 8.3),

$$\begin{cases} \dot{x}_1 = x_1 - \dfrac{x_1 x_2}{1 + \alpha x_1}, \\ \dot{x}_2 = -\gamma x_2 + \dfrac{x_1 x_2}{1 + \alpha x_1} - \delta x_2^2, \end{cases}$$

has Hopf and fold bifurcation curves in the $(\alpha, \delta)$-plane that touch at a Bogdanov-Takens point. Prove, at least for $\gamma = 1$, that this codim 2 bifurcation is nondegenerate.

(b) Show that a predator-prey system analyzed by Bazykin, Berezovskaya, Denisov & Kuznetsov [1981],

$$\begin{cases} \dot{x}_1 = x_1 - \dfrac{x_1 x_2}{(1 + \alpha_1 x_1)(1 + \alpha_2 x_2)}, \\ \dot{x}_2 = -\gamma x_2 + \dfrac{x_1 x_2}{(1 + \alpha_1 x_1)(1 + \alpha_2 x_2)}, \end{cases}$$

exhibits different types of nondegenerate Bogdanov-Takens bifurcations at a point on the $\alpha$-plane, depending on whether $\gamma > 1$ or $\gamma < 0$. (*Hint:* Introduce new variables

$$y_k = \frac{x_k}{1 + \alpha_k x_k}, \quad k = 1, 2,$$

or multiply the equations by $(1 + \alpha_1 x_1)(1 + \alpha_2 x_2)$ to get an orbitally equivalent polynomial system.)

Could you also analyze the case $\gamma = 1$? (*Hint:* For $\gamma = 1$ the system is invariant under the transformation $(x_1, x_2) \mapsto (x_2, x_1)$, $t \mapsto -t$, along its Hopf curve $\alpha = \beta$. Therefore, the Hopf bifurcation is degenerate and the system has a family of closed orbits around a center.)

**(9) (Normal form for a codim 3 bifurcation)** The system

$$\begin{cases} \dot{x}_1 = x_2, \\ \dot{x}_2 = \alpha_1 + \alpha_2 x_1 + \alpha_3 x_2 + b x_1^3 + d x_1 x_2 + e x_1^2 x_2, \end{cases}$$

where the coefficients satisfy $b \neq 0$, $d \neq 0$, and $d^2 + 8b \neq 0$, is a normal form for a degenerate Bogdanov-Takens bifurcation when condition $(BT.2)$ is violated (see Bazykin, Kuznetsov & Khibnik [1985, 1989], Dumortier, Roussarie, Sotomayor & Żołądek [1991]). Consider the "focus case,"

$$d > 0, \; e < 0, \; b < -\frac{d^2}{8}.$$

(a) Obtain the phase portrait of the normal form at $\alpha = 0$. How many equilibria can the system have for $\alpha \neq 0$, and what are their possible types?

(b) Derive the equations for the fold and Hopf bifurcation surfaces in the parameter space $(\alpha_1, \alpha_2, \alpha_3)$ and sketch them in a graph. Verify that the fold bifurcation surface has a cusp singularity line. Indicate the number and stability of equilibria in the different parameter regions.

(c) Compute the curve corresponding to the Bogdanov-Takens bifurcation in the system. Check that the fold and the Hopf surfaces are tangent along the Bogdanov-Takens curve. Guess how the homoclinic bifurcation surface bounded by the BT-curve is shaped.

(d) Find a line on the Hopf surface along which the first Lyapunov coefficient vanishes. Compute the second Lyapunov coefficient at this line and check that it is nonzero near the codim 3 point at the origin. Guess the location of the cycle fold bifurcation surface.

(e) Draw the intersections of the obtained surfaces with a small sphere centered at the origin in $\alpha$-space by projecting the two hemispheres onto the plane $(\alpha_1, \alpha_3)$. Show that there must be a "big" homoclinic loop bifurcation in the parameter portrait.

(f) Explain why the resulting parameter portrait is very similar to that of Bazykin's predator-prey system from Example 8.3, and find this codim 3 bifurcation $(\lambda_1 = \lambda_2 = b_{20} = 0)$ in the parameter space of that system.

**(10) (Bogdanov-Takens bifurcation revisited)**

(a) *Takens-Bogdanov equivalence.* Check that the change of coordinates

$$\begin{aligned} y_1 &= \frac{4a^2}{b}\left(x_1 + \frac{\alpha_2}{2a}\right), \\ y_2 &= \frac{8a^3}{b^2}\left(x_2 + \alpha_2 x_1 + a x_1^2\right), \end{aligned}$$

transforms a Takens normal form for the double-zero bifurcation

$$\begin{cases} \dot{x}_1 = x_2 + \alpha_2 x_1 + a x_1^2, \\ \dot{x}_2 = \alpha_1 + b x_1^2, \end{cases}$$

where $ab \neq 0$, into a system that is orbitally equivalent (after a possible reverse of time) to one of the Bogdanov forms:

$$\begin{cases} \dot{y}_1 = y_2, \\ \dot{y}_2 = \beta_1 + \beta_2 y_1 + y_1^2 \pm y_1 y_2 + O(\|y\|^3), \end{cases}$$

for

$$\beta_1 = \frac{16a^4}{b^3}\left(\alpha_1 + \frac{b\alpha_2^2}{4a^2}\right), \quad \beta_2 = -\frac{4a}{b}\alpha_2$$

[Dumortier 1978].

(b) *Lyapunov coefficient near BT-point.* Compute the Lyapunov coefficient $l_1$ along the Hopf line $H$ in the Bogdanov normal form (8.51), and show that it is negative near the origin.

(c) *Saddle quantity near BT-point.* Compute the saddle quantity $\sigma_0 = \lambda_1 + \lambda_2$ of the saddle point in (8.51), and show that it is negative along the homoclinic bifurcation curve $P$.

**(11) (Fold-Hopf bifurcation in Rössler's prototype chaotic system)** Consider the system due to Rössler [1979]:

$$\begin{cases} \dot{x} = -y - z, \\ \dot{y} = x + ay, \\ \dot{z} = bx - cz + xz, \end{cases}$$

with parameters $(a, b, c) \in \mathbb{R}^3$.

(a) Show that the system possesses at most two equilibria, $O$ and $P$, and find their coordinates.

(b) Check that the equilibria $O$ and $P$ "collide" at the surface

$$T = \{(a, b, c) : \ c = ab\}$$

and that the coinciding equilibria have eigenvalues $\lambda_1 = 0, \ \lambda_{2,2} = \pm i\omega_0$ with $\omega_0 = \sqrt{2 - a^2} > 0$, if $(a, b, c) \in T$ and

$$b = 1, \quad a^2 < 2.$$

(c) Compute the Poincaré (8.63) and Gavrilov (8.71) normal forms of the system along the locus of fold-Hopf points, and find the corresponding $s$ and $\theta$ using formulas from Section 8.5.1. Verify that $s = 1, \ \theta < 0$, and decide which of the possible "canonical" bifurcation diagrams appears in the Rössler system. (*Warning:* Since the system always has an equilibrium at the origin, the fold-Hopf bifurcation is degenerate with respect to the parameters in the system. Therefore, its bifurcation curves are only "induced" by those of the normal form.)

(d) Check that $-2 < \theta < 0$, so that if a saddle-focus homoclinic orbit exists, it satisfies the Shil'nikov condition and therefore "strange" dynamics exists near the codim 2 bifurcation. (*Hint:* See Gaspard [1993] for detailed treatment.)

**(12) (Lorenz-84 model)** Consider the following system appearing in atmospheric studies (Lorenz [1984], Shil'nikov, Nicolis & Nicolis [1995])

$$\begin{cases} \dot{x} = -y^2 - z^2 - ax + aF, \\ \dot{y} = xy - bxz - y + G, \\ \dot{z} = bxy + xz - z, \end{cases}$$

where $(a, b, F, G)$ are parameters. Show that the system undergoes a fold-Hopf bifurcation at

$$F^* = \frac{3a^2 + 3a^2b^2 + 12ab^2 + 12b^2 + 4a}{4(a + ab^2 + 2b^2)},$$

$$G^* = \frac{\sqrt{a}(a^2 + a^2b^2 + 4ab^2 + 4b^2)}{4\sqrt{a + ab^2 + 2b^2}}.$$

(*Hint:* Use the fact that at the fold-Hopf point of a three-dimensional system both the trace and the determinant of the Jacobian matrix vanish.)

**(13) (Fold-Hopf bifurcation revisited)**

(a) Compute the first Lyapunov coefficient along the Hopf bifurcation curve in the truncated amplitude system (8.82) for $s\theta < 0$, and verify the expression given in Section 8.5.2.

(b) Consider the case $s = 1$, $\theta < 0$, when a heteroclinic cycle is possible. Let $\lambda_1 < 0 < \lambda_2$ and $\mu_1 < 0 < \mu_2$ be the eigenvalues of the saddles $E_1$ and $E_2$ at the $\xi$-axis, respectively. Prove that, if it exists, the heteroclinic cycle is unstable from the inside whenever

$$\frac{\lambda_2\mu_2}{\lambda_1\mu_1} > 1,$$

and verify this inequality for the truncated system (8.82).

(c) Obtain the critical phase portraits of system (8.82) at $\beta_1 = \beta_2 = 0$. (*Hint:* See Wiggins [1990, pp. 331–337].)

(d) Explain why the truncated system (8.82) cannot have periodic orbits if $s\theta > 0$ and prove that in this case it is locally topologically equivalent near the origin to the quadratic system

$$\begin{cases} \dot{\xi} = \beta_1 + \xi^2 + s\rho^2, \\ \dot{\rho} = \rho(\beta_2 + \theta\xi). \end{cases}$$

**(14) (Codim 2 normal forms as perturbations of Hamiltonian systems; read Appendix A first)**

(a) Prove that $Q(\frac{1}{6}) = \frac{1}{7}$, where

$$Q(h) = \frac{I_2(h)}{I_1(h)}$$

is defined in Appendix A, which deals with the Bogdanov-Takens bifurcation. (*Hint:* There are two equivalent approaches to the problem:

(i) Check that the homoclinic orbit to the saddle in (A.6), which is given by $H(\zeta) = \frac{1}{6}$, intersects the $\zeta_1$-axis at $\zeta_1 = -\frac{1}{2}$. Express $\zeta_2$ along the upper part of the orbit as a function of $\zeta_1$, and evaluate the resulting integrals over the range $-\frac{1}{2} \le \zeta_1 \le 1$.

(ii) The solution of (A.6) starting at the horizontal axis and corresponding to the homoclinic orbit can be written explicitly as

$$\zeta_1(t) = 1 - \frac{6}{(e^{\frac{t}{2}} + e^{-\frac{t}{2}})^2}, \quad \zeta_2(t) = 6\,\frac{(e^{\frac{t}{2}} - e^{-\frac{t}{2}})}{(e^{\frac{t}{2}} + e^{-\frac{t}{2}})^3}.$$

Therefore, the integrals $I_1\left(\frac{1}{6}\right)$ and $I_2\left(\frac{1}{6}\right)$ reduce to certain standard integrals over $-\infty < t < +\infty$.)

(b) Consider the amplitude system (8.82) for the fold-Hopf bifurcation:

$$\begin{cases} \dot{\xi} = \beta_1 + \xi^2 + s\rho^2, \\ \dot{\rho} = \rho(\beta_2 + \theta\xi + \xi^2), \end{cases}$$

with $s\theta < 0$ ("difficult case").

(i) Show that the following singular rescaling and nonlinear time reparametrization:

$$\xi = \delta x, \quad \rho = \delta y, \quad dt = \frac{y^q}{\delta} d\tau,$$

where $q$ is some real number, bring the system into the form

$$\begin{cases} \dot{x} = y^q(-s + x^2 + sy^2), \\ \dot{y} = y^q(\theta xy + \delta(\alpha y + x^2 y)), \end{cases}$$

where $\alpha$ and $\delta$ should be considered as new parameters related to the original ones by the formulas

$$\beta_1 = -s\delta^2, \quad \beta_2 = \alpha\delta^2.$$

(ii) Prove that for

$$q + 1 = -\frac{2}{\theta}$$

the rescaled system with $\delta = 0$,

$$\begin{cases} \dot{x} = y^q(-s + x^2 + sy^2), \\ \dot{y} = y^q \theta xy, \end{cases}$$

is Hamiltonian with the Hamilton function

$$H(x, y) = \frac{\theta}{2} y^{q+1} \left( s - x^2 + \frac{sy^2}{\theta - 1} \right) \quad \text{if } \theta \neq 1,$$

or

$$H(x, y) = \frac{s - x^2}{2y^2} + s \ln y \quad \text{if } \theta = 1.$$

(Since $s\theta < 0$, $\theta = 1$ implies $s = -1$.) Draw the level curves of the Hamilton function $H$ for different combinations of $s$ and $\theta$ with $s\theta < 0$.

(iii) Take $s = 1$ and $\theta < 0$. Following the ideas presented in Appendix A, show that in this case the heteroclinic connection in the perturbed $(x, y)$-system happens if

$$\alpha = -\frac{1}{3} K(0),$$

where the function $K(h)$ is defined for $h \in \left[ \frac{\theta^2}{2(\theta - 1)}, 0 \right]$ by the ratio

$$K(h) = \frac{I_3(h)}{I_1(h)},$$

where

$$I_k(h) = \int_{H(x,y)=h} y^q x^k \, dy, \quad k = 1, 3.$$

Compute $K(0)$ as a function of $\theta$. Derive from this information the approximation of the heteroclinic curve $P$ in the original amplitude system on the $(\beta_1, \beta_2)$-plane.

(iv) Prove monotonicity of $K(h)$ with respect to $h$, thus establishing the uniqueness of the limit cycle in the truncated normal form. (*Hint:* See Chow, Li & Wang [1989*b*, 1989*a*].)

(c) Consider the truncated amplitude system (8.115) for the "difficult" case of the Hopf-Hopf bifurcation

$$\begin{cases} \dot{\xi}_1 = \xi_1(\mu_1 + \xi_1 - \theta\xi_2 + \Theta\xi_2^2), \\ \dot{\xi}_2 = \xi_2(\mu_2 + \delta\xi_1 - \xi_2 + \Delta\xi_1^2), \end{cases}$$

with $(\theta, \delta)$ belonging to one of the cases I, II, or VI (when the Hopf bifurcation is possible, see Section 8.6.2).

(i) Show that by introduction of new parameters $\alpha > 0$ and $\beta$, which parametrize the neighborhood of the Hopf curve,

$$\mu_1 = -\alpha, \ \mu_2 = \frac{\delta - 1}{\theta - 1}\alpha + \alpha\beta,$$

and by a singular rescaling,

$$\xi_1 = \alpha x, \quad \xi_2 = \alpha y, \quad t = \frac{1}{\alpha}\tau,$$

we transform the normal form into the system

$$\begin{aligned} \dot{x} &= x(-1 + x - \theta y + \alpha\Theta y^2), \\ \dot{y} &= y\left(\frac{\delta - 1}{\theta - 1} + \beta + \delta x - y + \alpha\Delta x^2\right). \end{aligned}$$

(ii) Check that the system corresponding to $\alpha = \beta = 0$,

$$\begin{aligned} \dot{x} &= x(-1 + x - \theta y), \\ \dot{y} &= y\left(\frac{\delta - 1}{\theta - 1} + \delta x - y\right), \end{aligned}$$

is orbitally equivalent for $x, y > 0$ to a Hamiltonian system with the Hamilton function

$$H = \frac{1}{q}x^p y^q \left(-1 + x + \frac{\theta - 1}{\delta - 1}y\right),$$

where

$$p = \frac{1 - \delta}{\theta\delta - 1}, \quad q = \frac{1 - \theta}{\theta\delta - 1}.$$

(*Hint:* The time reparametrization factor is $\gamma(x, y) = x^{p-1}y^{q-1}$.)

(iii) Sketch phase portraits of the Hamiltonian system

$$\begin{cases} \dot{x} = \dfrac{\partial H}{\partial y}, \\ \dot{y} = -\dfrac{\partial H}{\partial x}, \end{cases}$$

for cases I, II, and VI. Verify that in case VI ($p, q > 0$) closed level curves $H = \text{const}$ fill a triangle bounded by a heteroclinic cycle $H = 0$.

(iv) Consider case VI. Following the method of Appendix A, show that the rescaled system with $\alpha, \beta \neq 0$ and the reparametrized time has a heteroclinic cycle if

$$\beta = -R(0)\alpha + O(\alpha^2),$$

where

$$R(h) = \frac{I_1(h)}{I_0(h)},$$

and

$$I_1(h) = \int_{H \le h} \left(p\Theta x^{p-1}y^{q+1} + q\Delta x^{p+1}y^{q-1}\right) dx\, dy,$$

$$I_0(h) = \int_{H \le h} q x^{p-1} y^{q-1} dx\, dy.$$

(*Hint:* Use Stokes's formula and take into account that the divergence of a Hamiltonian vector field is equal to zero.)

(v) Compute $R(0)$ in terms of $\theta, \delta, \Theta$, and $\Delta$, and derive the quadratic approximation of the heteroclinic curve $Y$. Verify the representation for $Y$ given in Section 8.6.2.

(vi) Show that the Hopf bifurcation curve $H$ and the heteroclinic curve $Y$ in the truncated amplitude system (8.115) have at least quadratic tangency at $\mu_1 = \mu_2 = 0$.

**(15) (Poincaré normal form for Hopf-Hopf bifurcation)**

(a) Derive formulas (8.90)–(8.93) for the critical resonant coefficients of the Hopf-Hopf bifurcation using one of the available computer algebra systems.

(*Hint:* The following sequence of MAPLE commands solves the problem:

```
> readlib(mtaylor);
> readlib(coeftayl);
```

These commands load the procedures **`mtaylor`** and **`coeftayl`**, which compute the truncated multivariate Taylor series expansion and its individual coefficients, respectively, from the MAPLE library.

```
> P:=mtaylor(sum(sum(sum(sum(
>                g[j,k,l,m]*z^j*z1^k*u^l*u1^m,
>             j=0..3),k=0..3),l=0..3),m=0..3),
>    [z,z1,u,u1],4);
> Q:=mtaylor(sum(sum(sum(sum(
>                h[j,k,l,m]*z^j*z1^k*u^l*u1^m,
>             j=0..3),k=0..3),l=0..3),m=0..3),
>    [z,z1,u,u1],4);
> P1:=mtaylor(sum(sum(sum(sum(
>                g1[j,k,l,m]*z1^j*z^k*u1^l*u^m,
>             j=0..3),k=0..3),l=0..3),m=0..3),
>     [z,z1,u,u1],4);
> Q1:=mtaylor(sum(sum(sum(sum(
>                h1[j,k,l,m]*z1^j*z^k*u1^l*u^m,
>             j=0..3),k=0..3),l=0..3),m=0..3),
>     [z,z1,u,u1],4);
> for jj from 0 to 1 do
>     for kk from 0 to 1 do
>         for ll from 0 to 1 do
>             for mm from 0 to 1 do
>                 if jj+kk+ll+mm < 2 then
>                    g[jj,kk,ll,mm]:=0; h[jj,kk,ll,mm]:=0;
>                    g1[jj,kk,ll,mm]:=0; h1[jj,kk,ll,mm]:=0;
>                 fi;
>             od;
```

```
>         od;
>     od;
> od;
> g[1,0,0,0]:=I*omega; g1[1,0,0,0]:=-I*omega;
> h[0,0,1,0]:=I*Omega; h1[0,0,1,0]:=-I*Omega;
```

By this we have specified the right-hand sides of (8.88) and their conjugate expressions at the critical parameter value; **z,z1,u,u1** stand for $z_1, \bar{z}_1, z_2, \bar{z}_2$, respectively, while **omega** and **Omega** correspond to $\omega_1$ and $\omega_2$, respectively.

```
> R:=I*omega*v+G[2,1,0,0]*v^2*v1 + G[1,0,1,1]*v*w*w1;
> S:=I*Omega*w+H[1,1,1,0]*v*v1*w + H[0,0,2,1]*w^2*w1;
```

These are the specifications of the right-hand sides of the normalized system (8.89) (up to and including order 3), where **v,v1,w,w1** represent $w_1, \bar{w}_1, w_2, \bar{w}_2$, respectively.

```
> VV:=mtaylor(sum(sum(sum(sum(
>                  V[j,k,l,m]*z^j*z1^k*u^l*u1^m,
>             j=0..3),k=0..3),l=0..3),m=0..3),
>    [z,z1,u,u1],4);
> WW:=mtaylor(sum(sum(sum(sum(
>                  W[j,k,l,m]*z^j*z1^k*u^l*u1^m,
>             j=0..3),k=0..3),l=0..3),m=0..3),
>    [z,z1,u,u1],4);
> VV1:=mtaylor(sum(sum(sum(sum(
>                  V1[j,k,l,m]*z1^j*z^k*u1^l*u^m,
>             j=0..3),k=0..3),l=0..3),m=0..3),
>    [z,z1,u,u1],4);
> WW1:=mtaylor(sum(sum(sum(sum(
>                  W1[j,k,l,m]*z1^j*z^k*u1^l*u^m,
>             j=0..3),k=0..3),l=0..3),m=0..3),
>     [z,z1,u,u1],4);
> for j from 0 to 1 do
>     for k from 0 to 1 do
>         for l from 0 to 1 do
>             for m from 0 to 1 do
>                 if j+k+l+m < 2 then
>                    V[j,k,l,m]:=0; V1[j,k,l,m]:=0;
>                    W[j,k,l,m]:=0; W1[j,k,l,m]:=0;
>                 fi;
>             od;
>         od;
>     od;
> od;
> V[1,0,0,0]:=1; V[2,1,0,0]:=0; V[1,0,1,1]:=0;
> W[0,0,1,0]:=1; W[1,1,1,0]:=0; W[0,0,2,1]:=0;
> V1[1,0,0,0]:=1; V1[2,1,0,0]:=0; V1[1,0,1,1]:=0;
> W1[0,0,1,0]:=1; W1[1,1,1,0]:=0; W1[0,0,2,1]:=0;
```

By these commands the transformation (and its conjugate) that bring the system into the normal form is defined. Its coefficients have to be found.

```
> V_z:=diff(VV,z); V_z1:=diff(VV,z1);
> V_u:=diff(VV,u); V_u1:=diff(VV,u1);
> W_z:=diff(WW,z); W_z1:=diff(WW,z1);
> W_u:=diff(WW,u); W_u1:=diff(WW,u1);
```

The partial derivatives of the normalizing transformation are computed.

```
> D_1:=R-(V_z*P+V_z1*P1+V_u*Q+V_u1*Q1);
> D_2:=S-(W_z*P+W_z1*P1+W_u*Q+W_u1*Q1);
```

Conditions D_1 = 0 and D_2 = 0 are equivalent to the requirement that the specified transformation does the normalization. Now we should express D_1 and D_2 in terms of z,z1,u, and u1. This is achieved by

```
> v:=VV; v1:=VV1;
> w:=WW; w1:=WW1;
```

Now we can expand D_1 and D_2 as Taylor series with respect to z,z1,u,u1:

```
> DD_1:=mtaylor(D_1,[z,z1,u,u1],4);
> DD_2:=mtaylor(D_2,[z,z1,u,u1],4);
```

Everything is prepared to find the quadratic coefficients of the transformation. This can be done by equating the corresponding Taylor coefficients in DD_1 and DD_2 to zero, and solving the resulting equation for V[j,k,l,m] and W[j,k,l,m] with j+k+l+m=2:

```
> for j from 0 to 2 do
>     for k from 0 to 2 do
>         for l from 0 to 2 do
>             for m from 0 to 2 do
>                 if j+k+l+m=2 then
>                    V[j,k,l,m]:=solve(
>          coeftayl(DD_1,[z,z1,u,u1]=[0,0,0,0],[j,k,l,m])=0,
>                    V[j,k,l,m]);
>                    W[j,k,l,m]:=solve(
>          coeftayl(DD_2,[z,z1,u,u1]=[0,0,0,0],[j,k,l,m])=0,
>                    W[j,k,l,m]);
>                 fi;
>             od;
>         od;
>     od;
> od;
```

Finally, we are able to find the coefficients of the resonant terms:

```
> G[2,1,0,0]:=solve(
> coeftayl(DD_1,[z,z1,u,u1]=[0,0,0,0],[2,1,0,0])=0,G[2,1,0,0]);
> G[1,0,1,1]:=solve(
```

```
> coeftayl(DD_1,[z,z1,u,u1]=[0,0,0,0],[1,0,1,1])=0,G[1,0,1,1]);
> H[0,0,2,1]:=solve(
> coeftayl(DD_2,[z,z1,u,u1]=[0,0,0,0],[0,0,2,1])=0,H[0,0,2,1]);
> H[1,1,1,0]:=solve(
> coeftayl(DD_2,[z,z1,u,u1]=[0,0,0,0],[1,1,1,0])=0,H[1,1,1,0]);
```

The problem is solved.)

(b) Extend the described program to obtain the coefficients of the fifth-order resonant terms.

## 8.9 Appendix A: Limit cycles and homoclinic orbits of Bogdanov normal form

Consider the normal form for the Bogdanov-Takens bifurcation with $s = -1$:

$$\begin{cases} \dot{\xi}_1 = \xi_2, \\ \dot{\xi}_2 = \beta_1 + \beta_2\xi_1 + \xi_1^2 - \xi_1\xi_2. \end{cases} \tag{A.1}$$

**Theorem 8.9** *There is a unique, smooth curve $P$ corresponding to a saddle homoclinic bifurcation in the system (A.1) that originates at $\beta = 0$ and has the local representation*

$$P = \left\{(\beta_1, \beta_2) : \ \beta_1 = -\frac{6}{25}\beta_2^2 + o(\beta_2^2), \ \beta_2 < 0\right\}.$$

*Moreover, for $\|\beta\|$ small, system (A.1) has a unique and hyperbolic stable cycle for parameter values inside the region bounded by the Hopf bifurcation curve $H$ and the homoclinic bifurcation curve $P$, and no cycles outside this region.*

**Outline of the proof:**

*Step 1* (*Shift of coordinates*). The cycles and homoclinic orbits can exist only if there are two equilibria in (A.1). Thus, restrict attention to a parameter region to the left of the fold curve $T$ (see Section 8.4.2). Translate the origin of the coordinate system to the left (antisaddle) equilibrium $E_1$ of (A.1):

$$\begin{cases} \xi_1 = \eta_1 + \eta_1^0, \\ \xi_2 = \eta_2, \end{cases}$$

where

$$\eta_1^0 = -\frac{\beta_2 + \sqrt{\beta_2^2 - 4\beta_1}}{2}$$

is the $\eta_1$-coordinate of $E_1$. This obviously gives

$$\begin{cases} \dot{\eta}_1 = \eta_2, \\ \dot{\eta}_2 = \eta_1(\eta_1 - \nu) - (\eta_1^0\eta_2 + \eta_1\eta_2), \end{cases} \tag{A.2}$$

where

$$\nu = \sqrt{\beta_2^2 - 4\beta_1}$$

is the *distance* between the left equilibrium $E_1$ and the right equilibrium $E_2$.

*Step 2* (*Blowing-up*). Perform a *singular rescaling* that makes the distance between the equilibria equal to 1, *independent* of the parameters, and introduce a new time:

$$\eta_1 = \frac{\zeta_1}{\nu}, \quad \eta_2 = \frac{\zeta_2}{\nu^{3/2}}, \quad t = \frac{\tau}{\nu^{1/2}}. \tag{A.3}$$

This rescaling reduces (8.55) to

$$\begin{cases} \dot{\zeta}_1 = \zeta_2, \\ \dot{\zeta}_2 = \zeta_1(\zeta_1 - 1) - (\gamma_1\zeta_2 + \gamma_2\zeta_1\zeta_2), \end{cases} \tag{A.4}$$

where the dots mean derivatives with respect to the new time $\tau$ and

$$\begin{cases} \gamma_1 = \eta_1^0\nu^{-1/2}, \\ \gamma_2 = \nu^{1/2}. \end{cases} \tag{A.5}$$

Only *nonnegative* values of $\gamma_2$ should be considered. Clearly, $\gamma \to 0$ as $\beta \to 0$ inside the two-equilibrum region. The rescaling (A.3) acts as a "microscope" that blows up a neighborhood of the origin $\eta = 0$ (notice the difference in the expanding strength in the $\eta_1$- and $\eta_2$-directions). System (A.4) is orbitally equivalent to a system *induced* by (A.2) with the help of (A.5) (see Chapter 2 for the definitions). Studing the limit cycles and homoclinic orbits of (A.4) for $\gamma \neq 0$ provides the complete information on these objects in (A.2).

*Step 3* (*Hamiltonian properties*). For $\|\gamma\|$ small, system (A.4) can be viewed as a perturbation of the system

$$\begin{cases} \dot{\zeta}_1 = \zeta_2, \\ \dot{\zeta}_2 = \zeta_1(\zeta_1 - 1). \end{cases} \tag{A.6}$$

This system is a *Hamiltonian system*,

$$\begin{cases} \dot{\zeta}_1 = \dfrac{\partial H(\zeta)}{\partial \zeta_2}, \\ \dot{\zeta}_2 = -\frac{\partial H(\zeta)}{\partial \zeta_1}, \end{cases}$$

with the Hamilton (energy) function

$$H(\zeta) = \frac{\zeta_1^2}{2} + \frac{\zeta_2^2}{2} - \frac{\zeta_1^3}{3}. \tag{A.7}$$

The Hamiltonian $H$ is constant along orbits of (A.6), that is, $\dot{H} = 0$, and these orbits therefore are (oriented) level curves $H(\zeta) = \text{const}$ of the Hamiltonian of (A.7) (see **Fig. 8.33**). System (A.6) has two equilibria,

$$S_0 = (0,0), \quad S_1 = (0,1),$$

corresponding to the equilibria $E_1$ and $E_2$ of (A.1). The equilibrium $S_0$ is a *center* surrounded by closed orbits, while the equilibrium $S_1$ is a (neutral) saddle. The values of the Hamiltonian at these equilibria are

$$H(S_0) = 0, \quad H(S_1) = \frac{1}{6}.$$

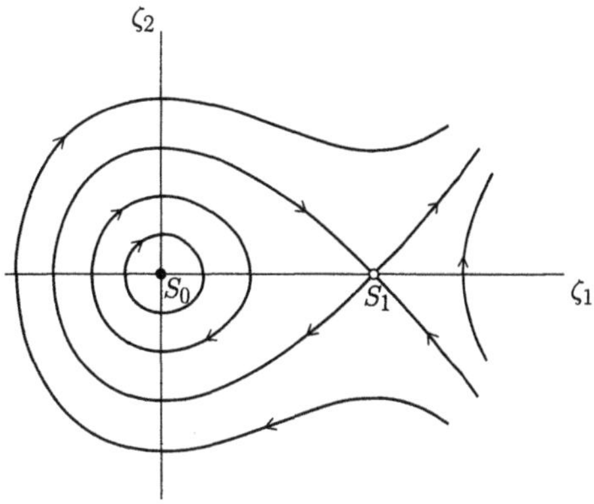


**Fig. 8.33.** Phase portrait of the Hamiltonian system (A.6).

The saddle separatrices are described by the level curve $H(\zeta) = \frac{1}{6}$. Two of them, located to the left of the saddle, form a homoclinic orbit bounding the family of closed orbits around $S_0$. The segment $0 \leq \zeta_1 \leq 1$ of the horizontal axis between $S_0$ and $S_1$ can be parametrized by $h \in \left[0, \frac{1}{6}\right]$ if we take as $h$ the value of the Hamiltonian function $H(\zeta_1, 0)$, which is monotone for $\zeta_1 \in [0, 1]$.

*Step 4 (Definition of the split function).* Consider now system (A.4) for small but nonzero $\|\gamma\|$ when it is no longer Hamiltonian. Notice that $S_0$ and $S_1$ are the equilibria for (A.4) for all $\gamma$. Since system (A.6) is highly structurally unstable, the topology of the phase portrait of (A.4) for $\gamma \neq 0$ is totally different from that for (A.6): The family of the closed orbits disappears and the saddle separatrices usually split.

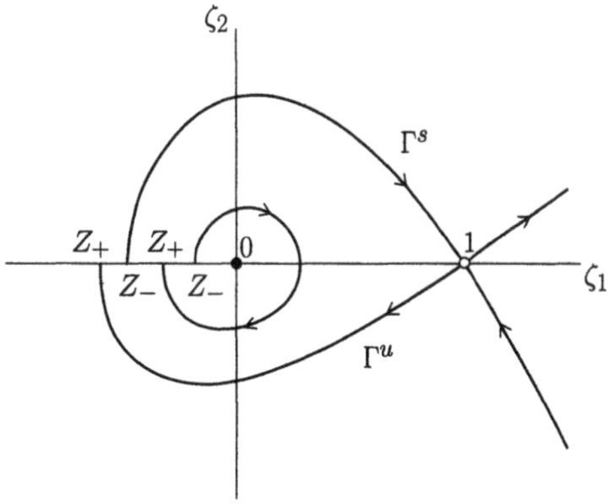


**Fig. 8.34.** Definition of the split function.

Using the introduced parametrization of the segment of the $\zeta_1$-axis between $S_0$ and $S_1$, take a point within this segment with $h \in \left(0, \frac{1}{6}\right)$ and consider an orbit $\Gamma$ of (A.4) passing through this point (see **Fig. 8.34**). Any such orbit will intersect the horizontal axis (at least) once more, in both forward and backward times. Denote these intersection points by $Z_+$ and $Z_-$, respectively. Now define an *orbit split function* $\Delta(h,\gamma)$ by taking the difference between the Hamiltonian values at the points $Z_-$ and $Z_+$:

$$\Delta(h,\gamma) = H(Z_-) - H(Z_+). \tag{A.8}$$

Extend this function to the end points of the segment by taking $\Delta(0,\gamma) = 0$ and using the same formula (A.8) for $h = \frac{1}{6}$, only now considering $Z_+$ and $Z_-$ as points of the intersection of the unstable and stable separatrices $\Gamma^u$ and $\Gamma^s$ of the saddle $S_1$ with the horizontal axis (see **Fig. 8.33**). Thus, $\Delta(\frac{1}{6},\gamma)$ is a *separatrix split function* as defined in Chapter 6. The function $\Delta$ is smooth in its domain of definition. In the Hamiltonian case, $\gamma = 0$, $\Delta(h,0) = 0$ for all $h \in \left[0, \frac{1}{6}\right]$ since the Hamiltonian is constant of motion. The equation

$$\Delta\left(\frac{1}{6},\gamma\right) = 0,$$

with the constraint $\gamma_2 \geq 0$, defines a curve $\mathcal{P}$ on the $(\gamma_1,\gamma_2)$-plane starting at the origin along which the system (A.4) has a *homoclinic orbit*. Similarly, the equation

$$\Delta(h,\gamma) = 0,$$

with $h \in (0, \frac{1}{6})$, specifies a curve $\mathcal{L}_h$ in the upper parameter half-plane $\gamma_2 > 0$ at which (A.4) has a *cycle* passing through a point between $S_0$ and $S_1$ corresponding to $h$.

*Step 5 (Approximation of the split function).* For $\gamma \neq 0$, the Hamiltonian $H(\zeta)$ varies along orbits of (A.4):

$$\dot{H} = \frac{\partial H}{\partial \zeta_1}\dot{\zeta}_1 + \frac{\partial H}{\partial \zeta_2}\dot{\zeta}_2 = -(\gamma_1\zeta_2^2 + \gamma_2\zeta_1\zeta_2^2).$$

Therefore,

$$\Delta(h,\gamma) = \int_{t_{Z_+}}^{t_{Z_-}} \dot{H}\, dt = \gamma_1 \int_\Gamma \zeta_2\, d\zeta_1 + \gamma_2 \int_\Gamma \zeta_1\zeta_2\, d\zeta_1, \tag{A.9}$$

where the orientation of $\Gamma$ is given by the direction of increasing time. Clearly, for $h = \frac{1}{6}$, the integrals should be interpreted as sums of the corresponding integrals along the separatrices $\Gamma^{u,s}$. Formula (A.9) is exact but involves the orbit(s) $\Gamma(\Gamma^{u,s})$ of (A.4), which we do not know explicitly. However, for small $\|\gamma\|$, the orbits of (A.4) deviate only slightly from the closed orbit (or the separatrix) of (A.6), and the integrals can be uniformly approximated by those taken along $H(\zeta) = h$:

$$\Delta(h,\gamma) = \gamma_1 \int_{H(\zeta)=h} \zeta_2\, d\zeta_1 + \gamma_2 \int_{H(\zeta)=h} \zeta_1\zeta_2\, d\zeta_1 + o(\|\gamma\|).$$

Denote the integrals involved in the last equation by

$$I_1(h) = \int_{H(\zeta)=h} \zeta_2\, d\zeta_1 = \int_{H(\zeta)\leq h} d\zeta_2\, d\zeta_1 \geq 0 \tag{A.10}$$

and

$$I_2(h) = \int_{H(\zeta)=h} \zeta_1\zeta_2\, d\zeta_1. \tag{A.11}$$

These integrals are certain *elliptic* integrals.

*Step 6* (*Uniqueness of the limit cycle in (A.4)*). By the Implicit Function Theorem, the curves $\mathcal{L}_h$ and $\mathcal{P}$ (introduced in Step 4) exist and have the representation

$$\gamma_1 = -\frac{I_2(h)}{I_1(h)}\gamma_2 + o(|\gamma_2|), \quad \gamma_2 \geq 0,$$

for $h \in (0, \frac{1}{6})$ and $h = \frac{1}{6}$, respectively. While $h$ varies from $h = 0$ to $h = \frac{1}{6}$, the curve $\mathcal{L}_h$ moves in the $(\gamma_1, \gamma_2)$-plane from the vertical half-axis $\{\gamma : \gamma_1 = 0, \gamma_2 \geq 0\}$ (since $I_2(0) = 0$) to the homoclinic curve $\mathcal{P}$. If this motion is *monotonous* with $h$, it will guarantee the uniqueness of the cycle of (A.4) for parameter values between the vertical half-axis and $\mathcal{P}$. The absence of cycles for all other parameter values $\gamma$ is obvious since any closed orbit of (A.4) must cross the segment between $S_0$ and $S_1$. Thus, the monotonicity of the function

$$Q(h) = \frac{I_2(h)}{I_1(h)} \tag{A.12}$$

for $h \in \left[0, \frac{1}{6}\right]$ is sufficient to prove the uniqueness of cycles. The function $Q(h)$ is a smooth function, $Q(0) = 0$, and

$$Q\left(\frac{1}{6}\right) = \frac{1}{7}$$

(see Exercise 14(a)). Meanwhile, the last equation leads to the following characterization for the homoclinic curve $\mathcal{P}$ in (A.4):

$$\mathcal{P} = \left\{(\gamma_1, \gamma_2) : \gamma_1 = -\frac{1}{7}\gamma_2 + o(|\gamma_2|), \ \gamma_2 \geq 0\right\}. \tag{A.13}$$

The graph of the function $Q(h)$ computed numerically is presented in **Fig. 8.35**. It is clearly monotonous. This fact can be proved without computers. Namely, the following lemma holds.

**Lemma 8.17** *$Q'(h) > 0$ for $h \in \left[0, \frac{1}{6}\right]$.* □

**Proof of Lemma 8.17:**

**Proposition 8.1 (Picard-Fuchs equations)** *The integrals $I_1(h)$ and $I_2(h)$ satisfy the system of differential equations*

$$\begin{cases} h\left(h - \frac{1}{6}\right)\dot{I}_1 = \left(\frac{5}{6}h - \frac{1}{6}\right)I_1 + \frac{7}{36}I_2, \\ h\left(h - \frac{1}{6}\right)\dot{I}_2 = -\frac{1}{6}hI_1 + \frac{7}{6}hI_2. \end{cases} \tag{A.14}$$

**Proof:**

Take the equation $H(\zeta) = h$ for the closed orbit of (A.6) corresponding to a value of $h \in (0, \frac{1}{6})$ (see (A.7)):

$$\frac{\zeta_1^2}{2} + \frac{\zeta_2^2}{2} - \frac{\zeta_1^3}{3} = h. \tag{A.15}$$

Considering $\zeta_2$ as a function of $\zeta_1$ and $h$, by differentiating (A.12) with respect to $h$ we obtain

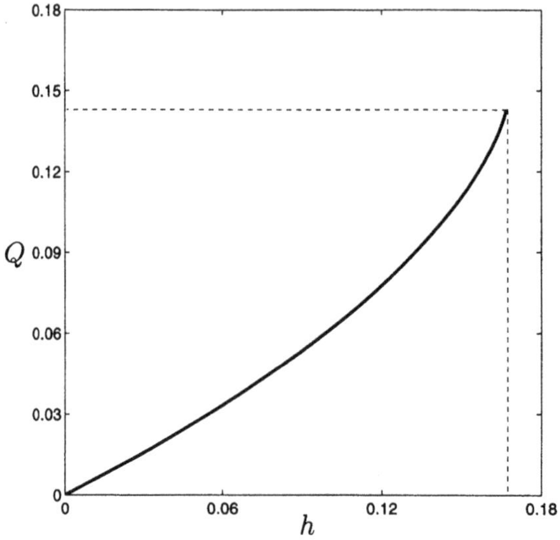


**Fig. 8.35.** Graph of the function $Q(h)$.

$$\zeta_2 \frac{\partial \zeta_2}{\partial h} = 1.$$

Thus,

$$\frac{dI_1}{dh} = \int_{H(\zeta)=h} \frac{d\zeta_1}{\zeta_2}, \tag{A.16}$$

and

$$\frac{dI_2}{dh} = \int_{H(\zeta)=h} \frac{\zeta_1 d\zeta_1}{\zeta_2}. \tag{A.17}$$

On the other hand, differentiating (A.15) with respect to $\zeta_1$ yields

$$\zeta_1 + \zeta_2 \frac{\partial \zeta_2}{\partial \zeta_1} - \zeta_1^2 = 0.$$

Multiplying the last equation by $\zeta_1^m \zeta_2^{-1}$ and integrating by parts, we get the following identity, which will actually be used only for $m = 0, 1$, and 2:

$$\int_{H(\zeta)=h} \frac{\zeta_1^{m+2} d\zeta_1}{\zeta_2} = \int_{H(\zeta)=h} \frac{\zeta_1^{m+1} d\zeta_1}{\zeta_2} - m \int_{H(\zeta)=h} \zeta_1^{m-1} \zeta_2 \, d\zeta_1. \tag{A.18}$$

Using the definitions (A.10) and (A.11) of the integrals, and the identities (A.16), (A.17), and (A.18) for $m = 1$ and 0, we get

$$\begin{aligned} h\frac{dI_1}{dh} &= h \int_{H(\zeta)=h} \frac{d\zeta_1}{\zeta_2} \\ &= \frac{1}{2} \int_{H(\zeta)=h} \frac{\zeta_1^2 \, d\zeta_1}{\zeta_2} + \frac{1}{2} \int_{H(\zeta)=h} \zeta_2 \, d\zeta_1 - \frac{1}{3} \int_{H(\zeta)=h} \frac{\zeta_1^3 \, d\zeta_1}{\zeta_2} \\ &= \frac{1}{2} I_1 + \frac{1}{2} \int_{H(\zeta)=h} \frac{\zeta_1^2 \, d\zeta_1}{\zeta_2} - \frac{1}{3} \int_{H(\zeta)=h} \frac{\zeta_1^3 \, d\zeta_1}{\zeta_2} \end{aligned}$$

$$= \frac{1}{2}I_1 + \frac{1}{6}\int_{H(\zeta)=h} \frac{\zeta_1^2\, d\zeta_1}{\zeta_2} + \frac{1}{3}I_1$$

$$= \frac{5}{6}I_1 + \frac{1}{6}\int_{H(\zeta)=h} \frac{\zeta_1\, d\zeta_1}{\zeta_2}$$

$$= \frac{5}{6}I_1 + \frac{1}{6}\frac{dI_2}{dh}.$$

Similarly, but with $m = 2, 1, 0$ in $(A.18)$, we have the following chain:

$$h\frac{dI_2}{dh} = h\int_{H(\zeta)=h} \frac{\zeta_1\, d\zeta_1}{\zeta_2}$$

$$= \frac{1}{2}\int_{H(\zeta)=h} \frac{\zeta_1^3\, d\zeta_1}{\zeta_2} + \frac{1}{2}\int_{H(\zeta)=h} \zeta_1\zeta_2\, d\zeta_1 - \frac{1}{3}\int_{H(\zeta)=h} \frac{\zeta_1^4\, d\zeta_1}{\zeta_2}$$

$$= \frac{1}{2}I_2 + \frac{1}{6}\int_{H(\zeta)=h} \frac{\zeta_1^3\, d\zeta_1}{\zeta_2} + \frac{2}{3}\int_{H(\zeta)=h} \frac{\zeta_1\, d\zeta_1}{\zeta_2}$$

$$= \frac{7}{6}I_2 + \frac{1}{6}\int_{H(\zeta)=h} \frac{\zeta_1^2\, d\zeta_1}{\zeta_2} - \frac{1}{6}\int_{H(\zeta)=h} \zeta_2\, d\zeta_1$$

$$= \frac{7}{6}I_2 - \frac{1}{6}I_1 + \frac{1}{6}\frac{dI_2}{dh}.$$

Taking into account the final results, we arrive at (A.14). □

**Proposition 8.2 (Riccati equation)** *The function $Q(h)$ defined by* (A.12) *satisfies the Riccati equation*

$$h\left(h - \frac{1}{6}\right)\dot{Q} = -\frac{7}{36}Q^2 + \left(\frac{h}{3} + \frac{1}{6}\right)Q - \frac{h}{6}. \tag{A.19}$$

**Proof:**

Indeed, using (A.14),

$$h\left(h - \frac{1}{6}\right)\dot{Q} = h\left(h - \frac{1}{6}\right)\left(\frac{\dot{I}_2}{I_1} - \frac{I_2\dot{I}_1}{I_1^2}\right)$$

$$= -\frac{h}{6}\frac{I_1}{I_1} + \frac{7h}{6}\frac{I_2}{I_1} - \frac{I_2}{I_1}\left(\left(\frac{5h}{6} - \frac{1}{6}\right)\frac{I_1}{I_1} + \frac{7}{36}\frac{I_2}{I_1}\right)$$

$$= -\frac{h}{6} + \frac{7h}{6}Q - Q\left(\frac{5h}{6} - \frac{1}{6} + \frac{7}{36}Q\right)$$

$$= -\frac{7}{36}Q^2 + \left(\frac{h}{3} + \frac{1}{6}\right)Q - \frac{h}{6}. \; \square$$

Substituting $Q(h) = \alpha h + O(h^2)$ into the Riccati equation (A.19) immediately gives $\alpha = \frac{1}{2}$, or

$$\dot{Q}(0) = \frac{1}{2} > 0.$$

The following proposition is also a direct consequence of (A.19).

**Proposition 8.3** *For all $h \in (0, \frac{1}{6})$, one has $0 \le Q(h) \le \frac{1}{7}$.*

**Proof:**

The function $Q(h)$ is positive for small $h > 0$. Suppose that $\bar{h} \in (0, \frac{1}{6})$ is the first intersection of the graph of $Q(h)$ with the $h$-axis, that is, $Q(\bar{h}) = 0$ and $Q(h) > 0$ for all $h \in (0, \bar{h})$. Then

$$\bar{h}\left(\bar{h} - \frac{1}{6}\right)\dot{Q}(\bar{h}) = -\frac{\bar{h}}{6} < 0.$$

Therefore, $\dot{Q}(\bar{h}) > 0$, which is a contradiction.

Now suppose that $\bar{h} \in (0, \frac{1}{6})$ is the first intersection of the graph of $Q(h)$ with the line $Q = \frac{1}{7}$, which means $Q(\bar{h}) = \frac{1}{7}$. Then

$$\bar{h}\left(\bar{h} - \frac{1}{6}\right)\dot{Q}(\bar{h}) = \frac{5}{42}\left(\frac{1}{6} - \bar{h}\right) > 0,$$

which implies $\dot{Q}(\bar{h}) < 0$, a contradiction. The proposition is proved. $\square$

Compute, finally, the second derivative of $Q(h)$ at a point $0 < \bar{h} < \frac{1}{6}$, where $\dot{Q}$ is supposed to vanish, $\dot{Q}(\bar{h}) = 0$. We have

$$\bar{h}\left(\bar{h} - \frac{1}{6}\right)\ddot{Q}(\bar{h})\bigg|_{\dot{Q}(\bar{h})=0} = \frac{1}{3}\left(Q - \frac{1}{2}\right) < 0.$$

Thus, $\ddot{Q}(\bar{h}) > 0$ at any point where $\dot{Q}(\bar{h}) = 0$ (i.e., all extrema are maximum points). This implies that such points do not exist since $Q(0) = 0$ and $Q(\frac{1}{6}) = \frac{1}{7} = \max_{0 \le h \le \frac{1}{6}} Q(h)$. Therefore, $\dot{Q}(h) > 0$ for $h \in [0, \frac{1}{6}]$. Lemma 8.17 is thus proved.

Lemma 8.17 provides the uniqueness of the cycle in (A.4).

**Remark:**

Actually, the lemma also gives the *hyperbolicity* of the cycle, because one can show that $Q'(h) > 0$ implies that the logarithm of the multiplier

$$\ln \mu = -\gamma_2 I_1(h) Q'(h) + o(\gamma_2) < 0,$$

for small $\gamma_2 > 0$, and, therefore, the multiplier $\mu$ of the limit cycle satisfies $0 < \mu < 1$. $\diamondsuit$

*Step 7* (*Return to the original parameters*). To apply the obtained results to the original system (A.2) (and, thus to (A.1)), we have to study in more detail the map $\gamma = \gamma(\beta)$ given by (A.5). This map is defined in the region

$$\{\beta : 4\beta_1 \le \beta_2^2\} \in \mathbb{R}^2,$$

bounded by the fold curve $T$ and located to the left of this curve. It maps this region homeomorphically onto the upper half-plane

$$\{\gamma : \gamma_1 \ge 0\} \in \mathbb{R}^2.$$

The inverse map is somewhat easy to study since it is smooth. From (A.5) we have

$$-\frac{\beta_2}{2} - \frac{\gamma_2^2}{2} = \gamma_1\gamma_2,$$
$$\beta_2^2 - 4\beta_1 = \gamma_2^4.$$

By the Inverse Function Theorem, these equations define a smooth function $\gamma(\beta)$. One can check that this function has the expansion

$$\begin{cases} \beta_1 = \gamma_1\gamma_2^2(\gamma_1+\gamma_2) + o(\|\gamma\|^4), \\ \beta_2 = -\gamma_2(2\gamma_1+\gamma_2) + o(\|\gamma\|^2). \end{cases} \quad (A.20)$$

The map (A.20) maps the vertical half-axis of the $(\gamma_1, \gamma_2)$-plane corresponding to $h = 0$ into the Hopf bifurcation line $H$ in the $(\beta_1, \beta_2)$-plane (as one could expect a priori since, as $h \to 0$, the cycle in (A.4) shrinks to $S_0$). The homoclinic curve $\mathcal{P}$ given by (A.20) is mapped by (A.11) into the curve

$$P = \left\{(\beta_1, \beta_2) : \beta_1 = -\frac{6}{25}\beta_2^2 + o(\beta_2^2),\ \beta_2 < 0\right\}$$

from the statement of Theorem 8.9. The cycle in (A.2) is unique and hyperbolic within the region bounded by $H$ and $P$. This completes the outline of the proof of Theorem 8.9. □

## 8.10 Appendix B: Bibliographical notes

The equilibrium structure of a two-parameter system near a triple equilibrium point (*cusp bifurcation*) has been known for a long time (see the survey by Arnol'd [1984]). Since any scalar system can be written as a gradient system $\dot{x} = -V_x(x, \alpha)$, the topological normal form for the cusp bifurcation naturally appeared in the list of seven elementary catastrophes by Thom [1972], along with the fold (codim 1) and swallow-tail (codim 3) singularities and their universal unfoldings.

Generic two-parameter bifurcation diagrams near a point where the first Lyapunov coefficient $l_1(0)$ vanishes were first obtained by Bautin [1949]. Therefore, we call this bifurcation the *Bautin bifurcation.* Serebriakova [1959] computed the second Laypunov coefficient $l_2(0)$ in terms of real Taylor coefficients of a general system and derived asymptotic formulas for the cycle-fold bifurcation curve near the Bautin point. Hassard, Kazarinoff & Wan [1981] present an expression for $c_2(0)$ in terms for the complex Taylor coefficients $g_{kl}(0)$ and some intermediate expressions. A compact formula expressing $l_2(0)$ directly in terms of $g_{kl}(0)$, which is equivalent to our (8.23), can be found in Bautin & Shil'nikov [1980] with a reference to Schuko [1968]. The modern treatments of the Bautin bifurcation based on ideas from singularity theory and Poincaré normal forms are due to Arnol'd [1972] and Takens [1973]. Takens has also studied higher degeneracies at the Hopf bifurcation with codim $> 2$ (e.g., the case $l_2(0) = 0$). A nice presentation of Bautin and related bifurcations can be found in [Shilnikov et al. 2001, Section 11.5].

The main features of the bifurcation diagram near an equilibrium with a double zero eigenvalue were known to mathematicians of the Andronov school in the late 1960s and were routinely used in the analysis of concrete models (see, e.g. Bautin & Leontovich [1976, pp. 183 – 186]). However, the complete picture, including the uniqueness of the limit cycle, is due to Bogdanov [1976*b*] (his results were announced by Arnold in 1972) and Takens [1974*b*]. Their analysis is based on the Pontryagin [1934] technique for locating limit cycles in dissipatively perturbated Hamiltonian systems and on nontrivial estimates of certain resulting elliptic integrals. Dumortier

& Rousseau [1990] found that one can establish the uniquness of the cycle in the Bogdanov-Takens normal form avoiding elliptic integrals by applying a theorem due to Coppel concerning Liénard planar systems. Although rather simple, this approach does not provide the local approximation of the homoclinic bifurcation curve. Annabi, Annabi & Dumortier [1992] proved that the conjugating homeomorphism depends continuously on the parameters. Our presentation of the double-zero bifurcation (Section 8.4 and Appendix A), closely follows the original Bogdanov papers. However, the proof of Lemma 8.17 in Appendix A, provided to the author by S. van Gils, is much simpler than Bogdanov's original proof. Let us also point out that in the otherwise perfect book by Arrowsmith & Place [1990], the Bogdanov-Takens bifurcation is called a "cusp" bifurcation because of the peculiar shape of the critical phase portrait, while the cusp bifurcation itself is lost.

The analysis of the last two codim 2 bifurcations, namely, the *fold-Hopf* and *Hopf-Hopf* cases, is more recent. Their study was initialized in the late 1970s, and early 1980s by Gavrilov [1978, 1980], and by Langford [1979], Keener [1981], and Guckenheimer [1981]. However, even the analysis of the truncated amplitude equations appeared to be difficult, and some hypothethes formulated in early papers and in the first edition of Guckenheimer & Holmes [1983] were proved wrong by subsequent analysis. The least trivial part of the analysis, namely proving the limit cycle uniqueness, was performed by Żołądek [1984, 1987]. His proofs were later considerably simplified by Carr, Chow & Hale [1985], van Gils [1985], Cushman & Sanders [1985], Chow, Li & Wang [1989*b*, 1989*a*], among others. Detailed uniqueness proofs for all the codim 2 cases are presented by Chow, Li & Wang [1994]. The study of nonsymmetric general perturbations of the truncated normal forms is far from complete. Let us refer only to Broer & Vegter [1984], Kirk [1991, 1993], and Gaspard [1993], where the problem of Shil'nikov type saddle-focus homoclinic orbits is analyzed.

Our presentation of the fold-Hopf bifurcation is based on the paper by Gavrilov [1978]. Lemma 8.10 was formulated by Gavrilov without stating the nondegeneracy conditions, as well as without explicit formulas for the normal form coefficients. The Poincaré normal form coefficients for the fold-Hopf case are also derived by Gamero, Freire & Rodríguez-Luis [1993] and by Wang [1993]. Our exposition of the Hopf-Hopf bifurcation also follows the general approach by Gavrilov [1980]; however, we both derive a different normal form and formulate explicitly the relevant nondegeneracy conditions. Notice that the normal form for the Hopf-Hopf case we use is different but equivalent to that studied by Żołądek [1987].

Explicit computational formulas for the normal form coefficients for all codim 2 equilibrium bifurcations in $n$-dimensional systems have been derived by Kuznetsov [1999] using the reduction/normalization technique by Coullet & Spiegel [1983]. Some of these coefficients were obtained earlier by Kurakin & Judovich [1986], who gave explicit criteria for stability of equilibria in $n$-dimensional ODEs in some critical cases. The fifth-order coefficients for the Hopf-Hopf case can be found in [Kuznetsov 1999].

All codim 3 local bifurcations possible in generic three-parameter *planar* systems have been identified and studied. The relevant normal forms are presented and discussed in Bazykin, Kuznetsov & Khibnik [1985, 1989], while the proofs can be found in Takens [1974*b*], Berezovskaya & Khibnik [1985], Dumortier, Roussarie & Sotomayor [1987], and Dumortier et al. [1991]. These codim 3 bifurcations are the following: (i) swallow-tail bifurcation; (ii) degenerate Bautin (Takens-Hopf) bifurcation; (iii) degenerate Bogdanov-Takens bifurcation with a double equilibrium; and (iv) degenerate Bogdanov-Takens bifurcation with a triple equilibrium (see Exercise 9).

# 9

# Two-Parameter Bifurcations of Fixed Points in Discrete-Time Dynamical Systems

This chapter is devoted to the study of generic bifurcations of fixed points of two-parameter maps. First we derive a list of such bifurcations. As for the final two bifurcations in the previous chapter, the description of the majority of these bifurcations is incomplete in principle. For all but two cases, only *approximate* normal forms can be constructed. Some of these normal forms will be presented in terms of associated planar continuous-time systems whose evolution operator $\varphi^1$ approximates the map in question (or an appropriate iterate of the map). We present bifurcation diagrams of the approximate normal forms in minimal dimensions and discuss their relationships with the original maps. In general $n$-dimensional situation, these results should be applied to a map restricted to the center manifold. We give explicit computational formulas for the critical normal form coefficients of the restricted map in most of the codim 2 cases.

## 9.1 List of codim 2 bifurcations of fixed points

Consider a two-parameter discrete-time dynamical system

$$x \mapsto f(x, \alpha), \tag{9.1}$$

where $x = (x_1, x_2, \dots, x_n)^T \in \mathbb{R}^n$, $\alpha = (\alpha_1, \alpha_2)^T \in \mathbb{R}^2$, and $f$ is sufficiently smooth in $(x, \alpha)$. Suppose that at $\alpha = \alpha^0$ system (9.1) has a fixed point $x = x^0$ for which the fold, flip, or Neimark-Sacker bifurcation condition is satisfied. Then, as in the continuous-time case, generically, there is a *bifurcation curve* $\mathcal{B}$ on the $(\alpha_1, \alpha_2)$-plane along which the system has a fixed point exhibiting the relevant bifurcation. More precisely, the fixed-point equation

$$f(x, \alpha) - x = 0$$

and a bifurcation condition

$$\psi(x, \alpha) = 0,$$

which is imposed on the eigenvalues (multipliers) of the Jacobian matrix[1] $f_x$ evaluated at $(x, \alpha)$, define a curve $\Gamma$ in the $(n+2)$-dimensional space $\mathbb{R}^{n+2}$ endowed with the coordinates $(x, \alpha)$ (see Fig. 9.1). Each point $(x^0, \alpha^0) \in \Gamma$

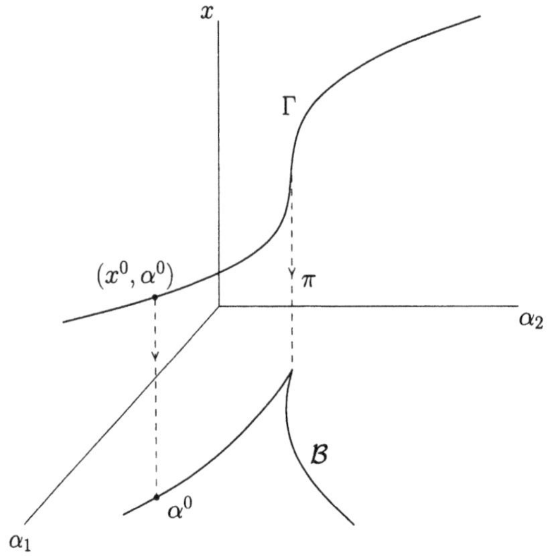


**Fig. 9.1.** A bifurcation curve $\Gamma$ and its projection $\mathcal{B}$.

corresponds to a fixed point $x^0$ of system (9.1) satisfying the relevant bifurcation condition for the parameter value $\alpha^0$. The *standard projection,*

$$\pi : (x, \alpha) \mapsto \alpha,$$

maps $\Gamma$ onto the bifurcation curve $\mathcal{B} = \pi\Gamma$ in the parameter plane.

**Example 9.1 (Fold and flip bifurcation curves)** Assume that at $\alpha = \alpha^0 = (\alpha_1^0, \alpha_2^0)^T$ system (9.1) has an equilibrium $x = x^0$ with a multiplier $\mu = 1$. As in Example 4.1 in Chapter 4, consider the system of nonlinear equations

$$\begin{cases} f(x, \alpha) - x = 0, \\ \det(f_x(x, \alpha) - I_n) = 0. \end{cases} \tag{9.2}$$

This is a system of $n+1$ equations for the $n+2$ variables $(x, \alpha_1, \alpha_2)$. Generically, it defines a smooth one-dimensional manifold (curve) $\Gamma \subset \mathbb{R}^{n+2}$ passing through the point $(x^0, \alpha_1^0, \alpha_2^0)$. As in Chapter 8, "generically" means that the rank of the Jacobian matrix of (9.2) is maximal (i.e., equal to $n+1$). For example, if the genericity conditions for the fold bifurcation are satisfied at $\alpha^0$ with respect to $\alpha_1$ or $\alpha_2$ (see Chapter 4), this condition is fulfilled.

[1] Actually, the function $\psi$ can be expressed directly in terms of the Jacobian matrix itself for all three codim 1 bifurcations.

Each point $(x, \alpha) \in \Gamma$ defines a fixed point $x$ of system (9.1) with a multiplier $\mu = 1$ for the parameter value $\alpha$. The standard projection maps $\Gamma$ onto a curve $\mathcal{B} = \pi\Gamma$ in the parameter plane. On this curve the fold bifurcation takes place.

A similar construction can be carried out for the flip bifurcation. In this case, system (9.2) should be substituted by the system

$$\begin{cases} f(x, \alpha) - x = 0, \\ \det(f_x(x, \alpha) + I_n) = 0. \end{cases} \tag{9.3}$$

System (9.3) is again a system of $n + 1$ equations in the $(n + 2)$-dimensional space, defining a curve $\Gamma \in \mathbb{R}^{n+2}$, under the maximal rank condition. Each point $(x, \alpha) \in \Gamma$ corresponds to a fixed point $x$ of system (9.1) with a multiplier $\mu = -1$ for the parameter value $\alpha$. The standard projection yields the flip bifurcation curve in the parameter plane $(\alpha_1, \alpha_2)$. $\Diamond$

**Example 9.2 (Neimark-Sacker bifurcation curve for planar maps)** Consider a *planar* system

$$x \mapsto f(x, \alpha), \quad x = (x_1, x_2)^T \in \mathbb{R}^2, \ \alpha = (\alpha_1, \alpha_2)^T \in \mathbb{R}^2, \tag{9.4}$$

having at $\alpha = \alpha^0 = (\alpha_1^0, \alpha_2^0)^T$ a fixed point $x^0 = (x_1^0, x_2^0)^T$ with a pair of (nonreal) multipliers on the unit circle: $\mu_{1,2} = e^{\pm i\theta_0}, 0 < \theta_0 < \pi$. Now consider the following system of three scalar equations in $\mathbb{R}^4$ with coordinates $(x_1, x_2, \alpha_1, \alpha_2)$:

$$\begin{cases} f(x, \alpha) - x = 0, \\ \det f_x(x, \alpha) - 1 = 0 \end{cases} \tag{9.5}$$

(notice the difference from (9.2); this construction has been already used in Chapter 4, Example 4.1). Clearly, $(x^0, \alpha^0)$ satisfies (9.5) since the determinant equals the product of the multipliers: $\det f_x = \mu_1\mu_2 = 1$. It can be shown that the Jacobian matrix of (9.5) has maximal rank (equal to 3) at $(x^0, \alpha^0)$ if the equilibrium $x^0$ exhibits a generic Neimark-Sacker bifurcation in the sense of Chapter 4 at $\alpha^0$. Therefore, system (9.5) defines a curve $\Gamma$ in $\mathbb{R}^4$ passing through $(x^0, \alpha^0)$. Each point on the curve specifies a fixed point of (9.4) with $\mu_{1,2} = e^{\pm i\theta_0}$, as long as the multipliers remain nonreal. The standard projection of $\Gamma$ onto the $(\alpha_1, \alpha_2)$-plane gives the Neimark-Sacker bifurcation boundary $\mathcal{B} = \pi\Gamma$.

Notice that the second equation in (9.5) can be satisfied by a fixed point with *real* multipliers

$$\mu_1 = \tau, \ \mu_2 = \frac{1}{\tau},$$

for which, obviously, $\mu_1\mu_2 = 1$. If $\tau \neq 1$, the fixed point is called a *neutral saddle*. The above construction allows for a generalization to higher dimensions (see Chapter 10). $\Diamond$

Let the parameters $(\alpha_1, \alpha_2)$ be varied simultaneously to track a bifurcation curve $\Gamma$ (or $\mathcal{B}$). Then, the following events might happen at some parameter value:

(i) extra multipliers can approach the unit circle, thus changing the dimension of the center manifold $W^c$;

(ii) some of the nondegeneracy conditions for the codim 1 bifurcation can be violated.

At nearby parameter values we can expect the appearance of new phase portraits of the system, which means a codim 2 bifurcation must occur. Let us recall the nondegeneracy conditions for the codim 1 bifurcations (see Chapter 4 for the details).

First, let us take a fold bifurcation curve $\mathcal{B}_T$. A typical point on this curve corresponds to a fixed point with a simple multiplier $\mu_1 = 1$ and no other multipliers on the unit circle. The restriction of (9.1) to a center manifold $W^c$, which is one-dimensional in this case, has the form

$$\xi \mapsto \xi + b\xi^2 + O(\xi^3). \tag{9.6}$$

By definition, the coefficient $b$ is nonzero at a nondegenerate fold bifurcation point.

If we follow a flip bifurcation curve $\mathcal{B}_F$, then a typical point corresponds to a fixed point with a simple multiplier $\mu_1 = -1$ and no other multipliers on the unit circle. The restriction of (9.1) to the one-dimensional center manifold $W^c$ at a nondegenerate flip point has (in an appropriate coordinate) the form

$$\xi \mapsto -\xi + c\xi^3 + O(\xi^4), \tag{9.7}$$

with $c \neq 0$.

Finally, while tracing a Neimark-Sacker bifurcation curve $\mathcal{B}_{NS}$, we typically have a fixed point with a simple pair of nonreal complex-conjugate multipliers $\mu_{1,2} = e^{\pm i\theta_0}$, which are the only multipliers on the unit circle. In this case, the center manifold $W^c$ is two-dimensional, and the system on this manifold can be written in complex notations as

$$z \mapsto ze^{i\theta_0}(1 + d_1|z|^2) + O(|z|^4), \tag{9.8}$$

where $d_1 \in \mathbb{C}^1$. The nondegeneracy conditions involved are of two types:

(i) *absence of "strong resonances"*:

$$e^{iq\theta_0} \neq 1, \quad q = 1, 2, 3, 4;$$

(ii) *"cubic nondegeneracy"*:

$$d = \text{Re } d_1 \neq 0.$$

Degenerate points of the following *eleven* types can be met in generic two-parameter discrete-time systems, while moving along codim 1 curves (see equations (9.6)–(9.8)):

(1) $\mu_1 = 1,\ b = 0$ (*cusp*);

(2) $\mu_1 = -1,\ c = 0$ (*generalized flip*);
(3) $\mu_{1,2} = e^{\pm i\theta_0},\ d = 0$ (*Chenciner bifurcation*);
(4) $\mu_1 = \mu_2 = 1$ (1:1 *resonance*);
(5) $\mu_1 = \mu_2 = -1$ (1:2 *resonance*);
(6) $\mu_{1,2} = e^{\pm i\theta_0},\ \theta_0 = \frac{2\pi}{3}$ (1:3 *resonance*);
(7) $\mu_{1,2} = e^{\pm i\theta_0},\ \theta_0 = \frac{\pi}{2}$ (1:4 *resonance*);
(8) $\mu_1 = 1,\ \mu_2 = -1$ (*fold-flip bifurcation*);
(9) $\mu_1 = 1,\ \mu_{2,3} = e^{\pm i\theta_0}$;
(10) $\mu_1 = -1,\ \mu_{2,3} = e^{\pm i\theta_0}$;
(11) $\mu_{1,2} = e^{\pm i\theta_0},\ \mu_{3,4} = e^{\pm i\theta_1}$.

We leave the reader to explain which of the cases we could meet while following each of the codim 1 bifurcation curves. The bifurcations listed are characterized by *two* independent conditions (i.e., have codim 2). There are no other codim 2 bifurcations in generic discrete-time systems. The rest of this chapter contains results concerning cases (1)–(8) in the least possible phase dimension ($n = 1$ for (1) and (2), and $n = 2$ for (3)–(8)). Notice that (1)–(7) are the only cases possible if the map (9.1) is a Poincaré map associated with a limit cycle in a three-dimensional autonomous system of ODEs or a period return map of a time-periodic planar system of ODEs. In these cases, there are only two multipliers $\mu_{1,2}$, and their product is always positive: $\mu_1\mu_2 > 0$. Although some results on cases (9)–(11) are known, we will not present them here.

Our treatment of the bifurcations will be organized similar to that of the codim 2 bifurcations in the continuous-time case and will proceed through studying truncated normal forms and discussing their relationship with the original maps in the minimal phase-space dimensions. In cases (4)–(8), it is convenient to present the approximating map for an appropriate iterate of the truncated normal form as the *unit-time shift* $\varphi^1$ under the flow $\varphi^t$ of a certain planar system of autonomous differential equations. Although this approximating map contains important information on the behavior of any generic system near the corresponding bifurcation, neither it nor the truncated normal form provides a topological normal form since taking into account terms of arbitrary high-order results in topologically nonequivalent bifurcation diagrams.

Similar to the continuous-time case, the obtained results can be applied to multidimensional systems with the help of the discrete-time versions of the Center Manifold Theorem and Shoshitaishvili's Theorem (see Chapter 5). To determine the bifurcation scenario for a given map (9.1), the corresponding critical normal form coefficients have to be computed. In Section 9.7 we derive such coefficients for most of the codim 2 bifurcations using a variant of the combined reduction/normalization technique employed in Section 8.7 of the previous chapter.

## 9.2 Cusp bifurcation

Suppose a smooth scalar map

$$x \mapsto f(x, \alpha), \quad x \in \mathbb{R}^1, \ \alpha \in \mathbb{R}^2, \tag{9.9}$$

has at $\alpha = 0$ a fixed point for which the cusp bifurcation conditions are satisfied. The following lemma can be proved by performing both a parameter-dependent shift of the coordinate and a scaling, and by introducing new parameters, as in Section 8.2 of Chapter 8.

**Lemma 9.1** *Suppose that a one-dimensional system*

$$x \mapsto f(x, \alpha), \quad x \in \mathbb{R}^1, \ \alpha \in \mathbb{R}^2,$$

*with $f$ smooth, has at $\alpha = 0$ the fixed point $x = 0$ for which the cusp bifurcation conditions hold:*

$$\mu = f_x(0,0) = 1, \ b = \frac{1}{2} f_{xx}(0,0) = 0.$$

*Assume that the following genericity conditions are satisfied:*

(C.1) $f_{xxx}(0,0) \neq 0$;
(C.2) $(f_{\alpha_1} f_{x\alpha_2} - f_{\alpha_2} f_{x\alpha_1})(0,0) \neq 0$.

*Then there are smooth invertible coordinate and parameter changes transforming the system into*

$$\eta \mapsto \eta + \beta_1 + \beta_2 \eta + s\eta^3 + O(\eta^4), \tag{9.10}$$

*where $s = \operatorname{sign} f_{xxx}(0,0) = \pm 1$.* □

We leave the proof to the reader.

Consider the *truncated normal form* corresponding to $s = -1$:

$$\eta \mapsto \eta + \beta_1 + \beta_2 \eta - \eta^3. \tag{9.11}$$

The equation for its fixed points,

$$\beta_1 + \beta_2 \eta - \eta^3 = 0, \tag{9.12}$$

coincides with that for the equilibria of the normal form for the cusp bifurcation in the continuous-time case. A fold bifurcation (fixed point "collision" and disappearance) happens on a semicubic parabola $T$ in the $(\beta_1, \beta_2)$-plane:

$$T = \left\{ (\beta_1, \beta_2) : \ 4\beta_2^3 - 27\beta_1^2 = 0 \right\}$$

(see **Fig. 9.2**). The curve $T$ has two branches, $T_1$ and $T_2$, which meet tangentially at the cusp point $(0,0)$. As in the continuous-time case, the resulting wedge divides the parameter plane into two regions. In region **1**, inside the

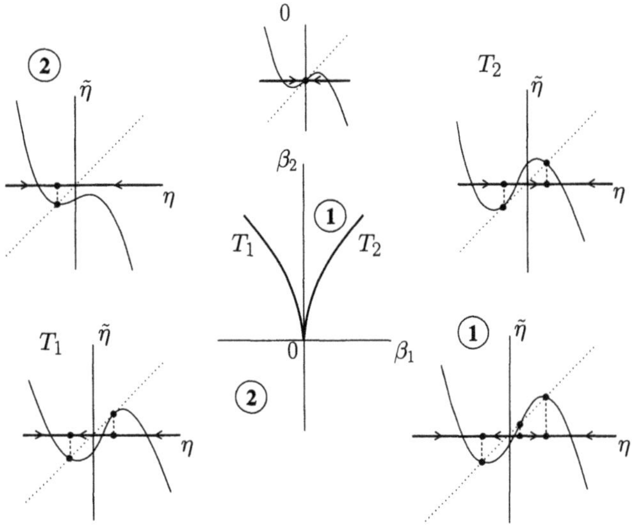


**Fig. 9.2.** Bifurcation diagram of the normal form (9.11).

wedge, there are three fixed points of (9.11), two stable and one unstable; while in region **2**, outside the wedge, there is a single fixed point, which is stable. A nondegenerate fold bifurcation (with respect to the parameter $\beta_1$) takes place if we cross either $T_1$ or $T_2$ away from the origin. If we approach the cusp point from inside region **1**, all three fixed points merge together into a *triple* root of (9.12).

The case $s = 1$ can be treated similarly. In this case, the truncated map typically has either one unstable fixed point or one stable and two unstable fixed points which can pairwise collide and disappear through fold bifurcations.

System (9.10) with the $O(\eta^4)$ terms truncated provides a *topological normal form* for the cusp bifurcation.

**Lemma 9.2** *The map*

$$\eta \mapsto G_\beta(\eta) = \eta + \beta_1 + \beta_2\eta \pm \eta^3 + O(\eta^4)$$

*is locally topologically equivalent near the origin to the map*

$$\eta \mapsto g_\beta(\eta) = \eta + \beta_1 + \beta_2\eta \pm \eta^3. \ \square$$

The lemma means (see Chapter 2) that there exists a local homeomorphism of the parameter plane $\varphi$, and a parameter-dependent homeomorphism $h_\beta$ of the one-dimensional phase space,[2] such that

[2] Recall, the homeomorphism $h_\beta$ is not assumed to depend continuously on the parameter $\beta$.

$$G_{\varphi(\beta)}(\eta) = (h_\beta^{-1} \circ g_\beta \circ h_\beta)(\eta),$$

in some neighborhood of $(\eta, \beta) = (0, 0)$. Proving that $G_\beta$ and $g_\beta$ have the same number of fixed points in corresponding (diffeomorphic) parameter regions is easy using the Implicit Function Theorem. The complete proof of the topological equivalence (i.e., the construction of $\varphi$ and $h_\beta$) is difficult and is omitted here. Summarizing, we can formulate the following theorem.

**Theorem 9.1 (Topological normal form for the cusp bifurcation)** *Any generic scalar two-parameter map*

$$x \mapsto f(x, \alpha),$$

*having at $\alpha = 0$ a fixed point $x = 0$ exhibiting the cusp bifurcation, is locally topologically equivalent near the origin to one of the normal forms*

$$\eta \mapsto \eta + \beta_1 + \beta_2 \eta \pm \eta^3. \ \square$$

Of course, "generic" here means "satisfying conditions (C.1) and (C.2)."

## 9.3 Generalized flip bifurcation

Consider a smooth scalar map

$$x \mapsto f(x, \alpha), \quad x \in \mathbb{R}^1, \ \alpha \in \mathbb{R}^2, \tag{9.13}$$

that has at $\alpha = 0$ a fixed point $x = 0$ with multiplier $\mu = f_x(0, 0) = -1$. As in Section 4.5 of Chapter 4, we can assume that $x = 0$ is a fixed point for all sufficiently small $\|\alpha\|$ and write

$$f(x, \alpha) = \mu(\alpha)x + a(\alpha)x^2 + b(\alpha)x^3 + e(\alpha)x^4 + g(\alpha)x^5 + O(x^6),$$

where $\mu(0) = -1$ and all the functions involved are smooth in $\alpha$. By performing a smooth change of coordinate:

$$x = y + \delta(\alpha)y^2 + \theta(\alpha)y^4, \tag{9.14}$$

where $\delta$ and $\theta$ are properly chosen smooth functions, we can transform (9.13) into

$$y \mapsto \mu(\alpha)y + c(\alpha)y^3 + d(\alpha)y^5 + O(y^6),$$

for some smooth $c$ and $d$. The functions $\delta(\alpha)$ and $\theta(\alpha)$ are selected exactly in order to annihilate the quadratic and fourth-order terms of the map. As seen in Chapter 4, setting

$$\delta(\alpha) = \frac{a(\alpha)}{\mu^2(\alpha) - \mu(\alpha)}$$

eliminates the quadratic term and results in the following expression for $c$:

$$c(\alpha) = b(\alpha) + \frac{2a^2(\alpha)}{\mu^2(\alpha) - \mu(\alpha)}. \tag{9.15}$$

If

$$c(0) = b(0) + a^2(0) = \frac{1}{4}[f_{xx}(0,0)]^2 + \frac{1}{6}f_{xxx}(0,0) \neq 0,$$

then, by definition, we have a nondegenerate flip bifurcation. At a generalized flip bifurcation point, we simultaneously have

$$\mu(0) = -1, \quad c(0) = 0;$$

thus the fifth-order coefficient $d$ enters the game. One can check that

$$d(0) = g(0) + 3a(0)e(0) - 2a^4(0) = \left(\frac{1}{120}f_{x^5} + \frac{1}{16}f_xf_{x^4} - \frac{1}{8}[f_x]^4\right)\bigg|_{(x,\alpha)=(0,0)}$$

*provided that* $c(0) = 0$. At this point it is useful to introduce new parameters. If the map

$$(\alpha_1, \alpha_2) \mapsto (\mu(\alpha) + 1, c(\alpha)),$$

where $c$ is given by (9.15), is regular at $\alpha = 0$, we can use

$$\begin{cases} \gamma_1 = -(\mu(\alpha) + 1), \\ \gamma_2 = c(\alpha), \end{cases}$$

as new parameters and consider $\alpha$ as a smooth function of $\gamma$. If $d(0) \neq 0$, then a nonsingular scaling of the coordinate and the new parameters brings the studied map into the form

$$\eta \mapsto -(1 + \beta_1)\eta + \beta_2\eta^3 + s\eta^5 + O(\eta^6),$$

for $s = \text{sign}\, d(0) = \pm 1$. Thus, the following lemma is proved.

**Lemma 9.3** *Suppose that a one-dimensional system*

$$x \mapsto f(x, \alpha), \quad x \in \mathbb{R}^1, \ \alpha \in \mathbb{R}^2,$$

*with $f$ smooth, has at $\alpha = 0$ the fixed point $x = 0$, and let the generalized flip bifurcation conditions hold:*

$$\mu = f_x(0,0) = -1, \quad c = \frac{1}{4}[f_{xx}(0,0)]^2 + \frac{1}{6}f_{xxx}(0,0) = 0.$$

*Assume that the following genericity conditions are satisfied:*

(GF.1) $D(0) = \left(\frac{1}{15}f_{x^5} + \frac{1}{2}f_xf_{x^4} - [f_x]^4\right)(0,0) \neq 0$;

(GF.2) *the map $\alpha \mapsto (\mu(\alpha) + 1, c(\alpha))^T$ is regular at $\alpha = 0$, where $c(\alpha)$ is given by* (9.15).

*Then there are smooth invertible coordinate and parameter changes transforming the system into*

$$\eta \mapsto -(1 + \beta_1)\eta + \beta_2\eta^3 + s\eta^5 + O(\eta^6), \tag{9.16}$$

*where* $s = \text{sign} D(0) = \pm 1$. □

System (9.16) without $O(\eta^6)$ terms is called the *truncated normal form* for the generalized flip bifurcation. Actually, it provides a topological normal form for the bifurcation.

Consider the normal form corresponding to $s = 1$:

$$\eta \mapsto g_\beta(\eta) = -(1+\beta_1)\eta + \beta_2\eta^3 + \eta^5. \tag{9.17}$$

Analyzing its *second* iterate,

$$g_\beta^2(\eta) = (1 + 2\beta_1 + \cdots)\eta - (2\beta_2 + \cdots)\eta^3 - (2 + \cdots)\eta^5 + \cdots,$$

we obtain the bifurcation diagram presented in **Fig. 9.3**. In region **1** the map

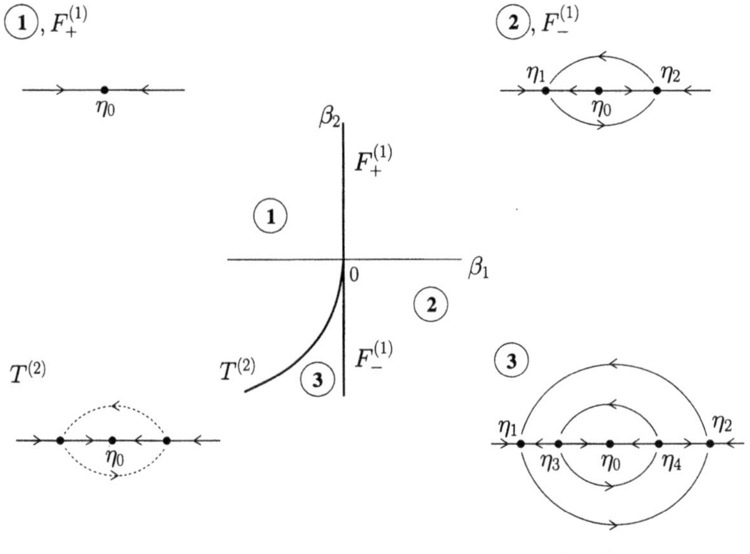


**Fig. 9.3.** Bifurcation diagram of the normal form (9.16) for $s = 1$.

(9.17) has a single stable fixed point $\eta = 0$ in a sufficiently small neighborhood of the origin. The iterates approach this point, "leap-frogging" around it. Crossing the upper half $F_+^{(1)}$ of the line

$$F^{(1)} = \{(\beta_1, \beta_2) : \beta_1 = 0\}$$

(corresponding to $\beta_2 > 0$) implies the flip bifurcation, leading to the creation of a *stable period-two* cycle (two distinct fixed points of $g_\beta^2$), while the fixed point at the origin becomes unstable (region **2**). Crossing the lower flip half-line $F_-^{(1)}$ generates an *unstable period-two cycle*, while the trivial fixed point regains stability. In region **3** two different period-two cycles coexist, a "big" stable one and a "small" unstable one. These two period-two cycles collide and disappear via the fold bifurcation of $g_\beta^2$ at the half-parabola:

$$T^{(2)} = \left\{ (\beta_1, \beta_2) : \beta_1 = -\frac{1}{4}\beta_2^2, \beta_2 < 0 \right\},$$

thus leaving a unique stable fixed point at the origin. In summary, the codim 2 point on the flip curve $F^{(1)}$ is the origin of an extra codim 1 bifurcation locus, namely, the fold bifurcation curve of period-two cycles $T^{(2)}$. The curve $T^{(2)}$ meets $F^{(1)}$ at $\beta = 0$, with a quadratic tangency.

The next lemma, which we give without proof, states that higher-order terms do not alter the picture obtained for (9.17).

**Lemma 9.4** *The map*

$$\eta \mapsto -(1 + \beta_1)\eta + \beta_2\eta^3 \pm \eta^5 + O(\eta^6)$$

*is locally topologically equivalent near the origin to the map*

$$\eta \mapsto -(1 + \beta_1)\eta + \beta_2\eta^3 \pm \eta^5. \;\Box$$

Now we can formulate the final theorem.

**Theorem 9.2 (Topological normal form for the generalized flip)** *Any generic scalar two-parameter map*

$$x \mapsto f(x, \alpha),$$

*having at $\alpha = 0$ a fixed point $x = 0$ exhibiting the generalized flip bifurcation, is locally topologically equivalent near the origin to one of the normal forms*

$$\eta \mapsto -(1 + \beta_1)\eta + \beta_2\eta^3 \pm \eta^5. \;\Box$$

**Remark:**

One might have noticed that there is a close similarity between the bifurcation diagrams of generalized flip and Bautin (generalized Hopf) bifurcations. This is not a coincidence since there is a deep analogy between flip and Hopf

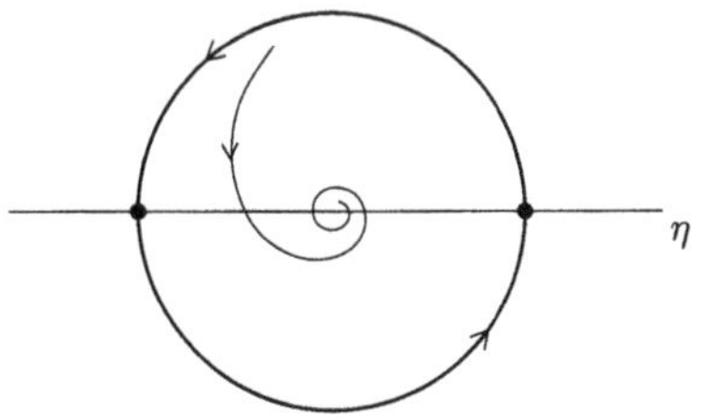


**Fig. 9.4.** A cycle corresponds to a period-two orbit of the Poincaré map.

bifurcations. For a Poincaré map defined on a cross-section *passing through* a focus, a limit cycle is a period-two orbit, thus, a Hopf bifurcation for a flow implies a flip bifurcation of the map so defined (see **Fig. 9.4**). ◊

## 9.4 Chenciner (generalized Neimark-Sacker) bifurcation

Consider a discrete-time system

$$x \mapsto f(x, \alpha), \quad x \in \mathbb{R}^2, \ \alpha \in \mathbb{R}^2, \tag{9.18}$$

with a smooth right-hand side $f$, having at $\alpha = 0$ a fixed point $x = 0$ for which the Neimark-Sacker bifurcation condition holds. That is, the multipliers of the fixed point are simple and are located on the unit circle $|\mu| = 1$:

$$\mu_{1,2} = e^{\pm i\theta_0}.$$

As seen in Chapter 4, system (9.18) can be written, for small $\|\alpha\|$, using a complex variable as

$$z \mapsto \mu(\alpha)z + g(z, \bar{z}, \alpha),$$

where $\mu$, $\omega$, and $g$ are smooth functions of their arguments,

$$\mu(\alpha) = r(\alpha)e^{i\theta(\alpha)},$$

$r(0) = 1$, $\theta(0) = \theta_0$, and formally

$$g(z, \bar{z}, \alpha) = \sum_{k+l \geq 2} \frac{1}{k!l!} g_{kl}(\alpha) z^k \bar{z}^l$$

for certain smooth complex-valued functions $g_{kl}(\alpha)$.

**Lemma 9.5 (Poincaré normal form)** *The map*

$$z \mapsto \mu(\alpha)z + \sum_{2 \leq k+l \leq 5} \frac{1}{k!l!} g_{kl}(\alpha) z^k \bar{z}^l + O(|z|^6), \tag{9.19}$$

*where* $\mu(\alpha) = r(\alpha)e^{i\theta(\alpha)}$, $r(0) = 1$, *and* $\theta_0 = \theta(0)$ *is such that*

(CH.0) $$e^{iq\theta_0} \neq 1, \quad q = 1, 2, \ldots, 6,$$

*can be transformed by an invertible smoothly parameter-dependent change of the complex coordinate:*

$$z = w + \sum_{2 \leq k+l \leq 5} \frac{1}{k!l!} h_{kl}(\alpha) w^k \bar{w}^l, \quad h_{21}(\alpha) = h_{32}(\alpha) = 0,$$

*for all sufficiently small* $\|\alpha\|$, *into the map*

$$\begin{aligned} w \mapsto &\ \mu(\alpha)w + c_1(\alpha)w|w|^2 + c_2(\alpha)w|w|^4 + O(|w|^6) \\ = &\ e^{i\theta(\alpha)}(r(\alpha) + d_1(\alpha))|w|^2 + d_2(\alpha)|w|^4)w + O(|w|^6). \ \square \end{aligned} \tag{9.20}$$

The lemma can be proved using the same method as for Lemma 4.7 in Chapter 4. By that lemma, we can assume that all the quadratic and non-resonant cubic terms in (9.19) have already been eliminated: $g_{20} = g_{11} = g_{02} = g_{30} = g_{12} = g_{21} = 0$ and $\frac{1}{2}g_{21} = c_1$. Then, by a proper selection of $h_{ij}$ with $i+j=4$, we can "kill" all the terms of order four in (9.19), keeping the coefficient of the resonant cubic term $c_1(\alpha)$ untouched but modifying the coefficients of the fifth- and higher-order terms. Finally, we can remove all the fifth-order terms except the *resonant* one shown in (9.20).

The coefficients $c_1(\alpha)$ and $c_2(\alpha)$ (as well as $d_1(\alpha)$ and $d_2(\alpha)$) are smooth complex-valued functions. Recall that $c_1(\alpha)$ can be computed by formula (4.25) from Chapter 4. We do not give the formula for $c_2(\alpha)$ here. In Section 9.7.3, the critical normal form coefficients of the map restricted to the center manifold at the Chenciner bifurcation will be derived for $n$-dimensional maps with $n \geq 2$. At a nondegenerate Neimark-Sacker bifurcation point, $d = \text{Re } d_1(0) = \text{Re}(e^{-i\theta_0}c_1(0)) \neq 0$.

Suppose that at $\alpha = 0$ we have simultaneously

$$\mu_{1,2} = e^{\pm i\theta_0}, \ d = \text{Re } d_1(0) = 0,$$

indicating that a Chenciner (generalized Neimark-Sacker) bifurcation takes place. In this case, a neighborhood of the point $\alpha = 0$ can be parametrized by new parameters, namely,

$$\begin{cases} \beta_1 = r(\alpha) - 1, \\ \beta_2 = \text{Re } d_1(\alpha), \end{cases}$$

provided that

(CH.1) *the map* $(\alpha_1, \alpha_2) \mapsto (r(\alpha) - 1, \text{Re } d_1(\alpha))$ *is regular at* $\alpha = 0$.

A zero value of $\beta_1$ corresponds to the Neimark-Sacker bifurcation condition, while the simultaneous vanishing of $\beta_2$ specifies a Chenciner point. Given (CH.1), one can express $\alpha$ in terms of $\beta$, thus obtaining the map

$$w \mapsto e^{i\theta(\beta)}(1 + \beta_1 + (\beta_2 + iD_1(\beta))|w|^2 + (D_2(\beta) + iE_2(\beta))|w|^4)w \\ + \Psi_\beta(w, \bar{w}), \tag{9.21}$$

where $\Psi_\beta = O(|w|^6)$; and $D_1(\beta) = \text{Im } d_1(\alpha(\beta))$, $D_2(\beta) = \text{Re } d_2(\alpha(\beta))$, and $E_2(\beta) = \text{Im } d_2(\alpha(\beta))$ are smooth real-valued functions of $\beta$. Here $\theta(\beta)$ is used instead of $\theta(\alpha(\beta))$ to save symbols. The map (9.21) is a "normal form" for the Chenciner bifurcation. Truncating $O(|w|^6)$ terms gives the map

$$w \mapsto e^{i\theta(\beta)}(1 + \beta_1 + (\beta_2 + iD_1(\beta))|w|^2 + (D_2(\beta) + iE_2(\beta))|w|^4)w. \tag{9.22}$$

Using polar coordinates $(\rho, \varphi)$, $z = \rho e^{i\varphi}$, we obtain the following representation of the *truncated* map (9.22):

$$\begin{cases} \rho \mapsto \rho(1 + \beta_1 + \beta_2\rho^2 + L_2(\beta)\rho^4) + \rho^6 R(\rho, \beta), \\ \varphi \mapsto \varphi + \theta(\beta) + \rho^2 Q(\rho, \beta), \end{cases} \tag{9.23}$$

where $L_2(\beta) = \frac{1}{2(1+\beta_1)} D_1^2(\beta) + D_2(\beta)$, while $R$ and $Q$ are smooth functions. Clearly, the first mapping in (9.23) is independent of $\varphi$ and can be studied separately.[3] The $\varphi$-map describes the rotation through a $\rho$-dependent angle that is approximately equal to $\theta(\beta)$. Any fixed point of the $\rho$-map corresponds to an *invariant circle* of the truncated normal form (9.22).

To say something definite about the $\rho$-map in (9.23), we have to assume that

$$\text{(CH.2)} \qquad L_2(0) = \frac{1}{2}\left[\mathrm{Im}(e^{-i\theta_0} c_1(0))\right]^2 + \mathrm{Re}(e^{-i\theta_0} c_2(0)) \neq 0.$$

This is an extra nondegeneracy condition for the Chenciner bifurcation.

Suppose, to begin with, that $L_2(0) < 0$. Then, there is a neighborhood of the origin in which the $\rho$-map has at most two positive fixed points of opposite stability (and the "outer" one of them is stable). These points branch from the trivial fixed point $\rho = 0$ at the curve

$$N = \{(\beta_1, \beta_2) : \beta_1 = 0\},$$

corresponding to a nondegenerate Neimark-Sacker bifurcation if $\beta_2 \neq 0$. The part $N_-$ of $N$ corresponding to $\beta_2 < 0$ gives rise to the creation of a stable fixed point when crossed from left to right. This is a supercritical Neimark-Sacker bifurcation producing a stable invariant circle. Crossing the upper part $N_+$ of $N$ at some $\beta_2 > 0$ in the opposite direction generates an unstable fixed point (invariant circle) via the subcritical Neimark-Sacker bifurcation. These two fixed points collide and disappear at the bifurcation curve

$$T_c = \left\{(\beta_1, \beta_2) : \beta_1 = \frac{1}{4L_2(0)}\beta_2^2 + o(\beta_2^2), \beta_2 > 0\right\},$$

resembling a half-parabola. For parameter values corresponding to the curve $T_c$, the truncated normal form (9.22) exhibits a collision and disappearance of two invariant circles. The resulting bifurcation picture illustrating the behavior of the truncated map is shown in **Fig. 9.5**. Notice that even for the truncated normal form, this picture captures only the existence of the invariant circles but not the orbit structure on them, which, generically, varies with the parameters and exhibits rational and irrational rotation numbers (see Chapter 7).

The case $L_2(0) > 0$ can be treated similarly. The only difference is that now the "outer" circle is unstable, while the "inner" one is stable.

A natural question is how much of the above picture remains if we consider the whole map (9.21) with generic higher-order terms. A short answer is that adding such terms results in topologically *nonequivalent* bifurcation diagrams. The Neimark-Sacker bifurcation curve is clearly independent of these terms.

[3] Actually, we have studied the similar map $g_\beta^2$ in Section 9.3 devoted to the generalized flip bifurcation.

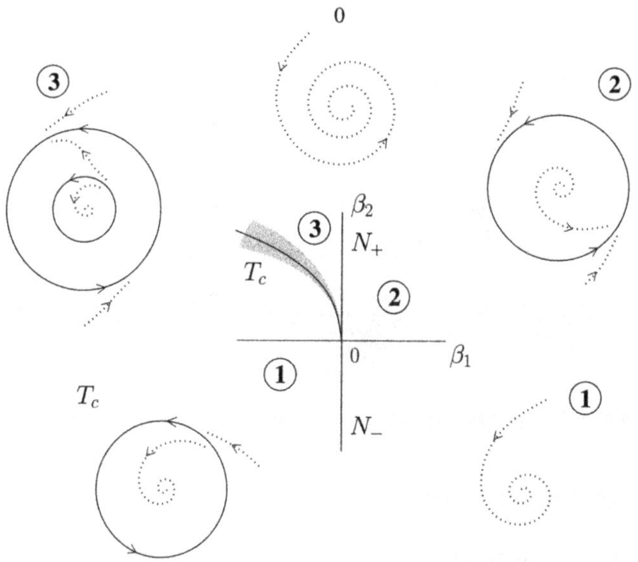


**Fig. 9.5.** Chenciner bifurcation in the truncated map (9.22) for $L_2(0) < 0$.

The nondegeneracy conditions for this bifurcation are also independent of these terms, provided that $\beta_2 \neq 0$, and thus remain valid for the map (9.21). Therefore, small closed invariant curves of corresponding stability do exist near the line $N$ for this map due to Theorem 4.6 (see Chapter 4). Moreover, these curves are smooth near $N$ and exist outside a narrow neighborhood of the collision curve $T_c$ as for the truncated map.

The situation is more complicated near the curve $T_c$. Generically, there is no single parameter curve corresponding to the collision of the closed invariant curves. Moreover, such invariant curves are not always present and might destruct, thus avoiding collision. It can be shown that there exists a parameter set in an exponentially narrow region containing $T_c$ for which the map (9.21) has a unique closed invariant curve, stable from the outside and unstable from the inside. Points of this set are limit points of an infinite number of parameter regions inside which there is a single (stable or unstable) invariant closed curve, and a sufficiently high iterate of the map (9.21) has saddle and stable (or unstable) fixed points (see **Fig. 9.6**). The stable and unstable invariant manifolds of the saddle points intersect and form a homoclinic structure, while the "inner" or "outer" closed invariant curve loses its smoothness and disappears without collision with its counterpart. Thus, there is an infinite number of appearing and disappearing high-periodic orbits of the map for parameter values near $T_c$. More details can be found in the literature cited in this chapter's appendix, but the complete picture seems to be unknown.

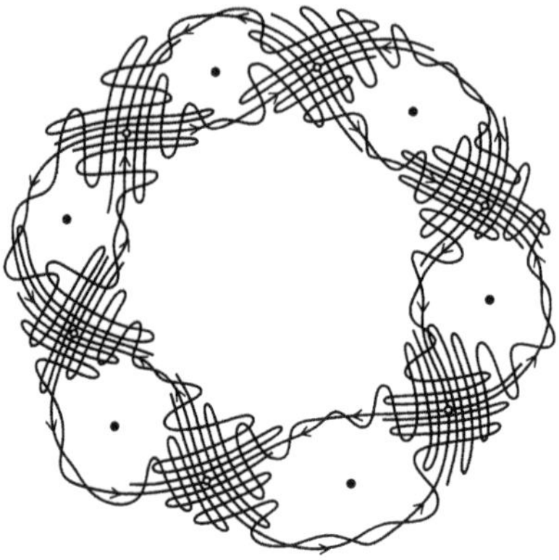

**Fig. 9.6.** Heteroclinic structure near a Chenciner bifurcation.

## 9.5 Strong resonances

### 9.5.1 Approximation by a flow

When dealing with strong resonances, we will repeatedly use the approximation of maps near their fixed points by shifts along the orbits of certain systems of autonomous ordinary differential equations. This allows us to predict *global* bifurcations of closed invariant curves happening in the maps near homo- and heteroclinic bifurcations of the approximating ODEs. Although the exact bifurcation structure is *different* for maps and approximating ODEs, their usage provides information that is hardly available by analysis of the maps alone. The planar case ($n = 2$) is sufficient for our purposes, but we shall give a general construction here.

Consider a map having a fixed point $x = 0$:

$$x \mapsto f(x) = Ax + f^{(2)}(x) + f^{(3)}(x) + \cdots, \quad x \in \mathbb{R}^n, \tag{9.24}$$

where $A$ is the Jacobian matrix, while each $f^{(k)}$ is a smooth polynomial vector-valued function of order $k$, $f^{(k)}(x) = O(\|x\|^k)$:

$$f_i^{(k)}(x) = \sum_{j_1+j_2+\cdots+j_n=k} b_{i,j_1j_2\cdots j_n}^{(k)} x_1^{j_1} x_2^{j_2} \cdots x_n^{j_n}.$$

Together with the map (9.24), consider a system of differential equations of the *same* dimension having an equilibrium at the point $x = 0$:

$$\dot{x} = F(x) = \Lambda x + F^{(2)}(x) + F^{(3)}(x) + \cdots, \quad x \in \mathbb{R}^n, \tag{9.25}$$

where $\Lambda$ is a matrix and the terms $F^{(k)}$ have the same properties as the corresponding $f^{(k)}$ above. Denote by $\varphi^t(x)$ the flow associated with (9.25). An interesting question is whether it is possible to construct a system (9.25) whose *unit-time shift* $\varphi^1$ along orbits coincides with (or, at least, approximates) the map $f$ given by (9.24).

**Definition 9.1** *The map* (9.24) *is said to be* approximated up to order $k$ *by system* (9.25) *if its Taylor expansion coincides with that of the unit-time shift along the orbits of* (9.25) *up to and including terms of order* $k$*:*

$$f(x) = \varphi^1(x) + O(\|x\|^{k+1}).$$

System (9.25) is called an *approximating system.*

Let us attack the approximation problem by constructing the Taylor expansion of $\varphi^t(x)$ with respect to $x$ at $x = 0$. This expansion can be obtained by *Picard iterations.* Namely, set

$$x^{(1)}(t) = e^{\Lambda t}x,$$

the solution of the linear equation $\dot{x} = \Lambda x$ with the initial data $x$, and define

$$x^{(k+1)}(t) = e^{\Lambda t}x + \int_0^t e^{\Lambda(t-\tau)} \left(F^{(2)}(x^{(k)}(\tau)) + \cdots + F^{(k+1)}(x^{(k)}(\tau))\right) d\tau. \tag{9.26}$$

A little thought shows that the $(k+1)$st iteration does not change $O(\|x\|^l)$ terms for any $l \leq k$. Substituting $t = 1$ into $x^{(k)}(t)$ provides the correct Taylor expansion of $\varphi^1(x)$ up to and including terms of order $k$:

$$\varphi^1(x) = e^{\Lambda}x + g^{(2)}(x) + g^{(3)}(x) + g^{(k)}(x) + O(\|x\|^{k+1}). \tag{9.27}$$

Now we can require the coincidence of the corresponding terms in (9.27) and (9.24):

$$e^{\Lambda} = A, \tag{9.28}$$

and

$$g^{(k)}(x) = f^{(k)}(x), \quad k = 2, 3, \ldots,$$

and then try to find $\Lambda$ and the coefficients of $g^{(k)}$ (and, eventually, the coefficients of $F^{(k)}$) in terms of those of $f^{(k)}$ (i.e., $b^{(k)}_{i,j_1 j_2 \cdots j_n}$). This is *not* always possible. It is easy to see that even the linear problem might cause difficulties. Take, for example, the planar linear map

$$f(x) = Ax, \quad A = \begin{pmatrix} -1 & 0 \\ 0 & 1 \end{pmatrix}.$$

There is no real matrix $\Lambda$ satisfying (9.28). Indeed, such a matrix must be diagonal, having the same eigenvectors as $A$,

$$\begin{pmatrix} \lambda_1 & 0 \\ 0 & \lambda_2 \end{pmatrix},$$

and one should thus have

$$e^{\lambda_1} = -1, \quad e^{\lambda_2} = 1.$$

The second equation gives $\lambda_2 = 0$, while the first one is unsolvable within real numbers. Therefore, the map $f$ cannot be approximated by a flow. Notice, however, that its *second* iterate,

$$f^2(x) = A^2 x, \quad A^2 = \begin{pmatrix} 1 & 0 \\ 0 & 1 \end{pmatrix},$$

allows for the approximation. Actually, $f^2(x) = x$ is the identity map, that is, a shift along orbits of the trivial equation $\dot{x} = 0$, with $\Lambda = 0$. One can prove that any map sufficiently close to the identity map can be approximated up to any order by a flow shift.

As we shall see later, it is possible that terms of order $k < l$, for some $l$ fixed, can be approximated by a system of differential equations, while those of order $l$ cannot. An example is provided by a planar map

$$f(x) = R_3 x + f^{(2)}(x) + \cdots, \quad x \in \mathbb{R}^2,$$

where $R_3$ is a matrix describing the planar rotation through the angle $\frac{2\pi}{3}$. The linear part of this map can be represented by the unit shift along the orbits of a linear system, while the quadratic terms $f^{(2)}(x)$ do not allow an approximation by a flow (see below). In this case, $l = 2$. Notice, however, that the *third* iteration $f^3$ can be approximated by a planar system of ODEs up to any order since it is close to the identity map near the origin.

It is worthwhile noticing that, even if a map allows for approximation by a flow up to any order, it may not be a unit-time shift of an ODE-system itself. This *flat* property is strongly related to the divergence of the approximating series and the appearance of *homo-* and *heteroclinic structures* for small perturbations of the maps. Examples will follow.

### 9.5.2 1:1 resonance

Consider a smooth planar map

$$x \mapsto f(x, \alpha), \quad x \in \mathbb{R}^2, \ \alpha \in \mathbb{R}^2. \tag{9.29}$$

Suppose that (9.29) has at $\alpha = 0$ a fixed point $x = 0$ with a double unit multiplier (1:1 resonance), $\mu_{1,2} = 1$. Write (9.29) at $\alpha = 0$ in the form

$$x \mapsto A_0 x + g(x), \tag{9.30}$$

where $A_0 = f_x(0,0)$, and $g(x) = f(x,0) - A_0 x = O(\|x\|^2)$ is smooth. Assume that there exist two real linearly independent vectors, $q_{0,1} \in \mathbb{R}^2$, such that

$$A_0 q_0 = q_0, \quad A_0 q_1 = q_1 + q_0. \tag{9.31}$$

The vector $q_0$ is the eigenvector of $A_0$ corresponding to the eigenvalue 1, while $q_1$ is the *generalized eigenvector* of $A_0$ corresponding to the same eigenvalue.[4]

---

[4] Note that this situation is more generic than the case where $A_0$ is *semisimple*; that is, having two independent eigenvectors $A_0 q_0 = q_0$, $A_0 q_1 = q_1$.

Moreover, there exist similar *adjoint* eigenvectors $p_{0,1} \in \mathbb{R}^2$ of the transposed matrix $A_0^T$:

$$A_0^T p_1 = p_1, \quad A_0^T p_0 = p_0 + p_1. \tag{9.32}$$

We can always select four vectors satisfying (9.31) and (9.32) such that

$$\langle p_0, q_0 \rangle = \langle p_1, q_1 \rangle = 1,$$

where $\langle \cdot, \cdot \rangle$ stands for the standard scalar product in $\mathbb{R}^2 : \langle x, y \rangle = x_1 y_1 + x_2 y_2$. The Fredholm Alternative Theorem implies

$$\langle p_0, q_1 \rangle = \langle p_1, q_0 \rangle = 0.$$

If $q_0$ and $q_1$ are selected, then any vector $x \in \mathbb{R}^2$ can be uniquely represented as

$$x = y_1 q_0 + y_2 q_1,$$

where new coordinates $(y_1, y_2)$ are given by

$$\begin{cases} y_1 = \langle p_0, x \rangle, \\ y_2 = \langle p_1, x \rangle. \end{cases} \tag{9.33}$$

In the coordinates $(y_1, y_2)$, the map (9.30) takes the form

$$\begin{pmatrix} y_1 \\ y_2 \end{pmatrix} \mapsto \begin{pmatrix} 1 & 1 \\ 0 & 1 \end{pmatrix} \begin{pmatrix} y_1 \\ y_2 \end{pmatrix} + \begin{pmatrix} \langle p_0, g(y_1 q_0 + y_2 q_1) \rangle \\ \langle p_1, g(y_1 q_0 + y_2 q_1) \rangle \end{pmatrix}. \tag{9.34}$$

Using the *same* coordinates $(y_1, y_2)$ for all $\alpha$ with $\|\alpha\|$ small, we can write the original map (9.29) as

$$\begin{pmatrix} y_1 \\ y_2 \end{pmatrix} \mapsto \begin{pmatrix} \langle p_0, f(y_1 q_0 + y_2 q_1, \alpha) \rangle \\ \langle p_1, f(y_1 q_0 + y_2 q_1, \alpha) \rangle \end{pmatrix}, \tag{9.35}$$

which reduces to system (9.30) for $\alpha = 0$. As in the analysis of the Bogdanov-Takens bifurcation in Chapter 8, expand the right-hand side of (9.35) as a Taylor series with respect to $y$ at $y = 0$,

$$\begin{cases} y_1 \mapsto y_1 + y_2 + a_{00}(\alpha) + a_{10}(\alpha) y_1 + a_{01}(\alpha) y_2 \\ \qquad + \frac{1}{2} a_{20}(\alpha) y_1^2 + a_{11}(\alpha) y_1 y_2 + \frac{1}{2} a_{02}(\alpha) y_2^2 + R_1(y, \alpha), \\ y_2 \mapsto y_2 + b_{00}(\alpha) + b_{10}(\alpha) y_1 + b_{01}(\alpha) y_2 \\ \qquad + \frac{1}{2} b_{20}(\alpha) y_1^2 + b_{11}(\alpha) y_1 y_2 + \frac{1}{2} b_{02}(\alpha) y_2^2 + R_2(y, \alpha), \end{cases} \tag{9.36}$$

where $a_{kl}(\alpha), b_{kl}(\alpha)$ and $R_{1,2}(y, \alpha) = O(\|y\|^3)$ are smooth functions of their arguments. Clearly,

$$a_{00}(0) = a_{10}(0) = a_{01}(0) = b_{00}(0) = b_{10}(0) = b_{01}(0) = 0.$$

The functions $a_{kl}(\alpha)$ and $b_{kl}(\alpha)$ can be expressed in terms of the right-hand side $f(x, \alpha)$ of (9.29) and the vectors $q_{0,1}, p_{0,1}$. For example,

$$a_{20}(\alpha) = \frac{\partial^2}{\partial y_1^2}\langle p_0, f(y_1 v_0 + y_2 v_1, \alpha)\rangle\bigg|_{y=0},$$
$$b_{20}(\alpha) = \frac{\partial^2}{\partial y_1^2}\langle p_1, f(y_1 v_0 + y_2 v_1, \alpha)\rangle\bigg|_{y=0},$$
$$b_{11}(\alpha) = \frac{\partial^2}{\partial y_1 \partial y_2}\langle p_1, f(y_1 v_0 + y_2 v_1, \alpha)\rangle\bigg|_{y=0}.$$

The following lemmas show that we can transform (9.36) into a simpler form by smooth invertible transformations that depend smoothly upon parameters. We begin with a result from linear algebra.

**Lemma 9.6 (Deformation of Jordan block)** *For any matrix*

$$L(\alpha) = \begin{pmatrix} \lambda + a_{10}(\alpha) & 1 + a_{01}(\alpha) \\ b_{10}(\alpha) & \lambda + b_{01}(\alpha) \end{pmatrix}$$

*with* $a_{10}(0) = a_{01}(0) = b_{10}(0) = b_{01}(0) = 0$, *there is a matrix* $P(\alpha)$, $P(0) = I$, *such that*

$$P^{-1}(\alpha)A(\alpha)P(\alpha) = \begin{pmatrix} \lambda & 1 \\ \varepsilon_1(\alpha) & \lambda + \varepsilon_2(\alpha) \end{pmatrix},$$

*where* $\varepsilon_1(0) = \varepsilon_2(0) = 0$.

**Proof:** The matrix

$$P(\alpha) = \begin{pmatrix} 1 + a_{01}(\alpha) & 0 \\ -a_{10}(\alpha) & 1 \end{pmatrix}$$

does the job, with

$$\varepsilon_1(\alpha) = b_{01}(\alpha) + a_{01}(\alpha)b_{10}(\alpha) - a_{10}(\alpha)b_{01}(\alpha), \quad \varepsilon_2(\alpha) = a_{10}(\alpha) + b_{01}(\alpha).$$

The matrix $P(\alpha)$ is as smooth in $\alpha$ as $L(\alpha)$. □

The nonsingular linear coordinate transformation

$$y = P(\alpha)u,$$

where $P$ is given above with $\lambda = 1$, reduces the map (9.36) for sufficiently small $\|\alpha\|$ into

$$\begin{cases} u_1 \mapsto u_1 + u_2 + g_{00}(\alpha) \\ \qquad + \frac{1}{2}g_{20}(\alpha)u_1^2 + g_{11}(\alpha)u_1u_2 + \frac{1}{2}g_{02}(\alpha)u_2^2 + S_1(u,\alpha), \\ u_2 \mapsto u_2 + h_{00}(\alpha) + \varepsilon_1(\alpha)u_1 + \varepsilon_2(\alpha)u_2 \\ \qquad + \frac{1}{2}h_{20}(\alpha)u_1^2 + h_{11}(\alpha)u_1u_2 + \frac{1}{2}h_{02}(\alpha)u_2^2 + S_2(u,\alpha), \end{cases} \tag{9.37}$$

where $S_{1,2}(u,\alpha) = O(\|u\|^3)$ and all introduced functions are smooth. We also have $\varepsilon_1(0) = \varepsilon_2(0) = 0$ and

$$\begin{pmatrix} g_{00}(\alpha) \\ h_{00}(\alpha) \end{pmatrix} = B^{-1}(\alpha)\begin{pmatrix} a_{00}(\alpha) \\ b_{00}(\alpha) \end{pmatrix},$$

so that $g_{00}(0) = h_{00}(0) = 0$, as well as $g_{kl}(0) = a_{kl}(0)$, $h_{kl}(0) = b_{kl}(0)$ for $k + l = 2$.

**Lemma 9.7 (Normal form map for 1:1 resonance)** *Assume that*

$$b_{20}(0) \neq 0.$$

*Then, there is a smooth invertible change of coordinates, smoothly depending on the parameters, that transforms* (9.37) *for all sufficiently small* $\|\alpha\|$ *into the map*

$$\begin{pmatrix} \xi_1 \\ \xi_2 \end{pmatrix} \mapsto \begin{pmatrix} \xi_1 + \xi_2 \\ \xi_2 + \nu_1(\alpha) + \nu_2(\alpha)\xi_2 + A(\alpha)\xi_1^2 + B(\alpha)\xi_1\xi_2 \end{pmatrix} + O(\|\xi\|^3) \quad (9.38)$$

*for smooth functions* $\nu_{1,2}(\alpha)$ *and* $A(\alpha), B(\alpha)$ *such that*

$$\nu_1(0) = \nu_2(0) = 0$$

*and*

$$A(0) = \frac{1}{2}b_{20}(0), \quad B(0) = a_{20}(0) + b_{11}(0). \quad (9.39)$$

**Proof:**

First, consider the case $\alpha = 0$, when (9.37) reduces to

$$\begin{cases} \tilde{u}_1 = u_1 + u_2 + \frac{1}{2}a_{20}u_1^2 + a_{11}u_1u_2 + \frac{1}{2}a_{02}u_2^2 + O(\|u\|^3), \\ \tilde{u}_2 = u_2 + \frac{1}{2}b_{20}u_1^2 + b_{11}u_1u_2 + \frac{1}{2}b_{02}u_2^2 + O(\|u\|^3), \end{cases} \quad (9.40)$$

where all $a_{ij}$ and $b_{ij}$ should be evaluated at $\alpha = 0$. Introduce a polynomial transformation

$$\begin{cases} u_1 = \xi_1 + \frac{1}{2}G_{20}\xi_1^2 + G_{11}\xi_1\xi_2, \\ u_2 = \xi_2 + \frac{1}{2}H_{20}\xi_1^2 + H_{11}\xi_1\xi_2, \end{cases} \quad (9.41)$$

where $G_{20}, G_{11}, H_{20}$, and $H_{11}$ are unknown coefficients. We select these coefficients so that (9.40) written in the $\xi$-coordinates will take the form

$$\begin{cases} \tilde{\xi}_1 = \xi_1 + \xi_2 + O(\|\xi\|^3), \\ \tilde{\xi}_2 = \xi_2 + A(0)\xi_1^2 + B(0)\xi_1\xi_2 + O(\|\xi\|^3). \end{cases} \quad (9.42)$$

Substituting (9.41) into (9.40), we express the components of $\tilde{u}$ in terms of those of $\xi$, while substituting (9.42) into

$$\begin{cases} \tilde{u}_1 = \tilde{\xi}_1 + \frac{1}{2}G_{20}\tilde{\xi}_1^2 + G_{11}\tilde{\xi}_1\tilde{\xi}_2, \\ \tilde{u}_2 = \tilde{\xi}_2 + \frac{1}{2}H_{20}\tilde{\xi}_1^2 + H_{11}\tilde{\xi}_1\tilde{\xi}_2, \end{cases} \quad (9.43)$$

which is equivalent to (9.41), we get another expression for the components of $\tilde{u}$. Comparing the coefficients of the $\xi_1^2$-, $\xi_1\xi_2$-, and $\xi_2^2$-terms in the expressions for $\tilde{u}_1$ and the coefficients of the $\xi_2^2$-term in $\tilde{u}_2$, we obtain the following linear system for the coefficients of the transformation (9.41):

$$\begin{pmatrix} 0 & 0 & -1 & 0 \\ 1 & 0 & 0 & -1 \\ 1 & 2 & 0 & 0 \\ 0 & 0 & 1 & 2 \end{pmatrix} \begin{pmatrix} G_{20} \\ G_{11} \\ H_{20} \\ H_{11} \end{pmatrix} = \begin{pmatrix} a_{20} \\ a_{11} \\ a_{02} \\ b_{02} \end{pmatrix}.$$

This system is nonsingular and has the unique solution

$$G_{20} = \frac{1}{2}a_{20} + a_{11} + \frac{1}{2}b_{02}, \tag{9.44}$$

$$G_{11} = -\frac{1}{4}a_{20} - \frac{1}{2}a_{11} + \frac{1}{2}a_{02} - \frac{1}{4}b_{02}, \tag{9.45}$$

$$H_{20} = -a_{20}, \tag{9.46}$$

$$H_{11} = \frac{1}{2}a_{20} + \frac{1}{2}b_{02}. \tag{9.47}$$

Taking into account (9.46), we obtain by comparing the remaining two quadratic terms in $\tilde{u}_2$ the expressions (9.39) for $A(0)$ and $B(0)$. Notice that we have not used the assumption $b_{20} \neq 0$ yet.

To prove the lemma for all $\alpha$ with sufficiently small $\|\alpha\|$, consider a parameter-dependent polynomial transformation

$$\begin{cases} u_1 = \xi_1 + \varepsilon_0(\alpha) + \frac{1}{2}G_{20}(\alpha)\xi_1^2 + G_{11}(\alpha)\xi_1\xi_2, \\ u_2 = \xi_2 + \delta_0(\alpha) + \delta_1(\alpha)\xi_1 + \delta_2(\alpha)\xi_2 + \frac{1}{2}H_{20}(\alpha)\xi_1^2 + H_{11}(\alpha)\xi_1\xi_2, \end{cases} \tag{9.48}$$

which reduces to (9.41) when $\alpha = 0$. The requirement that (9.37) in the $\xi$-coordinates has the form (9.38) translates into a system of algebraic equations

$$R_\alpha(\varepsilon_0, \delta_0, \delta_1, \delta_2, \nu_1, \nu_2, G_{20}, G_{11}, H_{20}, H_{11}, A, B) = 0,$$

where $R_\alpha : \mathbb{R}^{12} \to \mathbb{R}^{12}$ results from equating the corresponding Taylor coefficients. For the Jacobian matrix $J$ of this system evaluated at $\alpha = 0$, when

$$\varepsilon_0 = \delta_0 = \delta_1 = \delta_2 = \nu_1 = \nu_2 = 0$$

and the values of other variables are given by (9.44)–(9.47) and (9.39), we have $\det(J) = 8b_{20}(0) \neq 0$ by assumption. Therefore, the Implicit Function Theorem guarantees the local existence and smoothness of the coefficients of the transformation (9.48) and the map (9.38) as functions of $\alpha$. □

To use $(\nu_1, \nu_2)$ as new parameters, assume that the following transversality condition is satisfied:

(R1.0) $\det \left( \frac{\partial \nu}{\partial \alpha} \right) \Big|_{\alpha=0} \neq 0.$

Then, the Inverse Function Theorem allows one to consider $\alpha$ as a smooth function of $\nu$ for small $\|\nu\|$, such that $\alpha(0) = 0$. When $b_{20}(0) \neq 0$, condition (R1.0) is equivalent to the regularity of the map

$$T : \begin{pmatrix} u \\ \alpha \end{pmatrix} \mapsto \begin{pmatrix} G(u,\alpha) - u \\ \text{tr } G_u(u,\alpha) - 2 \\ \det\ G_u(u,\alpha) - 1 \end{pmatrix}$$

at the point $(u,\alpha) = (0,0)$. Here $G$ is the map defined by (9.37). Indeed, one can show by direct computation that at $(u,\alpha) = (0,0)$

$$\det\left(\frac{\partial \nu}{\partial \alpha}\right) = -\frac{1}{b_{20}(0)} \det\left(\frac{\partial T}{\partial(u,\alpha)}\right).$$

Moreover, since (9.37) and the original map (9.29) are related by a nonsingular linear transformation, one can check the regularity of the map

$$\begin{pmatrix} x \\ \alpha \end{pmatrix} \mapsto \begin{pmatrix} f(x,\alpha) - x \\ \text{tr } f_x(x,\alpha) - 2 \\ \det\ f_x(x,\alpha) - 1 \end{pmatrix}$$

at $(x,\alpha) = (0,0)$.

Using $(\nu_1,\nu_2)$ as the new parameters, we can write (9.38) in the form

$$N_\nu : \begin{pmatrix} \xi_1 \\ \xi_2 \end{pmatrix} \mapsto \begin{pmatrix} \xi_1 + \xi_2 \\ \xi_2 + \nu_1 + \nu_2\xi_2 + A_1(\nu)\xi_1^2 + B_1(\nu)\xi_1\xi_2 \end{pmatrix} + O(\|\xi\|^3), \quad (9.49)$$

where $A_1(\nu) = A(\alpha(\nu))$ and $B_1(\alpha) = B(\alpha(\nu))$ are smooth functions of $\nu$. Our next task is to approximate the normal form (9.49) by a flow.

Denote $a_0 = A_1(0) = A(0),\ b_0 = B_1(0) = B(0)$.

**Lemma 9.8** *The map* (9.49) *can be represented for all sufficiently small* $\|\nu\|$ *in the form*

$$N_\nu(\xi) = \varphi_\nu^1(\xi) + O(\|\nu\|^2) + O(\|\xi\|^2\|\nu\|) + O(\|\xi\|^3),$$

*where* $\varphi_\nu^t$ *is the flow of a smooth planar system*

$$\dot{\xi} = F(\xi,\nu), \quad \xi \in \mathbb{R}^2,\ \nu \in \mathbb{R}^2, \quad (9.50)$$

*with* $F(\xi,\nu) = F_0(\nu) + F_1(\xi,\nu) + F_2(\xi)$, *where*

$$F_0(\nu) = \begin{pmatrix} -\frac{1}{2}\nu_1 \\ \nu_1 \end{pmatrix},$$

$$F_1(\xi,\nu) = \begin{pmatrix} \xi_2 + \left(-\frac{1}{2}a_0 + \frac{1}{3}b_0\right)\nu_1\xi_1 + \left[\left(\frac{1}{5}a_0 - \frac{5}{12}b_0\right)\nu_1 - \frac{1}{2}\nu_2\right]\xi_2 \\ \left(\frac{2}{3}a_0 - \frac{1}{2}b_0\right)\nu_1\xi_1 + \left[\left(-\frac{1}{6}a_0 + \frac{1}{2}b_0\right)\nu_1 + \nu_2\right]\xi_2 \end{pmatrix},$$

*and*

$$F_2(\xi) = \begin{pmatrix} -\frac{1}{2}a_0\xi_1^2 + \left(\frac{2}{3}a_0 - \frac{1}{2}b_0\right)\xi_1\xi_2 + \left(-\frac{1}{6}a_0 + \frac{1}{3}b_0\right)\xi_2^2 \\ a_0\xi_1^2 + (-a_0 + b_0)\xi_1\xi_2 + \left(\frac{1}{6}a_0 - \frac{1}{2}b_0\right)\xi_2^2 \end{pmatrix}.$$

**Proof:**

Let us prove the lemma in detail for $\nu = 0$ by explicitly constructing $F(\xi, 0) = F_1(\xi, 0) + F_2(\xi)$ in (9.50).

Clearly, the linear part of (9.49) at $\nu = 0$,

$$\begin{pmatrix} \xi_1 \\ \xi_2 \end{pmatrix} \mapsto \begin{pmatrix} 1 & 1 \\ 0 & 1 \end{pmatrix} \begin{pmatrix} \xi_1 \\ \xi_2 \end{pmatrix},$$

is the unit-time shift along the orbits of the linear system

$$\begin{pmatrix} \dot{\xi}_1 \\ \dot{\xi}_2 \end{pmatrix} = \begin{pmatrix} 0 & 1 \\ 0 & 0 \end{pmatrix} \begin{pmatrix} \xi_1 \\ \xi_2 \end{pmatrix},$$

which is the linear part of the Bogdanov-Takens critical normal form. Introduce, therefore,

$$\Lambda_0 = \begin{pmatrix} 0 & 1 \\ 0 & 0 \end{pmatrix}.$$

Suppose that at $\nu = 0$ system (9.50) can be written as

$$\begin{cases} \dot{\xi}_1 = \xi_2 + \frac{1}{2}A_{20}\xi_1^2 + A_{11}\xi_1\xi_2 + \frac{1}{2}A_{02}\xi_2^2, \\ \dot{\xi}_2 = \frac{1}{2}B_{20}\xi_1^2 + B_{11}\xi_1\xi_2 + \frac{1}{2}B_{02}\xi_2^2, \end{cases} \tag{9.51}$$

where $A_{kl}$ and $B_{kl}$ with $k + l = 2$ are unknown coefficients to be defined. Let us perform two Picard iterations as described in Section 9.5.1. We have

$$\xi^{(1)}(\tau) = e^{\Lambda_0 \tau}\xi = \begin{pmatrix} \xi_1 + \tau\xi_2 \\ \xi_2 \end{pmatrix},$$

and using (9.26) we obtain

$$\xi^{(2)}(1) = \begin{pmatrix} \xi_1 + \xi_2 + \frac{1}{2}a_{20}\xi_1^2 + a_{11}\xi_1\xi_2 + \frac{1}{2}a_{02}\xi_2^2 \\ \xi_2 + \frac{1}{2}b_{20}\xi_1^2 + b_{11}\xi_1\xi_2 + \frac{1}{2}\xi_{02}\xi_2^2 \end{pmatrix},$$

where

$$\begin{aligned} a_{20} &= A_{20} + \frac{1}{2}B_{20}, \\ a_{11} &= A_{11} + \frac{1}{2}A_{20} + \frac{1}{6}B_{20} + \frac{1}{2}B_{11}, \\ a_{02} &= A_{02} + \frac{1}{3}A_{20} + A_{11} + \frac{1}{12}B_{20} + \frac{1}{3}B_{11} + \frac{1}{2}B_{02}, \\ b_{20} &= B_{20}, \\ b_{11} &= B_{11} + \frac{1}{2}B_{20}, \\ b_{02} &= B_{02} + \frac{1}{3}B_{20} + B_{11}. \end{aligned}$$

These quantities should be interpreted as quadratic Taylor coefficients of (9.49) so that

$$a_{20} = a_{11} = a_{02} = b_{02} = 0, \quad b_{20} = 2a_0, \quad b_{11} = b_0.$$

Solving the equations for $A_{20}, A_{11}, A_{02}, B_{20}, B_{11}$, and $B_{02}$, we get

$$\begin{aligned}
A_{20} &= a_{20} - \frac{1}{2}b_{20} \;=\; -a_0,\\
A_{11} &= a_{11} - \frac{1}{2}a_{20} + \frac{1}{3}b_{20} - \frac{1}{2}b_{11} \;=\; \frac{2}{3}a_0 - \frac{1}{2}b_0,\\
A_{02} &= a_{02} + \frac{1}{6}a_{20} - a_{11} - \frac{1}{6}b_{20} + \frac{2}{3}b_{11} - \frac{1}{2}b_{02} \;=\; -\frac{1}{3}a_0 + \frac{2}{3}b_0,\\
B_{20} &= b_{20} \;=\; 2a_0,\\
B_{11} &= b_{11} - \frac{1}{2}b_{20} \;=\; -a_0 + b_0,\\
B_{02} &= b_{02} + \frac{1}{6}b_{20} - b_{11} \;=\; \frac{1}{3}a_0 - b_0.
\end{aligned}$$

These formulas explicitly specify the quadratic part of system (9.51), whose time-one flow approximates (up to and including order 2) the map (9.49) at $\nu = 0$.

For $\nu \neq 0$ with small $\|\nu\|$, we construct $\varphi_\nu^t$ as the first two components of the flow

$$X \mapsto \phi^t(X) = \begin{pmatrix} \varphi_\nu^t(\xi) \\ \nu \end{pmatrix}, \quad X = \begin{pmatrix} \xi \\ \nu \end{pmatrix} \in \mathbb{R}^4,$$

generated by a 4-dimensional system with the parameters considered as constant variables:

$$\dot{X} = Y(X). \tag{9.52}$$

Here $Y(X) = JX + Y_2(X) + Y_3(X) + \cdots$, where

$$J = \begin{pmatrix} 0 & 1 & -\frac{1}{2} & 0 \\ 0 & 0 & 1 & 0 \\ 0 & 0 & 0 & 0 \\ 0 & 0 & 0 & 0 \end{pmatrix}, \quad Y_k(X) = \begin{pmatrix} Z_k(X) \\ 0 \end{pmatrix},$$

and each $Z_k$ is an order-$k$ homogeneous polynomial function from $\mathbb{R}^4$ to $\mathbb{R}^2$ with unknown coefficients. This trick is needed to assure that $X = 0$ is the equilibrium of (9.52), so that we can approximate the map

$$X \mapsto M(X) = \begin{pmatrix} N_\nu(\xi) \\ \nu \end{pmatrix}$$

near its fixed point $X = 0$ by the flow of (9.52), i.e.

$$M(X) = \phi^1(X) + O(\|X\|^3).$$

To find the vector field $Y$ explicitly, two Picard iterations for (9.52) are sufficient. We start with setting $X^{(1)}(t) = e^{Jt}X$. Then, the linear part of $M(X)$ coincides with $X^{(1)}(1) = e^J X$ (check!).

Since we know how the result $X^{(2)}(t)$ of the second Picard iteration should look, we set some coefficients of $Y_2$ equal to zero immediately:

$$Y_2(X) = \begin{pmatrix} A_{1010}\xi_1\nu_1 + A_{0110}\xi_2\nu_1 + A_{1001}\xi_1\nu 2 + A_{0101}\xi_2\nu_2 \\ B_{1010}\xi_1\nu_1 + B_{0110}\xi_2\nu_1 + B_{1001}\xi_1\nu_2 + B_{0101}\xi_2\nu_2 \\ 0 \\ 0 \end{pmatrix}$$
$$+ \begin{pmatrix} \frac{1}{2}A_{2000}\xi_1^2 + A_{1100}\xi_1\xi_2 + \frac{1}{2}A_{0200}\xi_2^2 \\ \frac{1}{2}B_{2000}\xi_1^2 + B_{1100}\xi_1\xi_2 + \frac{1}{2}B_{0200}\xi_2^2 \\ 0 \\ 0 \end{pmatrix}.$$

We leave the reader to complete the proof by computing

$$X^{(2)}(1) = e^J X + \int_0^t e^{J(1-\tau)} Y_2(X^{(1)}(\tau))d\tau$$

and comparing it with $M(X)$. Notice that $A_{2000} = A_{20}, A_{1100} = A_{11}$, and $A_{0200} = A_{02}$, while $B_{2000} = B_{20}, B_{1100} = B_{11}$, and $B_{0200} = B_{02}$, where $A_{jk}$ and $B_{jk}$ are as given above. □

The behavior of the approximate map $\varphi_\nu^1$ is described by the Bogdanov-Takens theory (see Chapter 8). As we know from Chapter 8, any generic system (9.50) is locally topologically equivalent to one of the normal forms

$$\begin{cases} \dot{\eta}_1 = \eta_2, \\ \dot{\eta}_2 = \beta_1 + \beta_2\eta_1 + \eta_1^2 + s\eta_1\eta_2, \end{cases} \tag{9.53}$$

where $s = \pm 1$. The genericity conditions of Theorem 8.4 include the nondegeneracy conditions (BT.0)–(BT.2) and transversality (BT.3) with respect to the parameters.

Condition (BT.3) for the system (9.50) is easy to verify. Indeed, the determinant of the Jacobian matrix of the map

$$\begin{pmatrix} \xi \\ \nu \end{pmatrix} \mapsto \begin{pmatrix} F(\xi, \nu) \\ \mathrm{tr}\ F_\xi(\xi, \nu) \\ \det F_\xi(\xi, \nu) \end{pmatrix}$$

evaluated at $\xi = \nu = 0$ is equal to $2a_0$. Therefore, we should assume that $2a_0 = B_{20} \neq 0$. Since eventually we want to apply all results to the original map that depends on $\alpha$, we have to assume that the transversality condition (R1.0) also holds.

The nondegeneracy condition (BT.0) holds automatically, while the conditions (BT.1) and (BT.2) for (9.50) at $\nu = 0$ (or, equivalently, for the system (9.51)) can be written as

$$A_{20} + B_{11} \neq 0, \quad B_{20} \neq 0.$$

Under these conditions,

$$s = \text{sign}[B_{20}(A_{20} + B_{11})].$$

Using the formulas derived above, we can express the quantities involved in the nondegeneracy conditions as

$$A_{20} + B_{11} = B(0) - 2A(0) = a_{20}(0) + b_{11}(0) - b_{20}(0), \quad B_{20} = 2A(0) = b_{20}(0).$$

Therefore the following nondegeneracy conditions should be imposed on the Taylor coefficients of the map (9.36) at the 1:1 resonance:

(R1.1) $a_{20}(0) + b_{11}(0) - b_{20}(0) \neq 0$;
(R1.2) $b_{20}(0) \neq 0$.

Notice that the second condition coincides with the assumption of Lemma 9.7 and simultaneously guarantees the transversality of the Bogdanov-Takens bifurcation. Thus, the expression for $s$ reads

$$s = \text{sign}[b_{20}(a_{20} + b_{11} - b_{20})](0) = \pm 1.$$

The bifurcation diagram of system (9.53) with $s = -1$ was presented in **Fig. 8.8**. That bifurcation diagram therefore describes the bifurcations of the approximate map $\varphi_\nu^1$ if we take into account the correspondence between $(\xi, \nu)$ and $(\eta, \beta)$. Equilibria correspond to fixed points, while limit cycles should be interpreted as closed invariant curves. Moving around the origin of the $\beta$-plane clockwise, we encounter the following bifurcations of the map $\varphi_\nu^1$. A pair of fixed points, a saddle and a stable one, appear when crossing the curve $T_-$ from region **1** to region **2**. The stable fixed point becomes unstable, producing a stable closed invariant curve along curve $H$, which now corresponds to a Neimark-Sacker bifurcation. The closed curve exists in region **3** and is destroyed via a homoclinic bifurcation at $P$. Finally, the saddle and unstable points collide and disappear via a fold bifurcation at $T_+$. Recall once more that all this happens in the approximate map.

What can be said about bifurcations of a generic map (9.29)? Certain features of the bifurcation diagram do survive, namely those related to *local* bifurcations. More precisely, there are bifurcation curves $\widetilde{T}$ and $\widetilde{H}$, corresponding to the curves $T$ and $H$, on which the map has a fold and a Neimark-Sacker bifurcation, respectively. These curves meet tangentially at the codim 2 point. A closed invariant curve bifurcating from the stable fixed point exists for parameter values close to $\widetilde{H}$.

However, even if a closed invariant curve exists for the original map, the orbit structure on it is generically different from that for $\varphi_\nu^1$. For the approximate map, all orbits in the curve are either periodic or dense, while for the original map, phase-locking phenomena occur that create and destroy stable and unstable long-period orbits as parameters vary inside region **3**. Actually, an infinite number of narrow phase-locking Arnold tongues are rooted at

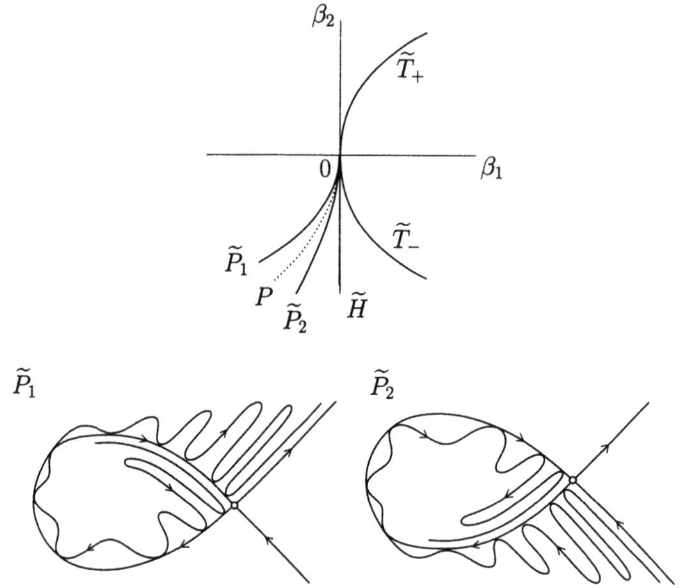


**Fig. 9.7.** Homoclinic tangencies along the curves $\widetilde{P}_1$ and $\widetilde{P}_2$.

the Neimark-Sacker curve $\widetilde{H}$. These tongues are delimited by fold bifurcation curves corresponding to a collision between stable and saddle periodic orbits. Some other bifurcations of the original map are not present in the approximating flow. These bifurcations take place near the homoclinic curve $P$ for the approximate map. The coincidence of the stable and unstable manifolds of the saddle at $P$ occurring for $\varphi_\nu^1$ is generically replaced by their transversal intersection, giving rise to a *homoclinic structure* implying the existence of an infinite number of saddle cycles (see Chapter 2). The transversal homoclinic structure exists in an exponentially narrow parameter region around $P$ bounded by two smooth bifurcation curves, $\widetilde{P}_1$ and $\widetilde{P}_2$, corresponding to *homoclinic tangencies* (see **Fig. 9.7** and Section 7.2.1 of Chapter 7). The fold curves delimiting the phase-locking tongues accumulate on $\widetilde{P}_{1,2}$ (see **Fig. 9.8**, where only one tongue is shown schematically). The complete picture includes other bifurcations (i.e., flips) and seems to be unknown (see the bibliographical notes in the appendix).

### 9.5.3 1:2 resonance

In this case we have a smooth planar map

$$x \mapsto f(x, \alpha), \quad x \in \mathbb{R}^2, \ \alpha \in \mathbb{R}^2,$$

having at $\alpha = 0$ the fixed point $x = 0$ with multipliers $\mu_{1,2} = -1$ (1:2 resonance). Since $\mu = 1$ is not an eigenvalue of its Jacobian matrix, we can

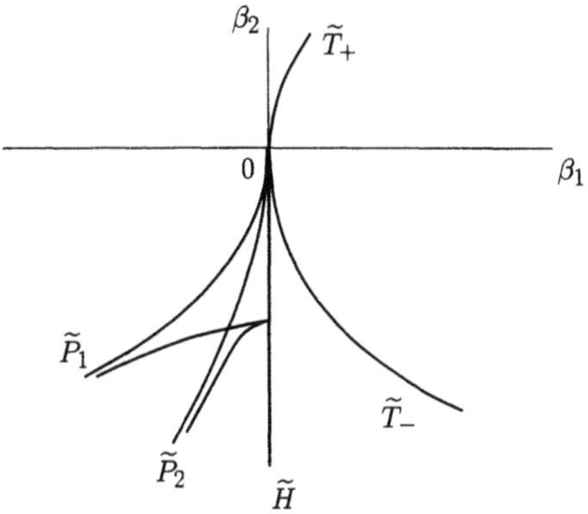


**Fig. 9.8.** An Arnold tongue rooted at a resonant Neimark-Sacker bifurcation point.

assume that $x = 0$ is a fixed point for all sufficiently small $\|\alpha\|$ and write our map as

$$x \mapsto A(\alpha)x + F(x, \alpha), \tag{9.54}$$

for some smooth $A(\alpha)$, $F(x, \alpha) = O(\|x\|^2)$. Let $A_0 = A(0)$, and assume that there exist two real linearly independent vectors, $q_{0,1} \in \mathbb{R}^2$:

$$A_0 q_0 = -q_0, \quad A_0 q_1 = -q_1 + q_0. \tag{9.55}$$

The vector $q_0$ is the eigenvector of $A_0$ corresponding to the eigenvalue $-1$, while $q_1$ is the generalized eigenvector of $A_0$ corresponding to the same eigenvalue. Moreover, there exist similar adjoint eigenvectors $p_{0,1} \in \mathbb{R}^2$ of the transposed matrix $A_0^T$:

$$A_0^T p_1 = -p_1, \quad A_0^T p_0 = -p_0 + p_1. \tag{9.56}$$

We can always select four vectors satisfying (9.55) and (9.56) such that

$$\langle q_0, p_0 \rangle = \langle q_1, p_1 \rangle = 1,$$

where $\langle \cdot, \cdot \rangle$ stands for the standard scalar product in $\mathbb{R}^2$ : $\langle x, y \rangle = x_1 y_1 + x_2 y_2$. The Fredholm Alternative Theorem implies

$$\langle q_1, p_0 \rangle = \langle q_0, p_1 \rangle = 0.$$

If $v_0$ and $v_1$ are selected, then any vector $x \in \mathbb{R}^2$ can be uniquely represented as

$$x = y_1 q_0 + y_2 q_1,$$

where new coordinates $(y_1, y_2)$ can be computed explicitly by

$$\begin{cases} y_1 = \langle p_0, x \rangle, \\ y_2 = \langle p_1, x \rangle. \end{cases} \tag{9.57}$$

In the coordinates $(y_1, y_2)$, the map (9.54) takes the form

$$\begin{pmatrix} y_1 \\ y_2 \end{pmatrix} \mapsto \begin{pmatrix} -1 + a_{10}(\alpha) & 1 + a_{01}(\alpha) \\ b_{10}(\alpha) & -1 + b_{01}(\alpha) \end{pmatrix} \begin{pmatrix} y_1 \\ y_2 \end{pmatrix} + \begin{pmatrix} g(y,\alpha) \\ h(y,\alpha) \end{pmatrix}, \tag{9.58}$$

where

$$g(y,\alpha) = \langle p_0, F(y_1 q_0 + y_2 q_1, \alpha)\rangle, \quad h(y,\alpha) = \langle p_1 F(y_1 q_0 + y_2 q_1, \alpha)\rangle,$$

and

$$a_{10}(\alpha) = \langle p_0, [A(\alpha) - A_0]q_0\rangle, \quad a_{01}(\alpha) = \langle p_0, [A(\alpha) - A_0]q_1\rangle,$$
$$b_{10}(\alpha) = \langle p_1, [A(\alpha) - A_0]q_0\rangle, \quad b_{01}(\alpha) = \langle p_1, [A(\alpha) - A_0]q_1\rangle.$$

Clearly,

$$a_{10}(0) = a_{01}(0) = b_{10}(0) = b_{01}(0) = 0.$$

The nonsingular linear coordinate transformation

$$y = P(\alpha)u,$$

where $P$ is given in Lemma 9.6 with $\lambda = -1$, reduces the map (9.58) for sufficiently small $\|\alpha\|$ into

$$\begin{pmatrix} u_1 \\ u_2 \end{pmatrix} \mapsto \begin{pmatrix} -1 & 1 \\ \nu_1(\alpha) & -1 + \nu_2(\alpha) \end{pmatrix} \begin{pmatrix} u_1 \\ u_2 \end{pmatrix} + P^{-1}(\alpha) \begin{pmatrix} g(P(\alpha)u,\alpha) \\ h(P(\alpha)u,\alpha) \end{pmatrix}, \tag{9.59}$$

where

$$\nu_1(\alpha) = b_{01}(\alpha) + a_{01}(\alpha)b_{10}(\alpha) - a_{10}(\alpha)b_{01}(\alpha), \quad \nu_2(\alpha) = a_{10}(\alpha) + b_{01}(\alpha).$$

Assume that

(R2.0) $\det\left(\frac{\partial \nu}{\partial \alpha}\right)\bigg|_{\alpha=0} \neq 0.$

Under this transversality assumption, we can use $\nu$ to parametrize a neighborhood of $\alpha = 0$:

$$\begin{cases} \beta_1 = \nu_1(\alpha), \\ \beta_2 = \nu_2(\alpha). \end{cases}$$

We can express $\alpha$ as a function of $\beta$ and write the map (9.59) as

$$\begin{pmatrix} u_1 \\ u_2 \end{pmatrix} \mapsto \begin{pmatrix} -1 & 1 \\ \beta_1 & -1 + \beta_2 \end{pmatrix} \begin{pmatrix} u_1 \\ u_2 \end{pmatrix} + \begin{pmatrix} G(u,\beta) \\ H(u,\beta) \end{pmatrix}, \tag{9.60}$$

where $G, H = O(\|u\|^2)$. Notice that $G(u,0) = g(u,0), H(u,0) = h(u,0)$.

Having performed these preliminary linear transformations, we can now make a nonlinear change of coordinates to simplify the map (9.60).

**Lemma 9.9 (Normal form map for 1:2 resonance)** *There is a smooth invertible change of coordinates, smoothly depending on the parameters, that transforms* (9.60) *into the map*

$$\begin{pmatrix} \xi_1 \\ \xi_2 \end{pmatrix} \mapsto \begin{pmatrix} -1 & 1 \\ \beta_1 & -1+\beta_2 \end{pmatrix} \begin{pmatrix} \xi_1 \\ \xi_2 \end{pmatrix} + \begin{pmatrix} 0 \\ C(\beta)\xi_1^3 + D(\beta)\xi_1^2\xi_2 \end{pmatrix} + O(\|\xi\|^4) \tag{9.61}$$

*for smooth functions* $C(\beta)$ *and* $D(\beta)$.

**Proof:**

Expand $G$ and $H$ into Taylor series with respect to $u$:

$$G(u,\beta) = \sum_{2\le j+k\le 3} g_{jk}(\beta)u_1^j u_2^k + O(\|u\|^4),$$

$$H(u,\beta) = \sum_{2\le j+k\le 3} h_{jk}(\beta)u_1^j u_2^k + O(\|u\|^4).$$

Let us try to find a transformation

$$\begin{cases} u_1 = \xi_1 + \displaystyle\sum_{2\le j+k\le 3} \phi_{jk}(\beta)\xi_1^j\xi_2^k, \\ u_2 = \xi_2 + \displaystyle\sum_{2\le j+k\le 3} \psi_{jk}(\beta)\xi_1^j\xi_2^k, \end{cases}$$

with

$$\phi_{03}(\beta) = \psi_{03}(\beta) = 0,$$

that annihilates all terms of order two and three except those displayed in (9.61).

First, fix $\alpha = 0$ and find the corresponding $\phi_{jk}(0), \psi_{jk}(0)$. This is sufficient to compute $C(0)$ and $D(0)$. To simplify notation, drop the argument of $\phi_{jk}$ and $\psi_{jk}$. In the coordinates $(\xi_1, \xi_2)$, the map (9.60) for $\beta = 0$ will have a form

$$\begin{cases} \tilde{\xi}_1 = -\xi_1 + \xi_2 + \displaystyle\sum_{2\le j+k\le 3} \gamma_{jk}\xi_1^j\xi_2^k + O(\|\xi\|^4), \\ \tilde{\xi}_2 = \quad \xi_2 + \displaystyle\sum_{2\le j+k\le 3} \sigma_{jk}\xi_1^j\xi_2^k + O(\|\xi\|^4), \end{cases}$$

where $\gamma_{jk}$ and $\sigma_{jk}$ are certain functions of $g_{jk}, h_{jk}$ and $\phi_{jk}, \psi_{jk}$.

By a proper selection of $\phi_{jk}, \psi_{jk}$ with $j+k=2$, we can eliminate all quadratic terms in the map $(\xi_1,\xi_2) \mapsto (\tilde{\xi}_1, \tilde{\xi}_2)$. Indeed, the vanishing of all these terms means

$$\gamma_{20} = \gamma_{11} = \gamma_{02} = \sigma_{20} = \sigma_{11} = \sigma_{02} = 0,$$

is equivalent to the linear algebraic system

$$\begin{pmatrix} 2 & 0 & 0 & -1 & 0 & 0 \\ -2 & 2 & 0 & 0 & -1 & 0 \\ 1 & -1 & 2 & 0 & 0 & -1 \\ 0 & 0 & 0 & 2 & 0 & 0 \\ 0 & 0 & 0 & -2 & 2 & 0 \\ 0 & 0 & 0 & 1 & -1 & 2 \end{pmatrix} \begin{pmatrix} \phi_{20} \\ \phi_{11} \\ \phi_{02} \\ \psi_{20} \\ \psi_{11} \\ \psi_{02} \end{pmatrix} = \begin{pmatrix} g_{20} \\ g_{11} \\ g_{02} \\ h_{20} \\ h_{11} \\ h_{02} \end{pmatrix}.$$

This system has a unique solution since its determinant obviously differs from zero. Then, we can use its solution

$$\phi_{20} = \frac{1}{2}g_{20} + \frac{1}{4}h_{20}, \tag{9.62}$$

$$\phi_{11} = \frac{1}{2}g_{20} + \frac{1}{2}g_{11} + \frac{1}{2}h_{20} + \frac{1}{4}h_{11}, \tag{9.63}$$

$$\phi_{02} = \frac{1}{4}g_{11} + \frac{1}{2}g_{02} + \frac{1}{8}h_{20} + \frac{1}{4}h_{11} + \frac{1}{4}h_{02}, \tag{9.64}$$

$$\psi_{02} = \frac{1}{2}h_{20}, \tag{9.65}$$

$$\psi_{11} = \frac{1}{2}h_{20} + \frac{1}{2}h_{11}, \tag{9.66}$$

$$\psi_{02} = \frac{1}{4}h_{11} + \frac{1}{2}h_{02}, \tag{9.67}$$

to eliminate all the quadratic $\xi$-terms at $\beta = 0$.

The next step is to "kill" cubic terms. Actually, it is possible to annihilate all but those terms displayed in (9.61). Namely, the vanishing conditions

$$\gamma_{21} = \gamma_{12} = \gamma_{03} = \gamma_{30} = \sigma_{12} = \sigma_{03} = 0$$

yield the linear system for $\phi_{jk}, \psi_{jk}$ of the form

$$\begin{pmatrix} 3 & 0 & 0 & -1 & 0 & 0 \\ -3 & 2 & 0 & 0 & -1 & 0 \\ 1 & -1 & 1 & 0 & 0 & -1 \\ 0 & 0 & 0 & 3 & 0 & 0 \\ 0 & 0 & 0 & -3 & 2 & 0 \\ 0 & 0 & 0 & 1 & -1 & 1 \end{pmatrix} \begin{pmatrix} \phi_{30} \\ \phi_{21} \\ \phi_{12} \\ \psi_{30} \\ \psi_{21} \\ \psi_{12} \end{pmatrix} = R[g,h],$$

where $R[g,h]$ is a certain vector-valued function of $g_{jk}, h_{jk}, j+k = 2,3$. This system is also solvable, thus specifying the cubic coefficients of the transformation. Explicit solution of this system (using (9.62)–(9.67)) leads to the following formulas for the coefficients in (9.61):

$$C(0) = h_{30}(0) + g_{20}(0)h_{20}(0) + \frac{1}{2}h_{20}^2(0) + \frac{1}{2}h_{20}(0)h_{11}(0) \tag{9.68}$$

and

$$D(0) = h_{21}(0) + 3g_{30}(0) + \frac{1}{2}g_{20}(0)h_{11}(0) + \frac{5}{4}h_{20}(0)h_{11}(0)$$
$$+ h_{20}(0)h_{02}(0) + 3g_{20}^2(0) + \frac{5}{2}g_{20}(0)h_{20}(0) + \frac{5}{2}g_{11}(0)h_{20}(0)$$
$$+ h_{20}^2(0) + \frac{1}{2}h_{11}^2(0). \tag{9.69}$$

Small $\|\beta\| \neq 0$ causes no difficulties since the above linear systems will be substituted by nearby linear systems. Because the original systems at $\beta = 0$ are regular, these new systems are uniquely solvable. We leave the details to the reader. □

**Remark:**

It is *insufficient* to make only the quadratic transformation with coefficients given by (9.62)–(9.67) to compute $D(0)$ since this quantity depends on $\psi_{30}(0)$. ◇

Denote the normal form map (9.61) by $\xi \mapsto \Gamma_\beta(\xi)$. Our next task would be to approximate this map by a flow. Clearly enough, this is not possible since its linear part for $\beta = 0$,

$$\begin{pmatrix} \xi_1 \\ \xi_2 \end{pmatrix} \mapsto \begin{pmatrix} -1 & 1 \\ 0 & -1 \end{pmatrix} \begin{pmatrix} \xi_1 \\ \xi_2 \end{pmatrix}$$

has negative eigenvalues. However, the *second* iterate,

$$\xi \mapsto \Gamma_\beta^2(\xi),$$

can be approximated by the unit-time shift of a flow. The map $\Gamma_\beta^2$ has the form

$$\begin{pmatrix} \xi_1 \\ \xi_2 \end{pmatrix} \mapsto \begin{pmatrix} 1+\beta_1 & -2+\beta_2 \\ -2\beta_1 + \beta_1\beta_2 & 1+\beta_1 - 2\beta_2 + \beta_2^2 \end{pmatrix} \begin{pmatrix} \xi_1 \\ \xi_2 \end{pmatrix} + \begin{pmatrix} V(\xi,\beta) \\ W(\xi,\beta) \end{pmatrix}, \tag{9.70}$$

where

$$V(\xi,\beta) = C(\beta)\xi_1^3 + D(\beta)\xi_1^2\xi_2$$

and

$$W(\xi,\beta) = (-2C(\beta) + \beta_1 D(\beta) + \beta_2 C(\beta))\xi_1^3$$
$$+ (3C(\beta) - 2D(\beta) - 2\beta_1 D(\beta) + \beta_2 D(\beta))\xi_1^2\xi_2$$
$$+ (-3C(\beta) + 2D(\beta) + \beta_1 D(\beta) - 2\beta_2 D(\beta))\xi_1\xi_2^2$$
$$+ (C(\beta) - D(\beta) + \beta_2 D(\beta))\xi_2^3 + O(\|\xi\|^4).$$

**Lemma 9.10** *The map* (9.70) *can be represented for all sufficiently small* $\|\beta\|$ *in the form*

$$\Gamma_\beta^2(\xi) = \varphi_\beta^1(\xi) + O(\|\xi\|^4),$$

*where* $\varphi_\beta^t$ *is the flow of a planar system*

$$\dot{\xi} = \Lambda_\beta \xi + U(\xi, \beta), \tag{9.71}$$

*where*

$$\Lambda_\beta = \begin{pmatrix} -\beta_1 & -2 - \frac{2}{3}\beta_1 - \beta_2 \\ -2\beta_1 & -\beta_1 - 2\beta_2 \end{pmatrix} + O(\|\beta\|^2),$$

*and* $U(\xi, \beta)$ *is a homogeneous cubic vector polynomial in* $\xi$.

**Proof:**

Start with $\beta = 0$. The linear part of (9.70) at $\beta = 0$,

$$\begin{pmatrix} \xi_1 \\ \xi_2 \end{pmatrix} \mapsto \begin{pmatrix} 1 & -2 \\ 0 & 1 \end{pmatrix} \begin{pmatrix} \xi_1 \\ \xi_2 \end{pmatrix},$$

is the unit-time shift along orbits of the planar linear system

$$\dot{\xi} = \Lambda_0 \xi,$$

where

$$\Lambda_0 = \begin{pmatrix} 0 & -2 \\ 0 & 0 \end{pmatrix}.$$

Suppose that the approximating cubic system $\dot{\xi} = \Lambda_0 \xi + U(\xi, 0)$ has the representation

$$\begin{cases} \dot{\xi}_1 = -2\xi_2 + A_{30}\xi_1^3 + A_{21}\xi_1^2\xi_2 + A_{12}\xi_1\xi_2^2 + A_{03}\xi_2^3, \\ \dot{\xi}_2 = B_{30}\xi_1^3 + B_{21}\xi_1^2\xi_2 + B_{12}\xi_1\xi_2^2 + B_{03}\xi_2^3. \end{cases} \tag{9.72}$$

Let us perform three Picard iterations (9.26) for (9.72). Since the system (9.72) has no quadratic terms, we have

$$\xi^{(1)}(\tau) = \xi^{(2)}(\tau) = e^{\Lambda_0 \tau}\xi = \begin{pmatrix} \xi_1 - 2\tau\xi_2 \\ \xi_2 \end{pmatrix}.$$

The third iteration yields

$$\xi^{(3)}(1) = \begin{pmatrix} \xi_1 - 2\xi_2 + a_{30}\xi_1^3 + a_{21}\xi_1^2\xi_2 + a_{12}\xi_1\xi_2^2 + a_{03}\xi_2^3 \\ \xi_2 + b_{30}\xi_1^3 + b_{21}\xi_1^2\xi_2 + b_{12}\xi_1\xi_2^2 + b_{03}\xi_2^3 \end{pmatrix},$$

where $a_{jk}, b_{jk}$ are expressed in terms of $A_{jk}, B_{jk}$ by the formulas

$$\begin{aligned}
a_{30} &= A_{30} - B_{30}, \\
a_{21} &= -3A_{30} + A_{21} + 2B_{30} - B_{21}, \\
a_{12} &= 4A_{30} - 2A_{21} + A_{12} - 2B_{30} + \frac{4}{3}B_{21} - B_{12}, \\
a_{03} &= -2A_{30} + \frac{4}{3}A_{21} - A_{12} + A_{03} + \frac{4}{5}B_{30} - \frac{2}{3}B_{21} + \frac{2}{3}B_{12} - B_{03}, \\
b_{30} &= B_{30}, \\
b_{21} &= -3B_{30} + B_{21}, \\
b_{12} &= 4B_{30} - 2B_{21} + B_{12}, \\
b_{03} &= -2B_{30} + \frac{4}{3}B_{21} - B_{12} + B_{03}.
\end{aligned}$$

Solving the above equations for $A_{jk}, B_{jk}$ and using (9.70), we obtain

$$\begin{aligned}
A_{30} &= a_{30} + b_{30} \;=\; -C(0),\\
A_{21} &= 3a_{30} + a_{21} + 4b_{30} + b_{21} \;=\; -2C(0) - D(0),\\
A_{12} &= 2a_{30} + 2a_{21} + a_{12} + 4b_{30} + \frac{8}{3}b_{21} + b_{12} \;=\; -C(0) - \frac{4}{3}D(0),\\
A_{03} &= \frac{2}{3}a_{21} + a_{12} + a_{03} + \frac{8}{15}b_{30} + \frac{4}{3}b_{21} + \frac{4}{3}b_{12} + b_{03}\\
&= -\frac{1}{15}C(0) - \frac{1}{3}D(0),\\
B_{30} &= b_{30} \;=\; -2C(0),\\
B_{21} &= 3b_{30} + b_{21} \;=\; -3C(0) - 2D(0),\\
B_{12} &= 2b_{30} + 2b_{21} + b_{12} \;=\; -C(0) - 2D(0),\\
B_{03} &= \frac{2}{3}b_{21} + b_{12} + b_{03} \;=\; -\frac{1}{3}D(0).
\end{aligned}$$

This proves the lemma and gives explicit formulas for the coefficients of the approximating system (9.71) at $\beta = 0$.

For small $\|\beta\| \neq 0$, we can verify the formula for the linear part of $\Lambda = \Lambda_\beta$ given in the lemma statement by first taking three terms in the expansion

$$e^{\Lambda} = I_2 + \sum_{k=1} \frac{1}{k!}\Lambda^k.$$

Then, Picard iterations produce expressions for $a_{jk}(\beta), b_{jk}(\beta)$ that, if we set $\beta = 0$, coincide with those obtained above. □

The map (9.71) can be simplified further.

**Lemma 9.11** *If we perform an invertible smoothly parameter-dependent transformation of variables, system* (9.71) *can be reduced to the form*

$$\begin{pmatrix} \dot{\eta}_1 \\ \dot{\eta}_2 \end{pmatrix} = \begin{pmatrix} 0 & 1 \\ \gamma_1 & \gamma_2 \end{pmatrix} \begin{pmatrix} \eta_1 \\ \eta_2 \end{pmatrix} + \begin{pmatrix} 0 \\ C_1\eta_1^3 + D_1\eta_1^2\eta_2 \end{pmatrix} + O(\|\eta\|^4), \qquad (9.73)$$

*for smooth functions* $\gamma_{1,2} = \gamma_{1,2}(\beta)$,

$$\begin{cases} \gamma_1(\beta) = 4\beta_1 + O(\|\beta\|^2), \\ \gamma_2(\beta) = -2\beta_1 - 2\beta_2 + O(\|\beta\|^2), \end{cases}$$

*and smooth* $C_1 = C_1(\beta), D_1 = D_1(\beta)$, *such that*

$$C_1(0) = 4C(0), \quad D_1(0) = -2D(0) - 6C(0).$$

**Proof:**

For $\beta = 0$, the desired transformation of (9.72) is given by

$$\begin{cases} \eta_1 = \xi_1 + \chi_{30}\xi_1^3 + \chi_{21}\xi_1^2\xi_2 + \chi_{12}\xi_1\xi_2^2, \\ \eta_2 = -2\xi_2 + \phi_{30}\xi_1^3 + \phi_{21}\xi_1^2\xi_2 + \phi_{12}\xi_1\xi_2^2, \end{cases}$$

where

$$\chi_{30} = \frac{1}{6}A_{21} + \frac{1}{12}B_{12}, \quad \chi_{21} = \frac{1}{4}A_{12} + \frac{1}{4}B_{03}, \quad \chi_{12} = \frac{1}{2}A_{03},$$

and

$$\phi_{30} = A_{30}, \quad \phi_{21} = -\frac{1}{2}B_{12}, \quad \phi_{12} = -B_{03}.$$

Making this transformation and using the above formulas for $B_{30}, B_{21}$, and $A_{30}$, one ends up with the coefficients $C_1(0)$ and $D_1(0)$ as in the lemma statement.

The case $\beta \neq 0$ is left to the reader. Notice that the map $\beta \mapsto \gamma$ is regular at $\beta = 0$, therefore $(\gamma_1, \gamma_2)$ can be used as new parameters near the origin of the parameter plane. □

By combining the two previous lemmas, we can formulate the following theorem.

**Theorem 9.3 (Normal form flow for 1:2 resonance)** *The second iterate of the smooth map,*

$$\begin{pmatrix} \xi_1 \\ \xi_2 \end{pmatrix} \mapsto \begin{pmatrix} -1 & 1 \\ \beta_1 & -1+\beta_2 \end{pmatrix} \begin{pmatrix} \xi_1 \\ \xi_2 \end{pmatrix} + \begin{pmatrix} 0 \\ C(\beta)\xi_1^3 + D(\beta)\xi_1^2\xi_2 \end{pmatrix} + O(\|\xi\|^4),$$

*can be represented for all sufficiently small $\|\beta\|$ in the form*

$$\xi \mapsto \varphi_\beta^1(\xi) + O(\|\xi\|^4),$$

*where $\varphi_\beta^t$ is the flow of a planar system, that is smoothly equivalent to the system*

$$\begin{pmatrix} \dot\eta_1 \\ \dot\eta_2 \end{pmatrix} = \begin{pmatrix} 0 & 1 \\ \gamma_1(\beta) & \gamma_2(\beta) \end{pmatrix} \begin{pmatrix} \eta_1 \\ \eta_2 \end{pmatrix} + \begin{pmatrix} 0 \\ C_1(\beta)\eta_1^3 + D_1(\beta)\eta_1^2\eta_2 \end{pmatrix}, \tag{9.74}$$

*where*

$$\begin{cases} \gamma_1(\beta) = 4\beta_1 + O(\|\beta\|^2), \\ \gamma_2(\beta) = -2\beta_1 - 2\beta_2 + O(\|\beta\|^2), \end{cases}$$

*and*

$$C_1(0) = 4C(0), \quad D_1(0) = -2D(0) - 6C(0). \ \square$$

Now consider the bifurcations of the approximating system (9.74), assuming the following nondegeneracy conditions:

$$C_1(0) \neq 0, \quad D_1(0) \neq 0.$$

These conditions can be expressed in terms of the normal map coefficients:

(R2.1) $C(0) \neq 0$;
(R2.2) $D(0) + 3C(0) \neq 0$;

and effectively verified for a given map using (9.68) and (9.69). We can also assume that

$$D_1(0) < 0;$$

otherwise, reverse time and substitute $\eta_1 \mapsto -\eta_1$. Under these assumptions, we can scale the variables, parameters, and time in (9.74), thus obtaining the system

$$\begin{cases} \dot{\zeta}_1 = \zeta_2, \\ \dot{\zeta}_2 = \varepsilon_1\zeta_1 + \varepsilon_2\zeta_2 + s\zeta_1^3 - \zeta_1^2\zeta_2, \end{cases} \tag{9.75}$$

where $s = \text{sign } C(0) = \pm 1$.

The bifurcation diagrams of (9.75) for $s = 1$ and $s = -1$ are presented in **Figs. 9.9** and **9.10**, respectively. The important feature of the approximating system is that it is *invariant* under the rotation through the angle $\pi$

$$\zeta \mapsto -\zeta.$$

Using the terminology of Chapter 7, we say that it is $\mathbb{Z}_2$-*symmetric*. The system always has the equilibrium

$$E_0 = (0, 0).$$

Two other possible equilibria are located on the horizontal axis $\zeta_2 = 0$,

$$E_{1,2} = \left(\mp\sqrt{-s\varepsilon_1}, 0\right),$$

and bifurcate simultaneously from the trivial one via a *pitchfork* bifurcation along the line

$$F^{(1)} = \{(\varepsilon_1, \varepsilon_2) : \ \varepsilon_1 = 0\}.$$

The nontrivial equilibria exist for $\varepsilon_1 < 0$ if $s = 1$, and for $\varepsilon_1 > 0$ if $s = -1$.

Consider the case $s = 1$ (**Fig. 9.9**). In region **1** there is a single trivial equilibrium $E_0$, which is a saddle. Crossing the lower branch of $F^{(1)}$ implies a pitchfork bifurcation generating a pair of symmetry-coupled saddles $E_{1,2}$, while the trivial equilibrium becomes a stable node. This node turns into a focus somewhere in region **2** and then loses its stability upon crossing the half-line

$$H^{(1)} = \{(\varepsilon_1, \varepsilon_2) : \ \varepsilon_2 = 0, \varepsilon_1 < 0\},$$

via a nondegenerate Hopf bifurcation. In region **3** a unique and stable limit cycle exists.[5] Crossing the curve

$$C = \left\{(\varepsilon_1, \varepsilon_2) : \ \varepsilon_2 = -\frac{1}{5}\varepsilon_1 + o(\varepsilon_1), \varepsilon_1 < 0\right\}$$

[5] In the terminology of Chapter 7, this cycle is an $S$-cycle.

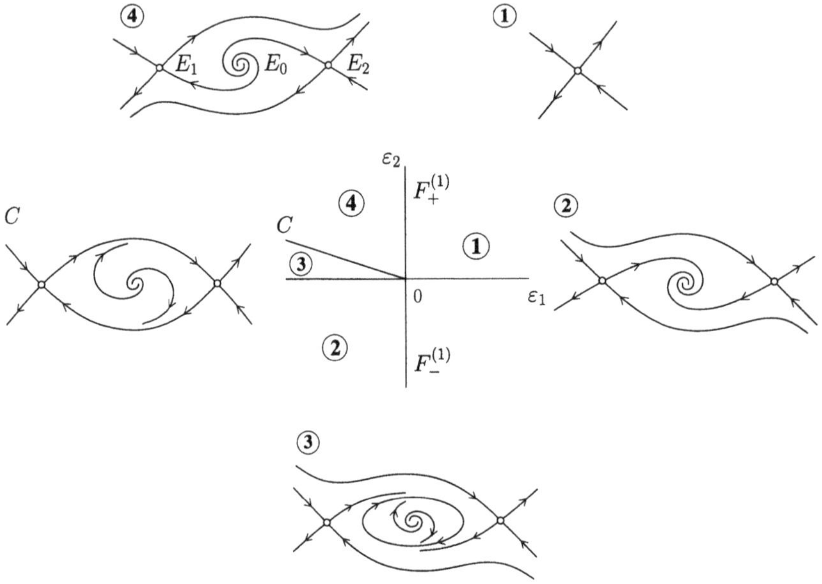


**Fig. 9.9.** Bifurcation diagram of the approximating system (9.75) for $s = 1$.

leads to the disappearance of the cycle through a heteroclinic bifurcation (see Exercise 5). Due to the $\mathbb{Z}_2$-symmetry, the heteroclinic orbits connecting the saddles $E_1$ and $E_2$ appear simultaneously, forming a *heteroclinic cycle* upon crossing $C$. In region **4** the totally unstable trivial equilibrium $E_0$ coexists with the saddles $E_{1,2}$. All three of these equilibria merge at the upper branch of the pitchfork bifurcation line $F^{(1)}$ as we return to region **1**. The most difficult fact to prove is the uniqueness of the limit cycle in region **3**. This can be done using singular rescaling, the Pontryagin method of Hamiltonian perturbations, and estimation of elliptic integrals in a similar way to that described in Appendix A to Chapter 8 for the Bogdanov-Takens bifurcation (see Exercise 5 for some of the details).

The bifurcation diagram in the case $s = -1$ is slightly more complicated (**Fig. 9.10**). In region **1** there is a single trivial equilibrium $E_0$ that is now a stable point (either a node or a focus). It undergoes a nondegenerate Hopf bifurcation on the half-line $H^{(1)}$ given above, giving rise to a stable limit cycle. Two unstable nodes (later becoming foci) branch from the trivial equilibrium when we cross the upper half-axis of $F^{(1)}$ from region **2** to region **3**. In region **3**, all three equilibria, $E_0, E_1$, and $E_2$, are located inside the surrounding "big" limit cycle that is still present. At the half-line,

$$H^{(2)} = \{(\varepsilon_1, \varepsilon_2) : \ \varepsilon_2 = \varepsilon_1, \ \varepsilon_1 > 0\},$$

the nontrivial foci $E_{1,2}$ simultaneously undergo Hopf bifurcations. These bifurcations lead to the appearance of two "small" unstable (symmetry-coupled)

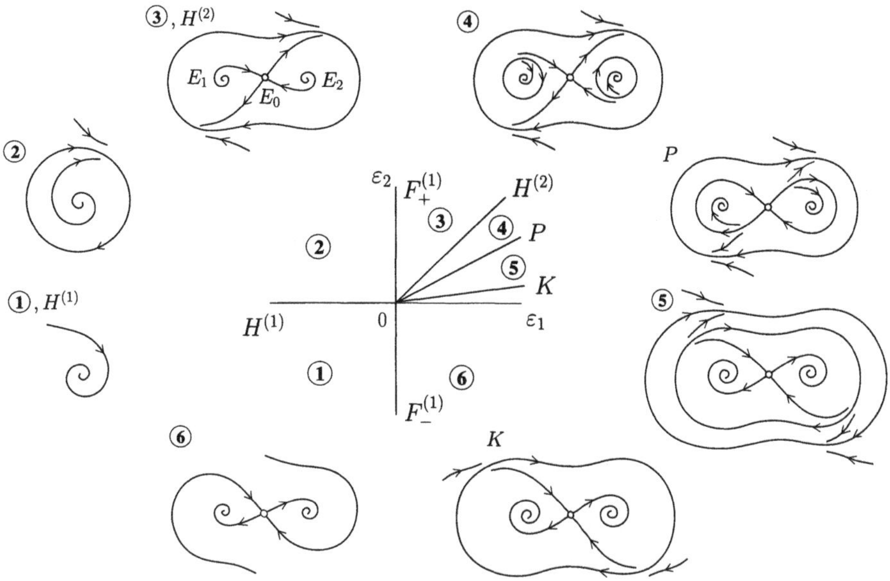


**Fig. 9.10.** Bifurcation diagram of the approximating system (9.75) for $s = -1$.

limit cycles around the nontrivial equilibria. The equilibria themselves become stable. Therefore, in region **4** we have *three* limit cycles: a "big" one and two "small" ones. Along the curve

$$P = \left\{ (\varepsilon_1, \varepsilon_2) : \ \varepsilon_2 = \frac{4}{5}\varepsilon_1 + o(\varepsilon_1) \ \varepsilon_1 > 0 \right\},$$

the "small" cycles disappear via a *symmetric figure-eight* homoclinic bifurcation (see Exercise 5). Along this curve, the saddle $E_0$ has two homoclinic orbits simultaneously. These orbits are transformed one into another by the rotation through $\pi$. The figure-eight is unstable from both inside and outside. Crossing the curve $P$ from region **4** to region **5** implies not only the destruction of the "small" cycles, but also the appearance of an extra unstable "big" limit cycle. Thus, in region **5**, we have two "big" cycles: The outer one is stable, while the inner one is unstable. These two "big" cycles collide and disappear along the curve

$$K = \{(\varepsilon_1, \varepsilon_2) : \ \varepsilon_2 = \kappa_0 \varepsilon_1 + o(\varepsilon_1), \ \varepsilon_1 > 0\},$$

where $\kappa_0 = 0.752\ldots$. This is a fold bifurcation of cycles. After the fold bifurcation, no limit cycles are left in the system. In region **6** we have three equilibria, the trivial saddle and two stable nontrivial foci/nodes. The nontrivial equilibria collide with the trivial one at the lower branch of the line $F^{(2)}$, as we return back to region **1**. We conclude the description of the bifurcation diagram in this case by pointing out that the most difficult part of the analysis is proving that there are no extra limit cycles in the system (see the bibliographical notes and Exercise 5).

We can now interpret the obtained results, first in terms of the approximating map $\varphi_\varepsilon^1$, and second in terms of the original map near the 1:2 resonance. Let $\varphi_\varepsilon^1$ be the unit-shift along orbits of the normal form (9.75). For this map, the equilibria become fixed points, the pitchfork bifurcation retains its sense, while the Hopf bifurcations turn into the Neimark-Sacker bifurcations since the limit cycles become closed invariant curves of corresponding stability. Recall now that the map $\varphi_\varepsilon^1$ approximates the *second* iterate $\Gamma_\beta^2$ of the original map $\Gamma_\beta$ near the 1:2 resonance. For the map $\Gamma_\beta$, the trivial fixed point of $\varphi_\varepsilon^1$ is a fixed point placed at the origin, while the nontrivial fixed points correspond to a single *period-two* orbit. Therefore, the pitchfork turns into a *period-doubling* bifurcation, which is natural to expect near a double multiplier $\mu_{1,2} = -1$. One can prove that bifurcation curves similar to the curves $F^{(1)}, H^{(1)}$, and $H^{(2)}$ exist for the corresponding $s$ in $\Gamma_\beta$.

On the contrary, curves analogous to the heteroclinic curve $C$, as well as to the homoclinic and cycle fold curves $P$ and $K$, do not exist for the original map with generic higher-order terms. As for the Chenciner bifurcation and the 1:1 resonance, complex bifurcation sets exist nearby. The "instant collisions" of the saddle invariant manifolds and closed invariant curves are substituted by infinite series of bifurcations in which *homoclinic structures* are involved (see **Fig. 9.11**(a,b)). Such structures imply the existence of long-period cycles

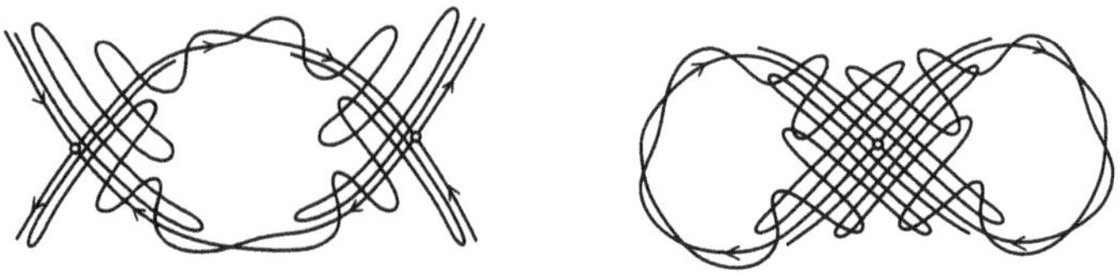

(a) (b)

**Fig. 9.11.** Homoclinic structures near 1:2 resonance: (a) $s = 1$; (b) $s = -1$.

appearing and disappearing via infinitely-many fold bifurcations as one crosses the corresponding bifurcation set. There is also an infinite cascade of flip bifurcations. The complete bifurcation picture is unknown.

**Remark:**

System (9.74) has a fundamental meaning if we consider it in the class of smooth planar $\mathbb{Z}_2$-symmetric systems, invariant under the rotation $x \mapsto \mathcal{I}x = -x$. Any such system has the form

$$\begin{cases} \dot{x}_1 = x_1 G_1(x_1^2, x_2^2, \alpha) + x_2 G_2(x_1^2, x_2^2, \alpha), \\ \dot{x}_2 = x_1 H_1(x_1^2, x_2^2, \alpha) + x_2 H_2(x_1^2, x_2^2, \alpha). \end{cases} \tag{9.76}$$

Assume that the origin $x = 0$ is an equilibrium with a double zero eigenvalue $\lambda_{1,2} = 0$ and that the Jacobian matrix has at least one nonzero element.

Similar to Lemma 9.11, one can show that any system (9.76) is smoothly orbitally equivalent to the system

$$\begin{cases} \dot{\eta}_1 = \eta_2 + \eta_1 \Psi_1(\eta_1^2, \eta_2^2, \beta) + \eta_2 \Psi_2(\eta_1^2, \eta_2^2, \beta), \\ \dot{\eta}_2 = \beta_1 \eta_1 + \beta_2 \eta_2 + C_1(\beta)\eta_1^3 + D_1(\beta)\eta_1^2 \eta_2 \\ \qquad + \eta_1 \Phi_1(\eta_1^2, \eta_2^2, \beta) + \eta_2 \Phi_2(\eta_1^2, \eta_2^2, \beta), \end{cases} \tag{9.77}$$

where $\Psi_k, \Phi_k = O(\|\eta\|^3), k = 1, 2$. It has been proved that, under conditions (R2.0)–(R2.2), system (9.77) is locally topologically equivalent to system (9.74). Moreover, the homeomorphism $h_\beta$ identifying the phase portraits can be selected to commute with $\mathcal{I}$:

$$h_\beta \circ \mathcal{I} = \mathcal{I} \circ h_\beta. \ \diamondsuit$$

### 9.5.4 1:3 resonance

Consider now the case when a planar smooth map

$$x \mapsto f(x, \alpha), \quad x \in \mathbb{R}^2, \ \alpha \in \mathbb{R}^2,$$

has at $\alpha = 0$ the fixed point $x = 0$ with simple multipliers $\mu_{1,2} = e^{\pm i\theta_0}$ for $\theta_0 = \frac{2\pi}{3}$. As with the 1:2 resonance, we can assume that $x = 0$ is the fixed point of the map for all sufficiently small $\|\alpha\|$ and write the map as

$$x \mapsto A(\alpha)x + F(x, \alpha),$$

where $F(x, \alpha) = O(\|x\|^2)$. Since the multipliers are simple, there is an eigenvector $q(\alpha) \in \mathbb{C}^2$:

$$A(\alpha)q(\alpha) = \mu(\alpha)q(\alpha),$$

for each $\|\alpha\|$ small, such that $\mu(0) = e^{i\theta_0}$. As usual, introduce the adjoint eigenvector $p(\alpha) \in \mathbb{C}^2$, satisfying

$$A^T(\alpha)p(\alpha) = \bar{\mu}(\alpha)p(\alpha),$$

which is normalized according to

$$\langle p, q \rangle = 1.$$

Now any vector $x \in \mathbb{R}^2$ can be represented in the form

$$x = zq + \bar{z}\bar{q},$$

and the studied map can be written in the complex form

$$z \mapsto \mu(\alpha)z + g(z, \bar{z}, \alpha), \tag{9.78}$$

where

$$g(z, \bar{z}, \alpha) = \langle p(\alpha), f(zq(\alpha) + \bar{z}\bar{q}(\alpha), \alpha) \rangle = \sum_{k+l \geq 2} \frac{1}{k!l!} g_{kl}(\alpha) z^k \bar{z}^l.$$

We can proceed in the same manner as when studying the Neimark-Sacker bifurcation in Chapter 4 (see Section 4.7).

**Lemma 9.12 (Normal form map for 1:3 resonance)** *The map* (9.78) *can be transformed by an invertible smooth and smoothly parameter-dependent change of variable, for all sufficiently small* $\|\alpha\|$, *into the form*

$$\zeta \mapsto \Gamma_\alpha(\zeta) = \mu(\alpha)\zeta + B(\alpha)\bar{\zeta}^2 + C(\alpha)\zeta|\zeta|^2 + O(|\zeta|^4), \tag{9.79}$$

*where*

$$B(\alpha) = \frac{g_{02}(\alpha)}{2}, \tag{9.80}$$

*and*

$$C(\alpha) = \frac{g_{20}(\alpha)g_{11}(\alpha)(2\mu(\alpha) + \bar{\mu}(\alpha) - 3)}{2(\bar{\mu}(\alpha) - 1)(\mu^2(\alpha) - \mu(\alpha))} + \frac{|g_{11}(\alpha)|^2}{1 - \bar{\mu}(\alpha)} + \frac{g_{21}(\alpha)}{2}. \tag{9.81}$$

**Proof:**

The proof is essentially contained in the proofs of Lemmas 4.5 and 4.6 in Chapter 4. As in Lemma 4.5, we can try to eliminate quadratic terms in (9.78) by the transformation

$$z = w + \frac{h_{20}}{2}w^2 + h_{11}w\bar{w} + \frac{h_{02}}{2}\bar{w}^2,$$

where $h_{kl} = h_{kl}(\alpha)$ are certain unknown functions. Exactly as in Lemma 4.5, we can annihilate the coefficients in front of the $w^2$- and $w\bar{w}$-terms by setting

$$h_{20} = \frac{g_{20}}{\mu^2 - \mu}, \quad h_{11} = \frac{g_{11}}{|\mu|^2 - \mu},$$

since the denominators are nonzero for all sufficiently small $\|\alpha\|$. However, an attempt to "kill" the $\bar{w}^2$-term of the resulting map by formally setting

$$h_{02} = \frac{g_{02}}{\bar{\mu}^2 - \mu}$$

(as in Lemma 4.5) fails spectacularly because

$$\bar{\mu}^2(0) - \mu(0) = e^{-2i\theta_0}(1 - e^{3i\theta_0}) = 0.$$

Thus, the $\bar{w}^2$-term cannot be removed by a transformation that depends smoothly on $\alpha$. Therefore, set $h_{02}(\alpha) = 0$, which gives

$$B(0) = \frac{g_{02}(0)}{2}. \tag{9.82}$$

The next step is to annihilate cubic terms. This can be done exactly as in Lemma 4.6 by performing an invertible smooth transformation

$$w = \zeta + \frac{h_{30}}{6}\zeta^3 + \frac{h_{12}}{2}\zeta\bar{\zeta}^2 + \frac{h_{03}}{6}\bar{\zeta}^3,$$

with smooth $h_{kl} = h_{kl}(\alpha)$. Since $\mu^2(0) \neq 1$ and $\mu^4(0) \neq 1$, the transformation removes all cubic terms in the resulting map, except the $\zeta|\zeta|^2$-term. To compute the coefficient $C(\alpha)$, we need only make the transformation

$$z = \zeta + \frac{1}{2}h_{20}(\alpha)\zeta^2 + h_{11}(\alpha)\zeta\bar{\zeta},$$

with $h_{20}$ and $h_{11}$ given above, and find the resulting coefficient in front of the $\zeta^2\bar{\zeta}$-term. This gives the expression (9.81). The critical value $C(0)$ is then provided by

$$C(0) = \frac{g_{20}(0)g_{11}(0)(1-2\mu_0)}{2(\mu_0^2 - \mu_0)} + \frac{|g_{11}(0)|^2}{1-\bar{\mu}_0} + \frac{g_{21}(0)}{2}, \tag{9.83}$$

where $\mu_0 = \mu(0) = e^{i\theta_0}$. Formula (9.83) can be obtained by merely substituting $g_{02} = 0$ into (4.20) from Chapter 4. The lemma is proved. □

We would like to approximate the map (9.79) by a flow. Its linear part

$$\zeta \mapsto \mu(\alpha)\zeta, \tag{9.84}$$

which describes the rotation through the angle $\frac{2\pi}{3}$ at $\alpha = 0$, provides no difficulties. Writing $\mu(\alpha)$ in the exponential form

$$\mu(\alpha) = e^{\varepsilon(\alpha)+i\theta(\alpha)},$$

where $\varepsilon(0) = 0$ and $\theta(0) = \theta_0 = \frac{2\pi}{3}$, we can see immediately that (9.84) is the unit-time shift along orbits of the linear equation

$$\dot{\zeta} = \lambda(\alpha)\zeta,$$

for $\lambda(\alpha) = \varepsilon(\alpha) + i\theta(\alpha)$. However, an attempt to approximate the quadratic term in $\Gamma_\alpha$ by a flow fails. Clearly, an approximating equation should have the form

$$\dot{\zeta} = \lambda\zeta + G_{02}\bar{\zeta}^2 + O(|\zeta|^3)$$

for some unknown function $G_{02} = G_{02}(\alpha)$. Perform two Picard iterations (9.26):

$$\zeta^{(1)}(\tau) = e^{\lambda\tau}\zeta,$$

$$\zeta^{(2)}(1) = e^{\lambda}\zeta + \frac{e^{2\bar{\lambda}}(e^{\lambda-2\bar{\lambda}} - 1)}{\lambda - 2\bar{\lambda}} G_{02}\bar{\zeta}^2.$$

The coefficient in front of $\bar{\zeta}^2$ vanishes at $\alpha = 0$ since

$$e^{\lambda(0)-2\bar{\lambda}(0)} = e^{3i\theta_0} = 1$$

at the 1:3 resonance. Therefore, we cannot approximate $\Gamma_0$ with $B(0) \neq 0$ by the unit-time shift of a flow. The same is true for its second iterate $\Gamma_\alpha^2$ (check!). Fortunately, the *third* iterate $\Gamma_\alpha^3$ allows for approximation by a flow.

**Lemma 9.13** *The third iterate of the map (9.79) can be represented for all sufficiently small* $\|\alpha\|$ *in the form*

$$\Gamma_\alpha^3(\zeta) = \varphi_\alpha^1 \zeta + O(|\zeta|^4),$$

*where* $\varphi_\alpha^t$ *is the flow of a planar system*

$$\dot{\zeta} = \omega(\alpha)\zeta + B_1(\alpha)\bar{\zeta}^2 + C_1(\alpha)\zeta|\zeta|^2, \tag{9.85}$$

*where* $\omega$, $B_1$, *and* $C_1$ *are smooth complex-valued functions of* $\alpha$, $\omega(0) = 0$, *and*

$$B_1(0) = 3\bar{\mu}_0 B(0), \tag{9.86}$$
$$C_1(0) = -3|B(0)|^2 + 3\mu_0^2 C(0), \tag{9.87}$$

*with* $\mu_0 = e^{i\theta_0}$, *and* $B(0)$ *and* $C(0)$ *are given by* (9.82) *and* (9.83), *respectively.*

**Proof:**

The third iterate of $\Gamma_\alpha$ has the form

$$\Gamma_\alpha^3(\zeta) = \mu^3\zeta + (\mu^2 + \bar{\mu}|\mu|^2 + \bar{\mu}^4)B\bar{\zeta}^2 + \left[2(|\mu|^2 + \bar{\mu}^3 + |\mu|^4)|B|^2 + \mu^2(1 + |\mu|^2 + |\mu|^4)C\right]\zeta|\zeta|^2 + O(|\zeta|^4).$$

Since $\mu^3(0) = 1$, we can represent $\mu^3$ near $\alpha = 0$ in the form

$$\mu^3(\alpha) = e^{\omega(\alpha)},$$

for some complex function $\omega(\alpha)$ such that $\omega(0) = 0$. This gives the linear term in (9.85).

For small $\|\alpha\|$ the map $\Gamma_\alpha^3$ is close to the *identity map* $\text{id}(\zeta) = \zeta$. As we have mentioned, such maps can always be approximated by flow shifts. To verify (9.86) and (9.88), let us first perform three Picard iterations for (9.85) at $\alpha = 0$:

$$\zeta^{(1)}(\tau) = \zeta,$$
$$\zeta^{(2)}(\tau) = \zeta + B_1(0)\bar{\zeta}^2\tau,$$
$$\zeta^{(3)}(1) = \zeta + B_1(0)\bar{\zeta}^2 + (|B_1(0)|^2 + C_1(0))\zeta|\zeta|^2 + O(|\zeta|^4).$$

Comparing the coefficients in $\zeta^{(3)}(1)$ with those in $\Gamma^3$ for $\alpha = 0$, one gets the expressions (9.86) and (9.87) by taking into account $\mu_0^3 = \bar{\mu}_0^3 = |\mu_0| = 1$. $\square$

Let us consider the real and imaginary parts of $\omega$ as new unfolding parameters $(\beta_1, \beta_2)$:

$$\omega(\alpha) = \beta_1(\alpha) + i\beta_2(\alpha).$$

We have

$$\begin{cases} \beta_1(\alpha) = 3\varepsilon(\alpha), \\ \beta_2(\alpha) = 3\theta(\alpha) \pmod{2\pi}. \end{cases}$$

Assuming

$$\text{(R3.0)} \qquad \det\left(\frac{\partial \beta}{\partial \alpha}\right) \neq 0,$$

at $\alpha = 0$, we can use $\beta$ to parametrize a neighborhood of the origin on the parameter plane and write (9.85) as

$$\dot{\zeta} = (\beta_1 + i\beta_2)\zeta + b_1(\beta)\bar{\zeta}^2 + c_1(\beta)\zeta|\zeta|^2, \tag{9.88}$$

for $b_1(\beta) = B_1(\alpha(\beta)), c_1(\beta) = C_1(\alpha(\beta))$. If the complex number

$$\text{(R3.1)} \qquad b_1(0) = B_1(0) \neq 0,$$

where $B_1(0)$ is given by (9.86), then we can scale (9.88) by taking

$$\zeta = \gamma(\beta)\eta, \ \gamma(\beta) \in \mathbb{C}^1,$$

with

$$\gamma(\beta) = \frac{1}{|b_1(\beta)|}\exp\left(i\frac{\arg b_1(\beta)}{3}\right).$$

The scaling results in

$$\dot{\eta} = (\beta_1 + i\beta_2)\eta + \bar{\eta}^2 + c(\beta)\eta|\eta|^2, \tag{9.89}$$

where

$$c(\beta) = \frac{c_1(\beta)}{|b_1(\beta)|^2}.$$

Writing (9.89) in polar coordinates $\eta = \rho e^{i\varphi}$, we obtain

$$\begin{cases} \dot{\rho} = \beta_1\rho + \rho^2\cos 3\varphi + a(\beta)\rho^3, \\ \dot{\varphi} = \beta_2 - \rho\sin 3\varphi + b(\beta)\rho^2, \end{cases} \tag{9.90}$$

with smooth real-valued functions $a(\beta) = \text{Re } c(\beta)$, $b(\beta) = \text{Im } c(\beta)$. We now introduce the final assumption concerning the approximating planar system, namely, we suppose

$$a(0) \neq 0,$$

which is equivalent to the nondegeneracy condition

$$\text{(R3.2)} \qquad \text{Re } C_1(0) \neq 0,$$

where $C_1(0)$ is given by (9.87).

Under the nondegeneracy conditions assumed, the bifurcation diagram of the approximating system (9.90) for $a(0) < 0$ is presented in **Fig. 9.12**. Notice that the system is *invariant* under the rotation $R_3$ through the angle $\varphi = \theta_0 = \frac{2\pi}{3}$. Such systems are called $\mathbb{Z}_3$-*symmetric*.

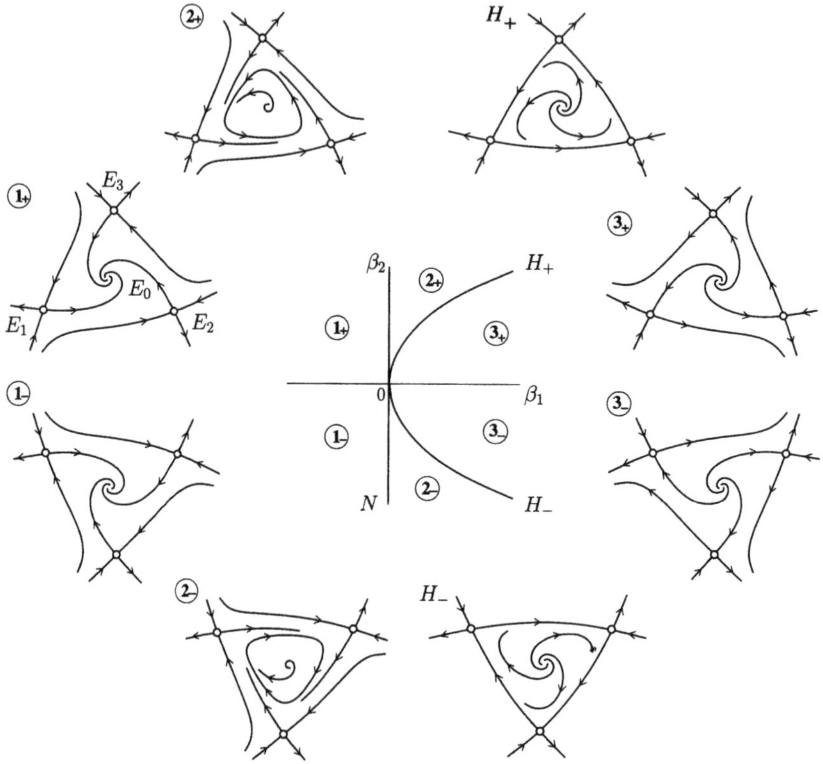


**Fig. 9.12.** Bifurcation diagram of the approximating system (9.89) for $a(0) = \text{Re}\ c(0) < 0$.

The system always has a trivial equilibrium $E_0$ with $\rho_0 = 0$. For all sufficiently small $\|\beta\| \neq 0$, there are also three nontrivial symmetric equilibria,

$$E_k = (\rho_s, \varphi_{s,k}), \quad k = 1, 2, 3,$$

all located on a circle of radius $r_s$,

$$r_s^2 = \beta_1^2 + \beta_2^2 + O(\|\beta\|^3),$$

and separated by the angle $\theta_0 = \frac{2\pi}{3}$ in $\varphi$-coordinate.

The nontrivial equilibria are always *saddles* and do not bifurcate for small $\|\beta\| \neq 0$. The trivial equilibrium is obviously stable for $\beta_1 < 0$ but becomes unstable for $\beta_1 > 0$ undergoing a supercritical Hopf bifurcation at $\beta_1 = 0$ due to the assumption $a(0) < 0$. Therefore, a unique and stable limit cycle appears in (9.90) if we cross the Hopf bifurcation line

$$N = \{(\beta_1, \beta_2) : \beta_1 = 0\}$$

from left to right at a point with $\beta_2 \neq 0$.

What happens with this limit cycle later? It can be proved that the system (9.90) can have no more than *one* limit cycle for all sufficiently small $\|\beta\|$, if the nondegeneracy conditions (R3.1) and (R3.2) are satisfied (see the bibliographical notes). One can show that there is a bifurcation curve

$$H = \left\{(\beta_1, \beta_2) : \beta_1 = -\frac{a}{2}\beta_2^2 + o(\beta_2^2)\right\}$$

at which the limit cycle disappears via a *heteroclinic* bifurcation (see Exercise 6). For parameter values on the curve $H$, the system has a heteroclinic cycle formed by coinciding stable and unstable separatrices of the nontrivial saddles. All three saddle connections happen simultaneously due to the symmetry. The heteroclinic cycle resembles a triangle and is stable from the inside. Thus, the limit cycle exists in two disjoint regions adjacent to the Hopf line $N$. We leave the reader to make a trip around the origin for the bifurcation diagram in **Fig. 9.12**, as well as consider the case $a(0) > 0$.

Let us briefly discuss the implications that result from the preceding analysis for the original map $\Gamma_\alpha$ near the 1:3 resonance. The map always has a trivial fixed point undergoing a nondegenerate Neimark-Sacker bifurcation on a bifurcation curve corresponding to the Hopf curve $N$ in the approximating system (9.90). The Neimark-Sacker bifurcation produces a closed invariant curve surrounding the trivial fixed point. For all parameter values close to the codim 2 point, the map $\Gamma_\alpha$ has a *saddle cycle of period three* corresponding to the three nontrivial saddle fixed points of $\Gamma_\alpha^3$, which, in turn, correspond to the saddles $E_k, k = 1, 2, 3$ of (9.90). Instead of the single heteroclinic bifurcation curve $H$, the map $\Gamma_\alpha$ with generic higher-order terms possesses a more complex bifurcation set. The stable and unstable invariant manifolds of the period-three cycle intersect transversally in an exponentially narrow parameter region forming a *homoclinic structure* (see **Fig. 9.13**). This region

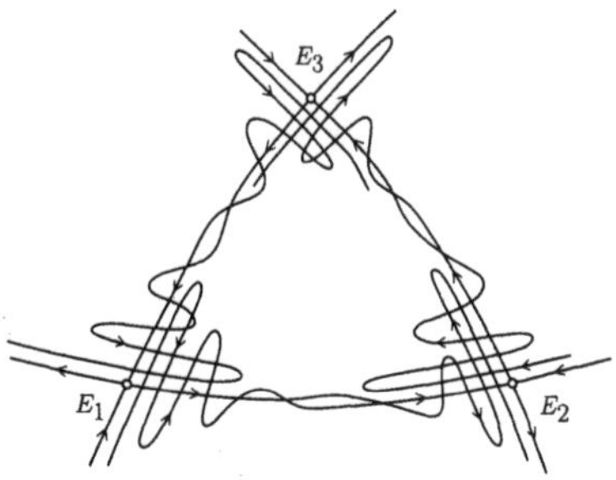


**Fig. 9.13.** Homoclinic structure near a 1:3 resonance.

is bounded by two smooth bifurcation curves $T_{1,2}$ at which the manifolds are tangent, obtaining a nontransversal homoclinic orbit. The intersection of these manifolds implies the existence of Smale horseshoes and, therefore, an infinite number of long-period orbits (see Chapter 1). These orbits appear and disappear via fold and flip bifurcations near the curves $T_{1,2}$. The closed invariant curve born at the Neimark-Sacker bifurcation loses its smoothness and is destroyed as it approaches the homoclinic structure. The complete picture is unknown.

**Remark:**

Adding any $\mathbb{Z}_3$-invariant higher-order terms to system (9.90) does not qualitatively change its bifurcation diagram. ◇

### 9.5.5 1:4 resonance

Consider a planar smooth map

$$x \mapsto f(x,\alpha), \quad x \in \mathbb{R}^2, \ \alpha \in \mathbb{R}^2,$$

having at $\alpha = 0$ the fixed point $x = 0$ with the simple multipliers $\mu_{1,2}(0) = \exp\left(\pm\frac{i\pi}{2}\right) = \pm i$ (1:4 resonance). Write the map, for sufficiently small $\|\alpha\|$, as

$$x \mapsto \Lambda(\alpha)x + F(x,\alpha),$$

where $F(x,\alpha) = O(\|x\|^2)$. Since the eigenvalue $\mu_1(0) = i$ of $\Lambda(0)$ is simple, there is a nearby smooth eigenvalue $\mu(\alpha)$, $\mu(0) = i$, of $\Lambda(\alpha)$ for all sufficiently small $\|\alpha\|$. As usual, introduce the corresponding ordinary and adjoint eigenvectors: $q(\alpha), p(\alpha) \in \mathbb{C}^2$,

$$\Lambda q = \mu q, \quad \Lambda^T p = \bar{\mu} p,$$

and normalize them according to

$$\langle p, q\rangle = 1.$$

Now, any vector $x \in \mathbb{R}^2$ can be represented in the form

$$x = zq + \bar{z}\bar{q},$$

and the map can be written in the complex form

$$z \mapsto \mu(\alpha)z + g(z,\bar{z},\alpha), \tag{9.91}$$

where

$$g(z,\bar{z},\alpha) = \langle p, f(zq(\alpha) + \bar{z}\bar{q}(\alpha),\alpha)\rangle = \sum_{k+l\geq 2} \frac{1}{k!l!} g_{kl}(\alpha) z^k \bar{z}^l.$$

By Lemma 4.5 from Chapter 4, we can make a smooth transformation, eliminating all quadratic terms in (9.91) for small $\|\alpha\|$ since $\mu_0 \neq 1$ and $\mu_0^3 \neq 1$, for $\mu_0 = i$. The transformation will change the cubic terms. Then, as in Lemma 4.6, we can try to annihilate as many cubic terms as possible. Denote the new complex coordinate by $\zeta$. Then, from the proof of Lemma 4.6, we immediately see that only two of the four cubic terms can be "killed" in the present case. Namely, attempts to remove $\zeta|\zeta|^2$- and $\bar{\zeta}^3$-terms fail because $|\mu_0|^2 = \mu_0^4 = 1$. Simple extra calculations prove the following lemma.

**Lemma 9.14 (Normal form map for 1:4 resonance)** *The map* (9.91) *can be transformed by an invertible smooth change of variable, smoothly depending on the parameters, for all sufficiently small* $\|\alpha\|$, *into the form*

$$\zeta \mapsto \Gamma_\alpha(\zeta) = \mu(\alpha)\zeta + C(\alpha)\zeta|\zeta|^2 + D(\alpha)\bar{\zeta}^3 + O(|\zeta|^4), \tag{9.92}$$

*where* $C$ *and* $D$ *are smooth functions of* $\alpha$ :

$$C(0) = \frac{1+3i}{4} g_{20}(0)g_{11}(0) + \frac{1-i}{2}|g_{11}(0)|^2 - \frac{1+i}{4}|g_{02}(0)|^2 + \frac{1}{2}g_{21}(0), \tag{9.93}$$

$$D(0) = \frac{i-1}{4} g_{11}(0)g_{02}(0) - \frac{1+i}{4} g_{02}(0)\bar{g}_{20}(0) + \frac{1}{6}g_{03}(0). \quad \square \tag{9.94}$$

The next aim is to approximate $\Gamma_\alpha$ by a flow. The linear part of (9.92),

$$\zeta \mapsto \mu(\alpha)\zeta, \tag{9.95}$$

is a rotation through the angle $\frac{\pi}{2}$ at $\alpha = 0$ and is easy to handle. Writing $\mu(\alpha)$ in the exponential form

$$\mu(\alpha) = e^{\varepsilon(\alpha)+i\theta(\alpha)},$$

where $\varepsilon(0) = 0$ and $\theta(0) = \frac{\pi}{2}$, we can immediately verify that (9.95) is the unit-time shift along orbits of the linear equation

$$\dot{\zeta} = \lambda(\alpha)\zeta,$$

for $\lambda(\alpha) = \varepsilon(\alpha) + i\theta(\alpha)$. However, if we are interested in nonlinear terms, only the *fourth iterate* of $\Gamma_\alpha$ allows for approximation by a flow.

**Lemma 9.15** *The fourth iterate of the map* (9.92) *can be represented, for all sufficiently small* $\|\alpha\|$, *in the form*

$$\Gamma_\alpha^4(\zeta) = \varphi_\alpha^1\zeta + O(|\zeta|^4),$$

*where* $\varphi_\alpha^t$ *is the flow of a planar system*

$$\dot{\zeta} = \omega(\alpha)\zeta + C_1(\alpha)\zeta|\zeta|^2 + D_1(\alpha)\bar{\zeta}^3, \tag{9.96}$$

*where* $\omega$, $C_1$, *and* $D_1$ *are smooth complex-valued functions of* $\alpha$, $\omega(0) = 0$, *and*

$$C_1(0) = -4iC(0), \quad D_1(0) = -4iD(0), \tag{9.97}$$

*with* $C(0)$ *and* $D(0)$ *given by* (9.93) *and* (9.94), *respectively.*

**Proof:**

The fourth iterate of $\Gamma_\alpha$ has the form

$$\Gamma_\alpha^4(\zeta) = \mu^4\zeta + \mu^3(|\mu|^2+1)(|\mu|^4+1)C\zeta|\zeta|^2 + (\mu+\bar{\mu}^3)(\mu^2+\bar{\mu}^6)D\bar{\zeta}^3 + O(|\zeta|^4).$$

Since $\mu^4(0) = 1$, we can represent $\mu^4$ near $\alpha = 0$ in the form

$$\mu^4(\alpha) = e^{\omega(\alpha)},$$

for some smooth complex-valued function $\omega(\alpha)$ such that $\omega(0) = 0$. This gives the linear term in (9.96).

For small $\|\alpha\|$ the map $\Gamma_\alpha^4$ is close to the *identity map* $\text{id}(\zeta) = \zeta$, and can therefore be approximated by the unit-time shift of a flow. To verify (9.97), let us first perform three Picard iterations for (9.96) at $\alpha = 0$:

$$\begin{aligned} \zeta^{(1)}(\tau) = \zeta^{(2)}(\tau) &= \zeta, \\ \zeta^{(3)}(1) = \zeta + C_1(0)\zeta|\zeta|^2 &+ D_1(0)\bar{\zeta}^3 + O(|\zeta|^4). \end{aligned}$$

Comparing the coefficients in $\zeta^{(3)}(1)$ with those in $\Gamma_\alpha^4$ for $\alpha = 0$, we get the expressions (9.97), after taking into account $\mu(0) = i$. □

As for the 1:3 resonance, consider the real and imaginary parts of $\omega$ as new unfolding parameters $(\beta_1, \beta_2)$:

$$\omega(\alpha) = \beta_1(\alpha) + i\beta_2(\alpha).$$

We have

$$\begin{cases} \beta_1(\alpha) = 4\varepsilon(\alpha), \\ \beta_2(\alpha) = 4\theta(\alpha) \pmod{2\pi}. \end{cases}$$

Assuming

(R4.0) $$\det\left(\frac{\partial\beta}{\partial\alpha}\right) \neq 0$$

at $\alpha = 0$, we can use $\beta$ to parametrize a neighborhood of the origin on the parameter plane and write (9.96) as

$$\dot{\zeta} = (\beta_1 + i\beta_2)\zeta + c_1(\beta)\zeta|\zeta|^2 + d_1(\beta)\bar{\zeta}^3, \tag{9.98}$$

for $c_1(\beta) = C_1(\alpha(\beta))$ and $d_1(\beta) = D_1(\alpha(\beta))$. If the complex number

(R4.1) $$d_1(0) = D_1(0) \neq 0,$$

where $D_1(0)$ is given by (9.97), then we can scale (9.98) by taking

$$\zeta = \gamma(\beta)\eta,\ \gamma(\beta) \in \mathbb{C}^1,$$

with

$$\gamma(\beta) = \frac{1}{\sqrt{|d_1(\beta)|}} \exp\left(i\frac{\arg d_1(\beta)}{4}\right).$$

The scaling results in

$$\dot{\eta} = (\beta_1 + i\beta_2)\eta + A(\beta)\eta|\eta|^2 + \bar{\eta}^3, \tag{9.99}$$

where

$$A(\beta) = \frac{c_1(\beta)}{|d_1(\beta)|}.$$

Notice that the system is *invariant* under the rotation $R_4$ through the angle $\theta_0 = \frac{\pi}{2}$, (i.e., the transformation $\eta \mapsto e^{i\theta_0}\eta$). Such systems are called $\mathbb{Z}_4$-*symmetric*. Perhaps, the symmetry of the system is more visible if we present it in polar coordinates $\eta = \rho e^{i\varphi}$:

$$\begin{cases} \dot{\rho} = \beta_1\rho + a(\beta)\rho^3 + \rho^3\cos 4\varphi, \\ \dot{\varphi} = \beta_2 + b(\beta)\rho^2 - \rho^2 \sin 4\varphi, \end{cases} \tag{9.100}$$

where $a(\beta) = \text{Re } A(\beta),\ b(\beta) = \text{Im } A(\beta)$.

The bifurcation analysis of system (9.99) is more complicated than that of the approximating systems in the previous sections and requires numerical techniques. The bifurcation diagram depends on $A = A(0) = (a(0), b(0))$, so the $(\text{Re } A, \text{Im } A)$-plane is divided into several regions with different bifurcation diagrams in the $(\beta_1, \beta_2)$-plane.[6] Fortunately, all the bifurcation diagrams are formed by straight lines originating at $\beta = 0$, thus we can completely describe them by making a roundtrip along the unit circle on the $\beta$-plane and detecting bifurcation points on it. This is equivalent to setting

$$\beta_1 + i\beta_2 = e^{i\alpha}, \quad \alpha \in [0, 2\pi),$$

and considering sequences of *one-parameter* bifurcations in (9.99) as $\alpha$ makes the complete circle. Crossing a boundary on the $A$-plane implies the appearance of new codim 1 bifurcations in the bifurcation sequence or the changing of their order. Therefore, these boundaries are the projections of codim 2 bifurcation curves in $(\alpha, A)$-space onto the $A$-plane. Unfortunately, only three of them correspond to bifurcations of equilibria and can be derived analytically. The others involve *degenerate heteroclinic bifurcations* and can be computed only numerically. As one can see, the boundaries on the $A$-plane are symmetric under reflections with respect to the coordinate axes. Thus, it is sufficient to study them in one quadrant of the $A$-plane. Assume

(R4.2) $$\text{Re } A \neq 0$$

and

(R4.3) $$\text{Im } A \neq 0,$$

and take the quadrant corresponding to

$$a = \text{Re } A < 0, \quad b = \text{Im } A < 0.$$

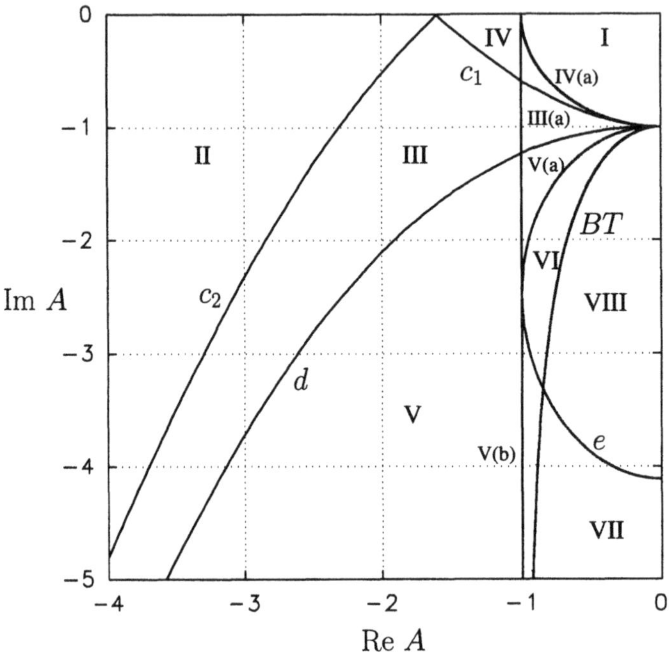


**Fig. 9.14.** Partitioning of the $A$-plane into regions with different bifurcation diagrams of (9.99).

The partitioning of the $A$-plane into regions with different bifurcation diagrams is given in **Fig. 9.14**. Some curves are known analytically, while others have been computed numerically (see the discussion ahead).

The system always has a trivial equilibrium $\eta = 0$, that is stable for $\beta_1 < 0$ and repelling for $\beta_1 > 0$ (see the first equation in (9.100)). At $\beta_1 = 0$ a Hopf bifurcation takes place; thus, a limit cycle appears/disappears when $\alpha$ passes through $\alpha = \pm\frac{\pi}{2}$, which is stable because we assumed $a < 0$. Notice that for $\beta_2 \neq 0$ the orbits spiral into or out of the origin depending on the sign of $\beta_2$ (see the second equation in (9.100)). For $\beta_2 = 0$ (i.e., $\alpha = 0$ and $\alpha = \pi$) the direction of rotation near the origin reverses.

Possible nontrivial equilibria $\eta = \rho e^{i\varphi}$ satisfy the complex equation

$$-\frac{e^{i\alpha}}{\rho^2} = A(\alpha) + e^{-4i\varphi},$$

which can be approximated for small $\|\beta\|$ by

$$-\frac{e^{i\alpha}}{\rho^2} = A + e^{-4i\varphi}, \tag{9.101}$$

[6] This resembles the situation with the Hopf-Hopf bifurcation in Chapter 8, where the bifurcation diagrams were different in different regions in the $(\theta, \delta)$-plane.

where $A = A(0)$. The left-hand side of (9.101) specifies (by fixing $\alpha$) a ray on the complex plane parametrized by $\rho > 0$, while the right-hand side defines a circle of unit radius centered at the point $A$, which is covered four times while $\varphi$ makes one full turn (see **Fig. 9.15**(a,b)). Any intersection $(\rho, \varphi)$ of

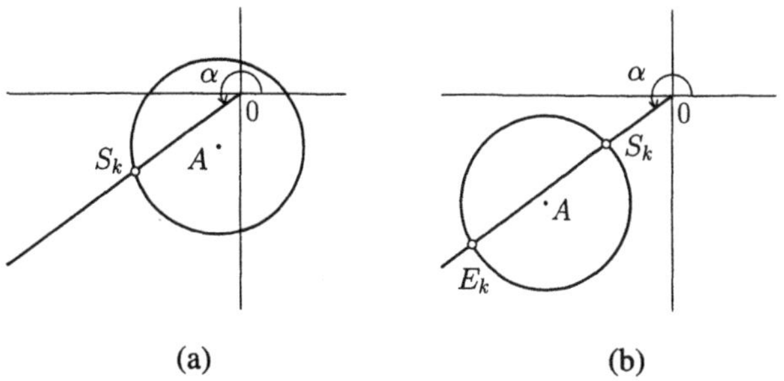


**Fig. 9.15.** Construction of the equilibria of (9.99): (a) $|A| < 1$; (b) $|A| > 1$.

the ray with the circle gives *four* symmetric equilibria of (9.99). Based on this geometrical construction, we can conclude that the system can have either none, four, or eight nontrivial equilibria.

Indeed, if $|A| < 1$, then the origin of the complex plane is located inside the circle; thus, any ray out of the origin has exactly one intersection with the circle (see **Fig. 9.15**(a)). This gives four symmetric equilibria $S_k, k = 1, 2, 3, 4$, in (9.99). As one can show in this case, the equilibria $S_k$ are *saddles* (see **Fig. 9.16**(a)). On the contrary, if $|A| > 1$, the origin of the complex plane is

**Fig. 9.16.** Equilibria of (9.99): (a) $|A| < 1$; (b) $|A| > 1$.

outside the circle. Therefore, a typical ray out of the origin has either none or two intersections with this circle (see **Fig. 9.15**(b)), giving in the latter case eight equilibria $S_k, E_k, k = 1, 2, 3, 4$, in (9.99). The equilibria $S_k$ closer to the origin are *saddles*, while the remote ones $E_k$ are *attractors* (see **Fig. 9.16**(b)) or *repellers*. As the ray rotates with $\alpha$ and becomes tangent to the circle, the equilibria $S_k$ and $E_k$ collide pairwise and disappear via fold bifurcations. Thus, for $|A| > 1$, there is an interval of $\alpha$-values within which the system has eight nontrivial equilibria. The interval of their existence is contained between the fold bifurcation values of the parameter $\alpha$.

Clearly, the case $|A| = 1$ is exceptional and should be avoided:

(R4.4) $$|A(0)| \neq 1.$$

Another exceptional case is when the circle is tangent to the imaginary axis of the complex plane, namely Re $A = -1$ within the considered quadrant (see **Fig. 9.17**). In this case, two different bifurcations happen simultaneously

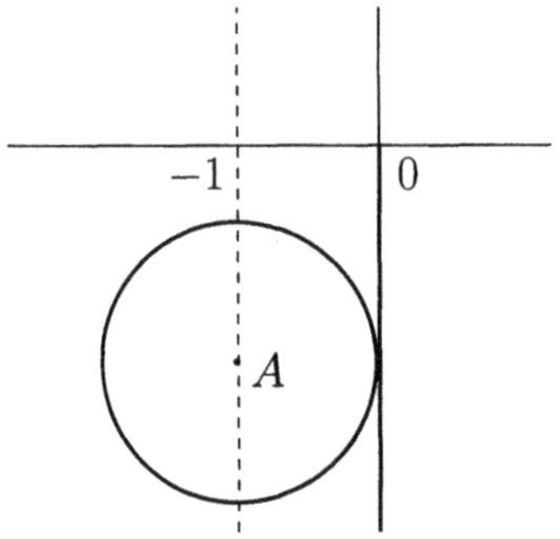


**Fig. 9.17.** Exceptional case Re $A = -1$.

at $\alpha = -\frac{\pi}{2}$: the Hopf bifurcation of the trivial equilibrium $E_0$ and the collision of the nontrivial ones $S_k, E_k$. Thus, suppose

(R4.5) $$|\text{Re } A(0)| \neq 1.$$

Depending on the value of $A$, the equilibria $E_k$ remain stable for all $\alpha$ or they change stability at a Hopf bifurcation. The curve in the $A$-plane separating these two cases is the projection of the Bogdanov-Takens bifurcation curve $BT$ in $(\alpha, A)$-space:

$$|b| = \frac{1+a^2}{\sqrt{1-a^2}},$$

where $a = \text{Re } A$, $b = \text{Im } A$. The curve $BT$ emanates from $-i$ and asymptotes to the line $a = -1$ within the selected quadrant of the $A$-plane (see **Fig. 9.14**). Above this curve the equilibria $E_k$ remain stable, while below the curve they exhibit a nondegenerate Hopf bifurcation. Therefore, assume the following nondegeneracy condition:

(R4.6) $$|\text{Im } A(0)| \neq \frac{1+(\text{Re } A(0))^2}{\sqrt{1-(\text{Re } A(0))^2}}.$$

Other possible codim 1 bifurcations involve limit cycles of (9.99). "Small" limit cycles born via Hopf bifurcations from the nontrivial equilibria $E_k$ die at homoclinic bifurcations when the separatrices of the saddles $S_k$ form *"small" homoclinic loops* (see **Fig. 9.18**(a)). The separatrices of the saddles $S_k$ can

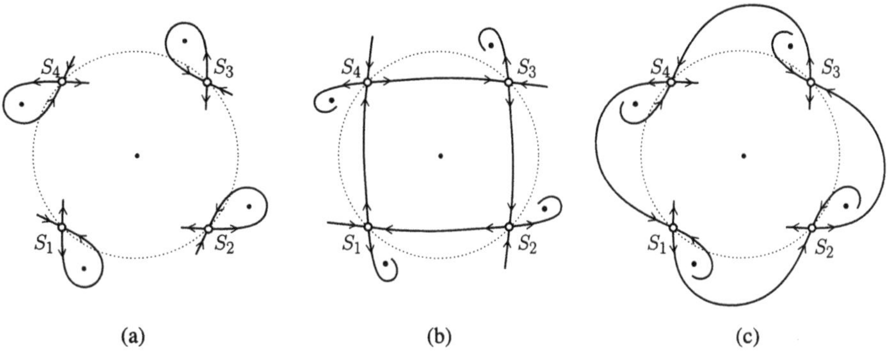


**Fig. 9.18.** (a) "Small" homoclinic loops; (b) "square" heteroclinic cycle; and (c) "clover" heteroclinic cycle.

also compose *heteroclinic cycles*. There are two possible types of these cycles, namely, a "square" heteroclinic cycle around the equilibrium $E_0$ (**Fig. 9.18**(b)) and a "clover" cycle surrounding all the equilibria $E_k, k = 0, 1, 2, 3, 4$ (**Fig. 9.18**(c)). All four connections appear simultaneously due to the $\mathbb{Z}_4$-symmetry. The bifurcation is similar to the standard homoclinic bifurcation studied in Chapter 6, where a single saddle was involved. Denote by $\sigma_0$ the saddle quantity of the saddle: $\sigma_0 = \text{tr } \Lambda(S_k)$. The "square" cycle is always stable from the inside ($\sigma_0 < 0$), while the "clover" one is stable or unstable from the outside depending on whether $\sigma_0 < 0$ or $\sigma_0 > 0$. When the parameter $\alpha$ passes a value corresponding to a "square" heteroclinic cycle with $\sigma_0 \neq 0$, a limit cycle of the relevant stability appears for nearby parameter values. This cycle surrounds the trivial equilibrium $E_0$ and has any existing nontrivial ones outside. Passing a value corresponding to a "clover" heteroclinic cycle with $\sigma_0 \neq 0$ brings in a limit cycle of the corresponding stability that surrounds all the equilibria of the system. If there is a limit cycle with no equilibrium outside, it is stable since Re $A < 0$. There is strong computer evidence that the system can have *at most* two "big" limit cycles surrounding all the nontrivial equilibria, and no more than one "small" limit cycle around each of the nontrivial equilibria. Another codim 1 bifurcation in which a "big" limit cycle can be involved is a *heteroclinic cycle* formed by the center manifolds of the *saddle-node equilibria* at the fold bifurcation (see **Fig. 9.19**). All four connections exist simultaneously due to the $\mathbb{Z}_4$-symmetry. This bifurcation is similar to the homoclinic orbit to a single saddle-node studied in Chapter 7.

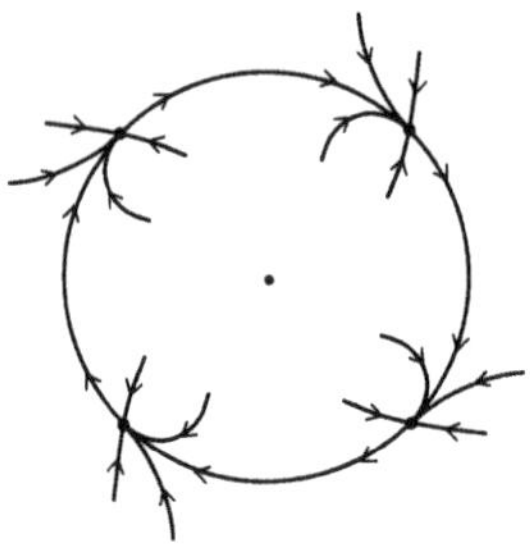

**Fig. 9.19.** Saddle-node heteroclinic connections.

When we cross a corresponding parameter value, a "big" limit cycle appears while the nontrivial equilibria disappear.

The remaining boundaries shown on the $A$-plane correspond to *degenerate heteroclinic bifurcations.* More precisely, in the $(\alpha, A)$-space there are curves corresponding to the presence of one of the heteroclinic cycles described above, having an *extra* degeneracy. There are three types of such codim 2 cases, defining three curves in **Fig. 9.14**. The "clover" heteroclinic cycle can involve saddles that all have $\sigma_0 = 0$ (*neutral saddles*). This implies that there can exist both stable and unstable "big" limit cycles, which can "collide." The corresponding boundary on the $A$-plane is marked by $e$. It looks like an ellipse passing through $A = -i$ if considered in the whole $A$-plane.[7] The other two boundaries correspond to *degenerate saddle-node connections* when the center manifold of a saddle-node (its "unstable separatrix") tends to another saddle-node along a noncenter direction (along the boundary of the "node sector"). There are two possible such degeneracies: a "square" (see **Fig. 9.20**(a)) or

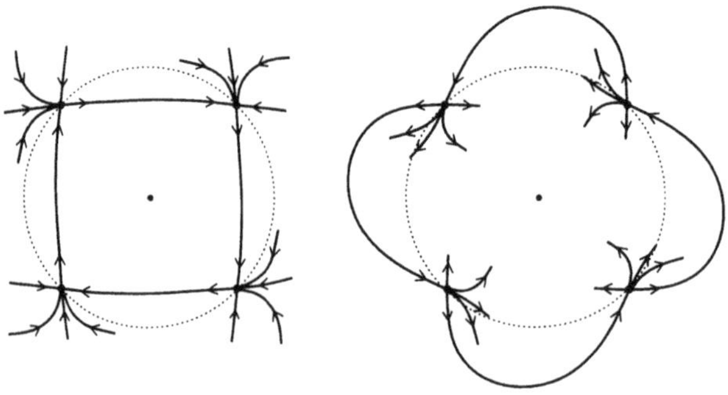

(a) (b)

**Fig. 9.20.** Degenerate saddle-node heteroclinic orbits: (a) "square"; (b) "clover".

[7] There is a symmetric one passing through $A = i$.

a "clover" (see **Fig. 9.20**(b)). The corresponding boundaries on the $A$-plane are marked by $c_{1,2}$ and $d$.

Analytically and numerically found boundaries divide the $A$-quadrant under consideration into twelve regions, denoted for historical reasons by I, II, III, III(a), IV, IV(a), V, V(a), V(b), VI, VII, and VIII. Each region is characterized by its own *bifurcation sequence* when $\alpha$ varies from, say, $\alpha = -\frac{\pi}{2}$ to $\alpha = \frac{3\pi}{2}$ increasing (counterclockwise). We have, therefore, three nondegeneracy conditions:

(R4.7) $$A \notin e,$$

(R4.8) $$A \notin c_{1,2},$$

and

(R4.9) $$A \notin d,$$

that have no analytical expression. Summarizing the previous discussion, we can encounter a number of the following codim 1 bifurcations, if $A$ is fixed *inside* one of the regions of the negative quadrant:

$H_0$ - *Hopf bifurcation of the trivial equilibrium* $E_0$. The first (at $\alpha = -\frac{\pi}{2}$) generates a stable limit cycle, while the second one (at $\alpha = \frac{\pi}{2}$) implies its disappearance.

$T$ - *tangent (fold) bifurcation of nontrivial equilibria.* At the corresponding parameter values eight nontrivial equilibrium points $E_k, S_k, k = 1,2,3,4$ appear/disappear. Actually, there are three possibilities for them to appear: either inside, on, or outside a "big" cycle if it exists when this bifurcation takes place. We will distinguish these possibilities by writing $T_{\text{in}}, T_{\text{on}}$, or $T_{\text{out}}$.

$H_1$ - *Hopf bifurcation of the nontrivial equilibria* $E_k, k = 1,2,3,4$. Four "small" limit cycles bifurcate from the nontrivial antisaddles $E_k$.

$L$ - *"small" homoclinic loop bifurcation.* The "small" cycles born via the Hopf bifurcation disappear via orbits homoclinic to the nontrivial saddles $S_k$.

$C_S$ - *"square" heteroclinic cycle.* A stable limit cycle bifurcates from the orbit.

$C_C$ - *"clover" heteroclinic cycle.* Depending on the sign of the saddle quantity $\sigma_0$, it generates either a stable or an unstable "big" limit cycle. We denote these cases by $C_C^-$ and $C_C^+$, respectively.

$F$ - *fold (tangent) bifurcation of "big" limit cycles.* Two "big" cycles, the outer of which is stable, collide and disappear.

The following symbolic sequences allow one to reconstruct completely the bifurcation diagrams of system (9.99) that are believed to exist in the corresponding regions on the $A$-plane. The numbers correspond to the phase portraits presented in **Figs. 9.21** and **9.22**, while the symbols over the arrows mean the bifurcations. In region I we start just before $\alpha = -\frac{\pi}{2}$ with only a stable trivial equilibrium $E_0$ and four symmetric nontrivial saddles $S_k, k = 1, 2, 3, 4$. In all the other regions, we start with a single globally stable trivial point $E_0$. The first bifurcation is always the supercritical Hopf bifurcation.

$$\mathrm{I} : 1 \xrightarrow{H_0} 2 \xrightarrow{C_S} 3(3') \xrightarrow{C_S} 2' \xrightarrow{H_0} 1'.$$

$$\mathrm{II} : 4 \xrightarrow{H_0} 5 \xrightarrow{T_{\mathrm{on}}} 10 \xrightarrow{T_{\mathrm{on}}} 5 \xrightarrow{H_0} 4'.$$

$$\mathrm{III} : 4 \xrightarrow{H_0} 5 \xrightarrow{T_{\mathrm{on}}} 10 \xrightarrow{C_S} 11 \xrightarrow{T_{\mathrm{out}}} 5 \xrightarrow{H_0} 4'.$$

$$\mathrm{III(a)} : 4 \xrightarrow{H_0} 5 \xrightarrow{T_{\mathrm{on}}} 10 \xrightarrow{C_S} 11 \xrightarrow{H_0} 12 \xrightarrow{T} 4'.$$

$$\mathrm{IV} : 4 \xrightarrow{H_0} 5 \xrightarrow{T_{\mathrm{out}}} 11 \xrightarrow{C_S} 10 \xrightarrow{C_S} 11 \xrightarrow{T_{\mathrm{out}}} 5 \xrightarrow{H_0} 4'.$$

$$\mathrm{IV(a)} : 4 \xrightarrow{H_0} 5 \xrightarrow{T_{\mathrm{out}}} 11 \xrightarrow{C_S} 10 \xrightarrow{C_S} 11 \xrightarrow{H_0} 12 \xrightarrow{T} 4'.$$

$$\mathrm{V} : 4 \xrightarrow{H_0} 5 \xrightarrow{T_{\mathrm{in}}} 8 \xrightarrow{C_C^-} 10 \xrightarrow{C_S} 11 \xrightarrow{T_{\mathrm{out}}} 5 \xrightarrow{H_0} 4'.$$

$$\mathrm{V(a,b)} : 4 \xrightarrow{H_0} 5 \xrightarrow{T_{\mathrm{in}}} 8 \xrightarrow{C_C^-} 10 \xrightarrow{C_S} 11 \xrightarrow{H_0} 12 \xrightarrow{T} 4'.$$

$$\mathrm{VI} : 4 \xrightarrow{H_0} 5 \xrightarrow{T_{\mathrm{in}}} 8 \xrightarrow{C_C^+} 9 \xrightarrow{F} 10 \xrightarrow{C_S} 11 \xrightarrow{H_0} 12 \xrightarrow{T} 4'.$$

$$\mathrm{VII} : 4 \xrightarrow{H_0} 5 \xrightarrow{T_{\mathrm{in}}} 6 \xrightarrow{H_1} 7 \xrightarrow{L} 8 \xrightarrow{C_C^-} 10 \xrightarrow{C_S} 11 \xrightarrow{H_0} 12 \xrightarrow{T} 4'.$$

$$\mathrm{VIII} : 4 \xrightarrow{H_0} 5 \xrightarrow{T_{\mathrm{in}}} 6 \xrightarrow{H_1} 7 \xrightarrow{L} 8 \xrightarrow{C_C^+} 9 \xrightarrow{F} 10 \xrightarrow{C_S} 11 \xrightarrow{H_0} 12 \xrightarrow{T} 4'.$$

A sequence of typical phase portraits in the simplest region I is given in **Fig. 9.21**. Notice that we have presented three pairs of topologically equivalent

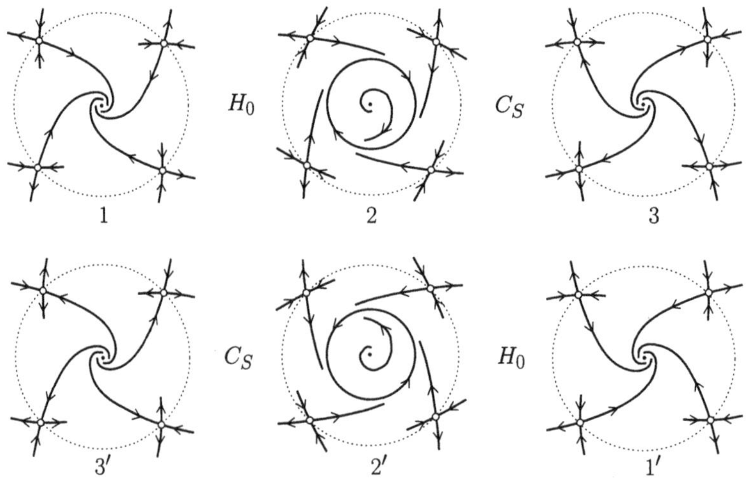


**Fig. 9.21.** Bifurcation sequence in region I.

phase portraits (1(1′), 2(2′), and 3(3′), which differ only in the direction of rotation around the origin) to facilitate the understanding of the sequence. A label between two phase portraits corresponds to the bifurcation transforming one into the other. The most complicated sequence of phase portraits corresponding to region VIII is depicted in **Fig. 9.22**. We recommend the reader

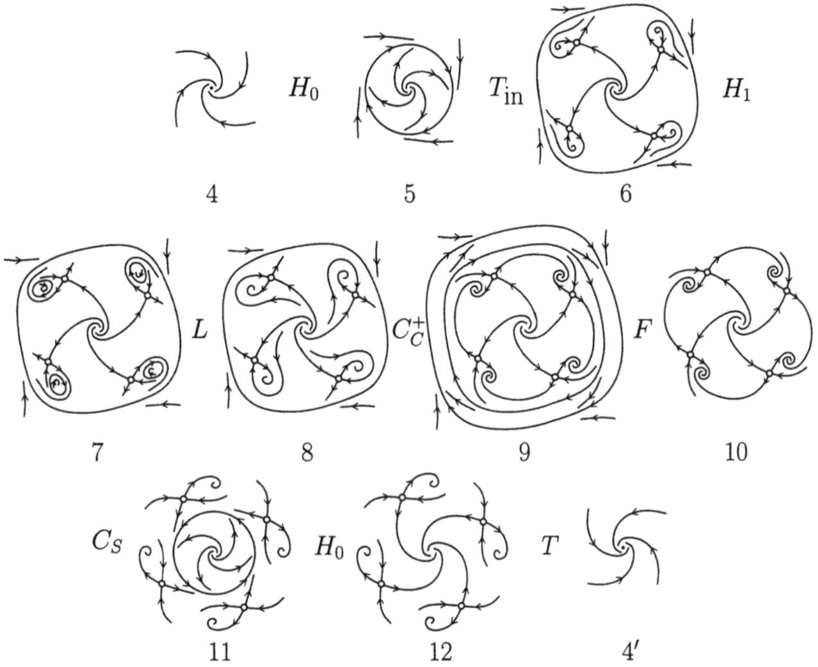


**Fig. 9.22.** Bifurcation sequence in region VIII.

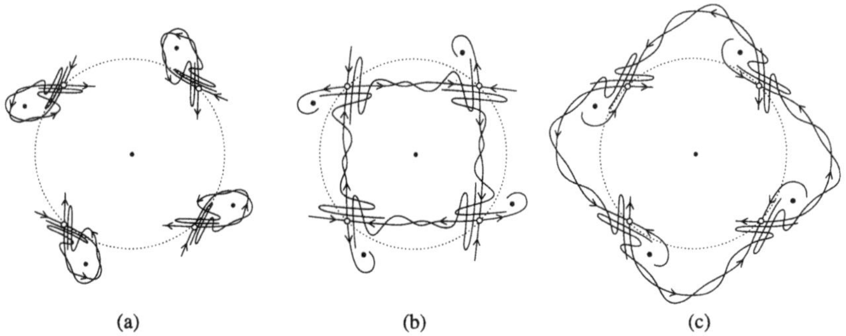


**Fig. 9.23.** Homoclinic structures near 1:4 resonance: (a) “small”; (b) “square”; and (c) “clover”.

to reconstruct all the other possible bifurcation sequences.

What do the obtained results imply for a generic map $\Gamma_\alpha$ exhibiting 1:4 resonance? Clearly, the trivial equilibrium of the approximating system corresponds to the trivial fixed point of the map, while four nontrivial $\mathbb{Z}_4$-coupled equilibria of the system correspond to a single period-four cycle of the original map. One can prove that tangent and Hopf bifurcations of the nontrivial equilibria give rise to tangent and Neimark-Sacker bifurcations of the nontrivial fixed points of the map. As usual, homo- and heteroclinic connections in the approximating system become heteroclinic structures of the map (see **Fig. 9.23**). They are formed by intersections of the stable and the unstable invariant manifolds of the nontrivial saddle period-four cycle. These structures imply the existence of an infinite number of periodic orbits. Closed invariant curves corresponding to limit cycles lose their smoothness and are destroyed, almost "colliding" with the saddle period-four cycle. Individual bifurcation sequences become dependent on $\|\beta\|$ and involve an infinite number of bifurcations. The complete details are likely to remain unknown forever!

## 9.6 Fold-flip bifurcation

Consider a smooth discrete-time planar dynamical system

$$x \mapsto f(x, \alpha), \quad x \in \mathbb{R}^2, \ \alpha \in \mathbb{R}^2, \tag{9.102}$$

having at $\alpha = 0$ a fixed point $x = 0$ with multipliers $\mu_1 = 1$ and $\mu_2 = -1$. These are two bifurcation conditions for a fold-flip bifurcation.

Expand $f$ in $x$ at $x = 0$ for any small $\|\alpha\|$:

$$f(x, \alpha) = \gamma(\alpha) + A(\alpha)x + R(x, \alpha),$$

where $\gamma(0) = 0$ and $R(x, \alpha) = O(\|x\|^2)$. The bifurcation conditions imply the existence of two eigenvectors, $q_{1,2}(\alpha) \in \mathbb{R}^2$, such that

$$A(\alpha)q_1(\alpha) = \lambda_1(\alpha)q_1(\alpha), \quad A(\alpha)q_2(\alpha) = \lambda_2(\alpha)q_2(\alpha),$$

where $\lambda_1(0) = 1$ and $\lambda_2(0) = -1$. Note that, due to the simplicity of the eigenvalues $\pm 1$ of $A(0)$, $\lambda_{1,2}$ depend smoothly on $\alpha$, and $q_{1,2}$ can also be chosen such that they are smooth functions of $\alpha$. Any $x \in \mathbb{R}^2$ can now be represented for all small $\|\alpha\|$ as

$$x = \xi_1 q_1(\alpha) + \xi_2 q_2(\alpha),$$

where $\xi = (\xi_1, \xi_2)^T \in \mathbb{R}^2$. One can compute the components of $\xi$ explicitly:

$$\xi_1 = \langle p_1(\alpha), x \rangle, \quad \xi_2 = \langle p_2(\alpha), x \rangle,$$

where

$$A^T(\alpha)p_1(\alpha) = \lambda_1(\alpha)p_1(\alpha), \quad A^T(\alpha)p_2(\alpha) = \lambda_2(\alpha)p_2(\alpha),$$

and $\langle p_1(\alpha), q_1(\alpha)\rangle = \langle p_2(\alpha), q_2(\alpha)\rangle = 1$. Since $\langle p_1(\alpha), q_2(\alpha)\rangle = \langle p_2(\alpha), q_1(\alpha)\rangle = 0$, the map $f$, when expressed in the $\xi$-coordinates, takes the form

$$\begin{pmatrix} \xi_1 \\ \xi_2 \end{pmatrix} \mapsto \begin{pmatrix} \sigma_1(\alpha) + \lambda_1(\alpha)\xi_1 + S_1(\xi, \alpha) \\ \sigma_2(\alpha) + \lambda_2(\alpha)\xi_2 + S_2(\xi, \alpha) \end{pmatrix}, \tag{9.103}$$

where

$$\sigma_k(\alpha) = \langle p_k(\alpha), \gamma(\alpha)\rangle, \quad S_k(\xi, \alpha) = \langle p_k(\alpha), R(\xi_1 q_1(\alpha) + \xi_2 q_2(\alpha), \alpha)\rangle$$

for $k = 1, 2$. Expanding $S_{1,2}(\xi, \alpha)$ further, we can write (9.103) as

$$\begin{pmatrix} \xi_1 \\ \xi_2 \end{pmatrix} \mapsto \begin{pmatrix} \sigma_1(\alpha) + \lambda_1(\alpha)\xi_1 + \displaystyle\sum_{i+j=2,3} \frac{1}{i!j!} g_{ij}(\alpha)\xi_1^i \xi_2^j \\ \sigma_2(\alpha) + \lambda_2(\alpha)\xi_2 + \displaystyle\sum_{i+j=2,3} \frac{1}{i!j!} h_{ij}(\alpha)\xi_1^i \xi_2^j \end{pmatrix} + O(\|\xi\|^4). \tag{9.104}$$

First consider map (9.104) for $\alpha = 0$ when $\sigma_1(0) = \sigma_2(0) = 0$ and $\lambda_1(0) = -\lambda_2(0) = 1$.

**Lemma 9.16 (Critical normal form)** *Suppose a smooth map $F_0 : \mathbb{R}^2 \to \mathbb{R}^2$ has the form*

$$\begin{pmatrix} \xi_1 \\ \xi_2 \end{pmatrix} \mapsto \begin{pmatrix} \xi_1 + \displaystyle\sum_{i+j=2,3} \frac{1}{i!j!} g_{ij}\xi_1^i \xi_2^j \\ -\xi_2 + \displaystyle\sum_{i+j=2,3} \frac{1}{i!j!} h_{ij}\xi_1^i \xi_2^j \end{pmatrix} + O(\|\xi\|^4) \tag{9.105}$$

*and $h_{11} \neq 0$. Then $F_0$ can be transformed by an invertible smooth change of variables into the form*

$$\begin{pmatrix} \eta_1 \\ \eta_2 \end{pmatrix} \mapsto \begin{pmatrix} \eta_1 + a(0)\eta_1^2 + b(0)\eta_2^2 + c(0)\eta_1^3 + d(0)\eta_1\eta_2^2 \\ -\eta_2 + \eta_1\eta_2 \end{pmatrix} + O(\|\eta\|^4), \tag{9.106}$$

*where*

$$a(0) = \frac{g_{20}}{2h_{11}}, \quad b(0) = \frac{1}{2}g_{02}h_{11}, \quad c(0) = \frac{1}{6h_{11}^2}\left(g_{30} + \frac{3}{2}g_{11}h_{20}\right), \tag{9.107}$$

$$\begin{aligned} d(0) = {} & \frac{3g_{02}(h_{02}h_{20} + 2h_{21} - 2g_{11}h_{20}) - g_{20}(3h_{02}^2 + 2h_{03})}{12h_{11}} - \frac{1}{2}g_{11}^2 + \frac{1}{2}g_{12} \\ & + \frac{1}{4}g_{11}h_{02} - \frac{1}{2}h_{02}^2 - \frac{1}{3}h_{03}. \end{aligned} \tag{9.108}$$

**Proof:**

*Step 1* (*Quadratic terms*) Applying to (9.105) a polynomial coordinate transformation

$$\begin{cases} \xi_1 = \eta_1 + \frac{1}{2}G_{20}\eta_1^2 + G_{11}\eta_1\eta_2 + \frac{1}{2}G_{02}\eta_2^2, \\ \xi_2 = \eta_2 + \frac{1}{2}H_{20}\eta_1^2 + H_{11}\eta_1\eta_2 + \frac{1}{2}H_{02}\eta_2^2, \end{cases} \tag{9.109}$$

we obtain

$$\begin{aligned} \eta_1 &\mapsto \eta_1 + \tfrac{1}{2}g_{20}\eta_1^2 + (g_{11} + 2G_{11})\eta_1\eta_2 + \tfrac{1}{2}g_{02}\eta_2^2 + \cdots, \\ \eta_2 &\mapsto -\eta_2 + \tfrac{1}{2}(h_{20} - 2H_{20})\eta_1^2 + h_{11}\eta_1\eta_2 + \tfrac{1}{2}(h_{02} - 2H_{02})\eta_2^2 + \cdots, \end{aligned}$$

where dots stand for higher-order terms. By setting

$$G_{11} = -\frac{1}{2}g_{11}, \quad H_{20} = \frac{1}{2}h_{20}, \quad H_{02} = \frac{1}{2}h_{02}, \tag{9.110}$$

we eliminate as many quadratic terms as possible. The remaining quadratic terms are called *resonant.*

*Step 2* (*Cubic terms*) Assume now that *Step 1* is already done, so that (9.105) has only resonant quadratic and all cubic terms. Consider a polynomial transformation

$$\begin{cases} \xi_1 = \eta_1 + \frac{1}{6}G_{30}\eta_1^3 + \frac{1}{2}G_{21}\eta_1^2\eta_2 + \frac{1}{2}G_{12}\eta_1\eta_2^2 + \frac{1}{6}G_{03}\eta_2^3, \\ \xi_2 = \eta_2 + \frac{1}{6}H_{30}\eta_1^3 + \frac{1}{2}H_{21}\eta_1^2\eta_2 + \frac{1}{2}H_{12}\eta_1\eta_2^2 + \frac{1}{6}H_{03}\eta_2^3. \end{cases} \tag{9.111}$$

Obviously, it does not change the quadratic terms. After this transformation, we get

$$\begin{aligned} \eta_1 \mapsto \eta_1 &+ \tfrac{1}{2}g_{20}\eta_1^2 + \tfrac{1}{2}g_{02}\eta_2^2 \\ &+ \tfrac{1}{6}g_{30}\eta_1^3 + \tfrac{1}{2}(g_{21} + 2G_{21})\eta_1^2\eta_2 + \tfrac{1}{2}g_{12}\eta_1\eta_2^2 + \tfrac{1}{6}(g_{03} + 2G_{03})\eta_2^3 + \cdots \end{aligned}$$

and

$$\begin{aligned} \eta_2 \mapsto -\eta_2 &+ h_{11}\eta_1\eta_2 \\ &+ \tfrac{1}{6}(h_{30} - 2H_{30})\eta_1^3 + \tfrac{1}{2}h_{21}\eta_1^2\eta_2 + \tfrac{1}{2}(h_{12} - 2H_{12})\eta_1\eta_2^2 + \tfrac{1}{6}h_{03}\eta_2^3 + \cdots. \end{aligned}$$

By setting

$$G_{21} = -\frac{1}{2}g_{21}, \quad G_{03} = -\frac{1}{2}g_{03}, \quad H_{30} = \frac{1}{2}h_{30}, \quad H_{12} = \frac{1}{2}h_{12},$$

we eliminate four cubic terms. The remaining cubic terms are also called *resonant.* They are not altered by (9.111).

*Step 3* (*Hyper-normalization*) The coefficients $H_{11}, G_{20}$, and $G_{02}$ of (9.109) do not affect the quadratic terms of (9.105) but alter its cubic terms. Taking into account (9.110) while computing the cubic terms of the transformed map, we obtain

$$\eta_1 \mapsto \eta_1 + \tfrac{1}{2}g_{20}\eta_1^2 + \tfrac{1}{2}g_{02}\eta_2^2 + \tfrac{1}{6}\left(g_{30} + \tfrac{3}{2}g_{11}h_{20}\right)\eta_1^3$$
$$+ \tfrac{1}{2}\left(2g_{02}H_{11} - g_{02}G_{20} + (g_{20} + 2h_{11})G_{02} + \tfrac{1}{2}g_{11}h_{02} + g_{12} - g_{11}^2\right)\eta_1\eta_2^2 + \cdots$$

and

$$\eta_2 \mapsto -\eta_2 + h_{11}\eta_1\eta_2 + \tfrac{1}{2}\left(g_{20}H_{11} + h_{11}G_{20} - g_{11}h_{20} + \tfrac{1}{2}h_{02}h_{20} + h_{21}\right)\eta_1^2\eta_2$$
$$+\tfrac{1}{6}\left(3g_{02}H_{11} + 3G_{02}h_{11} + h_{03} + \tfrac{3}{2}h_{02}^2\right)\eta_2^3 + \cdots,$$

where only the resonant cubic terms are displayed. Thus, we can try to eliminate three altered terms by selecting $H_{11}, G_{20}$, and $G_{02}$. This requires solving the linear system

$$\begin{pmatrix} 2g_{02} & -g_{02} & 2h_{11} + g_{20} \\ g_{20} & h_{11} & 0 \\ 3g_{02} & 0 & 3h_{11} \end{pmatrix} \begin{pmatrix} H_{11} \\ G_{20} \\ G_{02} \end{pmatrix} = \begin{pmatrix} -\frac{1}{2}g_{11}h_{02} - g_{12} + g_{11}^2 \\ g_{11}h_{20} - \frac{1}{2}h_{02}h_{20} - h_{21} \\ -h_{03} - \frac{3}{2}h_{02}^2 \end{pmatrix}.$$

Its matrix has zero determinant. However, using the nondegeneracy condition $h_{11} \neq 0$, we can eliminate the resonant cubic terms in the second component of the normal form. Thus, we set

$$H_{11} = 0 \tag{9.112}$$

and from the above linear system obtain

$$G_{02} = -\frac{1}{h_{11}}\left(\frac{1}{3}h_{03} + \frac{1}{2}h_{02}^2\right), \quad G_{20} = \frac{1}{h_{11}}\left(g_{11}h_{20} - h_{21} - \frac{1}{2}h_{02}h_{20}\right). \tag{9.113}$$

*Step 4* (*Final transformation*) Transform now the original map (9.105) using (9.109) with the coefficients defined in *Step 1* and *Step 3*. This results in a map with resonant quadratic terms, nonresonant cubic terms, and only two remaining resonant cubic terms in the first component. Transformation (9.111) from *Step 2* then allows elimination of all nonresonant cubic terms while keeping unchanged all remaining quadratic and cubic resonant terms. Finally, perform the linear scaling

$$\eta_1 \mapsto \frac{\eta_1}{h_{11}}$$

to put the coefficient in front of $\eta_1\eta_2$ in the second component equal to one. This results in the expressions (9.107) and (9.108) for the critical normal form coefficients. □

**Theorem 9.4 (Parameter-dependent normal form)** *Consider a planar map depending on two parameters*

$$\xi \mapsto F(\xi, \alpha), \quad \xi \in \mathbb{R}^2, \alpha \in \mathbb{R}^2,$$

*where* $F : \mathbb{R}^2 \times \mathbb{R}^2 \to \mathbb{R}^2$ *is smooth and such that*

(F.1): $F_0 : \xi \mapsto F_0(\xi) = F(\xi, 0)$ *satisfies the assumptions of* Lemma 9.16;

(F.2): *The map* $T : \mathbb{R}^2 \times \mathbb{R}^2 \to \mathbb{R}^2 \times \mathbb{R} \times \mathbb{R}$ *defined by*

$$\begin{pmatrix} \xi \\ \alpha \end{pmatrix} \mapsto T(\xi, \alpha) = \begin{pmatrix} F(\xi, \alpha) - \xi \\ \det F_\xi(\xi, \alpha) + 1 \\ \mathrm{tr}\ F_\xi(\xi, \alpha) \end{pmatrix} \tag{9.114}$$

*is regular at* $(\xi, \alpha) = (0, 0)$.

*Then* $F$ *can be transformed by an invertible smooth change of variables, smoothly depending on the parameters, and a smooth invertible change of parameters, for all sufficiently small* $\|\alpha\|$, *into the form*

$$\begin{pmatrix} \eta_1 \\ \eta_2 \end{pmatrix} \mapsto \begin{pmatrix} \beta_1 + (1 + \beta_2)\eta_1 + a(\beta)\eta_1^2 + b(\beta)\eta_2^2 + c(\beta)\eta_1^3 + d(\beta)\eta_1\eta_2^2 \\ -\eta_2 + \eta_1\eta_2 \end{pmatrix} + O(\|\eta\|^4), \tag{9.115}$$

*where all coefficients are smooth functions of* $\beta$ *whose values at* $\beta_1 = \beta_2 = 0$ *are given by* (9.107) *and* (9.108).

**Proof:**

We want to put (9.104) in the form (9.115) by means of a smooth coordinate transformation that depends smoothly on the parameters. Consider the change of variables

$$\begin{cases} \xi_1 = \eta_1 + \varepsilon_0(\alpha) + \varepsilon_1(\alpha)\eta_2 + \frac{1}{2}G_{20}(\alpha)\eta_1^2 + G_{11}(\alpha)\eta_1\eta_2 + \frac{1}{2}G_{02}(\alpha)\eta_2^2 + \\ \qquad \frac{1}{2}G_{21}(\alpha)\eta_1^2\eta_2 + \frac{1}{6}G_{03}(\alpha)\eta_2^3, \\ \xi_2 = \eta_2 + \delta_0(\alpha) + \delta_1(\alpha)\eta_1 + \frac{1}{2}H_{20}(\alpha)\eta_1^2 + \frac{1}{2}H_{02}(\alpha)\eta_2^2 + \\ \qquad \frac{1}{6}H_{30}(\alpha)\eta_2^3 + \frac{1}{2}H_{12}(\alpha)\eta_1\eta_2^2, \end{cases} \tag{9.116}$$

where all coefficients are as yet unknown smooth functions of $\alpha$ such that $\varepsilon_i(0) = \delta_i(0) = 0$ for $i = 0, 1$. Obviously, for $\alpha = 0$ (9.116) reduces to the transformation introduced in *Step 4* of the proof of Lemma 9.16 just before the final scaling.

Require now that the Taylor expansion of (9.104) in the $\eta$-coordinates takes the form

$$\begin{pmatrix} \eta_1 \\ \eta_2 \end{pmatrix} \mapsto \begin{pmatrix} \mu_1(\alpha) + (1 + \mu_2(\alpha))\eta_1 \\ -\eta_2 \end{pmatrix} + \begin{pmatrix} A(\alpha)\eta_1^2 + B(\alpha)\eta_2^2 + C(\alpha)\eta_1^3 + D(\alpha)\eta_1\eta_2^2 \\ E(\alpha)\eta_1\eta_2 \end{pmatrix} + O(\|\eta\|^4),$$

where $\mu_1(0) = \mu_2(0) = 0$. After all substitutions, this requirement translates into a system of algebraic equations

$$Q_\alpha(\varepsilon_0, \varepsilon_1, \delta_0, \delta_1, \beta_1, \beta_2, G_{20}, G_{11}, G_{02}, G_{21}, G_{03}, H_{20}, H_{02}, H_{30}, H_{12}, A, B, C, D, E) = 0,$$

where $Q_\alpha : \mathbb{R}^{20} \to \mathbb{R}^{20}$ results from equating the corresponding Taylor coefficients. For the Jacobian matrix $J$ of this system evaluated at $\alpha = 0$, when

$$\varepsilon_0 = \varepsilon_1 = \delta_0 = \delta_1 = \beta_1 = \beta_2 = 0,$$

we have $\det(J) = -147456\ h_{11}^3 \neq 0$. Therefore, the Implicit Function Theorem guarantees the local existence and smoothness of the coefficients of the transformation (9.116) as functions of $\alpha$.

The scaling

$$x_1 \mapsto \frac{x_1}{E(\alpha)},\quad \beta_1 \mapsto \frac{\beta_1}{E(\alpha)},$$

where $E(\alpha) = h_{11}(0) + O(\|\alpha\|)$, gives finally (9.115). Obviously, the critical coefficients are the same as in Lemma 9.16.

Moreover, one can show that

$$\det\left(\frac{\partial \beta}{\partial \alpha}\right)\bigg|_{\alpha=0} = \frac{1}{4h_{11}(0)} \det\left(\frac{\partial T}{\partial(\eta, \alpha)}\right)\bigg|_{\eta=\alpha=0},$$

where $T$ is the map (9.114) written in the $(\eta, \alpha)$-coordinates, i.e., for the map (9.104). Thus, if $h_{11}(0) \neq 0$, the regularity of (9.114) at the origin is equivalent to that of the map $\alpha \mapsto \beta$, and we can use $\beta_1$ and $\beta_2$ as the new parameters. □

Discard the $O(\|\eta\|^4)$-terms in (9.115) to obtain the *truncated normal form*

$$\begin{pmatrix} x_1 \\ x_2 \end{pmatrix} \mapsto \begin{pmatrix} \beta_1 + (1+\beta_2)x_1 + a(\beta)x_1^2 + b(\beta)x_2^2 + c(\beta)x_1^3 + d(\beta)x_1x_2^2 \\ -x_2 + x_1x_2 \end{pmatrix}, \tag{9.117}$$

where $(x_1, x_2)$ are used again instead of $(\eta_1, \eta_2)$. Denoting this map by $x \mapsto N_\beta(x)$, we see that

$$RN_\beta(x) = N_\beta(Rx),$$

where

$$R = \begin{pmatrix} 1 & 0 \\ 0 & -1 \end{pmatrix},\quad R^2 = I_2. \tag{9.118}$$

This implies that phase portraits of (9.117) are invariant under the reflection in the $x_1$-axis.

Denote the critical values of the normal form coefficients by

$$a_0 = a(0),\ b_0 = b(0),\ c_0 = c(0),\ d_0 = d(0).$$

To study bifurcations of (9.117), we will approximate this map by a flow. As in the study of the strong resonance 1:2, one could consider the second iterate $N_\beta^2$ and approximate it by the unit-time shift along orbits of a planar continuous-time system. However, it is more convenient to look for a flow that approximates the *composition* of $N_\beta$ with the reflection $x \mapsto Rx$, where $R$ is the matrix defined by (9.118).

**Theorem 9.5** *The truncated normal form* (9.117) *satisfies*

$$RN_\beta(x) = \varphi_\beta^1(x) + O(\|\beta\|^2) + O(\|x\|^2\|\beta\|) + O(\|x\|^4), \tag{9.119}$$

*where* $\varphi_\beta^t$ *is the flow of the planar system*

$$\begin{cases} \dot{x}_1 = \beta_1 + (-a_0\beta_1 + \beta_2)\,x_1 + a_0x_1^2 + b_0x_2^2 + d_1x_1^3 + d_2x_1x_2^2, \\ \dot{x}_2 = \frac{1}{2}\beta_1x_2 - x_1x_2 + d_3x_1^2x_2 + d_4x_2^3, \end{cases} \tag{9.120}$$

*with*

$$d_1 = c_0 - a_0^2,\ d_2 = d_0 - a_0b_0 + b_0,\ d_3 = \frac{1}{2}(a_0 - 1),\ d_4 = \frac{1}{2}b_0.$$

**Proof:**

As in the analysis of 1:1 resonance, we shall construct $\varphi_\beta^t$ as the first two components of the flow

$$X \mapsto \phi^t(X) = \begin{pmatrix} \varphi_\beta^t(x) \\ \beta \end{pmatrix},\quad X = \begin{pmatrix} x \\ \beta \end{pmatrix} \in \mathbb{R}^4,$$

generated by a 4-dimensional system with the parameters considered as constant variables:

$$\dot{X} = Y(X) = JX + Y_2(X) + Y_3(X) + \cdots,\ X \in \mathbb{R}^4. \tag{9.121}$$

Here

$$J = \begin{pmatrix} 0 & 0 & 1 & 0 \\ 0 & 0 & 0 & 0 \\ 0 & 0 & 0 & 0 \\ 0 & 0 & 0 & 0 \end{pmatrix},\quad Y_k(X) = \begin{pmatrix} Z_k(X) \\ 0 \end{pmatrix},$$

where each $Z_k$ is an order-$k$ homogeneous polynomial function from $\mathbb{R}^4$ to $\mathbb{R}^2$ with unknown coefficients. Define

$$M(X) = \begin{pmatrix} N_\beta(x) \\ \beta \end{pmatrix}$$

and introduce the $4 \times 4$ block-diagonal matrix

$$S = \begin{pmatrix} R & 0 \\ 0 & I_2 \end{pmatrix},$$

where $R$ is given in (9.118). We look now for a vector field $Y$ such that $SM(X) = \phi^1(X) + O(\|X\|^4)$.

To find the vector field $Y$ explicitly, perform three Picard iterations for (9.121). We start with setting $X^{(1)}(t) = e^{Jt}X$. Then, clearly, the linear part of $SM(X)$ coincides with $X^{(1)}(1)$.

Since we know how the result $X^{(2)}(t)$ of the second Picard iteration should look, we set some coefficients of $Y_2$ equal to zero immediately:

$$Y_2 = \begin{pmatrix} A_{10}\beta_1 x_1 + A_{01}\beta_2 x_1 + \frac{1}{2}A_{20}x_1^2 + \frac{1}{2}A_{02}x_2^2 \\ B_{11}x_1x_2 + B_{10}\beta_1 x_2 \\ 0 \\ 0 \end{pmatrix}.$$

Then

$$X^{(2)}(t) = e^{Jt}X + \int_0^t e^{J(t-\tau)}Y_2(X^{(1)}(\tau))d\tau =$$

$$\begin{pmatrix} x_1 + t\beta_1 \\ x_2 \\ \beta_1 \\ \beta_2 \end{pmatrix} + \begin{pmatrix} (A_{10}t + \frac{1}{2}A_{20}t^2)\beta_1 x_1 + A_{01}\beta_2 x_1 t + \frac{1}{2}A_{20}x_1^2 t + \frac{1}{2}A_{02}x_2^2 t \\ B_{11}x_1x_2 t + (\frac{1}{2}B_{11}t^2 + B_{10}t)\beta_1 x_2 \\ 0 \\ 0 \end{pmatrix}$$
$$+ \quad O(\|\beta\|^2).$$

Comparing quadratic terms of $SM(X)$ and $X^{(2)}(1)$, we find

$$A_{10} = -a_0,\ A_{20} = 2a_0,\ A_{01} = 1,\ A_{02} = 2b_0,\ B_{10} = \frac{1}{2},\ B_{11} = -1.$$

Passing on to the cubic part, we remark that we are only interested in cubic terms in $x$. Therefore, we put

$$Y_3 = \begin{pmatrix} \sum_{i+j=3} \frac{1}{i!j!}A_{ij}x_1^i x_2^j \\ \sum_{i+j=3} \frac{1}{i!j!}B_{ij}x_1^i x_2^j \\ 0 \\ 0 \end{pmatrix}$$

and get

$$X^{(3)}(1) = e^J X + \int_0^1 e^{J(1-\tau)}\left[Y_2(X^{(2)}(\tau)) + Y_3(X^{(2)}(\tau))\right] d\tau =$$

$$\begin{pmatrix} x_1 + \beta_1 + \beta_2 x_1 + a_0 x_1^2 + b_0 x_2^2 \\ x_2 - x_1 x_2 \\ \beta_1 \\ \beta_2 \end{pmatrix} +$$

$$\begin{pmatrix} \left(\frac{1}{6}A_{30} + a_0^2\right)x_1^3 + \frac{1}{2}A_{21}x_1^2 x_2 + \left(a_0 b_0 - b_0 + \frac{1}{2}A_{12}\right)x_1 x_2^2 + \frac{1}{6}A_{03}x_2^3 \\ \frac{1}{6}B_{30}x_1^3 + \frac{1}{2}(-a_0 + 1 + B_{21})x_1^2 x_2 + \frac{1}{2}B_{12}x_1 x_2^2 + \frac{1}{2}\left(\frac{1}{3}B_{03} - b_0\right)x_2^3 \\ 0 \\ 0 \end{pmatrix}$$
$$+ \quad O(\|\beta\|^2) + O(\|x\|^2\|\beta\|).$$

Comparing cubic terms of $SM(X)$ and $X^{(3)}(1)$, we find the coefficients of $Y_3$:

$$A_{30} = 6(c_0 - a_0^2),\ A_{21} = 0,\ A_{12} = 2(d_0 - a_0 b_0 + b_0),\ A_{03} = 0,$$

$$B_{30} = 0,\ B_{21} = a_0 - 1,\ B_{12} = 0,\ B_{03} = 3b_0.$$

This gives (9.120). □

Let us describe bifurcations of the approximating system (9.120) in a small neighborhood of the origin for small parameter values (see **Figs. 9.25–9.28**). The bifurcation diagrams resemble those of the truncated amplitude system (8.82) for the fold-Hopf bifurcation studied in Chapter 8. Since the substitution $x_2 \mapsto -x_2$ leaves (9.120) invariant, its phase portraits are symmetric with respect to the $x_1$-axis and we sketch them only in the upper half-plane.

In all cases, the system has no more than three equilibria with $x_2 \geq 0$. If

(FF.1) $$a_0 \neq 0,$$

then two trivial equilibria with $x_2 = 0$ appear via a nondegenerate fold bifurcation on the curve

$$F = \left\{(\beta_1, \beta_2) : \beta_1 = \frac{\beta_2^2}{4a_0} + o(\beta_2^2)\right\}.$$

The curve $F$ has two branches, $F_+$ and $F_-$, corresponding to $\beta_2 > 0$ and $\beta_2 < 0$, respectively. Provided that

(FF.2) $$b_0 \neq 0,$$

one of the trivial equilibria undergoes a nondegenerate pitchfork bifurcation along the line

$$P = \{(\beta_1, \beta_2) : \beta_1 = 0\},$$

giving rise to a nontrivial equilibrium with $x_2 > 0$. Moreover, when $b_0 > 0$, the nontrivial equilibrium exhibits a Hopf bifurcation at the bifurcation curve

$$NS = \left\{(\beta_1, \beta_2) : \beta_2 = \frac{(2b_0 + d_0)}{b_0}\beta_1 + o(\beta_1),\ \beta_1 < 0\right\}.$$

One can check that the corresponding first Lyapunov coefficient

$$l_1 = c_{NS} C_+ + o(\beta_1),$$

where $C_+ > 0$ and

$$c_{NS} = 3b_0 c_0 - a_0(2a_0 b_0 + 3b_0 + d_0). \tag{9.122}$$

Therefore, the Hopf bifurcation is nondegenerate and generates a single limit cycle if

(FF.3) $$3b_0 c_0 - a_0(2a_0 b_0 + 3b_0 + d_0) \neq 0.$$

If $a_0$ and $b_0$ are both positive, then two trivial equilibria of (9.120) are saddles, which are always connected by a heteroclinic orbit along the $x_1$-axis. There exists another heteroclinic connection when the parameters belong to the curve

$$J = \left\{ (\beta_1, \beta_2) : \beta_2 = \frac{2a_0(a_0 b_0 + a_0 d_0 + d_0) + 3b_0(a_0 + c_0)}{b_0 a_0 (3 + 2a_0)} \beta_1 + o(\beta_1), \beta_1 < 0 \right\}. \tag{9.123}$$

What do these results mean for the *truncated* normal form map (9.117)? The orbits of the map jump from the lower to the upper half-plane and back. This can be easily understood from (9.119), which implies that (9.115) is approximated by the composition of the unit-time shift $\varphi^1_\beta$ along the orbits of (9.120) with the reflection (jump) $R$. Notice that the unit shift gives only an *approximation* of the truncated normal form composed with $R$. However, as usual, the trivial equilibria of (9.120) correspond to fixed points of (9.117). The nontrivial equilibrium corresponds to a period-2 cycle of the map, while the limit cycle of (9.117) corresponds to a closed invariant curve or a more complicated invariant set existing nearby. The fold bifurcation of trivial equilibria gives rise to a fold bifurcation of fixed points in the $x_1$-axis, the pitchfork bifurcation becomes a period-doubling (flip) bifurcation of the fixed points, and the Hopf bifurcation of the nontrivial equilibrium assures a Neimark-Sacker bifurcation of the period-2 cycle. One can show that the corresponding bifurcation curves have the same asymptotic expressions as for the bifurcations of equilibria. Moreover, the bifurcations are nondegenerate under the same conditions. The presence of the $J$-curve for (9.120) implies for the map (9.117) the existence of two curves, along which heteroclinic tangencies occur (see curves $h_{1,2}$ in **Fig. 9.24**). Between these two curves, a heteroclinic structure exists.

Let us now go around the origin in the $(\mu_1, \mu_2)$-plane and discuss the bifurcations of the truncated normal form map (9.117). We have four cases with two subcases in the first two of them. These are determined by the signs of $a_0$, $b_0$, and $c_{NS}$.

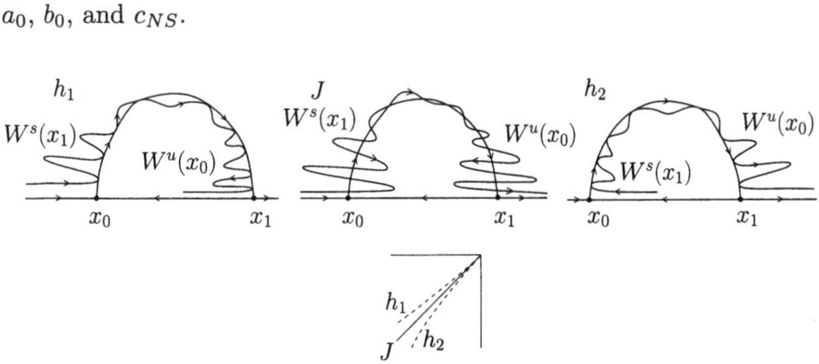


**Fig. 9.24.** Heteroclinic tangencies appearing in the lines $h_{1,2}$ together with a transversal heteroclinic structure between them.

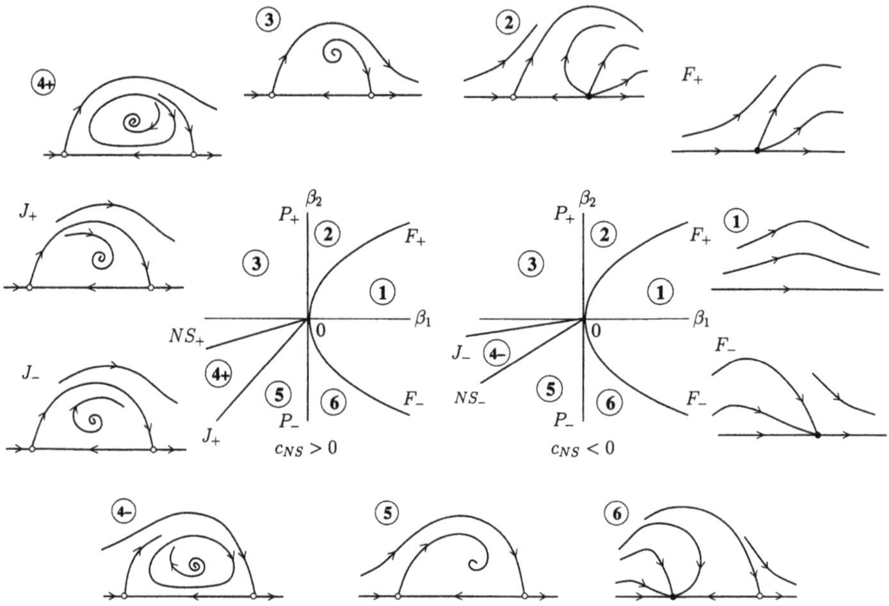


**Fig. 9.25.** Case 1: $a_0 > 0, b_0 > 0$.

Case 1 $(a_0 > 0, b_0 > 0)$: In region **1** (**Fig. 9.25**) orbits merely jump to the right. Crossing $F_+$ implies the appearance of two fixed points on the horizontal axis. In **2** one of these fixed points is totally unstable, while the other is a saddle. While crossing curve $P_+$ from **2** to **3**, the unstable fixed point becomes a saddle and an unstable period-2 cycle appears. If $c_{NS} > 0$, an unstable invariant curve "around" the period-2 cycles appears via the Neimark-Sacker bifurcation on $NS_+$ when we go from **3** to **4+**. The invariant curve disappears through a series of bifurcations associated with the heteroclinic bifurcations near $J_+$ if we come to **5**. If $c_{NS} < 0$, a stable closed invariant curve emerges in **4−** through a series of bifurcations associated with the heteroclinic structure. This stable invariant curve exists until we cross $NS_-$, where the stable period-2 cycle becomes attracting in **5**. Next we cross $P_-$ and the period-2 cycle disappears, leaving us with a stable fixed point and a saddle in **6**. These two collide if we return back to **1**.

Case 2 $(a_0 < 0, b_0 > 0)$: Fix a phase domain near the origin. Now we start with the two fixed points, one stable and one unstable, on the axis in region **1** (**Fig. 9.26**). Then, crossing the flip curve $P_+$ to **2**, one fixed point exhibits a period doubling and a period two cycle appears. The fixed points on the horizontal axis collide at the curve $F_+$, which separates region **2** from region **3**, where a stable period-2 cycle exists. If $c_{NS} > 0$, then an unstable invariant curve appears when we cross the Neimark-Sacker bifurcation curve $NS_+$. This invariant curve grows until it blows up and disappears from the selected fixed phase domain at some curve $B_+$. Actually, the invariant curve can lose its

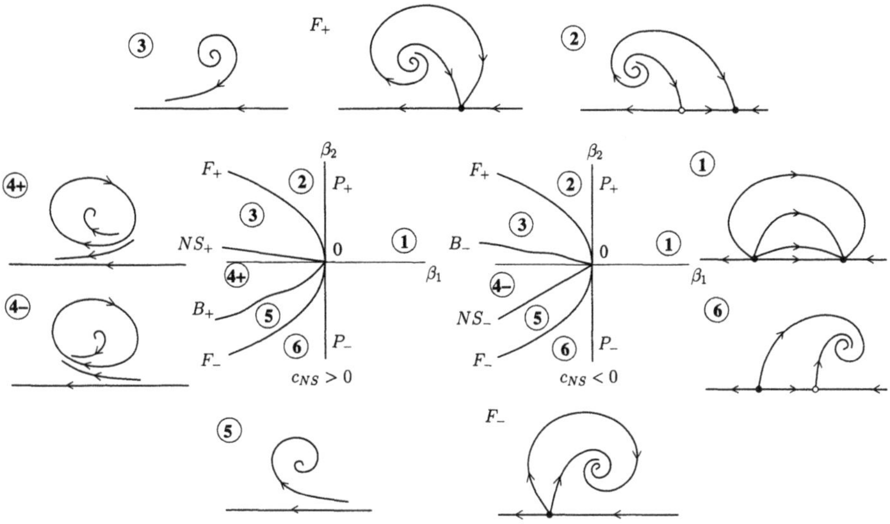


**Fig. 9.26.** Case 2: $a_0 < 0, b_0 > 0$.

smoothness and disappear before touching the boundary of the domain. If $c_{NS} < 0$, then we first encounter the "boundary bifurcation" curve $B_-$, where a big stable invariant curve appears in our fixed phase domain. The transition from **4**$-$ to **5** destroys the curve via the Neimark-Sacker bifurcation. Finally, crossing of the fold curve $F_-$ produces two fixed points in **6** and through the flip bifurcation on $P_-$ the period-two cycle disappears again as we go back to **1**.

Case 3 ($a_0 > 0, b_0 < 0$): We start with a period-2 saddle cycle in **1** (**Fig. 9.27**). Entering **2** through the fold curve $F_+$ creates two fixed points on the horizontal axis, a saddle and a repelling one. Then, while crossing the flip curve $P_+$ to **3**, the period-2 cycle is destroyed and we get two saddles on the $x_1$-axis. Passing $P_-$ one saddle becomes stable and a period-2 cycle in **4** is created. Finally, the fixed points on the horizontal axis collide on $F_-$ and we are in region **1** again.

Case 4 ($a_0 < 0, b_0 < 0$): Starting in region **1** (**Fig. 9.28**) we have, as in case 3, a period-2 saddle cycle but also a stable and an unstable fixed point on the $x_1$-axis. The unstable one becomes a saddle when we enter **2** through the $P_+$ curve. Then nothing special appears except for a "saddle-like flow" in region **3** after the saddle and the stable point collided on $F_+$. Going from **3** to **4** we get a saddle and an unstable point through the fold bifurcation on the curve $F_-$. We are back in **1**, when the flip bifurcation creates the period-two cycle on $P_-$.

The diagrams give a rather detailed description of the bifurcations of the truncated normal form (9.117). However, this description remains incomplete due to the presence of closed invariant curves and heteroclinic tangencies.

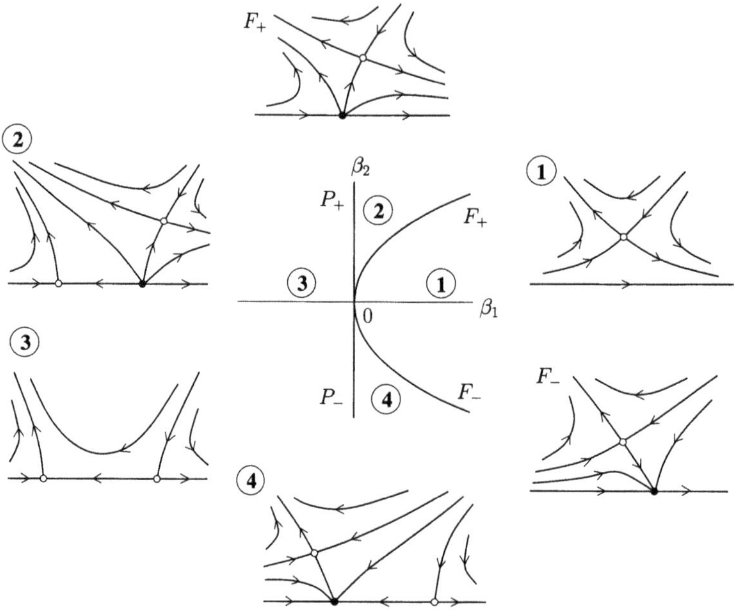


**Fig. 9.27.** Case 3: $a_0 > 0, b_0 < 0$.

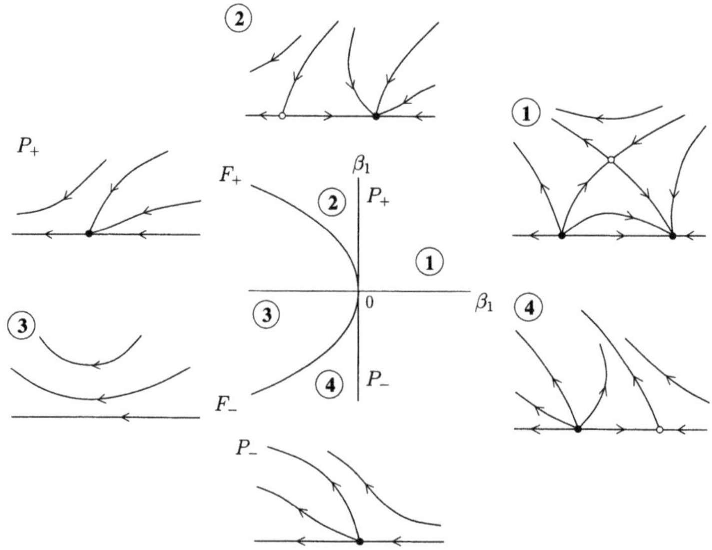


**Fig. 9.28.** Case 4: $a_0 < 0, b_0 < 0$.

Indeed, the rotation number on the closed invariant curve can change infinitely many times from rational to irrational and, moreover, the invariant curve can lose smoothness and disappear. Near a heteroclinic tangency, infinite series of bifurcations happen, including cascades of flips and folds.

Adding higher-order terms to the truncated normal form (9.117), i.e., restoring (9.115), complicates the bifurcation picture further. One can prove that for $\|\beta\|$ sufficiently small, the map (9.115) has the same bifurcations of fixed points and period-2 cycles as (9.117) with the same asymptotics for *arbitrary* higher-order terms. Therefore, we know what to expect locally. In particular, in cases 1 and 2 closed invariant curves appear. Moreover, the unit shift along the orbits of the vector field (9.120) composed with the reflection approximates (9.115) as good as (9.117). This implies that (9.115) in case 1 also has two bifurcation curves along which heteroclinic tangencies occur. Between these curves, a heteroclinic structure is present. Higher-order terms in (9.115) do affect these curves, but both remain tangent to the curve (9.123).

There are more differences between the phase portraits of (9.117) and a generic (9.115), which are related to other heteroclinic tangencies. For example, in the truncated normal form (9.117), the $x_1$-axis is always invariant. Therefore, in cases 1 and 3 we have the heteroclinic connections between the saddles located on the horizontal axis. However, generically, the higher-order terms in (9.115) break the reflectional symmetry, and the heteroclinic connection along the $x_1$-axis is lost. This allows for heteroclinic structures caused by intersections of the invariant manifolds of the saddles near the horizontal axis. These intersections can be either transversal (as in **Fig. 9.29**) or tangential. Therefore, in the first three cases, the bifurcation diagrams of (9.117) and a generic (9.115) are not locally topologically equivalent.

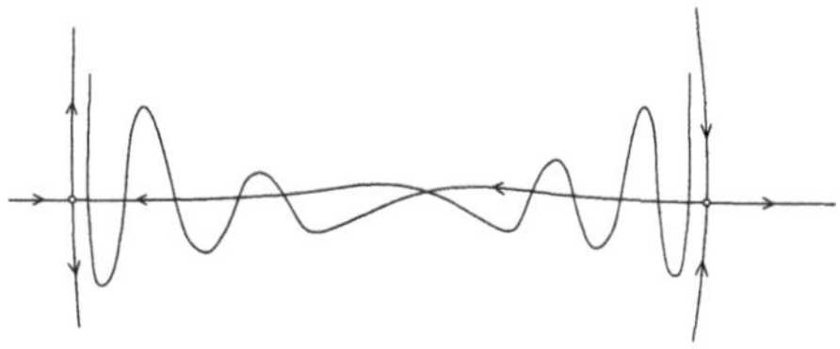

**Fig. 9.29.** A transversal heteroclinic structure near the horizontal axis.

## 9.7 Critical normal forms for $n$-dimensional maps

To apply the theory of the codim 2 bifurcations of fixed points to concrete maps, one needs to verify the nondegeneracy conditions at the bifurcation point, i.e., to compute the coefficients of the critical normal form on the center manifold up to a certain order. These coefficients could be found by first computing the Taylor expansion of the center manifold and then evaluating the corresponding normal form coefficients. This is how the explicit formulas (5.46), (5.49), and (5.54) for such coefficients at the codim 1 bifurcations of fixed points were found in Chapter 5. For codim 2 bifurcations, this approach is impractical.

In this section, we derive the critical normal form coefficients on the center manifold using an alternative "combined" normalization technique similar to that employed for the codim 2 bifurcations of equilibria in ODEs. We consider all codim 2 bifurcations of fixed points with *at most two critical multipliers.* Thus, all but cases (9), (10), and (11) listed in Section 9.1 will be covered.

The normalization method used below is very similar to that from Section 8.7. Suppose that a smooth map

$$x \mapsto F(x), \quad F : \mathbb{R}^n \to \mathbb{R}^n,$$

has a nonhyperbolic fixed point $x = 0$ and that the map obtained by restriction to the center manifold is transformed to a normal form

$$w \mapsto G(w), \quad G : \mathbb{R}^{n_c} \to \mathbb{R}^{n_c},$$

where $n_c$ is the number of the critical multipliers (counting multiplicity) or, equivalently, the dimension of the center manifold. Locally, the center manifold can be parametrized by $w \in \mathbb{R}^{n_c}$:

$$x = H(w), \quad H : \mathbb{R}^{n_c} \to \mathbb{R}^n.$$

Since the center manifold is invariant, we have the following *homological equation* for $H$ to satisfy:

$$H(G(w)) = F(H(w)). \tag{9.124}$$

As in Section 8.7 of Chapter 8, the Taylor coefficients of $G$ and $H$ can be found from (9.124) by an iterative procedure. Actually, the Taylor expansion of $H$ simultaneously defines the expansion of the center manifold, the normalizing transformation on it, and the coefficients of the resulting normal form. The resulting formulas involve only critical eigenvectors of the Jacobian matrix and its transpose, as well as the Taylor expansion (8.118) of the map $F$ at the critical fixed point in the original basis. The same "bordering technique" will be used to solve singular linear systems as in Section 8.7.

### 9.7.1 Cusp

In this case, there is only one critical multiplier $+1$ and the coefficient $b$ in (9.6) vanishes. The critical normal form for this bifurcation (cf. Section 9.2) can be written as

$$w \mapsto w + cw^3 + O(w^4),$$

where $w \in \mathbb{R}^1$ is a local coordinate along a one-dimensional center manifold represented by

$$H(w) = wq + \frac{1}{2}h_2w^2 + \frac{1}{6}h_3w^3 + O(w^4).$$

Here $h_i \in \mathbb{R}^n$ and

$$Aq = q.$$

We also introduce the adjoint eigenvector satisfying

$$A^Tp = p, \quad \langle p, q \rangle = 1.$$

Collecting the $w^2$-terms in (9.124), we find the equation

$$(A - I_n)h_2 = -B(q, q).$$

Since the matrix at the left-hand side is singular but the right-hand side satisfies Fredholm's solvability condition $\langle p, B(q, q)\rangle = 0$ due to (5.46), we use the bordering technique to solve for the unique $h_2 \in \mathbb{R}^n$ satisfying $\langle p, h_2 \rangle = 0$, which we denote $h_2 = -(A - I_n)^{INV}B(q, q)$. Next, we move on to the cubic terms and obtain

$$(A - I_n)h_3 = 6cq - C(q, q, q) - 3B(q, h_2).$$

This is another singular linear system. Using Fredholm's solvability condition and the normalization of $q$ with respect to $p$, we can express the critical coefficient $c$ as

$$c = \frac{1}{6}\langle p, C(q, q, q) - 3B(q, (A - I_n)^{INV}B(q, q))\rangle. \tag{9.125}$$

This expression coincides with (5.47) obtained in Chapter 5.

### 9.7.2 Generalized flip

In this case, there is only one critical multiplier $-1$ and the coefficient $c$ in (9.7) given by (5.49) vanishes. The critical normal form is

$$w \mapsto -w + dw^5 + O(w^6),$$

where $w \in \mathbb{R}^1$ is a properly selected local coordinate along the one-dimensional center manifold (see Section 9.3). The center manifold is defined by

$$H(w) = wq + \frac{1}{2}h_2w^2 + \frac{1}{6}h_3w^3 + \frac{1}{24}h_4w^4 + \frac{1}{120}h_5w^5 + O(w^6),$$

where $h_i \in \mathbb{R}^n$ and

$$Aq = -q.$$

We can find the adjoint eigenvector $p$ such that

$$A^Tp = -p, \quad \langle p, q \rangle = 1.$$

Collecting the $w^2$-terms in (9.124), we obtain

$$(A - I_n)h_2 = -B(q, q), \tag{9.126}$$

which is a nonsingular linear system that has the unique solution $h_2$. We continue with the $w^3$-terms to get

$$(A + I_n)h_3 = -C(q, q, q) - 3B(q, h_2). \tag{9.127}$$

This singular system is solvable since at the generalized flip

$$\langle p, C(q, q, q) + 3B(q, h_2)\rangle = 0$$

due to (5.49). Assume that $h_3$ is the unique solution of (9.127) satisfying $\langle p, h_3\rangle = 0$. The fourth-order terms in (9.124) give

$$(A - I_n)h_4 = -\left(4B(q, h_3) + 3B(h_2, h_2) + 6C(q, q, h_2) + D(q, q, q, q)\right). \tag{9.128}$$

This is a nonsingular system and thus we can solve for $h_4$. Finally, the critical coefficient $d$ appears in the fifth-order terms:

$$\begin{aligned}(A + I_n)h_5 = 120dq - (&5B(q, h_4) + 10B(h_2, h_3) + 10C(q, q, h_3) + \\ &15C(q, h_2, h_2) + 10D(q, q, q, h_2) + E(q, q, q, q, q)).\end{aligned}$$

The solvability of this singular system implies

$$\begin{aligned}d = \frac{1}{120}\langle p\ , &\ 5B(q, h_4) + 10B(h_2, h_3) + 10C(q, q, h_3) + \\ &15C(q, h_2, h_2) + 10D(q, q, q, h_2) + E(q, q, q, q, q)\rangle.\end{aligned}$$

In this formula, the vectors $h_2, h_3$, and $h_4$ are the solutions of the linear systems (9.126), (9.127), and (9.128), respectively.

### 9.7.3 Chenciner bifurcation

This bifurcation occurs when there is a pair of complex multipliers with $|\mu_{1,2}| = 1$ and the real part $d$ of the coefficient $d_1$ in (9.8) vanishes. It is also assumed that there are no other critical multipliers and that

$$e^{ik\theta_0} \neq 1 \text{ for } k = 1, 2, 3, 4, 5, 6.$$

The critical normal form (9.20) can be written as

$$w \mapsto e^{i\theta_0}w + c_1 w|w|^2 + c_2 w|w|^4 + O(|w|^6).$$

Here $w \in \mathbb{C}^1$ is a suitable local complex coordinate on the center manifold. The *first Lyapunov coefficient* $l_1 = \text{Re } d_1$, where $d_1 = e^{-i\theta_0}c_1$, vanishes at the Chenciner point. We choose complex eigenvectors satisfying

$$\begin{aligned} Aq = e^{i\theta_0}q, \quad A\bar{q} = e^{-i\theta_0}\bar{q}, \\ A^T p = e^{-i\theta_0}p, \quad A^T\bar{p} = e^{i\theta_0}\bar{p}, \end{aligned}$$

and such that $\langle p, q\rangle = 1$. The parametrization of the center manifold is now defined by

$$H(w, \overline{w}) = wq + \overline{w}\,\bar{q} + \sum_{1\le j+k\le 5} \frac{1}{j!k!} h_{jk} w^j \overline{w}^k + O(|w|^6),$$

with $h_{jk} \in \mathbb{C}^n$, $h_{kj} = \overline{h}_{jk}$.

Collecting the $w^j\bar{w}^k$-terms in (9.124) with $j+k=2$, we find the following nonsingular linear systems to be satisfied:

$$\begin{aligned} (A - e^{2i\theta_0}I_n)h_{20} &= -B(q,q), \\ (A - I_n)h_{11} &= -B(q,\bar{q}), \\ (A - e^{-2i\theta_0}I_n)h_{02} &= -B(\bar{q},\bar{q}). \end{aligned}$$

Collecting the $w^j\bar{w}^k$-terms in (9.124) with $j+k=3$ gives

$$(A - e^{3i\theta_0}I_n)h_{30} = -C(q,q,q) - 3B(q,h_{20}), \tag{9.129}$$

$$(A - e^{i\theta_0}I_n)h_{21} = 2c_1 q - C(q,q,\bar{q}) - B(\bar{q},h_{20}) - 2B(q,h_{11}), \tag{9.130}$$

$$(A - e^{-i\theta_0}I_n)h_{12} = 2\bar{c}_1\bar{q} - C(q,\bar{q},\bar{q}) - B(q,h_{02}) - 2B(\bar{q},h_{11}), \tag{9.131}$$

$$(A - e^{-3i\theta_0}I_n)h_{03} = -C(\bar{q},\bar{q},\bar{q}) - 3B(\bar{q},h_{02}). \tag{9.132}$$

The linear systems (9.129) and (9.132) are complex-conjugate and nonsingular. The systems (9.130) and (9.131) are also complex-conjugate but singular. As an intermediate result, we find from the solvability of (9.130) the expression

$$c_1 = \frac{1}{2}\langle p, C(q,q,\bar{q}) - B(\bar{q},(A - e^{2i\theta_0}I_n)^{-1}B(q,q)) - 2B(q,(A - I_n)^{-1}B(q,\bar{q}))\rangle,$$

which implies the formula (5.54) derived in Chapter 5. With this value of $c_1$, we can find from (9.130) the unique solution $h_{21}$ satisfying $\langle p, h_{21}\rangle = 0$ using the bordering technique.

The $w^j\bar{w}^k$-terms with $j+k=4$ lead to the nonsingular linear systems

$$\begin{aligned} (A - e^{4i\theta_0}I_n)h_{40} = &- [D(q,q,q,q) + 6C(q,q,h_{20}) + 3B(h_{20},h_{20}) + 4B(q,h_{30})], \\ (A - e^{2i\theta_0}I_n)h_{31} = &- [D(q,q,q,\bar{q}) + 3C(q,q,h_{11}) + 3C(q,\bar{q},h_{20}) + 3B(q,h_{21}) \\ &+ 3B(h_{11},h_{20}) + B(\bar{q},h_{30})] + 6c_1 h_{20} e^{i\theta_0}, \\ (A - I_n)h_{22} = &- [D(q,q,\bar{q},\bar{q}) + C(q,q,h_{02}) + C(\bar{q},\bar{q},h_{20}) + 4C(q,\bar{q},h_{11}) \\ &+ B(h_{20},h_{02}) + 2B(h_{11},h_{11}) + 2B(q,h_{12}) + 2B(\bar{q},h_{21})] \\ &+ 4h_{11}(c_1 e^{-i\theta_0} + \bar{c}_1 e^{i\theta_0}) \end{aligned}$$

$$\begin{aligned} (A - e^{-2i\theta_0}I_n)h_{13} = &- [D(q,\bar{q},\bar{q},\bar{q}) + 3C(\bar{q},\bar{q},h_{11}) + 3C(q,\bar{q},h_{02}) + 3B(\bar{q},h_{12}) \\ &+ 3B(h_{11},h_{02}) + B(q,h_{03})] + 6\bar{c}_1 h_{02} e^{-i\theta_0}, \\ (A - e^{-4i\theta_0}I_n)h_{04} = &- [D(\bar{q},\bar{q},\bar{q},\bar{q}) + 6C(\bar{q},\bar{q},h_{02}) + 3B(h_{02},h_{02}) + 4B(\bar{q},h_{03})]. \end{aligned}$$

The $w^3\bar{w}^2$-term in (9.124) gives the singular systems

$$
\begin{aligned}
(A - e^{i\theta_0} I_n) h_{32} = 12 c_2 q - \ & [E(q,q,q,\bar{q},\bar{q}) + D(q,q,q,h_{02}) + 6D(q,q,\bar{q},h_{11}) \\
& + 3D(q,\bar{q},\bar{q},h_{20}) + 3C(q,h_{20},h_{02}) + 6C(q,h_{11},h_{11}) \\
& + 3C(q,q,h_{12}) + 6C(q,\bar{q},h_{21}) + 6C(\bar{q},h_{11},h_{20}) \\
& + C(\bar{q},\bar{q},h_{30}) + 3B(h_{20},h_{12}) + 6B(h_{11},h_{21}) \\
& + 3B(q,h_{22}) + B(h_{02},h_{30}) + 2B(\bar{q},h_{31})] \\
& + 6h_{21}(2c_1 + \bar{c}_1 e^{2i\theta_0}).
\end{aligned}
$$

Note that $h_{21} = \bar{h}_{12}, h_{20} = \bar{h}_{02}$, and $c_1 e^{-i\theta_0} + \bar{c}_1 e^{i\theta_0} = 0$ at the Chenciner point. Also the vectors $h_{03}, h_{40}, h_{13}, h_{04}$, and $h_{32}$ need not be computed to find $c_2$. Further, only for $h_{21}$ do we need to use the bordering technique.

Finally, taking into account $\langle p, h_{21} \rangle = 0$, we obtain from the equation for $h_{32}$ the expression

$$
\begin{aligned}
c_2 = \frac{1}{12} \langle p\,,\, & E(q,q,q,\bar{q},\bar{q}) + D(q,q,q,h_{02}) + 6D(q,q,\bar{q},h_{11}) + 3D(q,\bar{q},\bar{q},h_{20}) \\
& + 3C(q,h_{20},h_{02}) + 6C(q,h_{11},h_{11}) + 3C(q,q,h_{12}) + 6C(q,\bar{q},h_{21}) \\
& + 6C(\bar{q},h_{11},h_{20}) + C(\bar{q},\bar{q},h_{30}) + 3B(h_{20},h_{12}) + 6B(h_{11},h_{21}) \\
& + 3B(q,h_{22}) + B(h_{02},h_{30}) + 2B(\bar{q},h_{31}) \rangle.
\end{aligned}
$$

### 9.7.4 Resonance 1:1

We have 1:1 resonance if there are two multipliers equal to 1 and no other critical multipliers exist. The critical normal form (9.38) for this bifurcation can be written as

$$
\begin{pmatrix} w_0 \\ w_1 \end{pmatrix} \mapsto \begin{pmatrix} w_0 + w_1 \\ w_1 + a w_0^2 + b w_0 w_1 \end{pmatrix} + O(\|(w\|^3),
$$

where $w = (w_0, w_1)^T \in \mathbb{R}^2$ provides a suitable local parametrization of the two-dimensional center manifold (see Section 9.5.2). We can find (generalized) eigenvectors of $A$ such that

$$
A q_0 = q_0, \quad A q_1 = q_1 + q_0,
$$

and similarly for the transposed matrix $A^T$

$$
A^T p_1 = p_1, \quad A^T p_0 = p_0 + p_1,
$$

so that $\langle p_0, q_0 \rangle = \langle p_1, q_1 \rangle = 1, \ \langle p_0, q_1 \rangle = \langle p_1, q_0 \rangle = 0$. Write the function parametrizing the center manifold as

$$
H(w_0, w_1) = w_0 q_0 + w_1 q_1 + \frac{1}{2} h_{20} w_0^2 + h_{11} w_0 w_1 + \frac{1}{2} h_{02} w_1^2 + O(\|w\|^3)
$$

with $h_{jk} \in \mathbb{R}^n$. Collecting the quadratic terms in (9.124), we obtain the singular linear systems

$$
\begin{aligned}
w_0^2 &: \ (A - I_n)h_{20} = -B(q_0, q_0) + 2aq_1, \\
w_0 w_1 &: \ (A - I_n)h_{11} = -B(q_0, q_1) + h_{20} + bq_1, \\
w_1^2 &: \ (A - I_n)h_{02} = -B(q_1, q_1) + 2h_{11} + h_{20}.
\end{aligned}
$$

The solvability of these singular linear systems requires their right-hand sides be orthogonal to $p_1$. The first equation immediately implies

$$a = \frac{1}{2}\langle p_1, B(q_0, q_0)\rangle.$$

Taking into account the identity $\langle p_1, h_{20}\rangle = -\langle p_0, B(q_0, q_0)\rangle$, we obtain from the second equation

$$b = \langle p_0, B(q_0, q_0)\rangle + \langle p_1, B(q_0, q_1)\rangle.$$

As in Section 8.7.4 of Chapter 8, we make the third system solvable by selecting a proper solution $h_{11}$ of the second system.

### 9.7.5 Resonance 1:2

Here we have two multipliers equal to $-1$ and no other critical multipliers. The critical normal form (9.61) can be written as

$$\begin{pmatrix} w_0 \\ w_1 \end{pmatrix} \mapsto \begin{pmatrix} -w_0 + w_1 \\ -w_1 + cw_0^3 + dw_0^2 w_1 \end{pmatrix} + O(\|(w_0, w_1)\|^4),$$

where $w = (w_0, w_1)^T \in \mathbb{R}^2$ is a suitable local parametrization of the critical center manifold (see Section 9.5.3). First we introduce the generalized eigenvectors of $A$ and $A^T$,

$$
\begin{aligned}
Aq_0 &= -q_0, & Aq_1 &= -q_1 + q_0, \\
A^T p_1 &= -p_1, & A^T p_0 &= -p_0 + p_1,
\end{aligned}
$$

satisfying the same normalization conditions as at 1:1 resonance. The expansion of the center manifold in this case should include cubic terms

$$H(w_0, w_1) = w_0 q_0 + w_1 q_1 + \sum_{1 \le i+j \le 3} \frac{1}{i!j!} h_{jk} w_0^k w_1^k + O(\|w\|^4)$$

with $h_{jk} \in \mathbb{R}^n$. Collecting the quadratic terms in (9.124), we get

$$
\begin{aligned}
w_0^2 &: \ (A - I_n)h_{20} = -B(q_0, q_0), \\
w_0 w_1 &: \ (A - I_n)h_{11} = -B(q_0, q_1) - h_{20}, \\
w_1^2 &: \ (A - I_n)h_{02} = -B(q_1, q_1) - 2h_{11} + h_{20}.
\end{aligned}
$$

Notice that the matrix $(A-I_n)$ is nonsingular since $A$ has only two eigenvalues $\lambda = -1$ on the unit circle. Therefore, one can solve these equations for $h_{20}, h_{11}$, and $h_{02}$ in the usual way.

From the cubic terms, we will find the equations

$$\begin{aligned} w_0^3 &: \ (A+I_n)h_{30} = 6cq_1 - 3B(q_0,h_{20}) - C(q_0,q_0,q_0), \\ w_0^2 w_1 &: \ (A+I_n)h_{21} = 2dq_1 + h_{30} - 2B(q_0,h_{11}) - B(q_1,h_{20}) - C(q_0,q_0,q_1), \\ w_0 w_1^2 &: \ (A+I_n)h_{12} = 2h_{21} - h_{30} - 2B(q_1,h_{11}) - B(q_0,h_{02}) - C(q_0,q_1,q_1), \\ w_1^3 &: \ (A+I_n)h_{03} = 3(h_{12}-h_{21}) + h_{30} - 3B(q_1,h_{02}) - C(q_1,q_1,q_1). \end{aligned}$$

Now we can easily determine the critical coefficient $c$:

$$c = \frac{1}{6}\langle p_1, C(q_0,q_0,q_0) + 3B(q_0,(I_n-A)^{-1}B(q_0,q_0))\rangle.$$

The equation for the critical coefficient $d$ involves the vector $h_{30}$. Taking the scalar product with $p_1$, we find from the equation at the $w_0^3$-term that $\langle p_1, h_{30}\rangle = -\langle p_0, 3B(q_0,h_{20}) + C(q_0,q_0,q_0)\rangle$. Then the solvability of the equation for $h_{21}$ implies

$$\begin{aligned} d = \frac{1}{2}[ & \langle p_1, 2B(q_0,h_{11}) + B(q_1,h_{20}) + C(q_0,q_0,q_1)\rangle \\ + & \ \langle p_0, 3B(q_0,h_{20}) + C(q_0,q_0,q_0)\rangle]. \end{aligned}$$

### 9.7.6 Resonance 1:3

Similar to the Chenciner bifurcation, the critical normal form (9.79) for the resonance 1:3 can be written in the complex form

$$w \mapsto e^{i\theta_0}w + b\bar{w}^2 + cw|w|^2 + O(|w|^4), \ \ \theta_0 = \frac{2\pi}{3},$$

where $w \in \mathbb{C}^1$ is a normalizing local coordinate on the center manifold (see Section 9.5.4). Now we select complex eigenvectors such that

$$\begin{aligned} Aq = e^{i\theta_0}q, \quad & A\bar{q} = e^{-i\theta_0}\bar{q}, \\ A^T p = e^{-i\theta_0}p, \quad & A^T\bar{p} = e^{i\theta_0}\bar{p}, \end{aligned}$$

and $\langle p,q\rangle = 1$. Introduce

$$H(w,\overline{w}) = wq + \overline{w}\,\bar{q} + \sum_{1\le j+k\le 3} \frac{1}{j!k!}h_{jk}w^j\overline{w}^k + O(|w|^4),$$

with $h_{jk} \in \mathbb{C}^n$, $h_{kj} = \overline{h}_{jk}$. The quadratic part of the homological equation (9.124) gives

$$\begin{aligned} w^2 &: \ (A - e^{2i\theta_0}I_n)h_{20} = \ 2\bar{b}\bar{q} - B(q,q), \\ w\bar{w} &: \ (A - I_n)h_{11} = -B(q,\bar{q}), \\ \bar{w}^2 &: \ (A - e^{-2i\theta_0}I_n)h_{02} = \ 2bq - B(\bar{q},\bar{q}). \end{aligned} \tag{9.133}$$

We remark that the first and third equations are conjugate, so that $\bar{h}_{20} = h_{02}$. Second, since $e^{i2\pi/3}$ is an eigenvalue of $A$, we have a singularity implying that $h_{02}$ should be found using a bordered system. As before, the solvability condition gives

$$b = \frac{1}{2}\langle p, B(\bar{q}, \bar{q})\rangle. \tag{9.134}$$

We only collect the $z^2\bar{z}$-terms since this is all we need to find the critical coefficient $c$,

$$(A - e^{i\theta_0} I_n)h_{21} = 2cq + e^{-i\theta_0}\bar{b}h_{02} - 2B(q, h_{11}) - B(\bar{q}, h_{20}) - C(q, q, \bar{q}).$$

From (9.134) it follows that $\langle p, h_{02}\rangle = 0$. Therefore we obtain the expression for $c$, which is similar to (5.54) if $b = 0$,

$$c = \frac{1}{2}\langle p,\ C(q, q, \bar{q}) \\ + 2B(q, (I_n - A)^{-1}B(q, \bar{q})) - B(\bar{q}, (e^{2i\theta_0} I_n - A)^{INV}(2\bar{b}\bar{q} - B(q, q)))\rangle.$$

### 9.7.7 Resonance 1:4

For this bifurcation the quadratic terms can be eliminated, but two cubic terms cannot be neglected. The critical normal form (9.94) can be written using a complex coordinate on the center manifold as

$$w \mapsto iw + cw|w|^2 + d\bar{w}^3 + O(|w|^4),\ w \in \mathbb{C}^1 \tag{9.135}$$

(see Section 9.5.5). As usual, we select complex eigenvectors such that

$$Aq = iq, \quad A\bar{q} = -i\bar{q}, \qquad A^T p = -ip, \quad A^T \bar{p} = i\bar{p}, \quad \langle p, q\rangle = 1.$$

Define the expansion of $H(w)$ as for the 1:3 resonance. The quadratic part of (9.124) then gives

$$\begin{aligned} w^2 &: \ (A + I_n)h_{20} = -B(q, q), \\ w\bar{w} &: \ (A - I_n)h_{11} = -B(q, \bar{q}), \\ \bar{w}^2 &: \ (A + I_n)h_{02} = -B(\bar{q}, \bar{q}). \end{aligned}$$

Since $\pm 1$ are not the eigenvalues of $A$, we can easily find $h_{20}, h_{11}$, and $h_{02}$. Now as above we only collect the coefficient in front of the resonant terms:

$$\begin{aligned} (A - iI_n)h_{21} &= 2cq - 2B(q, h_{11}) - B(\bar{q}, h_{20}) - C(q, q, \bar{q}), \\ (A - iI_n)h_{03} &= 6dq - 3B(\bar{q}, h_{02}) - C(\bar{q}, \bar{q}, \bar{q}). \end{aligned}$$

The solvability conditions imply

$$c = \frac{1}{2}\langle p, C(q, q, \bar{q}) + 2B(q, (I_n - A)^{-1}B(q, \bar{q})) - B(\bar{q}, (I_n + A)^{-1}B(q, q))\rangle$$

and

$$d = \frac{1}{6}\langle p, C(\bar{q}, \bar{q}, \bar{q}) - 3B(\bar{q}, (I_n + A)^{-1}B(\bar{q}, \bar{q}))\rangle.$$

### 9.7.8 Fold-flip

This bifurcation is characterized by two simple multipliers on the unit circle, one $+1$ and one $-1$. The matrix $A$ has the eigenvalues $\lambda_{1,2} = \pm 1$. Introduce the associated eigenvectors,

$$\begin{aligned} Aq_1 &= q_1, \quad A^T p_1 = p_1, \\ Aq_2 &= -q_2, \quad A^T p_2 = -p_2, \end{aligned}$$

such that $\langle p_1, q_1 \rangle = \langle p_2, q_2 \rangle = 1$. We have $\langle p_1, q_2 \rangle = \langle p_2, q_1 \rangle = 0$. The parametrization of the center manifold is given by

$$H(w) = w_1 q_1 + w_2 q_2 + \sum_{1 \le i+j \le 3} \frac{1}{i!j!} h_{jk} w_1^k w_2^k + O(\|w\|^4).$$

Assume that the restriction to the center manifold has been put into the normal form

$$G(w) = \begin{pmatrix} w_1 + \frac{1}{2}a_1 w_1^2 + \frac{1}{2}b_1 w_2^2 + \frac{1}{6}c_1 w_1^3 + \frac{1}{2}c_2 w_1 w_2^2 \\ -w_2 + e_1 w_1 w_2 + \frac{1}{2}c_3 w_1^2 w_2 + \frac{1}{6}c_4 w_2^3 \end{pmatrix} + O(\|w\|^4).$$

This is the resonant normal form appearing after *Step 2* in the proof of Lemma 9.16. Any map at the fold-flip bifurcation can be transformed by smooth invertible coordinate transformation to this form *without extra genericity assumptions.*

Collecting quadratic terms in (9.124), we find

$$\begin{aligned} (A - I_n)h_{20} &= a_1 q_1 - B(q_1, q_1), \\ (A + I_n)h_{11} &= e_1 q_2 - B(q_1, q_2), \\ (A - I_n)h_{02} &= b_1 q_1 - B(q_2, q_2), \end{aligned} \tag{9.136}$$

while the cubic coefficients are obtained from the singular systems

$$\begin{aligned} (A - I_n)h_{30} &= c_1 q_1 + 3a_1 h_{20} - 3B(q_1, h_{20}) - C(q_1, q_1, q_1), \\ (A - I_n)h_{12} &= c_2 q_1 + b_1 h_{20} - 2e_1 h_{02} - B(q_1, h_{02}) - 2B(q_2, h_{11}) - C(q_1, q_2, q_2), \\ (A + I_n)h_{21} &= c_3 q_2 + (2e_1 - a_1)h_{11} - 2B(q_1, h_{11}) - B(q_2, h_{20}) - C(q_1, q_1, q_2), \\ (A + I_n)h_{03} &= c_4 q_2 - 3b_1 h_{11} - 3B(q_2, h_{02}) - C(q_2, q_2, q_2). \end{aligned}$$

Since 1 and $-1$ are simple eigenvalues of $A$, the Fredholm solvability condition yields the critical quadratic coefficients

$$a_1 = \langle p_1, B(q_1, q_1) \rangle, \; b_1 = \langle p_1, B(q_2, q_2) \rangle, \; e_1 = \langle p_2, B(q_1, q_2) \rangle,$$

and using the bordering technique we find

$$\begin{aligned} h_{20} &= (A - I_n)^{INV} \left[ \langle p_1, B(q_1, q_1) \rangle q_1 - B(q_1, q_1) \right], \\ h_{11} &= (A + I_n)^{INV} \left[ \langle p_2, B(q_1, q_2) \rangle q_2 - B(q_1, q_2) \right], \\ h_{02} &= (A - I_n)^{INV} \left[ \langle p_1, B(q_2, q_2) \rangle q_1 - B(q_2, q_2) \right]. \end{aligned}$$

These vectors satisfy $\langle p_1, h_{20}\rangle = \langle p_1, h_{02}\rangle = \langle p_2, h_{11}\rangle = 0$. Then one can find the following expressions for the cubic coefficients:

$$\begin{aligned} c_1 &= \langle p_1, C(q_1,q_1,q_1) + 3B(q_1,h_{20})\rangle,\\ c_2 &= \langle p_1, C(q_1,q_2,q_2) + B(q_1,h_{02}) + 2B(q_2,h_{11})\rangle,\\ c_3 &= \langle p_2, C(q_1,q_1,q_2) + B(q_2,h_{20}) + 2B(q_2,h_{11})\rangle,\\ c_4 &= \langle p_2, C(q_2,q_2,q_2) + 3B(q_2,h_{02})\rangle. \end{aligned}$$

If we assume that $e_1 \neq 0$, the coefficients of the critical normal form (9.106) can be computed using (9.107) and (9.108) as

$$a(0) = \frac{a_1}{2e_1},\ b(0) = \frac{1}{2}b_1 e_1,\ c(0) = \frac{c_1}{6e_1^2},$$

$$d(0) = \frac{1}{2}c_2 + \frac{1}{2e_1}\left(b_1 c_3 - \frac{1}{3}c_4(a_1 + 2e_1)\right).$$

## 9.8 Codim 2 bifurcations of limit cycles

Consider a system of $(n+1)$ differential equations

$$\dot{y} = F(y,\alpha),\quad y \in \mathbb{R}^{n+1},\ \alpha \in \mathbb{R}^2, \tag{9.137}$$

where $F$ is smooth. Suppose system (9.137) has a *limit cycle* $L_\alpha$ with period $T$, and let

$$x \mapsto f(x,\alpha),\quad x \in \mathbb{R}^n,\ \alpha \in \mathbb{R}^2, \tag{9.138}$$

be the Poincaré map defined on an $n$-dimensional cross-section $\Sigma$ to the limit cycle. For simplicity assume that the cross-section is independent of the parameters. The cycle corresponds to a fixed point of the map $f$ (see Chapter 1). Therefore, any codim 2 bifurcation of the fixed point of the map will specify a certain codim 2 bifurcation of the limit cycle. We can use the bifurcation diagrams obtained in this chapter to deduce bifurcation pictures for the cycle bifurcations when the dimension of the center manifold of the Poincaré map is less than or equal to two.

Similar constructions allow us to apply the theory of codim 2 fixed-point bifurcations to codim 2 bifurcations of periodic solutions in nonautonomous time-periodic differential equations:

$$\dot{x} = f(x,t,\alpha),\quad x \in \mathbb{R}^n,\ \alpha \in \mathbb{R}^2, \tag{9.139}$$

where $f$ is periodic in $t$ with (minimal) period, say, $T_0 = 2\pi$. System (9.139) defines an autonomous system

$$\begin{cases} \dot{x} &= f(x, x_{n+1}, \alpha),\\ \dot{x}_{n+1} &= 1, \end{cases} \tag{9.140}$$

on the *cylinder* endowed with the coordinates $(x, x_{n+1} \bmod 2\pi)$. Isolated periodic solutions of (9.139) can be interpreted as limit cycles of (9.140) for which the hyperplane $\Sigma = \{x_{n+1} = 0\}$ is a cross-section. The Poincaré map (9.138) defined on this cross-section is simply the period return map for (9.139) (see Chapter 1).

In the cusp case (Section 9.2), the center manifold is one-dimensional. Each fixed point on this manifold corresponds to a limit cycle, while a collision of two points means a fold (tangent) bifurcation of limit cycles of different stability. Thus, there is generically a parameter region where system (9.137) has three nearby limit cycles that collide pairwise and disappear at the boundaries of this region (see **Fig. 9.2**).

For the generalized flip bifurcation (Section 9.3), the dimension of the center manifold still equals one. A cycle of period two on the manifold corresponds to a limit cycle that makes two turns around the "principal" cycle $L_\alpha$ before closure, thus having approximately double the period. Therefore, there is a region attached to the period-doubling bifurcation curve of the principal cycle where we have two cycles of different stability with approximately double the original period (see **Fig. 9.3**). At the boundary of the region, the cycles disappear via a fold bifurcation.

The Chenciner bifurcation (Section 9.4) implies that the center manifold for the Poincaré map is two-dimensional. A closed invariant curve on the manifold turns into an invariant two-dimensional *torus* of the relevant stability. There would be a region where two such tori coexist. However, they lose smoothness and destruct in a complex manner, almost colliding. Under parameter variation, an infinite number of long-period limit cycles appear and disappear via fold bifurcations both on and off the tori.

In all the strong resonance cases, the center manifold is two-dimensional. The central equilibrium in the phase portraits (see **Figs. 9.9**, **9.10**, **9.12**, and so forth) corresponds to the "principal" limit cycle, while nontrivial equilibria describe, actually, a *single* limit cycle making two, three, or four turns before closure (see **Fig. 9.30**, where a cycle with approximately *triple* period is shown). The closed invariant curves around the nontrivial equilibria correspond to a single invariant torus around the cycle with double, triple, and so on period. Stable and unstable invariant manifolds of saddle fixed points are the cross-sections of the corresponding manifolds of saddle limit cycles. Generically, they intersect transversally along orbits that are homo- or heteroclinic to the limit cycles. Near such structures, an infinite number of saddle limit cycles are present. This implies the presence of complex, "chaotic" dynamics near strong resonances. Consider, for example, the case 1:2 with $s = -1$ (see **Fig. 9.10**). A one-parameter system near this resonance can exhibit the following scenario. A stable limit cycle loses stability via a Neimark-Sacker bifurcation generating a smooth invariant torus. The remaining unstable cycle inside the torus undergoes a period-doubling bifurcation, giving rise to a repelling cycle of approximately double the period. After the period doubling, the "primary" cycle is a saddle cycle. The torus becomes pinched along a parallel while its

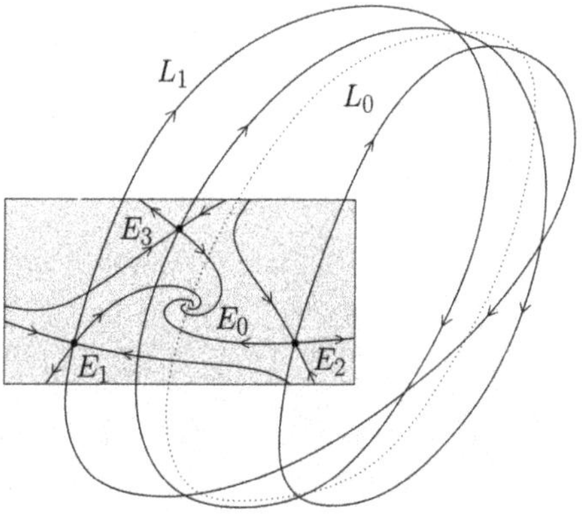


**Fig. 9.30.** A cycle of approximately triple period near 1:3 resonance.

meridian approaches the figure-eight shape. Then, the torus destructs near a homoclinic structure formed by the stable and the unstable invariant manifolds of the "primary" cycle. As a result, the orbits loop around one half and then the other half of the destroyed torus, jumping randomly from one side to the other. A "chaotic attractor" appears in which long-period stable cycles are embedded.

For the fold-flip bifurcation, the center manifold is also two-dimensional. The equilibria in the horizontal axis in **Figs. 9.25–9.28** correspond to the "principal" limit cycles, while a nontrivial equilibrium represents, actually, a limit cycle making *two turns* before closure. Similarly, the closed invariant curve corresponds to a "double torus" that makes two turns before closure. All remarks on the intersection of the invariant manifolds and the torus destruction are applicable in this case as well.

**Example 9.3 (Strong resonances in a periodically forced predator-prey system)** Consider the following time-periodic system of two differential equations [Kuznetsov, Muratori & Rinaldi 1992]:

$$\begin{cases} \dot{x}_1 = rx_1(1-x_1) - \dfrac{cx_1x_2}{\alpha(t)+x_1}, \\ \dot{x}_2 = -dx_2 + \dfrac{cx_1x_2}{\alpha(t)+x_1}, \end{cases}$$

where

$$\alpha(t) = b(1+\varepsilon\sin(t)).$$

Here $r, b, c, d$, and $\varepsilon$ are parameters, while $x_1$ and $x_2$ are scaled population numbers. For $\varepsilon = 0$, the system reduces to an autonomous predator-prey system studied in Chapter 3 (Example 3.1). As we have seen in that chapter, the system can have a stable limit cycle appearing via a Hopf bifurcation. The time-periodic function $\alpha(t)$ describes the influence of seasonal variability of the environment on the population dynamics. The time is scaled to make a year $2\pi$ in length. The parameter $0 \le \varepsilon \le 1$ measures the seasonal variation in the predator functional response. As we have already pointed out, the study of the model is equivalent to the study of the planar *first return map*:

$$f : x(0) \mapsto x(2\pi). \tag{9.141}$$

Fixed points of (9.141) correspond to $2\pi$-periodic solutions of the model, while period-$k$ points define *subharmonics* of period $2k\pi$ ($k$ years). Closed invariant curves correspond to *quasiperiodic* or *long-periodic (locked)* solutions, while irregular invariant sets describe *chaotic oscillations*. The map $f(x)$ depends on the parameters and exhibits several of the codim 2 bifurcations studied earlier. The following results were obtained with the help of a numerical continuation technique described in Chapter 10 and are based on the theory presented in this chapter. Let us fix

$$c = 2, \quad d = 1,$$

and analyze the bifurcation diagram in $(\varepsilon, b, r)$-space. The phenomena we shall see also persist for other parameter combinations.

To begin with, set

$$r = 1,$$

leaving only two parameters ($\varepsilon$ and $b$) active. Curves related to bifurcations of fixed points and period-two orbits are presented in **Fig. 9.31**. Let us trace step by step how this diagram was constructed.

For $\varepsilon = 0$ (no seasonal variation), the map $f$ is a $2\pi$-shift along the orbits of the autonomous system studied in Example 3.1 of Chapter 3. Its fixed points are trivial $2\pi$-solutions corresponding to the equilibria of the autonomous system. That system has a Hopf bifurcation at

$$b_H = \frac{c-d}{c+d}$$

($b_H = \frac{1}{3}$ for the selected parameter values). A stable limit cycle bifurcates for $b < b_H$. For the $2\pi$-shift map this means a Neimark-Sacker bifurcation generating a stable closed invariant curve. Therefore, the point

$$H = (0, b_H)$$

in the $(\varepsilon, b)$-plane can serve as an initial point for the continuation of a Neimark-Sacker bifurcation *curve* $h^{(1)}$ in these two parameters. Indeed, this curve is rooted at a point $H$ on the $b$-axis (see **Fig. 9.31**) and is transverse to this axis. Crossing $h^{(1)}$ at a typical point implies the generation of a small closed invariant curve, namely the appearance of small-amplitude oscillations around the "basic" $2\pi$-periodic solution corresponding to the nontrivial equilibrium of the constant-parameter model.[8]

[8] If the unforced system has a hyperbolic equilibrium, then the map $f$ has a nearby hyperbolic fixed point for small $\varepsilon$. This point corresponds to a $2\pi$-periodic solution of the forced system.

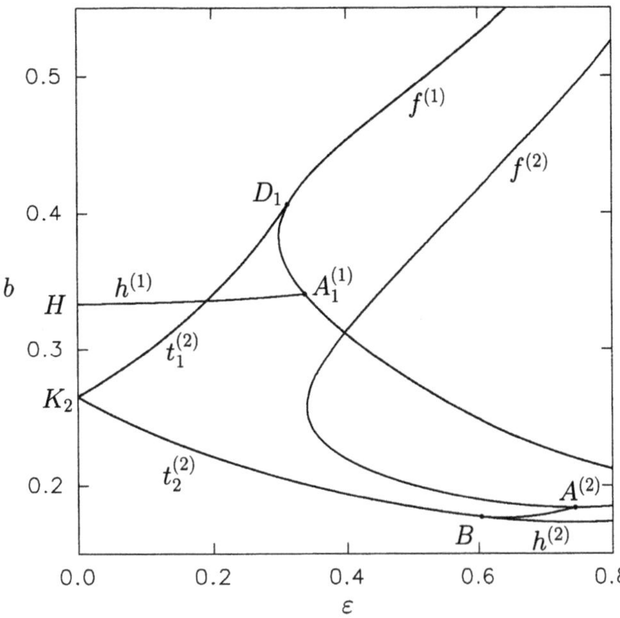


**Fig. 9.31.** Bifurcation curves on the $(\varepsilon, b)$-plane for $r = 1$.

While continue the curve $h^{(1)}$, from left to right on the $(\varepsilon, b)$-plane, both multipliers $\mu_{1,2}^{(1)}$ of the fixed point vary smoothly and become equal to $-1$ as the terminal point $A_1^{(1)}$ is reached. This is a codim 2 bifurcation – 1:2 *resonance*. As one can check numerically, the normal form coefficients satisfy

$$D_1(0) < 0, \; C_1(0) > 0,$$

therefore $s = 1$, and the bifurcation diagram of the system near $A_1^{(1)}$ is approximated by that depicted in **Fig. 9.9**. Thus, there is a single bifurcation curve related to fixed points or period-two orbits passing through the point $A_1^{(1)}$, specifically, the bifurcation curve $f^{(1)}$ along which the "basic" fixed point exhibits a flip (period-doubling) bifurcation. The fixed point has a simple multiplier $\mu_1^{(1)} = -1$ along this curve away from $A_1^{(1)}$. The two branches of $f^{(1)}$ have been obtained by numerical continuation starting from $A_1^{(1)}$ in the two possible directions. Crossing $f^{(1)}$ from the left to the right just above point $A_1^{(1)}$ results in the disappearance of a saddle period-two orbit while the stable period-one fixed point becomes a saddle (cf. **Fig. 9.9**). On the contrary, crossing the curve $f^{(1)}$ in the same direction just below point $A_1^{(1)}$ means the disappearance of a saddle period-two cycle while the repelling fixed point bifurcates into the saddle.

The analysis of the flip bifurcation on the upper branch of $f^{(1)}$ shows that there is another codim 2 bifurcation point $D_1$ at which the cubic normal form coefficient $c(0)$ vanishes and a nondegenerate *generalized flip bifurcation* occurs since $d(0) \neq 0$ (see Section 9.3). Thus, crossing the part of $f^{(1)}$ located above the point $D_1$ leads to a standard supercritical flip bifurcation generating a stable period-two cycle. Moreover, there is a tangent bifurcation curve $t_1^{(2)}$, at which two period-two cycles collide and disappear, that originates at point $D_1$ (see **Fig. 9.31**). A cycle of period two with a multiplier $\mu_1^{(2)} = 1$ exists along this curve. The curve $t_1^{(2)}$ terminates at a point $K_2$ on the $b$-axis where the limit cycle of the constant-parameter system has period $2 \cdot 2\pi$. The point $K_2$ is the end point of another branch $t_2^{(2)}$ of the tangent bifurcation curve for period-two cycles. As on the branch $t_1^{(2)}$, two period-two cycles appear as we cross $t_2^{(2)}$ near $K_2$, one of which is stable and the other of which is a saddle.

Following the lower branch $t_2^{(2)}$ of the tangent bifurcation curve, we encounter a point $B$ where the period-two cycle has two multipliers $\mu_{1,2}^{(2)} = 1$. This is a 1:1 *resonance* for the period-two cycle (see Section 9.5.2). A Neimark-Sacker bifurcation curve $h^{(2)}$ originates at this point. Along this curve the period-two cycle has a pair of the multipliers $\mu_{1,2}^{(2)}$ on the unit circle. The curve $h^{(2)}$ terminates at a point $A^{(2)}$ where the multipliers $\mu_{1,2}^{(2)}$ are both equal to $-1$ (1:2 *resonance*). A curve $f^{(2)}$ corresponding to a flip bifurcation of the period-two cycle goes through point $A^{(2)}$, which is also the root of a Neimark-Sacker bifurcation curve $h^{(4)}$ (not shown in **Fig. 9.31**) since for this 1:2 resonance we have $s = -1$. We will return to the discussion of the bifurcations near $A^{(2)}$ later.

For the current values of the parameters $(c, d, r)$, the Hopf bifurcation in the unperturbed system with $\varepsilon = 0$ generates a limit cycle with a period that is smaller than the seasonal period ($2\pi$). As $b$ was decreased, the cycle period grew and reached $2 \cdot 2\pi$ at the point $K_2$. Notice that the period of the cycle born via the Hopf bifurcation depends on the parameters, in particular on the value of $r$. Thus, now take a generic value or $r$, say

$$r = 0.73,$$

for which the period of the stable limit cycle appearing in the autonomous system through the Hopf bifurcation is *greater* than $2\pi$ (the forcing period). The resulting bifurcation curves on the $(\varepsilon, b)$-plane are presented in **Fig. 9.32**.

There is still a Neimark-Sacker bifurcation curve $h^{(1)}$ rooted at the Hopf point $H$ on the $b$-axis. As in the previous case, this curve terminates at a point $A_2^{(1)}$ corresponding to a 1:2 resonance, and a flip bifurcation curve $f^{(1)}$ passes through this point. However,

$$D_1(0) < 0, \ C_1(0) < 0,$$

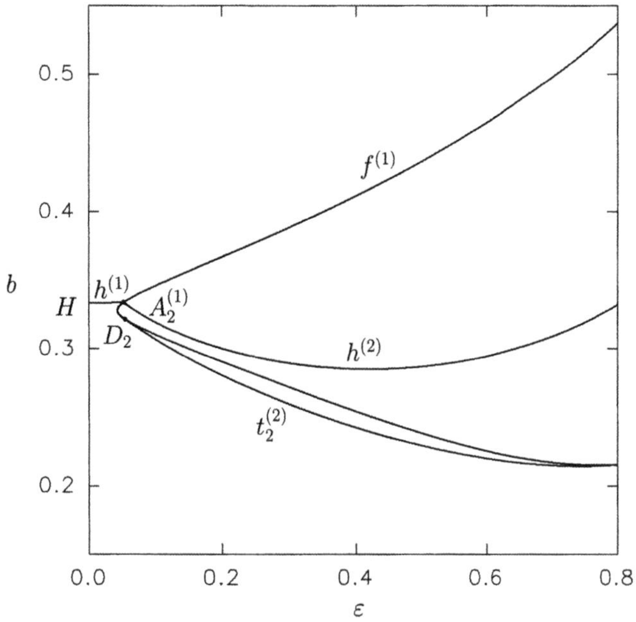


**Fig. 9.32.** Bifurcation curves on the $(\varepsilon, b)$-plane for $r = 0.73$.

in this case ($s = -1$). The theory of the 1:2 resonance suggests that there must be a curve $h^{(2)}$ originating at $A_2^{(1)}$ along which the period-two cycle has a pair of the multipliers $\mu_{1,2}^{(2)}$ on the unit circle (Neimark-Sacker bifurcation). Such a curve indeed exists and can be computed using the methods of Chapter 10 (see **Fig. 9.13**). In accordance with the theory of 1:2 resonance, a *subcritical* Neimark-Sacker bifurcation occurs along this curve. There is now only one branch $t_2^{(2)}$ of the tangent bifurcation curve terminating at a generalized flip point $D_2$ on the curve $f^{(1)}$.

How can we reconcile **Figs. 9.31** and **9.32**? Recall that there is a value $r = r_2$ ($r_2 = 0.75$) separating the two cases at which the asymptotic period of the limit cycle at the Hopf bifurcation for $\varepsilon = 0$ is *twice* as much as the forcing period. This is the 1:2 *resonant Hopf bifurcation* that has been analyzed theoretically by Gambaudo [1985] in the limit of small forcing, $\varepsilon \ll 1$ (see also the bibliographical notes). In a space of three parameters unfolding the bifurcation, with the forcing amplitude as the vertical axis, we have the remarkable bifurcation structure sketched in **Fig. 9.33**. There is a *conical* surface $f^{(1)}$ rooted at the resonant Hopf point at which a flip takes place. It is "cut" by a surface corresponding to the condition

$$\mu_1^{(1)} \mu_2^{(1)} = 1.$$

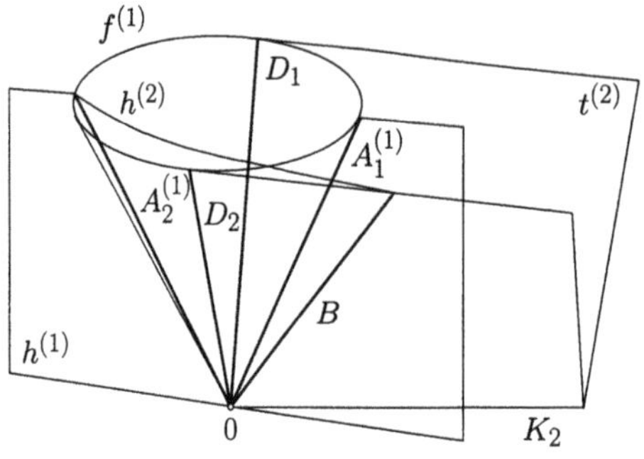


**Fig. 9.33.** Bifurcation structure near 1:2 resonant Hopf bifurcation.

The part of this surface located outside the cone, $h^{(1)}$, corresponds to the Neimark-Sacker bifurcation. The intersection curves $A_1^{(1)}$ and $A_2^{(1)}$ approaching the resonant Hopf point correspond to the 1:2 resonance (a double multiplier at $-1$) with opposite $s = \pm 1$. One of these curves $(A_2^{(1)})$ is a boundary of another codim 1 surface, namely $h^{(2)}$, on which a Neimark-Sacker bifurcation for the period-two cycle occurs. The surface $h^{(2)}$ is also bounded by a 1:1 resonance curve $B$ located on a surface of tangent bifurcation of period-two cycles $t^{(2)}$. This latter surface is bounded by two curves of generalized flip bifurcations $D_{1,2}$. **Fig. 9.34** gives a cross-section of the parameter space $(r, b, \varepsilon)$ by the plane $\varepsilon = 0.4$.

It agrees perfectly with the theory. The bifurcation diagrams shown in **Figs. 9.31**, **9.32**, and **9.34** are obviously not complete, because bifurcations of periodic orbits of period greater than or equal to three, as well as bifurcation sets related to homoclinic structures, are not presented. However, even those incomplete diagrams provide interesting information on the predator-prey model dynamics under periodic forcing, in particular, predicting the coexistence of various types of attractors.

Bifurcations far from the resonant Hopf point (e.g., the flip bifurcation of the period-two cycle on $f^{(2)}$) are not predicted by local analysis. Actually, there are accumulating sequences of flip bifurcation curves $f^{(2k)}$ leading to a Feigenbaum cascade of period doublings. Such cascades result in strange attractors in some regions of the parameter space, such as that presented in **Fig. 9.35**(a). Another scenario leading to chaos in the model is the destruction of a closed invariant curve. The closed curve can lose its smoothness and turn into an irregular invariant set near a homoclinic structure formed by the intersection of the stable and unstable manifolds of the period-two saddle cycle (see **Fig. 9.35**(b)).

Finally, let us look closer at the bifurcation diagram near the point $A^{(2)}$ in **Fig. 9.31** (when $r = 1$). This reveals an interesting phenomenon, namely the *accumulation of* 1:2 *resonances* (see **Fig. 9.36**). As we have already mentioned, there is an accumulating sequence of flip bifurcation curves on the $(\varepsilon, b)$-plane:

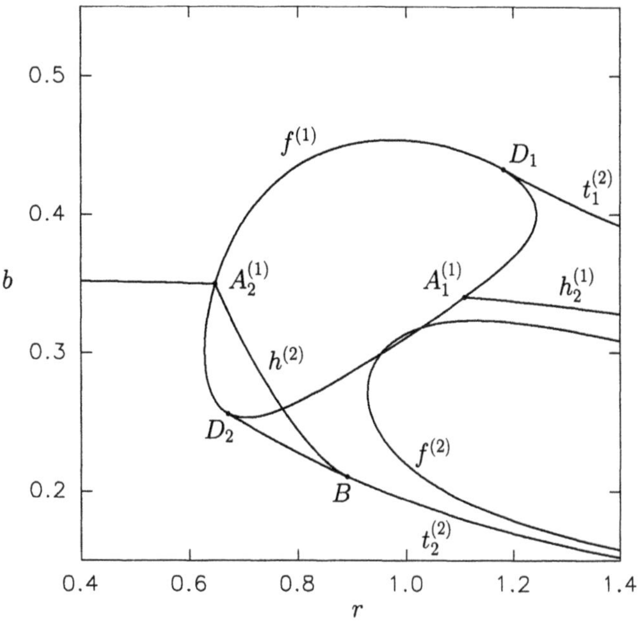


**Fig. 9.34.** Bifurcation curves on the $(r, b)$-plane for $\varepsilon = 0.4$.

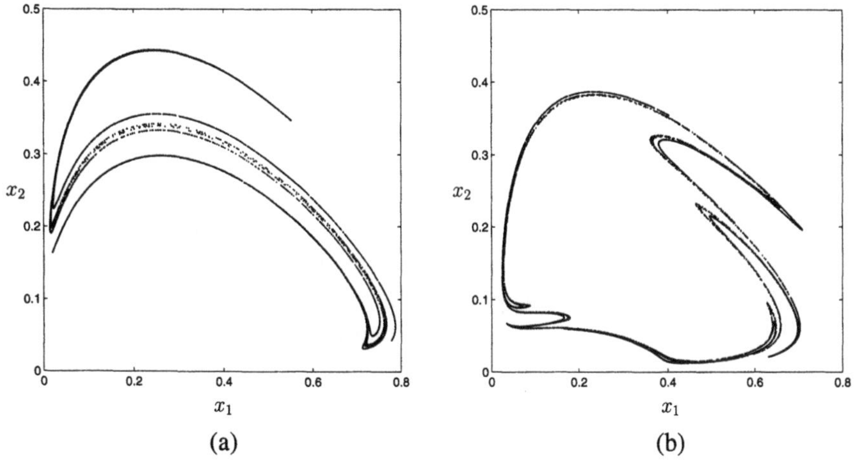


**Fig. 9.35.** Strange attractors of the Poincaré map.

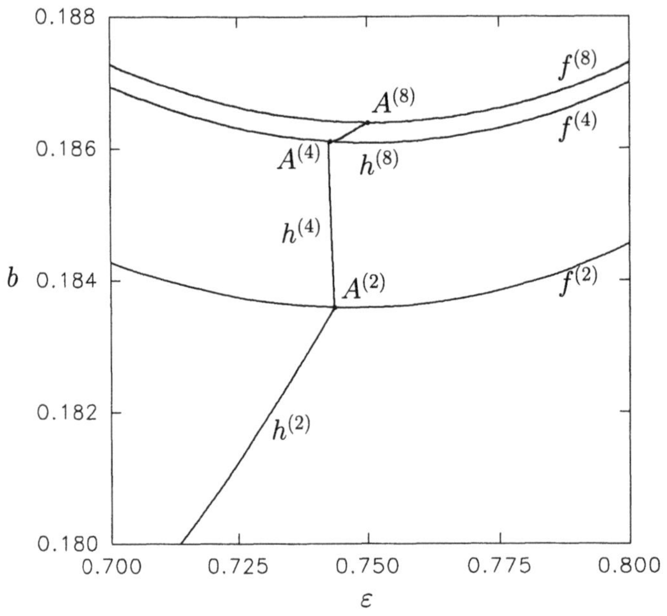


**Fig. 9.36.** Accumulation of 1:2 resonances.

$$f^{(2)}, f^{(4)}, f^{(8)}, \ldots .$$

Only the first three of them are shown in the figure since they get closer and closer according to Feigenbaum's universality. Crossing a curve $f^{(2k)}$ implies the appearance of stable period-$2k$ cycles while the previously stable cycle of period-$k$ becomes unstable (a supercritical flip). On each computed flip curve $f^{(2k)}$ there is a codim 2 point $A^{(2k)}$ of the 1:2 *resonance*. For corresponding parameter values we have a cycle of period $2k$, such that the $2k$th iterate $f^{2k}$ of the first return map (9.141) has a fixed point (cycle of period $2k$ for $f$) with multipliers $\mu_{1,2}^{(2k)} = -1$. Apparently, all the points $A^{(2k)}$ have the same topological type corresponding to the normal form flow (9.75) with $s = -1$. This means that there are Neimark-Sacker bifurcation curves $h^{(2k)}$ *and* $h^{(4k)}$ emanating from each $A^{(2k)}$. The computations and some theoretical arguments show that, in fact, the points $A^{(2k)}$ are *connected* by a sequence of Neimark-Sacker bifurcation curves. Therefore, this example illustrates how theoretical knowledge of the bifurcation structures associated with higher-codimension bifurcations guides numerical analysis and helps us to understand the behavior of a model. ◇

## 9.9 Exercises

**(1) (Strong resonances in adaptive control)** Consider the map

$$\begin{pmatrix} x \\ y \\ z \end{pmatrix} \mapsto \begin{pmatrix} y \\ bx + k + yz \\ z - \dfrac{ky}{c+y^2}(bx + k + zy - 1) \end{pmatrix},$$

which describes an adaptively controlled system [Frouzakis, Adomaitis & Kevrekidis 1991].

(a) Show that the fixed point $(x_0, y_0, z_0) = (1, 1, 1 - b - k)$ exhibits the 1:2 resonance at a point on the $(k, b)$-plane if the parameter $c$ is fixed. (*Hint:* Find a common point of the flip and the Neimark-Sacker bifurcation lines computed in Exercise 9 in Chapter 5.)

(b) Prove that we would encounter 1:3 and 1:4 resonances while moving along the Neimark-Sacker bifurcation line if $k$ decreases from the 1:2 resonant value to zero.

**(2) (Strong 1:4 resonance)** Verify that the map

$$\begin{pmatrix} x_1 \\ x_2 \end{pmatrix} \mapsto \begin{pmatrix} x_2 \\ \frac{3}{4} - x_1^2 \end{pmatrix}$$

[Hale & Koçak 1991] has a fixed point with multipliers $\mu_{1,2} = \pm i$, and compute the coefficients $C(0)$ and $D(0)$ of the normal form (9.92). (*Hint:* Do not forget to shift the fixed point to the origin!)

Which bifurcations do you expect in a generic two-parameter perturbation of this map?

**(3) (Degenerate 1:2 resonance in Hénon map)** Find parameter values at which the Hénon map (see Example 4.1 in Chapter 4 and Exercise 9 to Chapter 5)

$$\begin{pmatrix} x \\ y \end{pmatrix} \mapsto \begin{pmatrix} y \\ \alpha - \beta x - y^2 \end{pmatrix}$$

has a fixed point with a double multiplier $-1$. Compute the normal form coefficients $C(0)$ and $D(0)$ and check that the nondegeneracy condition (R2.2) does not hold. (*Hint:* For $\beta = 1$ the map is *area-preserving* since $\det J(x, y) = 1$, where $J(x, y)$ is its Jacobian evaluated at point $(x, y)$. Therefore, it cannot have any attracting or repelling fixed points and closed curves.)

**(4) (Map versus flow at a 1:2 resonance)** Consider the truncated normal form map (9.61) for the 1:2 resonance,

$$\begin{pmatrix} \xi_1 \\ \xi_2 \end{pmatrix} \mapsto \begin{pmatrix} -1 & 1 \\ \beta_1 & -1 + \beta_2 \end{pmatrix} \begin{pmatrix} \xi_1 \\ \xi_2 \end{pmatrix} + \begin{pmatrix} 0 \\ C\xi_1^3 + D\xi_1^2\xi_2 \end{pmatrix}, \tag{E.1}$$

and the approximating ODE system (9.141),

$$\begin{pmatrix} \dot{\xi}_1 \\ \dot{\xi}_2 \end{pmatrix} = \begin{pmatrix} 0 & 1 \\ \beta_1 & \beta_2 \end{pmatrix} \begin{pmatrix} \xi_1 \\ \xi_2 \end{pmatrix} + \begin{pmatrix} 0 \\ C\xi_1^3 + D\xi_1^2\xi_2 \end{pmatrix}. \tag{E.2}$$

Assume, for simplicity, that $C$ and $D$ are parameter-independent.

(a) Compare the equations for the Neimark-Sacker bifurcation of the trivial fixed point in (E.1) with the Hopf bifurcation of the trivial equilibrium in (E.2). Check that the transformation $(\beta_1, \beta_2) \mapsto (4\beta_1, -2\beta_1 - 2\beta_2)$ (cf. Lemma 9.11) maps the former bifurcation line into the latter.

(b) Compute the cubic normal form coefficient $c_1$ for the Neimark-Sacker bifurcation (see formula (4.26) in Chapter 4) in (E.1), and evaluate $d = \mathrm{Re}(e^{-i\theta_0} c_1) \neq 0$ near the origin. Verify that the nondegeneracy condition $d \neq 0$ is equivalent to condition (R2.2):

$$D + 3C \neq 0.$$

(c) Compute the Lyapunov coefficient $l_1$ along the Hopf bifurcation line in (E.2) near the origin. Then check that the nondegeneracy condition $l_1 \neq 0$ is equivalent to the condition

$$D \neq 0.$$

(d) Explain the difference in the results of (b) and (c) but their agreement with Lemma 9.11.

**(5) (Hetero- and homoclinic bifurcations in the approximating system at a 1:2 resonance)** Consider system (9.75):

$$\begin{cases} \dot{\zeta}_1 = \zeta_2, \\ \dot{\zeta}_2 = \varepsilon_1 \zeta_1 + \varepsilon_2 \zeta_2 + s\zeta_1^3 - \zeta_1^2 \zeta_2, \end{cases}$$

where $s = \pm 1$.

(a) Find a singular rescaling of the variables, time, and parameters that place the nontrivial equilibria (if they exist) at the points

$$\mathcal{E}_{1,2} = (\mp 1, 0),$$

and write the rescaled system in the form

$$\begin{cases} \dot{\eta}_1 = \eta_2, \\ \dot{\eta}_2 = s\eta_1(-1 + \eta_1^2) + \gamma_2 \eta_2 - \gamma_1 \eta_1^2 \eta_2, \end{cases}$$

where $(\gamma_1, \gamma_2)$ are new parameters. Derive the relationships between the new and old parameters:

$$\gamma_1 = \sqrt{-s\varepsilon_1}, \quad \gamma_2 = \frac{\varepsilon_2}{\sqrt{-s\varepsilon_1}}.$$

(b) Consider the system that results after substituting $\gamma_1 = \gamma_2 = 0$ into the previous equations, and verify that it is Hamiltonian with the Hamiltonian function

$$H_s(\eta) = \frac{s\eta_1^2}{2} + \frac{\eta_2^2}{2} - \frac{s\eta_1^4}{4}.$$

Draw the constant-level curves $H_s(\eta) = h$ for the cases $s = \pm 1$.

(c) Take the Hamiltonian system with $\gamma = 0$. Verify that, for $s = 1$,

$$\eta_1(t) = \frac{e^{\sqrt{2}t} - 1}{e^{\sqrt{2}t} + 1}, \quad \eta_2(t) = \frac{2\sqrt{2}e^{\sqrt{2}t}}{(e^{\sqrt{2}t} + 1)^2},$$

correspond to the upper heteroclinic orbit $H_1(\eta) = \frac{1}{4}$ connecting the saddles $\mathcal{E}_{1,2}$. Check that, for $s = -1$,

$$\eta_1(t) = \frac{2\sqrt{2}e^t}{1 + e^{2t}}, \quad \eta_2(t) = \frac{2\sqrt{2}e^t(1 - e^{2t})}{(1 + e^{2t})^2},$$

give an explicit solution corresponding to the right orbit $H_{-1}(\eta) = 0$ homoclinic to the saddle $\mathcal{E}_0$ at the origin.

(d) Following the strategy presented in Appendix A to Chapter 8, compute the derivative of the Hamiltonian $H_s$ along orbits of the system with $\gamma \neq 0$, and show that:

(i) for $s = 1$, the heteroclinic connection curve $C$ has the representation $\gamma_2 = k_1\gamma_1 + O(\gamma_1^2)$, with

$$k_1 = \frac{I_2(\frac{1}{4})}{I_1(\frac{1}{4})} = \frac{1}{5},$$

where $I_{1,2}(h)$ are defined by

$$I_1(h) = \int_{H_s(\eta)=h} \eta_2 \, d\eta_1, \quad I_2(h) = \int_{H_s(\eta)=h} \eta_1^2\eta_2 \, d\eta_1.$$

(ii) for $s = -1$, the "figure-eight" homoclinic curve $P$ has the representation $\gamma_2 = k_2\gamma_1 + O(\gamma_1^2)$, where

$$k_2 = \frac{I_2(0)}{I_1(0)} = \frac{4}{5}$$

with the above-defined integrals.
(*Hint:* To compute the integrals along the hetero- and homoclinic orbits, convert them into integrals over time and use the explicit solutions obtained in step (c).)

(e) Map the bifurcation curves obtained in step (d) onto the plane of the original parameters $(\varepsilon_1, \varepsilon_2)$, and verify the expressions for the bifurcation curves $C$ and $P$ given in Section 9.5.3.

(f) Prove that the integral ratio is a *monotone* function of $h$ in the case $s = 1$ but has a *unique minimum* for $s = -1$. (*Hint:* See Carr [1980], for example). What do these facts mean in terms of limit cycles of (9.75)?

**(6) (Heteroclinic bifurcations in the approximating system at a 1:3 resonance)** Consider the approximating system for the 1:3 resonance,

$$\dot{\eta} = (\beta_1 + i\beta_2)\eta + \bar{\eta}^2 + c\eta|\eta|^2.$$

(a) Analyze the system with the cubic term truncated, that is,

$$\dot{\eta} = (\beta_1 + i\beta_2)\eta + \bar{\eta}^2.$$

(i) Show that it has three nontrivial saddle equilibria placed at the vertices of an equilateral triangle.

(ii) Prove that for $\beta_1 = 0$ the system is Hamiltonian, and find the Hamilton function in polar coordinates. Draw its constant-level curves and verify that the orbits connecting the saddles at $\beta_1 = 0$ are straight lines.

(b) Perform a singular scaling of the system with the cubic term for small $\beta_1$ transforming it into a small perturbation of the quadratic Hamiltonian system studied in step (a). Compute the derivative of the Hamiltonian along orbits of the perturbed system, introduce a separatrix split function, and verify the asymptotic expression for the heteroclinic cycle curve $H$ given in Section 9.5.4. (*Hint:* See Horozov [1979] or Chow et al. [1994].)

**(7) (Equilibria of the approximating system at the 1:4 resonance)**
(a) Consider a linear planar system written in the complex form

$$\dot{z} = Pz + Q\bar{z},$$

where $P, Q \in \mathbb{C}^1$. Prove that the type of equilibrium at the origin is independent of the argument of $Q$. Show that the point is a saddle for $|P| < |Q|$, a focus for $|\text{Im } P| > |Q|$, and a node for $|\text{Im } P| < |Q| < |P|$. Verify also that the focus is stable for Re $P < 0$ and unstable for Re $P > 0$. (*Hint:* See Arnol'd [1983, p. 305].)

(b) Using step (a), prove that the nontrivial equilibria of the approximating system for the 1:4 resonance (9.99) are saddles if $|A| < 1$. Check that, for $|A| > 1$, the nontrivial equilibria of (9.99) with the smaller modulus are saddles, while those with the larger modulus are antisaddles.

(c) Find the condition on $A$ corresponding to the Bogdanov-Takens bifurcation of the nontrivial equilibria (i.e., when there are four nontrivial equilibria with a double zero eigenvalue), and verify that it coincides with that given in Section 9.5.5.

**(8) (Fold-flip bifurcation in the generalized Hénon map)**
Consider the planar map

$$\begin{pmatrix} x \\ y \end{pmatrix} \mapsto \begin{pmatrix} y \\ \mu - \varepsilon x - y^2 + Rxy \end{pmatrix},$$

where $(\alpha, \beta)$ are parameters and $R \neq 0$ is a constant (cf. Exercise **(3)**). This map appears in the study of codim 2 homoclinic tangencies [Gonchenko & Gonchenko 2000, Gonchenko, Gonchenko & Tatjer 2002].

(a) Show that at $(\mu, \varepsilon) = (0, -1)$ the map has a fixed point with eigenvalues $\lambda_{1,2} = \pm 1$.

(b) Prove that the corresponding fold-flip bifurcation is nondegenerate if $R \neq 1$, i.e., verify conditions (FF.1)–(FF.3) from Section 9.7. (*Hint*: First make the transformation $(x, y) = (\xi_1 + \xi_2, \xi_1 - \xi_2)$.)

## 9.10 Appendix: Bibliographical notes

Most of the results presented in this chapter have been known to specialists since the early 1960s although exact formulations and proofs appeared much later.

The *cusp* bifurcation of fixed points is parallel to that for equilibria. Theorem 9.1 is explicitly formulated in Arnol'd, Afraimovich, Il'yashenko & Shil'nikov [1994] (Russian original of 1986) together with a theorem on the generalized flip bifurcation.

The *generalized flip* bifurcation is briefly described by Holmes & Whitley [1984] and treated in detail together with higher-order flip degeneracies by Peckham & Kevrekidis [1991]. In particular, they derived the nondegeneracy condition equivalent to (GF.1) in Section 9.3.

The analysis of the *generalized Neimark-Sacker* bifurcation is due to Chenciner [1981, 1982, 1983*a*, 1983*b*, 1985*a*, 1985*b*, 1988]. A readable introduction to this bifurcation is given by Arrowsmith & Place [1990].

The study of *strong resonances* goes back to Melnikov [1962] and Sacker [1964]. The modern theory of strong resonances is due to Arnol'd [1977, 1983] and Takens [1974*a*]. It is presented in the textbooks by Arnol'd [1983] and Arrowsmith & Place [1990] (see also Arnol'd, Afraimovich, Il'yashenko & Shil'nikov [1994] and Chow, Li & Wang [1994]). The complete analysis of bifurcations in the approximating systems is performed for 1:1 resonance by Bogdanov [1975, 1976*b*, 1976*a*]; for 1:2 resonance by Takens [1974*a*], Holmes & Rand [1978], Carr [1981] and Horozov [1979]. The proof in the case 1:3 is also due to Horozov [1979]. Notice that the asymptotic expression for the heteroclinic bifurcation curve $H$ for this resonance in Arrowsmith & Place [1990] lacks the factor $\frac{1}{2}$.

The 1:4 resonance is the most complicated since even the analysis of the approximating system requires numerical computations. Most results on this resonance were obtained or predicted by Arnol'd [1977, 1983]. Analytical consideration of a neighborhood of the axis Re $A = 0$ (except $A = \pm i$) based on Hamiltonian perturbations is performed by Neishtadt [1978]. Analytical treatment of limit cycles in other regions on the $A$-plane (including the whole region Re $A < -1$) is made by Wan [1978*a*], Cheng [1990], Cheng & Sun [1992], and Zegeling [1993]. All known boundaries on the $A$-plane corresponding to degenerate heteroclinic bifurcations have been computed by Berezovskaya & Khibnik [1979, 1981], who also give possible bifurcation sequences of the phase portraits. Krauskopf [1994*a*] has generated computer pictures of these phase portraits. He studied the bifuircation diagram of the approximating system using the different scaling: $\dot{z} = e^{i\alpha}z + e^{i\varphi}z|z|^2 + b\bar{z}^3$ and computed its three-dimensional parametric portrait in the $(b, \varphi, \alpha)$-space [Krauskopf 1994*b*, Krauskopf 1997].

The standard approach (due to Arnol'd [1977]) to the study of the strong resonances starts with an autonomous system of differential equations[9] having a limit cycle whose multipliers satisfy $\mu^q = 1$, with $q = 1, 2, 3$, or 4. In proper coordinates, the system near the cycle is equivalent to a $2\pi$-periodic, nonautonomous system of the dimension subtracted by 1. By nonlinear transformations with $2\pi q$-periodic coefficients, this system can be reduced to an autonomous "principal part" plus higher-order terms with $2\pi q$-periodic coefficients. A shift along the orbits of this principal part approximates the $q$th iterate of the Poincaré map associated with the cycle. The construction is presented by Arnol'd [1983], Arrowsmith & Place [1990], and, in particular detail, by Iooss & Adelmeyer [1992]. It results in a $\mathbb{Z}_q$-symmetric autonomous system depending on parameters. We adopt a more elementary approach due to Neimark [1972] based on Picard iterations (see also Moser [1968] for the Hamiltonian case). It allows us to derive the approximating systems and the nondegeneracy conditions involved *directly* in terms of Taylor coefficients of the original maps. Notice that in some books it is wrongly assumed that the coefficients of the approximating systems coincide with those of the normalized maps.

The results on the *fold-flip* bifurcation are more recent. A connection between the corresponding bifurcation of limit cycles and bifurcations of the truncated amplitude system appearing in the analysis of the fold-Hopf bifurcation of ODEs was first indicated in [Arnol'd et al. 1994]. For maps, this bifurcation has been specially treated by Gheiner [1994] and Kuznetsov, Meijer & van Veen [2004]. The latter reference contains an example of the fold-flip bifurcation of a limit cycle in a simplified model of the atmospheric circulation.

The relationships between approximating systems and original maps are discussed in the books by Arnol'd [1983], Arrowsmith & Place [1990], and Arnol'd et al. [1994], among others. Important analytical and numerical results on closed invariant curves and homoclinic structures arising near codim 2 bifurcations of fixed points can be found in the cited papers by Chenciner and in those by Broer, Roussarie & Simó [1993, 1996], and Arrowsmith, Cartwright, Lansbury & Place [1993].

---

[9] It is sufficient to study three-dimensional systems.

Codim 2 bifurcations of fixed points (periodic orbits) have been found numerically in periodically driven models arising from economics [Ghezzi & Kuznetsov 1994], ecology [Kuznetsov et al. 1992], biotechnology [Pavlou & Kevrekidis 1992], engineering (Taylor & Kevrekidis [1991, 1993], Kuznetsov & Piccardi [1994*b*]), and epidemiology [Kuznetsov & Piccardi 1994*a*], and the list is growing rapidly. The normal form theory of the resonantly forced Hopf bifurcation has been developed by Gambaudo [1985], Bajaj [1986], Namachchivaya & Ariaratnam [1987], and Vance & Ross [1991].

Section 9.7 is based on [Kuznetsov & Meijer 2003].

# 10

# Numerical Analysis of Bifurcations

In this chapter we shall describe some of the basic techniques used in the numerical analysis of dynamical systems. We assume that low-level numerical routines like those for solving linear systems, finding eigenvectors and eigenvalues, and performing numerical integration of ODEs are known to the reader. Instead we focus on algorithms that are more specific to bifurcation analysis, specifically those for the location of equilibria (fixed points) and their continuation with respect to parameters, and for the detection, analysis, and continuation of bifurcations. Special attention is given to location and continuation of limit cycles and their associated bifurcations, as well as to continuation of homoclinic orbits. We deal mainly with the continuous-time case and give only brief remarks on discrete-time systems. Appendix A summarizes simple estimates of convergence of Newton-like methods. Appendix B gives some background information on the bialternate matrix product used to detect Hopf and Neimark-Sacker bifurcations. Appendix C presents numerical methods for detection of higher-order homoclinic bifurcations. The bibliographical notes in Appendix D include references to standard noninteractive software packages and interactive programs available for continuation and bifurcation analysis of dynamical systems. Actually, the main goal of this chapter is to provide the reader with an understanding of the methods implemented in widely used software for dynamical systems analysis.

Given a system of autonomous ODEs depending on parameters, our ultimate goal is to obtain its *bifurcation diagram* (i.e., to divide the parameter space into regions within which the system has topologically equivalent phase portraits and to describe how these portraits are transformed at the bifurcation boundaries). As we have seen in previous chapters, this task might be impossible since the bifurcation diagram can have an infinite number of complex-shaped regions. But even in the simplest cases, we have to rely on a computer to obtain information on the structure of the bifurcation boundaries of a dynamical system. This is particularly true for bifurcations of limit cycles and homoclinic orbits since there are only a few artificial examples of nonlinear systems allowing for closed-form analytical solutions. Even the analysis of equilibria (fixed points) in multidimensional systems is practically impossible without numerical calculations.

## 10.1 Numerical analysis at fixed parameter values

Consider a continuous-time system without parameters or with all parameters fixed at some values:

$$\dot{x} = f(x), \quad x \in \mathbb{R}^n, \tag{10.1}$$

where $f$ is sufficiently smooth. The analysis of system (10.1) means the construction of its phase portrait, that is, location of equilibria and limit cycles, studying the orbit structure near these phase objects and determining the global behavior of the system. We will discuss mainly *local* aspects of the problem, giving only a few remarks on global issues.

### 10.1.1 Equilibrium location

The analysis of system (10.1) starts with determining its equilibria, or the solutions of the system

$$f(x) = 0, \quad x \in \mathbb{R}^n. \tag{10.2}$$

Note that fixed points of a map $x \mapsto g(x)$ satisfy a system of the same type, namely: $f(x) = g(x) - x = 0$. A solution of (10.2) might be known analytically. This is especially likely if the system is *algebraic* with low-order polynomial components $f_i(x), i = 1, \dots, n$. In this case, we can apply algebraic results (see the bibliographical notes in Appendix D) to determine the number of solutions and (sometimes) find them explicitly. However, even polynomial systems of relatively low order do not have simple explicit solutions. Thus, numerical methods become unavoidable.

If the system has a *stable* equilibrium $x = x^0$, then we can find it to within a desired accuracy by numerical integration of (10.1) starting at a point $x$ within the basin of attraction of $x_0$ since

$$\lim_{t \to \infty} \|\varphi^t(x) - x^0\| = 0.$$

Here and throughout this chapter we use the norm $\|x\|^2 = \langle x, x \rangle = x^T x$, where $T$ means transpose. If the equilibrium $x^0$ is totally unstable (repelling), we can reverse time and repeat the procedure, thus converging backward to $x^0$ in the original time.

In general, for an equilibrium that is neither stable nor repelling, the location problem can only be solved provided the position of $x^0$ is known approximately. This is not unreasonable to assume, because one typically starts with an equilibrium found analytically or by integration at fixed parameter values and then "continues" it by small stepwise variations of parameters. The standard procedure that generates a sequence of points $\{x^{(i)}\}_{i=0}^{\infty}$ that converges, under very general conditions, to the equilibrium, is *Newton's method.*

## Newton's method

Denote by $A(x)$ the Jacobian matrix $f_x$ of (10.2) evaluated at a point $x$. Suppose $x^{(j)}$ is already close to $x^0$. Replace the left-hand side of (10.2) near $x^{(j)}$ by its linear part:

$$f(x^{(j)}) + A(x^{(j)})(x - x^{(j)}) \approx 0.$$

If the matrix $A(x^{(j)})$ is invertible, this linear system will have the solution

$$x = x^{(j)} - A^{-1}(x^{(j)})f(x^{(j)}),$$

which we could expect to be closer to $x^0$ than $x^{(j)}$ (see **Fig. 10.1** for the scalar case $n = 1$). Let $x^{(0)}$ be a given *initial point* near the equilibrium $x^0$. Inspired by the heuristic arguments above, *define* Newton iterations by the recurrence relation

$$x^{(j+1)} = x^{(j)} + \eta^{(j)}, \quad j = 0, 1, 2, \dots, \tag{10.3}$$

where the *displacement* $\eta^{(j)} \in \mathbb{R}^n$ is the solution of the linear system

$$A(x^{(j)})\eta^{(j)} = -f(x^{(j)}). \tag{10.4}$$

Notice that we need not invert the matrix $A(x^{(j)})$ to compute $x^{(j+1)}$; instead, only a solution to (10.4) is required. If the Jacobian matrix has a special structure, it is useful to take it into account when solving (10.4). Clearly, if iterations (10.3), (10.4) converge to some $x^0$, then $x^0$ is a solution of (10.2), or an equilibrium point of (10.1).

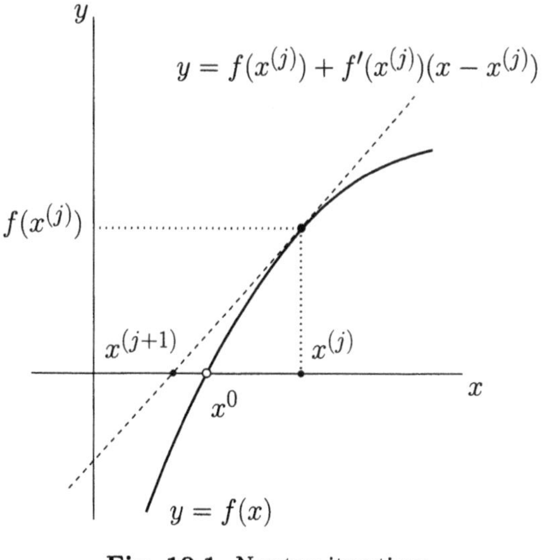


**Fig. 10.1.** Newton iterations.

**Theorem 10.1** *Suppose system* (10.1) *is smooth and has an equilibrium* $x = x^0$ *with no zero eigenvalue of the Jacobian matrix* $f_x(x^0)$. *Then there is a neighborhood* $U$ *of* $x^0$ *such that the Newton iterations* (10.3), (10.4) *converge to* $x^0$ *from any initial point* $x^{(0)} \in U$. *Moreover,*

$$\|x^{(j+1)} - x^0\| \leq \kappa_0 \|x^{(j)} - x^0\|^2, \quad j = 0, 1, 2, \ldots,$$

*for some* $\kappa_0 > 0$, *uniformly for* $x^{(0)} \in U$. □

This theorem follows from the Kantorovich theorem, which also provides more accurate error estimates (see Appendix A).

**Remarks:**

(1) Notice that the convergence is *independent* of the stability of the equilibrium. The absence of zero eigenvalues is equivalent to the invertibility of the Jacobian matrix $A(x)$ at the equilibrium point $x^0$. This can also be formulated as the condition for the map $f$ to be *regular* at $x^0$, or the matrix $A(x^0)$ to have maximal rank (equal to $n$).

(2) The convergence of the iterations is very rapid. The estimate in the theorem means that the error is roughly squared from iteration to iteration. Such a convergence is called *quadratic.* ◇

### 10.1.2 Modified Newton's methods

If we have no explicit formula for the Jacobian matrix, then the most expensive part of each Newton iteration is the numerical differentiation of $f(x)$, by finite differences, for example. Thus, several modifications of Newton's method have been proposed, all of which are aimed at reducing the number of the function calculations per iteration.

#### Newton-chord

Since $x^{(0)}$ is supposed to be close to $x^0$, we might not recompute the Jacobian matrix $A$ at each obtained point $x^{(j)}$ but use the initial matrix $A(x^{(0)})$ for all iterations. This simple idea leads to the *Newton-chord* iteration formula

$$x^{(j+1)} = x^{(j)} + \eta^{(j)}, \quad j = 0, 1, 2, \ldots, \tag{10.5}$$

where the displacement $\eta^{(j)}$ is now defined by

$$A(x^{(0)})\eta^{(j)} = -f(x^{(j)}). \tag{10.6}$$

This method also converges to $x^0$, though less rapidly.

**Theorem 10.2** *Under the conditions of* Theorem 10.1, *there is a neighborhood* $U$ *of* $x^0$ *such that Newton-chord iterations* (10.5),(10.6) *converge to* $x^0$ *from any initial point* $x^{(0)} \in U$. *However,*

$$\|x^{(j+1)} - x^0\| \leq \kappa_1 \|x^{(j)} - x^0\|, \quad j = 0, 1, 2, \ldots,$$

*for some* $0 < \kappa_1 < 1$, *uniformly for* $x^{(0)} \in U$. □

**Remarks:**

(1) The convergence of the Newton-chord method is *linear.*

(2) The idea of the proof is to write (10.5), (10.6) as a discrete-time dynamical system in $\mathbb{R}^n$,

$$x^{(j+1)} = g(x^{(j)}),$$

and then verify that the map $g : \mathbb{R}^n \to \mathbb{R}^n$ is a *contraction* in a sufficiently small ball around $x^0$. The Contraction Mapping Principle then guarantees convergence to the equilibrium and gives the error estimate. $\diamondsuit$

## Broyden update

A useful method called the *Broyden update* is a compromise between "true" Newton iterations and the Newton-chord method.

The idea is that we can use two successive iteration points and function values at these points to *update* the matrix involved in the computation of the displacement (i.e., make it closer to the Jacobian matrix at the next step). To understand the method, consider the $j$th iteration step of a Newton-like method. We have

$$x^{(j+1)} = x^{(j)} + \eta.$$

Suppose that a nonzero displacement $\eta$ is parallel to the $x_1$-axis:

$$\eta = \begin{pmatrix} \delta \\ 0 \\ \vdots \\ 0 \end{pmatrix}.$$

Let us try to use the coordinates of $x^{(j)}, x^{(j+1)}$ and the respective function values $f(x^{(j)}), f(x^{(j+1)})$ to approximate the Jacobian matrix at $x^{(j+1)}$, taking into account that $|\delta|$ is small. Clearly, we do not have enough data to approximate all the entries of the Jacobian matrix. However, *one* column can be updated. Indeed, for the first column,

$$A_{k1}(x^{(j+1)}) \approx \frac{f_k(x^{(j+1)}) - f_k(x^{(j)})}{\delta} = \frac{f_k(x^{(j+1)})}{\|\eta\|^2}\eta_1 - \frac{f_k(x^{(j)})}{\eta_1},$$

where $k = 1, 2, \ldots, n$, and $A_{kl}$ stands for the $kl$th element of $A$, while $\eta_1$ is the first component of $\eta$. Recalling that $\eta$ has to be found from

$$A(x^{(j)})\eta = -f(x^{(j)}),$$

and using the particular form of $\eta$, we obtain

$$A_{k1}(x^{(j)}) = -\frac{f_k(x^{(j)})}{\eta_1}, \quad k = 1, 2, \ldots, n.$$

Therefore, the following approximate relation holds between the "old" and "new" first-column elements:

$$A_{k1}(x^{(j+1)}) \approx A_{k1}(x^{(j)}) + \frac{f_k(x^{(j+1)})}{\|\eta\|^2}\eta_1, \quad k = 1, 2, \ldots, n.$$

We can formally write

$$A_{kl}(x^{(j+1)}) \approx A_{kl}(x^{(j)}) + \frac{f_k(x^{(j+1)})}{\|\eta\|^2}\eta_l$$

for $k, l = 1, 2, \ldots, n$, assuming that all columns except the first are left unmodified ($\eta_l = 0$ for $l = 2, 3, \ldots, n$). The last expression suggests the formula

$$A^{(j+1)} = A^{(j)} + \frac{f(x^{(j+1)})}{\|\eta^{(j)}\|^2}[\eta^{(j)}]^T \tag{10.7}$$

to update the matrix, where $T$ means transpose. It gives a reasonable update of the corresponding column of the Jacobian matrix if $\eta$ is parallel to one of the coordinate axis. For this reason, it is called a *rank-one* update.

Having in mind the above heuristics, define (*Newton-*)*Broyden iterations* as follows. Let $A^{(0)} = A(x^{(0)})$ be the Jacobian matrix at the initial point $x^{(0)}$. The *Broyden iterates* are defined by

$$x^{(j+1)} = x^{(j)} + \eta^{(j)}, \quad j = 0, 1, 2, \ldots, \tag{10.8}$$

where the displacement $\eta^{(j)}$ is the solution of the linear system

$$A^{(j)}\eta^{(j)} = -f(x^{(j)}), \tag{10.9}$$

and where the matrix $A^{(j+1)}$ to be used in the next iteration is given by (10.7).

The resulting method has better convergence than the Newton-chord modification. Under certain conditions (see the literature cited in Appendix D), Broyden iterations (10.7)–(10.9) converge to the equilibrium $x^0$ of system (10.2) *superlinearly*; namely, the following property holds:

$$\frac{\|x^{(j+1)} - x^0\|}{\|x^{(j)} - x^0\|} \to 0, \quad \text{as } j \to \infty.$$

**Remarks:**

(1) There is no reason to expect that $A^{(j)}$ converges to the Jacobian matrix $A(x^0)$ at the equilibrium $x^0$, even if the Broyden iterations converge to $x^0$ as $j \to \infty$. Thus we normally cannot use the finally obtained matrix $A^{(j)}$ as the Jacobian matrix, for example, to compute the eigenvalues of $x^0$.

(2) As we have pointed out, if the Jacobian matrix $A(x)$ has a certain special (i.e., band) structure, we usually try to use it in solving the linear system for $\eta^{(j)}$. Notice, however, that the Broyden update (10.7) may not preserve this special structure. ◊

### Convergence criteria

To terminate any of the described iterations when the required accuracy is achieved (or no convergence at all happens), some *convergence criteria* must be specified. We have two measures of convergence: the norm of the displacement $\|\eta^{(j)}\|$ and the norm of the function $\|f(x^{(j)})\|$ at the $j$th iteration. If the iterations converge, then both norms must tend to zero. However, one could easily construct examples for which small $\|\eta^{(j)}\|$ at finite $j$ does not imply that the corresponding $x^{(j)}$ satisfies (10.2) with reasonable accuracy. Therefore, the following combined criterion,

$$\|\eta^{(j)}\| < \varepsilon_x \quad \text{and} \quad \|f(x^{(j)})\| < \varepsilon_f,$$

prove to be most reliable. Here $\varepsilon_x$ and $\varepsilon_f$ are user-defined tolerances.

## 10.1.3 Equilibrium analysis

### Stability

After an equilibrium $x^0$ has been found with the desired accuracy, the next task is to analyze the phase portrait nearby, in particular, to determine whether $x^0$ is stable or unstable.

In the generic case, the stability of $x^0$ is determined by its eigenvalues, that is, the roots of the *characteristic polynomial*

$$p(\lambda) = \det(A - \lambda I_n).$$

In most existing eigenvalue solvers this polynomial is never constructed explicitly. Instead, certain transformations are used to bring $A$ into (block) diagonal form. From such a form the eigenvalues are easily extractable. Then, the number of the eigenvalues in the left (right) half-plane gives the dimension of the stable (unstable) manifold of $x^0$. The absence of eigenvalues with $\text{Re}\,\lambda \geq 0$ means (exponential) stability (see Chapter 2). Actually, to determine stability, we do not need to compute the eigenvalues at all. There are methods to check stability by computing the determinants of certain matrices whose elements are constructed from those of matrix $A$ or in terms of the coefficients of $p(\lambda)$ by simple rules (see below).

### Local approximation of invariant manifolds

Let $x^0 = 0$ be a saddle equilibrium of (10.1) with $n_-$ eigenvalues in the left half-plane and $n_+$ eigenvalues in the right half-plane, $n_- + n_+ = n$. In the eigenbasis of the matrix $A = f_x(x^0)$, (10.1) can be written as

$$\begin{cases} \dot{u} = Bu + g(u,v), \\ \dot{v} = Cv + h(u,v), \end{cases} \tag{10.10}$$

where $u \in \mathbb{R}^{n_+}$, $v \in \mathbb{R}^{n_-}$, and the $n_+ \times n_+$ matrix $B$ has only eigenvalues with Re $\lambda > 0$, while the $n_- \times n_-$ matrix $C$ has only eigenvalues satisfying Re $\lambda < 0$. The functions $g, h = O(\|(u,v)\|^2)$ are smooth. Locally, the stable and the unstable invariant manifolds of $x^0$ can be represented as the graphs of smooth functions,

$$W^s(0) = \{(u,v):\ u = U(v)\},\quad W^u(0) = \{(u,v):\ v = V(u)\},$$

where $U : \mathbb{R}^{n_-} \to \mathbb{R}^{n_+}$, $U(0) = 0$, $U_v(0) = 0$, $V : \mathbb{R}^{n_+} \to \mathbb{R}^{n_-}, V(0) = 0$, and $V_u(0) = 0$. Coefficients of the Taylor expansions of the functions $U$ and $V$ can be found by the method of unknown coefficients, similar to those of the center manifold in Chapter 5. Using the projection technique, we can avoid the transformation of (10.1) into the eigenform (10.10) and instead work in the original basis. Let us illustrate this technique by computing quadratic approximation of the manifolds in the case when $\dim W^u(0) = \operatorname{codim} W^s(0) = 1$ (i.e., $n_+ = 1,\ n_- = n - 1$).

Consider the system

$$\dot{x} = Ax + F(x), \tag{10.11}$$

where $A$ has one positive eigenvalue $\lambda > 0$ and $(n-1)$ eigenvalues with negative real part. Let $q \in \mathbb{R}^n$ be the eigenvector corresponding to $\lambda$,

$$Aq = \lambda q,$$

and let $p \in \mathbb{R}^n$ denote the adjoint eigenvector corresponding to the same eigenvalue:

$$A^T p = \lambda p.$$

Normalize these vectors by setting

$$\langle q, q\rangle = \langle p, q\rangle = 1,$$

where $\langle p, q\rangle = p^T q$ is the standard scalar product in $\mathbb{R}^n$. Now any vector $x \in \mathbb{R}^n$ can be uniquely represented in the form

$$x = \xi q + y,$$

where $\xi \in \mathbb{R}$ and $y \in \mathbb{R}^n$ are given by

$$\begin{cases} \xi = \langle p, x\rangle, \\ y = x - \langle p, x\rangle q. \end{cases}$$

The vector $y \in \mathbb{R}^n$ belongs to the eigenspace $T^s$ corresponding to all nonpositive eigenvalues, i.e., $\langle p, y\rangle = 0$. System (10.11) takes the form

$$\begin{cases} \dot{\xi} = \lambda\xi + G(\xi, y), \\ \dot{y} = Ay + H(\xi, y), \end{cases} \tag{10.12}$$

where

$$G(\xi, y) = \langle p, F(\xi q + y) \rangle = \langle y, Qy \rangle + \cdots,$$

with some symmetric $n \times n$ matrix $Q = Q^T$, while

$$H(\xi, y) = F(\xi q + y) - \langle p, F(\xi q + y) \rangle q = r\xi^2 + \cdots,$$

with $r \in T^s$, $\langle p, r \rangle = 0$. The dots in the last two formulas stand for all undisplayed terms of order two and higher, which are irrelevant in the following.

The unstable manifold $W^u(0)$ is locally represented by the function

$$y = Y(\xi) = s\xi^2 + O(\xi^3),$$

where $s \in T^s \subset \mathbb{R}^n$ is an unknown vector. Clearly, $Y(\xi)$ satisfies the equation (invariance condition)

$$AY + H(\xi, Y) - (\lambda\xi + G(\xi, Y))Y_\xi = 0.$$

Substituting the expression for $Y(\xi)$ and collecting $\xi^2$-terms, we get the equation from which $s$ can be found:

$$(2\lambda I_n - A)s = r.$$

The stable manifold $W^s(0)$ is given by

$$\xi = X(y) = \langle y, Ry \rangle + O(\|y\|^3),$$

where $y \in T^s$, and $R = R^T$ is a symmetric matrix to be defined. The function $X(y)$ satisfies the equation

$$\lambda X + G(X, y) - \langle X_y, Ay + H(X, y) \rangle = 0.$$

Quadratic terms provide the equation to be solved for $R$:

$$\langle y, (\lambda R + Q - 2A^T R)y \rangle = 0,$$

where $Q = Q^T$ is defined above. Therefore, the symmetric matrix $R$ can be found by solving the matrix equation

$$(A^T - \lambda I_n)R + RA = Q.$$

We leave the reader to check that both this equation has a unique symmetric solution, as well as to express $r$ and $Q$ in terms of the second partial derivatives of $F(x)$.

**Remarks:**

(1) Tangent or quadratic approximations of the invariant manifolds $W^{s,u}$ near the saddle point can be used in an attempt to compute them globally by integration forward or backward in time. This method works if the manifold is one-dimensional and system (10.1) is not stiff. It is particularly well suited for planar systems.

(2) Approximations of the stable and unstable manifolds near a saddle fixed point of a map can be obtained in a similar way. $\diamondsuit$

### 10.1.4 Location of limit cycles

Finding limit cycles of (10.1) is obviously a more complicated problem than locating its equilibria. There is no regular way to solve this problem in general. If the system possesses a stable cycle $L_0$, we can try to find it by numerical integration (simulation). If the initial point for the integration belongs to the basin of attraction of $L_0$, the computed orbit will converge to $L_0$ in forward time. Such a trick will fail to locate a saddle cycle, even if we reverse time.

For a general cycle, we might state the problem *locally*, by assuming that the position of the cycle is known approximately and then seeking to locate it more accurately. Such a setting arises naturally if the system depends on parameters. Then, we might know a cycle at certain parameter values and wish to "continue" it with respect to some parameter by making small steps. In this continuation process, a hyperbolic cycle will vary continuously, and its position at the "previous" parameter value provides a good approximation of the cycle at the "next" parameter value.

The cycle period $T_0$ is usually unknown. It is convenient to formulate the problem of cycle location as a periodic *boundary-value problem (BVP)* on a fixed interval. Specifically, consider $T_0$ as a parameter and introduce the system

$$\frac{du}{d\tau} = T_0 f(u), \tag{10.13}$$

which differs from (10.1) by the time-scaling factor $T_0$, and where the new time is called $\tau$. Clearly, a solution $u(\tau)$ of (10.13) with some $T_0$ fixed satisfying the *periodic boundary conditions*

$$u(0) = u(1) \tag{10.14}$$

corresponds to a $T_0$-periodic solution of (10.1). However, condition (10.14) does not define the periodic solution uniquely. Indeed, any time shift of a solution to the periodic BVP (10.13), (10.14) is another solution. Thus, an extra *phase condition* has to be appended to the problem (10.13), (10.14) in order to "select" a solution among all those corresponding to the cycle:

$$\Psi[u] = 0, \tag{10.15}$$

where $\Psi[u]$ is a scalar *functional* defined on periodic solutions. There are several ways to set up a phase condition (10.15).

The condition

$$\Psi[u] = g(u(0)) = 0, \tag{10.16}$$

where $g(x)$ is some smooth scalar function, selects a solution passing at $\tau = 0$ through a point on the surface

$$\Sigma_0 = \{u \in \mathbb{R}^n : g(u) = 0\}.$$

If $v(\tau)$ is a smooth vector-valued function with period one, then

$$\Psi[u] = \langle u(0) - v(0), \dot{v}(0) \rangle = 0 \tag{10.17}$$

specifies a solution $u(\tau)$ passing at $\tau = 0$ through the hyperplane orthogonal to the closed curve $\{u : u = v(\tau), \tau \in [0,1]\}$ at $\tau = 0$. The "reference" solution $v$ is assumed to be known. If we continue a limit cycle with respect to some parameter, $v(\tau)$ can be viewed as a solution corresponding to the limit cycle at the "previous" parameter values.

However, the most reliable phase condition is provided by

$$\Psi[u] = \int_0^1 \langle u(\tau), \dot{v}(\tau) \rangle \, d\tau = 0, \tag{10.18}$$

where $v(\tau)$ is the reference period-one solution. Condition (10.18) is called the *integral phase condition*. It is a necessary condition for a local minimum of the distance

$$\rho(\sigma) = \int_0^1 \|u(\tau + \sigma) - v(\tau)\|^2 \, d\tau$$

between $u$ and $v$ with respect to possible time shifts $\sigma$ (Exercise 3).

To solve the periodic BVP (10.13)–(10.15), no matter which phase condition is chosen, we have to reduce it to a finite-dimensional problem. There are several methods for such a *discretization*. Let us briefly describe the most frequently used of them.

### Shooting and multiple shooting

Let $\psi_{T_0}^{\tau}(u)$ be the solution of (10.13) at time $\tau$, with initial point $u$, which we should be able to compute numerically by some ODE-solver. Then, the problem (10.13), (10.14), (10.16) is equivalent to solving the system

$$\begin{cases} \psi_{T_0}^{1}(u^0) - u^0 = 0, \\ g(u^0) = 0, \end{cases} \tag{10.19}$$

where $u^0 = u(0)$. This system is a system of $n+1$ scalar equations[1] for $n+1$ unknowns, the components of $u^0$ and $T_0$, that we might try to solve by a Newton-like method. The Jacobian matrix of system (10.19) can be obtained by numerical differencing or via the variational equations. If the cycle has no unit multiplier $\mu = 1$ and the surface $\Sigma_0$ is a transversal cross-section to the cycle, then the Newton iterations converge to the cycle solution, starting from any sufficiently close approximation. Obviously, solving (10.19) means finding a point $u^0$ on the cross-section $\Sigma_0$ that is a fixed point of the associated Poincaré map, as well as the return time $T_0$ (see **Fig. 10.2**). The accuracy of the method is determined by the global integration error.

**Remark:**

[1] The first $n$ of them are defined *numerically*.

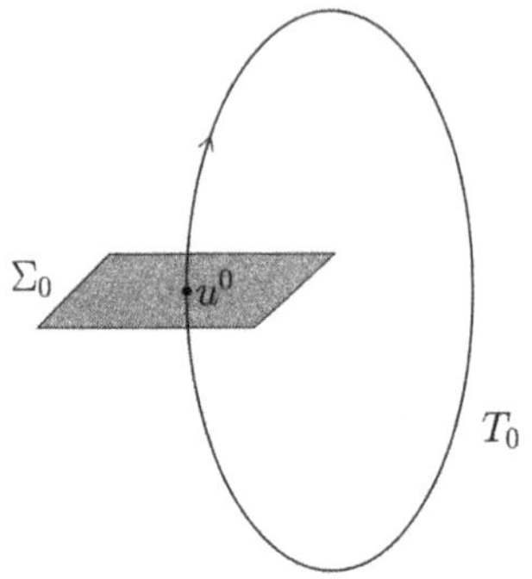


**Fig. 10.2.** Locating a cycle by shooting.

Notice that the multipliers $\mu_1, \mu_2, \ldots, \mu_{n-1}$ of the cycle $L_0$ can be found as the eigenvalues of the matrix

$$\left. \frac{\partial \psi_{T_0}^1(u)}{\partial u} \right|_{u=u^0}$$

if we discount its eigenvalue $\lambda_n = 1$, which is always present. If the Newton iterations converge, this matrix can be easily extracted from the Jacobian matrix of (10.19) at the last Newton iteration. ♢

The above *simple shooting* fails in many cases, particularly if the cycle under consideration is of saddle type. Then, a numerically obtained solution of (10.13) can be very different from the exact one $\psi_{T_0}^{\tau}(u)$ due to the strong error growth in the unstable directions. One might attempt to reduce the influence of this divergence by dividing the unit interval $[0, 1]$ into $N$ (nonequal) subintervals, hoping that the error would not grow much while integrating within each of them. To be more precise, introduce a mesh

$$0 = \tau_0 < \tau_1 < \cdots < \tau_N = 1,$$

and denote $\Delta_j = \tau_{j+1} - \tau_j$, $j = 0, 1, \ldots, N-1$. Let $u^j = u(\tau_j)$ be the (unknown) solution values at the mesh points. Then, $u^j, j = 0, 1, \ldots, N-1$, and $T_0$ can be found from the system

$$\begin{cases} \psi_{T_0}^{\Delta_0}(u^0) - u^1 = 0, \\ \psi_{T_0}^{\Delta_1}(u^1) - u^2 = 0, \\ \cdots \\ \psi_{T_0}^{\Delta_{N-1}}(u^{N-1}) - u^0 = 0, \\ g(u^0) = 0. \end{cases} \tag{10.20}$$

This is a system of $nN + 1$ scalar equations for the $n$ components of each $u^j$, $j = 0, 1, \ldots, N-1$, and the scaling factor (period) $T_0$. Essentially, this method is equivalent to the representation of the Poincaré map $P : \Sigma_0 \to \Sigma_0$ by a *composition* of $N - 1$ maps

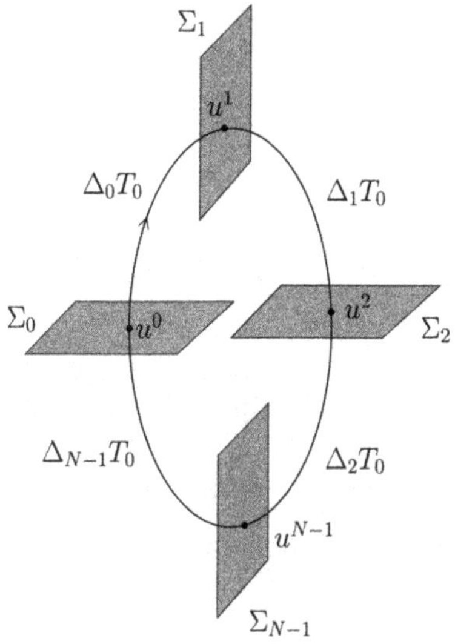


**Fig. 10.3.** Locating a cycle by multiple shooting.

$$P = P_{N-1} \circ \cdots \circ P_1 \circ P_0,$$

where

$$P_j : \Sigma_j \to \Sigma_{j+1}, \quad j = 0, 1, \ldots, N-1,$$

are *correspondence maps* along orbits of (10.13) between each pair of successive cross-sections from

$$\Sigma_0, \Sigma_1, \ldots, \Sigma_{N-1}, \Sigma_N = \Sigma_0$$

(see **Fig. 10.3**). Such a method is called *multiple shooting* and has better numerical stability than simple shooting, although it may fail due to the same reasons. Usually, when applying multiple shooting, one takes relatively few mesh points (small $N$) to avoid numerical difficulties in solving the multidimensional linear problems arising in the Newton-like method.

## Finite differences

Now take the partitioning of $[0, 1]$ by a sufficiently large number of mesh points. Then $\Delta_j$ will be small and the derivative $\dot{u}$ at point $\tau_j$ can be approximated by finite differences, for example,

$$\dot{u}(\tau_j) \approx \frac{u^{j+1} - u^{j-1}}{\tau_{j+1} - \tau_{j-1}}, \quad j = 1, 2, \ldots, N-1.$$

Then, BVP (10.13)–(10.15) can be approximated by the system of equations

$$\left\{ \begin{aligned} u^{j+1} - u^{j-1} - (\tau_{j+1} - \tau_{j-1}) T_0 f\left(\tfrac{1}{2}(u^{j+1} + u^j)\right) &= 0, \\ u^N - u^0 &= 0, \\ \psi[u^0, u^1, \ldots, u^N] &= 0, \end{aligned} \right. \qquad (10.21)$$

where $j = 1, 2, \ldots, N-1$, and $\psi[\cdot]$ is a proper discretization of the phase condition (10.15). For example, the phase condition (10.16) will read $\psi = g(u^0) = 0$. System (10.21) can be written (by eliminating $u^N$) as an equivalent system of $nN+1$ scalar equations for $nN$ components of $u^j$, $j = 0, 1, \ldots, N-1$, and $T_0$, and then solved by Newton's method provided that the usual regularity and sufficiently close initial data are given. It should be pointed out that Jacobian matrices arising in the Newton iterations for (10.21) have a special (band) structure that can be used to solve the corresponding linear problems efficiently.

The accuracy of the approximation of a smooth solution $u(\tau)$ of (10.13)–(10.15) by that of (10.21) can be estimated by

$$\|u(\tau_j) - u^j\| = O(h^2),$$

as $h = \max_{0 \le j \le N-1} \Delta_j \to 0$. Thus, to achieve reasonable accuracy of the solution, we have to take a sufficiently large $N$.

## Orthogonal collocation

Consider the BVP (10.13), (10.14) with the integral phase condition (10.18), and introduce once more the partitioning of the interval $[0,1]$ by $N-1$ mesh points,

$$0 = \tau_0 < \tau_1 < \cdots < \tau_N = 1,$$

but now let us seek to approximate the solution by a piecewise-differentiable continuous function that is a *vector-polynomial* $u^{(j)}(\tau)$ of maximal degree $m$ within each subinterval $[\tau_j, \tau_{j+1}]$, $j = 0, 1, \ldots, N-1$. "Collocation" consists of requiring the approximate solution to satisfy *exactly* system (10.13) at $m$ *collocation points* within each subinterval:

$$\tau_j < \zeta_{j,1} < \zeta_{j,2} < \cdots < \zeta_{j,m} < \tau_{j+1}.$$

Namely, we require

$$\left.\frac{du^{(j)}}{d\tau}\right|_{\tau=\zeta_{j,i}} = T_0 f(u^{(j)}(\zeta_{j,i})), \tag{10.22}$$

for $i = 1, \ldots, m$, $j = 0, 1, \ldots, N-1$.

It is convenient to characterize each polynomial $u^{(j)}(\tau)$, $j = 0, 1, \ldots, N-1$, by vectors

$$u^{j,k} = u^{(j)}(\tau_{j,k}) \in \mathbb{R}^n, \ k = 0, 1, \ldots, m,$$

which represent the (unknown) solution at the equidistant points

$$\tau_j = \tau_{j,0} < \tau_{j,1} < \tau_{j,2} < \cdots < \tau_{j,m} = \tau_{j+1},$$

given by

$$\tau_{j,k} = \tau_j + \frac{k}{m}(\tau_{j+1} - \tau_j), \quad j = 0, 1, \ldots, N-1, \ k = 0, 1, \ldots, m.$$

Note that by continuity $u^{j-1,m} = u^{j,0}$ for $j = 1, 2, \ldots, N-1$. Then, the polynomial $u^{(j)}(\tau)$ can be represented via the interpolation formula

$$u^{(j)}(\tau) = \sum_{i=0}^{m} u^{j,i} l_{j,i}(\tau), \tag{10.23}$$

where $l_{j,i}(\tau)$ are the *Lagrange basis polynomials*:

$$l_{j,i}(\tau) = \prod_{k=0, k \neq i}^{m} \frac{\tau - \tau_{j,k}}{\tau_{j,i} - \tau_{j,k}}.$$

These polynomials satisfy

$$l_{j,i}(\tau_{j,k}) = \begin{cases} 1 \text{ if } i = k, \\ 0 \text{ if } i \neq k \end{cases}$$

(check this!), which justifies (10.23).

Equations (10.22) now can be treated as equations for $u^{j,i}$. The periodicity condition (10.14) and the phase condition (10.18) can also be substituted by their discrete counterparts,

$$u^{0,0} = u^{N-1,m} \tag{10.24}$$

and

$$\sum_{j=0}^{N-1} \sum_{i=0}^{m} \omega_{j,i} \langle u^{j,i}, \dot{v}^{j,i} \rangle = 0, \tag{10.25}$$

where $\dot{v}^{j,i}$ are the values of the derivative of the reference periodic solution at the points $\tau_{j,i}$, and $\omega_{j,i}$ are the *Lagrange quadrature coefficients*. Equations (10.22), (10.24), and (10.25) compose a system of $nmN+n+1$ scalar equations for the unknown components of $u^{j,i}$ and the period $T_0$. The number of equations is equal to that of the unknowns. The resulting finite-dimensional system can be efficiently solved by a Newton-like method if we take into account the block structure of the Jacobian matrix.[2]

The most delicate problem in the described method, which we have not touched yet, is how to select the collocation points $\{\zeta_{j,i}\}$. We can try to tune their position to minimize the approximation error. As it can be proved, the optimal choice is to place them at the *Gauss points*, which are the roots $\zeta_{j,i}$, $i = 1, 2, \ldots, m$, of the $m$th degree *Legendre polynomial* relative to the subinterval $[\tau_j, \tau_{j+1}]$. These roots are well known numerically (see Exercise 4) on the standard interval $[-1, 1]$ and can be easily translated to each $[\tau_j, \tau_{j+1}]$. The Legendre polynomials compose the orthogonal system on the interval,

[2] One can also extract the cycle *multipliers* from this matrix.

giving the name to the method. Collocation at Gauss points leads to an extremely high accuracy of the approximation of a smooth solution of (10.13), (10.14), (10.18) by that of (10.22), (10.24), (10.25) at the *mesh points*, namely

$$\|u(\tau_j) - u^{j,0}\| = O(h^{2m}),$$

as $h = \max_{0 \leq j \leq N-1} \Delta_j \to 0$. The orthogonal collocation method has proved to be the most reliable among those implemented for limit cycle location.

As we shall see, the presented discretization methods are easily adaptable for automatic numerical continuation of limit cycles under parameter variation.

**Remark:**

Location of period-$k$ orbits of a discrete-time system $x \mapsto f(x)$ can be carried out by applying a Newton-like method to the system

$$f^k(x) - x = 0,$$

where $f^k$ is the $k$th iterate of the map $f$. $\diamondsuit$

## 10.2 One-parameter bifurcation analysis

Now consider a continuous-time system depending upon one parameter:[3]

$$\dot{x} = f(x, \alpha), \quad x \in \mathbb{R}^n, \ \alpha \in \mathbb{R}^1, \tag{10.26}$$

where $f$ is a smooth function of $(x, \alpha)$. The bifurcation analysis of the system means the construction of its one-parameter bifurcation diagram, in particular, studying the dependence of equilibria and limit cycles on the parameter, as well as locating and analyzing their bifurcations.

### 10.2.1 Continuation of equilibria and cycles

Equilibrium points of (10.26) satisfy

$$f(x, \alpha) = 0, \tag{10.27}$$

that is, a system of $n$ scalar equations in $\mathbb{R}^{n+1}$ endowed with the coordinates $(x, \alpha)$. As we have mentioned earlier, generically, (10.27) defines a smooth one-dimensional manifold (*curve*) $M$ in $\mathbb{R}^{n+1}$. Computation of this *equilibrium curve* gives the dependence of an equilibrium of (10.26) on the parameter $\alpha$ (see **Fig. 10.4**).

The problem of computing the curve $M$ is a specific case of the general (finite-dimensional) *continuation problem* – that means finding a curve in $\mathbb{R}^{n+1}$ defined by $n$ equations:

[3] The other parameters, if present, are assumed to be fixed.

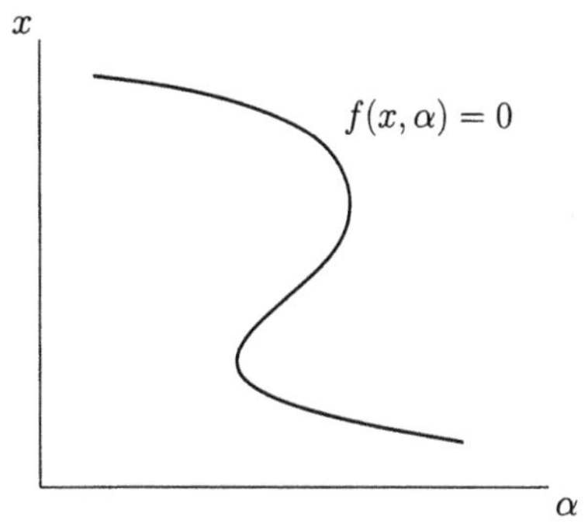


**Fig. 10.4.** Equilibrium curve.

$$F(y) = 0, \quad F : \mathbb{R}^{n+1} \to \mathbb{R}^n. \tag{10.28}$$

By the Implicit Function Theorem, system (10.28) locally defines a smooth curve $M$ passing through a point $y^0$ satisfying (10.28), provided that rank $J = n$, where $J = F_y(y^0)$ is the Jacobian matrix of (10.28) at $y^0$ (*regularity*). In (10.27), $y = (x, \alpha)$, and the regularity condition is definitely satisfied at hyperbolic equilibria, as well as at generic fold points.

The numerical solution of the continuation problem (10.28) means computing a *sequence* of points,

$$y^1, y^2, y^3, \ldots,$$

approximating the curve $M$ with desired accuracy. An *initial* point $y^0$, which is sufficiently close to $M$ (or belongs to it), from which the sequence can be generated in one of the two possible directions, is assumed to be known. In the equilibrium case, this point $y^0 = (x^0, \alpha^0)$ usually corresponds to an equilibrium $x^0$ of (10.26) found at some fixed parameter value $\alpha^0$ by one of the methods described in the previous section.

Most of the continuation algorithms used in bifurcation analysis implement *predictor-corrector methods* and include three basic steps performing repeatedly:

(i) *prediction of the next point*;
(ii) *correction*;
(iii) *step-size control.*

## Prediction

Suppose that a regular point $y^j$ in the sequence has been found. Then, the next point in the sequence can be guessed using the *tangent prediction*

$$\tilde{y}^{j+1} = y^j + h_j v^j, \tag{10.29}$$

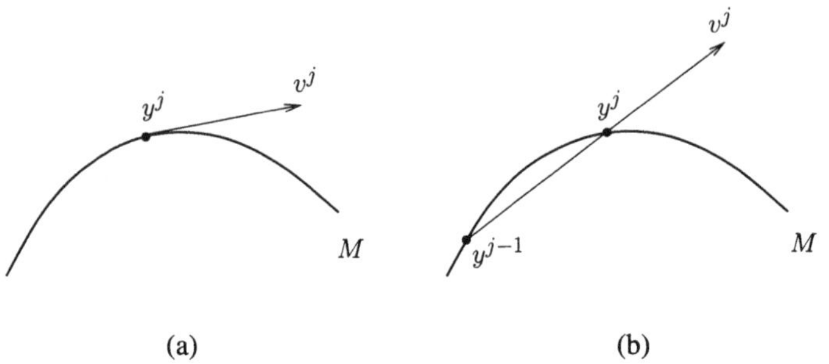


**Fig. 10.5.** (a) Tangent and (b) secant predictions.

where $h_j$ is the current *step size*, and $v^j \in \mathbb{R}^{n+1}$ is the normalized tangent vector to the curve $M$ at the point $y^j$, $\|v^j\| = 1$ (see **Fig. 10.5**(a)). If $y(s)$ is a parametrization of the curve near $y^j$, say, by the arclength with $y(0) = y^j$, then substituting $y = y(s)$ into (10.28) and computing the derivative with respect to $s$ give

$$J(y^j)v^j = 0 \tag{10.30}$$

since $v^j = \dot{y}(0)$ (a dot now denotes differentiation with respect to $s$). Here $J(y^j)$ is the Jacobian matrix of (10.28) evaluated at $y^j$,

$$J(y^j) = \left.\frac{\partial F}{\partial y}\right|_{y=y^j}.$$

System (10.30) has a unique solution (to within a scalar multiple) since rank $J(y^j) = n$ by the assumption of regularity.[4] To compute the tangent vector from (10.30), we have to fix its norm. The simplest way to do this is to preset some $v_{i_0} = 1$, solve the system for the other components, and then normalize the resulting vector, taking care to preserve the proper direction along the curve. An index $i_0$ that guarantees the solvability of the linear system for the other components is granted to exist. Equivalently, we could solve the $(n+1)$-dimensional *appended system*

$$\begin{pmatrix} J \\ (v^{j-1})^T \end{pmatrix} v^j = \begin{pmatrix} 0 \\ 1 \end{pmatrix},$$

where $v^{j-1} \in \mathbb{R}^{n+1}$ is the tangent vector at the previous point $y^{j-1}$ on the curve. This system is nonsingular for a regular curve $M$ if the points $y^j$ and $y^{j-1}$ are sufficiently close. The solution vector $v^j$ is tangent to the curve at $y^j$ and satisfies the normalization

$$\langle v^{j-1}, v^j \rangle = 1,$$

[4] Essentially, the regularity of $M$ is equivalent to the existence of a unique tangent direction at each point $y \in M$.

thus preserving the direction along the curve.

Another popular prediction method is the *secant prediction.* It requires *two* previous (distinct) points on the curve, $y^{j-1}$ and $y^j$. Then the prediction is given by (10.29), where now

$$v^j = \frac{y^{j-1} - y^j}{\|y^{j-1} - y^j\|} \tag{10.31}$$

(see **Fig. 10.5**(b)). The method cannot be applied at the first point $y^1$ on the curve.

## Correction

Having predicted a point $\widetilde{y}^{j+1}$ presumably close to the curve, we need to locate the next point $y^{j+1}$ on the curve to within a specified accuracy. This *correction* is usually performed by some Newton-like iterations. However, the standard Newton iterations can only be applied to a system in which the number of equations is equal to that of the unknowns. Thus, an extra scalar condition

$$g^j(y) = 0$$

has to be appended to (10.28) in order to apply Newton's method to the system

$$\begin{cases} F(y) = 0, \\ g^j(y) = 0. \end{cases} \tag{10.32}$$

Geometrically this means that we look for an intersection of the curve $M$ with some surface near $\widetilde{y}^{j+1}$. It is natural to assume that the prediction point $\widetilde{y}^{j+1}$ belongs to this surface (i.e., $g^j(\widetilde{y}^{j+1}) = 0$). There are several different ways to specify the function $g^j(y)$.

(*Natural continuation*) The simplest way is to take a hyperplane passing through the point $\widetilde{y}^{j+1}$ that is orthogonal to one of the coordinate axes, namely, set

$$g^j(y) = y_{i_0} - \widetilde{y}_{i_0}^{j+1}. \tag{10.33}$$

Clearly, the best choice of $i_0$ is provided by the index of the component of $v^j$ with the maximum absolute value (see **Fig. 10.6**(a)). In this case, the coordinate $y_{i_0}$ is locally the most rapidly changing along $M$. Obviously, the index $i_0$ may differ from point to point.

(*Pseudo-arclength continuation*) Another possibility is to select a hyperplane passing through the point $\widetilde{y}^{j+1}$ that is orthogonal to the vector $v^j$ (see **Fig. 10.6**(b)):

$$g^j(y) = \langle y - \widetilde{y}^{j+1}, v^j \rangle = \langle y - y^j, v^j \rangle - h_j. \tag{10.34}$$

If the curve is regular (rank $J(y) = n$) and the step size $h_j$ is sufficiently small, one can prove that the Newton iterations for (10.32) will converge to a

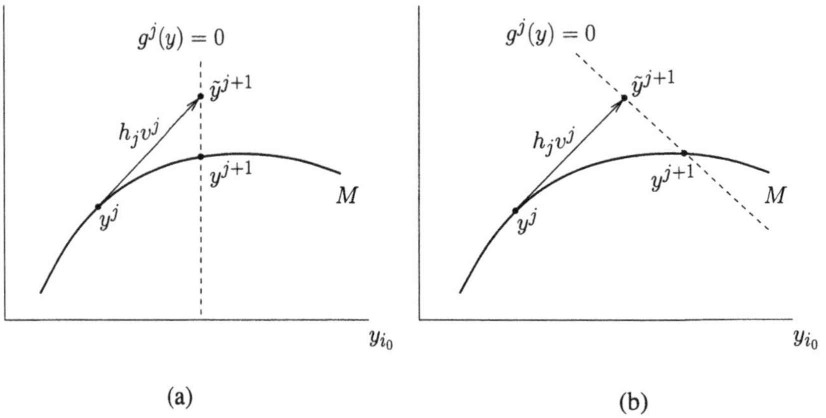


**Fig. 10.6.** (a) Natural and (b) pseudo-arclength continuation.

point $y^{j+1}$ on the curve $M$ from the predicted point $\tilde{y}^{j+1}$ for both methods. Notice that we can extract the matrix $J(y^{j+1})$ needed in (10.30) from the computed Jacobian matrix of (10.32) at the last Newton iteration. Moreover, the index $i_0$ used at the current step of natural continuation can be employed to determine the next tangent vector $v^{j+1}$ from (10.30) upon setting $v_{i_0}^{j+1} = 1$.

(*Moore-Penrose continuation*) The function $g^j(y)$ in (10.32) may be adapted in the course of the Newton iterations. For example, the plane in which the current iteration happens can be made orthogonal to a normalized vector $V^k$ satisfying $J(Y^{k-1})V^k = 0$, where $Y^{k-1}$ is the point obtained at the previous Newton iteration, so that

$$g_k^j(y) = \langle y - Y^{k-1}, V^k \rangle$$

(see **Fig. 10.7**). For the first iteration, $Y^0 = y^j + hv^j, V^1 = v^j$. Notice that vector $V^k$ with $k \geq 2$ is tangent to the "perturbed curve" $F(Y) = F(Y^{k-1})$ at the point $Y^{k-1}$ on it. One can prove local convergence of such modified iterations under the regularity condition. Exercise 13 explains the relation of this continuation method with the *Moore-Penrose matrix inverse*.

### Step-size control

There are many sophisticated algorithms to control the step size $h_j$; however, the simplest *convergence-dependent control* has proved to be reliable and easily implementable. That is, we decrease the step size and repeat the corrections if no convergence occurs after a prescribed number of iterations; we increase the step size $h^{j+1}$ with respect to $h^j$ if the convergence requires only a few iterations; and we keep the current step $h^{j+1} = h^j$ if the convergence happens after a "moderate" number of iterations.

**Remarks:**

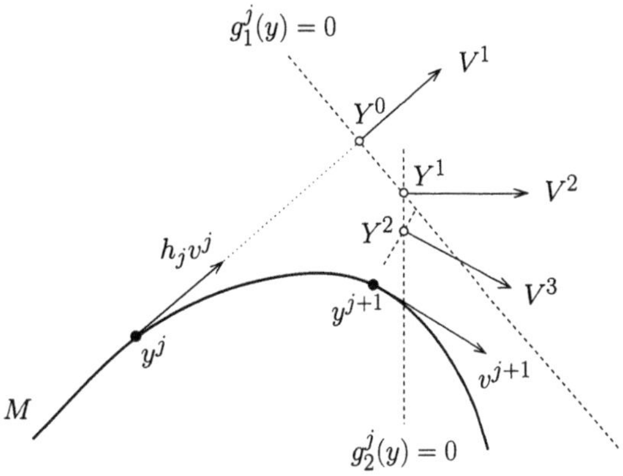


**Fig. 10.7.** Moore-Penrose continuation.

(1) The continuation methods described above are obviously applicable to compute *fixed-point curves* of a discrete-time system

$$x \mapsto f(x, \alpha), \quad x \in \mathbb{R}^n, \ \alpha \in \mathbb{R}^1.$$

Indeed, the corresponding equation

$$f(x, \alpha) - x = 0$$

has the form (10.28).

(2) The continuation of limit cycles in one-parameter systems also reduces to a continuation problem (10.28) by performing a discretization of the periodic boundary-value problem for the cycle continuation:

$$\begin{cases} \dot{u}(\tau) - Tf(u(\tau), \alpha) = 0, \\ u(1) - u(0) = 0, \\ \int_0^1 \langle u(\tau), \dot{u}_0(\tau) \rangle \, d\tau = 0, \end{cases}$$

where the last equation is the integral phase condition with a reference solution $u_0(\tau)$. The resulting finite-dimensional system will have the form

$$\mathcal{F}(\xi, \alpha) = 0, \quad \mathcal{F} : \mathbb{R}^{N+1} \to \mathbb{R}^N,$$

where $\xi \in \mathbb{R}^N$ contains the discretization data corresponding to $u(\cdot)$ *and* the cycle period $T$, while $\alpha \in \mathbb{R}^1$ is a free system parameter. For example, $\xi \in \mathbb{R}^N$ can be composed of the interpolation coefficients $\{u^{j,i}\}$ appearing in the orthogonal collocation and $T$. This continuation problem defines a curve in the direct product of the $\xi$-space and the parameter axis. The reference periodic solution $u_0$ involved in the phase condition is usually taken to be a solution corresponding to the cycle obtained at a previous step along the curve. ◇

### 10.2.2 Detection and location of codim 1 bifurcations

In continuous-time systems, there are two generic codim 1 bifurcations that can be detected along the equilibrium curve: the fold and the Hopf bifurcations.

The main idea of the following is to define some scalar smooth functions that have regular zeros at the bifurcation points. Such functions are called *test* or *bifurcation functions.* A bifurcation point is said to be *detected* between two successive points $y^k$ and $y^{k+1}$ on the curve

$$F(y) = 0, \quad F : \mathbb{R}^{n+1} \to \mathbb{R}^n,$$

if a test function $\psi = \psi(y)$ has the opposite signs at these points

$$\psi(y^k)\psi(y^{k+1}) < 0. \tag{10.35}$$

Then one may try to *locate* a point where $\psi$ vanishes more accurately in the same manner as a regular point, namely, by applying a Newton-like method to the system

$$\begin{cases} F(y) = 0, \\ \psi(y) = 0, \end{cases} \tag{10.36}$$

with the initial point $y^{(0)} = y^k$, for example. Clearly, to apply Newton's method, the test function $\psi$ has to be defined and differentiable in a neighborhood of the equilibrium curve. It might happen that the test function is such that the Jacobian matrix of (10.36) is *singular* at the point on the curve where $\psi = 0$. In this case, standard Newton's method is inappropriate; instead we can implement the one-dimensional secant method to locate $\psi = 0$ along the curve.

Let us describe two simple test functions to detect and locate fold and Hopf bifurcation points in continuous-time systems. Consider the equilibrium curve

$$f(x, \alpha) = 0, \quad x \in \mathbb{R}^n, \ \alpha \in \mathbb{R}^1, \tag{10.37}$$

corresponding to system (10.26).

#### Fold detection and location

Clearly, the function

$$\psi_t(x, \alpha) = \lambda_1(x, \alpha)\lambda_2(x, \alpha) \cdots \lambda_n(x, \alpha), \tag{10.38}$$

where $\lambda_j(x, \alpha)$ are the eigenvalues of the Jacobian matrix $A(x, \alpha) = f_x(x, \alpha)$, is a test function for the fold bifurcation. It is smooth and has a regular zero at a generic fold bifurcation point (check!). Moreover, the Jacobian matrix of (10.36) with $y = (x, \alpha)$ is nonsingular at such a point, so Newton's method can be applied to locate it. The Newton iterations, if converged, provide the coordinates of the equilibrium with zero eigenvalue and the critical parameter value at which it exists.

The test function (10.38) formally requires us to know all the eigenvalues of $A$ for its computation. These eigenvalues can be found numerically at each point on the curve by one of the standard routines. However, there is an obvious way to avoid eigenvalue computation. Notice that the product in the right-hand side of (10.38) is merely the determinant of $A(x, \alpha)$, which can be efficiently computed without dealing with its eigenvalues (by Gaussian elimination, for example). Thus, a popular test function for the fold is

$$\psi_t(x, \alpha) = \det \left( \frac{\partial f(x, \alpha)}{\partial x} \right). \tag{10.39}$$

Another way to detect a fold point is to monitor for *extremum points* with respect to the parameter on the equilibrium curve. Clearly, at a generic fold point the $\alpha$-component of the tangent vector $v$ changes sign.

### Hopf detection and location

Consider the function

$$\psi_H(x, \alpha) = \prod_{i>j} (\lambda_i(x, \alpha) + \lambda_j(x, \alpha)), \tag{10.40}$$

where the previous notation is used. This function vanishes at a Hopf bifurcation point, where there is a pair of multipliers $\lambda_{1,2} = \pm i\omega_0$. Clearly, $\psi_H = 0$ also if there is a pair of *real* eigenvalues

$$\lambda_1 = \kappa, \ \lambda_2 = -\kappa.$$

Thus we have to be careful to ignore such points when looking for Hopf bifurcations. The function $\psi_H$ is real and smooth and has a regular zero at a generic Hopf bifurcation point. Moreover, the Jacobian matrix of (10.36) is nonsingular at such a point, and Newton's method can be applied.

As in the previous case, there is a way to avoid explicit computation of all the eigenvalues of $A$, although it is slightly more difficult. As noted in Chapter 8, the *bialternate product* can be used to compute $\psi_H$.

Let $A$ and $B$ be $n \times n$ matrices with elements $\{a_{ij}\}$ and $\{b_{ij}\}$, respectively, $1 \le i, j \le n$. Set $m = \frac{1}{2}n(n-1)$.

**Definition 10.1** *The* bialternate product *of $A$ and $B$ is an $m \times m$ matrix, denoted $A \odot B$, whose rows are labeled by the multi-index $(p, q)$ $(p = 2, 3, \ldots, n;$ $q = 1, 2, \ldots, p-1)$, and whose columns are labeled by the multiindex $(r, s)$ $(r = 2, 3, \ldots, n; s = 1, 2, \ldots, r-1)$, and the elements are given by*

$$(A \odot B)_{(p,q),(r,s)} = \frac{1}{2} \left\{ \begin{vmatrix} a_{pr} & a_{ps} \\ b_{qr} & b_{qs} \end{vmatrix} + \begin{vmatrix} b_{pr} & b_{ps} \\ a_{qr} & a_{qs} \end{vmatrix} \right\}.$$

| | | | | | |
|---|---|---|---|---|---|
| (2,1),(2,1) | (2,1),(3,1) | (2,1),(3,2) | (2,1),(4,1) | (2,1),(4,2) | (2,1),(4,3) |
| (3,1),(2,1) | (3,1),(3,1) | (3,1),(3,2) | (3,1),(4,1) | (3,1),(4,2) | (3,1),(4,3) |
| (3,2),(2,1) | (3,2),(3,1) | (3,2),(3,2) | (3,2),(4,1) | (3,2),(4,2) | (3,2),(4,3) |
| (4,1),(2,1) | (4,1),(3,1) | (4,1),(3,2) | (4,1),(4,1) | (4,1),(4,2) | (4,1),(4,3) |
| (4,2),(2,1) | (4,2),(3,1) | (4,2),(3,2) | (4,2),(4,1) | (4,2),(4,2) | (4,2),(4,3) |
| (4,3),(2,1) | (4,3),(3,1) | (4,3),(3,2) | (4,3),(4,1) | (4,3),(4,2) | (4,3),(4,3) |

**Fig. 10.8.** Labeling of the elements of the bialternate product matrix for $n = 4$.

**Fig. 10.8** illustrates the labeling of the elements of the $6 \times 6$ matrix $A \otimes B$, when $A$ and $B$ are two $4 \times 4$ matrices. A motivation for this definition is given in Appendix B. The following classical theorem, which is proven in Appendix B, explains the importance of the bialternate product.

**Theorem 10.3 (Stéphanos [1900])** *Let $A$ be an $n \times n$ matrix with eigenvalues $\lambda_1, \lambda_2, \dots, \lambda_n$. Then*

(i) *$A \odot A$ has eigenvalues $\lambda_i \lambda_j$;*
(ii) *$2A \odot I_n$ has eigenvalues $\lambda_i + \lambda_j$;*

*where $i = 2, 3, \dots, n$; $j = 1, 2, \dots, i-1$, and $I_n$ is the $n \times n$ unit matrix.* □

Therefore, the test function (10.40) can be expressed as

$$\psi_H(x, \alpha) = \det\left(2\frac{\partial f(x, \alpha)}{\partial x} \odot I_n\right). \tag{10.41}$$

The definition of the bialternate product leads to the following formula for $2A \odot I_n$:

$$(2A \odot I_n)_{(p,q),(r,s)} = \begin{vmatrix} a_{pr} & a_{ps} \\ \delta_{qr} & \delta_{qs} \end{vmatrix} + \begin{vmatrix} \delta_{pr} & \delta_{ps} \\ a_{qr} & a_{qs} \end{vmatrix},$$

where $\delta_{ij}$ is the Kronecker delta.[5] Computing the determinants, we get

$$(2A \odot I_n)_{(p,q),(r,s)} = \begin{cases} -a_{ps} & \text{if } r = q, \\ a_{pr} & \text{if } r \neq p \text{ and } s = q, \\ a_{pp} + a_{qq} & \text{if } r = p \text{ and } s = q, \\ a_{qs} & \text{if } r = p \text{ and } s \neq q, \\ -a_{qr} & \text{if } s = p, \\ 0 & \text{otherwise.} \end{cases}$$

Thus, the computation of the elements of $2A \odot I_n$ can be efficiently programmed given the matrix $A$.

**Example 10.1** If $n = 3$, the matrix $B = 2A \odot I_3$ is the $3 \times 3$ matrix

[5] $\delta_{ii} = 1$, $\delta_{ij} = 0$ for $i \neq j$.

$$B = \begin{pmatrix} a_{22}+a_{11} & a_{23} & -a_{13} \\ a_{32} & a_{33}+a_{11} & a_{12} \\ -a_{31} & a_{21} & a_{33}+a_{22} \end{pmatrix}. \ \diamond$$

Let us construct test functions to locate codim 1 bifurcations in discrete-time systems. Suppose that we have a map

$$x \mapsto f(x,\alpha), \ \ x \in \mathbb{R}^n, \ \alpha \in \mathbb{R}^1,$$

and that we continue its fixed-point curve

$$f(x,\alpha) - x = 0.$$

Let $\mu_1, \mu_2, \ldots, \mu_n$ be multipliers of the Jacobian matrix $A = f_x$ evaluated at $(x, \alpha)$. The following test functions will obviously locate fold, flip, and Neimark-Sacker bifurcations, respectively:

$$\varphi_t = \prod_{i=1}^{n} (\mu_i - 1), \tag{10.42}$$

$$\varphi_f = \prod_{i=1}^{n} (\mu_i + 1), \tag{10.43}$$

$$\varphi_{NS} = \prod_{i>j} (\mu_i \mu_j - 1). \tag{10.44}$$

To detect a true Neimark-Sacker bifurcation, we must check that $\varphi_{NS} = 0$ is caused by the presence of *nonreal* multipliers with the unit product: $\mu_i \mu_j = 1$. Similarly to the continuous-time cases, we can express the test functions (10.42)–(10.44) in terms of the Jacobian matrix itself. Namely,

$$\varphi_t = \det(A - I_n), \tag{10.45}$$

$$\varphi_f = \det(A + I_n), \tag{10.46}$$

and

$$\varphi_{NS} = \det(A \odot A - I_m), \tag{10.47}$$

where $I_k$ is the $k \times k$ unit matrix, $m = \frac{1}{2}n(n-1)$. The last formula follows from statement (i) of Theorem 10.3. Using the definition of the bialternate product, we get

$$(A \odot A)_{(p,q),(r,s)} = \begin{vmatrix} a_{pr} & a_{ps} \\ a_{qr} & a_{qs} \end{vmatrix}.$$

### 10.2.3 Analysis of codim 1 bifurcations

To apply the theory developed in Chapters 3, 4, and 5 to detected bifurcation points, we have to verify that the appropriate genericity conditions are satisfied. The verification of the transversality with respect to the parameter is rather straightforward and will not be considered. Instead, we focus on the computation of relevant normal form coefficients at the critical parameter value, i.e., checking the nondegeneracy conditions.

Assume that a bifurcation point $(x^0, \alpha^0)$ has been detected on the equilibrium curve.

### Fold bifurcation

As we showed in Chapter 5 (see Section 5.4.1), the restriction of system (10.26) to its one-dimensional center manifold at a fold bifurcation point can be written as

$$\dot{u} = bu^2 + O(u^3),$$

where

$$b = \frac{1}{2}\langle p, B(q,q)\rangle$$

(see formula (5.23)). Here $q, p \in \mathbb{R}^n$ are eigenvectors satisfying

$$Aq = 0, \quad A^T p = 0,$$

which are normalized with respect to each other according to

$$\langle p, q\rangle = 1.$$

To specify the vectors completely, assume that

$$\langle q, q\rangle = 1.$$

The bilinear function $B : \mathbb{R}^n \times \mathbb{R}^n \to \mathbb{R}^n$ is defined by

$$B_i(x,y) = \sum_{j,k=1}^{n} \left.\frac{\partial^2 f_i(\xi, \alpha^0)}{\partial \xi_j \partial \xi_k}\right|_{\xi = x^0} x_j y_k, \; i = 1, 2, \ldots, n. \tag{10.48}$$

We can find $b$ by computing only one second-order derivative of a scalar function. Indeed, one can easily check that

$$B(q,q) = \left.\frac{d^2}{d\tau^2} f(x^0 + \tau q, \alpha^0)\right|_{\tau=0},$$

and, therefore,

$$\langle p, B(q,q)\rangle = \left.\frac{d^2}{d\tau^2}\langle p, f(x^0 + \tau q, \alpha^0)\rangle\right|_{\tau=0}. \tag{10.49}$$

The second derivative in (10.49) can be approximated numerically using finite differences, for example:

$$\langle p, B(q,q)\rangle = \frac{1}{h^2}\left[\langle p, f(x^0+hq,\alpha^0)\rangle + \langle p, f(x^0-hq,\alpha^0)\rangle\right] + O(h^2),$$

where $h$ is an *increment* and the equilibrium condition $f(x^0,\alpha^0)=0$ is taken into account. Notice that all the objects involved in (10.49) are expressed in the original coordinates. Thus, computation of $b$ is reduced to finding the eigenvectors $q$ and $p$, their normalization, and applying formula (10.49) or its finite-difference analogue. If $b \neq 0$, a generic fold bifurcation happens, given the transversality with respect to the parameter.

## Hopf bifurcation

According to formula (5.39) from Section 5.4.1 of Chapter 5, the first Lyapunov coefficient determining whether the Hopf bifurcation is sub- or supercritical is given by

$$\begin{aligned} l_1(0) = \frac{1}{2\omega_0}\mathrm{Re}\,&\left[\langle p, C(q,q,\bar{q})\rangle - 2\langle p, B(q, A^{-1}B(q,\bar{q}))\rangle\right. \\ &+ \left.\langle p, B(\bar{q}, (2i\omega_0 I_n - A)^{-1}B(q,q))\rangle\right], \end{aligned} \tag{10.50}$$

where $q$ and $p$ satisfy

$$Aq = i\omega_0 q, \;\; A^T p = -i\omega_0 p,$$

and are normalized by setting

$$\langle p, q\rangle = 1.$$

Assume also that

$$\langle q, q\rangle = 1, \;\; \langle \mathrm{Re}\, q, \mathrm{Im}\, q\rangle = 0.$$

The bilinear function $B(x,y)$ is given by (10.48), while the function $C : \mathbb{R}^n \times \mathbb{R}^n \times \mathbb{R}^n \to \mathbb{R}^n$ is defined by

$$C_i(x,y,z) = \sum_{j,k,l=1}^{n} \left.\frac{\partial^3 f_i(\xi,\alpha^0)}{\partial\xi_j\partial\xi_k\partial\xi_l}\right|_{\xi=x^0} x_j y_k z_l, \;\; i = 1,2,\ldots,n. \tag{10.51}$$

If the programming language you use supports complex arithmetic, formula (10.50) can be implemented directly with the partial derivatives in (10.48) and (10.51) evaluated numerically. However, there is a way to make computations real and to avoid computing all second- and third-order partial derivatives of $f$ at $(x^0,\alpha^0)$ involved in (10.48) and (10.51).

First, note that we can easily evaluate the multilinear functions $B(x,y)$ and $C(x,y,z)$ on any set of *coinciding* real vector arguments by computing certain *directional derivatives*. Indeed, the vector $B(v,v)$ for $v \in \mathbb{R}^n$ can be computed by the formula

$$B(v,v) = \left. \frac{d^2}{d\tau^2} f(x^0 + \tau v, \alpha^0) \right|_{\tau=0}, \tag{10.52}$$

which we have already used. The vector $C(v,v,v)$ for $v \in \mathbb{R}^n$ can be calculated by a similar formula:

$$C(v,v,v) = \left. \frac{d^3}{d\tau^3} f(x^0 + \tau v, \alpha^0) \right|_{\tau=0}. \tag{10.53}$$

We can approximate these derivatives with respect to the scalar variable $\tau$ via finite differences, for example,

$$B(v,v) = \frac{1}{h^2}\left[f(x^0 + hv, \alpha^0) + f(x^0 - hv, \alpha^0)\right] + O(h^2)$$

for (10.52), and

$$\begin{aligned} C(v,v,v) = \frac{1}{8h^3} & \left[f(x^0 + 3hv, \alpha^0) - 3f(x^0 + hv, \alpha^0)\right. \\ + & \left. 3f(x^0 - hv, \alpha^0) - f(x^0 - 3hv, \alpha^0)\right] + O(h^2) \end{aligned}$$

for (10.53), where $h \ll 1$ is a small increment. Therefore, let us rewrite (10.50) in terms of expressions computable via (10.52) and (10.53).

Denote the real and imaginary parts of the eigenvector $q$ by $q_R$ and $q_I$, respectively,

$$q = q_R + iq_I, \quad q \in \mathbb{C}^n, \ q_R, q_I \in \mathbb{R}^n.$$

Then we have

$$\begin{aligned} B(q,q) &= B(q_R,q_R) - B(q_I,q_I) + 2iB(q_R,q_I), \\ B(q,\bar{q}) &= B(q_R,q_R) + B(q_I,q_I), \\ C(q,q,\bar{q}) &= C(q_R,q_R,q_R) + C(q_R,q_I,q_I) + iC(q_R,q_R,q_I) + iC(q_I,q_I,q_I). \end{aligned}$$

Notice that the vectors $B(q_R,q_R), B(q_I,q_I), C(q_R,q_R,q_R)$, and $C(q_I,q_I,q_I)$ can be directly computed by (10.52) and (10.53), while $B(q_R,q_I), C(q_R,q_I,q_I)$, and $C(q_R,q_R,q_I)$ require more treatment. Essentially, it is enough to be able to compute the multilinear functions $B(v,w)$ and $C(v,v,w)$ for real vectors $v, w \in \mathbb{R}^n$ using (10.52) and (10.53).

The identities

$$\begin{aligned} B(v+w, v+w) &= B(v,v) + 2B(v,w) + B(w,w), \\ B(v-w, v-w) &= B(v,v) - 2B(v,w) + B(w,w) \end{aligned}$$

allow us to express the vector $B(v,w)$ as

$$B(v,w) = \frac{1}{4}\left[B(v+w, v+w) - B(v-w, v-w)\right]$$

(*"polarization identity"*). The right-hand side can be computed using (10.52).

Similarly, the identities

$$C(v+w,v+w,v+w) = C(v,v,v) + 3C(v,v,w) + 3C(v,w,w) + C(w,w,w)$$

and

$$C(v-w,v-w,v-w) = C(v,v,v) - 3C(v,v,w) + 3C(v,w,w) - C(w,w,w)$$

lead to the following expression for $C(v,v,w)$:

$$C(v,v,w) = \frac{1}{6}(C(v+w,v+w,v+w) - C(v-w,v-w,v-w)) - \frac{1}{3}C(w,w,w),$$

with the right-hand side computable by (10.53).

The linear system

$$Ar = B(q,\bar{q})$$

involves only real quantities and can be rewritten as

$$Ar = B(q_R,q_R) + B(q_I,q_I).$$

Its solution $r$ is a real vector. The scalar product $\langle p, B(q,r)\rangle$ from (10.50) can be transformed as

$$\langle p, B(q,r)\rangle = \langle p_R, B(q_R,r)\rangle + i\langle p_R, B(q_I,r)\rangle - i\langle p_I, B(q_R,r)\rangle + \langle p_I, B(q_I,r)\rangle,$$

so that

$$\mathrm{Re}\langle p, B(q,r)\rangle = \langle p_R, B(q_R,r)\rangle + \langle p_I, B(q_I,r)\rangle.$$

All the bilinear functions inside the scalar products to the right are of the form $B(v,w)$ and are thus reducible via the polarization identity to quantities computable by (10.52).

The linear complex system

$$(2i\omega_0 I_n - A)s = B(q,q)$$

is equivalent to the real system of double the dimension,

$$\begin{cases} -As_R - 2\omega_0 s_I = B(q_R,q_R) - B(q_I,q_I), \\ 2\omega_0 s_R - As_I = 2B(q_R,q_I), \end{cases}$$

where $s_R$ and $s_I$ are real and imaginary parts of $s$. Then the scalar product $\langle p, B(\bar{q},s)\rangle$ from (10.50) can be written as

$$\begin{aligned} \langle p, B(\bar{q},s)\rangle = & \langle p_R, B(q_R,s_R)\rangle + i\langle p_R, B(q_R,s_I)\rangle \\ & - i\langle p_R, B(q_I,s_R)\rangle + \langle p_R, B(q_I,s_I)\rangle \\ & - i\langle p_I, B(q_R,s_R)\rangle + \langle p_I, B(q_R,s_I)\rangle \\ & - \langle p_I, B(q_I,s_R)\rangle - i\langle p_I, B(q_I,s_I)\rangle, \end{aligned}$$

so that

$$\mathrm{Re}\langle p, B(\bar{q}, s)\rangle = \langle p_R, B(q_R, s_R)\rangle + \langle p_R, B(q_I, s_I)\rangle + \langle p_I, B(q_R, s_I)\rangle - \langle p_I, B(q_I, s_R)\rangle,$$

where all the bilinear functions are of the form $B(v, w)$ and are thus reducible to (10.52).

The scalar product $\langle p, C(q, q, \bar{q})\rangle$ from (10.48) can be expressed as

$$\begin{aligned}\langle p, C(q, q, \bar{q})\rangle = {} & \langle p_R, C(q_R, q_R, q_R)\rangle + \langle p_R, C(q_R, q_I, q_I)\rangle \\ & + i\langle p_R, C(q_R, q_R, q_I)\rangle + i\langle p_R, C(q_I, q_I, q_I)\rangle \\ & - i\langle p_I, C(q_R, q_R, q_R)\rangle - i\langle p_I, C(q_R, q_I, q_I)\rangle \\ & + \langle p_I, C(q_R, q_R, q_I)\rangle + \langle p_I, C(q_I, q_I, q_I)\rangle,\end{aligned}$$

which gives

$$\begin{aligned}\mathrm{Re}\langle p, C(q, q, \bar{q})\rangle = {} & \langle p_R, C(q_R, q_R, q_R)\rangle + \langle p_R, C(q_R, q_I, q_I)\rangle \\ & + \langle p_I, C(q_R, q_R, q_I)\rangle + \langle p_I, C(q_I, q_I, q_I)\rangle.\end{aligned}$$

This expression can be transformed into the formula

$$\begin{aligned}\mathrm{Re}\langle p, C(q, q, \bar{q})\rangle = {} & \frac{2}{3}\langle p_R, C(q_R, q_R, q_R)\rangle + \frac{2}{3}\langle p_I, C(q_I, q_I, q_I)\rangle \\ & + \frac{1}{6}\langle p_R + p_I, C(q_R + q_I, q_R + q_I, q_R + q_I)\rangle \\ & + \frac{1}{6}\langle p_R - p_I, C(q_R - q_I, q_R - q_I, q_R - q_I)\rangle\end{aligned}$$

involving only directional derivatives of the form (10.53).

Let us summarize the steps needed to compute $l_1(0)$.

*Step 0.* Evaluate the Jacobian matrix $A = f_x(x^0, \alpha^0)$ of (10.26) at the equilibrium $x^0$ exhibiting the Hopf bifurcation at the critical parameter value $\alpha^0$.
*Step 1.* Find four vectors $q_R, q_I, p_R, p_I \in \mathbb{R}^n$ satisfying the systems

$$\begin{cases} Aq_R + \omega_0 q_I = 0, \\ -\omega_0 q_R + Aq_I = 0, \end{cases}$$

and

$$\begin{cases} A^T p_R - \omega_0 p_I = 0, \\ \omega_0 p_R + A^T p_I = 0, \end{cases}$$

and normalize them according to

$$\langle q_R, q_R\rangle + \langle q_I, q_I\rangle = 1, \quad \langle q_R, q_I\rangle = 0,$$

$$\langle p_R, q_R\rangle + \langle p_I, q_I\rangle = 1, \quad \langle p_R, q_I\rangle - \langle p_I, q_R\rangle = 0.$$

*Step 2.* Compute the following vectors by the directional differentiation:

$$a = \left. \frac{d^2}{d\tau^2} f(x^0 + \tau q_R, \alpha^0) \right|_{\tau=0},$$
$$b = \left. \frac{d^2}{d\tau^2} f(x^0 + \tau q_I, \alpha^0) \right|_{\tau=0}$$

and

$$c = \left. \frac{1}{4} \frac{d^2}{d\tau^2} \left[ f(x^0 + \tau(q_R + q_I), \alpha^0) - f(x^0 + \tau(q_R - q_I), \alpha^0) \right] \right|_{\tau=0}.$$

*Step 3.* Solve the linear systems for $r$ and $(s_R, s_I)$:

$$Ar = a + b,$$

and

$$\begin{cases} -As_R - 2\omega_0 s_I = a - b, \\ 2\omega_0 s_R - As_I = 2c. \end{cases}$$

*Step 4.* Compute the following numbers:

$$\sigma_1 = \left. \frac{1}{4} \frac{d^2}{d\tau^2} \langle p_R, f(x^0 + \tau(q_R + r), \alpha^0) - f(x^0 + \tau(q_R - r), \alpha^0) \rangle \right|_{\tau=0},$$

$$\sigma_2 = \left. \frac{1}{4} \frac{d^2}{d\tau^2} \langle p_I, f(x^0 + \tau(q_I + r), \alpha^0) - f(x^0 + \tau(q_I - r), \alpha^0) \rangle \right|_{\tau=0},$$

and evaluate their sum

$$\Sigma_0 = \sigma_1 + \sigma_2.$$

*Step 5.* Compute the numbers

$$\delta_1 = \left. \frac{1}{4} \frac{d^2}{d\tau^2} \langle p_R, f(x^0 + \tau(q_R + s_R), \alpha^0) - f(x^0 + \tau(q_R - s_R), \alpha^0) \rangle \right|_{\tau=0},$$

$$\delta_2 = \left. \frac{1}{4} \frac{d^2}{d\tau^2} \langle p_R, f(x^0 + \tau(q_I + s_I), \alpha^0) - f(x^0 + \tau(q_I - s_I), \alpha^0) \rangle \right|_{\tau=0},$$

$$\delta_3 = \left. \frac{1}{4} \frac{d^2}{d\tau^2} \langle p_I, f(x^0 + \tau(q_R + s_I), \alpha^0) - f(x^0 + \tau(q_R - s_I), \alpha^0) \rangle \right|_{\tau=0},$$

$$\delta_4 = \left. \frac{1}{4} \frac{d^2}{d\tau^2} \langle p_I, f(x^0 + \tau(q_I + s_R), \alpha^0) - f(x^0 + \tau(q_I - s_R), \alpha^0) \rangle \right|_{\tau=0},$$

and evaluate

$$\Delta_0 = \delta_1 + \delta_2 + \delta_3 - \delta_4.$$

*Step 6.* Compute the numbers

$$\gamma_1 = \frac{d^3}{d\tau^3}\langle p_R, f(x^0 + \tau q_R, \alpha^0)\rangle\bigg|_{\tau=0},$$

$$\gamma_2 = \frac{d^3}{d\tau^3}\langle p_I, f(x^0 + \tau q_I, \alpha^0)\rangle\bigg|_{\tau=0},$$

$$\gamma_3 = \frac{d^3}{d\tau^3}\langle p_R + p_I, f(x^0 + \tau(q_R + q_I), \alpha^0)\rangle\bigg|_{\tau=0},$$

$$\gamma_4 = \frac{d^3}{d\tau^3}\langle p_R - p_I, f(x^0 + \tau(q_R - q_I), \alpha^0)\rangle\bigg|_{\tau=0},$$

and take

$$\Gamma_0 = \frac{2}{3}(\gamma_1 + \gamma_2) + \frac{1}{6}(\gamma_3 + \gamma_4).$$

*Step 7.* Finally, compute

$$l_1(0) = \frac{1}{2\omega_0}\,(\Gamma_0 - 2\Sigma_0 + \Delta_0).$$

If $l_1(0) \neq 0$ and the eigenvalues cross the imaginary axis with nonzero velocity, a unique limit cycle appears under variation of $\alpha$. To complete the Hopf bifurcation analysis, derive an algorithm allowing us to start the *continuation* of the bifurcating limit cycle away from the Hopf point by one of the methods described above. Therefore, we have to obtain a periodic solution from which we can initialize the cycle continuation. Recall that the cycles near a generic Hopf point form a paraboloid-like surface in the phase-parameter space that is *tangent* to the plane

$$x = zq + \bar{z}\bar{q},$$

where $z \in \mathbb{C}^1$, and $q$ is the properly normalized critical eigenvector. The linear part of the system restricted to the center manifold is simply

$$\dot{z} = i\omega_0 z,$$

having the solution $z(t) = z_0 e^{i\omega_0 t}$. Therefore, the linear approximation of the bifurcating limit cycle with the "amplitude" $z_0 = \varepsilon$ is given by

$$x_\varepsilon(t) = (e^{i\omega_0 t}q + e^{-i\omega_0 t}\bar{q})\varepsilon = 2\varepsilon\ \mathrm{Re}[e^{i\omega_0 t}q],$$

where $\varepsilon \in \mathbb{R}^1$ is a user-defined small number. The resulting starting solution can be presented in real form as

$$x_\varepsilon(t) = 2\varepsilon(q_R \cos\omega_0 t - q_I \sin\omega_0 t), \tag{10.54}$$

where $q_R, q_I \in \mathbb{R}^n$ are the real and imaginary parts of the complex eigenvector $q$. One can derive a quadratic approximation to the cycle near the Hopf bifurcation; however, formula (10.54) is accurate enough to start the cycle continuation in most cases.

### 10.2.4 Branching points

In general, the equilibrium curve (10.27) can have branching points.

**Definition 10.2** *A point* $y^*$ *is called a* branching (branch) point *for the continuation problem*

$$F(y) = 0, \quad F : \mathbb{R}^{n+1} \to \mathbb{R}^n, \tag{10.55}$$

*if* $F(y^*) = 0$ *and there are at least two different smooth curves satisfying* (10.55) *and passing through* $y^*$.

Two examples of branching are presented in **Fig. 10.9** for $n = 1$. We

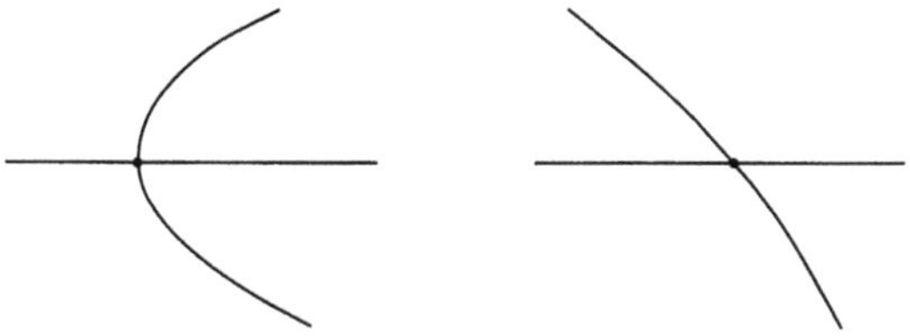

**Fig. 10.9.** Branching points.

can consider the curves in the figure as equilibrium curves of certain scalar systems depending on one parameter; $y = (x, \alpha)$.

**Remarks:**

(1) A branching point may appear if the curve makes a loop and intersects itself at a point $y^*$. That is why branching points are often called *self-crossing* points.

(2) Sometimes, branching points are merely called *bifurcation* points. The latter term is appropriate if we study equilibria only but is misleading in general. Indeed, neither nondegenerate fold nor Hopf bifurcation points are branching points of an equilibrium curve. ◊

The reader should understand that the appearance of branching points is a *nongeneric phenomenon*: A generic curve (10.55) (with rank $F_y = n$ along the curve) has no branching points by the Implicit Function Theorem. Moreover, if the continuation problem (10.55) has a branching point, $C^1$-close problems will, generically, have no branching points at all (see **Fig. 10.10**).

However, branching points appear easily if a certain *symmetry* is present. Indeed, as we have seen in Chapter 7, the *pitchfork* bifurcation, which is a branching point, happens generically in $\mathbb{Z}_2$-invariant systems. Another important example is provided by the *flip* bifurcation in discrete-time systems. Suppose that the fixed point $x^{(1)}(\alpha)$ of the map

$$x \mapsto f(x, \alpha) = f_\alpha(x), \quad x \in \mathbb{R}^n, \ \alpha \in \mathbb{R}^1,$$

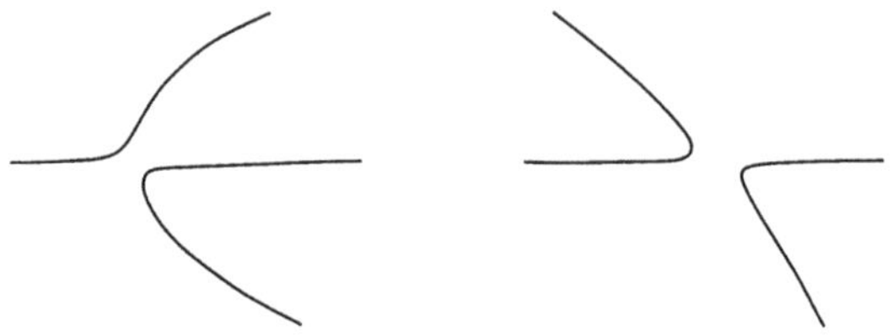

**Fig. 10.10.** Branching points disappear under generic perturbations.

exhibits a generic flip bifurcation at $\alpha = 0$. Denote by $x_{1,2}^{(2)}(\alpha)$ the points forming the period-two cycle bifurcating from $x^{(1)}(\alpha)$ at $\alpha = 0$. Since $x^{(1)}(\alpha)$ is a fixed point of the *second iterate* $f_\alpha^2$ of the map $f_\alpha$, the flip point $(x^{(1)}(0), 0)$ is a branching point for the continuation problem for cycles of period two:

$$f_\alpha^2(x) - x = 0, \quad x \in \mathbb{R}^n, \ \alpha \in \mathbb{R}^1.$$

Therefore, we need to be able to detect and locate branching points, while continuing a curve defined by (10.55), and to start the continuation of other branches emanating from each located point. Let us treat these problems. The following discussion involves quadratic terms of $F$. Let $y^* = 0$ be a branching point for (10.55). Write the Taylor expansion of $F(y)$ at $y = 0$ as

$$F(y) = J(0)y + \frac{1}{2}B(y, y) + O(\|y\|^3),$$

where $J(y) = F_y(y)$ is the $n \times (n+1)$ Jacobian matrix of (10.55), and $B : \mathbb{R}^{n+1} \times \mathbb{R}^{n+1} \to \mathbb{R}^n$ is the bilinear part of $F(y)$ at the branching point:

$$B_i(x, y) = \sum_{j,k=1}^{n+1} \left. \frac{\partial^2 F_i(\xi)}{\partial \xi_j \partial \xi_k} \right|_{\xi = x^0} x_j y_k, \ i = 1, 2, \ldots, n. \tag{10.56}$$

Let $y(s)$ be a smooth branch passing through the point $y = 0$ and parametrized by the arclength $s$ such that $y(0) = 0$. Denote by $v$ the vector tangent to this branch at $y = 0 : v = \dot{y}(0) \in \mathbb{R}^{n+1}$. As we have already pointed out, the vector $v$ satisfies the equation

$$J(0)v = 0 \tag{10.57}$$

(which results from $F(y(s)) = 0$ by differentiating with respect to $s$ and evaluating the result at $s = 0$). Denote by $\mathcal{K}$ the null-space of $J(0)$ composed of all vectors satisfying (10.57):

$$\mathcal{K} = \{v \in \mathbb{R}^{n+1} : J(0)v = 0\}.$$

All vectors tangent to branches passing through the point $y = 0$ belong to $\mathcal{K}$. If $y = 0$ were a regular point where rank $J(0) = n$, then $\dim \mathcal{K} = 1$. If

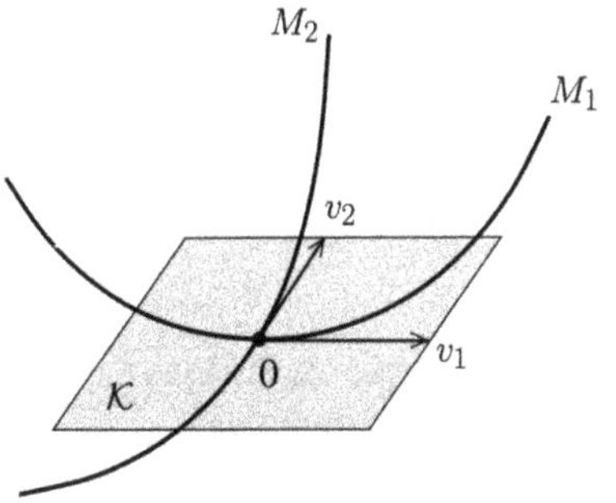


**Fig. 10.11.** Two intersecting curves passing through a branching point.

two branches of (10.55) intersect transversally at $y = 0$, $\dim \mathcal{K} = 2$, because their respective tangent vectors at the branching point, $v_1$ and $v_2$, both satisfy (10.57) (see **Fig. 10.11**).

Consider only the case when $\mathcal{K}$ is spanned by two linear independent vectors $q_1$ and $q_2$, and derive another equation which the tangent vectors satisfy. Differentiating $F(y(s)) = 0$ *twice* with respect to $s$ at $s = 0$, we obtain

$$J(0)\ddot{y}(0) + B(\dot{y}(0), \dot{y}(0)) = 0,$$

or, equivalently,

$$J(0)\ddot{y}(0) + B(v, v) = 0. \tag{10.58}$$

Consider now the transposed $(n+1) \times n$ matrix $J^T(0)$. Since $\dim \mathcal{K} = 2$, there is a unique vector (up to scalar multiple) $\varphi \in \mathbb{R}^n$ satisfying

$$J^T(0)\varphi = 0. \tag{10.59}$$

Computing the scalar product of the left-hand side of (10.58) with $\varphi$ in $\mathbb{R}^n$ and taking into account that

$$\langle \varphi, J(0)\ddot{y}(0) \rangle = \langle J^T(0)\varphi, \ddot{y}(0) \rangle = 0,$$

we obtain

$$\langle \varphi, B(v, v) \rangle = 0. \tag{10.60}$$

The left-hand side is a quadratic form defined on vectors $v \in \mathcal{K} \subset \mathbb{R}^n$ given explicitly by

$$\langle \varphi, B(v, v) \rangle = \sum_{i=1}^{n} \sum_{j,k=1}^{n+1} \varphi_i \left. \frac{\partial^2 F_i(\xi)}{\partial \xi_j \partial \xi_k} \right|_{\xi=0} v_j v_k.$$

Recall that any vector $v \in \mathcal{K}$ can be represented as

$$q = \beta_1 q_1 + \beta_2 q_2, \quad \beta = (\beta_1, \beta_2) \in \mathbb{R}^2.$$

Thus, the quadratic form $\langle \varphi, B(v,v)\rangle$ can be considered as a quadratic form on the $(\beta_1, \beta_2)$-plane:

$$b(\beta) = \langle \varphi, B(\beta_1 q_1 + \beta_2 q_2, \beta_1 q_1 + \beta_2 q_2)\rangle = b_{11}\beta_1^2 + 2b_{12}\beta_1\beta_2 + b_{22}\beta_2^2,$$

with

$$b_{ij} = \langle \varphi, B(q_i, q_j)\rangle, \ i,j = 1,2.$$

**Definition 10.3 (Simple branching points)** *A branching point is called* simple *if the following two conditions hold:*

(i) $\dim \mathcal{K} = 2$;
(ii) $b_{12}^2 - b_{11}b_{22} > 0$.

The second condition implies that the matrix of the quadratic form

$$\begin{pmatrix} b_{11} & b_{12} \\ b_{12} & b_{22} \end{pmatrix}$$

has one positive and one negative eigenvalue.

The following lemma completely characterizes simple branching points.

**Lemma 10.1** *Let $y^* = 0$ be a simple branching point of* (10.55). *Then there exist exactly two smooth curves satisfying* (10.55) *that pass through $y^*$. Moreover, their tangent vectors $v_k$ at $y^*$ are linearly independent and both satisfy* (10.60). □

Equation (10.60) allows us to compute the direction $v_2$ of the second branch, provided that the vector $v_1$ that is tangent to the first solution curve at the branching point of (10.55) is known. Indeed, set $q_1 = v_1$ and let $q_2 \in \mathcal{K}$ be a vector that is orthogonal to $q_1$ : $\langle q_2, q_1\rangle = 0$. Then

$$v_k = \beta_1^{(k)} q_1 + \beta_2^{(k)} q_2, \ \ k = 1,2,$$

for $\beta_1^{(1)} = 1$, $\beta_2^{(1)} = 0$. Therefore, a solution[6] to the *"algebraic branching equation"*

$$b_{11}\beta_1^2 + 2b_{12}\beta_1\beta_2 + b_{22}\beta_2^2 = 0$$

is known. This implies $b_{11} = 0$. Thus, the direction $(\beta_1^{(2)}, \beta_2^{(2)})$ of the second branch satisfies the equation

$$2b_{12}\beta_1^{(2)} + b_{22}\beta_2^{(2)} = 0,$$

where necessarily $b_{12} \neq 0$ due to the simplicity of the branching point. Thus,

$$\beta_1^{(2)} = -\frac{b_{22}}{2b_{12}}\beta_2^{(2)}.$$

The preceding results solve the problem of branch switching at a branching point. However, we should be able to detect such points and locate them accurately. The following lemma provides a test function to locate a branching point while continuing a curve defined by (10.55).

[6] Actually, a line $(\gamma, 0)$ of solutions.

**Lemma 10.2** *Let $y_{(1)}(s)$ correspond to a curve defined by* (10.55). *Consider the* $(n+1)\times(n+1)$ *matrix*

$$D(s) = \begin{pmatrix} J(y_{(1)}(s)) \\ \dot{y}_{(1)}^T(s) \end{pmatrix}, \tag{10.61}$$

*where $J(y)$ is the Jacobian matrix of* (10.55). *Then the scalar function*

$$\psi(s) = \det D(s) \tag{10.62}$$

*has a regular zero at $s = 0$.*

**Remark:**

The matrix $D$ differs slightly from the Jacobian matrix involved in the Newton corrections in the pseudo-arclength continuation method (10.34) for solving (10.55). If the iterations converge to a point on the curve, the function $\psi$ can be easily approximated at this point by the determinant of the Newton matrix and compared with that computed at the previous point. $\diamondsuit$

**Proof:**

The matrix $D(0)$ has a simple zero eigenvalue. Indeed,

$$D(0)q_2 = 0, \tag{10.63}$$

where $q_2$ is defined above as being orthogonal to $q_1 = v_1$ within the plane $\mathcal{K}$ at the branching point. Therefore, zero is an eigenvalue of $D(0)$. Moreover, $D(0)q_1 \neq 0$, so the zero eigenvalue is simple. Thus, there is a simple, parameter-dependent eigenvalue $\lambda(s)$ of $D(s)$,

$$D(s)u(s) = \lambda(s)u(s),$$

for all $|s|$ sufficiently small, such that $\lambda(0) = 0,\ u(0) = q_2$. Differentiating the above equation with respect to $s$ at $s = 0$ gives

$$\begin{pmatrix} B(q_1, q_2) \\ \ddot{y}_{(1)}^T(0) \end{pmatrix} + \begin{pmatrix} J(0) \\ q_1^T \end{pmatrix} \dot{u}(0) = \dot{\lambda}(0)q_2. \tag{10.64}$$

Notice now that a zero eigenvector $p$ of the transposed matrix $D^T(0)$,

$$D^T(0)p = 0,$$

has the form

$$p = \begin{pmatrix} \varphi \\ 0 \end{pmatrix},$$

where $\varphi$ was defined above by $J^T(0)\varphi = 0$ (see (10.59)). Moreover, $\langle p, q_2\rangle \neq 0$ due to the simplicity of $\lambda = 0$. Taking the scalar product of both sides of (10.64) with $p$, we get

$$\dot{\lambda}(0) = \frac{\langle \varphi, B(q_1, q_2) \rangle}{\langle p, q_2 \rangle} = \frac{b_{12}}{\langle p, q_2 \rangle} \neq 0$$

since $b_{12}$ must be nonzero at the simple branching point. Therefore, while $s$ passes the value $s = 0$, the simple real eigenvalue $\lambda(s)$ of $D(s)$ changes sign regularly, as does the determinant of $D(s)$. □

Therefore, $\psi(s)$ given by (10.62) can be monitored along the curve (10.55) to detect branching points. If such a point is detected, we can locate it, for example, by the secant method:

$$s^{j+1} = s^j - \frac{s^j - s^{j-1}}{\psi(s^j) - \psi(s^{j-1})} \psi(s^j), \; j = 1, 2, \ldots . \tag{10.65}$$

To perform branch switching at a located simple branch point, we need to find the vectors $q_1, q_2$, and $\varphi$ and evaluate $b_{22}$ and $b_{12}$ numerically. The vector $q_1$ can be interpolated from the tangent vectors at two points between which the branching point was detected; the vectors $q_2$ and $\varphi$ can be computed by solving the homogeneous linear systems (10.59) and (10.63). Finally, the scalar product of $\varphi$ with the function $B$ evaluated at $q_{1,2}$ can be computed by the directional derivative technique presented in the previous subsection. We leave the details to the reader.

**Remarks:**

(1) There is another effective technique to switch branches, called the *homotopy method.* It is based on the observation that generic perturbations of the continuation problem

$$F(y) = 0$$

destroy the intersection of the branches, making them all smooth (see **Fig. 10.10**). Thus, one can introduce an artificial small perturbation $\varepsilon \in \mathbb{R}^n$ and consider the continuation problem

$$F(y) - \varepsilon = 0.$$

Taking different $\varepsilon$ with $\|\varepsilon\|$ small, we can try to switch branches by computing a branch of the perturbed problem, starting at a point on the original branch of the unperturbed problem. Upon switching, we can turn off the perturbation by setting $\varepsilon$ back to zero, and then use the obtained point as the initial point for continuation of the new branch. This method works well in relatively low dimensions.

(2) Locating a branching point with the secant iterations (10.65) might be difficult since the domain of convergence for the Newton (or Newton-like) corrections used to compute the points on the curve (10.55) shrinks as one approaches the branching point. This difficulty can be avoided by considering *extended systems.* For example, a simple branch point $(x_0, \alpha_0)$ of the equilibrium curve (10.27) corresponds to a regular solution $(x, \alpha, \beta, p) = (x_0, \alpha_0, 0, p_0)$ to the defining system

$$\begin{cases} f(x,\alpha) + \beta p = 0, \\ f_x^T(x,\alpha)p \ \ = 0, \\ \langle p, f_\alpha(x,\alpha)\rangle = 0, \\ \langle p,p\rangle - 1 \ \ = 0, \end{cases} \tag{10.66}$$

where $p \in \mathbb{R}^n$ and $\beta \in \mathbb{R}^1$ are extra unknowns. Therefore, the standard Newton's method can be applied directly to to (10.66) to locate a simple branching point. ◇

## 10.3 Two-parameter bifurcation analysis

Here we start with a smooth system depending on two parameters:

$$\dot{x} = f(x,\alpha), \quad x \in \mathbb{R}^n, \ \alpha \in \mathbb{R}^2. \tag{10.67}$$

The aim of the system analysis now is to construct its two-parameter bifurcation diagram. The diagram generically includes curves corresponding to codim 1 bifurcations of equilibria, limit cycles, and homoclinic orbits. At isolated points on these curves, codim 2 bifurcations occur. As we saw in Chapters 8 and 9, such points are common points for several different codim 1 boundaries. Thus, the problem is to continue codim 1 bifurcation curves; to detect, locate, and analyze codim 2 singularities on them; and then to switch bifurcation branches at these points. We will mainly treat the continuation of codim 1 bifurcations, giving only a few remarks on the location and analysis of codim 2 bifurcations.

### 10.3.1 Continuation of codim 1 bifurcations of equilibria and fixed points

As discussed in Chapter 8, if system (10.67) has, at $\alpha = \alpha^0$, an equilibrium exhibiting a codim 1 bifurcation, then, generically, there is a bifurcation curve $\mathcal{B}$ in the $\alpha$-plane along which the system has an equilibrium demonstrating the relevant bifurcation. The curve $\mathcal{B}$ can be computed as a projection to the $\alpha$-plane of a certain curve $\Gamma$, defined in a space of larger dimension. Thus, we have to specify a continuation problem for $\Gamma$, that is, define functions determining the curve in this space.

#### Minimally augmented systems

This type of continuation problems for codim 1 bifurcations of equilibria has been introduced in Chapter 8 (see Section 8.1.1). In this approach we merely append the relevant test function to the equilibrium equation, thus obtaining a system of $n+1$ equations in an $(n+2)$-dimensional space with coordinates $(x,\alpha)$. More precisely, if $A(x,\alpha)$ is the Jacobian matrix of $f$ in (10.67) evaluated at $(x,\alpha)$, then we get the continuation problem

$$\begin{cases} f(x,\alpha) = 0, \\ \det A(x,\alpha) = 0, \end{cases} \tag{10.68}$$

for the fold bifurcation, and the continuation problem

$$\begin{cases} f(x,\alpha) = 0, \\ \det(2A(x,\alpha) \odot I_n) = 0, \end{cases} \tag{10.69}$$

for the Hopf bifurcation, where $\odot$ stands for the bialternate product. Each system is a system of $n+1$ scalar equations with $n+2$ scalar variables and is called the *minimally augmented system* since the dimension of the resulting continuation problem is enlarged by one with respect to the equilibrium continuation. Clearly, if a bifurcation point was detected while continuing an equilibrium and located as a zero of the corresponding test function $\psi_t$ or $\psi_H$, then we have all the necessary initial data to start the continuation of the bifurcation curve defined by (10.68) or (10.69) immediately.

**Lemma 10.3** *If* $(x, \alpha)$ *is a point corresponding to any generic* codim 1 *or* 2 *equilibrium bifurcation of* (10.67), *except the Hopf-Hopf singularity, then rank* $J = n+1$, *where* $J$ *is the Jacobian matrix of the corresponding minimally augmented system* (10.68) *or* (10.69). *A generic Hopf-Hopf point is a simple branching point for* (10.69). □

This lemma allows us to apply the standard predictor-corrector continuation technique to continue the bifurcation curves $\Gamma$ defined by (10.68) or (10.69). The standard projection of the computed curves onto the $(\alpha_1, \alpha_2)$-plane provides the parametric boundaries corresponding to the relevant bifurcations.

However, the defining system (10.68) (or (10.69)) has the following disadvantage: In general, it is impossible to express explicitly its Jacobian matrix $J$ in terms of the partial derivatives of $f(x, \alpha)$ since the determinant is involved in the test function. Thus, we have to rely on numerical differentiation while continuing the curve $\Gamma$, even if the derivatives of $f$ with respect to $(x, \alpha)$ are known analytically. To overcome this difficulty, we substitute the test function in (10.68) (or (10.69)) by a function $g(x, \alpha)$ that vanishes together with the corresponding test function but whose derivatives can be expressed analytically.

In the fold case, instead of (10.68), we introduce a modified minimally augmented system

$$\begin{cases} f(x,\alpha) = 0, \\ g(x,\alpha) = 0, \end{cases} \tag{10.70}$$

where $g = g(x, \alpha)$ is computed as the last component of the solution vector to the $(n+1)$-dimensional *bordered system*:

$$\begin{pmatrix} A(x,\alpha) & p_0 \\ q_0^T & 0 \end{pmatrix} \begin{pmatrix} w \\ g \end{pmatrix} = \begin{pmatrix} 0 \\ 1 \end{pmatrix}, \tag{10.71}$$

where $q_0, p_0 \in \mathbb{R}^n$ are some vectors. We have used a similar system in Chapter 5 to compute the quadratic approximation to the center manifold at the fold bifurcation (see equation (5.29)). If the vector $q_0$ is close to the null-vector of $A(x, \alpha)$ and the vector $p_0$ is close to the null-vector of $A^T(x, \alpha)$, the matrix

$$M = \begin{pmatrix} A(x, \alpha) & p_0 \\ q_0^T & 0 \end{pmatrix}$$

is nonsingular at $(x, \alpha)$ and (10.71) has the unique solution. In practical computations, $q_0$ and $p_0$ are the eigenvectors of $A$ and $A^T$, respectively, at the previous point on the curve. For $g = 0$, system (10.71) implies

$$Aw = 0, \quad \langle q_0, w \rangle = 1,$$

meaning that $w$ is a scaled null-vector of $A(x, \alpha)$ and $\det A(x, \alpha) = 0$ as in (10.68). In fact, $g$ is proportional to $\det A(x, \alpha)$. Indeed, by Cramer's rule

$$g(x, \alpha) = \frac{\det A(x, \alpha)}{\det M(x, \alpha)}.$$

The derivatives of $g$ with respect to $(x, \alpha)$ can be computed by differentiating (10.71). Let $z$ denote a component of $x$ or $\alpha$. Then,

$$\begin{pmatrix} A(x, \alpha) & p_0 \\ q_0^T & 0 \end{pmatrix} \begin{pmatrix} w_z \\ g_z \end{pmatrix} + \begin{pmatrix} A_z(x, \alpha) & 0 \\ 0 & 0 \end{pmatrix} \begin{pmatrix} w \\ g \end{pmatrix} = \begin{pmatrix} 0 \\ 0 \end{pmatrix}$$

and $(w_z, g_z)^T$ can be found by solving the system

$$\begin{pmatrix} A(x, \alpha) & p_0 \\ q_0^T & 0 \end{pmatrix} \begin{pmatrix} w_z \\ g_z \end{pmatrix} = - \begin{pmatrix} A_z(x, \alpha) w \\ 0 \end{pmatrix}. \tag{10.72}$$

This system has the same matrix $M$ as (10.71), while the right-hand side involves the known vector $w$ and the derivative $A_z$ of the Jacobian matrix $A$. The derivative $g_z$ can be expressed explicitly, if we introduce the solution $(v, h)^T$ to the *transposed system*

$$M^T \begin{pmatrix} v \\ h \end{pmatrix} = \begin{pmatrix} 0 \\ 1 \end{pmatrix}.$$

Multiplying (10.72) from the left by $(v^T, h)$ and taking into account that $(v^T, h)M = (0, 1)$, gives

$$g_z = -\langle v, A_z(x, \alpha) w \rangle.$$

In the Hopf case, the modified minimally augmented system looks exactly as (10.70), where now the function $g = g(x, \alpha)$ is computed by solving the bordered system

$$\begin{pmatrix} 2A(x,\alpha)\odot I_n & P_0 \\ Q_0^T & 0 \end{pmatrix} \begin{pmatrix} W \\ g \end{pmatrix} = \begin{pmatrix} 0 \\ 1 \end{pmatrix}. \tag{10.73}$$

This system is $(m+1)$-dimensional, where $2m = n(n-1)$, and is nonsingular if the vectors $Q_0, P_0 \in \mathbb{R}^m$ are the null-vectors of $2A \odot I_n$ and $(2A \odot I_n)^T$, respectively, at a nearby generic point on the Hopf curve.[7] The partial derivatives $g_z$ can be expressed in terms of $A_z$ as in the fold case.

Similar treatment can be given to codim 1 bifurcations of discrete-time systems. Consider a smooth two-parameter map

$$x \mapsto f(x,\alpha), \quad x \in \mathbb{R}^n, \ \alpha \in \mathbb{R}^2. \tag{10.74}$$

The system

$$\begin{cases} f(x,\alpha) - x = 0, \\ \det(A(x,\alpha) - I_n) = 0, \end{cases} \tag{10.75}$$

can be used to continue the fold bifurcation. The system

$$\begin{cases} f(x,\alpha) - x = 0, \\ \det(A(x,\alpha) + I_n) = 0, \end{cases} \tag{10.76}$$

is applicable to continue the flip bifurcation. Finally, the system

$$\begin{cases} f(x,\alpha) - x = 0, \\ \det\left(A(x,\alpha) \odot A(x,\alpha) - I_m\right) = 0, \end{cases} \tag{10.77}$$

allows for the continuation of the Neimark-Sacker bifurcation. A lemma similar to Lemma 10.3 can be formulated in the discrete-time case as well. We can also replace the systems (10.75)–(10.77) by modified minimally augmented systems whose last equation is defined by solving certain bordered systems.

**Remark:**

While continuing a Hopf curve defined by (10.69) (or (10.70) and (10.73)), we can pass a point where this curve loses its interpretation as a curve of Hopf bifurcations, turning instead into a curve corresponding to an equilibrium with real eigenvalues $\lambda_1 + \lambda_2 = 0$ ("neutral saddle"). This clearly happens at a Bogdanov-Takens point. A similar phenomenon takes place on a Neimark-Sacker curve at points of 1:1 and 1:2 resonances. At such points the Neimark-Sacker bifurcation curve turns into a nonbifurcation curve corresponding to a fixed point with real multipliers $\mu_1\mu_2 = 1$. $\Diamond$

### Standard augmented systems

If allowed to enlarge the dimension of the continuation problem by more than one, various defining systems can be proposed to compute codim 1 bifurcation curves.

[7] Note that (10.73) is singular at a Hopf-Hopf point regardless of the choice of $Q_0$ and $P_0$.

The system of $2n+1$ scalar equations for the $2n+2$ components of $(x, q, \alpha)$,

$$\begin{cases} f(x, \alpha) = 0, \\ A(x, \alpha)q = 0, \\ \langle q, q_0 \rangle - 1 = 0, \end{cases} \tag{10.78}$$

can be used to compute the fold bifurcation curve. Here, $q_0 \in \mathbb{R}^n$ is a *reference vector* that is not orthogonal to the null-space of $A$. In practical computations, $q_0$ is usually the null-vector of $A$ at the previously found point on the curve. If $\Gamma$ is a curve defined by (10.78), then its standard projection onto the parameter plane gives the fold bifurcation boundary $\mathcal{B}$. Indeed, if $(x^0, q^0, \alpha^0) \in \mathbb{R}^{2n+2}$ is a solution to (10.78) for a fixed $q_0$, then system (10.67) has at the parameter value $\alpha^0$ an equilibrium $x^0$ having a zero eigenvalue with the eigenvector $q^0$ normalized by $\langle q^0, q_0 \rangle = 1$. To start the continuation of $\Gamma$, we obviously need not only an approximation to the critical equilibrium and the critical parameter values, but also the null-vector $q_0$.

**Remark:**
The last equation in (10.78) can be substituted by the standard normalization condition $\langle q, q \rangle - 1 = 0$. $\Diamond$

Next, consider the following system of $3n+2$ scalar equations for the $3n+2$ components of $(x, v, w, \alpha)$ and $\omega$:

$$\begin{cases} f(x, \alpha) = 0, \\ A(x, \alpha)v + \omega w = 0, \\ A(x, \alpha)w - \omega v = 0, \\ \langle v, v_0 \rangle + \langle w, w_0 \rangle - 1 = 0, \\ \langle v, w_0 \rangle - \langle v_0, w \rangle = 0, \end{cases} \tag{10.79}$$

which is the real form of the complex system defining the Hopf bifurcation:

$$\begin{cases} f(x, \alpha) = 0, \\ A(x, \alpha)q - i\omega q = 0, \\ \langle q, q_0 \rangle_{\mathbb{C}^n} - 1 = 0. \end{cases}$$

Here $q = v + iw \in \mathbb{C}^n$ is the critical complex eigenvector and $q_0 = v_0 + iw_0 \in \mathbb{C}^n$ is a vector that is *not orthogonal* to the critical complex eigenspace corresponding to $\pm i\omega$; $\langle q, q_0 \rangle_{\mathbb{C}^n} = \bar{q}^T q_0$. As in the fold case, $q_0$ is usually the eigenvector $q = v + iw$ at the previously found point on the curve. The system (10.79) specifies a curve $\Gamma$ whose projection onto the $(\alpha_1, \alpha_2)$-plane gives the Hopf bifurcation boundary. If $(x^0, v^0, w^0, \alpha^0, \omega_0)$ is a point on $\Gamma$, then (10.67) has, at $\alpha^0$, the equilibrium $x^0$; the Jacobian matrix $A$ evaluated at this equilibrium has a pair of purely imaginary eigenvalues $\pm i\omega_0$, while $q = v^0 + iw^0$ is the corresponding normalized complex eigenvector. To start the continuation of a curve $\Gamma$, defined by the problem (10.79), from a Hopf point detected on an equilibrium curve, we have to compute additionally the Hopf frequency $\omega_0$ and two real vectors $v_0$ and $w_0$.

**Lemma 10.4** *The Jacobian matrix of the augmented system* (10.78) *has rank* $2n+1$ *at generic fold, Bogdanov-Takens, or cusp bifurcation points of* (10.67), *while the Jacobian matrix of the augmented system* (10.79) *has rank* $3n+2$ *at generic Hopf, Bautin, fold-Hopf, or Hopf-Hopf bifurcation points of* (10.67). □

The lemma allows us to use the standard continuation method to compute fold and Hopf bifurcation curves via (10.78) and (10.79), given a sufficiently good initial guess. Notice that the Jacobian matrices of (10.78) and (10.79) can easily be constructed in terms of the partial derivatives of $f(x, \alpha)$.

**Remark:**

The dimension of the augmented system (10.79) for the Hopf continuation can be reduced by the elimination of $w$, i.e., replacing the second and third equations in (10.79) with their implication

$$A^2 v + \omega^2 v = 0,$$

and considering the following augmented system of $2n+2$ scalar equations for the $2n+3$ variables $(x, v, \alpha, \kappa)$:

$$\left\{\begin{aligned} f(x,\alpha) &= 0, \\ \left[A^2(x,\alpha) + \kappa I_n\right] v &= 0, \\ \langle v, v\rangle - 1 &= 0, \\ \langle v, l_0\rangle &= 0, \end{aligned}\right. \tag{10.80}$$

where the reference vector $l_0 \in \mathbb{R}^n$ is not orthogonal to the real two-dimensional eigenspace of $A$ corresponding to the eigenvalues $\lambda_1 + \lambda_2 = 0$, $\lambda_1\lambda_2 = \kappa$. A solution to (10.80) with $\kappa > 0$ corresponds to the Hopf bifurcation point with $\omega^2 = \kappa$, while that with $\kappa < 0$ specifies a *neutral saddle* with two real eigenvalues $\lambda_{1,2} = \pm\sqrt{-\kappa}$. Contrary to (10.79), the system (10.80) is also regular at the Bogdanov-Takens point that separates these two cases and where $\omega^2 = \kappa = 0$. ◇

We leave the reader to explain why the following augmented systems,

$$\left\{\begin{aligned} f(x,\alpha) - x &= 0, \\ A(x,\alpha)q - q &= 0, \\ \langle q, q_0\rangle - 1 &= 0, \end{aligned}\right. \tag{10.81}$$

$$\left\{\begin{aligned} f(x,\alpha) - x &= 0, \\ A(x,\alpha)q + q &= 0, \\ \langle q, q_0\rangle - 1 &= 0, \end{aligned}\right. \tag{10.82}$$

and

$$\left\{\begin{aligned} f(x,\alpha) - x &= 0, \\ A(x,\alpha)v - v\cos\theta + w\sin\theta &= 0, \\ A(x,\alpha)w - v\sin\theta - w\cos\theta &= 0, \\ \langle v, v_0\rangle + \langle w, w_0\rangle - 1 &= 0, \\ \langle v, w_0\rangle - \langle v_0, w\rangle &= 0, \end{aligned}\right. \tag{10.83}$$

where $q_0, v_0, w_0 \in \mathbb{R}^n$ are proper reference vectors, are suitable for continuation of the fold, flip, and Neimark-Sacker bifurcations, respectively, in discrete-time dynamical systems. Note that (10.83) can be substituted by a smaller system similar to (10.80):

$$\begin{cases} f(x,\alpha) = 0, \\ \left[A^2(x,\alpha) - 2\cos\theta\, A(x,\alpha) + I_n\right] v = 0, \\ \langle v, v\rangle - 1 = 0, \\ \langle v, l_0\rangle = 0, \end{cases}$$

where $l_0 \in \mathbb{R}^n$ is not orthogonal to the real two-dimensional eigenspace of $A$ corresponding to the critical eigenvalues $\mu_{1,2} = e^{\pm i\theta}$.

## 10.3.2 Continuation of codim 1 limit cycle bifurcations

Continuation of codim 1 bifurcations of limit cycles of (10.67) is a more delicate problem than that for equilibria. If the studied system is not very stiff, we can compute the Poincaré map associated with the cycle and its Jacobian matrix by numerical integration and then apply the above-mentioned continuation methods for the fixed-point bifurcations. This method works satisfactorily in many cases; however, it fails if the cycle has multipliers with $|\mu| \gg 1$ or $|\mu| \ll 1$. In such situations, the BVP approach has proved to be more reliable. Namely, we can construct a *boundary-value problem* whose solutions will define the relevant bifurcation curve. Then, we can discretize the resulting BVP and apply the standard continuation technique, given the usual regularity conditions and good initial data.

The treatment of the fold case is relatively easy since the bifurcation implies a fold singularity of the corresponding boundary-value problem for the cycle continuation:

$$\begin{cases} \dot{u}(\tau) - Tf(u(\tau),\alpha) = 0, \\ u(1) - u(0) = 0, \\ \int_0^1 \langle u(\tau), \dot{u}^0(\tau)\rangle \, d\tau = 0, \end{cases} \qquad (10.84)$$

where the last equation is the integral phase condition (10.18) with a reference periodic solution $u^0(\tau)$. Generically, near the fold point the BVP (10.84) has two solutions that collide and disappear at the critical parameter values. As in the finite-dimensional case, this happens when the *linearized* BVP has a nontrivial solution (null-function). The linearization of (10.84) with respect to $(u(\cdot), T)$ around the periodic solution has the form

$$\begin{cases} \dot{v}(\tau) - Tf_x(u(\tau),\alpha)v(\tau) - \sigma f(u(\tau),\alpha) = 0, \\ v(1) - v(0) = 0, \\ \int_0^1 \langle v(\tau), \dot{u}^0(\tau)\rangle \, d\tau = 0, \end{cases}$$

and its nontrivial solution $(v(\cdot), \sigma)$ can be scaled to satisfy

$$\int_0^1 \langle v(\tau), v(\tau)\rangle \, d\tau + \sigma^2 = 1.$$

Thus the following periodic BVP for the functions $u(\tau), v(\tau)$ defined on $[0, 1]$ and scalar variables $T$ and $\sigma$:

$$\left\{\begin{aligned} \dot{u}(\tau) - Tf(u(\tau), \alpha) &= 0, \\ u(1) - u(0) &= 0, \\ \textstyle\int_0^1 \langle u(\tau), \dot{u}^0(\tau)\rangle \, d\tau &= 0, \\ \dot{v}(\tau) - Tf_x(u(\tau), \alpha)v(\tau) - \sigma f(u(\tau), \alpha) &= 0, \\ v(1) - v(0) &= 0, \\ \textstyle\int_0^1 \langle v(\tau), \dot{u}^0(\tau)\rangle \, d\tau &= 0, \\ \textstyle\int_0^1 \langle v(\tau), v(\tau)\rangle \, d\tau + \sigma^2 - 1 &= 0, \end{aligned}\right. \tag{10.85}$$

can be used for the continuation of the generic fold bifurcation of cycles. An appropriate discretization of this problem will have the form (cf. (10.78))

$$\left\{\begin{aligned} \mathcal{F}(\xi, \alpha) &= 0, \\ \mathcal{F}_\xi(\xi, \alpha)\eta &= 0, \\ \langle \eta, \eta\rangle - 1 &= 0, \end{aligned}\right. \tag{10.86}$$

where $\xi \in \mathbb{R}^N$ is a finite-dimensional approximation to $(u(\cdot), T)$, while $\eta$ is a finite-dimensional approximation to $(v(\cdot), \sigma)$. Here $\langle \cdot, \cdot \rangle$ is a scalar product in $\mathbb{R}^N$. Notice that (10.86) can be derived directly from the discretization $\mathcal{F}(\xi, \alpha) = 0$ of (10.84).

The flip (period-doubling) case is simpler. Introduce a vector-valued function $v(\tau)$ and consider the following *nonperiodic* BVP on the interval $[0, 1]$:

$$\left\{\begin{aligned} \dot{u}(\tau) - Tf(u(\tau), \alpha) &= 0, \\ u(1) - u(0) &= 0, \\ \textstyle\int_0^1 \langle u(\tau), \dot{u}^0(\tau)\rangle \, d\tau &= 0, \\ \dot{v}(\tau) - Tf_x(u(\tau), \alpha)v(\tau) &= 0, \\ v(1) + v(0) &= 0, \\ \textstyle\int_0^1 \langle v(\tau), v(\tau)\rangle \, d\tau - 1 &= 0. \end{aligned}\right. \tag{10.87}$$

As in the fold case, the first three equations specify the standard periodic BVP (10.84) for a limit cycle of (10.67). The fourth equation is the linearization of (10.67) (*variational equation*) around the periodic solution $u(\tau)$. The last equation provides a normalization to $v(\tau)$, while the boundary condition

$$v(1) = -v(0)$$

corresponds to the flip bifurcation. Indeed, if $(u^0(\tau), v^0(\tau), T_0)$ is a solution to (10.87) at $\alpha^0$, then the scaled system $\dot{u} = T_0 f(u, \alpha)$ has a limit cycle with period one satisfying the phase condition; moreover, the Jacobian matrix $\Lambda$ of the associated Poincaré map has a multiplier $\mu = -1$ since $\Lambda v(0) =$

$v(1) = -v(0)$. A discretization of (10.87) can be used to compute generic flip bifurcation curves of (10.67).

Finally, consider the continuation of the Neimark-Sacker bifurcation. Here we introduce a complex eigenfunction $w(\tau)$ and a scalar variable $\theta$ parametrizing the critical multipliers $\mu_{1,2} = e^{\pm i\theta}$. The boundary-value problem to continue the Neimark-Sacker bifurcation will read as follows:

$$\left\{\begin{aligned} \dot{u}(\tau) - Tf(u(\tau),\alpha) &= 0,\\ u(1) - u(0) &= 0,\\ \int_0^1 \langle u(\tau), \dot{u}^0(\tau)\rangle\, d\tau &= 0,\\ \dot{w}(\tau) - Tf_x(u(\tau),\alpha)w(\tau) &= 0,\\ w(1) - e^{i\theta}w(0) &= 0,\\ \int_0^1 \langle w(\tau), w^0(\tau)\rangle_{\mathbb{C}^n}\, d\tau - 1 &= 0, \end{aligned}\right. \tag{10.88}$$

where $u^0 : [0,1] \to \mathbb{R}^n$ and $w^0 : [0,1] \to \mathbb{C}^n$ are the reference functions (cf. (10.83)). The system (10.88) can be written in the real form, discretized, and used to continue generic Neimark-Sacker bifurcations.

**Remark:**

The presented BVP problems to continue fold, flip, and Neimark-Sacker bifurcations of cycles involve double or triple the number of the differential equations used in the cycle continuation: They are called *extended augmented BVPs*. It is possible to derive *minimally augmented BVPs* to continue these bifurcations, using a *bordered system* similar to that in the finite-dimensional case. Let us illustrate this approach for the flip bifurcation. The continuation of the flip bifurcation curve in two parameters can be reduced to the continuation of a solution of the following periodic boundary-value problem on the interval $[0,1]$:

$$\left\{\begin{aligned} \dot{u}(\tau) - Tf(u(\tau),\alpha) &= 0,\\ u(1) - u(0) &= 0,\\ \int_0^1 \langle u(\tau), \dot{u}^0(\tau)\rangle\, d\tau &= 0,\\ G[u,T,\alpha] &= 0, \end{aligned}\right. \tag{10.89}$$

where the value of the functional $G$ is computed from the linear BVP for $(v(\cdot), G)$ with given *bordering functions* $\varphi^0, \psi^0$ and factor $T$:

$$\left\{\begin{aligned} \dot{v}(\tau) - Tf_x(u(\tau),\alpha)v(\tau) + G\varphi^0(\tau) &= 0,\\ v(1) + v(0) &= 0,\\ \int_0^1 \langle \psi^0(\tau), v(\tau)\rangle\, d\tau &= 1. \end{aligned}\right. \tag{10.90}$$

The functions $\varphi^0$ and $\psi^0$ are selected to make (10.90) uniquely solvable. The first three equations in (10.89) specify the standard periodic BVP (10.84) defining a limit cycle of (10.67). The fourth equation is equivalent to the flip bifurcation condition. Indeed, if $G = 0$, the first equation in (10.84) reduces to the variation equation around the periodic solution $u(\tau)$. The last equation in (10.90) provides a normalization to the variational solution $v(\tau)$, while the boundary condition $v(1) = -v(0)$ corresponds to the multiplier $\mu = -1$ at the flip bifurcation. To apply the standard continuation technique, we have to replace (10.89) and (10.90) by their finite-dimensional approximations. It is also possible to compute efficiently the derivatives of $G$ with respect to $u, T$, and $\alpha$.

A similar approach is applicable to the continuation of the fold and Neimark-Sacker bifurcations. For example, (10.89), where $G$ is now computed from the linear BVP for $(v(\cdot), G, S)$ :

$$\begin{cases} \dot{v}(\tau) - Tf_x(u(\tau),\alpha)v(\tau) - Sf(u(\tau),\alpha) + G\varphi^0(\tau) = 0, \\ v(1) - v(0) = 0, \\ \int_0^1 \langle f(x(\tau),\alpha), v(\tau)\rangle \, d\tau = 0, \\ \int_0^1 \langle \psi^0(\tau), v(\tau)\rangle \, d\tau + S = 1, \end{cases} \tag{10.91}$$

can be used to continue the fold bifurcation of cycles in two parameters (cf., (10.85). The continuation of the Neimark-Sacker bifurcation is more involved (see references in Appendix D). ◊

### 10.3.3 Continuation of codim 1 homoclinic orbits

In this section we deal with the continuation of orbits homoclinic to a hyperbolic equilibrium or a nonhyperbolic equilibrium with a simple zero eigenvalue. As we saw in Chapters 6 and 7, the presence of such an orbit $\Gamma_0$ is a codim 1 phenomenon: In a generic two-parameter system (10.67) it exists along a curve in the $(\alpha_1, \alpha_2)$-plane. The problem then is to continue this curve, provided that an initial point on it is known together with the corresponding homoclinic solution.

Suppose that at $\alpha_0$, system (10.67) has a *hyperbolic equilibrium* $x_0$. Denote the Jacobian matrix $f_x(x,\alpha)$ by $A(x,\alpha)$. Assume, therefore, that $A(x_0,\alpha_0)$ has $n_+$ unstable eigenvalues $\lambda_i$, $i = 1, 2, \dots, n_+$ and $n_-$ stable eigenvalues $\mu_i$, $i = 1, 2, \dots, n_-$, such that $n_- + n_+ = n$ and

$$\text{Re } \mu_{n_-} \leq \cdots \leq \text{Re } \mu_2 \leq \text{Re } \mu_1 < 0 < \text{Re } \lambda_1 \leq \text{Re } \lambda_2 \leq \cdots \leq \text{Re } \lambda_{n_+}.$$

A homoclinic solution $x(t)$ of (10.67) satisfies the condition

$$x(t) \to x_0 \text{ as } t \to \pm\infty, \tag{10.92}$$

where $x_0$ is an equilibrium point,

$$f(x_0, \alpha) = 0. \tag{10.93}$$

Notice that conditions (10.92) and (10.93) do not specify the homoclinic solution completely. Indeed, any time shift of a solution to (10.67), (10.92), and (10.93) is still a homoclinic solution. Thus, a condition is required to fix the phase, similar to the situation for limit cycles. As for cycles, the following *integral phase condition* can be used

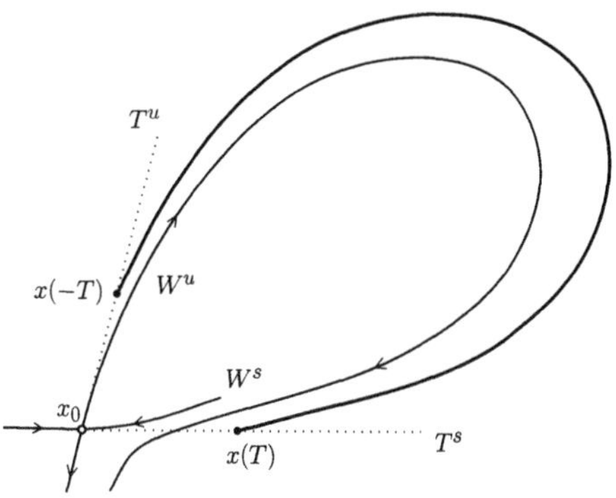


**Fig. 10.12.** Projection boundary conditions when $x_0$ is hyperbolic: $x(-T) - x_0 \in T^u$, $x(T) - x_0 \in T^s$.

$$\int_{-\infty}^{+\infty} \langle x(t) - x^0(t), \dot{x}^0(t)\rangle \, dt = 0, \qquad (10.94)$$

where $x^0(t)$ is a reference solution that is assumed to be known. Equation (10.94) is a necessary condition for a local minimum of the $L_2$-distance between $x$ and $x^0$ over time shifts. As usual, in the continuation setting, $x^0$ is the homoclinic solution obtained at the previously found point on the curve.

The problem (10.67), (10.92)–(10.94) defined on an infinite time interval has to be approximated by truncation to a finite interval, say $[-T, T]$, and specification of suitable boundary conditions. For example, we can place the solution at the two end points in the stable and unstable eigenspaces of $A(x_0, \alpha)$ which provide linear approximations to $W^{s,u}(x_0)$ (see **Fig. 10.12**).[8] This can be achieved via replacing (10.92) by the *projection boundary conditions*:

$$L_s(x_0, \alpha)(x(-T) - x_0) = 0, \quad L_u(x_0, \alpha)(x(T) - x_0) = 0. \qquad (10.95)$$

Here $L_s(x_0, \alpha)$ is a $(n_- \times n)$ matrix whose rows form a basis for the stable eigenspace of $A^T(x_0, \alpha)$. Accordingly, $L_u(x_0, \alpha)$ is a $(n_+ \times n)$ matrix, such that its rows form a basis for the unstable eigenspace of $A^T(x_0, \alpha)$. For example, if $n = 3$ and the matrix $A$ has real eigenvalues $\mu_2 < \mu_1 < 0 < \lambda_1$ (the saddle case), then $L_u = (p_1^u)^T$, where $p_1^u$ is the eigenvector of $A^T$ corresponding to $\lambda_1 : A^T p_1^u = \lambda_1 p_1^u$, while

$$L_s = \begin{pmatrix} (p_1^s)^T \\ (p_2^s)^T \end{pmatrix},$$

where $A^T p_k^s = \mu_k p_k^s$, $k = 1, 2$. There is a method to construct $L_s$ and $L_u$ so that they will depend smoothly on $\alpha$ (see the bibliographical notes).

[8] Notice that generically (10.67) has an orbit satisfying these boundary conditions for parameter values that are *close but not equal* to the parameter values at which the homoclinic orbit is present.

Finally, truncate the phase condition (10.94) to the interval $[-T, T]$,

$$\int_{-T}^{+T} \langle x(t) - x^0(t), \dot{x}^0(t) \rangle \, dt = 0. \tag{10.96}$$

Collecting the above equations gives the BVP for homoclinic continuation:

$$\left\{ \begin{aligned} f(x_0, \alpha) &= 0, \\ \dot{x}(t) - f(x(t), \alpha) &= 0, \\ L_s(x_0, \alpha)(x(-T) - x_0) &= 0, \\ L_u(x_0, \alpha)(x(T) - x_0) &= 0, \\ \textstyle\int_{-T}^{T} \langle x(t) - x^0(t), \dot{x}^0(t) \rangle \, dt - 1 &= 0, \end{aligned} \right. \tag{10.97}$$

It has been proved (see the bibliographical notes) that the existence of a *regular* homoclinic orbit in (10.67) implies the existence of a solution to the truncated BVP (10.97). Furthermore, as $T \to \infty$, the solution to (10.97) converges to the homoclinic solution restricted to an appropriate finite interval. The rate of convergence is exponential for both parameter values and solutions.

One can attempt to solve (10.97) by shooting (i.e., computing orbits on $W^s(x_0)$ and $W^u(x_0)$ and estimating the distance between these manifolds in order to make it zero by varying a parameter), however, better results can be achieved by approximating it with the help of the orthogonal collocation method (see Section 10.1.4). The resulting system defines a curve in a finite-dimensional space that can be continued by the standard technique; the projection of the curve onto the $(\alpha_1, \alpha_2)$-plane gives an approximation to the homoclinic bifurcation curve.

To start the procedure, we have to know an *initial homoclinic solution.* This solution might be obtained by shooting or by following a periodic orbit to a large period (since, according to the Shil'nikov theorems formulated in Chapter 6, such long-period cycles exist near homoclinic orbits). There are other possibilities for starting the continuation, including switching to the homoclinic curve at a Bogdanov-Takens bifurcation.

It is convenient to scale the time interval in (10.97) to $[0, 1]$ and consider instead the BVP

$$\left\{ \begin{aligned} f(x_0, \alpha) &= 0, \\ \dot{u}(\tau) - 2Tf(u(\tau), \alpha) &= 0, \\ L_s(x_0, \alpha)(u(0) - x_0) &= 0, \\ L_u(x_0, \alpha)(u(1) - x_0) &= 0, \\ \textstyle\int_0^1 \langle u(\tau) - u^0(\tau), \dot{u}^0(\tau) \rangle \, d\tau - 1 &= 0, \end{aligned} \right.$$

where $u^0$ is the reference solution on the unit interval.

**Remark:**

If the unstable manifold of $x_0$ is *one-dimensional*, the boundary-value problem can be slightly simplified. Namely, suppose that there is always only one eigenvalue with positive real part: $\lambda_1 > 0$. Then, the left-hand boundary condition can be written explicitly as

$$x(-T) = x_0 + \varepsilon q_1^u, \tag{10.98}$$

where $q_1$ is the eigenvector of the Jacobian matrix: $A(x_0, \alpha)q_1^u = \lambda_1 q_1^u$, and $\varepsilon > 0$ is a small user-defined constant. Thus, no integral phase condition is required. At the right-hand end point we have a single scalar equation

$$\langle p_1^u, x(T) - x_0 \rangle = 0,$$

where $p_1^u$ is the adjoint eigenvector: $A^T(x_0, \alpha)p_1^u = \lambda_1 p_1^u$. Notice that the eigenvectors $q_1^u$ and $p_1^u$ can be computed via continuation by appending their defining equations and normalization conditions to the continuation problem. The point $x(-T)$ given by (10.98) can also be used as a starting point to integrate $W^u(x_0)$ numerically in order to find a starting homoclinic orbit by shooting.[9] ◇

If $x_0$ has a simple zero eigenvalue (i.e., $x_0$ is a *saddle-node* or *saddle-saddle*), the projections (10.95) give only $n_- + n_+ = n - 1$ boundary conditions. In other words, the boundary conditions (10.95) place the solution at the two end points in the center-unstable and center-stable eigenspaces of $A(x_0, \alpha)$, if present (see **Fig. 10.13**). Thus, an extra equation is required, namely the

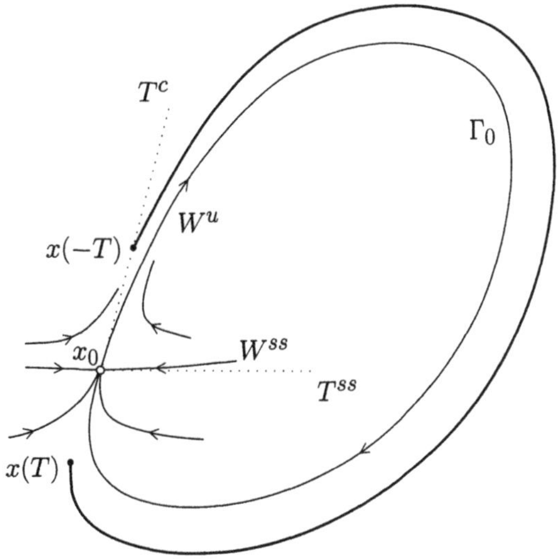


**Fig. 10.13.** Projection boundary condition when $x_0$ is a saddle-node: $x(-T) - x_0 \in T^c$; no conditions are imposed on $x(T)$.

[9] Higher-order approximations of $W^u$, such as the second-order one derived in Section 10.1.3, can also be used.

one defining the fold bifurcation, e.g.,

$$\det A(x_0, \alpha) = 0. \tag{10.99}$$

It can be proved (see the bibliographical notes) that a solution to (10.97), (10.99) exists and converges to the homoclinic solution as $T \to \infty$, provided a generic saddle-node homoclinic bifurcation (see Chapter 7) happens in (10.67). The convergence rate remains exponential for the parameters but is only $O(T^{-2})$ for the homoclinic solutions. Clearly, the projection of the curve defined by (10.97), (10.99) into the $(\alpha_1, \alpha_2)$-plane coincides with the fold bifurcation boundary

$$\begin{cases} f(x_0, \alpha) = 0, \\ \det A(x_0, \alpha) = 0, \end{cases}$$

that can be computed as described in Section 10.3.1, but solving (10.97), (10.99) also gives the homoclinic solution.

### 10.3.4 Detection, location, and analysis of codim 2 bifurcations

While following a bifurcation curve corresponding to a codim 1 bifurcation, we should be able to detect and locate possible codim 2 bifurcations. Thus, special test functions have to be derived. Here we will focus on detecting codim 2 equilibrium bifurcations. Test functions to detect codim 2 homoclinic bifurcations are discussed in Appendix C.

Notice that according to Lemmas 10.3 and 10.4, the continuation procedures for codim 1 curves will allow us to compute through higher-order singularities. If a singularity is caused by an extra linear degeneracy (i.e., $\dim W^c$ changes), the problem is most easily solved within the minimally augmented approach. Suppose we continue a fold bifurcation curve using the augmented system (10.68). Then, monitoring the test function $\psi_H$ for the Hopf bifurcation (10.41) along this curve allows us to detect both Bogdanov-Takens and fold-Hopf bifurcations. Similarly, evaluation of the fold test function $\psi_t$ given by (10.39) along a Hopf curve (10.69) provides an alternative way to detect the same bifurcations. As pointed out, the Hopf-Hopf bifurcation point can be detected as a branching point for the Hopf continuation problem (10.69). We can also detect this singularity by monitoring the Hopf test function $\psi_H$ restricted to the complementary eigenspace.

To detect singularities due to nonlinear terms, we have to compute the corresponding normal form coefficients. Since certain adjoint eigenvectors are required for these computations, we can extend the standard augmented system by appending equations for properly normalized adjoint eigenvectors. For example, to detect a cusp bifurcation on an equilibrium fold curve, we have to monitor the scalar product $\langle p, B(q,q) \rangle$, where $p$ is a zero eigenvector of the transposed Jacobian matrix. We can obtain the normalized adjoint eigenvector $p$ as a part of the continuation of the fold curve in the $(x, \alpha, q, p, \varepsilon)$-space using the defining system

$$\begin{cases} f(x,\alpha) = 0, \\ A(x,\alpha)q = 0, \\ A^T(x,\alpha)p - \varepsilon p = 0, \\ \langle q,q\rangle - 1 = 0, \\ \langle p,p\rangle - 1 = 0, \end{cases} \tag{10.100}$$

where $\varepsilon$ is an artificial parameter that is equal to zero along the fold curve. Note that a Bogdanov-Takens singularity can be detected while following the curve (10.100) as a regular zero of the test function $\psi_{BT} = \langle p, q\rangle$ (prove!). A similar system can be derived for the continuation of the Hopf bifurcation with simultaneous computation of the normalized adjoint eigenvector (its real and imaginary parts), that is required for the evaluation of the first Lyapunov coefficient $l_1$ and the detection of its zero (a Bautin point). To detect a Bogdanov-Takens point along the Hopf curve defined by the augmented system (10.80) we can monitor the function $\widetilde{\psi}_{BT} = \kappa$.

Appending the test function to be monitored to the appropriate continuation problem, we can locate the corresponding codim 2 bifurcation point via some iterative method. By freeing a third parameter, we could continue codim 2 points in three parameters (see the bibliographical notes). All the constructions presented can be carried out for discrete-time systems as well. We leave this as an exercise for the reader.

When a codim 2 equilibrium or fixed point is located, it should be analyzed. That is, the critical normal form coefficients of the ODEs and maps restricted to the corresponding center manifolds have to be computed as explained in Chapters 8 and 9. Normal form coefficients for the Bautin, Hopf-Hopf, and other codim 2 bifurcations depend on the fourth- and fifth-order partial derivatives of the system's right-hand side at the critical point (see Sections 8.7 and 9.7). Similar to the fold and Hopf cases, their computation can be reduced to the evaluation of certain directional derivatives. Indeed, the following *polarization identities* hold:

$$C(u,v,w) = \frac{1}{24}\Big[\, C(u+v+w,...) - C(u+v-w,...) - C(u-v+w,...) + C(u-v-w,...)\Big],$$

$$\begin{aligned} D(u,v,w,y) = \frac{1}{192}\Big[\, & D(u+v+w+y,...) - D(u+v+w-y,...) \\ &+ D(u+v-w-y,...) - D(u+v-w+y,...) \\ &- D(u-v+w+y,...) + D(u-v+w-y,...) \\ &- D(u-v-w-y,...) + D(u-v-w+y,...)\Big], \end{aligned}$$

$$E(u,v,w,y,z) = \frac{1}{1920}\Big[E(u+v+w+y+z,...) - E(u+v+w+y-z,...)$$

$$
\begin{aligned}
&+E(u+v+w-y-z,...)-E(u+v+w-y+z,...)\\
&-E(u+v-w+y+z,...)+E(u+v-w+y-z,...)\\
&-E(u+v-w-y-z,...)+E(u+v-w-y+z,...)\\
&-E(u-v+w+y+z,...)+E(u-v+w+y-z,...)\\
&-E(u-v+w-y-z,...)+E(u-v+w-y+z,...)\\
&+E(u-v-w+y+z,...)-E(u-v-w+y-z,...)\\
&+E(u-v-w-y-z,...)-E(u-v-w-y+z,...)\Big].
\end{aligned}
$$

Here the dots stand for repeating arguments. These formulas can be used together with the standard finite-difference approximations of the fourth- and fifth-order directional derivatives.

We finish this section by discussing the problem of branch switching at codim 2 points. We consider only equilibrium bifurcations of continuous-time systems. There is no branch-switching problem at a generic cusp point since there is only a single fold curve passing through without any geometric singularity. In contrast, a Hopf-Hopf point provides a branching point for the minimally augmented system. Thus, we might use the branching technique developed in the previous section to initialize the other Hopf branch. Switching to the Hopf curve from the fold curve at a generic Bogdanov-Takens or fold-Hopf point causes no difficulties, because these points are regular points on the respective curves: A codim 2 point can itself be used as the initial points. As we saw in Chapter 8, certain bifurcation curves at which limit cycle and homoclinic bifurcations take place emanate from some codim 2 points. Switching to such branches needs special techniques (see the bibliographical notes).

## 10.4 Continuation strategy

The analysis of a given dynamical system requires a clear strategy. Such a strategy is provided by bifurcation theory. Formally, while performing the analysis, we always continue certain curves in some spaces and monitor several test functions to detect and locate special points on these curves. Theoretical analysis of bifurcations suggests which higher-codimension bifurcations can be expected along the traced curve and therefore which test functions have to be monitored to detect and locate these points. It also predicts which bifurcation curves of the same codimension originate at a detected point. Having located a special point, we can either switch to one of these emanating curves or "activate" one more parameter to continue, if possible, the located point in more parameters. Of course, we can also merely continue the original branch further beyond the special point.

The analysis of any system usually starts with locating at least one equilibrium at some fixed parameter values. Then, we "activate" one of the system parameters and continue the obtained equilibrium with respect to this parameter. During continuation, some fold or Hopf points may be detected. Generically, these points are nondegenerate. Thus, a limit cycle bifurcates from the Hopf point in the direction determined by the sign of the first Lyapunov coefficient. Cycle continuation can be initialized from the Hopf point, and its possible codim 1 bifurcations detected and located as a parameter is varied. Switching to a cycle of double period can be done at flip points, for instance.

By freeing the second parameter and using one of the augmented systems, we can compute bifurcation curves in the plane defined by these two active parameters. Often, detected codim 2 points connect originally disjoint codim 1 points. For example, two fold points found in the one-parameter analysis can belong to the same fold bifurcation curve passing through a cusp point as two parameters vary. Thus, higher-order bifurcations play the role of "organizing centers" for nearby bifurcation diagrams. This role is even more prominent since certain codim 1 limit cycle bifurcation curves can originate at codim 2 points. The simplest examples are the fold curve for cycles originating at a generic Bautin point and the saddle homoclinic bifurcation curve emanating from a generic Bogdanov-Takens point. Their continuation can be started from these points. A more difficult problem is to start global bifurcation curves from fold-Hopf and Hopf-Hopf bifurcation points.

The continuation of codim 1 bifurcations of limit cycles usually reveals their own codim 2 bifurcation points (strong resonances, and so forth). Overlapping the obtained bifurcation boundaries for equilibria, cycles, and homoclinic orbits provides certain knowledge about the bifurcation diagram of the system and might give some insights on other bifurcations in which more complex invariant sets – for example, tori – are involved. Though well formalized, the analysis of a specific dynamical system will always be an art in which interactive computer tools (see the bibliographical notes) are a necessity.

## 10.5 Exercises

Most of the following exercises require the use of a computer and, desirably, some of the software tools mentioned in the bibliographical notes.

**(1) (Feigenbaum's universal map via Newton's method)** (Read Appendix A to Chapter 4 before attempting.)

(a) Assume that the fixed point $\varphi(x)$ of the doubling operator

$$(Tf)(x) = -\frac{1}{a}f(f(-ax)), \quad a = -f(1),$$

has the polynomial form

$$\varphi_0(x) = 1 + b_1x^2 + b_2x^4$$

with some unknown coefficients $b_{1,2}$. Substitute $\varphi_0$ into the fixed-point equation $\varphi - T\varphi = 0$ and truncate it by neglecting all $o(x^4)$ terms. The coefficients of the $x^2$- and $x^4$-terms define a polynomial system of two equations for $(b_1, b_2)$. Solve this system numerically by Newton's iterations starting from $(b_1^{(0)}, b_2^{(0)}) = (-1.5, 0.0)$. Verify that the iterations converge to

$$(b_1, b_2) \approx (-1.5222, 0.1276),$$

which is a good approximation to the true coefficient values.

(b) Now take

$$\varphi_0(x) = 1 + b_1 x^2 + b_2 x^4 + b_3 x^6,$$

which gives the next approximation to $\varphi$, and repeat the procedure, now dropping $o(x^6)$ terms. (*Hint*: A symbolic manipulation program may be useful.) Explain why introducing the term $b_3 x^6$ *changes* the resulting values of the coefficients $b_{1,2}$.

(c) Describe an algorithm allowing us to compute the approximation of $\varphi$ to within a given accuracy. Could a basis other than $\{1, x^2, x^4, \ldots\}$ be used, and would this give any advantages? (*Hint:* See Babenko & Petrovich [1983, 1984].)

(d) How would one approximate the Feigenbaum constant?

**(2) (Broyden versus Newton)** (Dennis & Schnabel [1983].) The system of two equations,

$$\begin{cases} x_1^2 + x_2^2 - 2 = 0, \\ e^{x_1 - 1} + x_2^3 - 2 = 0, \end{cases}$$

has the solution $x_1^0 = x_2^0 = 1$.

(a) Program Newton's method to solve the system using an analytically derived expression for the Jacobian matrix $A(x)$. Implement also the Broyden method for this system with $A^{(0)} = A(x^{(0)})$, where $x^{(0)}$ is the initial point for both methods.

(b) Compare experimentally the number of iterations required by each method to locate the specified solution with the accuracy $10^{-13}$, starting from the same initial point $x_1^{(0)} = 1.5$, $x_2^{(0)} = 2$.

(c) Make the same comparison between the Broyden method and the Newton-chord method with the matrix $A(x^{(0)})$ used in all iterations.

(d) Modify the program to compute approximately the Jacobian matrix in Newton's method by finite differences with increment $10^{-6}$. Compare the number of iterations and the number of right-hand side computations needed by the resulting algorithms to converge to the solution from the same initial point, with the same accuracy as in step (b).

**(3) (Integral phase condition)** Prove that the condition

$$\int_0^1 \langle u(\tau), \dot{v}(\tau) \rangle \, d\tau = 0$$

is a necessary condition for the $L_2$-distance between two (smooth) 1-periodic functions $u$ and $v$,

$$\rho(\sigma) = \int_0^1 \|u(\tau + \sigma) - v(\tau)\|^2 \, d\tau,$$

to achieve a local minimum with respect to possible shifts $\sigma$.

**(4) (Gauss points)**

(a) Apply the Gramm-Schmidt orthogonalization procedure:

$$\psi_0 = \varphi_0, \ \psi_n = \varphi_n - \sum_{j=0}^{n-1} \frac{\langle \varphi_n, \psi_j \rangle}{\langle \psi_j, \psi_j \rangle} \psi_j, \ n = 1, 2, \ldots,$$

with the scalar product

$$\langle f, g \rangle = \int_{-1}^{+1} f(x) g(x) \, dx, \quad f, g \in C[-1, 1],$$

to the set of functions $\varphi_j(x) = x^j$, $j = 0, 1, \ldots$. The resulting orthogonal polynomials are the *Legendre polynomials.* Verify that

$$\begin{aligned}
\psi_0(x) &= 1, \\
\psi_1(x) &= x, \\
\psi_2(x) &= \frac{1}{2}(3x^2 - 1), \\
\psi_3(x) &= \frac{1}{2}(5x^2 - 3x), \\
\psi_4(x) &= \frac{1}{8}(35x^4 - 30x^2 + 3), \\
\psi_5(x) &= \frac{1}{8}(63x^5 - 70x^3 + 15x).
\end{aligned}$$

(b) Find, by Newton's method, all roots (*Gauss points*) of the above polynomials $\psi_j(x)$, for $j = 1, 2, \ldots, 5$, with accuracy $\varepsilon = 10^{-13}$.

**(5) (Branching point)**
(a) Prove that $(0, 0)$ and $(1, 1)$ are simple branching points of the continuation problem

$$f(x_1, x_2) = x_1^2 - x_1 x_2 - x_1 x_2^2 + x_2^3 = 0,$$

and compute vectors tangent to the branches passing through them.

(b) Evaluate the test function (10.62) along the equilibrium branch $x_2 = x_1$, and check that it changes sign at the branching points.

(c) Continue a branch of the perturbed problem $f(x) - \varepsilon = 0$ passing near the point $x = (-2, -2)$ for several small $|\varepsilon|$. Use the results to continue the second branch of the original problem.

**(6) (Generalized flip continuation)** Specify an extended system that allows for the continuation of a generalized flip bifurcation of the discrete-time system

$$x \mapsto f(x, \alpha), \quad x \in \mathbb{R}^n, \ \alpha \in \mathbb{R}^3.$$

**(7) (Fixed points and periodic orbits of the Hassel-Lawton-May model)** Consider the recurrence relation

$$x_{k+1} = \frac{r x_k}{(1 + x_k)^\beta},$$

where $x_k$ is the density of a population at year $k$, and $r$ and $\beta$ are growth parameters.

(a) Write the model as a one-dimensional dynamical system

$$y \mapsto R + y - e^b \ln(1 + e^y), \tag{E.1}$$

by introducing the new variable and parameters: $y = \ln x, R = \ln r, b = \ln \beta$.

(b) Set $R = 3.0, b = 0.1$ and compute an orbit of (E.1) starting at $y_0 = 0$. Observe the convergence of the orbit to a stable fixed point $y^{(0)} \approx 2.646$.

(c) Continue the obtained fixed point $y^{(0)}$ with respect to the parameter $b$ within the interval $0 \leq b \leq 3$. Detect a supercritical flip bifurcation at $b_1 \approx 1.233$.

(d) Switch to the period-two cycle bifurcating at the flip point and continue it with respect to $b$ until the next period doubling happens at $b_2 \approx 2.1937$. Verify that $b_1$ is a branching point for the period-two cycle.

(e) Switch to the period-four cycle bifurcating at the flip point and continue it with respect to $b$ until the next period doubling happens at $b_4 \approx 2.5691$.

(f) Continue the flip bifurcations for period-one, -two, and -four cycles in two parameters $(b, R)$ within the region $0 \leq b \leq 3, 0 \leq R \leq 10.0$. Verify the Feingenbaum universality. Where could chaos be expected?

**(8) (Arnold tongue in the perturbed delayed logistic map)** Consider the following recurrence relation (see Example 7.2 in Chapter 7):

$$x_{k+1} = rx_k(1 - x_{k-1}) + \varepsilon, \tag{E.2}$$

where $x_k$ is the density of a population at year $k$, $r$ is the growth rate, and $\epsilon$ is the migration rate.

(a) Introduce $y_k = x_{k-1}$ and rewrite (E.2) as a planar dynamical system

$$\begin{pmatrix} x \\ y \end{pmatrix} \mapsto \begin{pmatrix} rx(1-y)+\varepsilon \\ x \end{pmatrix}. \tag{E.3}$$

(b) Set $r = 1.9$, $\varepsilon = 0$ and iterate an orbit of (E.3) starting at $(x_0, y_0) = (0.5, 0.2)$ until it converges (approximately) to a fixed point.

(c) Activate the parameter $r$ and continue the located fixed point. Check that a Neimark-Sacker bifurcation happens at $r = 2$, and verify that the critical multipliers have the form

$$\mu_{1,2} = e^{\pm i\theta_0}, \ \theta_0 = \frac{\pi}{3}.$$

Is it a strong resonance or not?

(d) Iterate the map (E.3) for $r = 2.01, 2.05, 2.10$, and $2.15$ using various initial data. Check that the iterations converge to closed invariant curves giving rise to quasiperiodic sequences $\{x_k\}$.

(e) Set $r = 2.177$ and iterate an orbit starting from the last point obtained in Step (d). Observe that the orbit converges to a cycle of period seven. Select a point on the cycle with maximal $x$-coordinate.

(f) Continue this period-seven cycle with respect to $r$ (see **Fig. 7.23** in Chapter 7). Interpret the resulting *closed* curve on the $(r, x)$-plane. (*Hint*: Each point of the period-seven orbit is a fixed point of the seventh iterate of the map (E.3).) Check that the cycle of period-seven exists inside the interval $r_1 < r < r_2$, where $r_1 \approx 2.176$ and $r_2 \approx 2.201$ are fold bifurcation points.

(g) Continue the period-seven cycle from Step (e) with respect to $\varepsilon$ and find its super-critical period doubling (flip) bifurcation at $\epsilon \approx 0.0365$.

(h) Continue the Neimark-Sacker bifurcation found in Step (c) on the $(r, \epsilon)$-plane and obtain the curve $h^{(1)}$ shown in **Fig. 7.24** in Chapter 7.

(i) Continue the fold for the period-seven orbits starting at two bifurcation points obtained in Step (f) varying $(r, \epsilon)$ (see the curves $t_{1,2}^{(7)}$ in **Fig. 7.24**). Which point on the Neimark-Sacker curve approaches the fold curves? Find the corresponding $\theta_0$. Is it in accordance with the theory of the Neimark-Sacker bifurcation and phase locking? Where is the period-six cycle?

(j) Continue the flip bifurcation curve $f^{(7)}$ starting at the point found in Step (g) on the $(r, \varepsilon)$-plane, thus delimiting the region of existence of the period-seven cycle (see **Fig. 7.24**). Verify that $f^{(7)}$ has two points in common (outside the region depicted in **Fig. 7.24**) with the fold bifurcation curves $t_{1,2}^{(7)}$, where the period-seven cycle has multipliers $\mu_{1,2}^{(7)} = \pm 1$ (a codim 2 bifurcation).

**(9) (Equilibria and limit cycles in a predator-prey model)** Consider the following system of ODEs [Bazykin 1985]:

$$\begin{cases} \dot{x} = x - \dfrac{xy}{1+\alpha x}, \\ \dot{y} = -\gamma y + \dfrac{xy}{1+\alpha x} - \delta y^2, \end{cases}$$

where $x$ and $y$ are prey and predator densities, respectively, $\alpha$ is a saturation parameter of the predator functional response, $\gamma$ describes the natural predator mortality, and $\delta$ is the predator competition rate for some external resources. When $\alpha = \delta = 0$ we obtain the classical Volterra system.

(a) Set $\alpha = 0, \gamma = 2, \delta = 0.5$, and numerically integrate an orbit of the system starting at $(x_0, y_0) = (1, 1)$. Verify that the orbit converges to the stable equilibrium $(x, y) = (2.5, 1.0)$. Try several different initial points.

(b) Continue the equilibrium found with respect to the parameter $\alpha$ and detect and locate a fold bifurcation point. Activate the parameter $\delta$ and compute the fold bifurcation curve $t$ in the $(\alpha, \delta)$-plane (see **Fig. 10.14**). Find a Bogdanov-Takens point $B = (\alpha_0, \delta_0) \approx (0.2808, 0.171)$ on the fold curve.

(c) Starting at the Bogdanov-Takens point, continue the Hopf bifurcation curve $h$ emanating from this point (see **Fig. 10.14**). Predict the stability of the limit cycle that appears upon crossing the Hopf curve near the point $B$.

(d) Choose a point on the Hopf bifurcation curve near its maximum with respect to the parameter $\delta$ at $(\alpha_1, \delta_1) \approx (0.199586, 0.2499)$. Decrease $\delta$ by a small increment (e.g., set $\delta = 0.229$) and find a stable limit cycle generated via the Hopf bifurcation by numerical integration of the system. Try several initial data not far from the cycle and check its stability. Determine (approximately) the period of the cycle.

(e) Continue the cycle from Step (d) with respect to the parameter $\delta$ in both possible directions. Monitor the period $T_0$ of the cycle as well as its multipliers. Notice that the period $T_0$ grows rapidly as $\delta$ approaches $\delta_0 \approx 0.177$ (see **Fig. 10.15**). Guess which bifurcation the cycle exhibits. Plot the cycle for increasing values of $T_0$, and trace the change in its shape as it approaches a homoclinic orbit.

(f) Continue the homoclinic bifurcation curve in the $(\alpha, \delta)$-plane and convince yourself that it tends to the Bogdanov-Takens point $B$.

**(10) (Lorenz-84 model)** Consider the following system [Lorenz 1984, Shil'nikov et al. 1995]:

$$\begin{cases} \dot{x} = -y^2 - z^2 - ax + aF, \\ \dot{y} = xy - bxz - y + G, \\ \dot{z} = bxy + xz - z, \end{cases}$$

where $(a, b, F, G)$ are parameters. Fix $a = 0.25, b = 4$.

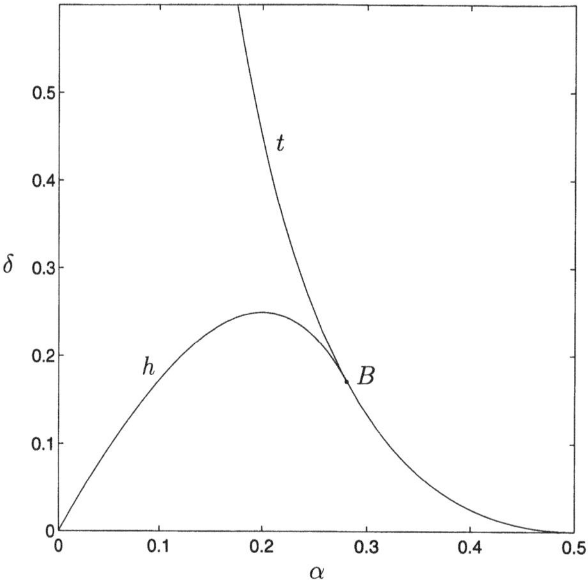


**Fig. 10.14.** Fold ($t$) and Hopf ($h$) bifurcation curves of the predator-prey model; the Bogdanov-Takens point is labled by $B$.

(a) Verify that the system exhibits a fold-Hopf bifurcation ($\lambda_1 = 0, \lambda_{2,3} = \pm i\omega_0$) at $ZH = (G_0, F_0) \approx (1.682969, 1.684052)$. (*Hint*: See Exercise 12 to Chapter 8.)

(b) Starting from the fold-Hopf point $ZH$, continue the fold ($t$) and the Hopf ($h$) bifurcation curves of the system. Find a cusp bifurcation point $C = (G_c, F_c) \approx (0.292, 0.466)$ on the fold curve (see **Fig. 10.16**).

**(11) (Limit cycles and heteroclinic orbits in a predator–double-prey system)** Consider the system

$$\begin{cases} \dot{x} = x(\alpha - x - 6y - 4z), \\ \dot{y} = y(\beta - x - y - 10z), \\ \dot{z} = -z(1 - 0.25x - 4y + z), \end{cases}$$

describing the dynamics of two prey populations affected by a predator (see, e.g., Bazykin [1985]).

(a) Fix $\alpha = 2.4, \beta = 1.77$. Integrate the system starting at $(x_0, y_0, z_0) = (0.9, 0.6, 0.001)$ and observe that the orbit converges toward a stable equilibrium $(x_0, y_0, z_0) \approx (0.6919, 0.228, 0.085)$.

(b) Continue the equilibrium from Step (a) with respect to the parameter $\beta$ and show that it undergoes a supercritical Hopf bifurcation at $\beta_H = 1.7638\ldots$.

(c) Take $\beta = 1.76$ and find a stable limit cycle in the system by numerical integration. Check that the period of the cycle $T_0 \approx 26.5$.

(d) Continue the limit cycle with respect to the parameter $\beta$ and monitor the dependence of its period $T_0$ upon $\beta$ (see **Fig. 10.17**). Plot the cycle for different values of $T_0$ and try to understand its limiting position as $T_0 \to \infty$ for $\beta \to \beta_0 \approx$

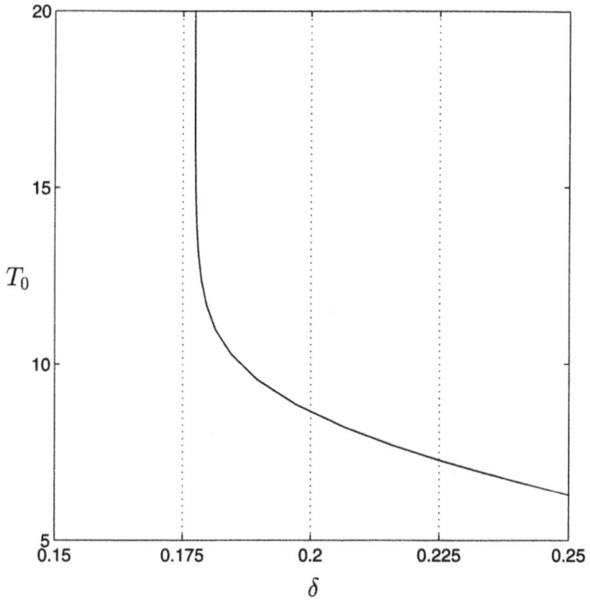


**Fig. 10.15.** Dependence of the cycle period $T_0$ on $\delta$; $\delta_0 \approx 0.177$ corresponds to the homoclinic bifurcation.

1.7353. (*Hint*: The limit cycle tends to a heteroclinic cycle composed of three orbits connecting saddle points; two of these orbits belong to the invariant coordinate planes and persist under parameter variation (see **Fig. 10.18**).)

(e) Analyze bifurcations detected on the cycle curve, and continue several flip and fold bifurcation curves varying the parameters $(\alpha, \beta)$.

**(12) (Periodically forced predator-prey system)** Reproduce the parametric portraits of the periodically forced predator-prey system presented in Example 9.3 of Chapter 9.

**(13) (Moore-Penrose continuation)**

**Definition 10.4** *Let $A$ be an $n \times (n+1)$ matrix of rank $n$. The* Moore-Penrose inverse *of $A$ is the $(n+1) \times n$ matrix $A^+ = A^T(AA^T)^{-1}$.*

(a) Consider the linear system

$$Ay = a, \quad y \in \mathbb{R}^{n+1}, \ a \in \mathbb{R}^n.$$

Prove that $y = A^+a$ is a solution to this system satisfying the orthogonality condition $\langle v, y \rangle = 0$, where $v \in \mathbb{R}^{n+1}$ is a nonzero vector such that $Av = 0$.

(b) Now consider a smooth continuation problem:

$$F(y) = 0, \quad F : \mathbb{R}^{n+1} \to \mathbb{R}^n.$$

Let $y^0 \in \mathbb{R}^{n+1}$ be a point sufficiently close to a regular point on the curve defined by the continuation problem. Denote $A(y) = F_y(y)$ and define the *Gauss-Newton corrections*

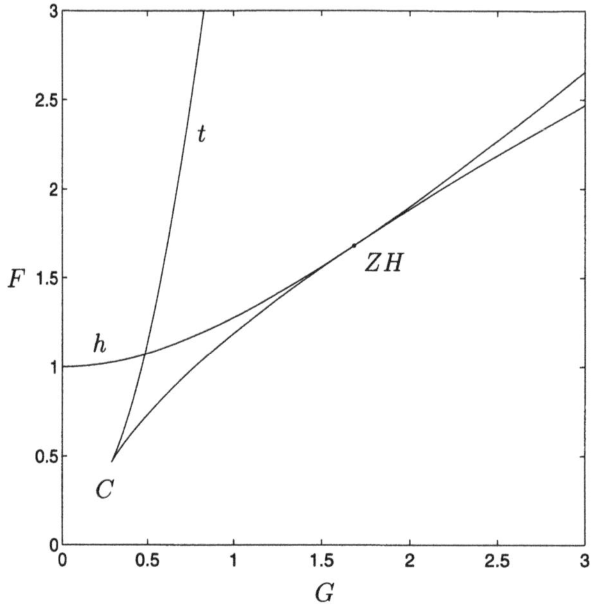


**Fig. 10.16.** Fold ($t$) and Hopf ($H$) bifurcation curves of the Lorenz-84 model; codim 2 bifurcation points: $ZH$ – fold-Hopf, $C$ – cusp.

$$y^{j+1} = y^j - A^+(y^j)F(y^j), \quad j = 0, 1, 2, \ldots.$$

Explain why these iterations are asymptotically equivalent to the geometric construction presented in **Fig. 10.7**.

(c) Devise an efficient implementation of the Moore-Penrose continuation, in which the computation of the tangent vector at the next point on the curve is incorporated into corrections. (*Hint*: See Allgower & Georg [1990].)

## 10.6 Appendix A: Convergence theorems for Newton methods

In this appendix we formulate without proof two basic theorems on the convergence of the Newton and Newton-chord methods. To give a theorem on the convergence of Newton's method, expand $f(\xi)$ near point $x$ as a Taylor series and explicitly write the quadratic terms:

$$f(\xi) = f(x) + A(x)(\xi - x) + \frac{1}{2}B(x; \xi - x, \xi - x) + O(\|\xi - x\|^3),$$

where

$$B_j(x; h, h) = \sum_{k,l=1,\ldots,n} \left. \frac{\partial^2 f_i(\xi)}{\partial \xi_k \partial \xi_l} \right|_{\xi = x} h_k h_l, \quad j = 1, 2, \ldots, n,$$

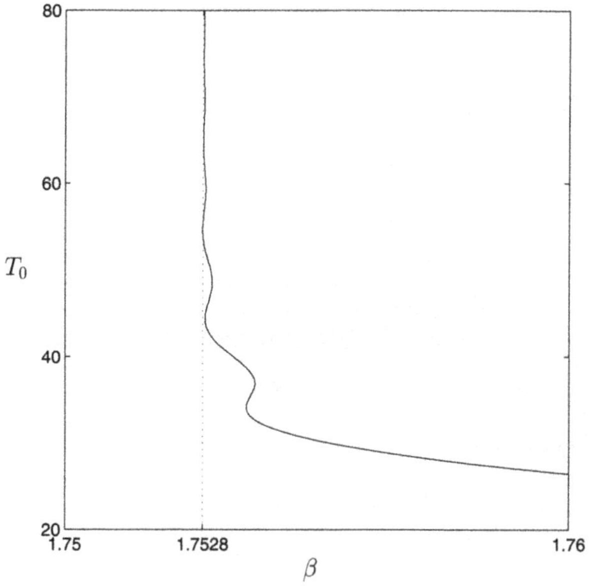


**Fig. 10.17.** Dependence of the cycle period $T_0$ upon $\beta$; $\beta_0 \approx 1.75353$ corresponds to the heteroclinic bifurcation.

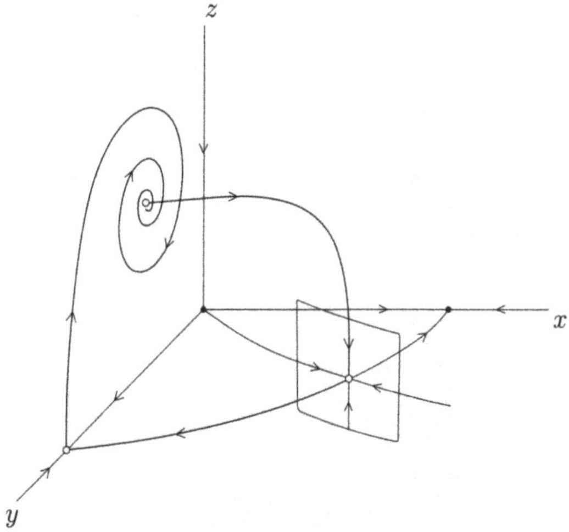


**Fig. 10.18.** Heteroclinic connection.

is the *bilinear* part of $f$ at $x$. Denote by $T(x, r)$ the ball bounded by a sphere of radius $r$ centered at the point $x \in \mathbb{R}^n$:

$$T(x, r) = \{\xi \in \mathbb{R}^n : \|\xi - x\| \leq r\}.$$

**Theorem 10.4 (Kantorovich)** *Let*

$$\sup_{\xi \in T(x^{(0)}, r_0)} \sup_{\|h\| \le 1} \|A^{-1}(x^{(0)}) B(\xi; h, h)\| = M < \infty,$$

*and let*

$$\delta = \|A^{-1}(x^{(0)}) f(x^{(0)})\|.$$

*Suppose that* $2\delta M < 1$ *and*

$$\delta \le \frac{r_0}{2}(1 + \sqrt{1 - 2\delta M}).$$

*Then, iterations* (10.3), (10.4) *converge to a solution* $x^0 \in T(x^{(0)}, r_0)$ *of system* (10.2), *moreover,*

$$\|x^{(j)} - x^0\| \le M^{-1} 2^{-j} (2\delta M)^{2^j}, \quad j = 0, 1, \ldots. \quad \square$$

**Theorem 10.5** *Under the conditions of* Theorem 10.4, *the Newton-chord method* (10.5), (10.6) *converges to a solution* $x^0 \in T(x^{(0)}, r_1)$, *of system* (10.2), *where* $r_1 = \min(r_0, M^{-1})$. *Moreover,*

$$\|x^{(j)} - x^0\| \le \delta(1 - 2\delta M)^{-1/2} \left[1 - (1 - 2\delta M)^{1/2}\right]^j, \quad j = 0, 1, \ldots. \quad \square$$

## 10.7 Appendix B: Bialternate matrix product

As we have already seen in Sections 10.2.2 and 10.3.1, the bialternate product plays a prominent role in the detection and continuation of Hopf and Neimark-Sacker bifurcations. Since this topic is not treated in standard courses on Linear Algebra, we prove here basic facts about the bialternate product, including Theorem 10.3.

Consider a $3 \times 3$ real matrix

$$A = \begin{pmatrix} a_{11} & a_{12} & a_{13} \\ a_{21} & a_{22} & a_{23} \\ a_{31} & a_{32} & a_{33} \end{pmatrix}.$$

Its eigenvalues $\lambda_1, \lambda_2$, and $\lambda_3$ satisfy the characteristic equation

$$\lambda^3 - \sigma\lambda^2 + \rho\lambda - \Delta = 0,$$

where $\sigma = \mathrm{tr}(A)$, $\Delta = \det(A)$, and

$$\rho = a_{11}a_{22} + a_{11}a_{33} - a_{12}a_{21} + a_{22}a_{33} - a_{13}a_{31} - a_{23}a_{32}.$$

If $\lambda_{1,2} = \pm i\omega$ with $\omega \ne 0$, then

$$\sigma\rho - \Delta = 0$$

(*Hurwitz's condition*). The left-hand side of this condition can be rewritten as the determinant of some $3 \times 3$ matrix whose elements are simple linear combinations of the elements $a_{ij}$ of the matrix $A$:

$$\sigma\rho - \Delta = \det \begin{pmatrix} a_{11} + a_{22} & a_{23} & -a_{13} \\ a_{32} & a_{11} + a_{33} & a_{12} \\ -a_{31} & a_{21} & a_{22} + a_{33} \end{pmatrix}.$$

The matrix on the right-hand side is $2A \odot I_3$, where $\odot$ is the *bialternate matrix product* formally introduced by Definition 10.1. How can one come to such a definition?

**Definition 10.5** *Index pairs* $(i, j), (m, n)$ *are listed in the* lexicographic order *if either* $i < m$ *or* ($i = m$ *and* $j < n$).

For example, the three pairs $(2,1), (3,1), (3,2)$ are listed in the lexicographic order.

**Definition 10.6** *The* wedge product *of two vectors from* $\mathbb{C}^n$,

$$v = (v_1, v_2, \ldots, v_n)^T, \quad w = (w_1, w_2, \ldots, w_n)^T,$$

*is a vector* $v \wedge w \in \mathbb{C}^m$, *where* $m = \frac{1}{2}n(n-1)$, *with the components*

$$(v \wedge w)_{(i,j)} = v_i w_j - v_j w_i, \quad n \geq i > j \geq 1,$$

*listed in the lexicographic order of their index pairs.*

The wedge product of two planar real vectors,

$$v = \begin{pmatrix} v_1 \\ v_2 \end{pmatrix}, \quad w = \begin{pmatrix} w_1 \\ w_2 \end{pmatrix},$$

is a real number $(v \wedge w)_{(2,1)} = v_2 w_1 - v_1 w_2$. Its absolute value equals the *area* of the parallelogram

$$\Pi = \{x \in \mathbb{R}^2 : x = \alpha v + \beta w, \ \alpha, \beta \in [0,1]\}.$$

The wedge product of two vectors from the space $\mathbb{R}^3$ is also related to a well-known object. When two real vectors

$$v = \begin{pmatrix} v_1 \\ v_2 \\ v_3 \end{pmatrix}, \quad w = \begin{pmatrix} w_1 \\ w_2 \\ w_3 \end{pmatrix}$$

are given, their wedge product

$$v \wedge w = \begin{pmatrix} (v \wedge w)_{(2,1)} \\ (v \wedge w)_{(3,1)} \\ (v \wedge w)_{(3,2)} \end{pmatrix} = \begin{pmatrix} v_2 w_1 - v_1 w_2 \\ v_3 w_1 - v_1 w_3 \\ v_3 w_2 - v_2 w_3 \end{pmatrix}$$

is a vector in $\mathbb{R}^3$. Denoting this vector by $u = v \wedge w$, we see that

$$\begin{pmatrix} u_3 \\ -u_2 \\ u_1 \end{pmatrix} = w \times v,$$

where $\times$ is the standard *cross product* of two vectors in $\mathbb{R}^3$.

We formulate the following two lemmas without proof.

**Lemma 10.5** *For any* $v, w, w^{1,2} \in \mathbb{C}^n$, *and* $\lambda \in \mathbb{C}$ :
(*i*) $v \wedge w = -w \wedge v$;
(*ii*) $v \wedge (\lambda w) = \lambda(v \wedge w)$;
(*iii*) $v \wedge (w^1 + w^2) = v \wedge w^1 + v \wedge w^2$. □

**Lemma 10.6** *If* $e^i \in \mathbb{C}^n$, $n \geq i \geq 1$, *form a basis in* $\mathbb{C}^n$, *then* $e^i \wedge e^j \in \mathbb{C}^m$, $n \geq i > j \geq 1$, *form a basis in* $\mathbb{C}^m$. □

If $e^i (n \geq i \geq 1)$ form the standard basis in $\mathbb{C}^n$, the resulting basis in $\mathbb{C}^m$ is also called *standard*.

**Lemma 10.7** *Fix a basis $e^i \in \mathbb{C}^n$, $n \geq i \geq 1$, in $\mathbb{C}^n$ and consider two vectors $v, w \in \mathbb{C}^n$,*

$$v = \sum_i v_i e^i, \quad w = \sum_j w_j e^j,$$

*and a vector $u \in \mathbb{C}^m$ :*

$$u = \sum_{n \geq i > j \geq 1} u_{(i,j)} e^i \wedge e^j.$$

*If $u = v \wedge w$, then $u_{(i,j)} = v_i w_j - v_j w_i$, $n \geq i > j \geq 1$.*

**Proof:** For $n \geq i, j \geq 1$:

$$\begin{aligned} v \wedge w = \left(\sum_i v_i e^i\right) \wedge \left(\sum_j w_j e^j\right) &= \sum_{i,j}^n v_i w_j \, (e^i \wedge e^j) \\ &= \sum_{i>j} v_i w_j \, (e^i \wedge e^j) + \sum_{j>i} v_i w_j \, (e^i \wedge e^j) \\ &= \sum_{i>j} v_i w_j \, (e^i \wedge e^j) - \sum_{j>i} v_i w_j \, (e^j \wedge e^i) \\ &= \sum_{i>j} v_i w_j \, (e^i \wedge e^j) - \sum_{i>j} v_j w_i \, (e^i \wedge e^j) \\ &= \sum_{i>j} (v_i w_j - v_j w_i)(e^i \wedge e^j) = \sum_{i>j} u_{(i,j)} e^i \wedge e^j. \quad \square \end{aligned}$$

Consider two linear transformations of $\mathbb{C}^n$,

$$v \mapsto Av, \quad w \mapsto Bw,$$

where $A$ and $B$ are $n \times n$ complex matrices with elements $a_{pq}$ and $b_{rs}$, respectively.

**Definition 10.7** *The transformation of $\mathbb{C}^m$ defined by*

$$(v \wedge w) \mapsto (A \odot B)(v \wedge w) = \frac{1}{2}(Av \wedge Bw - Aw \wedge Bv)$$

*is called the* bialternate product *of the above transformations.*

In particular,

$$(A \odot A)(v \wedge w) = Av \wedge Aw$$

and

$$(2A \odot I_n)(v \wedge w) = Av \wedge w + v \wedge Aw.$$

**Theorem 10.6** *The bialternate product is a linear transformation of $\mathbb{C}^m$. Its matrix $A \odot B$ in the standard basis $e^i \wedge e^j$, $n \geq i > j \geq 1$, has the elements*

$$(A \odot B)_{(p,q),(r,s)} = \frac{1}{2} \left\{ \begin{vmatrix} a_{pr} & a_{ps} \\ b_{qr} & b_{qs} \end{vmatrix} + \begin{vmatrix} b_{pr} & b_{ps} \\ a_{qr} & a_{qs} \end{vmatrix} \right\},$$

*where $n \geq p > q \geq 1$ and $n \geq r > s \geq 1$.*

**Proof:**

$$\begin{aligned}
(A \odot B)(e^r \wedge e^s) &= \frac{1}{2}(Ae^r \wedge Be^s - Ae^s \wedge Be^r) \\
&= \frac{1}{2}\left[\sum_p a_{pr}e^p \wedge \sum_q b_{qs}e^q - \sum_p a_{ps}e^p \wedge \sum_q b_{qr}e^q\right] \\
&= \frac{1}{2}\left[\sum_{p,q} a_{pr}b_{qs}(e^p \wedge e^q) - \sum_{p,q} a_{ps}b_{qr}(e^p \wedge e^q)\right] \\
&= \frac{1}{2}\left[\sum_{p>q} a_{pr}b_{qs}(e^p \wedge e^q) - \sum_{p<q} a_{qr}b_{ps}(e^p \wedge e^q)\right. \\
&\qquad \left. - \sum_{p>q} a_{ps}b_{qr}(e^p \wedge e^q) + \sum_{p<q} a_{qs}b_{pr}(e_p \wedge e_q)\right] \\
&= \frac{1}{2}\sum_{p>q}(a_{pr}b_{qs} - a_{ps}b_{qr} + a_{qs}b_{pr} - a_{qr}b_{ps})(e^p \wedge e^q) \\
&= \frac{1}{2}\sum_{p>q}\left\{\begin{vmatrix} a_{pr} & a_{ps} \\ b_{qr} & b_{qs} \end{vmatrix} + \begin{vmatrix} b_{pr} & b_{ps} \\ a_{qr} & a_{qs} \end{vmatrix}\right\}(e^p \wedge e^q).
\end{aligned}$$

On the other hand,

$$(A \odot B)(e^r \wedge e^s) = \sum_{p>q}(A \odot B)_{(p,q),(r,s)}(e^p \wedge e^q). \quad \square$$

The statement of Theorem 10.6 was used as Definition 10.1. The expression $2A \odot I_n$ has a special name: It is called the *biproduct* of $A$. We have

$$(2A \odot I_n)_{(p,q),(r,s)} = \begin{cases} -a_{ps} & r = q, \\ a_{pr} & r \neq p \text{ and } s = q, \\ a_{pp} + a_{qq} & r = p \text{ and } s = q, \\ a_{qs} & r = p \text{ and } s \neq q, \\ -a_{qr} & s = p, \\ 0 & \text{otherwise.} \end{cases}$$

For $n = 2, 3$, and 4, this gives $2A \odot I_2 = a_{11} + a_{22}$,

$$2A \odot I_3 = \begin{pmatrix} a_{11} + a_{22} & a_{23} & -a_{13} \\ a_{32} & a_{11} + a_{33} & a_{12} \\ -a_{31} & a_{21} & a_{22} + a_{33} \end{pmatrix},$$

and

$$2A \odot I_4 = \begin{pmatrix} a_{11} + a_{22} & a_{23} & -a_{13} & a_{24} & -a_{14} & 0 \\ a_{32} & a_{11} + a_{33} & a_{12} & a_{34} & 0 & -a_{14} \\ -a_{31} & a_{21} & a_{22} + a_{33} & 0 & a_{34} & -a_{24} \\ a_{42} & a_{43} & 0 & a_{11} + a_{44} & a_{12} & a_{13} \\ -a_{41} & 0 & a_{43} & a_{21} & a_{22} + a_{44} & a_{23} \\ 0 & -a_{41} & -a_{42} & a_{31} & a_{32} & a_{33} + a_{44} \end{pmatrix}.$$

The following three lemmas follow directly from Definition 10.7.

**Lemma 10.8** *For any complex $n \times n$ matrices $A, B, B_{1,2}$, and $\lambda \in \mathbb{C}$:*
(i) $A \odot B = B \odot A$;
(ii) $A \odot (\lambda B) = \lambda(A \odot B)$;
(iii) $A \odot (B_1 + B_2) = A \odot B_1 + A \odot B_2$. □

**Lemma 10.9** *For any complex $n \times n$ matrices $A$ and $B$,*

$$(A \odot A)(B \odot B) = (AB \odot AB). \quad \square$$

**Lemma 10.10** *For any complex $n \times n$ matrix $A$ and any nonsingular complex $n \times n$ matrix $P$:*
(i) $(P \odot P)^{-1} = P^{-1} \odot P^{-1}$;
(ii) $(PAP^{-1}) \odot (PAP^{-1}) = (P \odot P)(A \odot A)(P \odot P)^{-1}$;
(iii) $2(PAP^{-1}) \odot I_n = (P \odot P)(2A \odot I_n)(P \odot P)^{-1}$. □

Now we can prove the central result of this appendix.

**Theorem 10.3 (Stéphanos [1900])** *If a complex $n \times n$ matrix $A$ has eigenvalues $\mu_1, \mu_2, \ldots, \mu_n$, then*
(i) $A \odot A$ *has eigenvalues* $\mu_i \mu_j$,
(ii) $2A \odot I_n$ *has eigenvalues* $\mu_i + \mu_j$,
*where* $n \geq i > j \geq 1$.

**Proof:**

Suppose that all $\mu_i$ are simple. Then the corresponding eigenvectors $v^i$, $n \geq i \geq 1$, compose a basis in $\mathbb{C}^n$. Thus $v^i \wedge v^j$, $n \geq i > j \geq 1$, form a basis in $\mathbb{C}^m$. In particular, they are all nonzero vectors.

Compute

$$(A \odot A)(v^i \wedge v^j) = Av^i \wedge Av^j = \mu_i v^i \wedge \mu_j v^j = \mu_i \mu_j (v^i \wedge v^j).$$

This means that $v^i \wedge v^j$ is an eigenvector of $A \odot A$ corresponding to the eigenvalue $\mu_i \mu_j$. Similarly

$$(2A \odot I_n)(v^i \wedge v^j) = Av^i \wedge v^j + v^i \wedge Av^j = \mu_i v^i \wedge v^j + \mu_j v^i \wedge v^j = (\mu_i + \mu_j)(v^i \wedge v^j).$$

This means that $v^i \wedge v^j$ is an eigenvector of $2A \odot I_n$ corresponding to the eigenvalue $\mu_i + \mu_j$.

When multiple eigenvalues are present, the theorem follows now from the above results and the continuity of the eigenvalues as the functions of the matrix elements. □

The next two lemmas are important in constructing test functions for codim 2 bifurcations along the Neimark-Sacker and Hopf bifurcation curves.

**Lemma 10.11** *Suppose that an $n \times n$ real matrix $A$ has a single pair of eigenvalues with $\mu_1 \mu_2 = 1$. Then*

$$\frac{\mu_1 + \mu_2}{2} = \frac{\langle v, v\rangle\langle w, Aw\rangle + \langle w, w\rangle\langle v, Av\rangle - \langle v, w\rangle\langle w, Av\rangle - \langle w, v\rangle\langle v, Aw\rangle}{\langle v, v\rangle\langle w, w\rangle - \langle v, w\rangle^2},$$

*where* $(A \odot A - I_m)(v \wedge w) = 0$. □

If $\mu_{1,2} = \cos\theta \pm i \sin\theta$, then $\frac{1}{2}(\mu_1 + \mu_2) = \cos\theta$. This function can be used to detect strong resonances along the Neimark-Sacker curve.

**Lemma 10.12** *Suppose that an $n \times n$ real matrix $A$ has a single pair of eigenvalues with $\lambda_1 + \lambda_2 = 0$. Then*

$$-\lambda_1 \lambda_2 = \frac{\langle v, Av\rangle\langle w, Aw\rangle - \langle w, Av\rangle\langle v, Aw\rangle}{\langle v, v\rangle\langle w, w\rangle - \langle v, w\rangle^2},$$

*where $(2A \odot I_n)(v \wedge w) = 0$.* □

If $\lambda_{1,2} = \pm i\omega$, then $-\lambda_1\lambda_2 = \omega^2$. When $\omega = 0$, a Bogdanov-Takens bifurcation is expected generically.

## 10.8 Appendix C: Detection of codim 2 homoclinic bifurcations

While following a codim 1 homoclinic orbit in two parameters, we can expect codim 2 singularities at certain points on the obtained curve. By definition, at such a point one of Shil'nikov's nondegeneracy conditions (see (H.0)–(H.3) in Section 6.4.3 of Chapter 6 and (SNH.1)–(SNH.3) in Section 7.1.2 of Chapter 7) is violated. First of all, the equilibrium $x_0$ to which the homoclinic orbits tend can lose hyperbolicity via a fold or Hopf bifurcation. These points are end points of regular homoclinic orbit loci. Curves corresponding to homoclinic orbits to saddle-node/saddle-saddle equilibria originate at the fold parameter values. There are other types of end points, namely, those corresponding to the "shrinking" of the homoclinic orbit to a point (as at the Bogdanov-Takens and other local codim 2 bifurcations) or the "breaking" of the orbit into parts by the appearance of a heteroclinic cycle formed by more than one orbit connecting several equilibria. We will not consider these cases, instead we focus on codim 1 cases where there is a unique homoclinic orbit to a hyperbolic or saddle-node equilibrium. The appearance of codim 2 homoclinic bifurcation points leads to dramatic implications for system dynamics which will not be discussed here (see Champneys & Kuznetsov [1994] for a review of two-parameter bifurcation diagrams respective to these points).

Codimension-two homoclinic bifurcations are detected along branches of codim 1 homoclinic curves by locating zeros of certain *test functions* $\psi_i$ defined in general for an appropriate truncated boundary-value problem. In the simplest cases, test functions are computable via eigenvalues of the equilibrium or their eigenvectors and from the homoclinic solution at the end points. In other cases we have to enlarge the boundary-value problem and simultaneously solve variational equations with relevant boundary conditions. A test function is said to be *well defined* if, for all sufficiently large $T > 0$, it is a smooth function along the solution curve of the truncated problem and has a regular zero approaching the critical parameter value as $T \to \infty$. In fact, in all cases presented below, we have the stronger property that the limit of the test function exists and gives a regular test function for the original problem on the infinite interval also.

Let us label the eigenvalues of $A(x_0, \alpha) = f_x(x_0, \alpha)$ with nonzero real part as in Section 10.3.3:

$$\text{Re } \mu_{n_-} \leq \cdots \leq \text{Re } \mu_1 < 0 < \text{Re } \lambda_1 \leq \cdots \leq \text{Re } \lambda_{n_+}.$$

In accordance to Chapter 6, the eigenvalues with zero real part are called *critical*, while the stable (unstable) eigenvalues with real part closest to zero are termed the *leading* stable (unstable) eigenvalues. In this appendix we assume that all eigenvalues and necessary eigenvectors of $A(x_0, \alpha)$ (and its transpose $A^T(x_0, \alpha)$) can be accurately computed along the homoclinic curve.

### 10.8.1 Singularities detectable via eigenvalues

The following test functions can be monitored along a homoclinic curve corresponding to a hyperbolic equilibrium to detect codim 2 singularities which are well defined for both the original and truncated (10.97) problems.

*Neutral saddle*:

$$\psi_1 = \mu_1 + \lambda_1.$$

*Double real stable leading eigenvalue*:

$$\psi_2 = \begin{cases} (\text{Re } \mu_1 - \text{Re } \mu_2)^2, \text{ Im } \mu_1 = 0, \\ -(\text{Im } \mu_1 - \text{Im } \mu_2)^2, \text{ Im } \mu_1 \neq 0. \end{cases}$$

*Double real unstable leading eigenvalue*:

$$\psi_3 = \begin{cases} (\text{Re } \lambda_1 - \text{Re } \lambda_2)^2, \text{ Im } \lambda_1 = 0, \\ -(\text{Im } \lambda_1 - \text{Im } \lambda_2)^2, \text{ Im } \lambda_1 \neq 0. \end{cases}$$

Notice that the regularity of $\psi_{2,3}$ follows from the fact that they represent the discriminant of the quadratic factor of the characteristic polynomial corresponding to this pair of eigenvalues.

*Neutral saddle, saddle-focus, or focus-focus*:

$$\psi_4 = \text{Re } \mu_1 + \text{Re } \lambda_1.$$

*Neutrally divergent saddle-focus*:

$$\psi_5 = \text{Re } \mu_1 + \text{Re } \mu_2 + \text{Re } \lambda_1, \ \psi_6 = \text{Re } \lambda_1 + \text{Re } \lambda_2 + \text{Re } \mu_1.$$

*Three leading eigenvalues*:

$$\psi_7 = \text{Re } \mu_1 - \text{Re } \mu_3, \ \psi_8 = \text{Re } \lambda_1 - \text{Re } \lambda_3.$$

In order to detect homoclinic orbits to nonhyperbolic equilibria while continuing a locus of hyperbolic homoclinics, the truncated problem (10.97) should be formulated in such a way that it can be continued *through* the degenerate point. To this end it is necessary to modify the eigenvalue labeling to label as $\mu_i$ the $n_-$ leftmost eigenvalues and as $\lambda_i$ the $n_+$ rightmost eigenvalues irrespective of their location with respect to the imaginary axis. This accordingly modifies the meaning of the terms "stable" and "unstable" in the definition of the projection matrices $L_{u,s}$ in (10.95). With this modification, we can simply define the following test functions.

*Nonhyperbolic equilibria*:

$$\psi_9 = \text{Re } \mu_1, \ \psi_{10} = \text{Re } \lambda_1.$$

A zero of $\psi_{9,10}$ corresponds to either a fold or Hopf bifurcation of the continued equilibrium $x_0$. In the first case the bifurcation is called a *noncentral saddle-node*

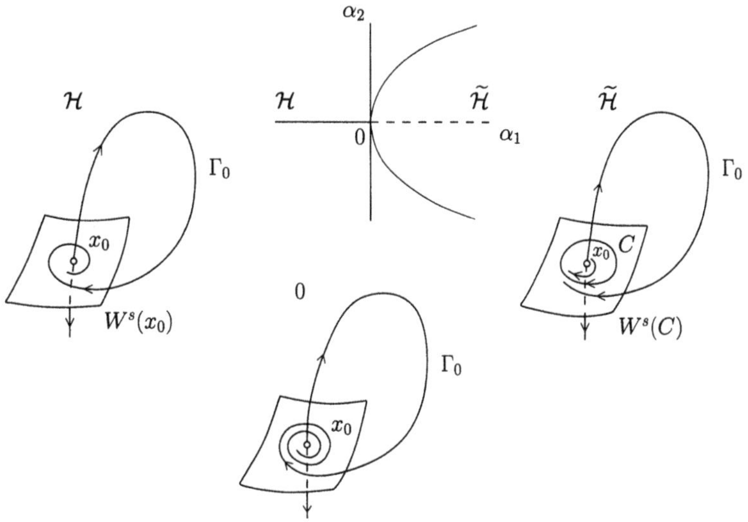


**Fig. 10.19.** Continuation through a Shil'nikov-Hopf with $n = 3$: The homoclinic locus is denoted by $\mathcal{H}$ and the point-to-periodic heteroclinic curve by $\widetilde{\mathcal{H}}$.

*homoclinic* bifurcation (see below), while the second one is usually referred to as

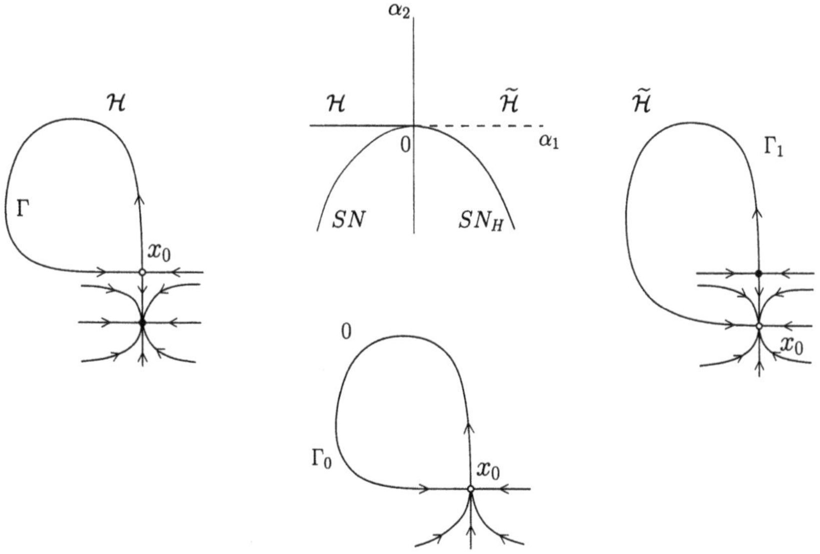


**Fig. 10.20.** Continuation through a noncentral saddle-node homoclinic bifurcation. The hyperbolic homoclinic curve is denoted by $\mathcal{H}$, the noncentral heteroclinic curve by $\widetilde{\mathcal{H}}$, and the curve of folds by $SN$. Along the right-hand branch $SN_H$ of the fold curve there also exists a central saddle-node homoclinic.

a *Shil'nikov-Hopf* bifurcation. Generically these singularities are end points of a locus of homoclinic orbits to hyperbolic equilibria. However, there exist continuous extensions of the solution curves of the truncated boundary-value problem (10.97) through both singularities. Beyond the Shil'nikov-Hopf codim 2 point, there exists a solution curve $\widetilde{\mathcal{H}}$ approximating a *heteroclinic connection* $\Gamma_1$ between $x_0$ and the limit cycle $C$ appearing via the Hopf bifurcation (see **Fig. 10.19**). A similar property holds for the saddle-node bifurcation. Suppose we continue a saddle homoclinic curve $\mathcal{H}$ toward a fold bifurcation of $x_0$, which means that an extra equilibrium approaches $x_0$. Assume that at the saddle-node point an unstable eigenvalue approaches zero. Beyond the codim 2 point, the truncated boundary-value problem (10.97) has a solution approximating a heteroclinic orbit connecting the two equilibria along the nonleading stable manifold and existing along $\widetilde{\mathcal{H}}$ (see **Fig. 10.20** for a planar illustration). More precisely, the continuation algorithm switches from the original equilibrium to the approaching one,[10] while the projection boundary conditions place the end points of the solution in the "stable" and "unstable" eigenspaces of the new $x_0$. Close to the codim 2 point, the latter gives a good approximation to the unstable eigenspace of the original equilibrium.

### 10.8.2 Orbit and inclination flips

Now consider test functions for two forms of *global* degeneracy along a curve of homoclinic orbits to a saddle, namely *orbit-* and *inclination-flip* bifurcations. Therefore we additionally assume that there are no eigenvalues of $A(x_0, \alpha)$ with zero real part, while the leading eigenvalues are real and simple, that is,

$$\text{Re}\ \mu_{n_-} \leq \cdots \leq \text{Re}\ \mu_2 < \mu_1 < 0 < \lambda_1 < \text{Re}\ \lambda_2 \leq \cdots \leq \text{Re}\ \lambda_{n_+}.$$

Then, it is possible to choose normalized eigenvectors $p_1^s$ and $p_1^u$ of $A^T(x_0, \alpha)$ depending smoothly on $(x_0, \alpha)$ and satisfying

$$A^T(x_0, \alpha)p_1^s = \mu_1 p_1^s, \quad A^T(x_0, \alpha)p_1^u = \lambda_1 p_1^u.$$

Here and in what follows the dependence on $x_0$ and $\alpha$ of eigenvalues and eigenvectors is not indicated, for simplicity. Accordingly, normalised eigenvectors $q_1^s$ and $q_1^u$ of $A(x_0, \alpha)$ are chosen depending smoothly on $(x_0, \alpha)$ and satisfying

$$A(x_0, \alpha)q_1^s = \mu_1 q_1^s, \quad A(x_0, \alpha)q_1^u = \lambda_1 q_1^u.$$

An orbit flip bifurcation occurs when the homoclinic orbit changes its direction of approach to the saddle between the two components of a leading eigenvector. The defining equation for the orbit-flip bifurcation (with respect to the stable manifold) can be written as

$$\lim_{t\to\infty} e^{-\mu_1 t} \langle p_1^s, x(t) - x_0 \rangle = 0, \qquad \text{(C.1)}$$

where $x(t)$ is a homoclinic solution and the growth of the exponential factor counterbalances the decay of $\|x(t) - x_0\|$ as $t \to \infty$. Similarly, the equation for the orbit-flip with respect to the unstable manifold is given by

$$\lim_{t\to-\infty} e^{-\lambda_1 t} \langle p_1^u, x(t) - x_0 \rangle = 0. \qquad \text{(C.2)}$$

[10] Beyond the critical point it is the approaching equilibrium that is labeled $x_0$.

At a point where either condition (C.1) or (C.2) is fulfilled, the homoclinic orbit tends to the saddle (in one time direction) along its nonleading eigenspace (see **Fig. 10.21** for an illustration in three dimensions). The truncated test functions which

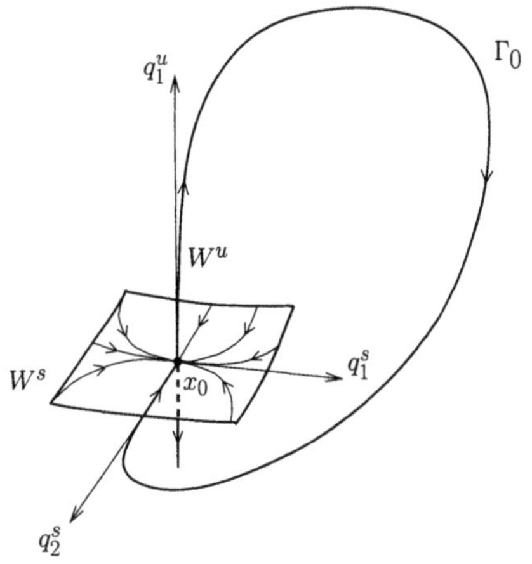


**Fig. 10.21.** Orbit flip with respect to the stable manifold in $\mathbb{R}^3$.

should be evaluated along the solution curve of (10.97) are therefore given by:

*Orbit-flip* (*with respect to the stable manifold*):

$$\psi_{11} = e^{-\mu_1 T} \langle p_1^s, x(+T) - x_0 \rangle.$$

*Orbit-flip* (*with respect to the unstable manifold*):

$$\psi_{12} = e^{\lambda_1 T} \langle p_1^u, x(-T) - x_0 \rangle.$$

The inclination-flip bifurcation is related to global twistedness of the stable and unstable manifolds $W^{s,u}(x_0)$ of the saddle $x_0$ around its homoclinic orbit. Recall from Chapter 6 that at each point $x(t)$ of the homoclinic orbit the sum of tangent spaces

$$Z(t) = X(t) + Y(t)$$

is defined, where

$$X(t) = T_{x(t)} W^s(x_0), \quad Y(t) = T_{x(t)} W^u(x_0).$$

Generically, codim $Z(t) = 1$, that is, $X(t) \cap Y(t) = \text{span}\{\dot{x}(t)\}$. In the three-dimensional case, the space $Z(t)$ merely coincides with the plane $X(t)$ tangent to $W^s(x_0)$ at a point $x(t)$ in $\Gamma$ (see **Fig. 10.22**). In order to describe the defining equations for the inclination-flip bifurcation we have to introduce the *adjoint variational problem*

$$\begin{cases} \dot{\varphi}(t) = -A^T(x(t), \alpha)\varphi(t), \\ \varphi(t) \to 0 \ \text{ as } \ t \to \pm\infty, \\ \int_{-\infty}^{\infty} \langle \varphi(t) - \varphi^0(t), \varphi^0(t) \rangle \, dt = 0, \end{cases} \tag{C.3}$$

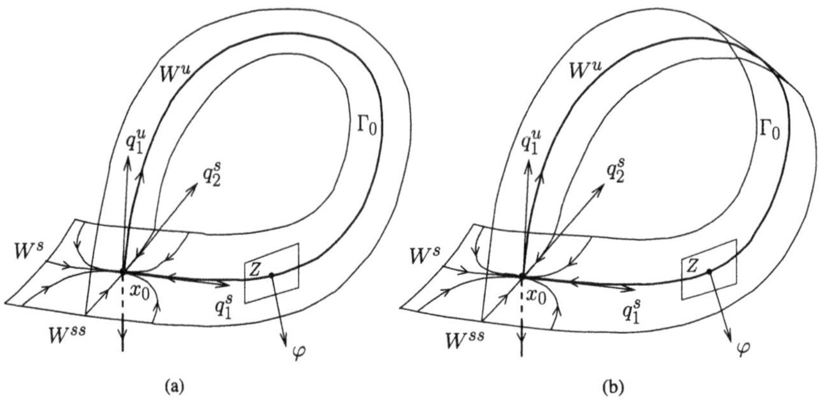


**Fig. 10.22.** (a) Simple and (b) twisted homoclinic orbits.

where $A(x,\alpha) = f_x(x,\alpha)$. The first equation in (C.3) is the adjoint variational equation introduced in Section 6.4.1 of Chapter 6. The integral phase condition with a reference vector-function $\varphi^0(\cdot)$ selects one solution out of the family $c\varphi(t)$ for $c \in \mathbb{R}^1$. The solution $\varphi(t)$ is orthogonal to the above defined subspace $Z(t)$ for each $t$, thus, its limit behavior as $t \to \pm\infty$ determines the twistedness of the space $Z(t)$ around the homoclinic orbit. Inclination-flip bifurcations occur at points along a homoclinic curve where this twistedness changes without an orbit flip occurring. The defining equations for the inclination-flip bifurcation with respect to the stable manifold are given by

$$\lim_{t\to-\infty} e^{\mu_1 t}\,\langle q_1^s, \varphi(t)\rangle = 0, \tag{C.4}$$

and with respect to the unstable manifold by

$$\lim_{t\to\infty} e^{\lambda_1 t}\,\langle q_1^u, \varphi(t)\rangle = 0, \tag{C.5}$$

where the exponential factors neutralize the decay of $\|\varphi(t)\|$ as $t \to \pm\infty$. If either (C.4) or (C.5) holds, the stable (unstable) manifolds of the saddle $x_0$ are neutrally twisted around the homoclinic orbit (see **Fig. 10.23** for a three-dimensional illustration).

Next we define $P_s(x_0,\alpha)$ to be the $(n_- \times n)$ matrix whose rows form a basis for the stable eigenspace of $A(x_0,\alpha)$. Similarly, $P_u(x_0,\alpha)$ is a $(n_+ \times n)$ matrix, such that its rows form a basis for the unstable eigenspace of $A(x_0,\alpha)$. Consider now replacing (C.3) by the truncated equations

$$\left\{\begin{aligned} \dot\varphi(t) + A^T(x(t),\alpha)\varphi(t) + \varepsilon f(x(t),\alpha) &= 0,\\ P_s(x_0,\alpha)\varphi(+T) &= 0,\\ P_u(x_0,\alpha)\varphi(-T) &= 0,\\ \textstyle\int_{-T}^{T}\langle \varphi(t)-\varphi^0(t), \varphi^0(t)\rangle\, dt &= 0. \end{aligned}\right. \tag{C.6}$$

Here, $\varepsilon \in \mathbb{R}^1$ is an artificial free parameter, which turns (C.6) into a well-posed boundary-value problem and remains almost zero along its solution curve. Evaluating the limits in (C.4) and (C.5) at $t = \pm T$, yields the following test functions.

*Inclination-flip* (*with respect to the stable manifold*):

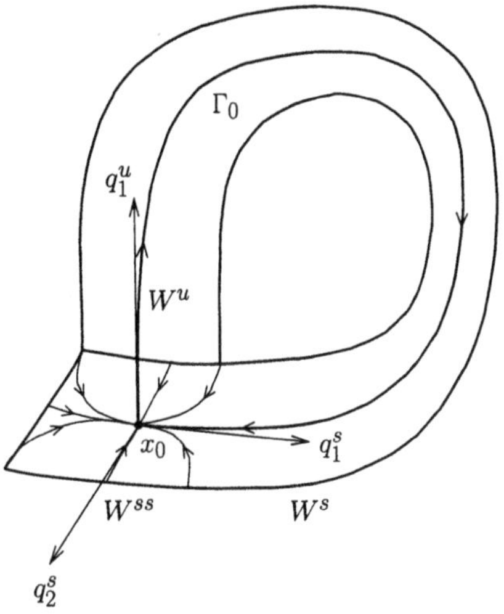


**Fig. 10.23.** Inclination flip with respect to the stable manifold in $\mathbb{R}^3$.

$$\psi_{13} = e^{-\mu_1 T} \langle q_1^s, \varphi(-T) \rangle.$$

*Inclination-flip (with respect to the unstable manifold)*:

$$\psi_{14} = e^{\lambda_1 T} \langle q_1^u, \varphi(+T) \rangle.$$

To evaluate practically the truncated test functions $\psi_{13,14}$ along the homoclinic curve, we have to continue the solution to the extended BVP composed of (10.97) and (C.6):

$$\left\{ \begin{aligned} f(x_0, \alpha) &= 0, \\ \dot{x}(t) - f(x(t), \alpha) &= 0, \\ L_s(x_0, \alpha)(x(-T) - x_0) &= 0, \\ L_u(x_0, \alpha)(x(T) - x_0) &= 0, \\ \int_{-T}^{T} \langle x(t) - x^0(t), \dot{x}^0(t) \rangle \, dt - 1 &= 0, \\ \dot{\varphi}(t) + A^T(x(t), \alpha)\varphi(t) + \varepsilon f(x(t), \alpha) &= 0, \\ P_s(x_0, \alpha)\varphi(+T) &= 0, \\ P_u(x_0, \alpha)\varphi(-T) &= 0, \\ \int_{-T}^{T} \langle \varphi(t) - \varphi^0(t), \varphi^0(t) \rangle \, dt &= 0. \end{aligned} \right.$$

This produces the (approximations to) homoclinic solution $x(t)$ and the bounded solution $\varphi(t)$ to the adjoint variational equation simultaneously.

The test functions $\psi_{11,12,13,14}$ are well defined for both the original and truncated boundary-value problems.

### 10.8.3 Singularities along saddle-node homoclinic curves

Suppose that a generic saddle-node homoclinic orbit is continued. Recall that in this case the truncated boundary-value problem is composed of equations (10.97) and (10.99). Let $p_0$ be a null-vector of $A^T(x_0, \alpha)$ normalized according to

$$\langle p_0, p_0 \rangle = 1$$

and differentiable along the saddle-node homoclinic curve. Then the following test

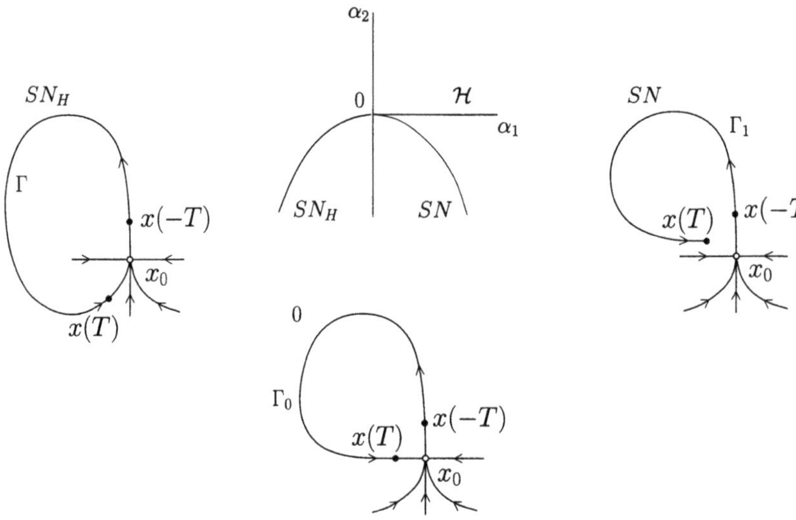


**Fig. 10.24.** Continuation through a noncentral saddle-node homoclinic bifurcation while following a central saddle-node homoclinic curve.

functions will detect *noncentral saddle-node homoclinic bifurcations*, where the closure of the homoclinic orbit becomes *nonsmooth* (i.e. the condition (SNH.1) of Section 7.1 in Chapter 7 is violated).

*Noncentral saddle-node homoclinic orbit:*

$$\psi_{15} = \frac{1}{T}\langle x(+T) - x_0), p_0 \rangle, \tag{C.7}$$

$$\psi_{16} = \frac{1}{T}\langle x(-T) - x_0), p_0 \rangle. \tag{C.8}$$

These functions measure the component of the one-dimensional center manifold in which the two end points of the approximate homoclinic orbit lie, and are well defined along the saddle-node bifurcation curve (see 10.24). Both $\psi_{15}$ and $\psi_{16}$ converge to smooth functions along the curve of central saddle-node homoclinic orbits as $T \to \infty$. Which of the test functions (C.7), (C.8) is annihilated is determined by whether the critical homoclinic orbit is a center-to-stable or unstable-to-center connection.

**Remarks:**

(1) The test functions $\psi_{9,10}$ and $\psi_{15,16}$ provide us with two different strategies for detecting noncentral saddle-node homoclinic orbits. This allows us to switch between the continuation of saddle and central saddle-node homoclinic orbits at such points.

(2) A *nontransverse saddle-node homoclinic* bifurcation (where the condition (SNH.3) from Section 7.1 is violated) can be detected as a limit point with respect to the parameter along a curve of central saddle-node homoclinic bifurcations. ◇

## 10.9 Appendix D: Bibliographical notes

The literature on numerical bifurcation analysis is vast and grows rapidly, together with computer software developed to support the analysis of dynamical systems. We can recommend the lectures by Beyn [1991] and the two tutorial papers by Doedel, Keller & Kernévez [1991*a*, 1991*b*] as good general introductions to numerical methods for dynamical systems (see also Guckenheimer & Worfolk [1993] and, in particular, Guckenheimer [2002]). More advanced readers will enjoy [Govaerts 2000]. For a comprehansive survey on the numerical bifurcation analysis, see [Beyn, Champneys, Doedel, Govaerts, Kuznetsov & Sandstede 2002].

The theory of Newton's method to locate solutions of nonlinear systems is fully presented in Kantorovich & Akilov [1964], for example. Rank-one updates to improve iteratively the quality of the Jacobian matrix approximation in Newton-like iterations were first introduced by Broyden [1965] using certain minimal conditions. A detailed convergence analysis of the Broyden method can be found in Dennis & Schnabel [1983].

An algorithm to compute the Taylor expansions of the stable and unstable invariant manifolds of a hyperbolic equilibrium was derived by Hassard [1980]. Our presentation in Section 10.1.3 that uses the projection technique is based on Kuznetsov [1983]. Computation of global two-dimensional invariant manifolds in ODEs (including stable and unstable invariant manifolds of equilibria and limit cycles, as well as invariant tori) has recently attracted much attention, see for example, Homburg, Osinga & Vegter [1995], Dieci & Lorenz[1995, 1997], Krauskopf & Osinga [1999], and Osinga [2003].

The integral phase condition for limit cycle computation was first proposed by Doedel [1981]. Computation of limit cycles by shooting and multiple shooting has been implemented by Khibnik [1979] and by Holodniok & Kubiček [1984*a*]. A combination of the shooting methods with automatic differentiation techniques has been studied by Guckenheimer & Meloon [2000]. Systematic usage of orthogonal collocation for computation of limit cycles is due to Doedel & Kernévez [1986] (see also Doedel, Jepson & Keller [1984]). General convergence theorems for collocation at Gaussian points have been established by de Boor & Swartz [1973]. An efficient algorithm, taking into account the special structure of matrices arising in Newton's method applied to the orthogonal collocation discretization, has been implemented by Doedel [1981]. It allows for the simultaneous computation of the cycle multipliers (see Fairgrieve & Jepson [1991] for an improvement).

Since equilibrium computation and many bifurcation problems can be reduced to continuation of implicitly defined curves, a number of continuation packages have been developed. They are all based on the predictor-corrector method with slight variations. Early standard codes to solve continuation problems were developed by Kubiček [1976] (DERPAR) and Balabaev & Lunevskaya [1978] (CURVE). Note that the Moore-Penrose inverse was implemented in CURVE to perform the corrections. Since then, several universal continuation codes have appeared. We mention PATH by Kaas-Petersen [1989], PITCON by Rheinboldt & Burkardt [1983], BEETLE by Nikolaev (see Khibnik, Kuznetsov, Levitin & Nikolaev [1993]), ALCON by Deuflhard, Fiedler & Kunkel [1987], several illustrative codes in the book by Allgower & Georg [1990], as well as the continuation segments of the bifurcation codes AUTO86/97 by Doedel & Kernévez [1986], BIFPACK by Seydel [1991], and CONTENT by Kuznetsov and Levitin [1997]. The survey paper by Allgower & Georg

[1993] contains some information on the availability of these and other continuation codes. The general theory of continuation is developed and presented by Keller [1977] (who introduced pseudo-arclength continuation), Rheinboldt [1986], Seydel [1988, 1991] and Allgower & Georg [1990].

The detection and location of bifurcation points based on monitoring the eigenvalues (multipliers) is implemented in AUTO86. Test functions to locate Hopf and Neimark-Sacker bifurcations based on Hurwitz determinants have been proposed by Khibnik [1990] and implemented in the code LINLBF, where they are also used for the continuation of the corresponding bifurcations by the minimally augmented approach. The detection of Hopf bifurcations with the help of the bialternate matrix product is essentially due to Fuller [1968] (who refers to even earlier contributions by Stéphanos [1900]). Jury & Gutman [1975] used the bialternate product to detect the Neimark-Sacker bifurcation of maps. This approach has been reintroduced within a bifurcation/continuation framework by Guckenheimer & Myers [1996] and Govaerts, Guckenheimer & Khibnik [1997], successfully applied to several problems, and implemented in CONTENT. Appendix B introduces the bialtretnate product in the most straightforward manner. An alternative approach based on the tensor matrix product can be found in [Govaerts 2000]. The regularity theorem for the resulting minimally augmented system is due to Guckenheimer, Myers & Sturmfels [1997]. Several test functions appropriate for large systems have been developed by Moore, Garret & Spence [1990] and Friedman [2001]. In the latter work, the continuation of low-dimensional invariant subspaces is used [Dieci & Friedman 2001]. Modified minimally augmented systems, where the defining function for the bifurcation is computed by solving a bordered linear system, were proposed by Griewank & Reddien [1984] for the fold continuation and by Govaerts et al. [1997] for the Hopf continuation. Other methods to continue Hopf bifurcation using bordered systems are discussed by Beyn [1991], Chu, Govaerts & Spence [1994], and Werner [1996]. General properties of bordered matrices were established by Govaerts & Pryce [1993] and presented in detail in [Govaerts 2000]. The location and continuation of fold and Hopf points via extended augmented systems has been considered by several authors, including Moore & Spence [1980], Rheinboldt [1982], Roose & Hlavaček [1985], and Holodniok & Kubiček [1984*b*]. These methods are implemented in AUTO86 (see Doedel, Keller & Kernévez [1991*a*]) and CONTENT. The solution of the linear systems arising at each Newton correction can be efficiently implemented by using their special structure.

Detection and analysis of branching points (called "bifurcation points") can be found in Allgower & Georg [1990]. Regular extended defining systems to locate the branching points are proposed by Moore [1980] and Mei [1989]. The program STAFF by Borisyuk [1981] (based on the program CURVE) detects branching points and supports branch switching. So do the programs AUTO86 and CONTENT, as well as several other bifurcation codes. Note that this feature allows switching to the continuation of the double-period cycle at a flip bifurcation point. AUTO86, CONTENT, and other programs can also start cycle continuation from a Hopf point.

Verification of the nondegeneracy condition for the fold bifurcation is supported by LINLBF, where formula (10.49) is used. Numerical computation of the first Lyapunov coefficient is performed by the code BIFOR2 by Hassard, Kazarinoff & Wan [1981] and by the corresponding part of the program LINLBF by Khibnik. Both programs use numerical differentiation and compute intermediate quadratic coefficients of the center manifold. The algorithm presented in Section 10.2.3 differs from those just mentioned in using the invariant expression for $l_1$ and requiring only the directional derivatives. It is implemented in CONTENT.

Two-parameter continuation of all equilibrium, fixed-point, and cycle codim 1 bifurcations, as well as many codim 2 bifurcations in three parameters, is supported by the program LINLBF and its variants (see Khibnik [1990], Khibnik, Kuznetsov, Levitin & Nikolaev [1993]). Codim 1 bifurcations of equilibria and cycles can be continued by AUTO86, CANDYS/QA [Feudel & Jansen 1992], CONTENT, and MATCONT [Dhooge, Govaerts & Kuznetsov 2003]. Moreover, CONTENT supports the three-parameter continuation of all codim 2 equilibrium bifurcations (see Govaerts, Kuznetsov & Sijnave [2000*a*, 2000*b*]). MATCONT implements the continuation of all limit cycle codim 1 bifurcations using bordered boundary-value problems as proposed in [Doedel, Govaerts & Kuznetsov 2003].

One way to locate a homoclinic orbit numerically is by the continuation of a limit cycle to large periods (see Doedel & Kernévez [1986]). A shooting technique for homoclinic orbit location and continuation was implemented into the code LOOPLN by Kuznetsov [1983] (see also Kuznetsov [1990, 1991]). Rodríguez-Luis, Freire & Ponce [1990] also developed a homoclinic continuation method based on shooting. Since these approaches have obvious limitations, boundary-value methods to locate and continue codim 1 homoclinic bifurcation have been proposed and analyzed by Hassard [1980], Miura [1982], Beyn [1990*b*, 1990*a*], Doedel & Friedman [1989], Friedman & Doedel [1991, 1993], Schecter [1993], Bai & Champneys [1996], Sandstede [1997*b*], and Demmel, Dieci & Friedman [2000]. Champneys & Kuznetsov [1994] and Champneys, Kuznetsov & Sandstede [1996] have extended these BVP methods to deal with codim 2 homoclinic bifurcations, including orbit and inclination flips (see Appendix C), and wrote a standard AUTO86 driver HomCont [Champneys, Kuznetsov & Sandstede 1995] for these problems. There is a way to obtain a good starting solution for the homoclinic continuation by a *homotopy method* (see Doedel, Friedman & Monteiro [1994], Champneys & Kuznetsov [1994], and Doedel, Friedman & Kunin [1997]). Beyn [1991, 1994] has developed a method to start the homoclinic curve from a Bogdanov-Takens bifurcation, while Gaspard [1993] proposed an algorithm to start such curves from a fold-Hopf point.

Numerical bifurcation analysis of maps was not considered in this chapter. Location, analysis, and continuation of fixed-point bifurcations are very similar to those for equilibria of ODEs (and are supported by AUTO97, LOCBIF, and CONTENT; see, for example, Govaerts, Kuznetsov & Sijnave [1999]). Other problems require special algorithms. Such algorithms were developed for the computation of the stable and unstable invariant manifolds of fixed points by You, Kostelich & Yorke [1991], Lou, Kostelich & Yorke [1992], Krauskopf & Osinga [1998] (and implemented in DYNAMICS [Nusse & Yorke 1998], and the latest version of DsTool [Back, Guckenheimer, Myers, Wicklin & Worfolk 1992]); for the continuation of homoclinic orbits and their tangencies by Beyn & Kleinkauf [1997]; and for the computation of normally hyperbolic invariant manifolds (i.e., closed invariant curves) by Broer, Osinga & Vegter [1997] and Edoh & Lorenz [2001].

Systematic bifurcation analysis requires repeated continuation of different phase objects, and detection and analysis of their singularities and branch switching. These computations produce a lot of numerical data that should be analyzed and finally presented in graphical form. Thus, continuation programs should not only be efficient numerically but should allow for interactive management and have a user-friendly graphics interface. The development of such programs is progressing rapidly. One of the most popular continuation/bifurcation programs, AUTO86, comes with a simple interactive graphics browser called PLAUT that allows for graphical presentation of computed data. There are versions of PLAUT for most of the widespread workstations, as well as a MATLAB version, `mplaut`,[11] written by O. De Feo. There were several attempts to improve the user interface of AUTO. A special interactive version of AUTO86 was developed at Princeton by Taylor & Kevrekidis [1990] for SGI workstations. The program XPPAUT for workstations and PCs is another example.[12]. It also performs simulations and computes one-dimensional invariant manifolds of equilibria (for a description of a recent version of XPPAUT, see Ermentrout [2002]). An interactive version, AUTO94, for UNIX workstations with X11 was also designed by E. Doedel, X. Wang, and T. Fairgrieve. This version has extended numerical capabilities, including the continuation of all codim 1 bifurcations of limit cycles and fixed points. The software was upgraded in 1997 and now supports the continuation of homoclinic orbits using HomCont. It is called AUTO97 [Doedel, Champneys, Fairgrieve, Kuznetsov, Sandstede & Wang 1997].[13] There is also a `C`-version, AUTO2000. For IBM-PC compatible computers, an interactive DOS program LOCBIF has been developed by Khibnik, Kuznetsov, Levitin & Nikolaev [1993]. The numerical part of the program is based on the noninteractive code LINLBF and allows for continuation of equilibrium, fixed-point, and limit cycle bifurcations up to codim 3. LOCBIF supported much of the continuation strategy described in Section 10.4 of this chapter.[14] A popular simulation program, DsTool[15] [Back et al. 1992], incorporates the numerical part of LOCBIF and emulates its interface.

The new generation of continuation/bifurcation software is represented by the interactive program CONTENT[16] developed by Yu.A. Kuznetsov and V.V. Levitin with contributions by O. De Feo, B. Sijnave, W. Govaerts, E. Doedel, and A.R. Skovoroda. It runs on most popular workstations under UNIX/X11/ (Open)Motif and on PCs under Linux or MS-Windows 95/NT/98/2000/ME/XP and supports numerical computation of orbits, continuation of equilibria (fixed points) and cycles, detection and normal form analysis of local bifurcations, their continuation in two and three parameters, and branch switching, as described in this chapter and in Kuznetsov, Levitin & Skovoroda [1996] and Govaerts, Kuznetsov & Sijnave [1999, 2000*a*, 2000*b*]. Two new software projects are based on CONTENT: (1) MATCONT [Dhooge et al. 2003], a MATLAB interactive toolbox for the continuation and bifurcation analysis of ODEs; (2) WEBCONT, an Internet client-server application originally developed by J. Val at the University of Amsterdam.

The development of computer algebra has an impact on dynamical system studies (see, e.g., a collection of papers edited by Tournier [1994] on computer algebra and differential equations). One can attempt to locate equilibria of a polynomial system using methods from commutative algebra implemented in popular symbolic manipulation systems such as MAPLE [Char, Geddes, Gonnet, Leong, Monagan &

[11] Available at `http://www.math.uu.nl/people/kuznet/cm`.

[12] XPPAUT is available via `http://www.math.pitt.edu/~bard/xpp/xpp.html`.

[13] AUTO97 is available via `http://indy.cs.concordia.ca/auto/main.html`.

[14] LOCBIF is freely available at `http://ftp.cwi.nl/pub/yuri/LOCBIF` but is no longer supported.

[15] DsTool is available via `http://www.cam.cornell.edu/~gucken/dstool`.

[16] CONTENT is available via `http://www.math.uu.nl/people/kuznet/CONTENT`.

Watt 1991*a*], [Char, Geddes, Gonnet, Leong, Monagan & Watt 1991*b*], Mathematica [Wolfram 1991], or REDUCE [Hearn 1993]. A very good presentation of the relevant notions (*ideals, varieties, Groëbner basis*, etc.) and algorithms is given by Cox, Little & O'Shea [1992]. Another important field of application of symbolic manipulations is the theory of normal forms (see, e.g., Chow, Drachman & Wang [1990], Sanders [1994], Murdock [2003]). Actually, the most complex expressions for normal form coefficients given in this book have been obtained with MAPLE. Finally, let us mention an interactive system SYMCON for continuation/bifurcation analysis of equilibria of symmetric systems of ODEs by Gatermann & Hohmann [1991], in which a combination of symbolic and numerical methods is implemented, and a variant of automatic differentiation implemented in CONTENT by Levitin [1995].

# A

# Basic Notions from Algebra, Analysis, and Geometry

In this appendix we summarize for the convenience of the reader some basic mathematical results that are assumed to be known in the main text. Of course, reading this appendix cannot substitute for a systematic study of the corresponding topics via standard textbooks.

## A.1 Algebra

### A.1.1 Matrices

Let $A$ be an $n \times m$ matrix with complex elements $a_{jk} \in \mathbb{C}^1$:

$$A = \begin{pmatrix} a_{11} & a_{12} & \cdots & a_{1m} \\ a_{21} & a_{22} & \cdots & a_{2m} \\ \cdots & & & \\ a_{n1} & a_{n2} & \cdots & a_{nm} \end{pmatrix},$$

and let $A^T$ denote its *transpose*:

$$A^T = \begin{pmatrix} a_{11} & a_{21} & \cdots & a_{n1} \\ a_{12} & a_{22} & \cdots & a_{n2} \\ \cdots & & & \\ a_{1m} & a_{2m} & \cdots & a_{nm} \end{pmatrix}.$$

The *product* of an $n \times m$ matrix $A$ and an $m \times l$ matrix $B$ is the $n \times l$ matrix $C = AB$ with the elements

$$c_{ij} = \sum_{k=1}^{m} a_{ik} b_{kj}, \quad i = 1, 2, \ldots, n, \quad j = 1, 2, \ldots, l.$$

The following property holds:

$$(AB)^T = B^T A^T.$$

The *determinant* of a square $n \times n$ matrix $A$ is the complex number defined by

$$\det A = \sum_{(i_1, i_2, \ldots, i_n) \in S_n} (-1)^{\delta(i_1, i_2, \ldots, i_n)} a_{i_1 1} a_{i_2 2} \cdots a_{i_n n},$$

where $S_n$ is the set of all permutations of $n$ indices, and $\delta = 0$ when the multiindex $(i_1, i_2, \ldots, i_n)$ can be obtained from the multi-index $(1, 2, \ldots, n)$ by an even number of one-step permutations; $\delta = 1$ otherwise. A square matrix $A$ is called *nonsingular* if $\det A \neq 0$. For a nonsingular matrix $A$, there is the *inverse* matrix $A^{-1}$, such that $AA^{-1} = A^{-1}A = I$, where $I$ is the unit (identity) $n \times n$ matrix

$$I_n = \begin{pmatrix} 1 & 0 & 0 & \cdots & 0 \\ 0 & 1 & 0 & \cdots & 0 \\ \cdots \\ 0 & 0 & \cdots & 0 & 1 \end{pmatrix}.$$

If $A$ and $B$ are two $n \times n$ matrices, then

$$\det(AB) = \det A \ \det B.$$

The order of the largest nonsingular square submatrix of an $n \times m$ matrix $A$ is called its *rank* and is denoted by $\operatorname{rank}(A)$.

The *trace* of a square $n \times n$ matrix $A$ is the sum of its diagonal elements:

$$\operatorname{tr} A = \sum_{i=1}^{n} a_{ii}.$$

The *sum* of two $n \times m$ matrices $A$ and $B$ is the $n \times m$ matrix $C = A + B$ with the elements

$$c_{ij} = a_{ij} + b_{ij}, \ i = 1, 2, \ldots, n, \ j = 1, 2, \ldots, m.$$

The product of a complex number $\lambda$ and an $n \times m$ matrix $A$ is the $n \times m$ matrix $B = \lambda A$ with the elements

$$b_{ij} = \lambda a_{ij}, \ i = 1, 2, \ldots, n, \ j = 1, 2, \ldots, m.$$

Consider a function $x \mapsto f(x)$ defined by a convergent series

$$f(x) = \sum_{k=0}^{\infty} f_k x^k$$

(*analytic function*). Given a square matrix $A$, we can introduce a square matrix $f(A)$ by

$$f(A) = \sum_{k=0}^{\infty} f_k A^k,$$

where $A^0 = I_n, \ A^k = AA^{k-1}, \ k = 1, 2, \ldots$. For example,

$$e^A = \sum_{k=0}^{\infty} \frac{1}{k!} A^k.$$

### A.1.2 Vector spaces and linear transformations

A complex $n \times 1$ matrix

$$v = (v_1, v_2, \dots, v_n)^T = \begin{pmatrix} v_1 \\ v_2 \\ \dots \\ v_n \end{pmatrix}$$

is called a *vector*. The set of all such vectors is a *linear space* that can be identified with $\mathbb{C}^n$. In this space, addition between two elements and multiplication of an element by a complex number are defined component-wise.

A subset $X \subset \mathbb{C}^n$ is called the *linear subspace* (*hyperplane*) of $\mathbb{C}^n$ if $x \in X$ and $y \in X$ imply $x + y \in X$ and $\lambda x \in X$ for any $\lambda \in \mathbb{C}^1$. A linear subspace $Z$ is called the *sum* of two linear subspaces $X$ and $Y$ if any vector $z \in Z$ can be represented as $z = x + y$ for some vectors $x \in X$ and $y \in Y$. Symbolically: $Z = X + Y$. If such a prepresentation is unique for each $z$, $Z$ is called the *direct sum* of $X$ and $Y$ and is denoted by $Z = X \oplus Y$.

Vectors $\{a^1, a^2, \dots, a^k\}$ from $\mathbb{C}^n$ are called *linearly independent* when

$$\alpha_1 a^1 + \alpha_2 a^2 + \dots + \alpha_k a^k = 0,$$

if and only if $\alpha_j = 0$ for all $j = 1, 2, \dots, k$. The set

$$L = \operatorname{span}\{a^1, a^2, \dots, a^k\} = \left\{ v \in \mathbb{C}^n : v = \sum_{i=1}^{k} \alpha_i a^i,\ \alpha_i \in \mathbb{C}^1 \right\}$$

is a linear subspace of $\mathbb{C}^n$. If $\{a^1, a^2, \dots, a^k\}$ are linearly independent, $\dim L = k$. A set of $n$ linearly independent vectors is called a *basis*. The set of unit vectors

$$e^1 = \begin{pmatrix} 1 \\ 0 \\ \dots \\ 0 \end{pmatrix}, e^2 = \begin{pmatrix} 0 \\ 1 \\ \dots \\ 0 \end{pmatrix}, \dots, e^n = \begin{pmatrix} 0 \\ 0 \\ \dots \\ 1 \end{pmatrix}$$

is the *standard basis* in $\mathbb{C}^n$. Any vector $v \in \mathbb{C}^n$ can be uniquely represented as

$$v = \begin{pmatrix} v_1 \\ v_2 \\ \dots \\ v_n \end{pmatrix} = v_1 e^1 + v_2 e^2 + \dots + v_n e^n.$$

If $\{\varepsilon^1, \varepsilon^2, \dots, \varepsilon^n\}$ is another basis, any vector $v \in \mathbb{C}^n$ can also be represented as

$$v = u_1 \varepsilon^1 + u_2 \varepsilon^2 + \dots + u_n \varepsilon^n,$$

where $u_k \in \mathbb{C}^1$ are *components* of $v$ in this basis. Denote $u = (u_1, u_2, \dots, u_n)^T$. Then

$$v = Cu,$$

where the $n \times n$ matrix $C$ is nonsingular and has the elements $c_{ij}$ that are the components of the basis vectors $\varepsilon_j$ in the standard basis

$$\varepsilon^j = c_{1j}e^1 + c_{2j}e^2 + \cdots + c_{nj}e^n, \quad j = 1, 2, \ldots, n.$$

An $n \times n$ matrix $A$ can be identified with a *linear transformation* of the space $\mathbb{C}^n$

$$v \mapsto Av.$$

In a basis $\{\varepsilon^1, \varepsilon^2, \ldots, \varepsilon^n\}$ this transformation will have the form

$$u \mapsto Bu,$$

where the matrix $B$ is given by

$$B = C^{-1}AC.$$

The matrices $A$ and $B$ are called *similar*. The ranks of similar matrices coincide.

### A.1.3 Eigenvectors and eigenvalues

A nonzero complex vector

$$v = (v_1, v_2, \ldots, v_n)^T \in \mathbb{C}^n$$

is called an *eigenvector* of an $n \times n$ matrix $A$ if

$$Av = \lambda v,$$

for some $\lambda \in \mathbb{C}^1$. The complex number $\lambda$ is called an *eigenvalue* of $A$ corresponding to the eigenvector $v$. The eigenvalues of $A$ are roots of the *characteristic polynomial*

$$h(\lambda) = \det(A - \lambda I_n),$$

and every root is an eigenvalue. Thus, there are $n$ eigenvalues if we count their multiplicities as the roots of $h(\lambda)$. The eigenvalues are continuous functions of matrix elements. The coefficients of the polynomial $h(\lambda)$ will not change if we replace $A$ by any similar to $A$ matrix. Since

$$h(\lambda) = (-1)^n\lambda^n + (-1)^{n-1}(\text{tr } A)\lambda^{n-1} + \cdots + \det \ A,$$

the determinants and traces of similar matrices coincide.

### A.1.4 Invariant subspaces, generalized eigenvectors, and Jordan normal form

A linear subspace $X \subset \mathbb{C}^n$ is called an *invariant subspace* of the matrix $A$ if $AX \subset X$, that is, if $w \in X$ implies $Aw \in X$.

If $\lambda$ is a root of the characteristic polynomial, then there is an invariant subspace (*eigenspace*) of $A$ that is spanned by the eigenvector $v \in \mathbb{C}^n$ associated with $\lambda$:

$$X = \{x \in \mathbb{C}^n : x = \omega v, \omega \in \mathbb{C}^1\}.$$

If $\lambda$ is a multiple root of the characteristic polynomial of multiplicity $m$, then one can find $1 \le l \le m$ linearly independent eigenvectors $v^1, v^2, \dots, v^l$, corresponding to $\lambda$. For each eigenvector $v^j$, there is a *maximal chain* of complex vectors $\{w^{(j,1)}, w^{(j,2)}, \dots, w^{(j,k_j)}\}$, such that

$$\begin{aligned} Aw^{(j,1)} &= \lambda w^{(j,1)}, \\ Aw^{(j,2)} &= \lambda w^{(j,2)} + w^{(j,1)}, \\ &\dots \\ Aw^{(j,k_j)} &= \lambda w^{(j,k_j)} + w^{(j,k_j-1)}. \end{aligned}$$

The chain can be composed of only one vector $w^{(j,1)}$, that is merely the eigenvector $v^j$. The vectors $w^{(j,k)}$ with $k \ge 2$ are called *generalized eigenvectors* of $A$ corresponding to the eigenvalue $\lambda$. They are not uniquely defined. The subspace

$$X = \{x \in \mathbb{C}^n : x = \omega_1 w^{(j,1)} + \omega_2 w^{(j,2)} + \dots + \omega_k w^{(j,k)},\ \omega_i \in \mathbb{C}^1\}$$

is an invariant subspace of $A$.

Eigenvectors and generalized eigenvectors corresponding to distinct eigenvalues are linearly independent. The vectors $\{w^{j,1}, w^{j,2}, \dots, w^{j,k_j}\}$ composing a chain corresponding to a multiple eigenvalue $\lambda$ are also linearly independent.

**Theorem A.1 (Jordan normal form)** *The space $\mathbb{C}^n$ can be decomposed into linear invariant subspaces of the matrix $A$ corresponding to its eigenvectors and generalized eigenvectors. In a basis given by all the eigenvectors and generalized eigenvectors, the matrix $A$ has a block-diagonal form with square blocks*

$$\begin{pmatrix} \lambda & 1 & 0 & \cdots & 0 \\ 0 & \lambda & 1 & \cdots & 0 \\ \cdots \\ 0 & \cdots & \cdots & \lambda & 1 \\ 0 & \cdots & \cdots & 0 & \lambda \end{pmatrix},$$

*whose dimension is equal to the length of the corresponding chain.* □

This form is called the *Jordan normal form* or *Jordan canonical form.* Notice that several Jordan blocks of dimension $m \geq 1$ can be associated with one eigenvalue $\lambda$. Their number $N(m, \lambda)$ can be computed by the formula

$$N(m, \lambda) = r_{m+1} - 2r_m + r_{m-1},$$

where $r_0 = n$ and $r_k = \text{rank}(A - \lambda I_n)^k$.

It also follows from Theorem A.1 that the product of all the eigenvalues of the matrix $A$ is equal to its determinant:

$$\det\ A = \lambda_1 \lambda_2 \cdots \lambda_n,$$

while their sum is equal to the trace:

$$\text{tr}\ A = \lambda_1 + \lambda_2 + \cdots + \lambda_n.$$

If the matrix $A$ is real, then it has linear invariant subspaces of $\mathbb{R}^n$ spanned by eigenvectors and generalized eigenvectors corresponding to its real eigenvalues and also by the real and imaginary parts of complex eigenvectors corresponding to its complex eigenvalues with, say, positive imaginary part. Such subspaces are called (*real*) generalized eigenspaces of $A$.

If $\lambda$ is an eigenvalue of $A$, then $\mu = f(\lambda)$ is an eigenvalue of $B = f(A)$, where $f$ is an analytic function.

### A.1.5 Fredholm Alternative Theorem

Let $A$ be a real $n \times m$ matrix and let $b \in \mathbb{R}^n$ be a real vector. The *null-space* of $A$ is the linear subspace of $\mathbb{R}^m$ composed of all vectors $x \in \mathbb{R}^m$ for which $Ax = 0$. The *range* of $A$ is the set of all $x \in \mathbb{R}^n$ for which $Ay = x$ for some $y \in \mathbb{R}^m$.

**Theorem A.2 (Fredholm Alternative Theorem)** *The equation $Ax = b$ has a solution if and only if $b^T v = 0$ for every vector $v \in \mathbb{R}^n$ satisfying $A^T v = 0$.* □

Notice that $b^T v = \sum_{j=1}^n b_j v_j$ is the standard *scalar product* in $\mathbb{R}^n$. The theorem means that the null-space of $A^T$ is the orthogonal complement of the range of $A$ and that together they span the whole $\mathbb{R}^n$. In other words, any vector $b \in \mathbb{R}^n$ can be uniquely decomposed as $b = b_r + b_0$, where $b_r$ is in the range of $A$, $b_0$ is in the null-space of $A^T$, and $b_r$ is orthogonal to $b_0$.

If $A$ is a complex matrix and $b$ is a complex vector, Theorem A.2 remains valid if we replace transposition by transposition composed with complex conjugation.

### A.1.6 Groups

A set $G$ is a *group* if a *product* "$\circ$": $G \times G \to G$ is defined which satisfies the following properties:

(i) $f \circ (g \circ h) = (f \circ g) \circ h$ for all $f, g, h \in G$;
(ii) there is a *unit* element $e \in G$ such that $g \circ e = e \circ g = g$, for all $g \in G$;
(iii) for each $g \in G$, there is a unique element $g^{-1} \in G$ such that $g^{-1} \circ g = g \circ g^{-1} = e$.

All real nonsingular $n \times n$ matrices with the matrix product and the unit matrix $I_n$ form the *general linear group* denoted by $GL(n)$. All $n \times n$ matrices satisfying $A^T A = I_n$ compose its *orthogonal subgroup* $O(n)$.

## A.2 Analysis

If $y = g(x)$, $g : \mathbb{R}^n \to \mathbb{R}^m$, and $z = f(y)$, $f : \mathbb{R}^m \to \mathbb{R}^k$, are two maps, then their *composition* $h = f \circ g$ is a map $z = h(x)$, $h : \mathbb{R}^n \to \mathbb{R}^k$, defined by the formula

$$h(x) = f(g(x)).$$

Let $f_y(y)$ denote the Jacobian matrix of $f$ evaluated at $y \in \mathbb{R}^m$:

$$f_y(y) = \left( \frac{\partial f_i(y)}{\partial y_j} \right),$$

where $i = 1, 2, \dots, k$, $j = 1, 2, \dots, m$. If we similarly define $h_x(x)$ and $g_x(x)$, then

$$h_x(x) = [f_y(y)]\big|_{y=g(x)} [g_x(x)]$$

(the *chain rule*).

### A.2.1 Implicit and Inverse Function Theorems

Consider a map

$$(x, y) \mapsto F(x, y),$$

where

$$F : \mathbb{R}^n \times \mathbb{R}^m \to \mathbb{R}^m,$$

is a smooth map defined in a neighborhood of $(x, y) = (0, 0)$ and such that $F(0, 0) = 0$. Let $F_y(0, 0)$ denote the matrix of first partial derivatives of $F$ with respect to $y$ evaluated at $(0, 0)$:

$$F_y(0,0) = \left( \frac{\partial F_i(x,y)}{\partial y_j} \right)\Bigg|_{(x,y)=(0,0)},$$

where $i, j = 1, 2, \dots, m$.

**Theorem A.3 (Implicit Function Theorem)** *If the matrix $F_y(0,0)$ is non-singular, then there is a unique smooth locally defined function $y = f(x)$,*

$$f : \mathbb{R}^n \to \mathbb{R}^m,$$

*such that*

$$F(x, f(x)) = 0,$$

*for all $x$ in some neighborhood of the origin of $\mathbb{R}^n$. Moreover,*

$$f_x(0) = -[F_y(0,0)]^{-1} F_x(0,0). \square$$

The degree of smoothness of the function $f$ is the same as that of $F$.

Consider now a map

$$y = g(x),$$

where

$$g : \mathbb{R}^n \to \mathbb{R}^n$$

is a smooth function defined in a neighborhood of $x = 0$ and satisfying $g(0) = 0$. The following theorem is a consequence of the Implicit Function Theorem.

**Theorem A.4 (Inverse Function Theorem)** *If the matrix $g_x(0)$ is non-singular, then there is a unique smooth locally defined function $x = f(y)$,*

$$f : \mathbb{R}^n \to \mathbb{R}^n,$$

*such that*

$$g(f(y)) = y$$

*for all $y$ in some neighborhood of the origin of $\mathbb{R}^n$.* $\square$

The function $f$ is called the *inverse function* for $g$ and is denoted by $f = g^{-1}$.

### A.2.2 Taylor expansion

Let $\Omega$ be a region in $\mathbb{R}^n$ containing the origin $x = 0$. Denote by $C^k(\Omega, \mathbb{R}^m)$ the set of maps (vector-valued functions) $y = f(x)$, $f : \Omega \to \mathbb{R}^m$, having continuously differentiable components up to and including order $k \geq 0$. If $f \in C^k(\Omega, \mathbb{R}^m)$ with a sufficiently large $k$, the function $f$ is called *smooth*. A $C^\infty$ function has continuous partial derivatives of any order. Any function $f \in C^k(\Omega, \mathbb{R}^m)$ can be represented near $x = 0$ in the form (*Taylor expansion*)

$$f(x) = \sum_{|i|=0}^{k} \frac{1}{i_1! i_2! \cdots i_n!} \left. \frac{\partial^{|i|} f(x)}{\partial x_1^{i_1} \partial x_2^{i_2} \cdots \partial x_n^{i_n}} \right|_{x=0} x_1^{i_1} x_2^{i_2} \cdots x_n^{i_n} + R(x),$$

where $|i| = i_1 + i_2 + \cdots + i_n$ and $R(x) = O(\|x\|^{k+1}) = o(\|x\|^k)$, namely,

$$\frac{\|R(x)\|}{\|x\|^k} \to 0$$

as $\|x\| \to 0$. Here $\|x\| = \sqrt{x^T x}$.

A $C^\infty$-function $f$ is called *analytic* near the origin if the corresponding *Taylor series*

$$\sum_{|i|=0}^{\infty} \frac{1}{i_1! i_2! \cdots i_n!} \frac{\partial^{|i|} f(x)}{\partial x_1^{i_1} \partial x_2^{i_2} \cdots \partial x_n^{i_n}}\bigg|_{x=0} x_1^{i_1} x_2^{i_2} \cdots x_n^{i_n}$$

converges to $f(x)$ at any point $x$ sufficiently close to $x = 0$.

### A.2.3 Metric, normed, and other spaces

A set $X$ is a *metric space* if a function $\rho : X \times X \to \mathbb{R}^1$ is defined such that:

(i) $\rho(x, y) = \rho(y, x)$ for all $x, y \in X$;
(ii) $\rho(x, y) \geq 0$ and $\rho(x, y) = 0$ if and only if $x = y$;
(iii) $\rho(x, y) \leq \rho(x, z) + \rho(z, y)$ for all $x, y, z \in X$.

The function $\rho$ is called a *metric* (or *distance*). A sequence $\{x_k\}_{k=1}^\infty$ of elements $x_k \in X$ has a *limit* $x^0 \in X$ (*convergent*) if for any $\varepsilon > 0$ there is an integer $N(\varepsilon)$ such that

$$\rho(x_k, x_0) < \varepsilon,$$

for all $k \geq N(\varepsilon)$. Notation: $x^0 = \lim_{k\to\infty} x_k$. A function $f : X \to X$ is *continuous* at $x^0$ if

$$\lim_{k\to\infty} f(x_k) = f(x^0),$$

for all sequences such that $\lim_{k\to+\infty} x_k = x^0$. A function $g : X \to X$ is called *Hölder-continuous* at $x^0$ if there exist a constant $L_0$ and an *index* $0 < \beta \leq 1$, such that

$$\rho(g(x), g(x_0)) \leq L_0[\rho(x, x_0)]^\beta$$

for all $x$ sufficiently close to $x_0$.

A set $S \subset X$ is *closed* if it contains the limits of all convergent sequences such that for any finite $k$, $x_k \in S$. A sequence $\{x_k\}_{k=1}^\infty$ of elements $x_k \in X$ is a *Cauchy sequence* if for any $\varepsilon > 0$ there is an integer $N(\varepsilon)$ such that for every $n, m \geq N$,

$$\rho(x_n, x_m) < \varepsilon.$$

If the sequence $\{x_k\}_{k=1}^\infty$ has a limit, it is a Cauchy sequence. If any Cauchy sequence has a limit in $X$, the space $X$ is called *complete.*

Let $X$ be a set of elements for which addition and multiplication by (complex) numbers satisfying standard axioms are defined. The set $X$ can consist of functions, for example, $X = C^k(\Omega, \mathbb{C}^m)$.

The set $X$ is a *normed space* if a function $\|\cdot\| : X \to \mathbb{R}^1$ is defined such that:

(i) $\|x\| \geq 0$ and $\|x\| = 0$ implies $x = 0$;
(ii) $\|\alpha x\| = |\alpha|\|x\|$ for any (complex) number $\alpha$;
(iii) $\|x + y\| \leq \|x\| + \|y\|$ for all $x, y \in X$.

The function $\|\cdot\|$ is called a *norm*. Any normed space is a metric space with the metric $\rho(x, y) = \|x - y\|$. If $X$ is complete in this metric, it is called a *Banach space*. The space of continuous functions $C^0(\Omega, \mathbb{C}^m)$ is a Banach in the norm

$$\|f\| = \max_{i=1,2,\ldots,m} \sup_{\xi \in \Omega} |f_i(\xi)|.$$

A set $S \subset X$ is *bounded* if $\|x\| < C$ with some $C > 0$ for all $x \in S$.

The set $X$ is a *space with a scalar product* if for each pair of $(x, y) \in X$ a complex number $\langle x, y \rangle$ (called the *scalar product*) is defined so that the following properties hold:

(i) $\langle x, y \rangle = \overline{\langle y, x \rangle}$ for all $x, y \in X$;
(ii) $\langle x, \alpha y \rangle = \alpha \langle x, y \rangle$ for all $x, y \in X$ and any complex number $\alpha$;
(iii) $\langle x + y, z \rangle = \langle x, z \rangle + \langle y, z \rangle$ for all $x, y, z \in X$;
(iv) $\langle x, x \rangle \geq 0$ and $\langle x, x \rangle = 0$ if and only if $x = 0$.

Any space $X$ with a scalar product is a normed space with $\|x\| = \sqrt{\langle x, x \rangle}$. If it is a Banach space in this norm, it is called a *Hilbert space*. The space $\mathbb{C}^n$ is a Hilbert space with the scalar product

$$\langle x, y \rangle = \bar{x}^T y = \sum_{k=1}^{n} \bar{x}_k y_k.$$

Thus, it is also a Banach space and a complete metric space. Note that

$$\langle x, Ay \rangle = \langle \bar{A}^T x, y \rangle,$$

for any $x, y \in \mathbb{C}^n$ and a complex matrix $A$. The space $C^0(\Omega, \mathbb{C}^m)$ is a space with the scalar product

$$\langle f, g \rangle = \int_{\Omega} \bar{f}^T(x) g(x)\, dx,$$

but is not a Hilbert space.

## A.3 Geometry

### A.3.1 Sets

To denote that $x$ is an element of a *set* $X$, we write $x \in X$. A set $A$ is a *subset* of $X$ ($A \subset X$) if $x \in A$ implies $x \in X$. If $A$ and $B$ are two sets, then the set $A \cup B$ consists of all elements that belong to either $A$ or $B$, while the set $A \cap B$ is composed of all elements that belong to both $A$ and $B$. The set consisting of all elements of $A$ which do not belong to $B$ is denoted by $A \setminus B$. The set of all ordered pairs $(a, b)$, such that $a \in A$ and $b \in B$, is called the *direct product* of two sets $A$ and $B$ and is denoted by $A \times B$.

The following notations are used for the standard sets:

$\mathbb{R}^1$: the set of all real numbers $-\infty < x < +\infty$; $\mathbb{R}^1_+$ denotes the set of all nonnegative real numbers $x \geq 0$;

$\mathbb{R}^n$: the direct product of $n$ sets $\mathbb{R}^1$; an element $x \in \mathbb{R}^n$ is considered as a vector (one-column matrix) $x = (x_1, x_2, \ldots, x_n)^T$;

$\mathbb{C}^1$: the set of all complex numbers $z = x + iy$, where $x, y \in \mathbb{R}^1$, $i^2 = -1$. Any $z \in \mathbb{C}^1$ can be represented as $z = \rho e^{i\varphi} = \rho(\cos\varphi + i\sin\varphi)$, where $\rho = |z| = \sqrt{x^2 + y^2}$ and $\varphi = \arg z$; $\bar{z} = x - iy$;

$\mathbb{C}^n$: the direct product of $n$ sets $\mathbb{C}^1$; an element $z \in \mathbb{C}^n$ is considered as a vector (one-column matrix) $z = (z_1, z_2, \ldots, z_n)^T$;

$\mathbb{Z}$: the set of all integer numbers $\{\ldots, -2, -1, 0, 1, 2, \ldots\}$; $\mathbb{Z}_+$ denotes the set of all nonnegative integers $k = 0, 1, 2, \ldots$;

$\mathbb{S}^1$: the unit circle: $\mathbb{S}^1 = \{x \in \mathbb{R}^2 : x_1^2 + x_2^2 = 1\}$;

$\mathbb{T}^2$: the two-torus: $\mathbb{T}^2 = \mathbb{S}^1 \times \mathbb{S}^1$.

### A.3.2 Maps

Let $X$ and $Y$ be two arbitrary sets. A (*single-valued*) *map* (or *function*)

$$f : X \to Y$$

is said to be defined from $X$ to $Y$ if, for any element $x \in X$, an element $y \in Y$ is specified. We write

$$y = f(x),$$

or

$$x \mapsto f(x).$$

A map $f : X \to X$ is called a *transformation* of $X$. A map $f : X \to Y$ can be defined only for elements of a subset $D \subset X$. In this case, $D$ is called the *domain of definition* of $f$.

The set $f(X_0)$ of all $y \in Y$ such that $y = f(x)$ for some $x \in X_0$ is called the *image* of $X_0 \subset X$. The image $f(X)$ is referred to as the *range* of $f$. The set $f^{-1}(Y_0)$ of all $x \in X$ such that $f(x) \in Y_0$ is called the *preimage* of $Y_0 \subset Y$.

A map $f$ is *invertible* if $f^{-1}(Y) = X$ and $f^{-1}(\{y\})$ consists precisely of one element $x \equiv f^{-1}(y)$ for any $y \in Y$. In this case, the *inverse* map $f^{-1} : Y \to X$ is defined, such that $f^{-1}(f(x)) = x$ for all $x \in X$, and $f(f^{-1}(y)) = y$ for all $y \in Y$.

### A.3.3 Manifolds

For our purposes, it is sufficient to consider the *manifold* $M \subset \mathbb{R}^n$ as a set of points in $\mathbb{R}^n$ that satisfy a system of $m$ scalar equations:

$$F(x) = 0,$$

where $F : \mathbb{R}^n \to \mathbb{R}^m$ for some $m \leq n$. The manifold $M$ is *smooth* (*differentiable*) if $F$ is smooth and the rank of the Jacobian matrix $F_x$ is equal to $m$ at each point $x \in M$. At each point $x$ of a smooth manifold $M$, an $(n-m)$-dimensional *tangent space* $T_xM$ is defined. This space consists of all vectors $v \in \mathbb{R}^n$ that can be represented as $v = \dot{\gamma}(0)$, where $\gamma : \mathbb{R}^1 \to M$ is a smooth *curve* on the manifold satisfying $\gamma(0) = x$. Alternatively, $T_xM$ can be characterized as the orthogonal complement to

$$\text{span}\{\nabla F_1, \nabla F_2, \ldots, \nabla F_m\},$$

where

$$\nabla F_k = \left(\frac{\partial F_k}{\partial x_1}, \frac{\partial F_k}{\partial x_2}, \ldots, \frac{\partial F_k}{\partial x_n}\right)^T, \quad k = 1, 2, \ldots, m,$$

are linear independent gradient vectors at point $x$. One can introduce $n - m$ coordinates near each point $x \in M$ by projecting to $T_xM$, so that a smooth manifold $M$ is locally equivalent to $\mathbb{R}^{n-m}$.

A *region* $\Omega \in \mathbb{R}^n$ is a closed set of points in $\mathbb{R}^n$ bounded by a piecewise smooth $(n-1)$-dimensional manifold.

# References


Afraimovich, V.S. & Shil'nikov, L.P. [1972], 'Singular trajectories of dynamical systems', *Uspekhi. Mat. Nauk* **27**, 189–190. In Russian.

Afraimovich, V.S. & Shil'nikov, L.P. [1974], 'The attainable transitions from Morse-Smale systems to systems with several periodic motions', *Math. USSR-Izv.* **8**, 1235–1270.

Afraimovich, V.S. & Shil'nikov, L.P. [1982], 'Bifurcation of codimension 1, leading to the appearance of a countable set of tori', *Dokl. Akad. Nauk SSSR* **262**, 777–780. In Russian.

Allgower, E. & Georg, K. [1990], *Numerical Continuation Methods: An Introduction*, Springer-Verlag, New York.

Allgower, E. & Georg, K. [1993], Continuation and path following, *in* 'Acta Numerica', Cambridge University Press, Cambridge, pp. 1–64.

Andronov, A.A. [1933], Mathematical problems of self-oscillation theory, *in* 'I All-Union Conference on Oscillations, November 1931', GTTI, Moscow-Leningrad, pp. 32–71. In Russian.

Andronov, A.A. & Leontovich, E.A. [1939], 'Some cases of the dependence of the limit cycles upon parameters', *Uchen. Zap. Gork. Univ.* **6**, 3–24. In Russian.

Andronov, A.A. & Pontryagin, L.S. [1937], 'Systèmes grossières', *C.R. (Dokl.) Acad. Sci. URSS (N.S.)* **14**, 247–251.

Andronov, A.A., Leontovich, E.A., Gordon, I.I. & Maier, A.G. [1973], *Theory of Bifurcations of Dynamical Systems on a Plane*, Israel Program for Scientific Translations, Jerusalem.

Annabi, H., Annabi, M. L. & Dumortier, F. [1992], Continuous dependence on parameters in the Bogdanov-Takens bifurcation, *in* 'Geometry and Analysis in Nonlinear Dynamics (Groningen, 1989)', Vol. 222 of *Pitman Research Notes in Mathematics Series*, Longman Scientific & Technical, Harlow, pp. 1–21.

Anosov, D.V. [1967], 'Geodesic flows on closed Riemannian manifolds of negative curvature', *Proc. Steklov Inst. Math.* **90**, 1–212. In Russian.

Arnol'd, V.I. [1972], 'Lectures on bifurcations in versal families', *Russian Math. Surveys* **27**, 54–123.

Arnol'd, V.I. [1973], *Ordinary Differential Equations*, MIT Press, Cambridge, MA.

Arnol'd, V.I. [1977], 'Loss of stability of self-induced oscillations near resonance, and versal deformations of equivariant vector fields', *Functional Anal. Appl.* **11**, 85–92.

Arnol'd, V.I. [1983], *Geometrical Methods in the Theory of Ordinary Differential Equations*, Springer-Verlag, New York.

Arnol'd, V.I. [1984], *Catastrophe Theory*, Springer-Verlag, New York.

Arnol'd, V.I., Afraimovich, V.S., Il'yashenko, Yu.S. & Shil'nikov, L.P. [1994], Bifurcation theory, *in* V.I. Arnol'd, ed., 'Dynamical Systems V. Encyclopaedia of Mathematical Sciences', Springer-Verlag, New York.

Arnol'd, V.I., Varchenko, A.N. & Guseĭn-Zade, S.M. [1985], *Singularities of Differentiable Maps I*, Birkhäuser, Boston, MA.

Aronson, D., Chory, M., Hall, G. & McGehee, R. [1982], 'Bifurcations from an invariant circle for two-parameter families of maps on the plane: A computer assisted study', *Comm. Math. Phys.* **83**, 303–354.

Arrowsmith, D. & Place, C. [1990], *An Introduction to Dynamical Systems*, Cambridge University Press, Cambridge.

Arrowsmith, D., Cartwright, J., Lansbury, A. & Place, C. [1993], 'The Bogdanov map: Bifurcations, mode locking, and chaos in a dissipative system', *Internat. J. Bifur. Chaos Appl. Sci. Engrg.* **3**, 803–842.

Auchmuty, J. & Nicolis, G. [1976], 'Bifurcation analysis of reaction–diffusion equations (III). Chemical oscillations', *Bull. Math. Biol.* **38**, 325–350.

Babenko, K.I. & Petrovich, V.Yu. [1983], Demonstrative computations on a computer, Preprint 83-133, Institute of Applied Mathematics, USSR Academy of Sciences, Moscow. In Russian.

Babenko, K.I. & Petrovich, V.Yu. [1984], 'Demonstrative calculations in the problem of existence of the solution of the doubling equation', *Soviet Math. Dokl.* **30**, 54–59.

Back, A., Guckenheimer, J., Myers, M., Wicklin, F. & Worfolk, P. [1992], 'DsTool: Computer assisted exploration of dynamical systems', *Notices Amer. Math. Soc.* **39**, 303–309.

Bai, F. & Champneys, A. [1996], 'Numerical detection and continuation of saddle-node homoclinic bifurcations of codimension one and two', *Dynam. Stability Systems* **11**, 325–346.

Bajaj, A. [1986], 'Resonant parametric perturbations of the Hopf bifurcation', *J. Math. Anal. Appl.* **115**, 214–224.

Balabaev, N.K. & Lunevskaya, L.V. [1978], Continuation of a curve in the $n$-dimensional space, FORTRAN Software Series, Vol. 1, Research Computing Centre, USSR Academy of Sciences, Pushchino, Moscow Region. In Russian.

Balakrishnan, A. [1976], *Applied Functional Analysis*, Springer-Verlag, Berlin.

Bautin, N.N. [1949], *Behavior of Dynamical Systems near the Boundaries of Stability Regions*, OGIZ Gostexizdat, Leningrad. In Russian.

Bautin, N.N. & Leontovich, E.A. [1976], *Methods and Rules for the Qualitative Study of Dynamical Systems on the Plane*, Nauka, Moscow. In Russian.

Bautin, N.N. & Shil'nikov, L.P. [1980], Supplement I: Behavior of dynamical systems close to the boundaries of domains of stability of equilibrium states and periodic motions ('dangerous' and 'safe' boundaries), *in* 'The Limit Cycle Bifurcation and Its Applications. Russian translation of the book by J.E. Marsden and M. McCracken', Mir, Moscow. In Russian.

Bazykin, A.D. [1974], Volterra system and Michaelis-Menten equation, *in* 'Problems of Mathematical Genetics', Novosibirsk State University, Novosibirsk, pp. 103–143. In Russian.

Bazykin, A.D. [1985], *Mathematical Biophysics of Interacting Populations*, Nauka, Moscow. In Russian (English translation: *Nonlinear Dynamics of Interacting Populations*, World Scientific Publishing Co. Inc., River Edge, NJ, 1998).

Bazykin, A.D. & Khibnik, A.I. [1981], 'On sharp excitation of self-oscillations in a Volterra-type model', *Biophysika* **26**, 851–853. In Russian.

Bazykin, A.D., Berezovskaya, F.S., Denisov, G.A. & Kuznetsov, Yu.A. [1981], 'The influence of predator saturation effect and competition among predators on predator-prey system dynamics', *Ecol. Modelling* **14**, 39–57.

Bazykin, A.D., Kuznetsov, Yu.A. & Khibnik, A.I. [1985], Bifurcation diagrams of planar dynamical systems, Research Computing Centre, USSR Academy of Sciences, Pushchino, Moscow Region. In Russian.

Bazykin, A.D., Kuznetsov, Yu.A. & Khibnik, A.I. [1989], *Portraits of Bifurcations: Bifurcation Diagrams of Dynamical Systems on the Plane*, Znanie, Moscow. In Russian.

Belitskii, G.R. [1973], 'Functional equations and conjugacy of local diffeomorphisms of a finite smoothness class', *Functional Anal. Appl.* **7**, 268–277.

Belitskii, G.R. [1979], *Normal Forms, Ivariants, and Local Mappings*, Naukova Dumka, Kiev. In Russian.

Belitskii, G.R. [2002], '$C^\infty$-normal forms of local vector fields. symmetry and perturbation theory', *Acta Appl. Math.* **70**, 23–41.

Belyakov, L.A. [1974], 'A certain case of the generation of periodic motion with homoclinic curves', *Math. Notes* **15**, 336–341.

Belyakov, L.A. [1980], 'The bifurcation set in a system with a homoclinic saddle curve', *Math. Notes* **28**, 910–916.

Belyakov, L.A. [1984], 'Bifurcations of systems with a homoclinic curve of the saddle-focus with a zero saddle value', *Math. Notes* **36**, 838–843.

Berezovskaya, F.S. & Khibnik, A.I. [1979], On the problem of bifurcations of self-oscillations close to a 1:4 resonance (investigation of a model equation), Research Computing Centre, USSR Academy of Sciences, Pushchino, Moscow Region. In Russian (English translation: *Selecta Math.* **13**, 1994, 197-215).

Berezovskaya, F.S. & Khibnik, A.I. [1981], 'On the bifurcations of separatrices in the problem of stability loss of auto-oscillations near 1:4 resonance', *J. Appl. Math. Mech.* **44**, 938–942.

Berezovskaya, F.S. & Khibnik, A.I. [1985], Bifurcations of a dynamical second-order system with two zero eigenvalues and additional degeneracy, *in* 'Methods of Qualitative Theory of Differential Eqiations', Gorkii State University, Gorkii, pp. 128–138. In Russian.

Beyn, W.-J. [1990*a*], Global bifurcations and their numerical computation, *in* D. Roose, B. De Dier & A. Spence, eds, 'Continuation and Bifurcations: Numerical Techniques and Applications (Leuven, 1989)', Vol. 313 of *NATO Adv. Sci. Inst. Ser. C Math. Phys. Sci.*, Kluwer Acad. Publ., Dordrecht, pp. 169–181.

Beyn, W.-J. [1990*b*], 'The numerical computation of connecting orbits in dynamical systems', *IMA J. Numer. Anal.* **9**, 379–405.

Beyn, W.-J. [1991], Numerical methods for dynamical systems, *in* W. Light, ed., 'Advances in Numerical Analysis, Vol. I, Nonlinear Partial Differential Equations and Dynamical Systems', Oxford University Press, Oxford, pp. 175–236.

Beyn, W.-J. [1994], 'Numerical analysis of homoclinic orbits emanating from a Takens-Bogdanov point', *IMA J. Numer. Anal.* **14**, 381–410.

Beyn, W.-J. & Kleinkauf, J.-M. [1997], 'The numerical computation of homoclinic orbits for maps', *SIAM J. Numer. Anal.* **34**, 1207–1236.

Beyn, W.-J., Champneys, A., Doedel, E., Govaerts, W., Kuznetsov, Yu.A. & Sandstede, B. [2002], Numerical continuation, and computation of normal forms, *in* B. Fiedler, ed., 'Handbook of Dynamical Systems, Vol. 2', Elsevier Science, Amsterdam, pp. 149–219.

Birkhoff, G. [1935], 'Nouvelles recherches sur les systèmes dynamiques', *Memoriae Pont. Acad. Sci. Novi. Lincaei, Ser. 3* **1**, 85–216.

Birkhoff, G. [1966], *Dynamical Systems*, With an addendum by Jurgen Moser. American Mathematical Society Colloquium Publications, Vol. IX, American Mathematical Society, Providence, RI.

Bogdanov, R.I. [1975], 'Versal deformations of a singular point on the plane in the case of zero eigenvalues', *Functional Anal. Appl.* **9**, 144–145.

Bogdanov, R.I. [1976*a*], Bifurcations of a limit cycle of a certain family of vector fields on the plane, *in* 'Proceedings of Petrovskii Seminar, Vol. 2', Moscow State University, Moscow, pp. 23–35. In Russian (English translation: *Selecta Math. Soviet.* **1**, 1981, 373-388).

Bogdanov, R.I. [1976*b*], The versal deformation of a singular point of a vector field on the plane in the case of zero eigenvalues, *in* 'Proceedings of Petrovskii Seminar, Vol. 2', Moscow State University, Moscow, pp. 37–65. In Russian (English translation: *Selecta Math. Soviet.* **1**, 1981, 389-421).

Borisyuk, R.M. [1981], Stationary solutions of a system of ordinary differential equations depending upon a parameter, FORTRAN Software Series, Vol. 6, Research Computing Centre, USSR Academy of Sciences, Pushchino, Moscow Region. In Russian.

Broer, H. & Vegter, G. [1984], 'Subordinate Šil'nikov bifurcations near some singularities of vector fields having low codimension', *Ergodic Theory Dynamical Systems* **4**, 509–525.

Broer, H., Osinga, H. & Vegter, G. [1997], 'Algorithms for computing normally hyperbolic invariant manifolds', *Z. Angew. Math. Phys.* **48**, 480–524.

Broer, H., Roussarie, R. & Simó, C. [1993], On the Bogdanov-Takens bifurcation for planar diffeomorphisms, *in* 'International Conference on Differential Equations, Vol. 1, 2 (Barcelona, 1991)', World Scientific, River Edge, NJ, pp. 81–92.

Broer, H., Roussarie, R. & Simó, C. [1996], 'Invariant circles in the Bogdanov-Takens bifurcation for diffeomorphisms', *Ergodic Theory Dynamical Systems* **16**, 1147–1172.

Broer, H., Simó, C. & Tatjer, J. [1998], 'Towards global models near homoclinic tangencies of dissipative diffeomorphisms', *Nonlinearity* **11**, 667–770.

Broyden, C. [1965], 'A class of methods for solving nonlinear simultaneous equations', *Math. Comp.* **19**, 577–593.

Butenin, N.V., Neimark, Ju.I. & Fufaev, N.A. [1976], *Introduction to the Theory of Nonlinear Oscillations*, Nauka, Moscow. In Russian (Spanish translation: *Introducción a la Teoría de Oscilaciones No Lineales*, Mir, Moscow, 1990).

Bykov, V.V. [1977], 'On the birth of periodic motions from a separatrix contour of a three-dimensional system', *Uspekhi Mat. Nauk* **32**, 213–214. In Russian.

Bykov, V.V. [1980], Bifurcations of dynamical systems close to systems with a seperatrix contour containing a saddle-focus, *in* 'Methods of Qualitative Theory of Differential Equations', Gorkii State University, Gorkii. In Russian.

Bykov, V.V. [1993], 'The bifurcations of separatrix contours and chaos', *Physica D* **62**, 290–299.

Bykov, V.V. [1999], 'On systems with separatrix contour containing two saddle-foci', *J. Math. Sci. (New York)* **95**, 2513–2522.

Carr, J. [1981], *Applications of Center Manifold Theory*, Springer-Verlag, New York.

Carr, J., Chow, S.-N. & Hale, J. [1985], 'Abelian integrals and bifurcation theory', *J. Differential Equations* **59**, 413–436.

Champneys, A. & Kuznetsov, Yu.A. [1994], 'Numerical detection and continuation of codimension-two homoclinic bifurcations', *Internat. J. Bifur. Chaos Appl. Sci. Engrg.* **4**, 795–822.

Champneys, A., Härterich, J. & Sandstede, B. [1996], 'A non-transverse homoclinic orbit to a saddle-node equilibrium', *Ergodic Theory Dynamical Systems* **16**, 431–450.

Champneys, A., Kuznetsov, Yu.A. & Sandstede, B. [1995], HomCont: An AUTO86 driver for homoclinic bifurcation analysis. Version 2.0, Report AM-R9516, Centrum voor Wiskunde en Informatica, Amsterdam.

Champneys, A., Kuznetsov, Yu.A. & Sandstede, B. [1996], 'A numerical toolbox for homoclinic bifurcation analysis', *Internat. J. Bifur. Chaos Appl. Sci. Engrg.* **6**, 867–887.

Char, B., Geddes, K., Gonnet, G., Leong, B., Monagan, M. & Watt, S. [1991*a*], *Maple V Language Reference Manual*, Springer-Verlag, New York.

Char, B., Geddes, K., Gonnet, G., Leong, B., Monagan, M. & Watt, S. [1991*b*], *Maple V Library Reference Manual*, Springer-Verlag, New York.

Chenciner, A. [1981], 'Courbes invariantes non normalement hyperboliques au voisinage d'une bifurcation de Hopf dégénérée de difféomorphismes de $\mathbf{R}^2$', *C. R. Acad. Sci. Paris Sér. I Math.* **292**, 507–510.

Chenciner, A. [1982], 'Points homoclines au voisinage d'une bifurcation de Hopf dégénérée de difféomorphismes de $\mathbf{R}^2$', *C. R. Acad. Sci. Paris Sér. I Math.* **294**, 269–272.

Chenciner, A. [1983*a*], Bifurcations de difféomorphismes de $\mathbf{R}^2$ au voisinage d'un point fixe elliptique, *in* G. Iooss, R. Helleman & R. Stora, eds, 'Chaotic Behavior of Deterministic Systems (Les Houches, 1981)', North-Holland, Amsterdam, pp. 273–348.

Chenciner, A. [1983*b*], 'Orbites périodiques et ensembles de Cantor invariants d'Aubry-Mather au voisinage d'une bifurcation de Hopf dégénérée de difféomorphismes de $\mathbf{R}^2$', *C. R. Acad. Sci. Paris Sér. I Math.* **297**, 465–467.

Chenciner, A. [1985*a*], 'Bifurcations de points fixes elliptiques. I. Courbes invariantes', *Inst. Hautes Études Sci. Publ. Math.* **61**, 67–127.

Chenciner, A. [1985*b*], 'Bifurcations de points fixes elliptiques. II. Orbites periodiques et ensembles de Cantor invariants', *Invent. Math.* **80**, 81–106.

Chenciner, A. [1988], 'Bifurcations de points fixes elliptiques. III. Orbites périodiques de "petites" périodes et élimination résonnante des couples de courbes invariantes', *Inst. Hautes Études Sci. Publ. Math.* **66**, 5–91.

Cheng, C.-Q. [1990], 'Hopf bifurcations in nonautonomous systems at points of resonance', *Sci. China Ser. A* **33**, 206–219.

Cheng, C.-Q. & Sun, Y.-S. [1992], 'Metamorphoses of phase portraits of vector fields in the case of symmetry of order 4', *J. Differential Equations* **95**, 130–139.

Chow, S.-N. & Hale, J. [1982], *Methods of Bifurcation Theory*, Springer-Verlag, New York.

Chow, S.-N. & Lin, X.-B. [1990], 'Bifurcation of a homoclinic orbit with a saddle-node equilibrium', *Differential Integral Equations* **3**, 435–466.

Chow, S.-N., Deng, B. & Fiedler, B. [1990], 'Homoclinic bifurcation at resonant eigenvalues', *J. Dynamics Differential Equations* **2**, 177–244.

Chow, S.-N., Drachman, B. & Wang, D. [1990], 'Computation of normal forms', *J. Comput. Appl. Math.* **29**, 129–143.

Chow, S.-N., Li, C. & Wang, D. [1989*a*], 'Erratum: "Uniqueness of periodic orbits of some vector fields with codimension two singularities"', *J. Differential Equations* **82**, 206.

Chow, S.-N., Li, C. & Wang, D. [1989*b*], 'Uniqueness of periodic orbits of some vector fields with codimension two singularities', *J. Differential Equations* **77**, 231–253.

Chow, S.-N., Li, C. & Wang, D. [1994], *Normal Forms and Bifurcations of Planar Vector Fields*, Cambridge University Press, Cambridge.

Chu, K., Govaerts, W. & Spence, A. [1994], 'Matrices with rank deficiency two in eigenvalue problems and dynamical systems', *SIAM J. Numer. Anal.* **31**, 524–539.

Coddington, E. & Levinson, N. [1955], *Theory of Ordinary Differential Equations*, McGraw-Hill, New York.

Coullet, P. & Eckmann, J.-P. [1980], *Iterated Maps on the Interval as a Dynamical System*, Birkhauser, Boston, MA.

Coullet, P. & Spiegel, E. [1983], 'Amplitude equations for systems with competing instabilities', *SIAM J. Appl. Math.* **43**, 776–821.

Cox, D., Little, J. & O'Shea, D. [1992], *Ideals, Varieties, and Algorithms*, Springer-Verlag, New York.

Cushman, R. & Sanders, J. [1985], 'A codimension two bifurcation with a third order Picard-Fuchs equation', *J. Differential Equations* **59**, 243–256.

de Boor, C. & Swartz, B. [1973], 'Collocation at Gaussian points', *SIAM J. Numer. Anal.* **10**, 582–606.

Demmel, J., Dieci, L. & Friedman, M. [2000], 'Computing connecting orbits via an improved algorithm for continuing invariant subspaces', *SIAM J. Sci. Comput.* **22**, 81–94.

Deng, B. [1989], 'The Šil'nikov problem, exponential expansion, strong $\lambda$-lemma, $C^1$-linearization, and homoclinic bifurcation', *J. Differential Equations* **79**, 189–231.

Deng, B. [1990], 'Homoclinic bifurcations with nonhyperbolic equilibria', *SIAM J. Math. Anal.* **21**, 693–720.

Deng, B. [1993*a*], 'Homoclinic twisting bifurcations and cusp horseshoe maps', *J. Dynamics Differential Equations* **5**, 417–467.

Deng, B. [1993*b*], 'On Šhil'nikov's homoclinic-saddle-focus theorem', *J. Differential Equations* **102**, 305–329.

Deng, B. [1994], 'Constructing homoclinic orbits and chaotic attractors', *Internat. J. Bifur. Chaos Appl. Sci. Engrg.* **4**, 823–841.

Deng, B. [1995], 'Constructing Lorenz type attractors through singular perturbations', *Internat. J. Bifur. Chaos Appl. Sci. Engrg.* **5**, 1633–1642.

Deng, B. & Sakamoto, K. [1995], 'Šil'nikov-Hopf bifurcations', *J. Differential Equations* **119**, 1–23.

Denjoy, A. [1932], 'Sur les courbes définies par les équations différentielles à la surface du tore', *J. Math.* **17(IV)**, 333–375.

Dennis, J. & Schnabel, R. [1983], *Numerical Methods for Unconstrained Optimization and Nonlinear Equations*, Prentice-Hall, Englewood Cliffs, NJ.

Deuflhard, P., Fiedler, B. & Kunkel, P. [1987], 'Efficient numerical pathfollowing beyond critical points', *SIAM J. Numer. Anal.* **24**, 912–927.

Dhooge, A., Govaerts, W. & Kuznetsov, Yu.A. [2003], 'MATCONT:A MATLAB package for numerical bifurcation analysis of ODEs', *ACM Trans. Math. Software* **29**, 141–164.

Dieci, L. & Friedman, M. [2001], 'Continuation of invariant subspaces', *Numer. Linear Algebra Appl.* **8**, 317–327.

Dieci, L. & Lorenz, J. [1995], 'Computation of invariant tori by the method of characteristics', *SIAM J. Numer. Anal.* **32**, 1436–1474.

Dieci, L. & Lorenz, J. [1997], 'Lyapunov-type numbers and torus breakdown: numerical aspects and a case study', *Numer. Algorithms* **14**, 79–102.

Diekmann, O. & van Gils, S. [1984], 'Invariant manifolds for Volterra integral equations of convolution type', *J. Differential Equations* **54**, 139–180.

Diekmann, O., van Gils, S., Verduyn Lunel, S. & Walther, H.-O. [1995], *Delay Equations: Functional, Complex, and Nonlinear Analysis*, Springer-Verlag, New York.

Diener, F. [1983], Quelques exemples de bifurcations et leurs canards, *in* 'Proceedings of the eleventh annual Iranian mathematics conference (Mashhad, 1980)', University of Mashhad, Mashhad, pp. 59–73.

Doedel, E. [1981], 'AUTO, a program for the automatic bifurcation analysis of autonomous systems', *Congr. Numer.* **30**, 265–384.

Doedel, E. & Friedman, M. [1989], 'Numerical computation of heteroclinic orbits', *J. Comput. Appl. Math.* **26**, 159–170.

Doedel, E. & Kernévez, J.-P. [1986], AUTO: Software for continuation problems in ordinary differential equations with applications, Applied Mathematics, California Institute of Technology, Pasadena, CA.

Doedel, E., Champneys, A., Fairgrieve, T., Kuznetsov, Yu.A., Sandstede, B. & Wang, X.-J. [1997], AUTO97: Continuation and bifurcation software for ordinary differential equations (with HomCont), Computer Science, Concordia University, Montreal.

Doedel, E., Friedman, M. & Kunin, B. [1997], 'Successive continuation for locating connecting orbits', *Numerical Algorithms* **14**, 103–124.

Doedel, E., Friedman, M. & Monteiro, A. [1994], 'On locating connecting orbits', *Appl. Math. Comput.* **65**, 231–239.

Doedel, E., Govaerts, W. & Kuznetsov, Yu.A. [2003], 'Computation of periodic solution bifurcations in ODEs using bordered systems', *SIAM J. Numer. Anal.* **41**, 401–435.

Doedel, E., Jepson, A. & Keller, H. [1984], Numerical methods for Hopf bifurcation and continuation of periodic solution paths, *in* R. Glowinski & J. Lions, eds, 'Computing Methods in Applied Science and Engineering VI', North-Holland, Amsterdam.

Doedel, E., Keller, H. & Kernévez, J.-P. [1991*a*], 'Numerical analysis and control of bifurcation problems: (I) Bifurcation in finite dimensions', *Internat. J. Bifur. Chaos Appl. Sci. Engrg.* **1**, 493–520.

Doedel, E., Keller, H. & Kernévez, J.-P. [1991*b*], 'Numerical analysis and control of bifurcation problems: (II) Bifurcation in infinite dimensions', *Internat. J. Bifur. Chaos Appl. Sci. Engrg.* **1**, 745–772.

Dulac, M. [1923], 'Sur les cycles limites', *Bull. Soc. Math. France* **51**, 45–188.

Dumortier, F. [1978], *Singularities of Vector Fields, Mathematical Monographs 32*, IMPA, Rio de Janeiro.

Dumortier, F. & Rousseau, R. [1990], 'Cubic Liénard equations with linear damping', *Nonlinearity* **3**, 1015–1039.

Dumortier, F., Roussarie, R. & Sotomayor, J. [1987], 'Generic 3-parameter families of vector fields on the plane, unfolding a singularity with nilpotent linear part. The cusp case of codimension 3', *Ergodic Theory Dynamical Systems* **7**, 375–413.

Dumortier, F., Roussarie, R., Sotomayor, J. & Żołądek, H. [1991], *Bifurcations of Planar Vector Fields: Nilpotent Singularities and Abelian Integrals*, Vol. 1480 of *Lecture Notes in Mathematics*, Springer-Verlag, Berlin.

Edoh, K. & Lorenz, J. [2001], 'Computation of Lyapunov-type numbers for invariant curves of planar maps', *SIAM J. Sci. Comput.* **23**, 1113–1134.

Ermentrout, B. [2002], *Simulating, Analyzing, and Animating Dynamical Systems: A Guide to XPPAUT for Researchers and Students*, Vol. 14 of *Software, Environments, and Tools*, SIAM, Philadelphia, PA.

Evans, J., Fenichel, N. & Feroe, J. [1982], 'Double impulse solutions in nerve axon equations', *SIAM J. Appl. Math.* **42**, 219–234.

Fairgrieve, T. & Jepson, A. [1991], 'O.K. Floquet multipliers', *SIAM J. Numer. Anal.* **28**, 1446–1462.

Feichtinger, G. [1992], 'Hopf bifurcation in an advertising diffusion model', *J. Econom. Behavior Organization* **17**, 401–411.

Feigenbaum, M. [1978], 'Quantitative universality for a class of nonlinear transformations', *J. Stat. Phys.* **19**, 25–52.

Fenichel, N. [1971], 'Persistence and smoothness of invariant manifolds for flows', *Indiana Univ. Math. J.* **21**, 193–226.

Feroe, J. [1981], 'Travelling waves of infinitely many pulses in nerve equations', *Math. Biosci.* **55**, 189–204.

Feudel, U. & Jansen, W. [1992], 'CANDYS/QA – A software system for qualitative analysis of nonlinear dynamical systems', *Internat. J. Bifur. Chaos Appl. Sci. Engrg.* **2**, 773–794.

Fiedler, B. [1988], *Global Bifurcations of Periodic Solutions with Symmetry*, Vol. 1309 of *Lecture Notes in Mathematics*, Springer-Verlag, Berlin.

FitzHugh, R. [1961], 'Impulses and physiological states in theoretical models of nerve membrane', *Biophys. J.* **1**, 445–446.

Friedman, M. [2001], 'Improved detection of bifurcations in large nonlinear systems via the continuation of invariant subspaces algorithm', *Internat. J. Bifur. Chaos Appl. Sci. Engrg.* **11**, 2277–2285.

Friedman, M. & Doedel, E. [1991], 'Numerical computation and continuation of invariant manifolds connecting fixed points', *SIAM J. Numer. Anal.* **28**, 789–808.

Friedman, M. & Doedel, E. [1993], 'Computational methods for global analysis of homoclinic and heteroclinic orbits: A case study', *J. Dynamics Differential Equations* **5**, 37–57.

Frouzakis, C., Adomaitis, R. & Kevrekidis, I. [1991], 'Resonance phenomena in an adaptively-controlled system', *Internat. J. Bifur. Chaos Appl. Sci. Engrg.* **1**, 83–106.

Fuller, A. [1968], 'Condition for a matrix to have only characteristic roots with negative real parts', *J. Math. Anal. Appl.* **23**, 71–98.

Gambaudo, J. [1985], 'Perturbation of a Hopf bifurcation by an external time-periodic forcing', *J. Differential Equations* **57**, 172–199.

Gamero, E., Freire, E. & Rodríguez-Luis, A. [1993], Hopf-zero bifurcation: normal form calculation and application to an electronic oscillator, *in* 'International Conference on Differential Equations, Vol. 1, 2 (Barcelona, 1991)', World Scientific, River Edge, NJ, pp. 517–524.

Gaspard, P. [1983], 'Generation of a countable set of homoclinic flows through bifurcation', *Phys. Lett. A* **97**, 1–4.

Gaspard, P. [1993], 'Local birth of homoclinic chaos', *Physica D* **62**, 94–122.

Gaspard, P., Kapral, R. & Nicolis, G. [1984], 'Bifurcation phenomena near homoclinic systems: a two-parameter analysis', *J. Stat. Phys.* **35**, 687–727.

Gatermann, K. & Hohmann, A. [1991], 'Symbolic exploitation of symmetry in numerical pathfollowing', *Impact Comput. Sci. Engrg.* **3**, 330–365.

Gavrilov, N.K. [1978], Bifurcations of an equilibrium with one zero and a pair of pure imaginary roots, *in* 'Methods of Qualitative Theory of Differential Equations', Gorkii State University, Gorkii. In Russian.

Gavrilov, N.K. [1980], Bifurcations of an equilibrium with two pairs of pure imaginary roots, *in* 'Methods of Qualitative Theory of Differential Equations', Gorkii State University, Gorkii, pp. 17–30. In Russian.

Gavrilov, N.K. & Shilnikov, A.L. [2000], Example of a blue sky catastrophe, *in* 'Methods of Qualitative Theory of Differential Equations and Related Topics', Vol. 200 of *Amer. Math. Soc. Transl. Ser. 2*, Amer. Math. Soc., Providence, RI, pp. 99–105.

Gavrilov, N.K. & Shilnikov, L.P. [1972], 'On three-dimensional systems close to systems with a structurally unstable homoclinic curve: I', *Math. USSR-Sb.* **17**, 467–485.

Gavrilov, N.K. & Shilnikov, L.P. [1973], 'On three-dimensional systems close to systems with a structurally unstable homoclinic curve: II', *Math. USSR-Sb.* **19**, 139–156.

Gheiner, J. [1994], 'Codimension-two reflection and nonhyperbolic invariant lines', *Nonlinearity* **7**, 109–184.

Ghezzi, L. & Kuznetsov, Yu.A. [1994], 'Strong resonances and chaos in a stock market model', *Internat. J. Systems Sci.* **11**, 1941–1955.

Glendinning, P. [1988], Global bifurcations in flows, *in* T. Bedford & J. Swift, eds, 'New Directions in Dynamical Systems', Vol. 127 of *London Math. Soc. Lecture Note Ser.*, Cambridge University Press, Cambridge, pp. 120–149.

Glendinning, P. & Sparrow, C. [1984], 'Local and global behaviour near homoclinic orbits', *J. Stat. Phys.* **35**, 645–696.

Golden, M. & Ydstie, B. [1988], 'Bifurcation in model reference adaptive control systems', *Systems Control Lett.* **11**, 413–430.

Golubitsky, M. & Schaeffer, D. [1985], *Singularities and Groups in Bifurcation Theory I*, Springer-Verlag, New York.

Golubitsky, M., Stewart, I. & Schaeffer, D. [1988], *Singularities and Groups in Bifurcation Theory II*, Springer-Verlag, New York.

Gonchenko, S.V. & Gonchenko, V.S. [2000], On Andronov-Hopf bifurcations of two-dimensional diffeomorphisms with homoclinic tangencies, Preprint 556, WIAAS, Berlin.

Gonchenko, S.V., Gonchenko, V.S. & Tatjer, J. [2002], Three-dimensional dissipative diffeomorphisms with codimension two homoclinic tangencies and generalized Hénon maps, *in* L. Lerman & L. Shilnikov, eds, 'Proceedings of the International Conference "Progress in Nonlinear Science" Dedicated to the 100th Anniversary of A.A. Andronov, Vol. I (Nizhny Novgorod, Russia, July 2001)', Institute of Applied Physics, Russian Academy of Sciences, Nizhny Novgorod, pp. 63–79.

Gonchenko, S.V., Turaev, D.V., Gaspard, P. & Nicolis, G. [1997], 'Complexity in the bifurcation structure of homoclinic loops to a saddle-focus', *Nonlinearity* **10**, 409–423.

Gonchenko, V.S., Kuznetsov, Yu.A. & Meijer, H.G.E. [2004], Generalized Hénon map and bifurcations of homoclinic tangencies, Preprint 1296, Department of Mathematics, Utrecht University.

Govaerts, W. [2000], *Numerical Methods for Bifurcations of Dynamical Equilibria*, SIAM, Philadelphia, PA.

Govaerts, W. & Pryce, J. [1993], 'Mixed block elimination for linear systems with wider borders', *IMA J. Numer. Anal.* **13**, 161–180.

Govaerts, W., Guckenheimer, J. & Khibnik, A.I. [1997], 'Defining functions for multiple Hopf bifurcations', *SIAM J. Numer. Anal.* **34**, 1269–1288.

Govaerts, W., Kuznetsov, Yu.A. & Sijnave, B. [1999], Bifurcations of maps in the software package CONTENT, *in* V. Ganzha, E. Mayr & E. Vorozhtsov, eds, 'Computer Algebra in Scientific Computing—CASC'99 (Munich)', Springer, Berlin, pp. 191–206.

Govaerts, W., Kuznetsov, Yu.A. & Sijnave, B. [2000*a*], Continuation of codimension-2 equilibrium bifurcations in CONTENT, *in* E. Doedel & L. Tuckerman, eds, 'Numerical Methods for Bifurcation Problems and Large-Scale Dynamical Systems (Minneapolis, 1997)', Vol. 119 of *IMA Vol. Math. Appl.*, Springer, New York, pp. 163–184.

Govaerts, W., Kuznetsov, Yu.A. & Sijnave, B. [2000*b*], 'Numerical methods for the generalized Hopf bifurcation', *SIAM J. Numer. Anal.* **38**, 329–346.

Griewank, A. & Reddien, G. [1984], 'Characterization and computation of generalized turning points', *SIAM J. Numer. Anal.* **21**, 176–184.

Grobman, D. [1959], 'Homeomorphisms of systems of differential equations', *Dokl. Akad. Nauk SSSR* **128**, 880–881. In Russian.

Gruzdev, V.G. & Neimark, Ju.I. [1975], A symbolic description of motion in the neighborhood of a not structurally stable homoclinic structure and of its change in transition to close systems, *in* 'System Dynamics, Vol. 8', Gorkii State University, Gorkii, pp. 13–33. In Russian.

Guckenheimer, J. [1981], On a codimension two bifurcation, *in* D. Rand & L. Young, eds, 'Dynamical Systems and Turbulence', Vol. 898 of *Lecture Notes in Mathematics*, Springer-Verlag, New York.

Guckenheimer, J. [2002], Numerical analysis of dynamical systems, *in* B. Fiedler, ed., 'Handbook of Dynamical Systems, Vol. 2', Elsevier Science, Amsterdam, pp. 345–390.

Guckenheimer, J. & Holmes, P. [1983], *Nonlinear Oscillations, Dynamical Systems and Bifurcations of Vector Fields*, Springer-Verlag, New York.

Guckenheimer, J. & Meloon, B. [2000], 'Computing periodic orbits and their bifurcations with automatic differentiation', *SIAM J. Sci. Comput.* **22**, 951–985.

Guckenheimer, J. & Myers, M. [1996], 'Computing Hopf bifurcations. II. Three examples from neurophysiology', *SIAM J. Sci. Comput.* **17**, 1275–1301.

Guckenheimer, J. & Worfolk, P. [1993], Dynamical systems: some computational problems, *in* D. Schlomiuk, ed., 'Bifurcations and Periodic Orbits of Vector Fields (Montreal, 1992)', Vol. 408 of *NATO Adv. Sci. Inst. Ser. C Math. Phys. Sci.*, Kluwer Acad. Publ., Dordrecht, pp. 241–278.

Guckenheimer, J., Myers, M. & Sturmfels, B. [1997], 'Computing Hopf bifurcations. I', *SIAM J. Numer. Anal.* **34**, 1–21.

Hadamard, J. [1901], 'Sur l'itération et let solutions asymptotiques des équations diffiérentialles', *Proc. Soc. Math. France* **29**, 224–228.

Hale, J. [1971], *Functional Differential Equations*, Springer-Verlag, New York.

Hale, J. [1977], *Theory of Functional Differential Equations*, Springer-Verlag, New York.

Hale, J. & Koçak, H. [1991], *Dynamics and Bifurcations*, Springer-Verlag, New York.

Hale, J. & Verduyn Lunel, S. [1993], *Introduction to Functional Differential Equations*, Springer-Verlag, New York.

Hartman, P. [1963], 'On the local linearization of differential equations', *Proc. Amer. Math. Soc.* **14**, 568–573.

Hartman, P. [1964], *Ordinary Differential Equations*, Wiley, New York.

Hassard, B. [1980], Computation of invariant manifolds, *in* P. Holmes, ed., 'New Approaches to Nonlinear Problems in Dynamics', SIAM, Philadelphia, PA, pp. 27–42.

Hassard, B., Kazarinoff, N. & Wan, Y.-H. [1981], *Theory and Applications of Hopf Bifurcation*, Cambridge University Press, London.

Hasselblatt, B. & Katok, A. [2003], *A First Course in Dynamics: With a Panorama of Recent Developments*, Cambridge University Press, New York.

Hastings, S. [1976], 'On the existence of homoclinic and periodic orbits for the FitzHugh-Nagumo equations', *Quart. J. Math. (Oxford)* **27**, 123–134.

Hearn, A. [1993], *REDUCE User's Manual, Version 3.5*, The RAND Corporation, Santa Monica.

Hénon, M. [1976], 'A two-dimensional mapping with a strange attractor', *Comm. Math. Phys.* **50**, 69–77.

Henry, D. [1981], *Geometric Theory of Semilinear Parabolic Equations*, Vol. 840 of *Lecture Notes in Mathematics*, Springer-Verlag, New York.

Hille, E. & Phillips, R. [1957], *Functional Analysis and Semigroups*, American Mathematical Society, Providence, RI.

Hirsch, M. & Smale, S. [1974], *Differential Equations, Dynamical Systems and Linear Algebra*, Academic Press, New York.

Hirsch, M., Pugh, C. & Shub, M. [1977], *Invariant Manifolds*, Vol. 583 of *Lecture Notes in Mathematics*, Springer-Verlag, Berlin.

Hirschberg, P. & Knobloch, E. [1993], 'Šil'nikov–Hopf bifurcation', *Physica D* **62**, 202–216.

Holling, C. [1965], 'The functional response of predators to prey density and its role in mimicry and population regulation', *Mem. Entomol. Soc. Canada* **45**, 5–60.

Holmes, P. & Rand, D. [1978], 'Bifurcations of the forced van der Pol oscillator', *Quart. Appl. Math.* **35**, 495–509.

Holmes, P. & Whitley, D. [1984], 'Bifurcations of one- and two-dimensional maps', *Philos. Trans. Roy. Soc. London, Ser. A* **311**, 43–102.

Holodniok, M. & Kubiček, M. [1984*a*], 'DERPER - An algorithm for the continuation of periodic solutions in ordinary differential equations', *J. Comput. Phys.* **55**, 254–267.

Holodniok, M. & Kubiček, M. [1984*b*], 'New algorithms for the evaluation of complex bifurcation points in ordinary differential equations. A comparative numerical study', *Appl. Math. Comput.* **15**, 261–274.

Homburg, A.J. [1993], Some global aspects of homoclinic bifurcations of vector fields, PhD thesis, Department of Mathematics, University of Groningen.

Homburg, A.J. [1996], 'Global aspects of homoclinic bifurcations of vector fields', *Mem. Amer. Math. Soc.* **121**(578), 1–128.

Homburg, A.J., Kokubu, H. & Krupa, M. [1994], 'The cusp horseshoe and its bifurcations in the unfolding of an inclination-flip homoclinic orbit', *Ergodic Theory Dynamical Systems* **14**, 667–693.

Homburg, A.J., Osinga, H. & Vegter, G. [1995], 'On the computation of invariant manifolds of fixed points', *Z. Angew. Math. Phys.* **46**, 171–187.

Hopf, E. [1942], 'Abzweigung einer periodischen Lösung von einer stationaren Lösung eines Differetialsystems', *Ber. Math.-Phys. Kl. Sachs, Acad. Wiss. Leipzig* **94**, 1–22.

Horozov, E. [1979], Versal deformations of equivariant vector fields for the cases of symmetry of order 2 and 3, *in* 'Proceedings of Petrovskii Seminar, Vol. 5', Moscow State University, Moscow, pp. 163–192. In Russian.

Ilyashenko, Yu. & Li, Weigu. [1999], *Nonlocal Bifurcations*, American Mathematical Society, Providence, RI.

Iooss, G. [1979], *Bifurcations of Maps and Applications*, North-Holland, Amsterdam.

Iooss, G. & Adelmeyer, M. [1992], *Topics in Bifurcation Theory and Applications*, World Scientific, Singapore.

Iooss, G. & Joseph, D. [1980], *Elementary Stability and Bifurcation Theory*, Springer-Verlag, New York.

Iooss, G., Arneodo, A., Coullet, P. & Tresser, C. [1981], Simple computation of bifurcating invariant circles for mappings, *in* D. Rand & L.-S. Young, eds, 'Dynamical Systems and Turbulence', Vol. 898 of *Lecture Notes in Mathematics*, Springer-Verlag, New York, pp. 192–211.

Irwin, M. [1980], *Smooth Dynamical Systems*, Academic Press, New York.

Jury, E. & Gutman, S. [1975], 'On the stability of the $A$ matrix inside the unit circle', *IEEE Trans. Automatic Control* **AC-20**, 533–535.

Kaas-Petersen, C. [1989], PATH - User's Guide, University of Leeds, Leeds.

Kantorovich, L.V. & Akilov, G.P. [1964], *Functional Analysis in Normed Spaces*, Pergamon Press, Oxford.

Katok, A. & Hasselblatt, B. [1995], *Introduction to the Modern Theory of Dynamical Systems*, Vol. 54 of *Encyclopedia of Mathematics and Its Applications*, Cambridge University Press, Cambridge.

Keener, J. [1981], 'Infinite period bifurcation and global bifurcation branches', *SIAM J. Appl. Math.* **41**, 127–144.

Keller, H. [1977], Numerical solution of bifurcation and nonlinear eigenvalue problems, *in* P. Rabinowitz, ed., 'Applications of Bifurcation Theory', Academic Press, New York, pp. 359–384.

Kelley, A. [1967], 'The stable, center stable, center, center unstable and unstable manifolds', *J. Differential Equations* **3**, 546–570.

Khibnik, A.I. [1979], Periodic solutions of a system of differential equations, FORTRAN Software Series, Vol. 5, Research Computing Centre, USSR Academy of Sciences, Pushchino, Moscow Region. In Russian.

Khibnik, A.I. [1990], LINLBF: A program for continuation and bifurcation analysis of equilibria up to codimension three, *in* D. Roose, B. De Dier & A. Spence, eds, 'Continuation and Bifurcations: Numerical Techniques and Applications (Leuven, 1989)', Vol. 313 of *NATO Adv. Sci. Inst. Ser. C Math. Phys. Sci.*, Kluwer Acad. Publ., Dordrecht, pp. 283–296.

Khibnik, A.I., Kuznetsov, Yu.A., Levitin, V.V. & Nikolaev, E.V. [1993], 'Continuation techniques and interactive software for bifurcation analysis of ODEs and iterated maps', *Physica D* **62**, 360–371.

Kielhöfer, H. [2004], *Bifurcation Theory: An Introduction with Applications to PDEs*, Springer-Verlag, New York.

Kirchgraber, U. & Palmer, K. [1990], *Geometry in the Neighborhood of Invariant Manifolds of Maps and Flows and Linearization*, Vol. 233 of *Pitman Research Notes in Mathematics Series*, Longman Scientific & Technical, Harlow.

Kirk, V. [1991], 'Breaking of symmetry in the saddle-node Hopf bifurcation', *Phys. Lett. A* **154**, 243–248.

Kirk, V. [1993], 'Merging of resonance tongues', *Physica D* **66**, 267–281.

Kisaka, M., Kokubu, H. & Oka, H. [1993*a*], 'Bifurcation to $n$-homoclinic orbits and $n$-periodic orbits in vector fields', *J. Dynamics Differential Equations* **5**, 305–358.

Kisaka, M., Kokubu, H. & Oka, H. [1993*b*], Supplement to homoclinic doubling bifurcation in vector fields, *in* R. Bamon, J. Labarca, J. Lewowicz & J. Palis, eds, 'Dynamical Systems', Longman, London, pp. 92–116.

Kolmogorov, A.N. [1957], Théorie générale des systèmes dynamiques et mécanique classique, *in* 'Proceedings of the International Congress of Mathematicians, Amsterdam, 1954, Vol. 1', Erven P. Noordhoff N.V., Groningen, pp. 315–333.

Krauskopf, B. [1994*a*], 'Bifurcation sequences at 1:4 resonance: an inventory', *Nonlinearity* **7**, 1073–1091.

Krauskopf, B. [1994*b*], 'The bifurcation set for the 1:4 resonance problem', *Experimental Mathematics* **3**, 107–128.

Krauskopf, B. [1997], 'Bifurcations at $\infty$ in a model for 1:4 resonance', *Ergodic Theory Dynamical Systems* **17**, 899–931.

Krauskopf, B. & Osinga, H. [1998], 'Growing 1D and quasi-2D unstable manifolds of maps', *J. Comput. Phys.* **146**, 404–419.

Krauskopf, B. & Osinga, H. [1999], 'Two-dimensional global manifolds of vector fields', *Chaos* **9**, 768–774.

Kubiček, M. [1976], 'Algorithm 502. Dependence of solutions of nonlinear systems on a parameter', *ACM Trans. Math. Software* **2**, 98–107.

Kurakin, L.G. & Judovich, V.I. [1986], 'Semi-invariant form of equilibrium stability criteria in critical cases', *J. Appl. Math. Mech.* **50**, 543–546.

Kuznetsov, Yu.A. [1983], One-dimensional invariant manifolds of saddles in ordinary differential equations depending upon parameters, FORTRAN Software Series, Vol. 8, Research Computing Centre, USSR Academy of Sciences, Pushchino, Moscow Region. In Russian.

Kuznetsov, Yu.A. [1984], Andronov-Hopf bifurcation in four-dimensional systems with circular symmetry, Research Computing Centre, USSR Academy of Sciences, Pushchino, Moscow Region. In Russian.

Kuznetsov, Yu.A. [1985], Auto-waves in reaction-diffusion systems with circular symmetry (center manifold approach), Research Computing Centre, USSR Academy of Sciences, Pushchino, Moscow Region. In Russian.

Kuznetsov, Yu.A. [1990], Computation of invariant manifold bifurcations, *in* D. Roose, B. De Dier & A. Spence, eds, 'Continuation and Bifurcations: Numerical Techniques and Applications (Leuven, 1989)', Vol. 313 of *NATO Adv. Sci. Inst. Ser. C Math. Phys. Sci.*, Kluwer Acad. Publ., Dordrecht, pp. 183–195.

Kuznetsov, Yu.A. [1991], Numerical analysis of the orientability of homoclinic trajectories, *in* R. Seydel, F. Schneider, T. Küpper & H. Troger, eds, 'Bifurcation and Chaos: Analysis, Algorithms, Applications (Würzburg, 1990)', Vol. 97 of *Internat. Ser. Numer. Math.*, Birkhäuser, Basel, pp. 237–242.

Kuznetsov, Yu.A. [1999], 'Numerical normalization techniques for all codim 2 bifurcations of equilibria in ODEs', *SIAM J. Numer. Anal.* **36**, 1104–1124.

Kuznetsov, Yu.A. & Meijer, H.G.E. [2003], Numerical normal forms for codim 2 bifurcations of fixed points with at most two critical eigenvalues, Preprint 1290, Department of Mathematics, Utrecht University.

Kuznetsov, Yu.A. & Panfilov, A.V. [1981], Stochastic waves in the FitzHugh-Nagumo system, Research Computing Centre, USSR Academy of Sciences, Pushchino, Moscow Region. In Russian.

Kuznetsov, Yu.A. & Piccardi, C. [1994*a*], 'Bifurcation analysis of periodic SEIR and SIR epidemic models', *J. Math. Biol.* **32**, 109–121.

Kuznetsov, Yu.A. & Piccardi, C. [1994*b*], 'Bifurcations and chaos in a periodically forced prototype adaptive control system', *Kybernetika* **30**, 121–128.

Kuznetsov, Yu.A. & Rinaldi, S. [1991], 'Numerical analysis of the flip bifurcation of maps', *Appl. Math. Comput.* **43**, 231–236.

Kuznetsov, Yu.A., De Feo, O. & Rinaldi, S. [2001], 'Belyakov homoclinic bifurcations in a tritrophic food chain model', *SIAM J. Appl. Math.* **62**, 462–487.

Kuznetsov, Yu.A., Levitin, V.V. & Skovoroda, A.R. [1996], Continuation of stationary solutions to evolution problems in CONTENT, Report AM-R9611, Centrum voor Wiskunde en Informatica, Amsterdam.

Kuznetsov, Yu.A., Meijer, H.G.E. & van Veen, L. [2004], 'The fold-flip bifurcation', *Internat. J. Bifur. Chaos Appl. Sci. Engrg.* **14**, to appear.

Kuznetsov, Yu.A., Muratori, S. & Rinaldi, S. [1992], 'Bifurcations and chaos in a periodic predator-prey model', *Internat. J. Bifur. Chaos Appl. Sci. Engrg.* **2**, 117–128.

Kuznetsov, Yu.A., Muratori, S. & Rinaldi, S. [1995], 'Homoclinic bifurcations in slow-fast second order systems', *Nonlinear Anal.* **25**, 747–762.

Lanford, O. [1980], 'A computer-assisted proof of the Feigenbaum conjectures', *Bull. Amer. Math. Soc.* **6**, 427–434.

Lanford, O. [1984], 'A shorter proof of the existence of the Feigenbaum fixed point', *Comm. Math. Phys.* **96**, 521–538.

Langford, W. [1979], 'Periodic and steady mode interactions lead to tori', *SIAM J. Appl. Math.* **37**, 22–48.

Lefever, R. & Prigogine, I. [1968], 'Symmetry-breaking instabilities in dissipative systems (II)', *J. Chem. Phys.* **48**, 1695–1700.

Levitin, V.V. [1995], Computation of functions and their derivatives in CONTENT, Report AM-R9512, Centrum voor Wiskunde en Informatica, Amsterdam.

Lin, X.-B. [1990], 'Using Melnikov's method to solve Silnikov's problems', *Proc. Roy. Soc. Edinburgh Sect. A* **116**, 295–325.

Lorenz, E. [1963], 'Deterministic non-periodic flow', *J. Atmos. Sci.* **20**, 130–141.

Lorenz, E. [1984], 'Irregularity: A fundamental property of the atmosphere', *Tellus A* **36**, 98–110.

Lou, Z., Kostelich, E. & Yorke, J. [1992], 'Erratum: "Calculating stable and unstable manifolds"', *Internat. J. Bifur. Chaos Appl. Sci. Engrg.* **2**, 215.

Lukyanov, V.I. [1982], 'Bifurcations of dynamical systems with a saddle-point separatrix loop', *Differential Equations* **18**, 1049–1059.

Lyapunov, A. [1892], *General Problem of Stability of Motion*, Mathematics Society of Kharkov, Kharkov.

Lyubich, M. [2000], 'The quadratic family as a qualitatively solvable model of chaos', *Notices Amer. Math. Soc.* **47**, 1042–1052.

Marsden, J. & McCracken, M. [1976], *Hopf Bifurcation and Its Applications*, Springer-Verlag, New York.

May, R. [1974], 'Biological populations with nonoverlapping generations: Stable points, stable cycles, and chaos', *Science* **17**, 645–647.

Maynard Smith, J. [1968], *Mathematical Ideas in Biology*, Cambridge University Press, London.

Medvedev, V.S. [1980], 'A new type of bifurcation on manifolds', *Mat. Sbornik* **113**, 487–492. In Russian.

Mei, Z. [1989], 'A numerical approximation for the simple bifurcation points', *Numer. Funct. Anal. Optimiz.* **10**, 383–400.

Melnikov, V.K. [1962], Qualitative description of resonance phenomena in nonlinear systems, P-1013, OIJaF, Dubna. In Russian.

Melnikov, V.K. [1963], 'On the stability of the center for time periodic perturbations', *Trans. Moscow Math. Soc.* **12**, 1–57.

Mishchenko, E.F. & Rozov, N.K. [1980], *Differential Equations with Small Parameters and Relaxation Oscillations*, Plenum, New York.

Miura, R. [1982], 'Accurate computation of the stable solitary wave for the FitzHugh-Nagumo equations', *J. Math. Biol.* **13**, 247–269.

Moon, F. & Rand, R. [1985], Parametric stiffness control of flexible structures, *in* 'Proceedings of the Workshop on Identification and Control of Flexible Space Structures, Vol. II', Jet Propulsion Laboratory Publication 85-29, Pasadena, CA, pp. 329–342.

Moore, G. [1980], 'The numerical treatment of non-trivial bifurcation points', *Numer. Funct. Anal. Optimiz.* **2**, 441–472.

Moore, G. & Spence, A. [1980], 'The calculation of turning points of nonlinear equations', *SIAM J. Numer. Anal.* **17**, 567–576.

Moore, G., Garret, T. & Spence, A. [1990], The numerical detection of Hopf bifurcation points, *in* D. Roose, B. De Dier & A. Spence, eds, 'Continuation and Bifurcations: Numerical Techniques and Applications (Leuven, 1989)', Vol. 313 of *NATO Adv. Sci. Inst. Ser. C Math. Phys. Sci.*, Kluwer Acad. Publ., Dordrecht, pp. 227–246.

Moser, J. [1968], 'Lectures on Hamiltonian systems', *Mem. Amer. Math. Soc.* **81**, 1–60.

Moser, J. [1973], *Stable and Random Motions in Dynamical Systems*, Princeton University Press, Princeton, NJ.

Murdock, J. [2003], *Normal Forms and Unfoldings for Local Dynamical Systems*, Springer-Verlag, New York.

Myrberg, P. [1962], 'Sur l'itération des polynomes réels quadratiques', *J. Math. Pures Appl.* **41**, 339–351.

Nagumo, J., Arimoto, S. & Yoshizawa, S. [1962], 'An active pulse transmission line simulating nerve axon', *Proc. IRE* **50**, 2061–2070.

Namachchivaya, S. & Ariaratnam, S. [1987], 'Periodically perturbed Hopf bifurcation', *SIAM J. Appl. Math.* **47**, 15–39.

Neimark, Ju.I. [1959], 'On some cases of periodic motions depending on parameters', *Dokl. Akad. Nauk SSSR* **129**, 736–739. In Russian.

Neimark, Ju.I. [1967], 'Motions close to doubly-asymptotic motion', *Soviet Math. Dokl.* **8**, 228–231.

Neimark, Ju.I. [1972], *The Method of Point Transformations in the Theory of Nonlinear Oscillations*, Nauka, Moscow. In Russian.

Neimark, Ju.I. & Shil'nikov, L.P. [1965], 'A case of generation of periodic motions', *Soviet Math. Dokl.* **6**, 305–309.

Neishtadt, A.I. [1978], 'Bifurcations of the phase pattern of an equation system arising in the problem of stability loss of self-oscillations close to 1 : 4 resonance', *J. Appl. Math. Mech.* **42**, 896–907.

Nemytskii, V.V. & Stepanov, V.V. [1949], *Qualitative Theory of Differential Equations*, GITTL, Moscow-Leningrad. In Russian.

Newhouse, S., Palis, J. & Takens, F. [1983], 'Bifurcations and stability of families of diffeomorphisms', *Inst. Hautes Études Sci. Publ. Math.* **57**, 5–71.

Nikolaev, E. [1995], 'Bifurcations of limit cycles of differential equations that admit involutory symmetry', *Mat. Sb.* **186**, 143–160.

Nikolaev, E.V. [1992], On bifurcations of closed orbits in the presence of involutory symmetry, Institute of Mathematical Problems of Biology, Russian Academy of Sciences, Pushchino, Moscow Region.

Nikolaev, E.V. [1994], Periodic motions in systems with a finite symmetry group, Institute of Mathematical Problems of Biology, Russian Academy of Sciences, Pushchino, Moscow Region.

Nikolaev, E.V. & Shnol, E.E. [1998*a*], 'Bifurcations of cycles in systems of differential equations with a finite symmetry group. I', *J. Dynam. Control Systems* **4**, 315–341.

Nikolaev, E.V. & Shnol, E.E. [1998*b*], 'Bifurcations of cycles in systems of differential equations with a finite symmetry group. II', *J. Dynam. Control Systems* **4**, 343–363.

Nitecki, Z. [1971], *Differentiable Dynamics*, MIT Press, Cambridge, MA.

Nozdrachova, V.P. [1982], 'Bifurcation of a noncoarse seperatrix loop', *Differential Equations* **18**, 1098–1104.

Nusse, H. & Yorke, J. [1998], *Dynamics: Numerical Explorations, 2nd ed.*, Springer-Verlag, New York.

Osinga, H. [2003], 'Nonorientable manifolds in threee-dimensional vector fields', *Internat. J. Bifur. Chaos Appl. Sci. Engrg.* **13**, 553–570.

Ovsyannikov, I.M. & Shil'nikov, L.P. [1987], 'On systems with a saddle-focus homoclinic curve', *Math. USSR-Sb.* **58**, 557–574.

Palis, J. & Pugh, C. [1975], Fifty problems in dynamical systems, *in* 'Dynamical Systems (Warwick, 1974)', Vol. 468 of *Lecture Notes in Mathematics*, Springer-Verlag, Berlin, pp. 345–353.

Palis, J. & Takens, F. [1993], *Hyperbolicity and Sensitive Chaotic Dynamics at Homoclinic Bifurcations: Fractal Dimensions and Infinitely Many Attractors*, Vol. 35 of *Cambridge Studies in Advanced Mathematics*, Cambridge University Press, Cambridge.

Palmer, K. [1984], 'Exponential dichotomies and transversal homoclinic points', *J. Differential Equations* **55**, 225–256.

Pavlou, S. & Kevrekidis, I. [1992], 'Microbial predation in a periodically operated chemostat: A global study of the interaction between natural and externally imposed frequencies', *Math. Biosci.* **108**, 1–55.

Peckham, B. & Kevrekidis, I. [1991], 'Period doubling with higher-order degeneracies', *SIAM J. Math. Anal.* **22**, 1552–1574.

Peixoto, M. [1962], 'Structural stability on two dimensional manifolds', *Topology* **1**, 101–120.

Perron, O. [1930], 'Die stabilitätsfrage bei differentialgleichungen', *Math. Z.* **32**, 703–728.

Petrovich, V.Yu. [1990], Numerical spectral analysis of the differential of the doubling operator by K.I. Babenko's method, Preprint 90-81, Institute of Applied Mathematics, USSR Academy of Sciences, Moscow. In Russian.

Pliss, V.A. [1964], 'A reduction principle in stability theory of motions', *Izv. Akad. Nauk SSSR, Ser. Mat.* **28**, 1297–1324.

Poincaré, H. [1879], *Sur les propriétés des fonctions définies par les équations aux différences partielles. Thèse*, Gauthier-Villars, Paris.

Poincaré, H. [1892,1893,1899], *Les Méthodes Nouvelles de la Méchanique Céleste*, Gauthier-Villars, Paris.

Pontryagin, L.S. [1934], 'On dynamical systems close to Hamiltonian systems', *J. Exptl. Theoret. Phys.* **4**, 234–238. In Russian.

Pontryagin, L.S. [1962], *Ordinary Differential Equations*, Pergamon Press, London.

Poston, T. & Stewart, I. [1978], *Catastrophe Theory and Its Applications*, Pitman, San Francisco.

Rheinboldt, W. [1982], 'Computation of critical boundaries on equilibrium manifolds', *SIAM J. Numer. Anal.* **19**, 653–669.

Rheinboldt, W. [1986], *Numerical Analysis of Parametrized Nonlinear Equations*, Wiley, New York.

Rheinboldt, W. & Burkardt, J. [1983], 'Algorithm 596: A program for a locally-parametrized continuation process', *ACM Trans. Math. Software* **9**, 236–241.

Richtmyer, R. [1978], *Principles of Advanced Mathematical Physics I*, Springer-Verlag, New York.

Richtmyer, R. [1981], *Principles of Advanced Mathematical Physics II*, Springer-Verlag, New York.

Ricker, W. [1954], 'Stock and recruitment', *J. Fish. Res. Board Canada* **11**, 559–663.

Robinson, J. [2001], *Infinite-dimensional Dynamical Systems: An Introduction to Dissipative Parabolic PDEs and the Theory of Global Attractors*, Cambridge University Press, Cambridge.

Rodríguez-Luis, A., Freire, E. & Ponce, E. [1990], A method for homoclinic and heteroclinic continuation in two and three dimensions, *in* D. Roose, B. De Dier & A. Spence, eds, 'Continuation and Bifurcations: Numerical Techniques and Applications (Leuven, 1989)', Vol. 313 of *NATO Adv. Sci. Inst. Ser. C Math. Phys. Sci.*, Kluwer Acad. Publ., Dordrecht, pp. 197–210.

Roose, D. & Hlavaček, V. [1985], 'A direct method for the computation of Hopf bifurcation points', *SIAM J. Appl. Math.* **45**, 897–894.

Roshchin, N.V. [1978], 'Unsafe stability boundaries of the Lorenz model', *J. Appl. Math. Mech.* **42**, 1038–1041.

Rössler, O. [1979], Continuous chaos—four prototype equations, *in* O. Gurel & O. Rossler, eds, 'Bifurcation Theory and Applications in Scientific Disciplines (Papers, Conf., New York, 1977)', New York Acad. Sci., pp. 376–392.

Ruelle, D. [1973], 'Bifurcation in the presence of a symmetry group', *Arch. Rational Mech. Anal.* **51**, 136–152.

Ruelle, D. & Takens, F. [1971], 'On the nature of turbulence', *Comm. Math. Phys.* **20**, 167–192.

Sacker, R. [1964], On invariant surfaces and bifurcation of periodic solutions of ordinary differential equations, Report IMM-NYU 333, New York University.

Sacker, R. [1965], 'A new approach to the perturbation theory of invariant surfaces', *Comm. Pure Appl. Math.* **18**, 717–732.

Salam, F. & Bai, S. [1988], 'Complicated dynamics of a prototype continuous-time adaptive control system', *IEEE Trans. Circuits and Systems* **35**, 842–849.

Sanders, J. [1982], 'Melnikov's method and averaging', *Celestial Mech.* **28**, 171–181.

Sanders, J. [1994], Versal normal form computation and representation theory, *in* E. Tournier, ed., 'Computer Algebra and Differential Equations', Cambridge University Press, Cambridge, pp. 185–210.

Sandstede, B. [1993], Verzweigungstheorie homokliner Verdopplungen, PhD thesis, Institut für Angewandte Analysis und Stochastik, Berlin.

Sandstede, B. [1997*a*], 'Constructing dynamical systems having homoclinic bifurcation points of codimension two', *J. Dynamics Differential Equations* **9**, 269–288.

Sandstede, B. [1997*b*], 'Convergence estimates for the numerical approximation of homoclinic solutions', *IMA J. Numer. Anal.* **17**, 437–462.

Sandstede, B. [2000], 'Center manifolds for homoclinic solutions', *J. Dynam. Differential Equations* **12**, 449–510.

Sandstede, B., Scheel, A. & Wulff, C. [1997], 'Dynamics of spiral waves on unbounded domains using center-manifold reductions', *J. Differential Equations* **141**, 122–149.

Schecter, S. [1993], 'Numerical computation of saddle-node homoclinic bifurcation points', *SIAM J. Numer. Anal.* **30**, 1155–1178.

Schuko, S.D. [1968], 'Derivation of the lyapunov coefficients on a digital computer', *Trudy Gorkii Inst. Inzh. Vodn. Transp.* **94**, 97–109. In Russian.

Serebriakova, N.N. [1959], 'On the behavior of dynamical systems with one degree of freedom near that point of the stability boundary where soft bifurcation turns into sharp', *Izv. Akad. Nauk SSSR.-Mech. Mash.* **2**, 1–10. In Russian.

Seydel, R. [1988], *From Equilibrium to Chaos: Practical Bifurcation and Stability Analysis*, Elsevier, New York.

Seydel, R. [1991], 'Tutorial on continuation', *Internat. J. Bifur. Chaos Appl. Sci. Engrg.* **1**, 3–11.

Shapiro, A.P. [1974], Mathematical models of competition, *in* 'Control and Information, Vol. 10', DVNC AN SSSR, Vladivostok, pp. 5–75. In Russian.

Shashkov, M.V. [1992], 'On bifurcations of separatrix contours with two saddles', *Internat. J. Bifur. Chaos Appl. Sci. Engrg.* **2**, 911–915.

Shil'nikov, A.L., Nicolis, G. & Nicolis, C. [1995], 'Bifurcation and predictability analysis of a low-order atmospheric circulation model', *Internat. J. Bifur. Chaos Appl. Sci. Engrg.* **5**, 1701–1711.

Shil'nikov, L.P. [1963], 'Some cases of generation of periodic motions from singular trajectories', *Mat. Sbornik* **61**, 443–466. In Russian.

Shil'nikov, L.P. [1965], 'A case of existence of a countable number of periodic motions', *Soviet Math. Dokl.* **6**, 163–166.

Shil'nikov, L.P. [1966], 'On the generation of a periodic motion from a trajectory which leaves and re-enters a saddle-saddle state of equilibrium', *Soviet Math. Dokl.* **7**, 1155–1158.

Shil'nikov, L.P. [1967*a*], 'The existence of a denumerable set of periodic motions in four-dimensional space in an extended neighborhood of a saddle-focus', *Soviet Math. Dokl.* **8**, 54–58.

Shil'nikov, L.P. [1967*b*], 'On a Poincaré–Birkhoff problem', *Math. USSR-Sb.* **3**, 353–371.

Shil'nikov, L.P. [1968], 'On the generation of periodic motion from trajectories doubly asymptotic to an equilibrium state of saddle type', *Math. USSR-Sb.* **6**, 427–437.

Shil'nikov, L.P. [1969], 'On a new type of bifurcation of multidimensional dynamical systems', *Sov. Math. Dokl.* **10**, 1368–1371.

Shil'nikov, L.P. [1970], 'A contribution to the problem of the structure of an extended neighborhood of a rough equilibrium state of saddle-focus type', *Math. USSR-Sb.* **10**, 91–102.

Shilnikov, L.P., Shilnikov, A.L., Turaev, D.V. & Chua, L. [1998], *Methods of Qualitative Theory in Nonlinear Dynamics. Part I*, World Scientific, Singapore.

Shilnikov, L.P., Shilnikov, A.L., Turaev, D.V. & Chua, L. [2001], *Methods of Qualitative Theory in Nonlinear Dynamics. Part II*, World Scientific, Singapore.

Shoshitaishvili, A.N. [1972], 'Bifurcations of topological type of singular points of vector fields that depend on parameters', *Functional Anal. Appl.* **6**, 169–170.

Shoshitaishvili, A.N. [1975], Bifurcations of topological type of a vector field near a singular point, *in* 'Proceedings of Petrovskii Seminar, Vol. 1', Moscow State University, Moscow, pp. 279–309. In Russian.

Smale, S. [1961], 'On gradient dynamical systems', *Ann. of Math.* **74**, 199–206.

Smale, S. [1963], Diffeomorphisms with many periodic points, *in* S. Carins, ed., 'Differential and Combinatorial Topology', Princeton University Press, Princeton, NJ, pp. 63–80.

Smale, S. [1966], 'Structurally stable systems are not dense', *Amer. J. Math.* **88**, 491–496.

Smale, S. [1967], 'Differentiable dynamical systems', *Bull. Amer. Math. Soc.* **73**, 747–817.

Stéphanos, C. [1900], 'Sur une extension du calcul des substitutions linéares', *J. Math. Pures Appl.* **6**, 73–128.

Sternberg, S. [1957], 'Local contractions and a theorem of Poincaré', *Amer. J. Math.* **79**, 809–824.

Szmolyan, P. [1991], 'Transversal heteroclinic and homoclinic orbits in singular perturbed problems', *J. Differential Equations* **92**, 252–281.

Takens, F. [1973], 'Unfoldings of certain singularities of vector fields: generalized Hopf bifurcations', *J. Differential Equations.* **14**, 476–493.

Takens, F. [1974*a*], 'Forced oscillations and bifurcations', *Comm. Math. Inst., Rijkuniversiteit Utrecht* **2**, 1–111. Reprinted in *Global Analysis of Dynamical Systems*, Instute of Physics, Bristol, 2001, pp. 1–61.

Takens, F. [1974*b*], 'Singularities of vector fields', *Inst. Hautes Études Sci. Publ. Math.* **43**, 47–100.

Taylor, M. & Kevrekidis, I. [1990], Interactive AUTO: A graphical interface for AUTO86, Department of Chemical Engineering, Princeton University.

Taylor, M. & Kevrekidis, I. [1991], 'Some common dynamic features of coupled reacting systems', *Physica D* **51**, 274–292.

Taylor, M. & Kevrekidis, I. [1993], 'Couple, double, toil and trouble: a computer assisted study of two coupled CSTRs', *Chem. Engng. Sci.* **48**, 1–86.

Temam, R. [1997], *Infinite-dimensional Dynamical Systems in Mechanics and Physics, 2nd edition*, Springer-Verlag, New York.

Thom, R. [1972], *Stabilité Structurelle et Morphogénèse*, Benjamin, New York.

Tournier, E., ed. [1994], *Computer Algebra and Differential Equations*, London Mathematical Society Lecture Note Series, Vol. 193, Cambridge University Press, Cambridge.

Tresser, C. [1984], 'About some theorems by L.P. Šil'nikov', *Ann. Inst. H. Poincaré Théor.* **40**, 441–461.

Turaev, D.V. [1991], On bifurcations of dynamical systems with two homoclinic curves of the saddle, PhD thesis, Gorkii State University. In Russian.

Turaev, D.V. & Shil'nikov, L.P. [1995], 'Blue sky catastrophes', *Dokl. Math.* **51**, 404–407.

Vaĭnberg, M.M. & Trenogin, V.A. [1974], *Theory of Branching of Solutions of Non-linear Equations*, Noordhoff International Publishing, Leyden.

van Gils, S. [1982], On a formula for the direction of Hopf bifurcation, Centre for Mathematics and Computer Science, Report TW/225.

van Gils, S. [1985], 'A note on "Abelian integrals and bifurcation theory"', *J. Differential Equations* **59**, 437–441.

van Gils, S. & Mallet-Paret, J. [1986], 'Hopf bifurcation and symmetry: Travelling and standing waves on the circle', *Proc. Roy. Soc. Edinburgh Sect. A* **104**, 279–307.

van Strien, S. [1979], 'Center manifolds are not $C^\infty$', *Math. Z.* **166**, 143–145.

van Strien, S. [1991], Interval dynamics, *in* E. van Groesen & E. de Jager, eds, 'Structures in Dynamics', Vol. 2 of *Studies in Mathematical Physics*, North-Holland, Amsterdam, pp. 111–160.

Vance, W. & Ross, J. [1991], 'Bifurcation structures of periodically forced oscillators', *Chaos* **1**, 445–453.

Vanderbauwhede, A. [1989], 'Centre manifolds, normal forms and elementary bifurcations', *Dynamics Reported* **2**, 89–169.

Volterra, V. [1931], *Lecònse sur la Théorie Mathematique de la Lutte pur la Via*, Gauthier-Villars, Paris.

Wan, Y.-H. [1978*a*], 'Bifurcation into invariant tori at points of resonance', *Arch. Rational Mech. Anal.* **68**, 343–357.

Wan, Y.-H. [1978*b*], 'Computations of the stability condition for the Hopf bifurcation of diffeomorphisms on $\mathbf{R}^2$', *SIAM J. Appl. Math.* **34**, 167–175.

Wang, D. [1993], A recursive formula and its application to computations of normal forms and focal values, *in* S.-T. Liao, T.-R. Ding & Y.-Q. Ye, eds, 'Dynamical Systems (Tianjin, 1990/1991)', Vol. 4 of *Nankai Ser. Pure Appl. Math. Theoret. Phys.*, World Sci. Publishing, River Edge, NJ, pp. 238–247.

Werner, B. [1996], 'Computation of Hopf bifurcations with bordered matrices', *SIAM J. Numer. Anal.* **33**, 435–455.

Whitley, D. [1983], 'Discrete dynamical systems in dimensions one and two', *Bull. London Math. Soc.* **15**, 177–217.

Wiggins, S. [1988], *Global Bifurcations and Chaos*, Springer-Verlag, New York.

Wiggins, S. [1990], *Introduction to Applied Non-linear Dynamical Systems and Chaos*, Springer-Verlag, New York.

Wolfram, S. [1991], *Mathematica: A System for Doing Mathematics by Computer*, Addison-Wesley, Redwood City, CA.

Yanagida, E. [1987], 'Branching of double pulse solutions from single pulse solutions in nerve axon equations', *J. Differential Equations* **66**, 243–262.

You, Z., Kostelich, E. J. & Yorke, J. [1991], 'Calculating stable and unstable manifolds', *Internat. J. Bifur. Chaos Appl. Sci. Engrg.* **1**, 605–623.

Zegeling, A. [1993], 'Equivariant unfoldings in the case of symmetry of order 4', *Serdica* **19**, 71–79.

Żołądek, H. [1984], 'On the versality of a family of symmetric vector-fields in the plane', *Math. USSR-Sb.* **48**, 463–498.

Żołądek, H. [1987], 'Bifurcations of certain family of planar vector fields tangent to axes', *J. Differential Equations* **67**, 1–55.

# Index

# Applied Mathematical Sciences

*(continued from page ii)*



*(continued on next page)*

# Applied Mathematical Sciences

*(continued from previous page)*

117. *Taylor:* Partial Differential Equations III, Nonlinear Equations.
118. *Godlewski/Raviart:* Numerical Approximation of Hyperbolic Systems of Conservation Laws.
119. *Wu:* Theory and Applications of Partial Functional Differential Equations.
120. *Kirsch:* An Introduction to the Mathematical Theory of Inverse Problems.
121. *Brokate/Sprekels:* Hysteresis and Phase Transitions.
122. *Gliklikh:* Global Analysis in Mathematical Physics: Geometric and Stochastic Methods.
123. *Le/Schmitt:* Global Bifurcation in Variational Inequalities: Applications to Obstacle and Unilateral Problems.
124. *Polak:* Optimization: Algorithms and Consistent Approximations.
125. *Arnold/Khesin:* Topological Methods in Hydrodynamics.
126. *Hoppensteadt/Izhikevich:* Weakly Connected Neural Networks.
127. *Isakov:* Inverse Problems for Partial Differential Equations.
128. *Li/Wiggins:* Invariant Manifolds and Fibrations for Perturbed Nonlinear Schrödinger Equations.
129. *Müller:* Analysis of Spherical Symmetries in Euclidean Spaces.
130. *Feintuch:* Robust Control Theory in Hilbert Space.
131. *Ericksen:* Introduction to the Thermodynamics of Solids, Revised ed.
132. *Ihlenburg:* Finite Element Analysis of Acoustic Scattering.
133. *Vorovich:* Nonlinear Theory of Shallow Shells.
134. *Vein/Dale:* Determinants and Their Applications in Mathematical Physics.
135. *Drew/Passman:* Theory of Multicomponent Fluids.
136. *Cioranescu/Saint Jean Paulin:* Homogenization of Reticulated Structures.
137. *Gurtin:* Configurational Forces as Basic Concepts of Continuum Physics.
138. *Haller:* Chaos Near Resonance.
139. *Sulem/Sulem:* The Nonlinear Schrödinger Equation: Self-Focusing and Wave Collapse.
140. *Cherkaev:* Variational Methods for Structural Optimization.
141. *Naber:* Topology, Geometry, and Gauge Fields: Interactions.
142. *Schmid/Henningson:* Stability and Transition in Shear Flows.
143. *Sell/You:* Dynamics of Evolutionary Equations.
144. *Nédélec:* Acoustic and Electromagnetic Equations: Integral Representations for Harmonic Problems.
145. *Newton:* The *N*-Vortex Problem: Analytical Techniques.
146. *Allaire*: Shape Optimization by the Homogenization Method.
147. *Aubert/Kornprobst:* Mathematical Problems in Image Processing: Partial Differential Equations and the Calculus of Variations.
148. *Peyret:* Spectral Methods for Incompressible Viscous Flow.
149. *Ikeda/Murota:* Imperfect Bifurcation in Structures and Materials: Engineering Use of Group-Theoretic Bifucation Theory.
150. *Skorokhod/Hoppensteadt/Salehi:* Random Perturbation Methods with Applications in Science and Engineering.
151. *Bensoussan/Frehse:* Regularity Results for Nonlinear Elliptic Systems and Applications.
152. *Holden/Risebro:* Front Tracking for Hyperbolic Conservation Laws.
153. *Osher/Fedkiw:* Level Set Methods and Dynamic Implicit Surfaces.
154. *Bluman/Anco:* Symmetry and Integration Methods for Differential Equations.
155. *Chalmond:* Modeling and Inverse Problems in Image Analysis.
156. *Kielhöfer:* Bifurcation Theory: An Introduction with Applications to PDEs.
157. *Kaczynski/Mischaikow/Mrozek:* Computational Homology.
158. *Oertel:* Prandtl's Essentials of Fluid Mechanics, 2nd ed.
159. *Ern/Guermond:* Theory and Practice of Finite Elements.

GPSR Compliance
The European Union's (EU) General Product Safety Regulation (GPSR) is a set of rules that requires consumer products to be safe and our obligations to ensure this.

If you have any concerns about our products, you can contact us on

ProductSafety@springernature.com

In case Publisher is established outside the EU, the EU authorized representative is:

Springer Nature Customer Service Center GmbH
Europaplatz 3
69115 Heidelberg, Germany

www.ingramcontent.com/pod-product-compliance
Ingram Content Group UK Ltd.
Pitfield, Milton Keynes, MK11 3LW, UK
UKHW020652210926
15297UKWH00010B/152
* 9 7 8 1 4 7 5 7 3 9 7 9 4 *